Introduction to Orthogonal Transforms

With Applications in Data Processing and Analysis

A systematic, unified treatment of orthogonal transform methods for signal processing, data analysis, and communications, this book guides the reader from mathematical theory to problem solving in practice. It examines each transform method in depth, emphasizing the common mathematical principles and essential properties of each method in terms of signal decorrelation and energy compaction. The different forms of Fourier transform, as well as the Laplace, Z-, Walsh–Hadamard, slant, Haar, Karhunen–Loève, and wavelet transforms, are all covered, with discussion of how these transform methods can be applied to real-world problems. Numerous practical examples and end-of-chapter problems, supported by online Matlab and C code and an instructor-only solutions manual, make this an ideal resource for students and practitioners alike.

Ruye Wang is a Professor in the Engineering Department at Harvey Mudd College. Previously a Principal Investigator at the Jet Propulsion Laboratory, NASA, his research interests include image processing, computer vision, machine learning, and remote sensing.

Introduction to Orthogonal Transforms

With Applications in Data Processing and Analysis

RUYE WANG
Harvey Mudd College, California, USA

CAMBRIDGE
UNIVERSITY PRESS

CAMBRIDGE
UNIVERSITY PRESS

University Printing House, Cambridge CB2 8BS, United Kingdom

One Liberty Plaza, 20th Floor, New York, NY 10006, USA

477 Williamstown Road, Port Melbourne, VIC 3207, Australia

314-321, 3rd Floor, Plot 3, Splendor Forum, Jasola District Centre, New Delhi - 110025, India

79 Anson Road, #06-04/06, Singapore 079906

Cambridge University Press is part of the University of Cambridge.

It furthers the University's mission by disseminating knowledge in the pursuit of education, learning and research at the highest international levels of excellence.

www.cambridge.org
Information on this title: www.cambridge.org/9780521516884

First published 2012

A catalogue record for this publication is available from the British Library

ISBN 978-0-521-51688-4 Hardback

Additional resources for this publication at www.cambridge.org/9780521516884

To my parents

Contents

Preface

What is the book all about?

When a straight line standing on a straight line makes the adjacent angles equal to one another, each of the equal angles is right, and the straight line standing on the other is called a *perpendicular* to that on which it stands.
— Euclid, *Elements, Book 1, definition 10*

This is Euclid's definition for "perpendicular", which is synonymous with the word "orthogonal" used in the title of this book. Although the meaning of this word has been generalized since Euclid's time to describe the relationship between two functions as well as two vectors, as what we will be mostly concerned with in this book, they are essentially no different from two perpendicular straight lines, as discussed by Euclid some 23 centuries ago.

Orthogonality is of important significance not only in geometry and mathematics, but also in science and engineering in general, and in data processing and analysis in particular. This book is about a set of mathematical and computational methods, known collectively as the orthogonal transforms, that enables us to take advantage of the orthogonal axes of the space in which the data reside. As we will see throughout the book, such orthogonality is a much desired property that can keep things untangled and nicely separated for ease of manipulation, and an orthogonal transform can rotate a signal, represented as a vector in a Euclidean space, or more generally Hilbert space, in such a way that the signal components tend to become, approximately or accurately, orthogonal to each other. Such orthogonal transforms, typified by the most well-known Fourier transform, lend themselves well to various data processing and analysis needs, and therefore are used in a wide variety of disciplines and areas, including both social and natural sciences and engineering. The book also covers the Laplace and z-transforms, which can be considered as the extended versions of the Fourier transform for continuous and discrete functions respectively, and the wavelet transforms which may not be strictly orthogonal but which are still closely related to those that are.

In the last few decades the scales of data collection across almost all fields have been increasing dramatically due mostly to the rapid advances in technologies. Consequently, how best to make sense of the fast accumulating data has become more challenging than ever. Wherever a large amount of data is collected, from stock market indices in economics to microarray data in bioinformatics, from seismic data in geophysics to audio and video data in communication and broadcasting engineering, there is always the need to process, analyze, and compress the data in some meaningful way for the purposes of effective and efficient data transmission, interpretation, and storage by various computational methods and algorithms. The transform methods discussed in this book can be used as a set of basic tools for the data processing and the subsequent analysis, such as data mining, knowledge discovery, and machine learning.

The specific purpose of each data processing and analysis task at hand may vary from case to case. From a set of given data, one may desire to remove a certain type of noise, extract particular kinds of features of interest, and/or reduce the quantity of the data without losing useful information for storage and transmission. On the other hand, many operations needed for achieving these very different goals may all be carried out using the same mathematical tool of orthogonal transform, by which the data are manipulated and represented in such a way that the desired results can be achieved effectively in the subsequent stage. To address all such needs, this book presents a thorough introduction to the mathematical background common to these transform methods, and provides a repertoire of computational algorithms for these methods.

The basic approach of the book is the combination of the theoretical derivation and practical implementation of each transform method considered. Certainly, many existing books touch upon the topics of both orthogonal and wavelet transforms, from either a mathematical or an engineering point of view. Some of them may concentrate on the theories of these methods, while others may emphasize their applications, but relatively few would guide the reader directly from the mathematical theories to the computational algorithms, and then to their applications to real data analysis, as this book intends to do. Here, deliberate efforts are made to bridge the gap between the theoretical background and the practical implementation, based on the belief that, to truly understand a certain method, one needs ultimately to be able to convert the mathematical theory into computer code for the algorithms to be actually implemented and tested. This idea has been the guiding principle throughout the writing of the book. For each of the methods covered, we will first derive the theory mathematically, then present the corresponding computational algorithm, and finally provide the necessary code segments in Matlab or C for the key parts of the algorithm. Moreover, we will also include some relatively simple application examples to illustrate the actual data-processing effects of the algorithm. In fact, every one of the transform methods considered in the book has been implemented in either Matlab and/or the C programming language and tested on real data. The complete programs are also made readily available on a website dedicated to the book at: www.cambridge.org/orthogonaltransforms. The reader is encouraged and expected to try these algorithms out by running the code on his/her own data.

Why orthogonal transforms?

The transform methods covered in the book are a collection of both old and new ideas ranging from the classical Fourier series expansion that goes back almost 200 years, to some relatively recent thoughts such as the various origins of the wavelet transform. While all of these ideas were originally developed with different goals and applications in mind, from solving the heat equation to the analysis of seismic data, they can all be considered to belong to the same family, based on the common mathematical framework they all share, and their similar applications in data processing and analysis. The discussions of specific methods and algorithms in the chapters will all be approached from such a unified point of view.

Before the specific discussion of each of the methods, let us first address a fundamental issue: why do we need to carry out an orthogonal transform to start with? A signal, as the measurement of a certain variable (e.g., the temperature of a physical process) tends to vary continuously and smoothly, as the energy associated with the physical process is most probably distributed relatively evenly in both space and time. Most such spatial or temporal signals are likely to be correlated, in the sense that, given the value of a signal at a certain point in space or time, one can predict with reasonable confidence that the signal at a neighboring point will take a similar value. Such everyday experience is due to the fundamental nature of the physical world governed by the principles of minimum energy and maximum entropy, in which any abruption and discontinuities, typically caused by an energy surge of some kind, are relatively rare and unlikely events (except in the microscopic world governed by quantum mechanics). On the other hand, from the signal processing viewpoint, the high signal correlation and even energy distribution are not desirable in general, as it is difficult to decompose such a signal, as needed in various applications such as information extraction, noise reduction, and data compression. The issue, therefore, becomes one of how the signal can be converted in such a way that it is less correlated and its energy less evenly distributed, and to what extent such a conversion can be carried out to achieve the goal.

Specifically, in order to represent, process, and analyze a signal, it needs to be decomposed into a set of components along a certain dimension. While a signal is typically represented by default as a continuous or discrete function of time or space, it may be desirable to represent it along some alternative dimension, most commonly (but not exclusively) frequency, so that it can be processed and analyzed more effectively and conveniently. Viewed mathematically, a signal is a vector in some vector space which can be represented by any of a set of different orthogonal bases all spanning the same space. Each representation corresponds to a different decomposition of the signal. Moreover, all such representations are equivalent, in the sense that they are related to each other by certain rotation in the space by which the total energy or information contained in the signal is conserved. From this point of view, all different orthogonal transform methods developed in the last 200 years by mathematicians, scientists, and engineers for various purposes can be unified to form a family of methods for the same general purpose.

While all transform methods are equivalent, as they all conserve the total energy or information of the signal, they can be very different in terms of how the total energy or information in the signal is redistributed among its components after the transform, and how much these components are correlated. If, after a properly chosen orthogonal transform, the signal is represented in such a way that its components are decorrelated and most of the signal information of interest is concentrated in a small subset of its components, then the remaining components could be neglected as they carry little information. This simple idea is essentially the answer to the question asked above about why an orthogonal transform is needed, and it is actually the foundation of the general orthogonal transform method for feature selection, data compression, and noise reduction. In a certain sense, once a proper basis of the space is chosen so that the signal is represented in such a favorable manner, the signal-processing goal is already achieved to a significant extent.

What is in the chapters?

The purpose of the first two chapters is to establish a solid mathematical foundation for the thorough understanding of the topics of the subsequent chapters, which each discuss a specific type of transform method. Chapter 1 is a brief summary of the basic concepts of signals and linear time-invariant (LTI) systems. For readers with an engineering background, much of this chapter may be a quick review that could be scanned through or even skipped. For others, this chapter serves as an introduction to the mathematical language by which the signals and systems will be described in the following chapters.

Chapter 2 sets up the stage for all transform methods by introducing the key concepts of the vector space, or more strictly speaking the Hilbert space, and the linear transformations in such a space. Here, a usual N-dimensional space can be generalized in several aspects: (1) the dimension N of the space may be extended to infinity, (2) a vector space may also include a function space composed of all continuous functions satisfying certain conditions, and (3) the basis vectors of a space may become uncountable. The mathematics needed for a rigorous treatment of these much-generalized spaces is likely to be beyond the comfort zone of most readers with a typical engineering or science background, and it is therefore also beyond the scope of this book. The emphasis of the discussion here is not mathematical rigor, but the basic understanding and realization that many of the properties of these generalized spaces are just the natural extensions of those of the familiar N-dimensional vector space. The purpose of such discussions is to establish a common foundation for all transform methods so that they can all be studied from a unified point of view, namely, that any given signal, either continuous or discrete, with either finite or infinite duration, can be treated as a vector in a certain space and represented differently by any of a variety of orthogonal transform methods, each corresponding to one of the orthogonal bases that span the space. Moreover, all of these different representations are related to each other by rotations in the space. Such basic ideas may also be extended to non-orthogonal (e.g., biorthogonal) bases that are used in wavelet transforms. All transform methods considered in later chapters will be studied in light of such a framework. Although it is highly recommended for the reader to at least read through the materials in the first two chapters, those who feel it is difficult to thoroughly follow the discussions could skip them and move on to the following chapters, as many of the topics could be studied relatively independently, and one can always come back to learn some of the concepts in the first two chapters when needed.

In Chapters 3 and 4 we study the classical Fourier methods for continuous and discrete signals respectively. Fourier's theory is mathematically beautiful and is referred to as "mathematical poem"; and it has great significance throughout a wide variety of disciplines, in practice as well as in theory. While the general topic of the Fourier transform is covered in a large number of textbooks in various fields, such as engineering, physics, and mathematics, a not-so-conventional approach is adopted here to treat all Fourier-related methods from a unified point of view. Specifically, the Fourier series expansion, the continuous- and discrete-time Fourier transforms (CTFT and DTFT), and the discrete Fourier transform (DFT) will be considered as four different variations of the same general Fourier transform, corresponding to the four combinations of the two basic categories of signals: continuous versus discrete, periodic versus non-periodic. By doing so, many of the dual and symmetrical relationships among these four different forms and between time and frequency domains of the Fourier transform can be much more clearly and conveniently presented and understood.

Chapter 5 discusses the Laplace and z-transforms. Strictly speaking, these transforms do not belong to the family of orthogonal transforms, which convert a one-dimensional (1-D) signal of time t into another 1-D function along a different variable, typically frequency f or angular frequency $\omega = 2\pi f$. Instead, the Laplace converts a 1-D continuous signal from the time domain into a function in a two-dimensional (2-D) complex plane $s = \sigma + j\omega$, and the z-transform converts a 1-D discrete signal from the time domain into a function in a 2-D complex plane $z = e^s$. However, as these transforms are respectively the natural extensions of the CTFT and DTFT, and are widely used in signal processing and system analysis, they are included in the book as two extra tools in our toolbox.

Chapter 6 discusses the Hartly and sine/cosine transforms, both of which are closely related to the Fourier transform. As real transforms, both Hartly and sine/cosine transforms have the advantage of reduced computational cost when compared with the complex Fourier transform. If the signal in question is real with zero imaginary part, then half of the computation in its Fourier transform is redundant and, therefore, wasted. However, this redundancy is avoided by a real transform such as the cosine transform, which is widely used for data compression, such as in the image compression standard JPEG.

Chapter 7 combines three transform methods, the Walsh-Hadamard, slant, and Haar transforms, all sharing some similar characteristics (i.e., the basis functions associated with these transforms all have square-wave-like waveforms). Moreover, as the Haar transform also possesses the basic characteristics of the wavelet transform method, it can also serve as a bridge between the two camps of the orthogonal transforms and the wavelet transforms, leading to a natural transition from the former to the latter.

In Chapter 8 we discuss the Karhunen-Loeve transform (KLT), which can be considered as a capstone of all previously discussed transform methods, and the associated principal component analysis (PCA), which is popularly used in many data-processing applications. The KLT is the optimal transform method among all orthogonal transforms in terms of the two main characteristics of the general orthogonal transform method, namely the compaction of signal energy and the decorrelation among all signal components. In this regard, all orthogonal transform methods can be compared against the optimal KLT for an assessment of their performances.

We next consider in Chapter 9 both the continuous- and discrete-time wavelet transforms (CTWT and DTWT), which differ from all orthogonal transforms discussed previously in two main aspects. First, the wavelet transforms are not strictly orthogonal, as the bases used to span the vector space and to represent a given signal may not be necessarily orthogonal. Second, the wavelet transform converts a 1-D time signal into a 2-D function of two variables, one for different levels of details or scales, corresponding to different frequencies in the Fourier transform, and the other for different temporal positions, which is totally absent in the Fourier or any other orthogonal transform. While redundancy is inevitably introduced into the 2-D transform domain by such a wavelet transform, the additional second dimension also enables the transform to achieve both temporal and frequency localities in signal representation at the same time (while all other transform methods can only achieve either one of the two localities). Such a capability of the wavelet transform is its main advantage over orthogonal transforms in some applications such as signal filtering.

Finally, in Chapter 10 we introduce the basic concept of multiresolution analysis (MRA) and Mallat's fast algorithm for the discrete wavelet transform (DWT), together with its filter bank implementation. Similar to the orthogonal transforms, this algorithm converts a discrete signal of size N into a set of DWT coefficients also of size N, from which the original signal can be perfectly reconstructed; i.e., there is no redundancy introduced by the DWT. However, different from orthogonal transforms, the DWT coefficients represent the signal with temporal as well as frequency (levels of details) localities, and can, therefore, be more advantageous in some applications, such as data compressions.

Moreover, some fundamental results in linear algebra and statistics are also summarized in the two appendices at the back of the book.

Who are the intended readers?

The book can be used as a textbook for either an undergraduate or graduate course in digital signal processing, communication, or other related areas. In such a classroom setting, all orthogonal transform methods can be systematically studied following a thorough introduction of the mathematical background common to these methods. The mathematics prerequisite is no more than basic calculus and linear algebra. Moreover, the book can also be used as a reference by practicing professionals in both natural and social sciences, as well as in engineering. A financial analyst or a biologist may need to learn how to effectively analyze and interpret his/her data, a database designer may need to know how to compress his data before storing them in the database, and a software engineer may need to learn the basic data-processing algorithms while developing a software tool in the field. In general, anyone who deals with a large quantity of data may desire to gain some basic knowledge in data-processing, regardless of his/her backgrounds and specialties. In fact the book has been developed with such potential readers in mind. Owing possibly to personal experience, I always feel that self-learning (or, to borrow a machine learning terminology, "unsupervised learning") is no less important than formal classroom learning. One may have been out of school for some years but still feel the need to update and expand one's knowledge. Such readers could certainly study whichever chapters of interest, instead of systematically reading through each chapter from beginning to end. They can also skip certain mathematical derivations, which are included in the book for completeness (and for those who feel comfortable only if the complete proof and derivations of all conclusions are provided). For some readers, neglecting much of the mathematical discussion for a specific transform method should be just fine if the basic ideas regarding the method and its implementation are understood. It is hoped that the book can serve as a toolbox, as well as a textbook, from which certain transform methods of interest can be learned and applied, in combination with the reader's expertise in his/her own field, to solving the specific data-processing/analysis problems at hand.

About the homework problems and projects

Understanding the transform methods and the corresponding computational algorithms is not all. Eventually they all need to be implemented and realized by either software or hardware, specifically by computer code of some sort. This is why the book emphasizes the algorithm and coding as well as theoretical derivation, and many homework problems and projects require certain basic coding skills, such as some knowledge in Matlab. However, being able to code is not expected of all readers. Those who may not need or wish to learn coding can by all means skip the sections in the text and those homework problems involving software programming. However, all readers are encouraged to at least run some of the Matlab functions provided to see the effects of the transform methods. (There are a lot of such Matlab m-files on the website of the book. In fact, all functions used to generate many of the figures in the book are provided on the site.) If a little more interested, the reader can read through the code to see how things are done. Of course, a step further is to modify the code and use different parameters and different datasets to better appreciate the various effects of the algorithms.

Back to Euclid

Finally, let us end by again quoting Euclid, this time, a story about him.

> A youth who had begun to study geometry with Euclid, when he had learned the first proposition, asked, "What do I get by learning these things?" So Euclid called a slave and said "Give him three pence, since he must make a gain out of what he learns."

Surely, explicit efforts are made in this book to discuss the practical uses of the orthogonal transforms and the mathematics behind them, but one should realize that, after all, the book is about a set of mathematical tools, just like those propositions in Euclid's geometry, out of learning which the reader may not be able to make a direct and immediate gain. However, in the end, it is the application of these tools toward solving specific problems in practice that will enable the reader to make a gain out of the book; much more than three pence, hopefully.

Acknowledgments

I am in debt to two of my colleagues, Professors John Molinder and Ellis Cumberbatch for their support and help with the book project. In addition to our discussions regarding some of the topics in the book, John provided the application example of orthogonal frequency division modulation discussed in section 5.8, together with the Matlab code that is used in a homework problem. Also, Ellis read through the first two chapters of the manuscript and made numerous suggestions for the improvement of the coverage of the topics in these two chapters. All such valuable help and support are greatly appreciated.

Notation

General notation

iff	if and only if
$j = \sqrt{-1} = e^{j\pi/2}$	imaginary unit
$\overline{u + jv} = u - jv$	complex conjugate of $u + jv$
$\text{Re}(u + jv) = u$	real part of $u + jv$
$\text{Im}(u + jv) = v$	imaginary part of $u + jv$
$\|u + jv\| = \sqrt{u^2 + v^2}$	magnitude (absolute value) of a complex value $u + jv$
$\angle(u + jv) = \tan^{-1}(v/u)$	phase of $u + jv$
$\boldsymbol{x}_{n \times 1}$	an n by 1 column vector
$\overline{\boldsymbol{x}}$	complex conjugate of $\boldsymbol{x}$
$\boldsymbol{x}^{\mathrm{T}}$	transpose of $\boldsymbol{x}$, a 1 by n row vector
$\boldsymbol{x}^* = \overline{\boldsymbol{x}}^{\mathrm{T}}$	conjugate transpose of matrix $\boldsymbol{A}$
$\|\|\boldsymbol{x}\|\|$	norm of vector $\boldsymbol{x}$
$\boldsymbol{A}_{m \times n}$	an m by n matrix of m rows and n columns
$\overline{\boldsymbol{A}}$	complex conjugate of matrix $\boldsymbol{A}$
$\boldsymbol{A}^{-1}$	inverse of matrix $\boldsymbol{A}$
$\boldsymbol{A}^{\mathrm{T}}$	transpose of matrix $\boldsymbol{A}$
$\boldsymbol{A}^* = \overline{\boldsymbol{A}}^{\mathrm{T}} = \overline{\boldsymbol{A}^{\mathrm{T}}}$	conjugate transpose of matrix $\boldsymbol{A}$
$\mathbb{N}$	set of all positive integers including 0
$\mathbb{Z}$	set of all real integers
$\mathbb{R}$	set of all real numbers
$\mathbb{C}$	set of all complex numbers
$\mathbb{R}^N$	N-dimensional Euclidean space
$\mathbb{C}^N$	N-dimensional unitary space
$\mathcal{L}^2$	space of all square-integrable functions
l^2	space of all square-summable sequences
$x(t)$	a function representing a continuous signal
$\boldsymbol{x} = [\dots, x[n], \dots]^{\mathrm{T}}$	a vector representing a discrete signal
$\dot{x}(t) = dx(t)/dt$	first order time derivative of $x(t)$
$\ddot{x}(t) = dx^2/dt^2$	second order time derivative of $x(t)$
f	frequency (cycle per unit time)
$\omega = 2\pi f$	angular frequency (radian per unit time)

Throughout the book, angular frequency ω will be used interchangeably with $2\pi f$, whichever is more convenient in the context of the discussion.

As a convention, a bold-faced lower case letter $\boldsymbol{x}$ is typically used to represent a vector, while a bold-faced upper case letter $\boldsymbol{A}$ represents a matrix, unless noted otherwise.

1 Signals and systems

In the first two chapters we will consider some basic concepts and ideas as the mathematical background for the specific discussions of the various orthogonal transform methods in the subsequent chapters. Here, we will set up a framework common to all such methods, so that they can be studied from a unified point of view. While some discussions here may seem mathematical, the emphasis is on the intuitive understanding, rather than the theoretical rigor.

1.1 Continuous and discrete signals

A physical signal can always be represented as a real- or complex-valued continuous function of time $x(t)$ (unless specified otherwise, such as a function of space). The continuous signal can be sampled to become a discrete signal $x[n]$. If the time interval between two consecutive samples is assumed to be $\triangle$, then the nth sample is

$$x[n] = x(t)\big|_{t=n\triangle} = x(n\triangle). \tag{1.1}$$

In either the continuous or discrete case, a signal can be assumed in theory to have infinite duration; i.e., $-\infty < t < \infty$ for $x(t)$ and $-\infty < n < \infty$ for $x[n]$. However, any signal in practice is finite and can be considered as the truncated version of a signal of infinite duration. We typically assume $0 \le t \le T$ for a finite continuous signal $x(t)$, and $1 \le n \le N$ (or sometimes $0 \le n \le N-1$ for certain convenience) for a discrete signal $x[n]$. The value of such a finite signal $x(t)$ is not defined if $t < 0$ or $t > T$; similarly, $x[n]$ is not defined if $n < 0$ or $n > N$. However, for mathematical convenience we could sometimes assume a finite signal to be periodic; i.e., $x(t+T) = x(t)$ and $x[n+N] = x[n]$.

A discrete signal can also be represented as a vector $\boldsymbol{x} = [\ldots, x[n-1], x[n], x[n+1], \ldots]^{\mathrm{T}}$ of finite or infinite dimensions composed of all of its samples or components as the vector elements. We will always represent a discrete signal as a column vector (transpose of a row vector).

Definition 1.1. *The discrete unit impulse or Kronecker delta function is defined as*

$$\delta[n] = \begin{cases} 1 & n = 0 \\ 0 & n \neq 0 \end{cases}. \tag{1.2}$$

Based on this definition, a discrete signal can be represented as

$$x[n] = \sum_{m=-\infty}^{\infty} x[m]\delta[n-m], \qquad (n = 0, \pm 1, \pm 2, \ldots). \tag{1.3}$$

This equation can be interpreted in two conceptually different ways.

- First, a discrete signal $x[n]$ can be decomposed into a set of unit impulses each at a different moment $n = m$ and weighted by the signal amplitude $x[m]$ at the moment, as shown in Fig. 1.1(a).
- Second, the Kronecker delta $\delta[n-m]$ acts as a filter that sifts out a particular value of the signal $x[n]$ at the moment $m = n$ from a sequence of signal samples $x[m]$ for all m. This is the *sifting property* of the Kronecker delta.

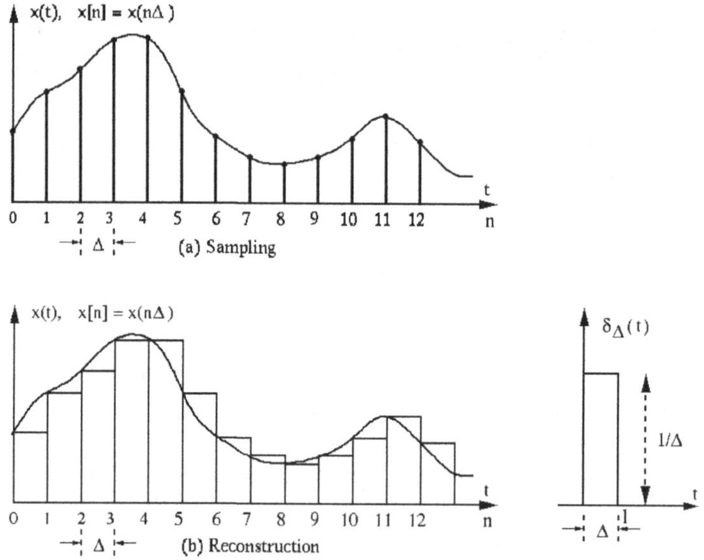

Figure 1.1 Sampling and reconstruction of a continuous signal.

In a similar manner, a continuous signal $x(t)$ can also be represented by its samples. We first define a unit square impulse function as

$$\delta_\triangle(t) = \begin{cases} 1/\triangle & 0 \le t < \triangle \\ 0 & \text{else} \end{cases}. \tag{1.4}$$

Note that the width and height of this square impulse are respectively $\triangle$ and $1/\triangle$; i.e, it covers a unit area $\triangle \times 1/\triangle = 1$, independent of the value of $\triangle$:

$$\int_{-\infty}^{\infty} \delta_{\triangle}(t)\, dt = \triangle \cdot 1/\triangle = 1. \tag{1.5}$$

Now a continuous signal $x(t)$ can be approximated as a sequence of square impulses $\delta_{\triangle}(t - n\triangle)$ weighted by the sample value $x[n]$ for the amplitude of the signal at the moment $t = n\triangle$:

$$x(t) \approx \hat{x}(t) = \sum_{n=-\infty}^{\infty} x\,[n]\, \delta_{\triangle}(t - n\triangle)\triangle. \tag{1.6}$$

This is shown in Fig. 1.1(b).

The approximation $\hat{x}(t)$ above will become a perfect reconstruction of the signal if we take the limit $\triangle \to 0$, so that the square impulse becomes a *continuous unit impulse* or *Dirac delta*:

$$\lim_{\triangle \to 0} \delta_{\triangle}(t) = \delta(t). \tag{1.7}$$

which is formally defined as

Definition 1.2. *The continuous unit impulse or Dirac delta function $\delta(t)$ is a function that has an infinite height but zero width at $t = 0$, and it covers a unit area; i.e., it satisfies the following two conditions:*

$$\delta(t) = \begin{cases} \infty & t = 0 \\ 0 & t \neq 0 \end{cases} \quad \textit{and} \quad \int_{-\infty}^{\infty} \delta(t)\, dt = \int_{0^-}^{0^+} \delta(t)\, dt = 1. \tag{1.8}$$

Now at the limit $\triangle \to 0$, the summation in the approximation of Eq. (1.6) above becomes an integral, the square impulse becomes a Dirac delta, and the approximation becomes a perfect reconstruction of the continuous signal:

$$x(t) = \lim_{\triangle \to 0} \sum_{n=-\infty}^{\infty} x[n]\delta_{\triangle}(t - n\triangle)\triangle = \int_{-\infty}^{\infty} x(\tau)\delta(t - \tau)\, d\tau. \tag{1.9}$$

In particular, when $t = 0$, Eq. (1.9) becomes

$$x(0) = \int_{-\infty}^{\infty} x(\tau)\delta(\tau)\, d\tau. \tag{1.10}$$

Equation (1.9) can be interpreted in two conceptually different ways.

- First, a continuous signal $x(t)$ can be decomposed into an uncountably infinite set of unit impulses each at a different moment $t = \tau$, weighted by the signal intensity $x(\tau)$ at the moment $t = \tau$.
- Second, the Dirac delta $\delta(\tau - t)$ acts as a filter that sifts out the value of $x(t)$ at the moment $\tau = t$ from a sequence of uncountably infinite signal samples. This is the *sifting property* of the Dirac delta.

Note that the discrete impulse function $\delta[n]$ has a unit height, while the continuous impulse function $\delta(t)$ has a unit area (product of height and width for time); i.e., the two types of impulses have different dimensions. The dimension of the discrete impulse is the same as that of the signal (e.g., voltage), while the dimension of the continuous impulse is the signal's dimension divided by time (e.g., voltage/time). In other words, $x(\tau)\delta(t-\tau)$ represents the density of the signal at $t = \tau$, only when integrated over time will the continuous impulse functions have the same dimension as the signal $x(t)$.

The results above indicate that a time signal, either discrete or continuous, can be decomposed in the time domain to become a linear combination, either a summation or an integral, of a set of time impulses (components), either countable or uncountable. However, as we will see in future chapters, the decomposition of the time signal is not unique. The signal can also be decomposed in different domains other than time, such as frequency, and the representations of the signal in different domains are related by certain orthogonal transformations.

1.2 Unit step and nascent delta functions

Here we define some important functions to be used frequently in the future. The *discrete unit step function* is defined as

Definition 1.3.

$$u[n] = \begin{cases} 1 & n \geq 0 \\ 0 & n < 0 \end{cases}. \tag{1.11}$$

The Kronecker delta can be obtained as the first-order difference of the unit step function:

$$\delta[n] = u[n] - u[n-1] = \begin{cases} 1 & n = 0 \\ 0 & n \neq 0 \end{cases}. \tag{1.12}$$

Similarly, in continuous case, the impulse function $\delta(t)$ is also closely related to the *continuous unit step function* (also called *Heaviside step function*) $u(t)$. To see this, we first consider a piece-wise linear function defined as

$$u_\triangle(t) = \begin{cases} 0 & t < 0 \\ t/\triangle & 0 \leq t < \triangle \\ 1 & t \geq \triangle \end{cases}. \tag{1.13}$$

Taking the time derivative of this function, we get the square impulse considered before in Eq. (1.4):

$$\delta_\triangle(t) = \frac{d}{dt} u_\triangle(t) = \begin{cases} 0 & t < 0 \\ 1/\triangle & 0 \leq t < \triangle \\ 0 & t \geq \triangle \end{cases}. \tag{1.14}$$

If we let $\triangle \to 0$, then $u_\triangle(t)$ becomes the *unit step function* $u(t)$ at the limit:

Definition 1.4.

$$u(t) = \lim_{\triangle \to 0} u_\triangle(t) = \begin{cases} 0 & t < 0 \\ 1/2 & t = 0 \\ 1 & t > 0 \end{cases}. \tag{1.15}$$

Here, we have defined $u(0) = 1/2$ at $t = 0$ for reasons to be discussed in the future. [1] Also, at the limit $\triangle \to 0$, $\delta_\triangle(t)$ becomes Dirac delta discussed above:

$$\delta(t) = \lim_{\triangle \to 0} \delta_\triangle(t) = \begin{cases} \infty & t = 0 \\ 0 & t \neq 0 \end{cases}. \tag{1.16}$$

Therefore, by taking the limit $\triangle \to 0$ on both sides of Eq. (1.14) we obtain a useful relationship between $u(t)$ and $\delta(t)$:

$$\frac{d}{dt} u(t) = \delta(t), \qquad u(t) = \int_{-\infty}^{t} \delta(\tau)\, d\tau. \tag{1.17}$$

This process is shown in the three cases for different values of $\triangle$ in Fig. 1.2.

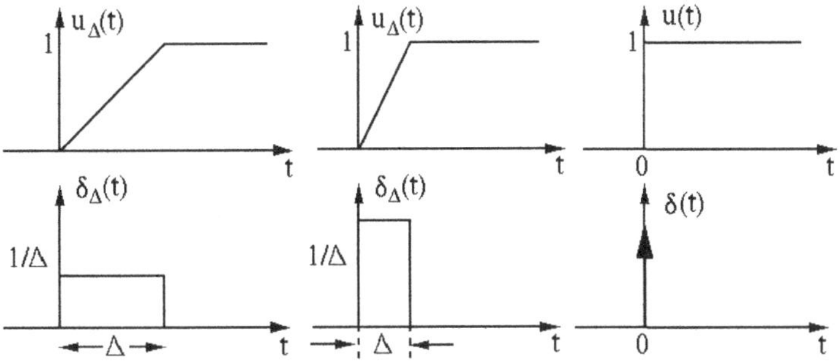

Figure 1.2 Generation of unit step and unit impulse. Three functions $u_\triangle(t)$ with different values of $\triangle$ together with their derivatives $\delta_\triangle(t)$ are shown. In particular, when $\delta \to 0$, these functions become $u(t)$ and $\delta(t)$, as shown on the right.

In addition to the square impulse $\delta_\triangle(t)$, the Dirac delta $\delta(t)$ can also be generated from a variety of different *nascent delta functions* at the limit when a certain parameter of the function approaches the limit of either zero or infinity. Consider, for example, the Gaussian function:

$$g(t) = \frac{1}{\sqrt{2\pi\sigma^2}} e^{-t^2/2\sigma^2}, \tag{1.18}$$

which is the probability density function of a normally distributed random variable t with zero mean and variance σ^2. Obviously the area underneath this

[1] Although in some of the literature it could be alternatively defined as either $u(0) = 0$ or $u(0) = 1$.

density function is always one, independent of σ:

$$\int_{-\infty}^{\infty} g(t)\, dt = \frac{1}{\sqrt{2\pi\sigma^2}} \int_{-\infty}^{\infty} e^{-t^2/2\sigma^2}\, dt = 1. \tag{1.19}$$

At the limit $\sigma \to 0$, this Gaussian function $g(t)$ becomes infinity at $t = 0$ but it is zero for all $t \neq 0$; i.e., it becomes the unit impulse function:

$$\lim_{\sigma \to 0} \frac{1}{\sqrt{2\pi\sigma^2}} e^{-t^2/2\sigma^2} = \delta(t). \tag{1.20}$$

The Gaussian functions with three different σ values are shown in Fig. 1.3.

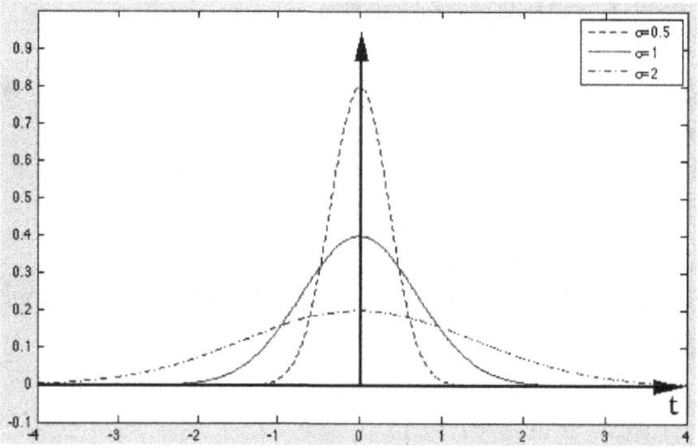

Figure 1.3 Gaussian functions with different σ values (0.5, 1, 2).

The argument t of a Dirac delta $\delta(t)$ may be scaled so that it becomes $\delta(at)$. In this case Eq. (1.10) becomes

$$\int_{-\infty}^{\infty} x(\tau)\delta(a\tau)\, d\tau = \int_{-\infty}^{\infty} x\left(\frac{u}{a}\right) \delta(u) \frac{1}{|a|}\, du = \frac{1}{|a|} x(0), \tag{1.21}$$

where we have defined $u = a\tau$. Comparing this result with Eq. (1.10), we see that

$$\delta(at) = \frac{1}{|a|}\delta(t); \quad \text{i.e.} \quad |a|\delta(at) = \delta(t). \tag{1.22}$$

For example, a delta function $\delta(f)$ of frequency f can also be expressed as a function of angular frequency $\omega = 2\pi f$ as

$$\delta(f) = 2\pi\delta(\omega). \tag{1.23}$$

More generally, the Dirac delta can also be defined over a function $f(t)$ of a variable, instead of a variable t. Now the Dirac delta becomes $\delta(f(t))$, which is zero except when $f(t_k) = 0$, where $t = t_k$ is one of the roots of $f(t)$. To see how such an impulse is scaled, consider the following integral:

$$\int_{-\infty}^{\infty} x(\tau)\delta(f(\tau))\, d\tau = \int_{-\infty}^{\infty} x(\tau)\delta(u)\, \frac{1}{|f'(\tau)|}\, du, \tag{1.24}$$

where we have changed the integral variable from τ to $u = f(\tau)$. If $\tau = \tau_0$ is the only root of $f(\tau)$; i.e., $u = f(\tau_0) = 0$, then the integral above becomes

$$\int_{-\infty}^{\infty} x(\tau)\delta(f(\tau))\, d\tau = \frac{x(\tau_0)}{|f'(\tau_0)|}. \tag{1.25}$$

If $f(\tau)$ has multiple roots τ_k, then we have

$$\int_{-\infty}^{\infty} x(\tau)\delta(f(\tau))\, d\tau = \sum_k \frac{x(\tau_k)}{|f'(\tau_k)|}. \tag{1.26}$$

This is the generalized sifting property of the impulse function. We can now express the delta function as

$$\delta(f(t)) = \sum_k \frac{\delta(t - t_k)}{|f'(\tau_k)|}, \tag{1.27}$$

which is composed of a set of impulses each corresponding to one of the roots of $f(t)$, weighted by the reciprocal of the absolute value of the derivative of the function evaluated at the root.

1.3 Relationship between complex exponentials and delta functions

Here we list a set of important formulas that will be used in the discussions of various forms of the Fourier transform in Chapters 3 and 4. These formulas show that the Kronecker and Dirac delta functions can be generated as the sum or integral of some forms of the general complex exponential function $e^{j2\pi ft} = e^{j\omega t}$. The proofs of these formulas are left as homework problems.

- I. Dirac delta as an integral of a complex exponential:

$$\begin{aligned}\int_{-\infty}^{\infty} e^{\pm j2\pi ft}\, dt &= \int_{-\infty}^{\infty} \cos(2\pi ft)\, dt \pm j\int_{-\infty}^{\infty} \sin(2\pi ft)\, dt \\ &= 2\int_{0}^{\infty} \cos(2\pi ft)\, dt = \delta(f) = 2\pi\delta(\omega).\end{aligned} \tag{1.28}$$

  Note that the integral of the odd function $\sin(2\pi ft)$ over all time $-\infty < t < \infty$ is zero, while the integral of the even function $\cos(2\pi ft)$ over all time is twice the integral over $0 < t < \infty$. Equation (1.28) can also be interpreted intuitively. The integral of any sinusoid over all time is always zero, except if $f = 0$ and $e^{\pm j2\pi ft} = 1$, then the integral becomes infinity. Alternatively, if we integrate the complex exponential with respect to frequency f, we get

$$\int_{-\infty}^{\infty} e^{\pm j2\pi ft}\, df = 2\int_{0}^{\infty} \cos(2\pi ft)\, df = \delta(t), \tag{1.29}$$

  which can also be interpreted intuitively as a superposition of uncountably infinite sinusoids with progressively higher frequency f. These sinusoids cancel each other at any time $t \neq 0$ except if $t = 0$ and $\cos(2\pi ft) = 1$ for all f, then their superposition becomes infinity.

- Ia. This formula is a variation of Eq. (1.28):

$$\int_0^{\infty} e^{\pm j2\pi ft}\, dt = \int_0^{\infty} e^{\pm j\omega t}\, dt = \frac{1}{2}\delta(f) \mp \frac{1}{j2\pi f} = \pi\delta(\omega) \mp \frac{1}{j\omega}. \quad (1.30)$$

Given the above, we can also get:

$$\int_{-\infty}^{0} e^{\pm j\omega t}\, dt = \int_0^{-\infty} e^{\pm j\omega t} d(-t) = \int_0^{\infty} e^{\mp j\omega t}\, dt$$
$$= \frac{1}{2}\delta(f) \pm \frac{1}{j2\pi f} = \pi\delta(\omega) \pm \frac{1}{j\omega}. \quad (1.31)$$

Adding the two equations above we get the same result as given in Eq. (1.28):

$$\int_{-\infty}^{\infty} e^{\pm j\omega t}\, dt = \int_{-\infty}^{0} e^{\pm j\omega t}\, dt + \int_0^{\infty} e^{\pm j\omega t}\, dt = \delta(f) = 2\pi\delta(\omega). \quad (1.32)$$

- II. Kronecker delta as an integral of a complex exponential:

$$\frac{1}{T}\int_T e^{\pm j2\pi kt/T}\, dt = \frac{1}{T}\int_T \cos(2\pi kt/T)\, dt \pm j\frac{1}{T}\int_T \sin(2\pi kt/T)\, dt$$
$$= \frac{1}{T}\int_T \cos(2\pi kt/T)\, dt = \delta[k]. \quad (1.33)$$

In particular, if $T = 1$ we have

$$\int_0^1 e^{\pm j2\pi kt}\, dt = \delta[k]. \quad (1.34)$$

- III. A train of Dirac deltas with period F as a summation of a complex exponential:

$$\frac{1}{F}\sum_{n=-\infty}^{\infty} e^{\pm j2\pi fn/F} = \frac{1}{F}\sum_{n=-\infty}^{\infty} \cos(2\pi fn/F) \pm j\frac{1}{F}\sum_{n=-\infty}^{\infty} \sin(2\pi fn/F)$$
$$= \frac{1}{F}\sum_{n=-\infty}^{\infty} \cos(2\pi fn/F) = \sum_{k=-\infty}^{\infty} \delta(f - kF) = \sum_{k=-\infty}^{\infty} 2\pi\delta(\omega - 2\pi kF). \quad (1.35)$$

In particular, if $F = 1$ we have

$$\sum_{n=-\infty}^{\infty} e^{\pm j2\pi fn} = \sum_{k=-\infty}^{\infty} \delta(f - k) = \sum_{k=-\infty}^{\infty} 2\pi\delta(\omega - 2\pi k). \quad (1.36)$$

- IIIa. This formula is a variation of Eq. (1.36):

$$\sum_{n=0}^{\infty} e^{\pm j2\pi fn} = \frac{1}{2}\sum_{k=-\infty}^{\infty} \delta(f - k) + \frac{1}{1 - e^{\pm j2\pi f}} = \sum_{k=-\infty}^{\infty} \pi\delta(\omega - 2\pi k) + \frac{1}{1 - e^{\pm j\omega}}. \quad (1.37)$$

Given the above, we can also get

$$\sum_{n=-\infty}^{-1} e^{\pm j2\pi fn} = \sum_{n=0}^{\infty} e^{\mp j2\pi fn} - 1 = \frac{1}{2}\sum_{k=-\infty}^{\infty} \delta(f-k) + \frac{1}{1-e^{\mp j2\pi f}} - 1$$
$$= \frac{1}{2}\sum_{k=-\infty}^{\infty} \delta(f-k) - \frac{1}{1-e^{\pm j2\pi f}}. \quad (1.38)$$

Adding the two equations above we get the same result as given in Eq. (1.36):

$$\sum_{n=-\infty}^{\infty} e^{\pm j2\pi fn} = \sum_{n=-\infty}^{-1} e^{\pm j2\pi fn} + \sum_{n=0}^{\infty} e^{\pm j2\pi fn}$$
$$= \sum_{k=-\infty}^{\infty} \delta(f-k) = 2\pi \sum_{k=-\infty}^{\infty} \delta(\omega - 2\pi k). \quad (1.39)$$

- IV. A train of Kronecker deltas with period N as a summation of complex exponential:

$$\frac{1}{N}\sum_{n=0}^{N-1} e^{\pm j2\pi nm/N} = \frac{1}{N}\sum_{n=0}^{N-1} \cos(2\pi nm/N) \pm \frac{j}{N}\sum_{n=0}^{N-1} \sin(2\pi nm/N)$$
$$= \frac{1}{N}\sum_{n=0}^{N-1} \cos(2\pi nm/N) = \sum_{k=-\infty}^{\infty} \delta[m-kN]. \quad (1.40)$$

1.4 Attributes of signals

A time signal can be characterized by the following parameters.

- The *energy* contained in a continuous signal $x(t)$ is

$$\mathcal{E} = \int_{-\infty}^{\infty} |x(t)|^2 \, dt, \quad (1.41)$$

or in a discrete signal $x[n]$, it is

$$\mathcal{E} = \sum_{n=-\infty}^{\infty} |x[n]|^2. \quad (1.42)$$

Note that $|x(t)|^2$ and $|x[n]|^2$ have different dimensions and they represent respectively the power and energy of the signal at the corresponding moment. If the energy contained in a signal is finite $\mathcal{E} < \infty$, then the signal is called an *energy signal* . A continuous energy signal is said to be *square-integrable*, and a discrete energy signal is said to be *square-summable*. All signals to be considered in the future, either continuous or discrete, will be assumed to be energy signals.

- The *average power* of the signal is

$$\mathcal{P} = \lim_{T\to\infty} \frac{1}{T} \int_0^{\mathrm{T}} |x(t)|^2 \, dt, \tag{1.43}$$

or for a discrete signal, it is

$$\mathcal{P} = \lim_{N\to\infty} \frac{1}{N} \sum_{n=1}^{N} |x[n]|^2. \tag{1.44}$$

If $\mathcal{E}$ of $x(t)$ is not finite but $\mathcal{P}$ is, then $x(t)$ is a *power signal.* Obviously, the average power of an energy signal is zero.
- The *cross-correlation* defined below measures the similarity between two signals as a function of the relative time shift:

$$\begin{aligned} r_{xy}(\tau) = x(t) \star y(t) &= \int_{-\infty}^{\infty} x(t)\, \overline{y}(t-\tau)\, dt = \int_{-\infty}^{\infty} x(t+\tau)\, \overline{y}(t)\, dt \\ &\neq \int_{-\infty}^{\infty} \overline{x}(t-\tau)\, y(t)\, dt = y(t) \star x(t) = r_{yx}(\tau). \end{aligned} \tag{1.45}$$

Note that the cross-correlation is not commutative. For a discrete signal, we have

$$r_{xy}[m] = x[n] \star y[n] = \sum_{n=-\infty}^{\infty} x[n]\, \overline{y}[n-m] = \sum_{n=-\infty}^{\infty} x[n+m]\, \overline{y}[n]. \tag{1.46}$$

In particular, when $x(t) = y(t)$ and $x[n] = y[n]$, the cross-correlation becomes the *autocorrelation*, which measures the self-similarity of the signal:

$$r_x(\tau) = \int_{-\infty}^{\infty} x(t)\overline{x}(t-\tau)\, dt = \int_{-\infty}^{\infty} x(t+\tau)\overline{x}(t)\, dt, \tag{1.47}$$

and

$$r_x[m] = \sum_{n=-\infty}^{\infty} x[n]\, \overline{x}[n-m] = \sum_{n=-\infty}^{\infty} x[n+m]\, \overline{x}[n]. \tag{1.48}$$

More particularly when $\tau = 0$ and $m = 0$ we have

$$r_x(0) = \int_{-\infty}^{\infty} |x(t)|^2\, dt \qquad r_x[0] = \sum_{n=-\infty}^{\infty} |x[n]|^2\, dt, \tag{1.49}$$

which represent the total energy contained in the signal.
- A random time signal $x(t)$ is also called a *stochastic process.* Its mean or expectation is (Appendix B):

$$\mu_x(t) = E[x(t)]. \tag{1.50}$$

The cross-covariance of two stochastic processes $x(t)$ and $y(t)$ is

$$\begin{aligned} Cov_{xy}(t,\tau) = \sigma_{xy}^2(t,\tau) &= E[(x(t) - \mu_x(t))\, (\overline{y}(\tau) - \overline{\mu}_y(\tau))] \\ &= E[x(t)\overline{y}(\tau)] - \mu_x(t)\overline{\mu}_y(\tau). \end{aligned} \tag{1.51}$$

- A stochastic process $x(t)$ can be truncated and sampled to become a random vector $\boldsymbol{x} = [x[1], \ldots, x[N]]^{\mathrm{T}}$. The mean or expectation of $\boldsymbol{x}$ is a vector:

$$\boldsymbol{\mu}_x = E[\boldsymbol{x}], \tag{1.52}$$

The nth element of $\boldsymbol{\mu}$ is $\mu[n] = E[x[n]]$. The cross-covariance of $\boldsymbol{x}$ and $\boldsymbol{y}$ is an N by N matrix:

$$\boldsymbol{\Sigma}_{xy} = E[(\boldsymbol{x} - \boldsymbol{\mu}_x)(\boldsymbol{y} - \boldsymbol{\mu}_y)^*] = E[\boldsymbol{x}\boldsymbol{y}^*] - \boldsymbol{\mu}_x\boldsymbol{\mu}_y^* \tag{1.53}$$

The mnth element of $\boldsymbol{\Sigma}_{xy}$ is

$$\sigma_{xy}^2[m, n] = E[(x[m] - \mu_x[m])(\overline{y}[n] - \overline{\mu}_y[n])] = E[x[m]\overline{y}[n]] - \mu_x[m]\overline{\mu}_y[n]. \tag{1.54}$$

In particular, when $x(t) = y(t)$ and $x[n] = y[n]$, the cross-covariance becomes autocovariance

$$\begin{aligned} Cov_x(t, \tau) = \sigma_x^2(t, \tau) &= E[(x(t) - \mu_x(t))\,(\overline{x}(\tau) - \overline{\mu}_x(\tau))] \\ &= E[x(t)\overline{x}(\tau)] - \mu_x(t)\overline{\mu}_x(\tau), \end{aligned} \tag{1.55}$$

and

$$\boldsymbol{\Sigma}_x = E[(\boldsymbol{x} - \boldsymbol{\mu}_x)(\boldsymbol{x} - \boldsymbol{\mu}_x)^*] = E[\boldsymbol{x}\boldsymbol{x}^*] - \boldsymbol{\mu}_x\boldsymbol{\mu}_x^*. \tag{1.56}$$

More particularly, when $t = \tau$ and $m = n$ we have

$$\sigma_x^2(t) = E[|x(t)|^2] - |\mu_x(t)|^2 \qquad \sigma_x^2[n] = E[|x[n]|^2] - |\mu_x[n]|^2. \tag{1.57}$$

We see that $\sigma_x^2(t)$ represents the average dynamic power of the signal $x(t)$, and $\sigma_x^2[n]$ represents the average dynamic energy contained in the nth signal component $x[n]$.

1.5 Signal arithmetics and transformations

Any of the arithmetic operations (addition/subtraction and multiplication/division) can be applied to two continuous signal $x(t)$ and $y(t)$, or two discrete signals $x[n]$ and $y[n]$, to produce a new signal $z(t)$ or $z[n]$:

- Scaling: $z(t) = ax(t)$ or $z[n] = ax[n]$.
- Addition/subtraction: $z(t) = x(t) \pm y(t)$ or $z[n] = x[n] \pm y[n]$.
- Multiplication: $z(t) = x(t)y(t)$ or $z[n] = x[n]y[n]$.
- Division: $z(t) = x(t)/y(t)$ or $z[n] = x[n]/y[n]$.

Note that these operations are actually applied to the amplitude values of the two signals $x(t)$ and $y(t)$ at each moment t, and the result becomes the value of $z(t)$ at the same moment; and the same is true for the operations on the discrete signals.

Moreover, a linear transformation in the general form of $y = ax + b = a(x + b/a)$ can be applied to the amplitude of a function $x(t)$ (vertical in the time plot) in two steps:

- Translation:
  $y(t) = x(t) + x_0$: the time function $x(t)$ is moved either upward if $x_0 > 0$ or downward if $x_0 < 0$.
- Scaling:
  $y(t) = ax(t)$: the time function $x(t)$ is either up-scaled if $|a| > 1$ or down-scaled if $|a| < 1$. $x(t)$ is also flipped vertically (upside-down) if $a < 0$.

The same linear transformation $y = ax + b$ can also be applied to the time argument t of the function $x(t)$ (horizontal in the time plot) as well as to its amplitude:

$$\tau = at + t_0 = a(t + t_0/a), \qquad y(\tau) = x(at + t_0) = x[a(t + t_0/a)]. \tag{1.58}$$

- Translation or shift:
  $y(t) = x(t + t_0)$ is translated by $|t_0|$ either to the right if $t_0 < 0$, or to the left if $t_0 > 0$.
- Scaling:
  $y(t) = x(at)$ is either compressed if $|a| > 1$, or expanded if $|a| < 1$. The signal is also reversed (flipped horizontally) in time if $a < 0$.

In general, the transformation in time $y(t) = x(at + t_0) = x(a(t + t_0/a))$ containing both translation and scaling can be carried out in either of the two methods.

1. A two-step process.
   - Step 1: define an intermediate signal $z(t) = x(t + t_0)$ due to translation.
   - Step 2: find the transformed signal $y(t) = z(at)$ due to time-scaling (containing time reversal if $a < 0$).

   The two steps can be carried out equivalently in reverse order.
   - Step 1: define an intermediate signal $z(t) = x(at)$ due to time-scaling (containing time reversal if $a < 0$).
   - Step 2: find the transformed signal $y(t) = z(t + t_0/a)$ due to translation.

   Note that the translation parameters (direction and amount) are different depending on whether the translation is carried out before or after scaling.
2. A two-point process:
   Evaluate $x(t)$ at two arbitrarily chosen time points $t = t_1$ and $t = t_2$ to get $v_1 = x(t_1)$ and $v_2 = x(t_2)$. Then $y(t) = x(at + t_0) = v_1$ when its argument is $at + t_0 = t_1$; i.e., when $t = (t_1 - t_0)/a$, and $y(t) = x(at + t_0) = v_2$ when $at + t_0 = t_1$, i.e., $t = (t_2 - t_0)/a$. As the transformation $at + t_0$ is linear, the value of $y(t)$ at any other time moment t can be found by linear interpolation based on these two points.

Example 1.1: Consider the transformation of a time signal

$$x(t) = \begin{cases} t & 0 < t < 2 \\ 0 & \text{else} \end{cases} . \tag{1.59}$$

- Translation: $y(t) = x(t+3)$ and $z(t) = x(t-1)$ are shown in Fig. 1.4(a).
- Expansion/compression: $y(t) = x(2t/3)$ and $z(t) = x(2t)$ are shown in Fig. 1.4(b).
- Time reversal: $y(t) = x(-t)$ and $z(t) = x(-2t)$ are shown in Fig. 1.4(c).
- Combination of translation, scaling, and reversal:

$$y(t) = x(-2t+3) = x\left[-2\left(t - \frac{3}{2}\right)\right] . \tag{1.60}$$

  - Method 1: based on the first expression $y(t) = x(-2t+3)$ we get (Fig. 1.4 (d)):

$$z(t) = x(t+3), \quad y(t) = z(-2t) \tag{1.61}$$

    Alternatively, based on the second expression $y(t) = x(-2(t-3/2))$ we get (Fig. 1.4 (e)):

$$z(t) = x(-2t), \quad y(t) = z\left(t - \frac{3}{2}\right) \tag{1.62}$$

  - Method 2: the signal has two break points at $t_1 = 0$ and $t_2 = 2$, correspondingly, the two break points for $y(t)$ can be found to be:

$$-2t + 3 = t_1 = 0 \quad \Longrightarrow \quad t = \frac{3}{2},$$
$$-2t + 3 = t_2 = 2 \quad \Longrightarrow \quad t = \frac{1}{2}.$$

    By linear interpolation based on these two points, the waveform of $y(t)$ can easily be obtained, which is the same as that obtained by the previous method shown in Fig. 1.4(d) and (e).

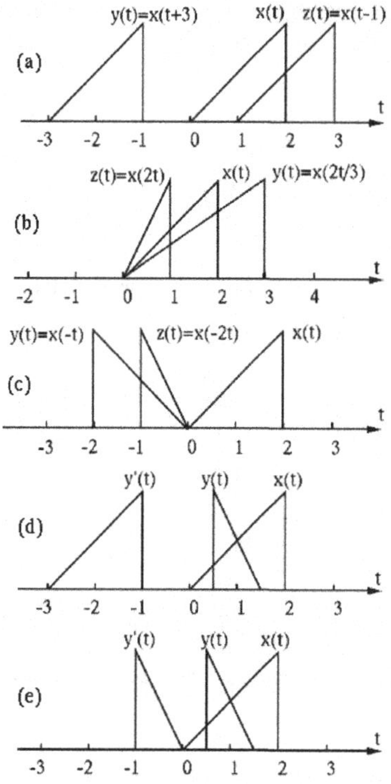

Figure 1.4 Transformation of continuous signal.

In the transformation of discrete signals, the expansion and compression for continuous signals are replaced respectively by *up-sampling* and *down-sampling.*

- **Down-sampling (decimation)**
  Keep every Nth sample and discard the rest. Signal size becomes $1/N$ of the original one:

$$x_{(N)}[n] = x[nN]. \tag{1.63}$$

  For example, if $N = 3$, $x_{(3)}[0] = x[0]$, $x_{(3)}[1] = x[3]$, $x_{(3)}[2] = x[6]$, ...
- **Up-sampling (interpolation by zero stuffing)**
  Insert $N-1$ zeros between every two consecutive samples $x[n]$ and $x[n+1]$. Signal size becomes N times the original one:

$$x^{(N)}[n] = \begin{cases} x[n/N] & n = 0, \pm N, \pm 2N, \dots \\ 0 & \text{else} \end{cases}. \tag{1.64}$$

  For example, if $N = 2$, $x^{(2)}[0] = x[0]$, $x^{(2)}[2] = x[1]$, $x^{(2)}[4] = x[2]$, ..., and $x[n] = 0$ for all other n.

Example 1.2: Given $x[n]$ as shown in Fig. 1.5(a), a transformation $y[n] = x[-n + 4]$, shown in Fig. 1.5(b), can be obtained based on two time points:

$$\begin{aligned} -n + 4 = 0 &\implies n = 4, \\ -n + 4 = 3 &\implies n = 1. \end{aligned} \tag{1.65}$$

The down- and up-sampling of the signal in Fig. 1.5(a) can be obtained from the following table and are shown in Fig. 1.5(c) and (d), respectively.

n	$\cdots$	-1	0	1	2	3	4	5	6	7	$\cdots$
$x[n]$	$\cdots$	0	1	2	3	4	0	0	0	0	$\cdots$
$x_{(2)}[n]$	$\cdots$	0	1	3	0	0	0	0	0	0	$\cdots$
$x^{(2)}[n]$	$\cdots$	0	1	0	2	0	3	0	4	0	$\cdots$

(1.66)

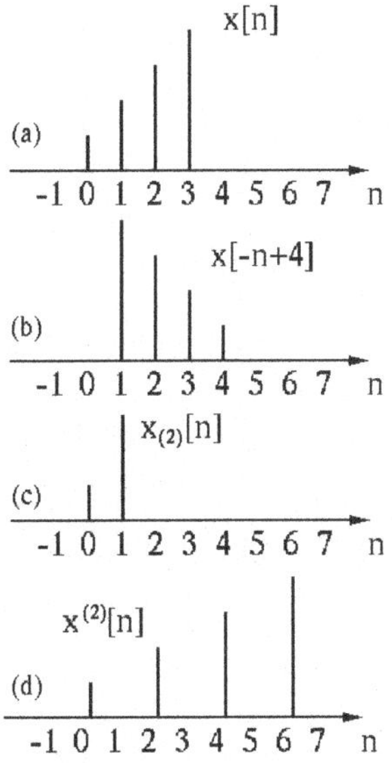

Figure 1.5 Transformation of discrete signal.

1.6 Linear and time-invariant systems

A generic system (electrical, mechanical, biological, economical, etc.) can be symbolically represented in terms of the relationship between its input $x(t)$ (stimulus, excitation) and output $y(t)$ (response, reaction):

$$\mathcal{O}[x(t)] = y(t), \tag{1.67}$$

where the symbol $\mathcal{O}[\ ]$ represents the operation applied by the system to its input. A system is *linear* if its input-output relationship satisfies both *homogeneity* and *superposition.*

- **Homogeneity**

$$\mathcal{O}\left[ax(t)\right] = a\mathcal{O}[x(t)] = ay(t), \tag{1.68}$$

- **Superposition**
  If $\mathcal{O}[x_n(t)] = y_n(t)$ $(n = 1, 2, \ldots, N)$, then

$$\mathcal{O}\left[\sum_{n=1}^{N} x_n(t)\right] = \sum_{n=1}^{N} \mathcal{O}[x_n(t)] = \sum_{n=1}^{N} y_n(t), \tag{1.69}$$

  or

$$\mathcal{O}\left[\int_{-\infty}^{\infty} x(t)\, dt\right] = \int_{-\infty}^{\infty} \mathcal{O}[x(t)]\, dt = \int_{-\infty}^{\infty} y(t)\, dt. \tag{1.70}$$

Combining these two properties, we have

$$\mathcal{O}\left[\sum_{n=1}^{N} a_n x_n(t)\right] = \sum_{n=1}^{N} a_n \mathcal{O}[x_n(t)] = \sum_{n=1}^{N} a_n y_n(t), \tag{1.71}$$

or

$$\mathcal{O}\left[\int_{-\infty}^{\infty} a(\tau)x(t,\tau)\, d\tau\right] = \int_{-\infty}^{\infty} a(\tau)\mathcal{O}[x(t,\tau)]\, d\tau = \int_{-\infty}^{\infty} a(\tau)y(t,\tau)\, d\tau. \tag{1.72}$$

A system is *time-invariant* if how it responds to the input does not change over time. In other words,

$$\text{if} \quad \mathcal{O}[x(t)] = y(t), \quad \text{then} \quad \mathcal{O}[x(t-\tau)] = y(t-\tau). \tag{1.73}$$

A *linear and time-invariant (LTI) system* is both linear and time-invariant.

As an example, we see that the response of an LTI system $y(t) = \mathcal{O}[x(t)]$ to $dx(t)/dt$ is $dy(t)/dt$:

$$\mathcal{O}\left[\frac{1}{\triangle}[x(t+\triangle) - x(t)]\right] = \frac{1}{\triangle}[y(t+\triangle t) - y(t)]. \tag{1.74}$$

Taking the limit $\triangle \to 0$, we get

$$\mathcal{O}\left[\frac{d}{dt}x(t)\right] = \mathcal{O}[\dot{x}(t)] = \frac{d}{dt}y(t) = \dot{y}(t). \tag{1.75}$$

Example 1.3: Determine if each of the following systems is linear.

- The input $x(t)$ is the voltage across a resistor R and the output $y(t)$ is the current through R:

$$y(t) = \mathcal{O}[x(t)] = \frac{x(t)}{R}. \tag{1.76}$$

  This is obviously a linear system.

- The input $x(t)$ is the voltage across a resistor R and the output $y(t)$ is the power consumed by R:

$$y(t) = \mathcal{O}[x(t)] = \frac{x^2(t)}{R}. \tag{1.77}$$

This is not a linear system.
- The input $x(t)$ is the voltage across a resistor R and a capacitor C in series and the output is the voltage across C:

$$RC\frac{d}{dt}y(t) + y(t) = \tau\frac{d}{dt}y(t) + y(t) = x(t). \tag{1.78}$$

where $\tau = RC$ is the *time constant* of the system. As the system is characterized by a linear, first-order ordinary differential equation (ODE), it is linear.
- A system produces its output $y(t)$ by adding a constant a to its input $x(t)$:

$$y(t) = \mathcal{O}[x(t)] = x(t) + a. \tag{1.79}$$

Consider

$$\begin{aligned} &\mathcal{O}[x_1(t) + x_2(t)] = x_1(t) + x_2(t) + a \\ \neq\ &\mathcal{O}[x_1(t)] + \mathcal{O}[x_2(t)] = x_1(t) + x_2(t) + 2a. \end{aligned} \tag{1.80}$$

This is not a linear system.
- The input $x(t)$ is the force f applied to a spring of length l_0 and spring constant k; the output is the length of the spring. According to Hooke's law, $\Delta l = -kf = -kx(t)$, we have

$$y(t) = l = l_0 + \Delta l = l_0 - kx(t). \tag{1.81}$$

This is not a linear system.
- As above, except the output $y(t) = l - l_0 = \Delta l$ is the displacement of the moving end of the spring:

$$y(t) = \Delta l = -kf = -kx(t). \tag{1.82}$$

This system is linear.

1.7 Signals through continuous LTI systems

If the input to an LTI system is an impulse $x(t) = \delta(t)$ at $t = 0$, then the response of the system, called the *impulse response function*, is

$$h(t) = \mathcal{O}[\delta(t)]. \tag{1.83}$$

We now show that, given the impulse response $h(t)$ of an LTI system, we can find its response to *any* input $x(t)$. First, according to Eq. (1.9), we can express the input as

$$x(t) = \int_{-\infty}^{\infty} x(\tau)\delta(t-\tau)\, d\tau. \tag{1.84}$$

As the system is linear and time-invariant; i.e., both Eqs. (1.72) and (1.73) hold, we have

$$\begin{aligned} y(t) &= \mathcal{O}[x(t)] = \mathcal{O}\left[\int_{-\infty}^{\infty} x(\tau)\delta(t-\tau)\, d\tau\right] \\ &= \int_{-\infty}^{\infty} x(\tau)\mathcal{O}[\delta(t-\tau)]\, d\tau = \int_{-\infty}^{\infty} x(\tau)h(t-\tau)]\, d\tau. \end{aligned} \tag{1.85}$$

This process is illustrated in Fig. 1.6.

The integration on the right-hand side of Eq. (1.85) is called the *continuous convolution* of $x(t)$ and $h(t)$, which is more generally defined as an operation of two continuous functions $x(t)$ and $y(t)$:

$$z(t) = x(t) * y(t) = \int_{-\infty}^{\infty} x(\tau)y(t-\tau)\, d\tau = \int_{-\infty}^{\infty} y(\tau)x(t-\tau)\, d\tau = y(t) * x(t). \tag{1.86}$$

Note that convolution is commutative; i.e., $x(t) * y(t) = y(t) * x(t)$. Also note that Eq. (1.9) can be written as $x(t) = x(t) * \delta(t)$, i.e., any function $x(t)$ convolved with $\delta(t)$ remains unchanged.

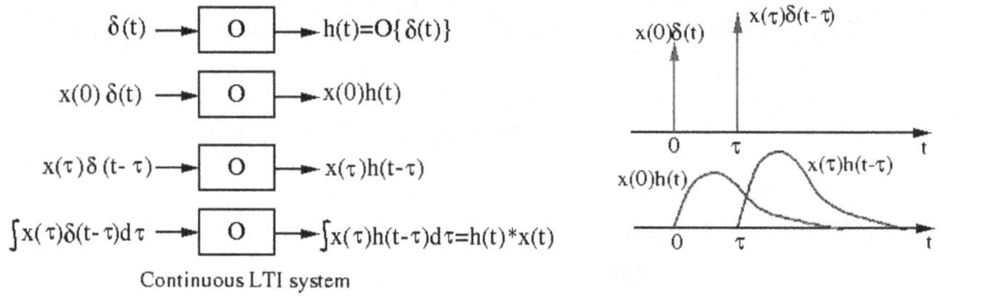

Figure 1.6 Response of a continuous LTI system.

In particular, if the input to an LTI system is a complex exponential function:

$$x(t) = e^{st} = e^{(\sigma+j\omega)t} = [\cos(\omega t) + j\, \sin(\omega t)]e^{\sigma t}, \tag{1.87}$$

where $s = \sigma + j\omega$ is a complex parameter, the corresponding output is

$$y(t) = \mathcal{O}[e^{st}] = \int_{-\infty}^{\infty} h(\tau)e^{s(t-\tau)}\, d\tau = e^{st}\int_{-\infty}^{\infty} h(\tau)e^{-s\tau}\, d\tau = H(s)e^{st} \tag{1.88}$$

where $H(s)$ is a constant (independent of the time variable t) defined as

$$H(s) = \int_{-\infty}^{\infty} h(\tau)e^{-s\tau}\, d\tau. \tag{1.89}$$

This is called the *transfer function (TF)* of the continuous LTI system, which is the *Laplace transform* of the impulse response function $h(t)$ of the system, to be discussed in Chapter 6. We can rewrite Eq. (1.88) as an *eigenequation*:

$$\mathcal{O}[e^{st}] = H(s)e^{st}, \tag{1.90}$$

where the constant $H(s)$ and the complex exponential e^{st} are, respectively, the *eigenvalue* and the corresponding *eigenfunction* of the LTI system, i.e., the response of the system to the complex exponential input e^{st} is equal to the input multiplied by a constant $H(s)$. Also note that the complex exponential e^{st} is the eigenfunction of *any* continuous LTI system, independent of its specific impulse response $h(t)$.

In particular, when $s = j\omega = j2\pi f$ ($\sigma = 0$), $H(s)$ becomes

$$H(j\omega) = \int_{-\infty}^{\infty} h(\tau)e^{-j\omega\tau}\, d\tau = \int_{-\infty}^{\infty} h(\tau)e^{-j2\pi f\tau}\, d\tau. \tag{1.91}$$

This is the *frequency response function (FRF)* of the system, which is the *Fourier transform* of the impulse response function $h(t)$, to be discussed in Chapter 3. Alternative notations such as $H(f)$ and $H(\omega)$ are also used for the FRF as a function of frequency f or angular frequency $\omega = 2\pi f$ in the various literature, depending on the convention adopted by the authors. We will use these notations interchangeably, whichever is most convenient and suitable in the specific discussion, as no confusion should be caused given the context.

Given the FRF $H(\omega)$ of a system, its response to a complex exponential $x(t) = e^{j\omega_0 t}$ with a specific frequency $\omega_0 = 2\pi f_0$ can be found by evaluating Eq. (1.88) at $s = j\omega_0$:

$$y(t) = \mathcal{O}[e^{j\omega_0 t}] = H(\omega_0)e^{j\omega_0 t} = H(f_0)e^{j2\pi f_0 t}. \tag{1.92}$$

Moreover, if an input $x(t)$ can be written as a linear combination of a set of complex exponentials:

$$x(t) = \sum_{k=-\infty}^{\infty} X[k]e^{jk\omega_0 t}, \tag{1.93}$$

where $X[k]$ is the weighting coefficient for the kth complex exponential of frequency $k\omega_0$, then, due to the linearity of the system, its output is

$$\begin{aligned} y(t) = \mathcal{O}[x(t)] &= \mathcal{O}\left[\sum_{k=-\infty}^{\infty} X[k]e^{jk\omega_0 t}\right] = \sum_{k=-\infty}^{\infty} X[k]\mathcal{O}[e^{jk\omega_0 t}] \\ &= \sum_{k=-\infty}^{\infty} X[k]H(k\omega_0)e^{jk\omega_0 t} = \sum_{k=-\infty}^{\infty} Y[k]e^{jk\omega_0 t}, \end{aligned} \tag{1.94}$$

where $Y[k] = X[k]H(k\omega_0)$ is the kth coefficient for the output.

The result can be further generalized to cover signals composed of uncountably infinite exponentials:

$$x(t) = \int_{-\infty}^{\infty} X(f)e^{j2\pi ft}\, df \tag{1.95}$$

where $X(f)$ is the weighting function for all exponentials with frequency in the range of $-\infty < f < \infty$. The output of the system is

$$\begin{aligned} y(t) &= \mathcal{O}[x(t)] = \mathcal{O}\left[\int_{-\infty}^{\infty} X(f)e^{j2\pi ft}\, df\right] = \int_{-\infty}^{\infty} X(f)\mathcal{O}[e^{j2\pi ft}]\, df \\ &= \int_{-\infty}^{\infty} X(f)H(f)e^{j2\pi ft}\, df = \int_{-\infty}^{\infty} Y(f)e^{j2\pi ft}\, df, \end{aligned} \tag{1.96}$$

where

$$Y(f) = X(f)H(f) \tag{1.97}$$

is the weighting function for the output.

In summary, the response $y(t)$ of an LTI system to an arbitrary input $x(t)$ can be obtained by two different but equivalent approaches. First, $y(t)$ can be obtained by the convolution in Eq. (1.85) based on the system's impulse response function $h(t)$. Second, $Y(f)$ can be obtained by the multiplication in Eq. (1.97) based on the system's frequency response function $H(f)$, when both the input and output are represented as a linear combination of a set of complex exponentials in Eqs. (1.95) and (1.96) in terms of $X(f)$ and $Y(f)$. This result is also an important conclusion of the continuous-time Fourier transform (CTFT) theory to be considered in Chapter 3.

An LTI system is *stable* if its response to any bounded input is also bounded:

$$\text{if} \quad |x(t)| < B_x \quad \text{then} \quad |y(t)| < B_y. \tag{1.98}$$

As the input and output of an LTI are related by convolution

$$y(t) = h(t) * x(t) = \int_{-\infty}^{\infty} h(\tau)x(t-\tau)\, d\tau, \tag{1.99}$$

we have

$$\begin{aligned} |y(t)| &= \left|\int_{-\infty}^{\infty} h(\tau)x(t-\tau)\, d\tau\right| \le \int_{-\infty}^{\infty} |h(\tau)||x(t-\tau)|\, d\tau \\ &< B_x \int_{-\infty}^{\infty} |h(\tau)| < B_y, \end{aligned} \tag{1.100}$$

which obviously requires

$$\int_{-\infty}^{\infty} |h(\tau)|\, d\tau < \infty. \tag{1.101}$$

In other words, if the impulse response function $h(t)$ of an LTI system is integrable, then the system is stable; i.e., Eq. (1.101) is the sufficient condition for an LTI system to be stable. We can show that this condition is also necessary; i.e., all stable LTI systems' impulse response functions are absolutely integrable.

An LTI system is *causal* if its output $y(t)$ only depends on the current and past input $x(t)$ (but not the future). If the system is initially at rest with zero output $y(t) = 0$ for $t < 0$, then its response $y(t) = h(t)$ to an impulse $x(t) = \delta(t)$

at moment $t = 0$ will be at rest before the moment $t = 0$; i.e., $h(t) = h(t)u(t)$. Its response to a general input $x(t)$ is

$$y(t) = h(t) * x(t) = \int_{-\infty}^{\infty} h(\tau)x(t-\tau)\,d\tau = \int_{0}^{\infty} h(\tau)x(t-\tau)\,d\tau. \tag{1.102}$$

Moreover, if the input begins at a specific moment; e.g., $t = 0$; i.e., $x(t) = x(t)u(t)$ and $x(t-\tau) = 0$ for $\tau > t$, then we have

$$y(t) = h(t) * x(t) = \int_{-\infty}^{\infty} h(\tau)x(t-\tau)\,d\tau = \int_{0}^{t} h(\tau)x(t-\tau)\,d\tau. \tag{1.103}$$

1.8 Signals through discrete LTI systems

Similar to the above discussion for continuous signals and systems, the following results can be obtained for discrete signals and systems. First, as shown in Eq. (1.3), any discrete signal can be written as

$$x[n] = \sum_{m=-\infty}^{\infty} x[m]\delta[n-m]. \tag{1.104}$$

Let the impulse response of a discrete LTI system be

$$h[n] = \mathcal{O}[\delta[n]], \tag{1.105}$$

then its response to the signal $x[n]$ is

$$\begin{aligned} y[n] &= \mathcal{O}[x[n]] = \mathcal{O}\left[\sum_{m=-\infty}^{\infty} x[m]\delta[n-m]\right] = \sum_{m=-\infty}^{\infty} x[m]\mathcal{O}[\delta[n-m]] \\ &= \sum_{m=-\infty}^{\infty} x[m]h[n-m] = \sum_{m=-\infty}^{\infty} x[n-m]h[m]. \end{aligned} \tag{1.106}$$

This process is illustrated in Fig. 1.7.

The last summation in Eq. (1.106) is called the *discrete convolution*, which is generally defined as an operation of two discrete functions $x[n]$ and $h[n]$

$$z[n] = x[n] * y[n] = \sum_{m=-\infty}^{\infty} x[m]y[n-m] = \sum_{m=-\infty}^{\infty} y[m]x[n-m] = y[n] * x[n]. \tag{1.107}$$

Note that convolution is commutative; i.e., $x[n] * y[n] = y[n] * x[n]$. Similar to the continuous case, if the system is causal and the input $x[n]$ is zero until $n = 0$, we have

$$y[n] = \sum_{m=0}^{n} x[m]h[n-m] = \sum_{m=0}^{n} x[n-m]h[m]. \tag{1.108}$$

Also note that Eq. (1.3) can be written as $x[n] = x[n] * \delta[n]$, i.e., any sequence $x[n]$ convolved with $\delta[n]$ remains unchanged.

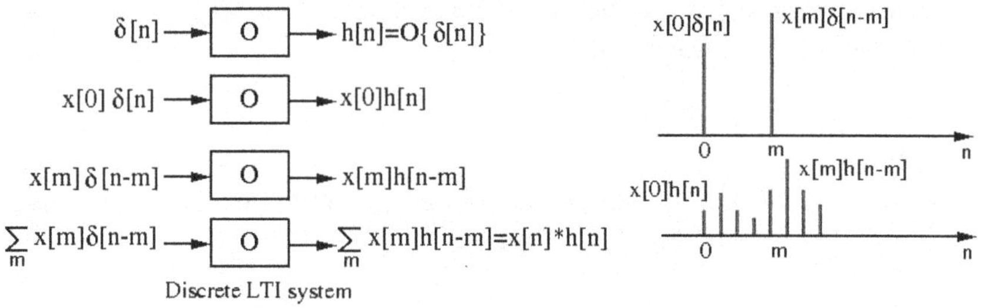

Figure 1.7 Response of a discrete LTI system.

In particular, if the input to an LTI system is a complex exponential function:

$$x[n] = e^{sn} = (e^s)^n = z^n, \tag{1.109}$$

where $s = \sigma + j\omega$ as defined above and z is defined as $z = e^s$, then according to Eq. (1.106), the corresponding output is

$$y[n] = \mathcal{O}[z^n] = \sum_{k=-\infty}^{\infty} h[k]z^{n-k} = z^n \sum_{k=-\infty}^{\infty} h[k]z^{-k} = H(z)z^n, \tag{1.110}$$

where $H(z)$ is a constant (independent of the time variable n) defined as

$$H(z) = \sum_{k=-\infty}^{\infty} h[k]z^{-k}. \tag{1.111}$$

This is called the *transfer function (TF)* of the discrete LTI system, which is the *Z-transform* of the impulse response $h[n]$ of the system, to be discussed in Chapter 6. We note that Eq. (1.110) is an eigenequation, where the constant $H(z)$ and the complex exponential z^n are, respectively, the eigenvalue and the corresponding eigenfunction of the LTI system. Also note that the complex exponential z^n is the eigenfunction of *any* discrete LTI system, independent of its specific impulse response $h[n]$. In particular, when $s = j\omega$ ($\sigma = 0$) and $z = e^s = e^{j\omega}$, $H(z)$ becomes

$$H(e^{j\omega}) = \sum_{k=-\infty}^{\infty} h[k]e^{-jk\omega} = \sum_{k=-\infty}^{\infty} h[k]e^{-j2k\pi f}. \tag{1.112}$$

This is the FRF of the system, which is the Fourier transform of the discrete impulse response function $h[n]$, to be discussed in Chapter 4. As in the continuous case, alternative notations such as $H(f)$ and $H(\omega)$ can also be used for the FRF as a function of frequency f or angular frequency $\omega = 2\pi f$ in the various literature, and we will use these notations interchangeably, whichever is most convenient and suitable in the specific discussion.

Given $H(e^{j\omega})$ of a discrete system, its response to a discrete input $x[n] = z^n = e^{j\omega_0 n}$ with a specific frequency $\omega_0 = 2\pi f_0$ can be found by evaluating Eq. (1.110) at $z = e^{j\omega_0}$:

$$y[n] = \mathcal{O}[e^{j\omega_0 n}] = H(e^{j\omega_0})e^{j\omega_0 n}. \tag{1.113}$$

Moreover, if the input $x[n]$ can be written as a linear combination of a set of complex exponentials:

$$x[n] = \sum_{k=0}^{N-1} X[k]e^{jk\omega_0 n/N}, \tag{1.114}$$

where $X[k]$ $(0 \leq k < N)$ are a set of constant coefficients, then, due to the linearity of the system, its output is

$$\begin{aligned} y[n] &= \mathcal{O}[x[n]] = \mathcal{O}\left[\sum_{k=0}^{N-1} X[k]e^{jk\omega_0 n/N}\right] = \sum_{k=0}^{N-1} X[k]\mathcal{O}[e^{jk\omega_0 n}] \\ &= \sum_{k=0}^{N-1} X[k]H(e^{jk\omega_0})e^{jk\omega_0 n} = \sum_{k=0}^{N-1} Y[k]e^{jk\omega_0 n}, \end{aligned} \tag{1.115}$$

where $Y[k] = X[k]H(e^{jk\omega_0})$ is the kth coefficient of the output. The result can be generalized to cover signals composed of uncountably infinite complex exponentials:

$$x[n] = \int_0^F X(f)e^{j2\pi fn/F}\, df, \tag{1.116}$$

where $X(f)$ is the weighting function for all exponentials with frequencies in the range of $0 < f < F$. The output of the system is

$$\begin{aligned} y[n] &= \mathcal{O}[x[n]] = \mathcal{O}\left[\int_0^F X(f)e^{j2\pi fn/F}\, df\right] = \int_0^F X(f)\mathcal{O}[e^{j2\pi fn/F}]\, df \\ &= \int_0^F X(f)H(e^{j2\pi fn/F})e^{j2\pi fn/F}\, df = \int_0^F Y(f)e^{j2\pi fn/F}\, df, \end{aligned} \tag{1.117}$$

where

$$Y(f) = X(f)H(e^{j2\pi fn/F}) \tag{1.118}$$

is the weighting function for the output.

In summary, the response $y[n]$ of a discrete LTI system to an arbitrary input $x[n]$ can be obtained by two different but equivalent approaches. First, $y[n]$ can be obtained by the convolution in Eq. (1.106) based on the system's impulse response function $h[n]$. Second, $Y(f)$ can be obtained by the multiplication in Eq. (1.118) based on the system's frequency response function $H(e^{j\omega})$, when both the input and output are represented as a linear combination of a set of complex exponentials in Eqs. (1.116) and (1.117). This result is also an important conclusion of the discrete-time Fourier transform(DTFT) theory to be considered in Chapter 4.

Similar to a stable continuous LTI system, a stable discrete LTI system's response to any bounded input is also bounded for all n:

$$\text{if } |x[n]| < B_x \quad \text{then} \quad |y[n]| < B_y. \tag{1.119}$$

As the output and input of an LTI is related by convolution

$$y[n] = h[n] * x[n] = \sum_{m=-\infty}^{\infty} h[m]x[n-m], \tag{1.120}$$

we have

$$\begin{aligned} |y[n]| &= \Big|\sum_{m=-\infty}^{\infty} h[m]x[n-m]\Big| \le \sum_{m=-\infty}^{\infty} |h[m]||x[n-n]| \\ &< B_x \sum_{m=-\infty}^{\infty} |h[m]|\, d\tau < B_y. \end{aligned} \tag{1.121}$$

which obviously requires

$$\sum_{m=-\infty}^{\infty} |h[m]| < \infty. \tag{1.122}$$

In other words, if the impulse response function $h[n]$ of an LTI system is absolutely summable, then the system is stable; i.e., Eq. (1.122) is the sufficient condition for an LTI system to be stable. We can show that this condition is also necessary; i.e., all stable LTI systems' impulse response functions are absolutely summable.

Also, a discrete LTI system is causal if its output $y[n]$ only depends on the current and past input $x[n]$ (but not the future). Assuming the system is initially at rest with zero output $y[n] = 0$ for $n < 0$, then its response $y[n] = h[n]$ to an impulse $x[n] = \delta[n]$ at moment $n = 0$ will be at rest before the moment $n = 0$; i.e., $h[n] = h[n]u[n]$. Its response to a general input $x[n]$ is

$$y[n] = h[n] * x[n] = \sum_{m=-\infty}^{\infty} h[m]x[n-m] = \sum_{m=0}^{\infty} h[m]x[n-m]. \tag{1.123}$$

Moreover, if the input begins at a specific moment; e.g., $n = 0$; i.e., $x[n] = x[n]u[n]$ and $x[n-m] = 0$ for $m > n$, then we have

$$y[n] = h[n] * x[n] = \sum_{m=-\infty}^{\infty} h[m]x[n-m] = \sum_{m=0}^{n} h[m]x[n-m]. \tag{1.124}$$

1.9 Continuous and discrete convolutions

The continuous and discrete convolutions defined respectively in Eqs. (1.86) and (1.107) are of great importance in the future discussions. Here, we further consider how these convolutions can be specifically carried out. First, we reconsider the continuous convolution

$$y(t) = x(t) * h(t) = \int_{-\infty}^{\infty} x(\tau)h(t-\tau)\, d\tau, \tag{1.125}$$

which can be carried out conceptually in the following three steps:

1. Find the time reversal of one of the two functions, say $h(\tau)$, by flipping it in time to get $h(-\tau)$.
2. Slide this flipped function along the τ axis to get $h(t-\tau)$ as the shift amount t goes from $-\infty$ to ∞.
3. For each shift amount t, find the integral of $x(\tau)h(t-\tau)$ over all τ, the area of overlap between $x(\tau)$ and $h(t-\tau)$, which is the convolution $y(t) = x(t) * h(t)$ at moment t.

This process is illustrated in the following example and in Fig. 1.8.

Although cross-correlation (Eq. (1.45)) and convolution are two different operations, they look similar and are closely related. If we flip one of the two functions in a convolution, it becomes the same as the cross-correlation.

$$x(t) * y(-t) = \int_{-\infty}^{\infty} x(\tau)y(\tau - t)\, d\tau = r_{xy}(t) = x(t) \star y(t). \tag{1.126}$$

In other words, if one of the signals $y(t) = y(-t)$ is even, then the two operations are the same $x(t) * y(t) = x(t) \star y(t)$.

Example 1.4: Let $x(t) = u(t)$ be the input to an LTI system with impulse response function $h(t) = e^{-at}u(t)$ (a first order system to be considered in Example 5.2), the output $y(t)$ of the system is

$$\begin{aligned} y(t) &= h(t) * x(t) = \int_0^t h(t-\tau)\, d\tau = \int_0^t e^{-a(t-\tau)}\, d\tau \\ &= \frac{1}{a}e^{-at}e^{a\tau}\Big|_0^t = \frac{1}{a}e^{-at}(e^{at}-1) = \frac{1}{a}(1-e^{-at}) \qquad (t>0). \end{aligned} \tag{1.127}$$

The result can be written as $h(t) = \frac{1}{a}(1-e^{-at})u(t)$ as it is zero when $t<0$. Alternatively, the convolution can also be written as

$$\begin{aligned} y(t) = x(t) * h(t) &= \int_{-\infty}^{\infty} h(\tau)x(t-\tau)\, d\tau = \int_0^t h(\tau)\, d\tau = \int_0^t e^{-a\tau}\, d\tau \\ &= -\frac{1}{a}e^{-a\tau}\Big|_0^t = \frac{1}{a}(1-e^{-at})u(t). \end{aligned} \tag{1.128}$$

Moreover, if the input is

$$x(t) = u(t) - u(t-\tau) = \begin{cases} 1 & 0 \le t < \tau \\ 0 & \text{else} \end{cases}. \tag{1.129}$$

Then, owing to the previous result and the linearity of the system, its output is

$$\begin{aligned} y(t) &= h(t) * [u(t) - u(t-\tau)] = h(t) * u(t) - h(t) * u(t-\tau) \\ &= \frac{1}{a}[(1-e^{-at})u(t) - (1-e^{-a(t-\tau)})u(t-\tau)]. \end{aligned} \tag{1.130}$$

This result is shown in Fig. 1.9

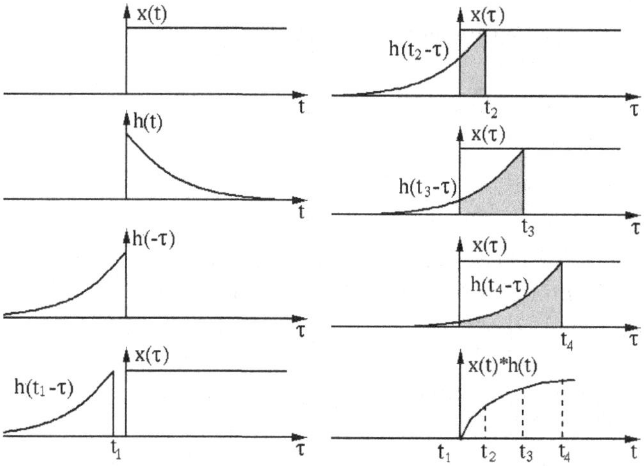

Figure 1.8 Continuous convolution. The three steps are shown top-down, then left to right. The shaded area represents the convolution evaluated at a specific time moment such as $t = t_2$, $t = t_3$, and $t = t_4$.

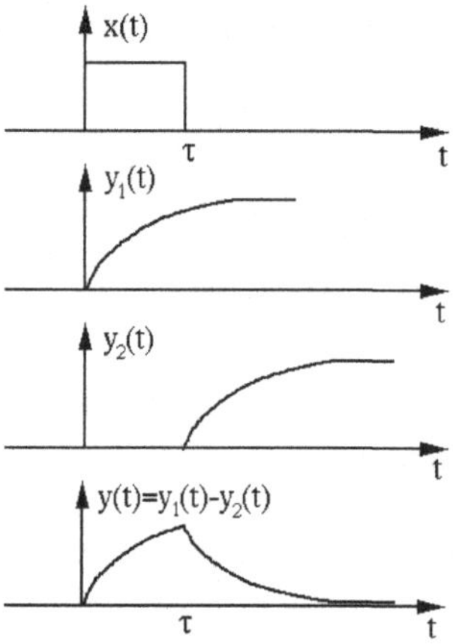

Figure 1.9 The linearity of convolution. Given $y_1(t) = h(t) * u(t)$ and $y_2(t) = h(t) * u(t - \tau)$, then $h(t) * [u(t) - u(t - \tau)] = y_1(t) - y_2(t)$.

Example 1.5: Let $x(t) = e^{-at}u(t)$ and $y(t) = e^{-bt}u(t)$, and both a and b are positive. We first find their convolution:

$$x(t) * y(t) = \int_{-\infty}^{\infty} x(\tau)y(t - \tau)\, d\tau. \tag{1.131}$$

As $y(t-\tau)$ can be written as

$$y(t-\tau) = e^{-b(t-\tau)} u(t-\tau) = \begin{cases} e^{-b(t-\tau)} & \tau < t \\ 0 & \tau > t \end{cases}, \tag{1.132}$$

we have

$$\begin{aligned} x(t) * y(t) &= \int_0^t e^{-at} e^{-b(t-\tau)} \, d\tau = e^{-bt} \int_0^t e^{-(a-b)\tau} \, d\tau = \frac{1}{a-b}(e^{-bt} - e^{-at}) \\ &= \frac{1}{b-a}(e^{-at} - e^{-bt}) = y(t) * x(t). \end{aligned}$$

Next we find the cross-correlation $x(t) \star y(t)$:

$$x(t) \star y(t) = \int_{-\infty}^{\infty} x(\tau) y(\tau - t) \, d\tau. \tag{1.133}$$

Consider two cases:

- When $t > 0$, the above becomes

$$\int_t^{\infty} e^{-a\tau} e^{-b(\tau-t)} \, d\tau = e^{bt} \int_t^{\infty} e^{-(a+b)\tau} \, d\tau = \frac{e^{-at}}{a+b} u(t). \tag{1.134}$$

- When $t < 0$, the above becomes

$$\int_0^{\infty} e^{-a\tau} e^{-b(\tau-t)} \, d\tau = e^{bt} \int_0^{\infty} e^{-(a+b)\tau} \, d\tau = \frac{e^{bt}}{a+b} u(-t). \tag{1.135}$$

Combining these two cases, we have

$$x(t) \star y(t) = \frac{1}{a+b} \begin{cases} e^{-at} u(t) & t > 0 \\ e^{bt} u(-t) & t < 0 \end{cases}. \tag{1.136}$$

We next consider the discrete convolution:

$$y[n] = x[n] * h[n] = \sum_{m=-\infty}^{\infty} x[m] h[n-m], \tag{1.137}$$

which can be carried out by the following three steps similar to those for the continuous convolution above:

1. Find the time reversal of one of the two functions, say, $h[m]$, by flipping it in time to get $h[-m]$;
2. Slide this flipped function along the m axis to get $h[n-m]$ as the shift amount n goes from $-\infty$ to ∞;
3. For each shift amount n, find the sum of $x[m]h[n-m]$ over all m, which is the convolution $y[n] = x[n] * h[n]$ at n.

This process is illustrated in the following example and in Fig. 1.10.

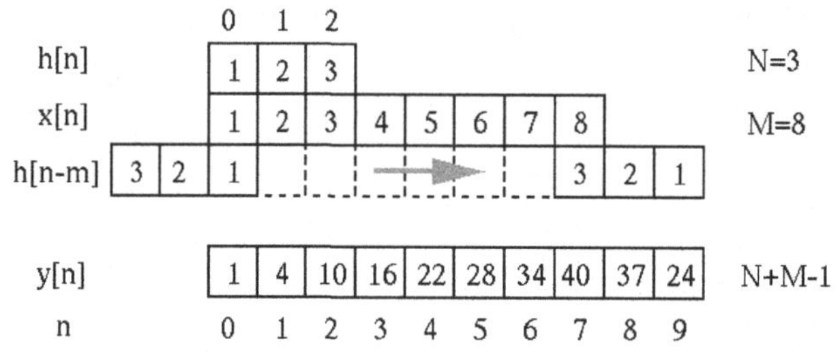

Figure 1.10 Discrete convolution.

Example 1.6: Eq. (1.139) shows the convolution of two finite discrete signals $x[n]$ of size $M = 8$ and $h[n]$ of size $N = 3$ where

$$\boldsymbol{x} = [1\ 2\ 3\ 4\ 5\ 6\ 7\ 8]^{\mathrm{T}}, \qquad \boldsymbol{h} = [1\ 2\ 3]^{\mathrm{T}}. \tag{1.138}$$

Note that $x[n] = 0$ outside the range of $0 \le n \le M - 1 = 7$ and $h[n] = 0$ outside the range of $0 \le n \le N - 1 = 2$. Consequently, in Eq. (1.137), $x[m] = 0$ outside the range $0 \le m \le M - 1 = 7$ and $h[n - m] = 0$ outside the range of $0 \le n - m \le N - 1 = 2$. Combining these two ranges for m and $n - m$, we get the range $0 \le m \le n \le N + m - 1 \le N + M - 2 = 9$ for the output $y[n]$; i.e., outside the range $0 \le n \le 9$, $y[n] = 0$.

m	···	−2	−1	0	1	2	3	4	5	6	7	8	9	10	···
$x[m]$	···	0	0	1	2	3	4	5	6	7	8	0	0	0	···
$h[-1-m]$	···	2	1												···
$h[0-m]$	···	3	2	1											···
$h[1-m]$	···		3	2	1										···
$h[2-m]$	···			3	2	1									···
$h[3-m]$	···				3	2	1								···
$h[4-m]$	···					3	2	1							···
$h[5-m]$	···						3	2	1						···
$h[6-m]$	···							3	2	1					···
$h[7-m]$	···								3	2	1				···
$h[8-m]$	···									3	2	1			···
$h[9-m]$	···										3	2	1		···
$h[10-m]$	···											3	2	1	···
$y[n]$	···	0	0	1	4	10	16	22	28	34	40	37	24	0	···

(1.139)

For example, when $n = 2$, we have

$$\begin{aligned} y[2] &= \sum_{m=-\infty}^{\infty} h[2-m]x[m] = h[2]x[0] + h[1]x[1] + h[0]x[2] \\ &= 3 \times 1 + 2 \times 2 + 1 \times 3 = 10. \end{aligned} \tag{1.140}$$

Example 1.7: Let $x[n] = u[n]$ be the input to a discrete LTI system with impulse response $h[n] = a^n u[n]$ ($|a| < 1$), the output $y[n]$ is the following convolution (illustrated in Fig. 1.11):

$$\begin{aligned} y[n] &= h[n] * x[n] = \sum_{m=-\infty}^{\infty} h[m]x[n-m] = \sum_{m=0}^{n} h[m] \\ &= \sum_{m=0}^{n} a^m = \frac{1-a^{n+1}}{1-a}. \end{aligned} \tag{1.141}$$

Alternatively, the convolution can also be written as

$$\begin{aligned} y[n] = x[n] * h[n] &= \sum_{m=-\infty}^{\infty} x[m]h[n-m] = \sum_{m=0}^{n} h[n-m] \\ &= a^n \sum_{m=0}^{n} a^{-m} = a^n \frac{1-a^{-(n+1)}}{1-a^{-1}} = \frac{1-a^{n+1}}{1-a}. \end{aligned} \tag{1.142}$$

If $a = 1/2$, then the output $y[n]$ is $[\ldots, 0, 1, 3/2, 7/4, 15/8, \ldots]$, and when $n \to \infty$, $y[n] \to 1/(1-a) = 2$, as shown in the bottom panel of Fig. 1.11.

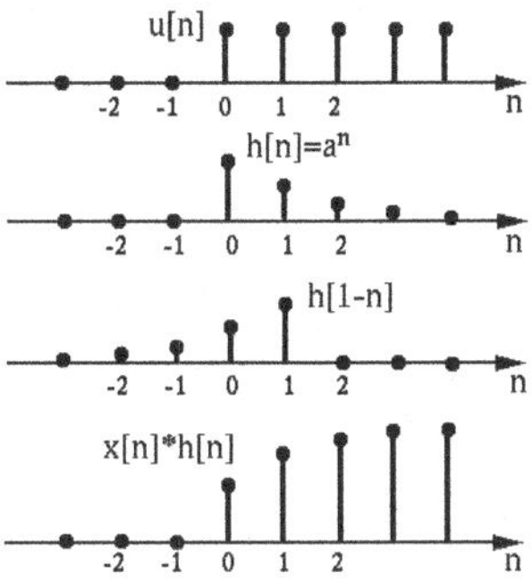

Figure 1.11 Discrete convolution of $u[n]$ and $a^n u[n]$.

1.10 Homework problems

1. Given the two square impulses as shown below:

$$r_a(t) = \begin{cases} 1 & |t| < a/2 \\ 0 & \text{else} \end{cases} \qquad r_b(t) = \begin{cases} 1 & |t| < b/2 \\ 0 & \text{else} \end{cases}, \tag{1.143}$$

where we assume $b > a$, find their convolution $x(t) = r_a(t) * r_b(t)$ in analytical form (piecewise functions; i.e., one expression for one particular time interval) as well as graphic form.

2. Given the triangle wave, an isosceles triangle

$$s_a(t) = \begin{cases} 1 + t/a & -a < t < 0 \\ 1 - t/a & 0 < t < a \end{cases}, \tag{1.144}$$

find the convolution $s_a(t) * s_a(t)$ in analytical form (piecewise function) as well as graphic form.

3. Prove the identity in Eq. (1.28):

$$\int_{-\infty}^{\infty} e^{\pm j2\pi ft}\, dt = \delta(f). \tag{1.145}$$

Hint: Follow these steps:

(a) Change the lower and upper integral limits to $-a/2$ and $a/2$, respectively, and show that this definite integral results in a sinc function $a\ \mathrm{sinc}(af)$ of frequency f with a parameter a. A sinc function is defined as $\mathrm{sinc}(x) = \sin(\pi x)/\pi x$, and $\lim_{x\to 0}\mathrm{sinc}(x) = 1$.

(b) Show that the following integral of this sinc function $a\ \mathrm{sinc}(af)$ is 1 (independent of a):

$$a\int_{-\infty}^{\infty} \mathrm{sinc}(af)\, df = 1, \tag{1.146}$$

based on the integral formula

$$\int_{0}^{\infty} \frac{\sin(x)}{x} dx = \frac{\pi}{2}. \tag{1.147}$$

(c) Let $a \to \infty$ and show that $a\ \mathrm{sinc}(af)$ approaches a unit impulse:

$$\lim_{a\to\infty} s(f,a) = \delta(f). \tag{1.148}$$

4. Prove the identity in Eq. (1.30):

$$\int_{0}^{\infty} e^{\pm j2\pi ft}\, dt = \frac{1}{2}\delta(f) \mp \frac{1}{j2\pi f} = \pi\delta(\omega) \mp \frac{1}{j\omega}. \tag{1.149}$$

Hint: Follow these steps:

(a) Introduce an extra term e^{-at} with a real parameter $a > 0$ so that the integrand becomes $e^{-(a+j\omega)t}$ and the integral can be carried out. Note that we cannot take the limit $a \to 0$ for the integral result due to the singularity at $f = 0$.

(b) Take the limit $a \to 0$ on the imaginary part, which is odd without singularity at $f = 0$.

(c) Take the limit on the real part, which is even with a singularity at $f = 0$. However, show this impulse is one half of Dirac delta as its integral over $-\infty < f < \infty$ is $1/2$. You may need to use this integral:

$$\int \frac{1}{a^2 + x^2} dx = \frac{1}{a}\tan^{-1}\left(\frac{x}{a}\right). \tag{1.150}$$

5. Prove the identity in Eq. (1.33):

$$\frac{1}{T}\int_{T} e^{\pm j2\pi kt/T}\, dt = \delta[k]. \tag{1.151}$$

Hint: Use Euler's formula to represent the integrand as

$$e^{\pm j2\pi kt/T} = \cos\left(\frac{2\pi t}{T/k}\right) \pm j\sin\left(\frac{2\pi t}{T/k}\right). \tag{1.152}$$

6. Prove the identity in Eq. (1.35):

$$\frac{1}{F}\sum_{k=-\infty}^{\infty} e^{\pm j2k\pi f/F} = \sum_{n=-\infty}^{\infty} \delta(f - nF). \tag{1.153}$$

Hint: Follow these steps:

(a) Find the summation of the following series:

$$\begin{aligned}\sum_{k=-\infty}^{\infty} (a\,e^{x})^k &= \sum_{k=0}^{\infty}(a\,e^{x})^k + \sum_{k=-\infty}^{0} (a\,e^{x})^k - 1 \\ &= \sum_{k=0}^{\infty}(a\,e^{x})^k + \sum_{k=0}^{\infty}(a\,e^{-x})^k - 1,\end{aligned} \tag{1.154}$$

based on the power series formula for $|a| < 1$:

$$\sum_{k=0}^{\infty}(a\,e^{x})^k = \frac{1}{1 - a\,e^{x}}. \tag{1.155}$$

(b) Show that when $a = 1$ the sum above is zero if $f \neq nF$ but infinity when $f = nF$, for any integer n; i.e., the sum is a train of impulses.

(c) Show that each impulse is a Dirac delta, a unit impulse, as its integral over the period of F with respect to f is 1. Here, the result of the previous problem may be needed.

7. Prove the identity in Eq. (1.37):

$$\begin{aligned}\sum_{m=0}^{\infty} e^{-j2\pi fm} &= \frac{1}{2}\sum_{k=-\infty}^{\infty} \delta(f - k) + \frac{1}{1 - e^{-j2\pi f}} \\ &= \sum_{k=-\infty}^{\infty} \pi\delta(\omega - 2k\pi) + \frac{1}{1 - e^{-j\omega}}.\end{aligned} \tag{1.156}$$

Hint: Follow these steps:

(a) Introduce an extra term a^n with a real parameter $0 < a < 1$ so that the summation term becomes $(a\,e^{-j\omega})^n$ and the summation can be carried out. Note that we cannot take the limit $a \to 1$ directly on the result due to the singularity at $f = k$ ($\omega = 2k\pi$) for any integer value of k.

(b) Take the limit $a \to 0$ on the imaginary part, which is odd without singularity at $f = k$.

(c) Take the limit on the real part, which is even with a singularity at $f = k$. However, show each impulse is one half of Dirac delta as its integral over $-1/2 < f - k < 1/2$ is $1/2$. You may need to use this integral:

$$\int \frac{dx}{a^2 + b^2 - 2ab\cos x} = \frac{2}{a^2 - b^2}\tan^{-1}\left[\frac{a+b}{a-b}\tan\left(\frac{x}{2}\right)\right]. \tag{1.157}$$

8. Prove the identity in Eq. (1.40):

$$\frac{1}{N}\sum_{n=0}^{N-1} e^{\pm j2\pi nm/N} = \sum_{k=-\infty}^{\infty} \delta[m-kN]. \tag{1.158}$$

Hint: Consider the summation on the left-hand side in the following two cases to show that:
(a) If $m = kN$ for any integer value of k, the summation is 1;
(b) If $m \neq kN$, the summation is 0, based on the formula of geometric series:

$$\sum_{n=0}^{N-1} x^n = \frac{1-x^N}{1-x}. \tag{1.159}$$

9. Consider the three signals $x(t)$, $y(t)$, and $z(t)$ in Fig. 1.12.

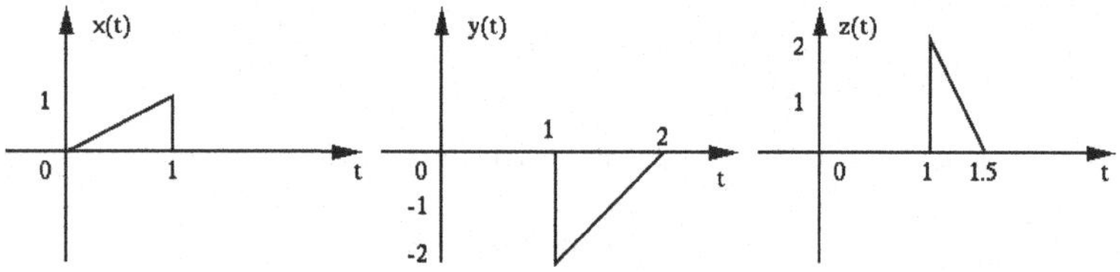

Figure 1.12 Orthogonal projection.

- Give the expressions for $y(t)$ in terms of $x(t)$.
- Give the expressions for $z(t)$ in terms of $x(t)$.
- Give the expressions for $y(t)$ in terms of $z(t)$.
- Give the expressions for $z(t)$ in terms of $y(t)$.

10. Let $\boldsymbol{x} = [1, 1, -1, -1, 1, 1, -1, -1]^{\mathrm{T}}$ be the input to an LTI system with impulse response $\boldsymbol{h} = [1, 2, 3]^{\mathrm{T}}$. Find the output $y[n] = h[n] * x[n]$. Write a Matlab program to confirm your result.
Note that given the input $x[n]$ and the corresponding output $y[n]$, it is difficult to find $h[n]$, similarly, given the output $y[n]$ and the impulse response $h[n]$, it is also difficult to find the input $x[n]$. As we will see later, such difficulties can be resolved by the Fourier transform method in the frequency domain.

11. The impulse response $h(t)$ of an LTI system is shown in Fig. 1.13, and the input signal is $x(t) = \sum_{k=-\infty}^{\infty} \delta(t - kT)$. Draw the system's response $y(t) = h(t) * x(t)$ when T takes each of these values: $T = 2$, $T = 1$, $T = 2/3$, $T = 1/2$, and $T = 1/3$ ($x(t)$ shown in the figure is only for the case when $T = 1$).

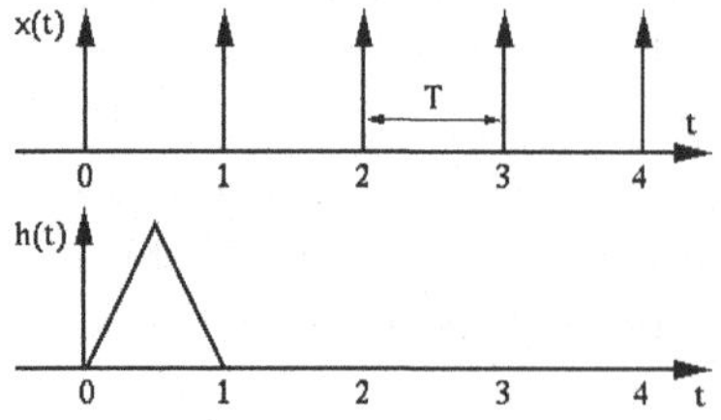

Figure 1.13 Impulse and input of an LTI system.

12. The impulse response of an LTI system is

$$h(t) = \begin{cases} 1 & 0 < t < T \\ 0 & \text{else} \end{cases}. \tag{1.160}$$

Find the response of the system to an input $x(t) = \cos(2\pi f t)$, and then write a Matlab program to confirm your result.

13. The impulse response of a discrete LTI system is $h[n] = a^n u[n]$ with $|a| < 1$ and the input is $x[n] = \cos(2\pi n f_0)$. Find the corresponding output $y[n] = h[n] * x[n]$.

<u>Hint:</u> When needed, any complex expression (such as $1/(1 - a\,e^{j2\pi f_0})$) can be represented in polar form $r\,e^{j\theta}$. But the magnitude r and angle θ need to be expressed in terms of the given parameters (such as a and f_0).

2 Vector spaces and signal representation

In this chapter we discuss some basic concepts of Hilbert space and the related operations and properties as the mathematical foundation for the topics of the subsequent chapters. Specifically, based on the concept of unitary transformation in a Hilbert space, all of the unitary transform methods to be specifically considered in the following chapters can be treated from a unified point of view: they are just a set of different rotations of the standard basis of the Hilbert space in which a given signal, as a vector, resides. By such a rotation the signal can be better represented in the sense that the various signal processing needs, such as noise filtering, information extraction and data compression, can all be carried out more effectively and efficiently.

2.1 Inner product space

2.1.1 Vector space

In our future discussion, any signal, either a continuous one represented as a time function $x(t)$, or a discrete one represented as a vector $\boldsymbol{x} = [\ldots, x[n], \ldots]^{\mathrm{T}}$, will be considered as a *vector* in a *vector space*, which is just a generalization of the familiar concept of N-dimensional (N-D) space, formally defined as below.

Definition 2.1. *A vector space is a set V with two operations of addition and scalar multiplication defined for its members, referred to as vectors.*

1. *Vector addition maps any two vectors $\boldsymbol{x}, \boldsymbol{y} \in V$ to another vector $\boldsymbol{x} + \boldsymbol{y} \in V$ satisfying the following properties:*
   - *Commutativity: $\boldsymbol{x} + \boldsymbol{y} = \boldsymbol{y} + \boldsymbol{x}$.*
   - *Associativity: $\boldsymbol{x} + (\boldsymbol{y} + \boldsymbol{z}) = (\boldsymbol{x} + \boldsymbol{y}) + \boldsymbol{z}$.*
   - *Existence of zero: there is a vector $\boldsymbol{0} \in V$ such that: $\boldsymbol{0} + \boldsymbol{x} = \boldsymbol{x} + \boldsymbol{0} = \boldsymbol{x}$.*
   - *Existence of inverse: for any vector $\boldsymbol{x} \in V$, there is another vector $-\boldsymbol{x} \in V$ such that $\boldsymbol{x} + (-\boldsymbol{x}) = \boldsymbol{0}$.*
2. *Scalar multiplication maps a vector $\boldsymbol{x} \in V$ and a real or complex scalar $a \in \mathbb{C}$ to another vector $a\boldsymbol{x} \in V$ with the following properties:*
   - $a(\boldsymbol{x} + \boldsymbol{y}) = a\boldsymbol{x} + a\boldsymbol{y}$.
   - $(a + b)\boldsymbol{x} = a\boldsymbol{x} + b\boldsymbol{x}$.
   - $ab\boldsymbol{x} = a(b\boldsymbol{x})$.
   - $1\boldsymbol{x} = \boldsymbol{x}$.

Listed below is a set of typical vector spaces for various types of signal of interest.

- N-D vector space $\mathbb{R}^N$ or $\mathbb{C}^N$
  This space contains all N-D vectors expressed as an N-tuple, an ordered list of N elements (or components):
$$\boldsymbol{x} = \begin{bmatrix} x[1] \\ x[2] \\ \vdots \\ x[N] \end{bmatrix} = [x[1], x[2], \ldots, x[N]]^{\mathrm{T}}, \tag{2.1}$$
  which can be used to represent a discrete signal containing N samples. We will always represent a vector as a column vector, or the transpose of a row vector. The space is denoted by either $\mathbb{C}^N$ if the elements are complex $x[n] \in \mathbb{C}$, or $\mathbb{R}^N$ if they are all real $x[n] \in \mathbb{R}$ $(n = 1, \ldots, N)$. Sometimes the N elements of a vector can be alternatively indexed by $n = 0, \ldots, N-1$ to gain some convenience, as can be seen in future chapters.
- A vector space can be defined to contain all $M \times N$ matrices composed of N M-D column vectors:
$$\boldsymbol{X} = [\boldsymbol{x}_1, \ldots, \boldsymbol{x}_N] = \begin{bmatrix} x[1,1] & x[1,2] & \cdots & x[1,N] \\ x[2,1] & x[2,2] & \cdots & x[2,N] \\ \vdots & \vdots & \ddots & \vdots \\ x[M,1] & x[M,2] & \cdots & x[M,N] \end{bmatrix}, \tag{2.2}$$
  where the nth column is an M-D vector $\boldsymbol{x}_n = [x[1,n], \ldots, x[M,n]]^{\mathrm{T}}$. Such a matrix can be converted to an MN-D vector by cascading all of the column (or row) vectors. A matrix $\boldsymbol{X}$ can be used to represent a 2-D signal, such as an image.
- l^2 space:
  The dimension N of $\mathbb{R}^N$ or $\mathbb{C}^N$ can be extended to infinity so that a vector in the space becomes a sequence $\boldsymbol{x} = [\ldots, x[n], \ldots]^{\mathrm{T}}$ for $0 \le n < \infty$ or $-\infty < n < \infty$. If all vectors are square-summable, the space is denoted by l^2. All discrete energy signals are vectors in l^2.
- $\mathcal{L}^2$ space:
  A vector space can also be a set of real or complex valued continuous functions $x(t)$ defined over either a finite range such as $0 \le t < T$, or an infinite range $-\infty < t < \infty$. If all functions are square-integrable, the space is denoted by $\mathcal{L}^2$. All continuous energy signals are vectors in $\mathcal{L}^2$.

Note that the term "vector", generally denoted by $\boldsymbol{x}$, may be interpreted in two different ways. First, in the most general sense, it represents a member of a vector space, such as any of the vector spaces considered above; e.g., a function $\boldsymbol{x} = x(t) \in \mathcal{L}^2$. Second, in a more narrow sense, it can also represent a tuple of N elements, an N-D vector $\boldsymbol{x} = [x[1], \ldots, x[N]]^{\mathrm{T}} \in \mathbb{C}^N$, where N may be infinity. It should be clear what a vector $\boldsymbol{x}$ represents from the context in our future discussion.

Definition 2.2. *The sum of two subspaces $S_1 \subset V$ and $S_2 \subset V$ of a vector space V is defined as*

$$S_1 + S_2 = \{\boldsymbol{s}_1 + \boldsymbol{s}_2 | \boldsymbol{s}_1 \in S_1, \boldsymbol{s}_2 \in S_2\}. \tag{2.3}$$

In particular, if S_1 and S_2 are mutually exclusive:

$$S_1 \cap S_2 = \emptyset, \tag{2.4}$$

then their sum $S_1 + S_2$ is called a direct sum, denoted by $S_1 \oplus S_2$. Moreover, if $S_1 \oplus S_2 = V$, then S_1 and S_2 form a direct sum decomposition of the vector space V, and S_1 and S_2 are said to be complementary. The direct sum decomposition of V can be generalized to include multiple subspaces:

$$V = \oplus_{n=1}^{N} S_n = S_1 \oplus \ldots \oplus S_N, \tag{2.5}$$

where all subspaces $S_n \subset V$ are mutually exclusive:

$$S_m \cap S_n = \emptyset, \qquad (m \neq n). \tag{2.6}$$

Definition 2.3. *Let $S_1 \subset V$ and $S_2 \subset V$ be subsets of V and $S_1 \oplus S_2 = V$. Then*

$$\boldsymbol{p}_{S_1,S_2}(\boldsymbol{s}_1 + \boldsymbol{s}_2) = \boldsymbol{s}_1, \qquad (\boldsymbol{s}_1 \in S_1,\ \boldsymbol{s}_2 \in S_2) \tag{2.7}$$

is called the projection of $\boldsymbol{s}_1 + \boldsymbol{s}_2$ onto S_1 along S_2.

2.1.2 Inner product space

Definition 2.4. *An inner product on a vector space V is a function that maps two vectors $\boldsymbol{x}, \boldsymbol{y} \in V$ to a scalar $\langle \boldsymbol{x}, \boldsymbol{y} \rangle \in \mathbb{C}$ and satisfies the following conditions:*

- *Positive definiteness:*

$$\langle \boldsymbol{x}, \boldsymbol{x} \rangle \geq 0, \qquad \langle \boldsymbol{x}, \boldsymbol{x} \rangle = 0 \quad \text{iff } \boldsymbol{x} = \boldsymbol{0}. \tag{2.8}$$

- *Conjugate symmetry:*

$$\langle \boldsymbol{x}, \boldsymbol{y} \rangle = \overline{\langle \boldsymbol{y}, \boldsymbol{x} \rangle}. \tag{2.9}$$

  If the vector space is real, the inner product becomes symmetric*:*

$$\langle \boldsymbol{x}, \boldsymbol{y} \rangle = \langle \boldsymbol{y}, \boldsymbol{x} \rangle. \tag{2.10}$$

- *Linearity in the first variable:*

$$\langle a\boldsymbol{x} + b\boldsymbol{y}, \boldsymbol{z} \rangle = a\langle \boldsymbol{x}, \boldsymbol{z} \rangle + b\langle \boldsymbol{y}, \boldsymbol{z} \rangle, \tag{2.11}$$

  where $a, b \in \mathbb{C}$. The linearity does not apply to the second variable:

$$\begin{aligned} \langle \boldsymbol{x}, a\boldsymbol{y} + b\boldsymbol{z} \rangle &= \overline{\langle a\boldsymbol{y} + b\boldsymbol{z}, \boldsymbol{x} \rangle} = \overline{a\langle \boldsymbol{y}, \boldsymbol{x} \rangle + b\langle \boldsymbol{z}, \boldsymbol{x} \rangle} \\ &= \overline{a}\langle \boldsymbol{x}, \boldsymbol{y} \rangle + \overline{b}\langle \boldsymbol{x}, \boldsymbol{z} \rangle \neq a\langle \boldsymbol{x}, \boldsymbol{y} \rangle + b\langle \boldsymbol{x}, \boldsymbol{z} \rangle, \end{aligned} \tag{2.12}$$

  unless the coefficients are real $a, b \in \mathbb{R}$. As a special case, when $b = 0$, we have

$$\langle a\boldsymbol{x}, \boldsymbol{y} \rangle = a\langle \boldsymbol{x}, \boldsymbol{y} \rangle, \qquad \langle \boldsymbol{x}, a\boldsymbol{y} \rangle = \overline{a}\langle \boldsymbol{x}, \boldsymbol{y} \rangle. \tag{2.13}$$

More generally we have

$$\left\langle \sum_n c_n \boldsymbol{x}_n, \boldsymbol{y} \right\rangle = \sum_n c_n \langle \boldsymbol{x}_n, \boldsymbol{y} \rangle, \qquad \left\langle \boldsymbol{x}, \sum_n c_n \boldsymbol{y}_n \right\rangle = \sum_n \bar{c}_n \langle \boldsymbol{x}, \boldsymbol{y}_n \rangle. \quad (2.14)$$

Definition 2.5. *A vector space with inner product defined is called an inner product space.*

In particular, when the inner product is defined, $\mathbb{C}^N$ is called a *unitary space* and $\mathbb{R}^N$ is called a *Euclidean space.* All vector spaces in the future discussion will be assumed to be inner product spaces. Some examples of the inner product are listed below:

- In an N-D vector space, the inner product, also called the *dot product*, of two vectors $\boldsymbol{x} = [x[1], \ldots, x[N]]^{\mathrm{T}}$ and $\boldsymbol{y} = [y[1], \ldots, y[N]]^{\mathrm{T}}$ is defined as

$$\langle \boldsymbol{x}, \boldsymbol{y} \rangle = \boldsymbol{x}^{\mathrm{T}} \overline{\boldsymbol{y}} = \boldsymbol{y}^* \boldsymbol{x} = [x[1], \ldots, x[N]] \begin{bmatrix} \overline{y}[1] \\ \vdots \\ \overline{y}[N] \end{bmatrix} = \sum_{n=1}^{N} x[n]\overline{y}[n], \quad (2.15)$$

  where $\boldsymbol{y}^* = \overline{\boldsymbol{y}}^{\mathrm{T}}$ is the conjugate transpose of $\boldsymbol{y}$.
- In a space of 2-D matrices containing $M \times N$ elements, the inner product of two matrices $\boldsymbol{X}$ and $\boldsymbol{Y}$ is defined as

$$\langle \boldsymbol{X}, \boldsymbol{Y} \rangle = \sum_{m=1}^{M} \sum_{n=1}^{N} x[m,n]\overline{y}[m,n]. \quad (2.16)$$

  This inner product is equivalent to Eq. (2.15) if we cascade the column (or row) vectors of $\boldsymbol{X}$ and $\boldsymbol{Y}$ to form two MN-D vectors.
- In a function space, the inner product of two function vectors $\boldsymbol{x} = x(t)$ and $\boldsymbol{y} = y(t)$ is defined as

$$\langle x(t), y(t) \rangle = \int_a^b x(t)\overline{y(t)}\, dt = \overline{\int_a^b \overline{x(t)} y(t)\, dt} = \overline{\langle y(t), x(t) \rangle}. \quad (2.17)$$

  In particular, Eq. (1.10) for the sifting property of the delta function $\delta(t)$ is an inner product:

$$\langle x(t), \delta(t) \rangle = \int_{-\infty}^{\infty} x(\tau)\delta(\tau)\, d\tau = x(0). \quad (2.18)$$

- The inner product of two random variables x and y can be defined as

$$\langle x, y \rangle = E[x\overline{y}]. \quad (2.19)$$

  If the two random variables have zero means; i.e., $\mu_x = E(x) = 0$ and $\mu_x = E(x) = 0$, the inner product above is also their covariance:

$$\sigma_{xy}^2 = E[(x - \mu_x)\overline{(y - \mu_y)}] = E(x\overline{y}) - \mu_x \overline{\mu}_y = E(x\overline{y}) = \langle x, y \rangle. \quad (2.20)$$

The concept of inner product is of essential importance based on which a whole set of other important concepts can be defined.

Definition 2.6. *If the inner product of two vectors* $\boldsymbol{x}$ *and* $\boldsymbol{y}$ *is zero,* $\langle \boldsymbol{x}, \boldsymbol{y} \rangle = 0$, *they are orthogonal (perpendicular) to each other, denoted by* $\boldsymbol{x} \perp \boldsymbol{y}$.

Definition 2.7. *The norm (or length) of a vector* $\boldsymbol{x} \in V$ *is defined as*

$$||\boldsymbol{x}|| = \sqrt{\langle \boldsymbol{x}, \boldsymbol{x} \rangle} = \langle \boldsymbol{x}, \boldsymbol{x} \rangle^{1/2}, \qquad or \qquad ||\boldsymbol{x}||^2 = \langle \boldsymbol{x}, \boldsymbol{x} \rangle. \tag{2.21}$$

The norm $||\boldsymbol{x}||$ is non-negative and it is zero if and only if $\boldsymbol{x} = \mathbf{0}$. In particular, if $||\boldsymbol{x}|| = 1$, then it is said to be *normalized* and becomes a *unit vector.* Any vector can be normalized when divided by its own norm: $\boldsymbol{x}/||\boldsymbol{x}||$. The vector norm squared $||\boldsymbol{x}||^2 = \langle \boldsymbol{x}, \boldsymbol{x} \rangle$ can be considered as the energy of the vector.

Specifically, in an N-D unitary space, the norm of a vector $\boldsymbol{x} = [x[1], \ldots, x[N]]^{\mathrm{T}} \in \mathbb{C}^N$ is

$$||\boldsymbol{x}|| = \sqrt{\langle \boldsymbol{x}, \boldsymbol{x} \rangle} = \sqrt{\boldsymbol{x}^{\mathrm{T}} \overline{\boldsymbol{x}}} = \left[\sum_{n=1}^{N} x[n]\overline{x}[n] \right]^{1/2} = \left[\sum_{n=1}^{N} |x[n]|^2 \right]^{1/2}. \tag{2.22}$$

The total energy contained in this vector is its norm squared:

$$\mathcal{E} = ||\boldsymbol{x}||^2 = \langle \boldsymbol{x}, \boldsymbol{x} \rangle = \sum_{n=1}^{N} |x[n]|^2. \tag{2.23}$$

This norm can be generalized to *p-norm* defined as

$$||\boldsymbol{x}||_p = \left[\sum_{n=1}^{N} |x[n]|^p \right]^{1/p}. \tag{2.24}$$

Particularly,

$$||\boldsymbol{x}||_1 = \sum_{n=1}^{N} |x[n]|, \qquad ||\boldsymbol{x}||_\infty = \max(|x[1]|, \ldots, |x[N]|). \tag{2.25}$$

The norm of a matrix $\boldsymbol{X}$ can be defined differently but here we will only consider the element-wise norm defined as

$$||\boldsymbol{X}||_p = \left[\sum_{n=1}^{N} |x[m][n]|^p \right]^{1/p}. \tag{2.26}$$

When $p = 2$, $||\boldsymbol{X}||_2^2$ can be considered as the total energy contained in the 2-D signal $\boldsymbol{X}$. We will always use this matrix norm in the future.

The concept of N-D unitary (or Euclidean) space can be generalized to an infinite-dimensional space, in which case the range of the summation will cover all real integers $\mathbb{Z}$ in the entire real axis $-\infty < n < \infty$. This norm exists only if the summation converges to a finite value; i.e., the vector $\boldsymbol{x}$ is an energy signal with finite energy:

$$\sum_{n=-\infty}^{\infty} |x[n]|^2 < \infty. \tag{2.27}$$

All such vectors $\boldsymbol{x}$ satisfying the above are square-summable and form the vector space denoted by $l^2(\mathbb{Z})$.

Similarly, in a function space, the norm of a function vector $\boldsymbol{x} = x(t)$ is defined as

$$||\boldsymbol{x}|| = \left[\int_a^b x(t)\overline{x(t)}\ dt\right]^{1/2} = \left[\int_a^b |x(t)|^2\ dt\right]^{1/2}, \tag{2.28}$$

where the lower and upper integral limits $a < b$ are two real numbers, which may be extended to all real values $\mathbb{R}$ in the entire real axis $-\infty < t < \infty$. This norm exists only if the integral converges to a finite value; i.e., $x(t)$ is an energy signal containing finite energy:

$$\int_{-\infty}^{\infty} |x(t)|^2\ dt < \infty. \tag{2.29}$$

All such functions $x(t)$ satisfying the above are square-integrable, and they form a function space denoted by $\mathcal{L}^2(\mathbb{R})$.

All vectors and functions in the future discussion are assumed to be square-summable/integrable; i.e., they represent energy signals containing finite amount of energy, so that these conditions do not need to be mentioned every time a signal vector is considered.

Theorem 2.1. *(The Cauchy-Schwarz inequality) The following inequality holds for any two vectors $\boldsymbol{x}, \boldsymbol{y} \in V$ in an inner product space V:*

$$|\langle \boldsymbol{x}, \boldsymbol{y}\rangle|^2 \ \le\ \langle \boldsymbol{x}, \boldsymbol{x}\rangle\langle \boldsymbol{y}, \boldsymbol{y}\rangle; \qquad \textit{i.e.,} \qquad 0 \le |\langle \boldsymbol{x}, \boldsymbol{y}\rangle| \le ||\boldsymbol{x}||\ ||\boldsymbol{y}||. \tag{2.30}$$

Proof: If either $\boldsymbol{x}$ or $\boldsymbol{y}$ is zero, we have $\langle \boldsymbol{x}, \boldsymbol{y}\rangle = 0$; i.e., Eq. (2.30) holds (an equality). Otherwise, we consider the following inner product:

$$\langle \boldsymbol{x} - \lambda\boldsymbol{y}, \boldsymbol{x} - \lambda\boldsymbol{y}\rangle = ||\boldsymbol{x}||^2 - \overline{\lambda}\langle \boldsymbol{x}, \boldsymbol{y}\rangle - \lambda\langle \boldsymbol{y}, \boldsymbol{x}\rangle + |\lambda|^2||\boldsymbol{y}||^2 \ge 0, \tag{2.31}$$

where $\lambda \in \mathbb{C}$ is an arbitrary complex number, which can be assumed to be:

$$\lambda = \frac{\langle \boldsymbol{x}, \boldsymbol{y}\rangle}{||\boldsymbol{y}||^2}, \quad \text{then} \quad \overline{\lambda} = \frac{\langle \boldsymbol{y}, \boldsymbol{x}\rangle}{||\boldsymbol{y}||^2}, \quad |\lambda|^2 = \frac{|\langle \boldsymbol{x}, \boldsymbol{y}\rangle|^2}{||\boldsymbol{y}||^4}. \tag{2.32}$$

Substituting these into Eq. (2.31), we get

$$||\boldsymbol{x}||^2 - \frac{|\langle \boldsymbol{x}, \boldsymbol{y}\rangle|^2}{||\boldsymbol{y}||^2} \ge 0; \quad \text{i.e.,} \quad |\langle \boldsymbol{x}, \boldsymbol{y}\rangle| \le ||\boldsymbol{x}||\ ||\boldsymbol{y}||. \tag{2.33}$$

Definition 2.8. *The angle between two vectors $\boldsymbol{x}$ and $\boldsymbol{y}$ is defined as*

$$\theta = \cos^{-1}\left(\frac{\langle \boldsymbol{x}, \boldsymbol{y}\rangle}{||\boldsymbol{x}||\ ||\boldsymbol{y}||}\right). \tag{2.34}$$

Now the inner product of $\boldsymbol{x}$ and $\boldsymbol{y}$ can also be written as

$$\langle \boldsymbol{x}, \boldsymbol{y} \rangle = ||\boldsymbol{x}|| \, ||\boldsymbol{y}|| \cos\theta. \tag{2.35}$$

In particular, if $\theta = 0$, then $\cos\theta = 1$, and $\boldsymbol{x}$ and $\boldsymbol{y}$ are collinear, and the inner product $\langle \boldsymbol{x}, \boldsymbol{y} \rangle = ||\boldsymbol{x}|| \, ||\boldsymbol{y}||$ in Eq. (2.30) is maximized. Else if $\theta = \pi/2$, then $\cos\theta = 0$, and $\boldsymbol{x}$ and $\boldsymbol{y}$ are orthogonal to each other, and the inner product $\langle \boldsymbol{x}, \boldsymbol{y} \rangle = 0$ is minimized.

Definition 2.9. *The orthogonal projection of a vector $\boldsymbol{x} \in V$ onto another vector $\boldsymbol{y} \in V$ is defined as*

$$\boldsymbol{p}_{\boldsymbol{y}}(\boldsymbol{x}) = \frac{\langle \boldsymbol{x}, \boldsymbol{y} \rangle}{||\boldsymbol{y}||} \frac{\boldsymbol{y}}{||\boldsymbol{y}||} = \frac{\langle \boldsymbol{x}, \boldsymbol{y} \rangle}{\langle \boldsymbol{y}, \boldsymbol{y} \rangle} \boldsymbol{y} = ||\boldsymbol{x}|| \cos\theta \frac{\boldsymbol{y}}{||\boldsymbol{y}||}, \tag{2.36}$$

where $\theta = \cos^{-1}[\langle \boldsymbol{x}, \boldsymbol{y} \rangle / (||\boldsymbol{x}|| \, ||\boldsymbol{y}||)]$ is the angle between the two vectors.

The projection $\boldsymbol{p}_{\boldsymbol{y}}(\boldsymbol{x})$ is a vector and its norm is a scalar denoted by:

$$p_{\boldsymbol{y}}(\boldsymbol{x}) = ||\boldsymbol{p}_{\boldsymbol{y}}(\boldsymbol{x})|| = \frac{\langle \boldsymbol{x}, \boldsymbol{y} \rangle}{||\boldsymbol{y}||} = ||\boldsymbol{x}|| \cos\theta, \tag{2.37}$$

which is sometimes also referred to as the scalar projection or simply projection. The projection $\boldsymbol{p}_{\boldsymbol{y}}(\boldsymbol{x})$ is illustrated in Fig. 2.1. In particular, if $\boldsymbol{y}$ is a unit (normalized) vector with $||\boldsymbol{y}|| = 1$, we have

$$\boldsymbol{p}_{\boldsymbol{y}}(\boldsymbol{x}) = \langle \boldsymbol{x}, \boldsymbol{y} \rangle \boldsymbol{y}, \qquad ||\boldsymbol{p}_{\boldsymbol{y}}(\boldsymbol{x})|| = \langle \boldsymbol{x}, \boldsymbol{y} \rangle. \tag{2.38}$$

In other words, the magnitude of the projection of $\boldsymbol{x}$ onto a unit vector is simply their inner product.

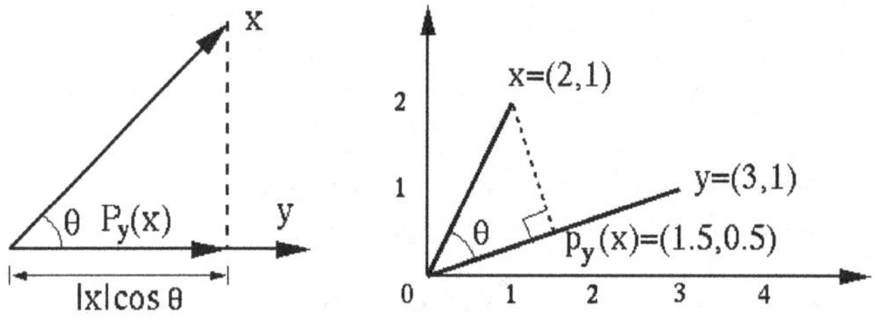

Figure 2.1 Orthogonal projection.

Example 2.1: Find the projection of $\boldsymbol{x} = [1, 2]^{\mathrm{T}}$ onto $\boldsymbol{y} = [3, 1]^{\mathrm{T}}$.

The angle between the two vectors is

$$\theta = \cos^{-1}\left(\frac{\langle \boldsymbol{x}, \boldsymbol{y} \rangle}{\sqrt{\langle \boldsymbol{x}, \boldsymbol{x} \rangle \langle \boldsymbol{y}, \boldsymbol{y} \rangle}}\right) = \cos^{-1}\left(\frac{5}{\sqrt{5 \times 10}}\right) = \cos^{-1} 0.707 = 45^\circ. \tag{2.39}$$

The projection of $\boldsymbol{x}$ on $\boldsymbol{y}$ is

$$\boldsymbol{p}_{\boldsymbol{y}}(\boldsymbol{x}) = \frac{\langle \boldsymbol{x}, \boldsymbol{y} \rangle}{\langle \boldsymbol{y}, \boldsymbol{y} \rangle} \boldsymbol{y} = \frac{5}{10} \begin{bmatrix} 3 \\ 1 \end{bmatrix} = \begin{bmatrix} 1.5 \\ 0.5 \end{bmatrix}. \tag{2.40}$$

The norm of the projection is $\sqrt{1.5^2 + 0.5^2} \approx 1.58$, which is of course the same as $||\boldsymbol{x}|| \cos\theta = \sqrt{5} \cos 45° \approx 1.58$. If $\boldsymbol{y}$ is normalized to become $\boldsymbol{z} = \boldsymbol{y}/||\boldsymbol{y}|| = [3, 1]/\sqrt{10}$, then the projection of $\boldsymbol{x}$ onto $\boldsymbol{z}$ can be simply obtained as their inner product:

$$p_{\boldsymbol{z}}(\boldsymbol{x}) = ||\boldsymbol{p}_{\boldsymbol{z}}(\boldsymbol{x})|| = \langle \boldsymbol{x}, \boldsymbol{z} \rangle = [1,\ 2] \begin{bmatrix} 3 \\ 1 \end{bmatrix} /\sqrt{10} = 5/\sqrt{10} \approx 1.58. \tag{2.41}$$

Definition 2.10. *Two subspaces $S_1 \subset V$ and $S_2 \subset V$ of an inner product space V are orthogonal, denoted by $S_1 \perp S_2$, if $\boldsymbol{s}_1 \perp \boldsymbol{s}_2$ for any $\boldsymbol{s}_1 \in S_1$ and $\boldsymbol{s}_2 \in S_2$. In particular, if one of the subsets contains only one vector $S_1 = \{\boldsymbol{s}_1\}$, then the vector is orthogonal to the other subset $\boldsymbol{s}_1 \perp S_2$.*

Definition 2.11. *The orthogonal complement of a subspace $S \subset V$ is the set of all vectors in V that are orthogonal to S:*

$$S^\perp = \{\boldsymbol{v} \in V \mid \boldsymbol{v} \perp S\} = \{\boldsymbol{v} \in V \mid \langle \boldsymbol{v}, \boldsymbol{u} \rangle = 0, \forall \boldsymbol{u} \in S\}. \tag{2.42}$$

Definition 2.12. *An inner product space V as the direct sum of N mutually orthogonal subspaces $S_k \subset V$ $(k = 1, \ldots, N)$ is called the orthogonal direct sum of these subspaces:*

$$V = S_1 \oplus \ldots \oplus S_k, \qquad \textit{with} \quad S_k \perp S_l \quad \textit{for all } k \neq l. \tag{2.43}$$

It can be shown that if $V = S_1 \oplus S_2$ and $S_1 \perp S_2$, then

$$S \cap S^\perp = \emptyset, \qquad \text{and} \qquad S \oplus S^\perp = V. \tag{2.44}$$

Definition 2.13. *Let $S \subset V$ and $S \oplus S^\perp = V$ and $\boldsymbol{s} \in S$, $\boldsymbol{r} \in S^\perp$. Then $\boldsymbol{p}_S(\boldsymbol{s} + \boldsymbol{r}) = \boldsymbol{s}$ is called the orthogonal projection of $\boldsymbol{s} + \boldsymbol{r}$ onto S.*

All of these definitions can be intuitively and trivially visualized in a 3-D space spanned by three perpendicular coordinates (x, y, z) representing three mutually orthogonal subspaces. The orthogonal direct sum of these subspaces is the 3-D space, and the orthogonal complement of the subspace in the x direction is the 2-D yz plane formed by coordinates y and z. The orthogonal projection of a vector $\boldsymbol{v} = [1, 2, 3]^{\mathrm{T}}$ onto the subspace in the x direction is $[1, 0, 0]^{\mathrm{T}}$, and its orthogonal projection onto the yz subspace is a 2-D vector $[0, 2, 3]^{\mathrm{T}}$.

Definition 2.14. *The distance between two vectors $\boldsymbol{x}, \boldsymbol{y}$ is*

$$d(\boldsymbol{x}, \boldsymbol{y}) = ||\boldsymbol{x} - \boldsymbol{y}||. \tag{2.45}$$

Theorem 2.2. *The distance satisfies the following three conditions:*

- *Non-negative:*

$$d(\boldsymbol{x},\boldsymbol{y}) \geq 0, \qquad d(\boldsymbol{x},\boldsymbol{y}) = 0 \quad \textit{iff} \quad \boldsymbol{x} = \boldsymbol{y}. \tag{2.46}$$

- *Symmetric:*

$$d(\boldsymbol{x},\boldsymbol{y}) = d(\boldsymbol{y},\boldsymbol{x}). \tag{2.47}$$

- *Triangle inequality:*

$$d(\boldsymbol{x},\boldsymbol{y}) \leq d(\boldsymbol{x},\boldsymbol{z}) + d(\boldsymbol{z},\boldsymbol{y}). \tag{2.48}$$

Proof: The first two conditions are self-evident based on the definition. We now show the third condition also holds by considering the following:

$$\begin{aligned} ||\boldsymbol{u}+\boldsymbol{v}||^2 &= \langle \boldsymbol{u}+\boldsymbol{v}, \boldsymbol{u}+\boldsymbol{v}\rangle = ||\boldsymbol{u}||^2 + \langle \boldsymbol{u},\boldsymbol{v}\rangle + \langle \boldsymbol{v},\boldsymbol{u}\rangle + ||\boldsymbol{v}||^2 \\ &= ||\boldsymbol{u}||^2 + 2\ \mathrm{Re}\langle \boldsymbol{u},\boldsymbol{v}\rangle + ||\boldsymbol{v}||^2 \leq ||\boldsymbol{u}||^2 + 2\,|\langle \boldsymbol{u},\boldsymbol{v}\rangle| + ||\boldsymbol{v}||^2 \\ &\leq ||\boldsymbol{u}||^2 + 2\,||\boldsymbol{u}||\,||\boldsymbol{v}|| + ||\boldsymbol{v}||^2 = (||\boldsymbol{u}|| + ||\boldsymbol{v}||)^2. \end{aligned} \tag{2.49}$$

The first $\leq$ sign above is due to the fact that the magnitude of a complex number is no less that its real part, and the second $\leq$ sign is simply the Cauchy-Schwarz inequality. Taking the square root on both sides, we get

$$||\boldsymbol{u}+\boldsymbol{v}|| \leq ||\boldsymbol{u}|| + ||\boldsymbol{v}||. \tag{2.50}$$

We further let $\boldsymbol{u} = \boldsymbol{x} - \boldsymbol{z}$ and $\boldsymbol{v} = \boldsymbol{z} - \boldsymbol{y}$, and the above becomes the triangle inequality:

$$||\boldsymbol{x}-\boldsymbol{y}|| \leq ||\boldsymbol{x}-\boldsymbol{z}|| + ||\boldsymbol{z}-\boldsymbol{y}||. \tag{2.51}$$

This is Eq. (2.48). Q.E.D.

Definition 2.15. *When distance is defined between any two vectors in a vector space, it is called a* metric space.

In a unitary space $\mathbb{C}^N$, the *Euclidean distance* between any two vectors $\boldsymbol{x}$ and $\boldsymbol{y}$ can be defined as the norm of the difference vector $\boldsymbol{x} - \boldsymbol{y}$:

$$d(\boldsymbol{x},\boldsymbol{y}) = ||\boldsymbol{x}-\boldsymbol{y}|| = \left(\sum_{n=1}^{N} |x[n] - y[n]|^2\right)^{1/2}. \tag{2.52}$$

This distance can be considered as a special case ($p = 2$) of the more general *p-norm distance* defined as

$$d_p(\boldsymbol{x},\boldsymbol{y}) = \left(\sum_{n=1}^{N} |x[n] - y[n]|^p\right)^{1/p}. \tag{2.53}$$

Other commonly used p-norm distances include

$$d_1(\boldsymbol{x},\boldsymbol{y}) = \sum_{n=1}^{N} |x[n]-y[n]| \tag{2.54}$$

$$d_\infty(\boldsymbol{x},\boldsymbol{y}) = \max(|x[1]-y[1]|,\dots,|x[N]-y[N]|). \tag{2.55}$$

In a function space, the p-norm distance between two functions $x(t)$ and $y(t)$ is similarly defined as

$$d_p(x(t),y(t)) = \left(\int_a^b |x(t)-y(t)|^p \; dt\right)^{1/p}. \tag{2.56}$$

In particular, when $p = 2$, we have

$$d_2(x(t),y(t)) = ||x(t)-y(t)|| = \left(\int_a^b |x(t)-y(t)|^2 \; dt\right)^{1/2}. \tag{2.57}$$

2.1.3 Bases of vector space

Definition 2.16. *In a vector space V, the subspace W of all linear combinations of a set of M vectors* $\boldsymbol{b}_k \in V$, $(k = 1,\dots,M)$ *is called the linear span of the vectors:*

$$W = \text{span}(\boldsymbol{b}_1,\dots,\boldsymbol{b}_M) = \left\{\sum_{k=1}^{M} c[k]\boldsymbol{b}_k \;\middle|\; c[k] \in \mathbb{C}\right\}. \tag{2.58}$$

Definition 2.17. *A set of linearly independent vectors that spans a vector space is called a basis of the space.*

The basis vectors are linearly independent; i.e., none of them can be represented as a linear combination of the rest. They are also complete; i.e., by including any additional vector in the basis it would no longer be linearly independent, and removing any of them would result in inability to represent certain vectors in the space. In other words, a basis is a minimum set of vectors capable of representing any vector in the space. Also, as any rotation of a given basis will result in a different basis, we see that there are infinitely many bases that all span the same space. This idea is of great importance in our future discussion.

For example, any vector $\boldsymbol{x} \in \mathbb{C}^N$ can be uniquely expressed as a linear combination of some N basis vectors $\boldsymbol{b}_k$:

$$\boldsymbol{x} = \sum_{k=1}^{N} c[k]\boldsymbol{b}_k. \tag{2.59}$$

Moreover, the concept of a finite N-D space spanned by a basis composed of N discrete (countable) linearly independent vectors can be generalized to a vector space V spanned by a basis composed of a family of uncountably infinite vectors

$\boldsymbol{b}(f)$. Any vector $\boldsymbol{x} \in V$ in the space can be expressed as a linear combination, an integral, of these basis vectors:

$$\boldsymbol{x} = \int_a^b c(f)\boldsymbol{b}(f)\, df. \tag{2.60}$$

We see that the index k for the summation in Eq. (2.59) is replaced by a continuous variable f for the integral, and the coefficient $c[k]$ is replaced by a continuous weighting function $c(f)$ for the uncountably infinite set of basis vectors $\boldsymbol{b}(f)$ with $a < f < b$. The significance of this generalization becomes clear during our future discussion of orthogonal transforms of continuous signals $x(t)$. An important issue is how to find the coefficients $c[k]$ or the weighting function $c(f)$, given the vector $\boldsymbol{x}$ and the basis $\boldsymbol{b}_k$ or $\boldsymbol{b}(f)$.

Consider specifically the case of an N-D unitary space $\mathbb{C}^N$ as an example. Let $\{\boldsymbol{b}_1, \ldots, \boldsymbol{b}_M\}$ be a basis consisting of M linearly independent N-D vectors. Then any vector $\boldsymbol{x} \in \mathbb{C}^N$ can be represented as a linear combination of these basis vectors:

$$\boldsymbol{x} = \begin{bmatrix} x[1] \\ \vdots \\ x[N] \end{bmatrix}_{N\times 1} = \sum_{k=1}^{M} c[k]\boldsymbol{b}_k = [\boldsymbol{b}_1, \ldots, \boldsymbol{b}_M]_{N\times M} \begin{bmatrix} c[1] \\ \vdots \\ c[M] \end{bmatrix}_{M\times 1} = \boldsymbol{B}\boldsymbol{c}, \tag{2.61}$$

where $\boldsymbol{B} = [\boldsymbol{b}_1, \ldots, \boldsymbol{b}_M]$ is an N by M matrix composed of the M N-D basis vectors as its columns, and the nth coefficient $c[n]$ is the nth element of an M-D vector $\boldsymbol{c} = [c[1], \ldots, c[M]]^{\mathrm{T}}$. This coefficient vector $\boldsymbol{c}$ can be found by solving the equation system in Eq. (2.61). For the solution to exist, the number of unknown coefficients must be no fewer than the number of constraining equations; i.e., $M \geq N$. On the other hand, as there can be no more than N independent basis vectors in this N-D space, we must also have $M \leq N$. Therefore there must be exactly $M = N$ vectors in a basis of an N-D space. In this case, $\boldsymbol{B}$ is an N by N square matrix with full rank (as all column vectors are independent); i.e., its inverse $\boldsymbol{B}^{-1}$ exists and the coefficients can be obtained by solving the system with N unknowns and N equations:

$$\boldsymbol{c} = \begin{bmatrix} c[1] \\ \vdots \\ c[N] \end{bmatrix} = [\boldsymbol{b}_1, \ldots, \boldsymbol{b}_N]^{-1} \begin{bmatrix} x[1] \\ \vdots \\ x[N] \end{bmatrix} = \boldsymbol{B}^{-1}\boldsymbol{x}. \tag{2.62}$$

The computational complexity to solve this system of N equations and N unknowns is $O(N^3)$.

Similarly, we may need to find the weighting function $c(f)$ in Eq. (2.60) in order to represent a vector $\boldsymbol{x}$ in terms of the basis $\boldsymbol{b}(f)$. However, solving this equation for $c(f)$ is not as trivial as solving Eq. (2.61) for $\boldsymbol{c}$ in the previous case of a vector space spanned by a finite and discrete basis. In the next subsection, this problem will be reconsidered when some additional condition is imposed on the basis $\boldsymbol{c}$ to make the problem easier to solve.

Example 2.2: A 2-D Euclidean $\mathbb{R}^2$ space can be spanned by two basis vectors $\boldsymbol{e}_1 = [1,0]^{\mathrm{T}}$ and $\boldsymbol{e}_2 = [0,1]^{\mathrm{T}}$, by which two vectors $\boldsymbol{a}_1 = [1,0]^{\mathrm{T}}$ and $\boldsymbol{a}_2 = [-1,2]^{\mathrm{T}}$ can be represented as

$$\boldsymbol{a}_1 = 1\boldsymbol{e}_1 + 0\boldsymbol{e}_2 = \begin{bmatrix} 1 \\ 0 \end{bmatrix}, \qquad \boldsymbol{a}_2 = -1\boldsymbol{e}_1 + 2\boldsymbol{e}_2 = \begin{bmatrix} -1 \\ 2 \end{bmatrix}. \tag{2.63}$$

As $\boldsymbol{a}_1$ and $\boldsymbol{a}_2$ are independent (as they are not collinear), they in turn form a basis of the space. Any given vector such as

$$\boldsymbol{x} = \begin{bmatrix} 1 \\ 2 \end{bmatrix} = 1\boldsymbol{e}_1 + 2\boldsymbol{e}_2 = 1\begin{bmatrix} 1 \\ 0 \end{bmatrix} + 2\begin{bmatrix} 0 \\ 1 \end{bmatrix} \tag{2.64}$$

can be expressed in terms of $\{\boldsymbol{a}_1, \boldsymbol{a}_2\}$ as

$$\boldsymbol{x} = \begin{bmatrix} 1 \\ 2 \end{bmatrix} = c[1]\boldsymbol{a}_1 + c[2]\boldsymbol{a}_2 = c[1]\begin{bmatrix} 1 \\ 0 \end{bmatrix} + c[2]\begin{bmatrix} -1 \\ 2 \end{bmatrix} = \begin{bmatrix} 1 & -1 \\ 0 & 2 \end{bmatrix}\begin{bmatrix} c[1] \\ c[2] \end{bmatrix}. \tag{2.65}$$

Solving this we get $c[1] = 2$ and $c[2] = 1$, so that $\boldsymbol{x}$ can be expressed by $\boldsymbol{a}_1$ and $\boldsymbol{a}_2$ as

$$\boldsymbol{x} = c[1]\boldsymbol{a}_1 + c[2]\boldsymbol{a}_2 = 2\begin{bmatrix} 1 \\ 0 \end{bmatrix} + 1\begin{bmatrix} -1 \\ 2 \end{bmatrix} = \begin{bmatrix} 1 \\ 2 \end{bmatrix}. \tag{2.66}$$

This example is illustrated in Fig. 2.2.

Figure 2.2 Different basis vectors of a 2-D space.

Example 2.3: The previous example in $\mathbb{R}^2$ can also be extended to a function space defined over $[0,2]$ spanned by two basis functions:

$$a_1(t) = \begin{cases} 1 & 0 \le t < 1 \\ 0 & 1 \le t < 2 \end{cases}, \qquad a_2(t) = \begin{cases} -1 & 0 \le t < 1 \\ 2 & 1 \le t < 2 \end{cases}. \tag{2.67}$$

A given time function $x(t)$ in the space

$$x(t) = \begin{cases} 1 & 0 \le t < 1 \\ 2 & 1 \le t < 2 \end{cases}. \tag{2.68}$$

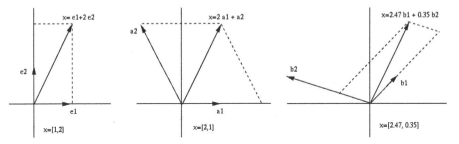

can be represented by the two basis functions as

$$x(t) = c[1]a_1(t) + c[2]a_2(t). \tag{2.69}$$

To obtain the coefficients $c[1]$ and $c[2]$, we first find the inner products of this equation with the following two functions:

$$e_1(t) = \begin{cases} 1 & 0 \le t < 1 \\ 0 & 1 \le t < 2 \end{cases}, \quad e_2(t) = \begin{cases} 0 & 0 \le t < 1 \\ 1 & 1 \le t < 2 \end{cases} \tag{2.70}$$

to get

$$\begin{aligned} \langle x(t), e_1(t) \rangle &= 1 = c[1]\langle a_1(t), e_1(t) \rangle + c[2]\langle a_2(t), e_1(t) \rangle = c[1] - c[2] \\ \langle x(t), e_2(t) \rangle &= 2 = c[1]\langle a_1(t), e_2(t) \rangle + c[2]\langle a_2(t), e_2(t) \rangle = 2c[2]. \end{aligned} \tag{2.71}$$

Solving this equation system, which is identical to that in the previous example, we get the same coefficients $c[1] = 2$ and $c[2] = 1$. Now $x(t)$ can be expressed as $x(t) = 2a_1(t) + a_2(t)$, as illustrated in Fig. 2.3.

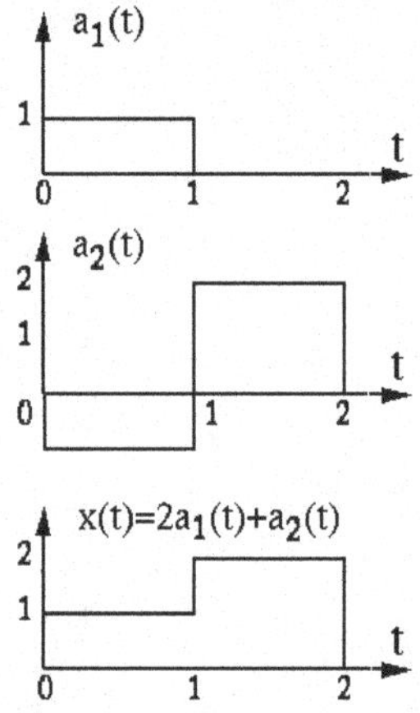

Figure 2.3 Representation of a time function by basis functions.

So far we have only considered inner product spaces of finite dimensions. Additional theory is needed to deal with spaces of infinite dimensions.

Definition 2.18.

- *In a metric space V, a sequence $\{\boldsymbol{x}_1, \boldsymbol{x}_2, \dots\}$ is a Cauchy sequence if for any $\epsilon > 0$ there exists an $N > 0$ such that for any $m, n > N$, $d(\boldsymbol{x}_m, \boldsymbol{x}_n) < \epsilon$.*
- *A metric space V is complete if every Cauchy sequence $\{\boldsymbol{x}_n\}$ in V converges to $\boldsymbol{x} \in V$:*

$$\lim_{m \to \infty} d(\boldsymbol{x}_m, \boldsymbol{x}) = \lim_{m \to \infty} ||\boldsymbol{x} - \boldsymbol{x}_m|| = 0. \tag{2.72}$$

In other words, for any $\epsilon > 0$, *there exists an* $N > 0$ *such that*

$$d(\boldsymbol{x}_m, \boldsymbol{x}) < \epsilon \quad \textit{if} \quad m > N. \tag{2.73}$$

- *A complete inner product space is a Hilbert space, denoted by* H.
- *Let* $\boldsymbol{b}_k$ *be a set of orthogonal vectors* ($k = 1, 2, \ldots$) *in* H, *and an arbitrary vector* $\boldsymbol{x}$ *is approximated in an* M*-D subspace by*

$$\hat{\boldsymbol{x}}_M = \sum_{k=1}^{M} c[k]\boldsymbol{b}_k. \tag{2.74}$$

If the least-squares error of this approximation $||\boldsymbol{x} - \hat{\boldsymbol{x}}_M||^2$ *converges to zero when* $M \to \infty$; *i.e.,*

$$\lim_{M\to\infty} ||\boldsymbol{x} - \hat{\boldsymbol{x}}_M||^2 = \lim_{M\to\infty} \left|\left| \boldsymbol{x} - \sum_{k=1}^{M} c[k]\boldsymbol{b}_k \right|\right|^2 = 0, \tag{2.75}$$

then this set of orthogonal vectors is said to be complete, called a complete orthogonal system, and the approximation converges to the given vector:

$$\lim_{M\to\infty} \sum_{k=1}^{M} c[k]\boldsymbol{b}_k = \sum_{k=1}^{\infty} c[k]\boldsymbol{b}_k = \boldsymbol{x}. \tag{2.76}$$

In the following, to keep the discussion generic, the lower and upper limits of a summation or an integral may not be always explicitly specified, as the summation or integral may be finite (e.g., from 1 to N) or infinite (e.g., from 0 or $-\infty$ to ∞), depending on each specific case.

2.1.4 Signal representation by orthogonal bases

As shown in Eqs. (2.59) and (2.60), a vector $\boldsymbol{x} \in V$ in a vector space can be represented as a linear combination of a set of linearly independent basis vectors, either countable like $\boldsymbol{b}_k$, or uncountable like $\boldsymbol{b}(f)$, that span the space V. However, it may not be always easy to find the weighting coefficients $c[k]$ or function $c(f)$. As shown in Eq. (2.62) for the simple case of the finite dimensional space $\mathbb{C}^N$, in order to obtain the coefficient vector $\boldsymbol{c}$, we need to find the inverse of the $N \times N$ matrix $\boldsymbol{B} = [\boldsymbol{b}_1, \ldots, \boldsymbol{b}_N]$, which may not be a trivial problem if N is large. Moreover, in the case of uncountable basis $\boldsymbol{b}(f)$ of Eq. (2.60), it is certainly not a trivial problem to find the coefficient function $c(f)$. However, as to be shown below, finding the coefficients $c[k]$ or weighting function $c(f)$ can become most straight forward if the basis is orthogonal.

Theorem 2.3. *Let* $\boldsymbol{x}$ *and* $\boldsymbol{y}$ *be any two vectors in a Hilbert space* H *spanned by a complete orthonormal system* $\{\boldsymbol{u}_k\}$ *satisfying*

$$\langle \boldsymbol{u}_k, \boldsymbol{u}_l \rangle = \delta[k - l]. \tag{2.77}$$

Then we have

1. *Series expansion:*

$$\boldsymbol{x} = \sum_k \langle \boldsymbol{x}, \boldsymbol{u}_k \rangle \boldsymbol{u}_k. \tag{2.78}$$

2. *Plancherel theorem:*

$$\langle \boldsymbol{x}, \boldsymbol{y} \rangle = \sum_k \langle \boldsymbol{x}, \boldsymbol{u}_k \rangle \overline{\langle \boldsymbol{y}, \boldsymbol{u}_k \rangle}. \tag{2.79}$$

3. *Parseval's theorem:*

$$\langle \boldsymbol{x}, \boldsymbol{x} \rangle = ||\boldsymbol{x}||^2 = \sum_k |\langle \boldsymbol{x}, \boldsymbol{u}_k \rangle|^2. \tag{2.80}$$

Here, the dimensionality of the space is not specified to keep the discussion more general.

Proof: As $\{\boldsymbol{u}_k\}$ is the basis of H, any $\boldsymbol{x} \in H$ can be written as

$$\boldsymbol{x} = \sum_k c[k] \boldsymbol{u}_k. \tag{2.81}$$

Taking an inner product with $\boldsymbol{u}_l$ on both sides we get

$$\langle \boldsymbol{x}, \boldsymbol{u}_l \rangle = \left\langle \sum_k c[k] \boldsymbol{u}_k, \boldsymbol{u}_l \right\rangle = \sum_k c[k] \langle \boldsymbol{u}_k, \boldsymbol{u}_l \rangle = \sum_k c[k] \delta[k-l] = c[l]. \tag{2.82}$$

We therefore have $c[k] = \langle \boldsymbol{x}, \boldsymbol{u}_k \rangle$ and

$$\boldsymbol{x} = \sum_k c[k] \boldsymbol{u}_k = \sum_k \langle \boldsymbol{x}, \boldsymbol{u}_k \rangle \boldsymbol{u}_k. \tag{2.83}$$

Here, $\boldsymbol{x}$ is expressed as the vector sum of its projections $\boldsymbol{p}_{\boldsymbol{u}_k}(\boldsymbol{x}) = \langle \boldsymbol{x}, \boldsymbol{u}_k \rangle \boldsymbol{u}_k$ onto each of the unit basis vectors $\boldsymbol{u}_k$ (Eq. (2.38)), and the scalar coefficient $c[k] = \langle \boldsymbol{x}, \boldsymbol{u}_k \rangle$ is the norm of the projection. Vector $\boldsymbol{y} \in H$ can also be written as

$$\boldsymbol{y} = \sum_l d[l] \boldsymbol{u}_l = \sum_l \langle \boldsymbol{y}, \boldsymbol{u}_l \rangle \boldsymbol{u}_l, \tag{2.84}$$

and we have

$$\begin{aligned} \langle \boldsymbol{x}, \boldsymbol{y} \rangle &= \left\langle \sum_k c[k] \boldsymbol{u}_k, \sum_l d[l] \boldsymbol{u}_l \right\rangle = \sum_k c[k] \sum_l \overline{d}[l] \langle \boldsymbol{u}_k, \boldsymbol{u}_l \rangle \\ &= \sum_k c[k] \sum_l \overline{d}[l] \delta[k-l] = \sum_k c[k] \overline{d}[k] \\ &= \sum_k \langle \boldsymbol{x}, \boldsymbol{u}_k \rangle \overline{\langle \boldsymbol{y}, \boldsymbol{u}_k \rangle} = \langle \boldsymbol{c}, \boldsymbol{d} \rangle, \end{aligned} \tag{2.85}$$

where $\boldsymbol{c} = [\ldots, c[k], \ldots]^{\mathrm{T}}$ and $\boldsymbol{d} = [\ldots, d[k], \ldots]^{\mathrm{T}}$ are the coefficient vectors of either finite or infinite dimensions. This is the Plancherel theorem. In particular, when $\boldsymbol{x} = \boldsymbol{y}$, we have

$$\langle \boldsymbol{x}, \boldsymbol{x} \rangle - ||\boldsymbol{x}||^2 = \sum_k |\langle \boldsymbol{x}, \boldsymbol{u}_k \rangle|^2 = \sum_k |c[k]|^2 = \langle \mathbf{c}, \mathbf{c} \rangle = ||\mathbf{c}||^2. \tag{2.86}$$

This is Parseval's theorem or identity. Q.E.D.

Eqs. (2.82) and (2.83) can be combined to form a pair of equations:

$$\boldsymbol{x} = \sum_k c[k]\boldsymbol{u}_k = \sum_k \langle \boldsymbol{x}, \boldsymbol{u}_k \rangle \boldsymbol{u}_k \tag{2.87}$$

$$c[k] = \langle \boldsymbol{x}, \boldsymbol{u}_k \rangle, \qquad \text{for all } k. \tag{2.88}$$

The first equation is the *generalized Fourier expansion*, which represents a given vector $\boldsymbol{x}$ as a linear combination of the basis $\{\boldsymbol{u}_k\}$, and the weighting coefficient $c[k]$ given in the second equation is the *generalized Fourier coefficient.*

The results above can be generalized to a vector space spanned by a basis composed of a continuum of uncountable orthogonal basis vectors $\boldsymbol{u}(f)$ satisfying:

$$\langle \boldsymbol{u}(f), \boldsymbol{u}(f') \rangle = \delta(f - f'). \tag{2.89}$$

Under this basis, any vector $\boldsymbol{x}$ in the space can be expressed as

$$\boldsymbol{x} = \int c(f)\boldsymbol{u}(f)\, df. \tag{2.90}$$

Same as Eq. (2.60), this equation also represents a given vector $\boldsymbol{x}$ in the space as a linear combination (an integral) of the basis function $\boldsymbol{u}(f)$, weighted by $c(f)$. However, different from the case in Eq. (2.60), here the weighting function $c(f)$ can be easily obtained due to the orthogonality of the basis $\boldsymbol{u}(f)$. Taking the inner product with $\boldsymbol{u}(f')$ on both sides of Eq. (2.90), we get

$$\begin{aligned} \langle \boldsymbol{x}, \boldsymbol{u}(f') \rangle &= \left\langle \int c(f)\boldsymbol{u}(f)\, df,\ \boldsymbol{u}(f') \right\rangle = \int c(f) \langle \boldsymbol{u}(f), \boldsymbol{u}(f') \rangle\, df \\ &= \int c(f)\delta(f - f')\, df = c(f'). \end{aligned} \tag{2.91}$$

We therefore have

$$c(f) = \langle \boldsymbol{x}, \boldsymbol{u}(f) \rangle, \tag{2.92}$$

representing the projection of $\boldsymbol{x}$ onto the unit basis vector $\boldsymbol{u}(f)$. Now Eq. (2.90) can also be written as

$$\boldsymbol{x} = \int c(f)\boldsymbol{u}(f)\, df = \int \langle \boldsymbol{x}, \boldsymbol{u}(f) \rangle \boldsymbol{u}(f)\, df. \tag{2.93}$$

Also, based on Eq. (2.90), we can easily show that Parseval's identity holds:

$$||\boldsymbol{x}||^2 = \langle \boldsymbol{x}, \boldsymbol{x} \rangle = \int c(f)\overline{c}(f)\, df = \langle c(f), c(f) \rangle = ||c(f)||^2. \tag{2.94}$$

As a specific example, space $\mathbb{C}^N$ can be spanned by N orthonormal vectors $\{\boldsymbol{u}_1, \ldots, \boldsymbol{u}_N\}$, where the kth basis vector is $\boldsymbol{u}_k = [u[1,k], \ldots, u[N,k]]^{\mathrm{T}}$, that satisfy:

$$\langle \boldsymbol{u}_k, \boldsymbol{u}_l \rangle = \boldsymbol{u}_k^{\mathrm{T}} \overline{\boldsymbol{u}}_l = \sum_{n=1}^{N} u[n,k]\overline{u}[n,l] = \delta[k-l]. \tag{2.95}$$

Any vector $\boldsymbol{x} = [x[1], \ldots, x[N]]^{\mathrm{T}} \in \mathbb{C}^N$ can be expressed as

$$\boldsymbol{x} = \sum_{k=1}^{N} c[k]\boldsymbol{u}_k = [\boldsymbol{u}_1, \ldots, \boldsymbol{u}_N] \begin{bmatrix} c[1] \\ \vdots \\ c[N] \end{bmatrix} = \boldsymbol{U}\boldsymbol{c}, \tag{2.96}$$

where $\boldsymbol{c} = [c[1], \ldots, c[N]]^{\mathrm{T}}$ and

$$\boldsymbol{U} = [\boldsymbol{u}_1, \ldots, \boldsymbol{u}_N] = \begin{bmatrix} u[1,1] & \ldots & u[1,N] \\ \vdots & \ddots & \vdots \\ u[N,1] & \ldots & u[N,N] \end{bmatrix}. \tag{2.97}$$

As the column (and row) vectors in $\boldsymbol{U}$ are orthogonal, it is a unitary matrix that satisfies $\boldsymbol{U}^{-1} = \boldsymbol{U}^*$; i.e., $\boldsymbol{U}\boldsymbol{U}^* = \boldsymbol{U}^*\boldsymbol{U} = \boldsymbol{I}$ (Eq. (A.51). To find the coefficient vector $\boldsymbol{c}$, we pre-multiply $\boldsymbol{U}^{-1} = \boldsymbol{U}^*$ on both sides of Eq. (2.96) and get:

$$\boldsymbol{U}^*\boldsymbol{x} = \boldsymbol{U}^{-1}\boldsymbol{x} = \boldsymbol{U}^{-1}\boldsymbol{U}\boldsymbol{c} = \boldsymbol{c}. \tag{2.98}$$

Equations (2.96) and (2.98) can be rewritten as a pair of transforms:

$$\begin{cases} \boldsymbol{c} = \boldsymbol{U}^*\boldsymbol{x} = \boldsymbol{U}^{-1}\boldsymbol{x} \\ \boldsymbol{x} = \boldsymbol{U}\boldsymbol{c} \end{cases}. \tag{2.99}$$

We see that the norm of $\boldsymbol{x}$ is conserved (Parseval's identity):

$$||\boldsymbol{x}||^2 = \langle \boldsymbol{x}, \boldsymbol{x} \rangle = \langle \boldsymbol{U}\boldsymbol{c}, \boldsymbol{U}\boldsymbol{c} \rangle = (\boldsymbol{U}\boldsymbol{c})^*\boldsymbol{U}\boldsymbol{c} = \boldsymbol{c}^*\boldsymbol{U}^*\boldsymbol{U}\boldsymbol{c} = \boldsymbol{c}^*\boldsymbol{c} = \langle \boldsymbol{c}, \boldsymbol{c} \rangle = ||\boldsymbol{c}||^2. \tag{2.100}$$

Equivalently, the coefficient $c[k]$ can also be found by an inner product with $\boldsymbol{u}_l$ on both sides of Eq. (2.96):

$$\langle \boldsymbol{x}, \boldsymbol{u}_l \rangle = \langle \sum_{k=1}^{N} c[k]\boldsymbol{u}_k, \boldsymbol{u}_l \rangle = \sum_{k=1}^{N} c[k]\langle \boldsymbol{u}_k, \boldsymbol{u}_l \rangle = \sum_{k=1}^{N} c[k]\delta[k-l] = c[l]. \tag{2.101}$$

Now the transform pair above can also be written as

$$c[k] = \langle \boldsymbol{x}, \boldsymbol{u}_k \rangle = \sum_{n=1}^{N} x[n]\overline{u}[n,k], \qquad k = 1, \ldots, N \tag{2.102}$$

$$\boldsymbol{x} = \sum_{k=1}^{N} c[k]\boldsymbol{u}_k = \sum_{k=1}^{N} \langle \boldsymbol{x}, \boldsymbol{u}_k \rangle \boldsymbol{u}_k. \tag{2.103}$$

The second equation can also be written in component form as

$$x[n] = \sum_{k=1}^{N} c[k]u[k,n], \qquad n = 1, \ldots, N. \tag{2.104}$$

Obviously, the N coefficients $c[k]$ $(k = 1, \ldots, N)$ can be obtained with computational complexity $O(N^2)$, in comparison with the complexity $O(N^3)$ needed to find $\boldsymbol{U}^{-1}$ in Eq. (2.62) when non-orthogonal basis $\boldsymbol{b}_k$ is used.

Consider another example of $\mathcal{L}^2$ space composed of all square-integrable functions defined over $a < t < b$, spanned by a set of orthonormal basis functions $u_k(t)$ satisfying:

$$\langle u_k(t), u_l(t) \rangle = \int_a^b u_k(t)\overline{u}_l(t)\, dt = \delta[k-l]. \tag{2.105}$$

Any $x(t)$ in the space can be written as

$$x(t) = \sum_k c[k]u_k(t). \tag{2.106}$$

Taking an inner product with $u_l(t)$ on both sides, we get

$$\langle x(t), u_l(t) \rangle = \sum_k c[k]\langle u_k(t), u_l(t) \rangle = \sum_k c[k]\delta[k-l] = c[l]; \tag{2.107}$$

i.e.,

$$c[k] = \langle x(t), u_k(t) \rangle = \int_a^b x(t)\overline{u}_k(t)\, dt. \tag{2.108}$$

which is the projection of $x(t)$ onto the unit basis function $\phi_k(t)$. Again we can easily get:

$$||x(t)||^2 = \langle x(t), x(t) \rangle = \int_a^b x(t)\overline{x}(t)\, dt = \sum_k |c[k]|^2 = ||\boldsymbol{c}||^2. \tag{2.109}$$

Since orthogonal bases are more advantageous than non-orthogonal ones, it is often desirable to convert a given non-orthogonal basis $\{\boldsymbol{a}_1, \ldots, \boldsymbol{a}_N\}$ into an orthogonal one $\{\boldsymbol{u}_1, \ldots, \boldsymbol{u}_N\}$ by the following *Gram-Schmidt orthogonalization process*:

- $\boldsymbol{u}_1 = \boldsymbol{a}_1$
- $\boldsymbol{u}_2 = \boldsymbol{a}_2 - P_{\boldsymbol{u}_1}\boldsymbol{a}_2$
- $\boldsymbol{u}_3 = \boldsymbol{a}_3 - P_{\boldsymbol{u}_1}\boldsymbol{a}_3 - P_{\boldsymbol{u}_2}\boldsymbol{a}_3$
- $\ldots$
- $\boldsymbol{u}_N = \boldsymbol{a}_N - \sum_{n=1}^{N-1} P_{\boldsymbol{u}_n}\boldsymbol{a}_N$.

Example 2.4: In Example 2.2, a vector $\boldsymbol{x} = [1, 2]^{\mathrm{T}}$ in a 2-D space is represented under a basis composed of $\boldsymbol{a}_1 = [1, 0]^{\mathrm{T}}$ and $\boldsymbol{a}_2 = [-1, 2]^{\mathrm{T}}$. Now we show that based on this basis an orthogonal basis can be constructed by the Gram-Schmidt orthogonalization process. In this case of $n = 2$, we have $\boldsymbol{u}_1 = \boldsymbol{a}_1 = [1, 0]^{\mathrm{T}}$, $P_{\boldsymbol{u}_1}\boldsymbol{a}_2 = [-1, 0]^{\mathrm{T}}$, and

$$\boldsymbol{u}_2 = \boldsymbol{a}_2 - P_{\boldsymbol{u}_1}\boldsymbol{a}_2 = \begin{bmatrix} -1 \\ 2 \end{bmatrix} - \begin{bmatrix} -1 \\ 0 \end{bmatrix} = \begin{bmatrix} 0 \\ 2 \end{bmatrix}. \tag{2.110}$$

We see that the new basis $\{\boldsymbol{u}_1, \boldsymbol{u}_2\}$ is indeed orthogonal as $\langle \boldsymbol{u}_1, \boldsymbol{u}_2 \rangle = 0$. Now the same vector $\boldsymbol{x} = [1, 2]^{\mathrm{T}}$ can be represented by the new orthogonal basis as

$$\boldsymbol{x} = \begin{bmatrix} 1 \\ 2 \end{bmatrix} = 1\boldsymbol{u}_1 + 1\boldsymbol{u}_2 = \begin{bmatrix} 1 \\ 0 \end{bmatrix} + \begin{bmatrix} 0 \\ 2 \end{bmatrix}. \tag{2.111}$$

In this particular case, both coefficients $c[1] = c[2] = 1$ happen to be 1, as illustrated in Fig. 2.4.

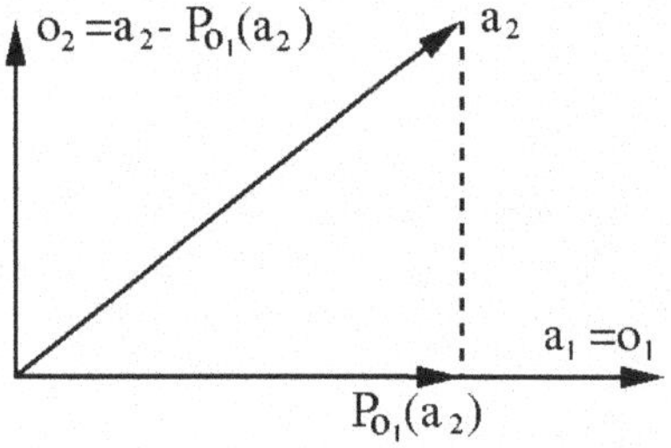

Figure 2.4 Gram-Schmidt orthogonalization.

2.1.5 Signal representation by standard bases

Here, we consider, as a special case of the orthogonal bases, the standard basis in the N-D space $\mathbb{R}^N$. When $N = 3$, a vector $\boldsymbol{v} = [x, y, z]^{\mathrm{T}}$ is conventionally represented as

$$\boldsymbol{v} = \begin{bmatrix} x \\ y \\ z \end{bmatrix} = x\boldsymbol{i} + y\boldsymbol{j} + z\boldsymbol{k} = x\begin{bmatrix} 1 \\ 0 \\ 0 \end{bmatrix} + y\begin{bmatrix} 0 \\ 1 \\ 0 \end{bmatrix} + z\begin{bmatrix} 0 \\ 0 \\ 1 \end{bmatrix}, \tag{2.112}$$

where $\boldsymbol{i} = [1, 0, 0]^{\mathrm{T}}$, $\boldsymbol{j} = [0, 1, 0]^{\mathrm{T}}$, and $\boldsymbol{k} = [0, 0, 1]^{\mathrm{T}}$ are the three standard (or canonical) basis vectors along each of the three mutually perpendicular axes. This standard basis $\{\boldsymbol{i}, \boldsymbol{j}, \boldsymbol{k}\}$ in $\mathbb{R}^3$ can be generalized to $\mathbb{R}^N$ spanned by a set of N standard basis vectors defined as

$$\boldsymbol{e}_1 = \begin{bmatrix} 1 \\ 0 \\ \vdots \\ 0 \end{bmatrix}, \quad \boldsymbol{e}_2 = \begin{bmatrix} 0 \\ 1 \\ \vdots \\ 0 \end{bmatrix}, \quad \cdots, \quad \boldsymbol{e}_N = \begin{bmatrix} 0 \\ \vdots \\ 0 \\ 1 \end{bmatrix}. \tag{2.113}$$

All components of the nth standard basis vector $\boldsymbol{e}_n$ are zero except the nth one which is 1; i.e., the mth component of the nth vector $\boldsymbol{e}_n$ is $e[m,n] = \delta[m-n]$. These standard basis vectors are indeed orthogonal as $\langle \boldsymbol{e}_m, \boldsymbol{e}_n \rangle = \delta[m-n]$ $(m, n = 1, \ldots, N)$, and they form an identity matrix $\boldsymbol{I} = [\boldsymbol{e}_1, \ldots, \boldsymbol{e}_N]$, which is a special unitary matrix satisfying $\boldsymbol{I}^* = \boldsymbol{I}^{-1} = \boldsymbol{I}^{\mathrm{T}} = \boldsymbol{I}$.

Given this standard basis in $\mathbb{R}^N$, a vector $\boldsymbol{x} = [x[1], \ldots, x[N]]^{\mathrm{T}}$ representing N samples of a time signal can be expressed as a linear combination of the N standard basis vectors:

$$\boldsymbol{x} = \sum_{n=1}^{N} x[n]\boldsymbol{e}_n = [\boldsymbol{e}_1, \ldots, \boldsymbol{e}_N]\boldsymbol{x} = \boldsymbol{I}\boldsymbol{x}, \tag{2.114}$$

and the mth component $x[m]$ of $\boldsymbol{x}$ is

$$x[m] = \sum_{n=1}^{N} x[n]e[m,n] = \sum_{n=1}^{N} x[n]\delta[m-n] \qquad m = 1, \ldots, N. \tag{2.115}$$

Comparing this equation with Eq. (1.3) in the previous chapter we see that they are actually in exactly the same form (except here the signal $\boldsymbol{x}$ has a finite number of N samples), indicating the fact that whenever a discrete time signal is given in the form of a vector $\boldsymbol{x} = [x[1], \ldots, x[N]]^{\mathrm{T}}$, it is represented implicitly by the standard basis; i.e., the signal is decomposed in time in terms of a set of components $x[m]$ each corresponding to a particular time segment $\delta[m-n]$ at $n = m$. However, while it may seem only natural and reasonable to decompose a signal into a set of time samples, or equivalently, to represent the signal vector by the standard basis, it is also possible, and sometime more beneficial, to decompose the signal into a set of components along some dimension other than time, or equivalently to represent the signal vector by an orthogonal basis which can be obtained by rotating the standard basis. This is an important point which is to be emphasized through out the book.

The concept of representing a discrete time signal $x[n]$ by the standard basis can be extended to the representation of a continuous time signal $x(t)$ $(0 < t < T)$. We first recall the unit square impulse function defined in Eq. (1.4):

$$\delta_\triangle(t) = \begin{cases} 1/\triangle & 0 \le t < \triangle \\ 0 & \text{else} \end{cases}, \tag{2.116}$$

based on which a set of basis functions $e_n(t) = \delta_\triangle(t - n\triangle)$ $(n = 0, \ldots, N-1)$ can be obtained by a translation of $n\triangle$ in time. These basis functions are obviously orthonormal:

$$\langle e_m(t), e_n(t) \rangle = \int_0^{\mathrm{T}} \delta_\triangle(t - m\triangle)\, \delta_\triangle(t - n\triangle)\, dt = \delta[m-n]. \tag{2.117}$$

Next, we sample the continuous time signal $x(t)$ with a sampling interval $\triangle = T/N$ to get a set of discrete samples $\{x[0], \ldots, x[N-1]\}$, and approximate the signal as

$$x(t) \approx \tilde{x}(t) = \sum_{n=0}^{N-1} x[n] e_n(t) = \sum_{n=0}^{N-1} x[n] \delta_\triangle(t - n\triangle)\, \triangle. \tag{2.118}$$

Here, $x[n]e_n(t)$ represents the nth segment of the signal over the time duration $n\triangle < t < (n+1)\triangle$, as illustrated in Fig. 2.5.

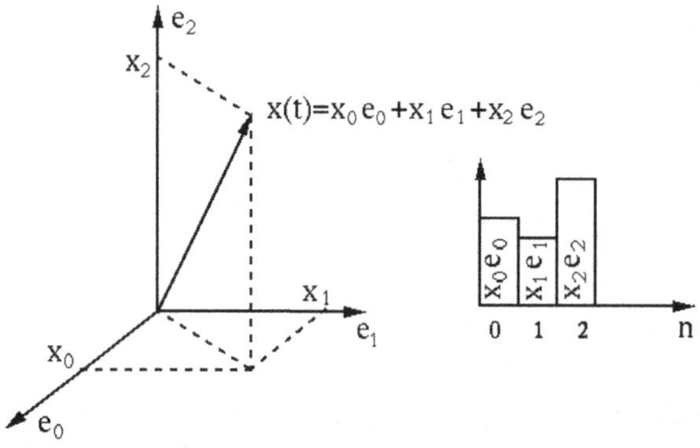

Figure 2.5 Vector representation of an N-D space ($N = 3$).

We see that each of these functions $e_n(t) = \delta_\triangle(t - n\triangle)$ represents a certain time segment, same as the standard basis $e[m, n] = \delta[m - n]$ in $\mathbb{C}^N$. However, we note that these functions $\delta_\triangle(t - n\triangle)$ do not form a basis that spans the space $\mathcal{L}^2$, as they are not complete; i.e., they can only approximate, but not precisely represent, a continuous function $x(t) \in \mathcal{L}^2$. This shortcoming can be overcome if we keep reducing the sampling interval $\triangle$ to get the Dirac delta at the limit $\triangle \to 0$:

$$\lim_{\triangle \to 0} \delta_\triangle(t) = \delta(t). \tag{2.119}$$

Now the summation in Eq. (2.118) becomes an integral, by which the function $x(t)$ can be precisely represented:

$$\lim_{\triangle \to 0} \tilde{x}(t) = \int x(\tau)\delta(t - \tau)\, d\tau = x(t). \tag{2.120}$$

This equation is actually the same as Eq. (1.9) in the previous chapter. Now we have obtained a continuum of uncountable basis functions $e_\tau(t) = \delta(t - \tau)$ (for all τ), which are complete as well as orthonormal; i.e., they form a standard basis of the function space $\mathcal{L}^2$, by which any continuous signal $x(t)$ can be represented, just as the standard basis $\boldsymbol{e}_n$ in $\mathbb{C}^N$ by which any discrete signal $x[n]$ can be represented.

Again, it may seem only natural to represent a continuous time signal $x(t)$ by the corresponding standard basis representing a sequence of time impulses $x(\tau)\delta(t - \tau)$. However, this is not the only way or the best way to represent the signal. The time signal can also be represented by a basis other than the standard basis $\delta(t-\tau)$, so that the signal is decomposed along some different dimension other than time. Such an alternative way of signal decomposition and representation may be desirable, as the signal can be more conveniently processed and analyzed, for whatever purpose of the signal processing task. This is actually the fundamental reason why different orthogonal transforms are developed, as will be discussed in detail in future chapters.

Fig. 2.6 illustrates the idea that any given vector $\boldsymbol{x}$ can be equivalently represented under different bases each corresponding to a different set of coefficients, such as the standard basis, an orthogonal basis (any rotated version of the standard basis), or an arbitrary basis not necessarily orthogonal at all. While non-orthogonal axes are never actually used, one always has many options in terms of what orthogonal basis to use.

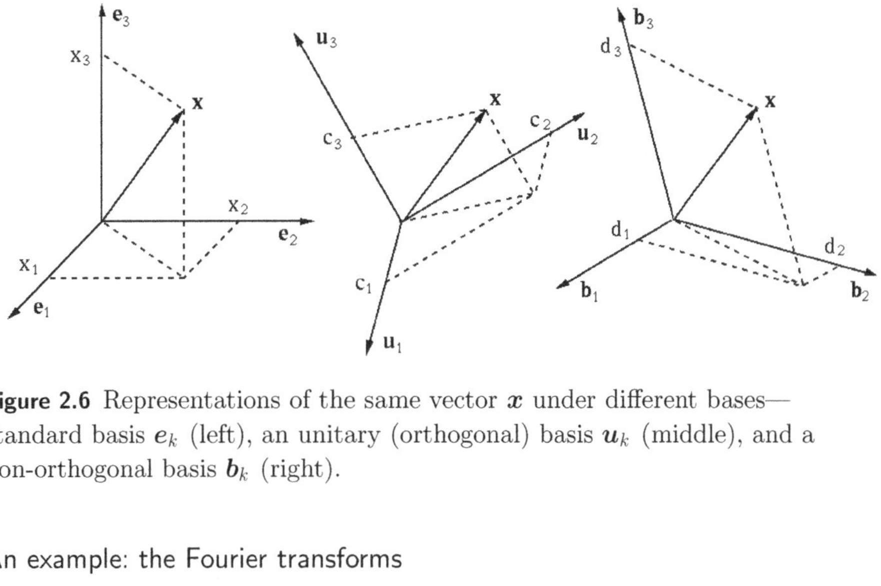

Figure 2.6 Representations of the same vector $\boldsymbol{x}$ under different bases—standard basis $\boldsymbol{e}_k$ (left), an unitary (orthogonal) basis $\boldsymbol{u}_k$ (middle), and a non-orthogonal basis $\boldsymbol{b}_k$ (right).

2.1.6 An example: the Fourier transforms

To illustrate how a vector can be represented by an orthogonal basis that spans the space, we consider the following four Fourier bases that span four different types of vector spaces for signals that are either continuous or discrete, of finite or infinite duration.

- $\boldsymbol{u}_k = [e^{j2\pi k0/N}, \ldots, e^{j2\pi k(N-1)/N}]^{\mathrm{T}}/\sqrt{N}$ $(k = 0, \ldots, N-1)$ form a set of N orthonormal basis vectors that span $\mathbb{C}^N$ (Eq. (1.40)):

$$\langle \boldsymbol{u}_k, \boldsymbol{u}_l \rangle = \frac{1}{N} \sum_{n=0}^{N-1} e^{j2\pi(k-l)n/N} = \delta[k-l]. \tag{2.121}$$

  Any vector $\boldsymbol{x} = [x[0], \ldots, x[N-1]]^{\mathrm{T}}$ in $\mathbb{C}^N$ can be expressed as

$$\boldsymbol{x} = \sum_{k=0}^{N-1} X[k]\boldsymbol{u}_k = \sum_{k=0}^{N-1} \langle \boldsymbol{x}, \boldsymbol{u}_k \rangle \boldsymbol{u}_k, \tag{2.122}$$

or in component form:

$$x[n] = \frac{1}{\sqrt{N}} \sum_{k=0}^{N-1} X[k] e^{j2\pi kn/N} \qquad 0 \le n \le N-1, \tag{2.123}$$

where the coefficient $X[k]$ is the projection of $\boldsymbol{x}$ onto $\boldsymbol{u}_k$:

$$X[k] = \langle \boldsymbol{x}, \boldsymbol{u}_k \rangle = \sum_{n=0}^{N-1} x[n]\overline{u}[n,k] = \frac{1}{\sqrt{N}} \sum_{n=0}^{N-1} x[n] e^{-j2\pi nk/N}. \tag{2.124}$$

- $\boldsymbol{u}(f) = [\ldots, e^{j2\pi mf/F}, \ldots]^{\mathrm{T}}/\sqrt{F}$ $(0 < f < F)$ form a set of uncountably infinite orthonormal basis vectors (of infinite dimensions) (Eq. (1.35)) that spans l^2 space of all square-summable vectors of infinite dimensions:

$$\langle \boldsymbol{u}_f, \boldsymbol{u}_{f'} \rangle = \frac{1}{F} \sum_{m=-\infty}^{\infty} e^{j2\pi(f-f')m/F} = \delta(f-f'). \tag{2.125}$$

Any vector $\boldsymbol{x} = [\ldots, x[n], \ldots]^{\mathrm{T}}$ in this space can be expressed as

$$\boldsymbol{x} = \int_{-\infty}^{\infty} X(f)\boldsymbol{u}(f)\, df = \int_{-\infty}^{\infty} \langle \boldsymbol{x}, \boldsymbol{u}(f) \rangle \boldsymbol{u}(f)\, df, \tag{2.126}$$

or in component form:

$$x[n] = \frac{1}{\sqrt{F}} \int_{-\infty}^{\infty} X(f) e^{j2\pi fn/F}\, df, \qquad -\infty < n < \infty, \tag{2.127}$$

where the coefficient function $X(f)$ is the projection of $\boldsymbol{x}$ onto $\boldsymbol{u}(f)$:

$$X(f) = \langle \boldsymbol{x}, \boldsymbol{u}(f) \rangle = \frac{1}{\sqrt{F}} \sum_{n=-\infty}^{\infty} x[n] e^{-j2\pi fn/F}. \tag{2.128}$$

- $u_k(t) = e^{j2\pi kt/T}/\sqrt{T}$ $(-\infty < k < \infty)$ form a set of infinite orthonormal basis functions (Eq. (1.33)) that spans the space of all square-integrable functions defined over $0 < t < T$:

$$\langle u_k(t), u_l(t) \rangle = \frac{1}{T} \int_0^{\mathrm{T}} e^{j2\pi(k-l)t/T}\, dt = \delta[k-l]. \tag{2.129}$$

Any function $x_T(t)$ in this space can be expressed as

$$x_T(t) = \sum_{k=-\infty}^{\infty} X[k] u_k(t) = \frac{1}{\sqrt{T}} \sum_{k=-\infty}^{\infty} X[k] e^{j2\pi kt/T}, \tag{2.130}$$

where the coefficient $X[k]$ is the projection of $x(t)$ onto the kth basis function $u_k(t)$:

$$X[k] = \langle x(t), u_k(t) \rangle = \int_{-\infty}^{\infty} x(t)\overline{u}_k(t)\, df = \frac{1}{\sqrt{T}} \int_{-\infty}^{\infty} x(t) e^{-j2\pi kt/T}\, dt. \tag{2.131}$$

- $u_f(t) = e^{j2\pi ft}$ $(-\infty < f < \infty)$ is a set of uncountably infinite orthonormal basis functions (Eq. (1.28)) that spans $\mathcal{L}^2$ space of all square-integrable functions defined over $-\infty < t < \infty$.

$$\langle u_f(t), u_{f'}(t)\rangle = \int_{-\infty}^{\infty} e^{j2\pi(f-f')t}\, dt = \delta(f-f'). \tag{2.132}$$

Any function $x(t)$ in this space can be expressed as

$$x(t) = \int_{-\infty}^{\infty} X(f) u_f(t)\, df = \int_{-\infty}^{\infty} X(f) e^{j2\pi ft}\, df, \tag{2.133}$$

where the coefficient function is the projection of $x(t)$ onto $u_f(t)$:

$$X(f) = \langle x(t), u_f(t)\rangle = \int_{-\infty}^{\infty} x(t)\overline{u}_f(t)\, df = \int_{-\infty}^{\infty} x(t) e^{-j2\pi ft}\, dt. \tag{2.134}$$

2.2 Unitary transformation and signal representation

2.2.1 Linear transformation

Definition 2.19.

- *Let V and W be two vector spaces. A transformation is a function or mapping $T: V \longrightarrow W$ that converts a vector $\boldsymbol{x} \in V$ to another vector $\boldsymbol{u} \in W$ denoted by: $T\boldsymbol{x} = \boldsymbol{u}$. If $W = V$, the linear transformation T is a linear operator.*
- *If the transformation is invertible; i.e., then a transformation that converts $\boldsymbol{u} \in W$ back to $\boldsymbol{x} \in V$ is an inverse transformation denoted by: $\boldsymbol{x} = T^{-1}\boldsymbol{u}$.*
- *An identity transformation maps a vector to itself: $I\boldsymbol{x} = \boldsymbol{x}$.*
  Obviously $TT^{-1} = T^{-1}T = I$ is an identity operator that maps a vector to itself:

$$\begin{aligned} TT^{-1}\boldsymbol{u} &= T(T^{-1}\boldsymbol{u}) = T\boldsymbol{x} = \boldsymbol{u} = I\boldsymbol{u}, \\ T^{-1}T\boldsymbol{x} &= T^{-1}(T\boldsymbol{x}) = T^{-1}\boldsymbol{u} = \boldsymbol{x} = I\boldsymbol{x}. \end{aligned} \tag{2.135}$$

- *A transformation T is linear if the following is true:*

$$T(a\boldsymbol{x} + b\boldsymbol{y}) = aT\boldsymbol{x} + bT\boldsymbol{y} \tag{2.136}$$

for any scalars $a, b \in \mathbb{C}$ and any vectors $\boldsymbol{x}, \boldsymbol{y} \in V$.

For example, the derivative and integral of a continuous function $x(t)$ are linear operators:

$$T_d x(t) = \frac{d}{dt}x(t) = \dot{x}(t), \qquad T_i x(t) = \int x(\tau)\, d\tau. \tag{2.137}$$

For another example, an M by N matrix $\boldsymbol{A}$ with its mnth element being $a[m,n] \in \mathbb{C}$ is a linear transformation $T_A : \mathbb{C}^N \longrightarrow \mathbb{C}^M$ that maps an N-D vector $\boldsymbol{x} \in \mathbb{C}^N$

to an M-D vector $\boldsymbol{y} \in \mathbb{C}^M$:

$$T_A \boldsymbol{x} = \boldsymbol{A}\boldsymbol{x} = \boldsymbol{y}, \tag{2.138}$$

or in component form:

$$\begin{bmatrix} y[1] \\ y[2] \\ \vdots \\ y[M] \end{bmatrix}_{M\times 1} = \begin{bmatrix} a[1,1] & a[1,2] & \cdots & a[1,N] \\ a[2,1] & a[2,2] & \cdots & a[2,N] \\ \vdots & \vdots & \ddots & \vdots \\ a[M,1] & a[M,2] & \cdots & a[M,N] \end{bmatrix}_{M\times N} \begin{bmatrix} x[1] \\ x[2] \\ \vdots \\ x[n] \end{bmatrix}_{N\times 1}. \tag{2.139}$$

If $M = N$, then $\boldsymbol{x}, \boldsymbol{y} \in \mathbb{C}^N$ and $\boldsymbol{A}$ becomes a linear operator.

However, note that the operation of translation $T_t \boldsymbol{x} = \boldsymbol{x} + \boldsymbol{t}$ is not a linear transformation:

$$T_t(a\boldsymbol{x} + b\boldsymbol{y}) = a\boldsymbol{x} + b\boldsymbol{y} + \boldsymbol{t} \neq aT_t\boldsymbol{x} + bT_t\boldsymbol{y} = a\boldsymbol{x} + b\boldsymbol{y} + (a+b)\boldsymbol{t}. \tag{2.140}$$

Definition 2.20.

- *For a linear transformation $T : V \to W$, if there is another transformation $T^* : W \to V$ so that*

$$\langle T\boldsymbol{x}, \boldsymbol{u} \rangle = \langle \boldsymbol{x}, T^*\boldsymbol{u} \rangle, \tag{2.141}$$

  for any $\boldsymbol{x} \in V$ and $\boldsymbol{u} \in W$, the T^* is called the Hermitian adjoint or simply adjoint of T.
- *If a linear operator $T : V \to V$ is its own adjoint; i.e.,*

$$\langle T\boldsymbol{x}, \boldsymbol{y} \rangle = \langle \boldsymbol{x}, T\boldsymbol{y} \rangle, \tag{2.142}$$

  for any $\boldsymbol{x}, \boldsymbol{y} \in V$, then T is called a self-adjoint or Hermitian transformation.

In the following, the terms "self-adjoint" and "Hermitian" are used interchangeably.

In particular, in the unitary space $\mathbb{C}^N$, let $\boldsymbol{B} = \boldsymbol{A}^*$ be the adjoint of matrix $\boldsymbol{A}$; i.e., $\langle \boldsymbol{A}\boldsymbol{x}, \boldsymbol{y} \rangle = \langle \boldsymbol{x}, \boldsymbol{B}\boldsymbol{y} \rangle$, then we have

$$\langle \boldsymbol{A}\boldsymbol{x}, \boldsymbol{y} \rangle = (\boldsymbol{A}\boldsymbol{x})^{\mathrm{T}} \overline{\boldsymbol{y}} = \boldsymbol{x}^{\mathrm{T}} \boldsymbol{A}^{\mathrm{T}} \overline{\boldsymbol{y}} = \langle \boldsymbol{x}, \boldsymbol{B}\boldsymbol{y} \rangle = \boldsymbol{x}^{\mathrm{T}} \overline{\boldsymbol{B}\boldsymbol{y}}. \tag{2.143}$$

Comparing the two sides, we get $\boldsymbol{A}^{\mathrm{T}} = \overline{\boldsymbol{B}}$; i.e., the adjoint matrix $\boldsymbol{B} = \boldsymbol{A}^* = \overline{\boldsymbol{A}}^{\mathrm{T}}$ is the conjugate transpose of $\boldsymbol{A}$:

$$\boldsymbol{A}^* = \overline{\boldsymbol{A}}^{\mathrm{T}}. \tag{2.144}$$

A matrix $\boldsymbol{A}$ is self-adjoint, or*Hermitian*, if $\boldsymbol{A} = \boldsymbol{A}^* = \overline{\boldsymbol{A}}^{\mathrm{T}}$; i.e.,

$$\langle \boldsymbol{A}x, \boldsymbol{y} \rangle = \langle \boldsymbol{x}, \boldsymbol{A}y \rangle. \tag{2.145}$$

In particular, when $\overline{\boldsymbol{A}} = \boldsymbol{A}$ is real, a self-adjoint matrix $\boldsymbol{A} = \boldsymbol{A}^* = \boldsymbol{A}^{\mathrm{T}}$ is symmetric. Note that we have always used $\boldsymbol{A}^*$ to denote the conjugate transpose of a matrix $\boldsymbol{A}$, which we now see is also the self-adjoint of $\boldsymbol{A}$, and the notation T^* is more generally used to denote the self-adjoint of any operator T.

In a function space, if T^* is the adjoint of a linear operator T, then the following holds:

$$\langle Tx(t), y(t)\rangle = \int Tx(t)\, \overline{y(t)}\, dt = \langle x(t), T^*y(t)\rangle = \int x(t)\, \overline{T^*y(t)}\, dt. \tag{2.146}$$

If $T = T^*$, it is a self-adjoint or Hermitian operator.

2.2.2 Eigenvalue problems

Definition 2.21. *If the application of an operator T to a vector $\boldsymbol{x} \in V$ results in another vector $\lambda\boldsymbol{x} \in V$, where $\lambda \in \mathbb{C}$ is a constant scalar:*

$$T\boldsymbol{x} = \lambda\boldsymbol{x}, \tag{2.147}$$

then the scalar λ is an eigenvalue of T and vector $\boldsymbol{x}$ is the corresponding eigenvector or eigenfunctions of T, and the equation above is called the eigenequation of the operator T. The set of all eigenvalues of an operator is called the spectrum of the operator.

Note that if $\boldsymbol{x}$ is an eigenvector of operator T then $-\boldsymbol{x}$ is also an eigenvector of T, as Eq. (2.147) is satisfied by either of the two vectors.

In a unitary space $\mathbb{C}^N$, an N by N matrix $\boldsymbol{A}$ is a linear operator and the associated eigenequation is

$$\boldsymbol{A}\boldsymbol{\phi}_n = \lambda_n\boldsymbol{\phi}_n \qquad n = 1, \ldots, N, \tag{2.148}$$

where λ_n and $\boldsymbol{\phi}_n$ are the nth eigenvalue and the corresponding eigenvector of $\boldsymbol{A}$, respectively.

In a function space, the nth-order differential operator $D^n = d^n/dt^n$ is a linear operator with the following eigenequation:

$$D^n\phi(t) = D^n\, e^{st} = \frac{d^n}{dt^n}\, e^{st} = s^n\, e^{st} = \lambda\phi(t), \tag{2.149}$$

where s is a complex scalar. Here, the $\lambda = s^n$ is the eigenvalue and the complex exponential $\phi(t) = e^{st}$ is the corresponding eigenfunction. More generally, we can write an Nth-order *linear constant coefficient differential equation (LCCDE)* as

$$\sum_{n=0}^{N} a_n \frac{d^n}{dt^n} y(t) = \left[\sum_{n=0}^{N} a_n D^n\right] y(t) = x(t), \tag{2.150}$$

where $\sum_{n=0}^{N} a_n D^n$ is a linear operator that is applied to function $y(t)$, representing the response of a linear system to an input $x(t)$. Obviously, the same complex exponential $\phi(t) = e^{st}$ is also the eigenfunction corresponding to the eigenvalue $\lambda = \sum_{k=0}^{n} a_k s^k$ of this operator.

Perhaps the most well-known eigenvalue problem in physics is the Schrödinger equation, which describes a particle in terms of its energy and the de Broglie wave function. Specifically, for a 1-D stationary single particle system, we have

$$\hat{\mathcal{H}}\psi(x) = \left[-\frac{\hbar^2}{2m}\frac{\partial^2}{\partial x^2} + V(x)\right]\psi(x) = \mathcal{E}\psi(x), \tag{2.151}$$

where

$$\hat{\mathcal{H}} = -\frac{\hbar^2}{2m}\frac{\partial^2}{\partial x^2} + V(x) \tag{2.152}$$

is the Hamiltonian operator, $\hbar$ is the Planck constant, m and $V(x)$ are the mass and potential energy of the particle, respectively. $\mathcal{E}$ is the eigenvalue of $\hat{\mathcal{H}}$, representing the total energy of the particle, and the wave function $\psi(x)$ is the corresponding eigenfunction, also called eigenstate, representing probability amplitude of the particle; i.e., $|\psi(x)|^2$ is the probability for the particle to be found at position x.

Theorem 2.4. *A self-adjoint operator has the following properties:*

1. *All eigenvalues are real.*
2. *The eigenvectors corresponding to different eigenvalues are orthogonal.*
3. *The family of all eigenvectors forms a complete orthogonal system.*

Proof: Let λ and μ be two different eigenvalues of a self-adjoint operator T, and $\boldsymbol{x}$ and $\boldsymbol{y}$ be the corresponding eigenvectors:

$$T\boldsymbol{x} = \lambda\boldsymbol{x}, \qquad T\boldsymbol{y} = \mu\boldsymbol{y}. \tag{2.153}$$

As $T = T^*$ is self-adjoint, we have

$$\langle T\boldsymbol{x}, \boldsymbol{y}\rangle = \langle \boldsymbol{x}, T\boldsymbol{y}\rangle. \tag{2.154}$$

Substituting $T\boldsymbol{x} = \lambda\boldsymbol{x}$ into Eq. (2.154) and letting $\boldsymbol{y} = \boldsymbol{x}$, we get

$$\langle \lambda\boldsymbol{x}, \boldsymbol{x}\rangle = \langle \boldsymbol{x}, \lambda\boldsymbol{x}\rangle; \quad \text{i.e.} \quad \lambda\langle \boldsymbol{x}, \boldsymbol{x}\rangle = \overline{\lambda}\langle \boldsymbol{x}, \boldsymbol{x}\rangle. \tag{2.155}$$

As in general $\langle \boldsymbol{x}, \boldsymbol{x}\rangle \neq 0$, we see that $\lambda = \overline{\lambda}$ is real. Next, we substitute $T\boldsymbol{x} = \lambda\boldsymbol{x}$ and $T\boldsymbol{y} = \mu\boldsymbol{y}$ into Eq. (2.154) and get:

$$\lambda\langle \boldsymbol{x}, \boldsymbol{y}\rangle = \overline{\mu}\langle \boldsymbol{x}, \boldsymbol{y}\rangle = \mu\langle \boldsymbol{x}, \boldsymbol{y}\rangle. \tag{2.156}$$

As in general $\lambda \neq \mu$, we get $\langle \boldsymbol{x}, \boldsymbol{y}\rangle = 0$; i.e., $\boldsymbol{x}$ and $\boldsymbol{y}$ are orthogonal. The proof of the third property is beyond the scope of the book and is therefore omitted. Q.E.D.

For example, the Hamiltonian operator $\hat{\mathcal{H}}$ in the Schrödinger equation is a self-adjoint operator with real eigenvalues $\mathcal{E}$ representing different energy levels corresponding to different eigenstates of the particle.

The third property in Theorem 2.4 indicates that the eigenvectors of a self-adjoint operator can be used as an orthogonal basis of a vector space, so that any vector in the space can be represented as a linear combination of these eigenvectors.

In space $\mathbb{C}^N$, let λ_k and $\boldsymbol{\phi}_k$ $(k = 1, \ldots, N)$ be the eigenvalues and the corresponding eigenvectors of a Hermitian matrix $\boldsymbol{A} = \boldsymbol{A}^*$, then its eigenequation can be written as

$$\boldsymbol{A}\boldsymbol{\phi}_k = \lambda_k \boldsymbol{\phi}_k, \qquad k = 1, \ldots, N. \tag{2.157}$$

We can further combine all N eigenequations to have

$$\boldsymbol{A}[\boldsymbol{\phi}_1, \ldots, \boldsymbol{\phi}_N] = [\boldsymbol{\phi}_1, \ldots, \boldsymbol{\phi}_N]\boldsymbol{\Lambda}, \qquad \text{or} \qquad \boldsymbol{A}\boldsymbol{\Phi} = \boldsymbol{\Phi}\boldsymbol{\Lambda}, \tag{2.158}$$

where matrices $\boldsymbol{\Phi}$ and $\boldsymbol{\Lambda}$ are defined as

$$\boldsymbol{\Phi} = [\boldsymbol{\phi}_1, \ldots, \boldsymbol{\phi}_N], \qquad \boldsymbol{\Lambda} = \begin{bmatrix} \lambda_1 & 0 & \cdots & 0 \\ 0 & \lambda_2 & \cdots & 0 \\ \vdots & \vdots & \ddots & \vdots \\ 0 & 0 & \cdots & \lambda_N \end{bmatrix}. \tag{2.159}$$

As $\boldsymbol{A}$ is a self-adjoint operator, its eigenvalues λ_k are real, and their corresponding eigenvectors $\boldsymbol{\phi}_k$ are orthogonal:

$$\langle \boldsymbol{\phi}_k, \boldsymbol{\phi}_l \rangle = \boldsymbol{\phi}_k^{\mathrm{T}} \overline{\boldsymbol{\phi}}_l = \delta[k-l], \tag{2.160}$$

and they form a complete orthogonal system to span the N-D unitary space. Also, $\boldsymbol{\Phi}$ is a unitary matrix satisfying

$$\boldsymbol{\Phi}^*\boldsymbol{\Phi} = \boldsymbol{I}, \qquad \text{or} \qquad \boldsymbol{\Phi}^* = \boldsymbol{\Phi}^{-1}. \tag{2.161}$$

The eigenequation in Eq. (2.158) can also be written in some other useful forms. First, pre-multiplying both sides of the equation by $\boldsymbol{\Phi}^{-1} = \boldsymbol{\Phi}^*$, we get

$$\boldsymbol{\Phi}^{-1}\boldsymbol{A}\boldsymbol{\Phi} = \boldsymbol{\Phi}^*\boldsymbol{A}\boldsymbol{\Phi} = \boldsymbol{\Lambda}; \tag{2.162}$$

i.e., the matrix $\boldsymbol{A}$ can be diagonalized by $\boldsymbol{\Phi}$. Alternatively, if we post-multiply both sides of Eq. (2.158) by $\boldsymbol{\Phi}^*$, we get

$$\boldsymbol{A} = \boldsymbol{\Phi}\boldsymbol{\Lambda}\boldsymbol{\Phi}^* = [\boldsymbol{\phi}_1, \boldsymbol{\phi}_2, \ldots, \boldsymbol{\phi}_N] \begin{bmatrix} \lambda_1 & 0 & \cdots & 0 \\ 0 & \lambda_2 & \cdots & 0 \\ \vdots & \vdots & \ddots & \vdots \\ 0 & 0 & \cdots & \lambda_N \end{bmatrix} \begin{bmatrix} \boldsymbol{\phi}_1^* \\ \boldsymbol{\phi}_2^* \\ \vdots \\ \boldsymbol{\phi}_N^* \end{bmatrix} = \sum_{k=1}^{N} \lambda_k \boldsymbol{\phi}_k \boldsymbol{\phi}_k^*; \tag{2.163}$$

i.e., the matrix $\boldsymbol{A}$ can be series expanded to become a linear combination of N *eigen-matrices* $\boldsymbol{\phi}_k \boldsymbol{\phi}_k^*$ $(k = 1, \ldots, N)$.

2.2.3 Eigenvectors of D^2 as Fourier basis

Here we consider a particular example of the self-adjoint operators, the second-order differential operator $D^2 = d^2/dt^2$ in $\mathcal{L}^2$-space, which is of important significance as its orthogonal eigenfunctions form the basis used in the Fourier transform.

First we show that D^2 is indeed a self-adjoint operator:

$$\langle D^2x(t), y(t)\rangle = \langle x(t), D^2y(t)\rangle, \tag{2.164}$$

where $x(t)$ and $y(t)$ are two functions defined over a certain time interval, such as $[0, T]$, and $D^2x(t) = \ddot{x}(t)$ is the second time derivative of function $x(t)$. Using integration by parts, we can show that this equation does hold:

$$\begin{aligned}\langle D^2x(t), y(t)\rangle &= \int_0^T \ddot{x}(t)\overline{y}(t)\,dt = \dot{x}(t)\overline{y}(t)\Big|_0^T - \int_0^T \dot{x}(t)\dot{\overline{y}}(t)\,dt \\ &= \dot{x}(t)\overline{y}(t)\Big|_0^T - x(t)\dot{\overline{y}}(t)\Big|_0^T + \int_0^T x(t)\ddot{\overline{y}}(t)\,dt = \langle x(t), D^2y(t)\rangle.\end{aligned} \tag{2.165}$$

Here, we have assumed all functions satisfy $x(0) = x(T)$, $\dot{x}(0) = \dot{x}(T)$, so that

$$\left[\dot{x}(t)\overline{y}(t) - x(t)\dot{\overline{y}}(t)\right]\Bigg|_0^T = 0. \tag{2.166}$$

Next, we find the eigenvalues and eigenfunctions of D^2 by solving this equation:

$$\begin{cases} D^2\phi(t) = \lambda\phi(t); \quad \text{i.e.} \quad \ddot{\phi}(t) - \lambda\phi(t) = 0 \\ \text{subject to:} \quad \phi(0) = \phi(T), \quad \dot{\phi}(0) = \dot{\phi}(T) \end{cases}. \tag{2.167}$$

Consider the following three cases:

1. $\lambda = 0$
   The equation becomes $\ddot{\phi}(t) = 0$ with solution $\phi(t) = c_1t + c_2$. Substituting this $\phi(t)$ into the boundary conditions, we have

   $$\phi(0) = c_2 = \phi(T) = c_1T + c_2. \tag{2.168}$$

   We get $c_1 = 0$ and the eigenfunction $\phi(t) = c_2$ is any constant.
2. $\lambda > 0$
   We assume $\phi(t) = e^{st}$ and substitute it into the equation to get

   $$(s^2 - \lambda)e^{st} = 0; \quad \text{i.e.} \quad s = \pm\sqrt{\lambda}. \tag{2.169}$$

   The solution is $\phi(t) = c\,e^{\pm\sqrt{\lambda}t}$. Substituting this into the boundary conditions, we have

   $$\phi(0) = c = \phi(T) = c\,e^{\pm\sqrt{\lambda}T}. \tag{2.170}$$

   Obviously, this equation holds only if $\lambda = 0$, as in the previous case.
3. $\lambda < 0$
   We assume $\lambda = -\omega^2$; i.e., $\sqrt{\lambda} = \pm j\omega$, and the solution is

   $$\phi(t) = c\,e^{\pm\sqrt{\lambda}t} = c\,e^{\pm j\omega t}. \tag{2.171}$$

   Substituting this into the boundary conditions we have

   $$\phi(0) = c = \phi(T) = c\,e^{\pm j\omega T}; \quad \text{i.e.} \quad e^{\pm j\omega T} = 1, \tag{2.172}$$

which can be solved to get

$$\omega T = 2k\pi; \quad \text{i.e.} \quad \omega = \frac{2k\pi}{T} = 2k\pi f_0 = k\omega_0, \qquad k = 0, \pm 1, \pm 2, \ldots, \tag{2.173}$$

where we have defined

$$f_0 = \frac{1}{T}, \quad \omega_0 = 2\pi f_0 = \frac{2\pi}{T}. \tag{2.174}$$

Now the eigenvalues and the corresponding eigenfunctions can be written as

$$\lambda_k = -\omega^2 = -(k\omega_0)^2 = -(2k\pi f_0)^2 = -(2k\pi/T)^2 \tag{2.175}$$

$$\phi_k(t) = c\, e^{\pm jk\omega_0} = c\, e^{\pm j2k\pi f_0} = c\, e^{\pm j2k\pi/T} \tag{2.176}$$

$$k = 0, \pm 1, \pm 2, \ldots.$$

In particular, when $k = 0$, we have $\lambda_k = 0$ and $\phi_0(t) = c$, which is the same as the first case above.

These eigenvalues and their corresponding eigenfunctions have the following properties:

- The eigenvalues are discrete, the gap between two consecutive eigenvalues is

$$\triangle\lambda_k = \lambda_{k+1} - \lambda_k. \tag{2.177}$$

- All eigenfunctions are also discrete with a frequency gap between two consecutive eigenfunctions:

$$\omega_0 = 2\pi f_0 = 2\pi/T. \tag{2.178}$$

- All eigenfunctions $\phi_k(t)$ are periodic with period T:

$$\phi_k(t+T) = e^{j2k\pi(t+T)/T} = e^{j2k\pi t/T} e^{j2k\pi} = e^{j2k\pi t/T} = \phi_k(t). \tag{2.179}$$

According to the properties of self-adjoint operators discussed above, the eigenfunctions $\phi_k(t)$ of D^2 form a complete orthogonal system. The orthogonality can be easily verified:

$$\langle \phi_k(t), \phi_l(t) \rangle = c^2 \int_0^{\mathrm{T}} e^{jk\omega_0 t} e^{-jl\omega_0 t}\, dt = c^2 \int_0^{\mathrm{T}} e^{j2\pi(k-l)t/T}\, dt$$
$$= c^2 \int_0^{\mathrm{T}} \cos\left(\frac{2\pi(k-l)t}{T}\right) dt + jc^2 \int_0^{\mathrm{T}} \sin\left(\frac{2\pi(k-l)t}{T}\right) dt = \begin{cases} T & k = l \\ 0 & k \neq l \end{cases}. \tag{2.180}$$

If we let $c = 1/\sqrt{T}$, then the eigenfunctions become

$$\phi_k(t) = \frac{1}{\sqrt{T}} e^{j2k\pi t/T} = \frac{1}{\sqrt{T}} e^{j2k\pi f_0 t}, \tag{2.181}$$

which are orthonormal:

$$\langle \phi_k(t), \phi_l(t) \rangle = \frac{1}{T} \int_0^{\mathrm{T}} e^{j2\pi(k-l)t/T}\, dt = \delta[k-l]. \tag{2.182}$$

This is actually Eq. (2.129). As a complete orthogonal system, these orthogonal eigenfunctions form a basis to span the function space over $[0, T]$; i.e., all periodic functions $x_T(t) = x_T(t+T)$ can be represented as a linear combination of these basis functions:

$$x_T(t) = \sum_{k=-\infty}^{\infty} X[k]\phi_k(t) = \sum_{k=-\infty}^{\infty} X[k]e^{j2k\pi f_0 t} = \sum_{k=-\infty}^{\infty} X[k]e^{jk\omega_0}, \tag{2.183}$$

where $X[k]$ $(k = 0, \pm 1, \pm 2, \ldots)$ are the coefficients given in Eq. (2.131). This is the Fourier expansion, to be discussed in detail in the next chapter.

The expansion of a non-periodic function can be similarly obtained if we let $T \to \infty$ so that at the limit a periodic function $x_T(t)$ becomes non-periodic, and the following will take place:

- The discrete variables $k\omega_0 = 2k\pi/T$ $(k = 0, \pm 1, \pm 2, \ldots)$ become a continuous variable $-\infty < \omega < \infty$.
- The gap between two consecutive eigenvalues becomes zero; i.e., $\triangle\lambda_k \to 0$, so the discrete eigenvalues $\lambda_k = -(2k\pi/T)^2$ become a continuous eigenvalue function $\lambda = -\omega^2$.
- The frequency gap ω_0 between two consecutive eigenfunctions becomes zero, so the discrete eigenfunctions $\phi_k(t) = e^{j2k\pi t/T}$ $(k = 0, \pm 1, \pm 2, \ldots)$ become a set of uncountable non-periodic eigenfunctions $\phi_f(t) = e^{j2\pi f t}$ for all $-\infty < f < \infty$.

We see that the same self-adjoint operator D^2 is now defined over a different interval $(-\infty, \infty)$ and correspondingly its eigenfunctions $\phi(t) = e^{j\omega t} = e^{j2\pi f t} = \phi(t, f)$ become a continuous function of f as well as t and they form a complete orthogonal system spanning the function space of all non-periodic functions:

$$\langle \phi_f(t), \phi_{f'}(t) \rangle = \int_{-\infty}^{\infty} e^{j2\pi(f-f')t}\, dt = \delta(f - f'). \tag{2.184}$$

This is actually Eq. (2.132). Now $\phi_f(t)$ becomes a set of uncountably infinite basis functions and any non-periodic square-integrable function $x(t)$ can be represented as

$$x(t) = \int_{-\infty}^{\infty} X(f)\phi_f(t)\, df = \int_{-\infty}^{\infty} X(f)e^{j2\pi f t}\, df, \tag{2.185}$$

where $X(f)$ is the weighting function given in Eq. (2.134). This is the Fourier transform, to be discussed in detail in the next chapter.

2.2.4 Unitary transformations

Definition 2.22. *A linear transformation $U : V \to W$ is a unitary transformation if it conserves inner products:*

$$\langle \boldsymbol{x}, \boldsymbol{y} \rangle = \langle U\boldsymbol{x}, U\boldsymbol{y} \rangle. \tag{2.186}$$

In particular, if the vectors are real with symmetric inner product $\langle \boldsymbol{x}, \boldsymbol{y} \rangle = \langle \boldsymbol{y}, \boldsymbol{x} \rangle$, then U is an orthogonal transformation.

Obviously, a unitary transformation also conserves any measurement based on the inner product, such as the norm of a vector, the distance and angle between two vectors, and the projection of one vector on another. Also, if in particular $\boldsymbol{x} = \boldsymbol{y}$, we have

$$\langle \boldsymbol{x}, \boldsymbol{x} \rangle = ||\boldsymbol{x}||^2 = \langle U\boldsymbol{x}, U\boldsymbol{x} \rangle = ||U\boldsymbol{x}||^2; \tag{2.187}$$

i.e., the unitary transformation conserves the vector norm (length). This is Parseval's identity for a generic unitary transformation $U\boldsymbol{x}$. Owing to this property, a unitary operation $R : V \to V$ can be intuitively interpreted as a rotation in space V. [1]

Theorem 2.5. *A linear transformation U is unitary if and only if its adjoint U^* is equal to its inverse U^{-1}:*

$$U^* = U^{-1}; \qquad i.e. \qquad U^*U = UU^* = I. \tag{2.188}$$

Proof: We let $U\boldsymbol{y} = \boldsymbol{d}$; i.e., $\boldsymbol{y} = U^{-1}\boldsymbol{d}$ in Eq. (2.186), and get

$$\langle U\boldsymbol{x}, \boldsymbol{d} \rangle = \langle \boldsymbol{x}, U^{-1}\boldsymbol{d} \rangle = \langle \boldsymbol{x}, U^*\boldsymbol{d} \rangle; \tag{2.189}$$

i.e., $U^{-1} = U^*$. Q.E.D.

Eq. (2.188) can be used as an alternative definition for the unitary operator.

In the generalized Fourier expansion in Eqs. (2.87) and (2.88) based on the Plancherel Theorem (Theorem. 2.3), the coefficient vector $\boldsymbol{c} = [\ldots, c[k], \ldots]^{\mathrm{T}}$ composed of $c[k] = \langle \boldsymbol{x}, \boldsymbol{u}_k \rangle$ can be considered as a transformation $\boldsymbol{c} = U\boldsymbol{x}$. Assuming another transformation $\boldsymbol{d} = U\boldsymbol{y}$, we get the Plancherel identity (Eq. (2.85)):

$$\langle \boldsymbol{x}, \boldsymbol{y} \rangle = \langle \boldsymbol{c}, \boldsymbol{d} \rangle = \langle U\boldsymbol{x}, U\boldsymbol{y} \rangle, \tag{2.190}$$

indicating that the inner product is conserved by U; i.e., the generalized Fourier expansion $\boldsymbol{c} = U\boldsymbol{x}$ is actually a unitary transformation. In particular when $\boldsymbol{y} = \boldsymbol{x}$, the above equation becomes Parseval's identity:

$$\langle \boldsymbol{x}, \boldsymbol{x} \rangle = ||\boldsymbol{x}||^2 = \langle U\boldsymbol{x}, U\boldsymbol{x} \rangle = \langle \boldsymbol{c}, \boldsymbol{c} \rangle = ||\boldsymbol{c}||^2. \tag{2.191}$$

When a unitary operator U is applied to an orthonormal basis $\{\boldsymbol{u}_k\}$, the basis is rotated to become another orthonormal basis $\{\boldsymbol{v}_k = U\boldsymbol{u}_k\}$ that spans the same space:

$$\langle \boldsymbol{v}_k, \boldsymbol{v}_l \rangle = \langle U\boldsymbol{u}_k, U\boldsymbol{u}_l \rangle = \langle \boldsymbol{u}_k, \boldsymbol{u}_l \rangle = \delta[k-l]. \tag{2.192}$$

Specially, when a unitary operator U is applied to the standard basis $\{\boldsymbol{e}_k\}$, this basis is rotated to become a unitary basis $\{\boldsymbol{u}_k = U\boldsymbol{e}_k\}$.

[1] Strictly speaking, a unitary transformation may also correspond to other norm-preserving operations such as reflection and inversion, which could all be treated as rotations in the most general sense.

2.2.5 Unitary transformations in N-D space

We consider specifically the unitary transformation in the N-D unitary space $\mathbb{C}^N$.

Definition 2.23. *A matrix $\boldsymbol{U}$ is unitary if it conserves inner products:*

$$\langle \boldsymbol{Ux}, \boldsymbol{Uy} \rangle = \langle \boldsymbol{x}, \boldsymbol{y} \rangle. \tag{2.193}$$

Theorem 2.6. *A matrix $\boldsymbol{U}$ is unitary if and only if $\boldsymbol{U}^*\boldsymbol{U} = \boldsymbol{I}$; i.e., the following two statements are equivalent:*

$$(a)\ \ \langle \boldsymbol{Ux}, \boldsymbol{Uy} \rangle = \langle \boldsymbol{x}, \boldsymbol{y} \rangle \tag{2.194}$$

$$(b)\ \ \boldsymbol{U}^*\boldsymbol{U} = \boldsymbol{U}\boldsymbol{U}^* = \boldsymbol{I}; \quad i.e., \quad \boldsymbol{U}^{-1} = \boldsymbol{U}^*. \tag{2.195}$$

Proof: We first show if (b) then (a):

$$\langle \boldsymbol{Ux}, \boldsymbol{Uy} \rangle = (\boldsymbol{Ux})^{\mathrm{T}}\overline{\boldsymbol{Uy}} = \boldsymbol{x}^{\mathrm{T}}\boldsymbol{U}^{\mathrm{T}}\overline{\boldsymbol{U}}\overline{\boldsymbol{y}} = \boldsymbol{x}^{\mathrm{T}}\boldsymbol{I}\overline{\boldsymbol{y}} = \langle \boldsymbol{x}, \boldsymbol{y} \rangle. \tag{2.196}$$

Next we show if (a) then (b). (a) can be written as

$$(\boldsymbol{Ux})^*\boldsymbol{Ux} = \boldsymbol{x}^*\boldsymbol{U}^*\boldsymbol{Ux} = \boldsymbol{x}^*\boldsymbol{x}; \tag{2.197}$$

i.e.,

$$\boldsymbol{x}^*(\boldsymbol{U}^*\boldsymbol{U} - \boldsymbol{I})\boldsymbol{x} = 0. \tag{2.198}$$

Since in general $\boldsymbol{x} \neq 0$, we must have $\boldsymbol{U}^*\boldsymbol{U} = \boldsymbol{I}$. Post-multiplying this equation by $\boldsymbol{U}^{-1}$, we get $\boldsymbol{U}^* = \boldsymbol{U}^{-1}$. Pre-multiplying this new equation by $\boldsymbol{U}$, we get $\boldsymbol{U}\boldsymbol{U}^* = \boldsymbol{I}$. Q.E.D.

As (a) and (b) in Theorem 2.6 are equivalent, either of them can be used as the definition of a unitary matrix. If a unitary matrix $\overline{\boldsymbol{U}} = \boldsymbol{U}$ is real; i.e., $\boldsymbol{U}^{-1} = \boldsymbol{U}^{\mathrm{T}}$, then it is called an *orthogonal matrix.*

A unitary matrix $\boldsymbol{U}$ has the following properties:

- Unitary transformation $\boldsymbol{Ux}$ conserves the vector norm; i.e., $||\boldsymbol{Ux}|| = ||\boldsymbol{x}||$ for any $\boldsymbol{x} \in \mathbb{C}^N$.
- All eigenvalues $\{\lambda_1, \ldots, \lambda_N\}$ of $\boldsymbol{U}$ have an absolute value of 1: $|\lambda_k| = 1$; i.e., they lie on the unit circle in the complex plain.
- The determinant of $\boldsymbol{U}$ has an absolute value of 1: $|det(\boldsymbol{U})| = 1$. This can be easily seen as $det(\boldsymbol{U}) = \prod_{k=1}^{N} \lambda_k$.
- All column (or row) vectors of $\boldsymbol{U} = [\boldsymbol{u}_1, \ldots, \boldsymbol{u}_N]$ are orthonormal:

$$\langle \boldsymbol{u}_k, \boldsymbol{u}_l \rangle = \delta[k - l]. \tag{2.199}$$

The last property indicates that the column (row) vectors $\{\boldsymbol{u}_k\}$ form an orthogonal basis that spans $\mathbb{C}^N$. Any vector $\boldsymbol{x} = [x[1], \ldots, x[N]]^{\mathrm{T}} \in \mathbb{C}^N$ represented by the standard basis $\boldsymbol{I} = [\boldsymbol{e}_1, \ldots, \boldsymbol{e}_N]$ as

$$\boldsymbol{x} = \begin{bmatrix} x[1] \\ \vdots \\ x[N] \end{bmatrix} = \sum_{n=1}^{N} x[n]\boldsymbol{e}_n = [\boldsymbol{e}_1, \ldots, \boldsymbol{e}_N] \begin{bmatrix} x[1] \\ \vdots \\ x[N] \end{bmatrix} = \boldsymbol{I}\boldsymbol{x} \tag{2.200}$$

can also be represented by the basis $\boldsymbol{U} = [\boldsymbol{u}_1, \ldots, \boldsymbol{u}_N]$ as

$$\boldsymbol{x} = \boldsymbol{I}\boldsymbol{x} = \boldsymbol{U}\boldsymbol{U}^*\boldsymbol{x} = \boldsymbol{U}\boldsymbol{c} = [\boldsymbol{u}_1, \ldots, \boldsymbol{u}_N] \begin{bmatrix} c[1] \\ \vdots \\ c[N] \end{bmatrix} = \sum_{k=1}^{N} c[k]\boldsymbol{u}_k, \tag{2.201}$$

where we have defined

$$\boldsymbol{c} = \begin{bmatrix} c[1] \\ \vdots \\ c[N] \end{bmatrix} = \boldsymbol{U}^*\boldsymbol{x} = \begin{bmatrix} \boldsymbol{u}_1^* \\ \vdots \\ \boldsymbol{u}_N^* \end{bmatrix} \boldsymbol{x}; \qquad \text{i.e.} \quad c[k] = \boldsymbol{u}_k^*\boldsymbol{x} = \langle \boldsymbol{x}, \boldsymbol{u}_k \rangle. \tag{2.202}$$

Combining the two equations we get

$$\begin{cases} \boldsymbol{c} = \boldsymbol{U}^*\boldsymbol{x}, \\ \boldsymbol{x} = \boldsymbol{U}\boldsymbol{c}. \end{cases} \tag{2.203}$$

This is the generalized Fourier transform in Eqs. (2.87) and (2.88), by which a vector $\boldsymbol{x}$ is rotated to become another vector $\boldsymbol{c}$.

This result can be extended to the continuous transformation first given in Eqs. (2.90) and (2.92) for signal vectors in the form of continuous functions. In general, corresponding to any given unitary transformation U, a signal vector $\boldsymbol{x} \in H$ can be alternatively represented by a coefficient vector $\boldsymbol{c} = U^*\boldsymbol{x}$ (where $\boldsymbol{c}$ can be either a set of discrete coefficients $c[k]$ or a continuous function $c(f)$). The original signal vector $\boldsymbol{x}$ can always be reconstructed from $\boldsymbol{c}$ by applying U on both sides of $\boldsymbol{c} = U^*\boldsymbol{x}$ to get $U\boldsymbol{c} = UU^*\boldsymbol{x} = I\boldsymbol{x} = \boldsymbol{x}$; i.e., we get a unitary transform pair in the most general form:

$$\begin{cases} \boldsymbol{c} = U^*\boldsymbol{x}, \\ \boldsymbol{x} = U\boldsymbol{c}. \end{cases} \tag{2.204}$$

The first equation is the forward transform that maps the signal vector $\boldsymbol{x}$ to a coefficient vector $\boldsymbol{c}$, while the second equation is the inverse transform by which the signal is reconstructed. In particular, when $U = I$ is an identity operator, both equations in Eq. (2.204) become an identity $\boldsymbol{x} = I\boldsymbol{x} = \boldsymbol{x}$; i.e., no transformation is carried out.

Previously we considered the rotation of a given vector $\boldsymbol{x}$ We next consider the rotation of the basis that spans the space. Specifically, let $\{\boldsymbol{a}_k\}$ be an arbitrary basis of $\mathbb{C}^N$ (not necessarily orthogonal), then any vector $\boldsymbol{x}$ can be represented in terms of a set of coefficients $c[k]$:

$$\boldsymbol{x} = \sum_{k=1}^{N} c[k] \boldsymbol{a}_k. \tag{2.205}$$

Rotating this vector by a unitary matrix $\boldsymbol{U}$, we get a new vector:

$$\boldsymbol{U}\boldsymbol{x} = \boldsymbol{U} \left[\sum_{k=1}^{N} c[k] \boldsymbol{a}_k \right] = \sum_{k=1}^{N} c[k] \boldsymbol{U} \boldsymbol{a}_k = \sum_{k=1}^{N} c[k] \boldsymbol{a}'_k = \boldsymbol{y}. \tag{2.206}$$

This equation indicates that vector $\boldsymbol{y}$ after the rotation can still be represented by the same set of coefficients $c[k]$, if the basis $\{\boldsymbol{a}_k\}$ is also rotated the same way to become $\boldsymbol{a}'_k = \boldsymbol{U}\boldsymbol{a}_k$, as illustrated in Fig. 2.7(a) for the 2-D case.

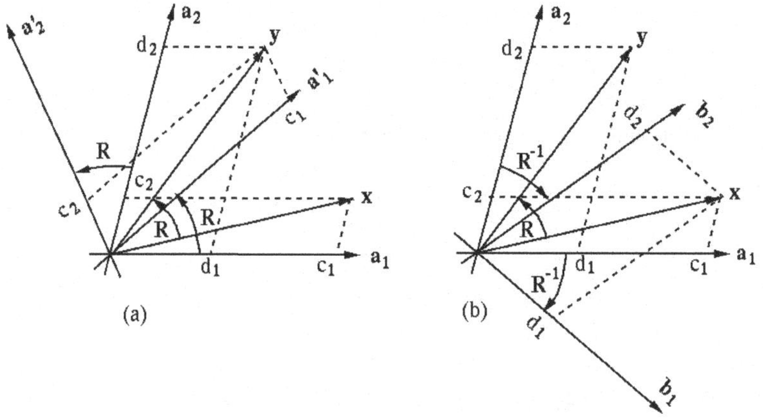

Figure 2.7 Rotation of vectors and bases.

Under the original basis $\{\boldsymbol{a}_k\}$, the rotated vector $\boldsymbol{y}$ can be represented in terms of a set of new coefficients $\{\ldots, d[k], \ldots\}$:

$$\boldsymbol{y} = \sum_{k=1}^{N} d[k] \boldsymbol{a}_k = [\boldsymbol{a}_1, \ldots, \boldsymbol{a}_N] \begin{bmatrix} d[1] \\ \vdots \\ d[N] \end{bmatrix}. \tag{2.207}$$

The N new coefficients $d[n]$ can be obtained by solving this linear equation system with N equations (with $O(N^3)$ complexity).

On the other hand, if we rotate $\boldsymbol{y}$ in the opposite direction by the inverse matrix $\boldsymbol{U}^{-1} = \boldsymbol{U}^*$, we of course get $\boldsymbol{x}$ back:

$$\boldsymbol{U}^{-1}\boldsymbol{y} = \boldsymbol{U}^{-1} \left[\sum_{k=1}^{N} d[k] \boldsymbol{a}_k \right] = \sum_{k=1}^{N} d[k] \boldsymbol{U}^{-1} \boldsymbol{a}_k = \sum_{k=1}^{N} d[k] \boldsymbol{b}_k, \tag{2.208}$$

where $\boldsymbol{b}_k = \boldsymbol{U}^{-1}\boldsymbol{a}_k = \boldsymbol{U}^*\boldsymbol{a}_k$ is the kth vector of a new basis obtained by rotating $\boldsymbol{a}_k$ of the old basis in the opposite direction. In fact, as

$$P_{\boldsymbol{a}_k}(\boldsymbol{y}) = \frac{\langle \boldsymbol{y}, \boldsymbol{a}_k \rangle}{||\boldsymbol{a}_k||} = \frac{\langle \boldsymbol{U}\boldsymbol{x}, \boldsymbol{U}\boldsymbol{b}_k \rangle}{||\boldsymbol{U}\boldsymbol{a}_k||} = \frac{\langle \boldsymbol{x}, \boldsymbol{b}_k \rangle}{||\boldsymbol{n}_k||} = P_{\boldsymbol{b}_k}(\boldsymbol{x}), \tag{2.209}$$

we see that the scalar projection of the new vector $\boldsymbol{y} = \boldsymbol{U}\boldsymbol{x}$ onto the old basis $\boldsymbol{a}_k$ is the same as that of the old vector $\boldsymbol{x}$ onto the new basis $\boldsymbol{b}_k = \boldsymbol{U}^{-1}\boldsymbol{a}_k$. In other words, a rotation of the vector is equivalent to a rotation in the opposite direction of the basis, as one would intuitively expect. This is illustrated in Fig. 2.7(b). A rotation in a 3-D space is illustrated in Fig. 2.8.

In summary, multiplication of a vector $\boldsymbol{x} \in \mathbb{C}^N$ by a unitary matrix corresponds to a rotation of the vector. The transformation pair in Eq. (2.203) can, therefore, be interpreted as a rotation of $\boldsymbol{x}$ to get the coefficients $\boldsymbol{U}^*\boldsymbol{x} = \boldsymbol{c}$, and a rotation of $\boldsymbol{c}$ in the opposite direction $\boldsymbol{x} = \boldsymbol{U}\boldsymbol{c}$ gets the original vector $\boldsymbol{x}$ back. Moreover, a different rotation $\boldsymbol{d} = \boldsymbol{V}^*\boldsymbol{x}$ by another unitary matrix $\boldsymbol{V}$ will result in a different set of coefficients $\boldsymbol{d}$, and these two sets of coefficients $\boldsymbol{c}$ and $\boldsymbol{d}$ are also related by a rotation corresponding to a unitary matrix $\boldsymbol{W} = \boldsymbol{V}^*\boldsymbol{U}$:

$$\boldsymbol{d} = \boldsymbol{V}^*\boldsymbol{x} = \boldsymbol{V}^*\boldsymbol{U}\boldsymbol{c} = \boldsymbol{W}\boldsymbol{c}. \tag{2.210}$$

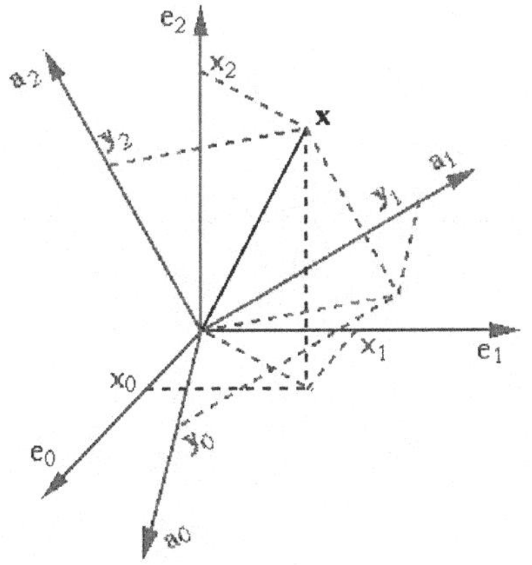

Figure 2.8 Rotation of coordinate system.

Example 2.5: In Example 2.2, a vector $\boldsymbol{x} = [1,\ 2]^{\mathrm{T}} = 1\boldsymbol{e}_1 + 2\boldsymbol{e}_2$ is represented under a basis composed of $\boldsymbol{a}_1 = [1, 0]^{\mathrm{T}}$ and $\boldsymbol{a}_2 = [-1, 2]^{\mathrm{T}}$:

$$\boldsymbol{x} = 1\boldsymbol{a}_1 + 2\boldsymbol{a}_2 = 2\begin{bmatrix} 1 \\ 0 \end{bmatrix} + 1\begin{bmatrix} -1 \\ 2 \end{bmatrix} = \begin{bmatrix} 1 \\ 2 \end{bmatrix}. \tag{2.211}$$

This basis $\{\boldsymbol{a}_1, \boldsymbol{a}_2\}$ can be rotated by $\theta = 45°$ by an orthogonal matrix

$$\boldsymbol{R} = \begin{bmatrix} \cos\theta & -\sin\theta \\ \sin\theta & \cos\theta \end{bmatrix} = 0.707 \begin{bmatrix} 1 & -1 \\ 1 & 1 \end{bmatrix} \tag{2.212}$$

to become a new basis $\{\boldsymbol{b}_1, \boldsymbol{b}_2\}$:

$$\boldsymbol{b}_1 = \boldsymbol{R}\boldsymbol{a}_1 = \boldsymbol{R}\begin{bmatrix}1\\0\end{bmatrix} = 0.707\begin{bmatrix}1\\1\end{bmatrix}, \quad \boldsymbol{b}_2 = \boldsymbol{R}\boldsymbol{a}_2 = \boldsymbol{R}\begin{bmatrix}-1\\2\end{bmatrix} = 0.707\begin{bmatrix}-3\\1\end{bmatrix}. \tag{2.213}$$

Under this new basis, $\boldsymbol{x}$ is represented as

$$\begin{aligned}\boldsymbol{x} &= c'[1]\boldsymbol{b}_1 + c'[2]\boldsymbol{b}_2 = c'[1]\,0.707\begin{bmatrix}1\\1\end{bmatrix} + c'[2]\,0.707\begin{bmatrix}-3\\1\end{bmatrix}\\ &= 0.707\begin{bmatrix}1 & -3\\1 & 1\end{bmatrix}\begin{bmatrix}c'[1]\\c'[2]\end{bmatrix} = \begin{bmatrix}1\\2\end{bmatrix}.\end{aligned} \tag{2.214}$$

Solving this, we get $c'[1] = 2.47$ and $c'[2] = 0.35$; i.e., $\boldsymbol{x} = 2.47\boldsymbol{b}_1 + 0.35\boldsymbol{b}_2$, as shown in Fig. 2.9. In this case, the coefficients $c'[1]$ and $c'[2]$ cannot be found as the projections of $\boldsymbol{x}$ onto basis vectors $\boldsymbol{b}_1$ and $\boldsymbol{b}_2$ as they are not orthogonal. We see that the same vector $\boldsymbol{x}$ can be equivalently represented by different bases:

$$\boldsymbol{x} = 1\boldsymbol{e}_1 + 2\boldsymbol{e}_2 = 2\boldsymbol{a}_1 + 1\boldsymbol{a}_2 = 2.47\boldsymbol{b}_1 + 0.35\boldsymbol{b}_2. \tag{2.215}$$

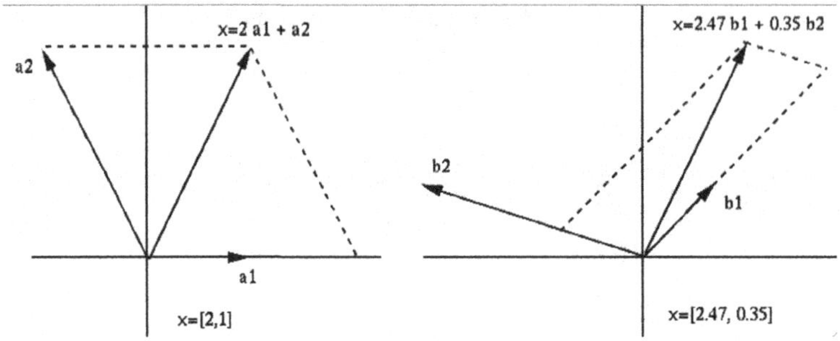

Figure 2.9 Rotation of a basis.

2.3 Projection theorem and signal approximation

2.3.1 Projection theorem and pseudo-inverse

A signal in a high-dimensional space (possibly infinite-dimensional) may need to be approximated in a lower dimensional subspace for various reasons, such as computational complexity reduction and data compression. Although a complete basis is necessary to represent any given vector in a vector space, it is still possible to approximate the vector in a subspace if a certain error is allowed. Also, a continuous function may not be accurately representable in a finite-dimensional space, but it may still be needed to approximate the function in such a space for a certain desired signal processing. The issue in such an approximation is how to minimize the error.

Let H be a Hilbert space (finite or infinite dimensional), and $U \subset H$ be an M-D subspace spanned by a set of M basis vectors $\{\boldsymbol{a}_1, \ldots, \boldsymbol{a}_M\}$ (not necessarily orthogonal), and assume a given vector $\boldsymbol{x} \in H$ is approximated by a vector $\hat{\boldsymbol{x}} \in U$:

$$\boldsymbol{x} \approx \hat{\boldsymbol{x}} = \sum_{k=1}^{M} c[k]\boldsymbol{a}_k. \tag{2.216}$$

An error vector is defined as

$$\tilde{\boldsymbol{x}} = \boldsymbol{x} - \hat{\boldsymbol{x}} = \boldsymbol{x} - \sum_{k=1}^{M} c[k]\boldsymbol{a}_k. \tag{2.217}$$

The least-squares error of the approximation is defined as

$$\varepsilon = ||\tilde{\boldsymbol{x}}||^2 = \langle \tilde{\boldsymbol{x}}, \tilde{\boldsymbol{x}} \rangle. \tag{2.218}$$

The goal is to find a set of coefficients $c[1], \ldots, c[M]$ so that the error ε is minimized.

Theorem 2.7. *(The projection theorem) The least-squares error $\varepsilon = ||\tilde{\boldsymbol{x}}||^2$ of the approximation by Eq. (2.216) is minimized if and only if the error vector $\tilde{\boldsymbol{x}} = \boldsymbol{x} - \hat{\boldsymbol{x}}$ is orthogonal to the subspace U:*

$$\tilde{\boldsymbol{x}} \perp U; \quad \textit{i.e.,} \quad \tilde{\boldsymbol{x}} \perp \boldsymbol{a}_k, \quad k = 1, \ldots, M. \tag{2.219}$$

Proof: Let $\hat{\boldsymbol{x}}$ and $\hat{\boldsymbol{x}}'$ be two vectors both in the subspace U, where $\hat{\boldsymbol{x}}'$ is arbitrary but $\hat{\boldsymbol{x}}$ is the projection of $\boldsymbol{x}$ onto U; i.e., $(\boldsymbol{x} - \hat{\boldsymbol{x}}) \perp U$. As $\hat{\boldsymbol{x}} - \hat{\boldsymbol{x}}'$ is also a vector in U, we have $(\boldsymbol{x} - \hat{\boldsymbol{x}}) \perp (\hat{\boldsymbol{x}} - \hat{\boldsymbol{x}}')$; i.e., $\langle \boldsymbol{x} - \hat{\boldsymbol{x}},\ \hat{\boldsymbol{x}} - \hat{\boldsymbol{x}}' \rangle = 0$. Now consider the error associated with $\hat{\boldsymbol{x}}'$:

$$\begin{aligned} &||\boldsymbol{x} - \hat{\boldsymbol{x}}'||^2 = ||\boldsymbol{x} - \hat{\boldsymbol{x}} + \hat{\boldsymbol{x}} - \hat{\boldsymbol{x}}'||^2 \\ &= ||\boldsymbol{x} - \hat{\boldsymbol{x}}||^2 + \langle \boldsymbol{x} - \hat{\boldsymbol{x}},\ \hat{\boldsymbol{x}} - \hat{\boldsymbol{x}}' \rangle + \langle \hat{\boldsymbol{x}} - \hat{\boldsymbol{x}}',\ \boldsymbol{x} - \hat{\boldsymbol{x}} \rangle + ||\hat{\boldsymbol{x}} - \hat{\boldsymbol{x}}'||^2 \\ &= ||\boldsymbol{x} - \hat{\boldsymbol{x}}||^2 + ||\hat{\boldsymbol{x}} - \hat{\boldsymbol{x}}'||^2 \geq ||\boldsymbol{x} - \hat{\boldsymbol{x}}||^2. \end{aligned} \tag{2.220}$$

We see that the error $||\boldsymbol{x} - \hat{\boldsymbol{x}}'||^2$ associated with $\hat{\boldsymbol{x}}'$ is always greater than the error $||\boldsymbol{x} - \hat{\boldsymbol{x}}||^2$ associated with $\hat{\boldsymbol{x}}$, unless $\hat{\boldsymbol{x}}' = \hat{\boldsymbol{x}}$; i.e., the error is minimized if and only if the approximation is $\hat{\boldsymbol{x}}$, the projection of $\boldsymbol{x}$ onto the subspace U. Q.E.D.

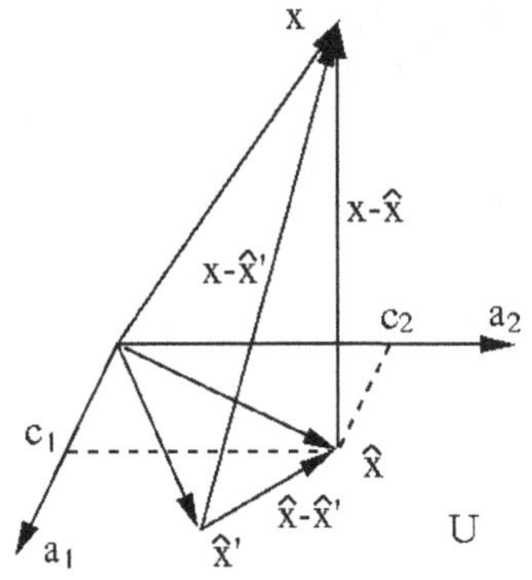

Figure 2.10 Projection theorem.

This theorem can be understood intuitively as shown in Fig. 2.10, where a vector $\boldsymbol{x}$ in a 3-D space is approximated by a vector $\hat{\boldsymbol{x}}$ in a 2-D subspace $\hat{\boldsymbol{x}} = c[1]\boldsymbol{a}_1 + c[2]\boldsymbol{a}_2$. The error $\varepsilon = ||\tilde{\boldsymbol{x}}||^2 = ||\boldsymbol{x} - \hat{\boldsymbol{x}}||^2$ is indeed minimum if $\boldsymbol{x} - \hat{\boldsymbol{x}}$ is orthogonal to the 2-D plane spanned by the basis vectors $\boldsymbol{a}_1$ and $\boldsymbol{a}_2$, as any other vector $\hat{\boldsymbol{x}}'$ in this plane would be associated with a larger error; i.e., the approximation $\hat{\boldsymbol{x}}$ is the *projection* of $\boldsymbol{x}$ onto the subspace U.

The coefficients corresponding to the optimal approximation can be found based on the projection theorem. As the minimum error vector $\tilde{\boldsymbol{x}} = \boldsymbol{x} - \hat{\boldsymbol{x}}$ has to be orthogonal to each of the basis vectors that span the subspace U, we have

$$\langle \tilde{\boldsymbol{x}}, \boldsymbol{a}_l \rangle = \langle \boldsymbol{x} - \sum_{k=1}^{M} c[k]\boldsymbol{a}_k,\ \boldsymbol{a}_l \rangle = \langle \boldsymbol{x}, \boldsymbol{a}_l \rangle - \sum_{k=1}^{M} c[k]\langle \boldsymbol{a}_k, \boldsymbol{a}_l \rangle = 0$$
$$l = 1, \ldots, M; \tag{2.221}$$

i.e.

$$\langle \boldsymbol{x}, \boldsymbol{a}_l \rangle = \sum_{k=1}^{M} c[k]\langle \boldsymbol{a}_k, \boldsymbol{a}_l \rangle, \qquad m = 1, \ldots, M. \tag{2.222}$$

These M equations can be written in matrix form:

$$\begin{bmatrix} \langle \boldsymbol{x}, \boldsymbol{a}_1 \rangle \\ \vdots \\ \langle \boldsymbol{x}, \boldsymbol{a}_M \rangle \end{bmatrix}_{M\times 1} = \begin{bmatrix} \langle \boldsymbol{a}_1, \boldsymbol{a}_1 \rangle & \cdots & \langle \boldsymbol{a}_M, \boldsymbol{a}_1 \rangle \\ \vdots & \ddots & \vdots \\ \langle \boldsymbol{a}_1, \boldsymbol{a}_M \rangle & \cdots & \langle \boldsymbol{a}_M, \boldsymbol{a}_M \rangle \end{bmatrix}_{M\times M} \begin{bmatrix} c[1] \\ \vdots \\ c[M] \end{bmatrix}_{M\times 1}. \tag{2.223}$$

Solving this system of M equations and M unknowns, we get the optimal coefficients $c[k]$ and the vector $\boldsymbol{x}$ can be approximated in the M-D subspace as shown in Eq. (2.216).

In particular, if the basis vectors of the Hilbert space are orthogonal; i.e., $\langle \boldsymbol{a}_k, \boldsymbol{a}_l \rangle = 0$ for all $k \neq l$, then all off-diagonal components of the $M \times M$ matrix in Eq. (2.223) are zero, and each of the coefficients can be obtained independently:

$$c[k] = \frac{\langle \boldsymbol{x}, \boldsymbol{a}_k \rangle}{\langle \boldsymbol{a}_k, \boldsymbol{a}_k \rangle} = \frac{\langle \boldsymbol{x}, \boldsymbol{a}_k \rangle}{||\boldsymbol{a}_k||^2}, \qquad k = 1, \dots, M. \tag{2.224}$$

Equation (2.216) now becomes

$$\hat{\boldsymbol{x}} = \sum_{k=1}^{M} c[k]\boldsymbol{a}_k = \sum_{k=1}^{M} \frac{\langle \boldsymbol{x}, \boldsymbol{a}_k \rangle}{||\boldsymbol{a}_k||^2} \boldsymbol{a}_k = \sum_{k=1}^{M} \boldsymbol{p}_{\boldsymbol{a}_k}(\boldsymbol{x}). \tag{2.225}$$

We see that $\hat{\boldsymbol{x}}$ is the vector sum of the projections of $\boldsymbol{x}$ onto each of the basis vectors $\boldsymbol{a}_k$ $(k = 1, \dots, M)$ of the subspace U. Moreover, if the basis is also normalized; i.e. $\langle \boldsymbol{a}_k, \boldsymbol{a}_l \rangle = \delta[k - l]$, then we have

$$c[k] = \langle \boldsymbol{x}, \boldsymbol{a}_k \rangle, \qquad k = 1, \dots, M, \tag{2.226}$$

and Eq. (2.216) becomes

$$\hat{\boldsymbol{x}} = \sum_{k=1}^{M} \langle \boldsymbol{x}, \boldsymbol{a}_k \rangle \boldsymbol{a}_k. \tag{2.227}$$

Consider for example the space $\mathbb{C}^N$ spanned by a basis $\{\boldsymbol{a}_1, \dots, \boldsymbol{a}_N\}$ (not necessarily orthogonal). We wish to express a given vector $\boldsymbol{x} \in \mathbb{C}^N$ in an M-D subspace spanned by M basis vectors $\{\boldsymbol{a}_1, \dots, \boldsymbol{a}_M\}$ as

$$\boldsymbol{x} = \begin{bmatrix} x[1] \\ \vdots \\ x[N] \end{bmatrix}_{N \times 1} = \sum_{k=1}^{M} c[k]\boldsymbol{a}_k = [\boldsymbol{a}_1, \cdots, \boldsymbol{a}_M]_{N \times M} \begin{bmatrix} c[1] \\ \vdots \\ c[M] \end{bmatrix}_{M \times 1} = \boldsymbol{A}\boldsymbol{c}. \tag{2.228}$$

This equation system is over-determined with only M unknowns $c[1], \dots, c[M]$ but $N > M$ equations. As the N by M non-square matrix $\boldsymbol{A}$ is not invertible, the system has no solution in general, indicating the impossibility of representing the N-D vector $\boldsymbol{x}$ in an M-D subspace. However, based on the projection theorem, we can find the optimal approximation of $\boldsymbol{x}$ in the M-D subspace by solving Eq. (2.223). In this case the inner products in the equation become $\langle \boldsymbol{x}, \boldsymbol{a}_k \rangle = \boldsymbol{a}_k^* \boldsymbol{x}$ and $\langle \boldsymbol{a}_k, \boldsymbol{a}_l \rangle = \boldsymbol{a}_l^* \boldsymbol{a}_k$, Eq. (2.223) can be written as

$$\boldsymbol{A}^* \boldsymbol{x} = \boldsymbol{A}^* \boldsymbol{A} \boldsymbol{c}, \tag{2.229}$$

where $\boldsymbol{A}^* \boldsymbol{A}$ is an M by M square matrix and, therefore, invertible. Premultiplying its inverse $(\boldsymbol{A}^{\mathrm{T}} \boldsymbol{A})^{-1}$ on both sides, we can find the optimal solution for $\boldsymbol{c}$ of the overdetermined equation system corresponding to the minimum least-squares error

$$\boldsymbol{c} = (\boldsymbol{A}^* \boldsymbol{A})^{-1} \boldsymbol{A}^* \boldsymbol{x} = \boldsymbol{A}^{-} \boldsymbol{x}, \tag{2.230}$$

where

$$\boldsymbol{A}^{-} = (\boldsymbol{A}^* \boldsymbol{A})^{-1} \boldsymbol{A}^* \tag{2.231}$$

is an M by N matrix, known as the *generalized inverse or pseudo-inverse* of the N by M matrix $\boldsymbol{A}$ (Appendix A), and we have $\boldsymbol{A}^{-}\boldsymbol{A} = \boldsymbol{I}$. [2] If all N basis vectors can be used, then $\boldsymbol{A}$ becomes an N by N square matrix and the pseudo-inverse becomes the regular inverse:

$$\boldsymbol{A}^{-} = \boldsymbol{A}^{-1}(\boldsymbol{A}^{*})^{-1}\boldsymbol{A}^{*} = \boldsymbol{A}^{-1}, \tag{2.233}$$

and the coefficients can be found simply by

$$\boldsymbol{c} = \boldsymbol{A}^{-1}\boldsymbol{x}. \tag{2.234}$$

If the basis is orthogonal; i.e., $\langle \boldsymbol{a}_k, \boldsymbol{a}_l \rangle = 0$ for all $k \neq l$, the M coefficients can be found as

$$c[k] = \frac{\langle \boldsymbol{x}, \boldsymbol{a}_k \rangle}{\langle \boldsymbol{a}_k, \boldsymbol{a}_k \rangle} = \frac{\langle \boldsymbol{x}, \boldsymbol{a}_k \rangle}{||\boldsymbol{a}_k||^2} \qquad k = 1, \ldots, M, \tag{2.235}$$

with complexity $O(M^2)$. Moreover, if the basis is orthonormal with $||\boldsymbol{a}_k||^2 = 1$, the coefficients become

$$c[k] = \langle \boldsymbol{x}, \boldsymbol{a}_k \rangle = \boldsymbol{a}_k^{*}\boldsymbol{x} \qquad k = 1, \ldots, M, \tag{2.236}$$

and the approximation becomes

$$\hat{\boldsymbol{x}}_M = \sum_{k=1}^{M} c[k]\boldsymbol{a}_k = \sum_{k=1}^{M} \langle \boldsymbol{x}, \boldsymbol{a}_k \rangle \boldsymbol{a}_k. \tag{2.237}$$

This is actually the unitary transformation in Eq. (2.203). We see that, under any orthonormal basis $\{\boldsymbol{a}_k\}$ of $\mathbb{C}^N$, a given vector $\boldsymbol{x}$ can always be optimally approximated in the M-D subspace ($M < N$) with least-squares error

$$\begin{aligned}
\varepsilon &= ||\tilde{\boldsymbol{x}}||^2 = \langle \boldsymbol{x} - \hat{\boldsymbol{x}}_M, \boldsymbol{x} - \hat{\boldsymbol{x}}_M \rangle \\
&= \langle \boldsymbol{x}, \boldsymbol{x} \rangle - \langle \boldsymbol{x}, \hat{\boldsymbol{x}}_M \rangle - \langle \hat{\boldsymbol{x}}_M, \boldsymbol{x} \rangle + \langle \hat{\boldsymbol{x}}_M, \hat{\boldsymbol{x}}_M \rangle \\
&= ||\boldsymbol{x}||^2 - \sum_{k=1}^{M} \langle \boldsymbol{x}, \boldsymbol{a}_k \rangle \bar{c}[k] - \sum_{k=1}^{M} c[k] \langle \boldsymbol{a}_k, \boldsymbol{x} \rangle + \sum_{k=1}^{M} |c[k]|^2 \\
&= ||\boldsymbol{x}||^2 - \sum_{k=1}^{M} |c[k]|^2 = \sum_{k=M+1}^{N} |c[k]|^2 \geq 0.
\end{aligned} \tag{2.238}$$

The last equation is due to Parseval's identity $||\boldsymbol{x}||^2 = ||\boldsymbol{c}||^2 = \sum_{k=1}^{N} |c[k]|^2$. When $M \to N$, the sequence $\hat{\boldsymbol{x}}_M$ converges to $\boldsymbol{x}$:

$$\lim_{M \to N} \hat{\boldsymbol{x}}_M = \lim_{M \to N} \sum_{k=1}^{M} c[k]\boldsymbol{a}_k = \sum_{k=1}^{N} c[k]\boldsymbol{a}_k = \boldsymbol{x}, \tag{2.239}$$

[2] The pseudo-inverse in Eq. (2.231) is for the case where $\boldsymbol{A}$ has more columns than rows ($M < N$ in this case). If $\boldsymbol{A}$ has more rows than columns ($M > N$ in this case), the pseudo-inverse becomes

$$\boldsymbol{A}^{-} = \boldsymbol{A}^{*}(\boldsymbol{A}\boldsymbol{A}^{*})^{-1}. \tag{2.232}$$

and Eq. (2.238) becomes

$$\lim_{M \to N} \varepsilon = ||\boldsymbol{x}||^2 - \sum_{k=1}^{N} |c[k]|^2 = 0. \tag{2.240}$$

This is, of course, Parseval's identity $||\boldsymbol{x}||^2 = ||\boldsymbol{c}||^2$.

Example 2.6: Consider a 3-D Euclidean space $\mathbb{R}^3$ spanned by a set of three linearly independent vectors:

$$\boldsymbol{a}_1 = \begin{bmatrix} 1 \\ 0 \\ 0 \end{bmatrix}, \quad \boldsymbol{a}_2 = \begin{bmatrix} 1 \\ 1 \\ 0 \end{bmatrix}, \quad \boldsymbol{a}_3 = \begin{bmatrix} 1 \\ 1 \\ 1 \end{bmatrix}. \tag{2.241}$$

We want to find two coefficients $c[1]$ and $c[2]$ so that a given vector $\boldsymbol{x} = [1, 2, 3]^{\mathrm{T}}$ can be optimally approximated as $\hat{\boldsymbol{x}} = c[1]\boldsymbol{a}_1 + c[2]\boldsymbol{a}_2$ in the 2-D subspace spanned by $\boldsymbol{a}_1$ and $\boldsymbol{a}_2$. First we construct a matrix composed of $\boldsymbol{a}_1$ and $\boldsymbol{a}_2$:

$$\boldsymbol{A} = [\boldsymbol{a}_1, \boldsymbol{a}_2] = \begin{bmatrix} 1 & 1 \\ 0 & 1 \\ 0 & 0 \end{bmatrix}. \tag{2.242}$$

Next, we find the pseudo-inverse of $\boldsymbol{A}$:

$$\boldsymbol{A}^{-} = (\boldsymbol{A}^{\mathrm{T}}\boldsymbol{A})^{-1}\boldsymbol{A}^{\mathrm{T}} = \begin{bmatrix} 1 & -1 & 0 \\ 0 & 1 & 0 \end{bmatrix}. \tag{2.243}$$

The two coefficients can then be obtained as

$$\boldsymbol{c} = \begin{bmatrix} c[1] \\ c[2] \end{bmatrix} = \boldsymbol{A}^{-}\boldsymbol{x} = \begin{bmatrix} 1 & -1 & 0 \\ 0 & 1 & 0 \end{bmatrix} \begin{bmatrix} 1 \\ 2 \\ 3 \end{bmatrix} = \begin{bmatrix} -1 \\ 2 \end{bmatrix}. \tag{2.244}$$

The optimal approximation, therefore, is

$$\hat{\boldsymbol{x}} = c[1]\boldsymbol{a}_1 + c[2]\boldsymbol{a}_2 = -1 \begin{bmatrix} 1 \\ 0 \\ 0 \end{bmatrix} + 2 \begin{bmatrix} 1 \\ 1 \\ 0 \end{bmatrix} = \begin{bmatrix} 1 \\ 2 \\ 0 \end{bmatrix}, \tag{2.245}$$

which is indeed the projection of $\boldsymbol{x} = [1, 2, 3]^{\mathrm{T}}$ onto the 2-D subspace spanned by $\boldsymbol{a}_1$ and $\boldsymbol{a}_2$.

Alternatively, if we want to approximate $\boldsymbol{x}$ by $\boldsymbol{a}_2$ and $\boldsymbol{a}_3$ as $\hat{\boldsymbol{x}} = c[2]\boldsymbol{a}_2 + c[3]\boldsymbol{a}_3$, we have

$$\boldsymbol{A} = [\boldsymbol{a}_2, \boldsymbol{a}_3] = \begin{bmatrix} 1 & 1 \\ 1 & 1 \\ 0 & 1 \end{bmatrix} \qquad \boldsymbol{A}^{-} = \frac{1}{2} \begin{bmatrix} 1 & 1 & -2 \\ 0 & 0 & 2 \end{bmatrix}, \tag{2.246}$$

and

$$\mathbf{c} = \mathbf{A}^{-}\mathbf{x} = \begin{bmatrix} -1.5 \\ 3 \end{bmatrix}, \qquad \hat{\mathbf{x}} = c[2]\mathbf{a}_2 + c[3]\mathbf{a}_3 = -1.5 \begin{bmatrix} 1 \\ 1 \\ 0 \end{bmatrix} + 3 \begin{bmatrix} 1 \\ 1 \\ 1 \end{bmatrix} = \begin{bmatrix} 1.5 \\ 1.5 \\ 3 \end{bmatrix}. \tag{2.247}$$

If all three basis vectors can be used, then the coefficients can be found as

$$\mathbf{c} = \mathbf{A}^{-1}\mathbf{x} = [\mathbf{a}_1, \mathbf{a}_2, \mathbf{a}_3]^{-1}\mathbf{x} = \begin{bmatrix} 1 & -1 & 0 \\ 0 & 1 & -1 \\ 0 & 0 & 1 \end{bmatrix} \begin{bmatrix} 1 \\ 2 \\ 3 \end{bmatrix} = \begin{bmatrix} -1 \\ -1 \\ 3 \end{bmatrix}, \tag{2.248}$$

and $\mathbf{x}$ can be precisely represented as

$$\mathbf{x} = c[1]\mathbf{a}_1 + c[2]\mathbf{a}_2 + c[3]\mathbf{a}_3 = \mathbf{A}\mathbf{c} = \begin{bmatrix} 1 \\ 2 \\ 3 \end{bmatrix}. \tag{2.249}$$

2.3.2 Signal approximation

As discussed above, a signal vector can be represented equivalently under different bases that span the space, in terms of the total energy (Parseval's equality). However, these representations may differ drastically in terms of how different types of information contained in the signal are concentrated in different signal components and represented by the coefficients. Sometimes certain advantages can be gained from one particular basis compared with another, depending on the specific application. In the following we consider two simple examples to illustrate such issues.

Example 2.7: Given a signal $x(t) = t$ defined over $0 \le t < 2$ (undefined outside the range), we want to optimally approximate it in a subspace spanned by the following two bases.

- First we use the standard functions $e_1(t)$ and $e_2(t)$:

$$\hat{x}(t) = c[1]e_1(t) + c[2]e_2(t), \tag{2.250}$$

where $e_1(t)$ and $e_2(t)$ are defined as

$$e_1(t) = \begin{cases} 1, & 0 \le t < 1 \\ 0, & 1 \le t < 2 \end{cases}, \qquad e_2(t) = \begin{cases} 0, & 0 \le t < 1 \\ 1, & 1 \le t < 2 \end{cases}. \tag{2.251}$$

These two basis functions are obviously orthonormal $\langle e_m(t), e_n(t)\rangle = \delta[m - n]$. Following the projection theorem, the coefficients $c[1]$ and $c[2]$ can be found by solving these two simultaneous equations (Eq. (2.222)):

$$c[1] \int_0^2 e_1(t)e_1(t)\,dt + c[2] \int_0^2 e_2(t)e_1(t)\,dt = \int_0^2 x(t)e_1(t)\,dt$$
$$c[1] \int_0^2 e_1(t)e_2(t)\,dt + c[2] \int_0^2 e_2(t)e_2(t)\,dt = \int_0^2 x(t)e_2(t)\,dt.$$

As $e_1(t)$ and $e_2(t)$ are orthonormal, the equation system becomes decoupled and the two coefficients $c[1]$ and $c[2]$ can be obtained independently as the projections of $x(t)$ onto each of the basis functions.

$$c[1] = \int_0^2 x(t)e_1(t)\,dt = \int_0^1 t\,dt = 0.5$$
$$c[2] = \int_0^2 x(t)e_2(t)\,dt = \int_1^2 t\,dt = 1.5. \tag{2.252}$$

Now the signal $x(t)$ can be approximated as

$$\hat{x}(t) = 0.5e_1(t) + 1.5e_2(t) = \begin{cases} 0.5, & 0 \le t < 1 \\ 1.5, & 1 \le t < 2 \end{cases}. \tag{2.253}$$

- Next, we use two different basis functions $u_1(t)$ and $u_2(t)$:

$$\hat{x}(t) = d[1]u_1(t) + d[2]u_2(t), \tag{2.254}$$

where

$$u_1(t) = \frac{1}{\sqrt{2}}[e_1(t) + e_2(t)] = \frac{1}{\sqrt{2}},$$
$$u_2(t) = \frac{1}{\sqrt{2}}[e_1(t) - e_2(t)] = \begin{cases} 1/\sqrt{2}, & 0 \le t < 1 \\ -1/\sqrt{2}, & 1 \le t < 2 \end{cases}.$$

Again, these two basis functions are orthonormal $\langle u_k(t), u_l(t)\rangle = \delta[k-l]$, and the two coefficients $d[1]$ and $d[2]$ can be obtained independently as

$$d[1] = \int_0^2 x(t)u_1(t)\,dt = \sqrt{2}, \qquad d[2] = \int_0^2 x(t)u_2(t)\,dt = -\frac{1}{\sqrt{2}}. \tag{2.255}$$

The approximation is

$$\hat{x}(t) = \sqrt{2}u_1(t) - \frac{1}{\sqrt{2}}u_2(t) = \begin{cases} 0.5, & 0 \le t < 1 \\ 1.5, & 1 \le t < 2 \end{cases}. \tag{2.256}$$

We see that the approximations based on these two different bases happen to be identical, as illustrated in Fig. 2.11.

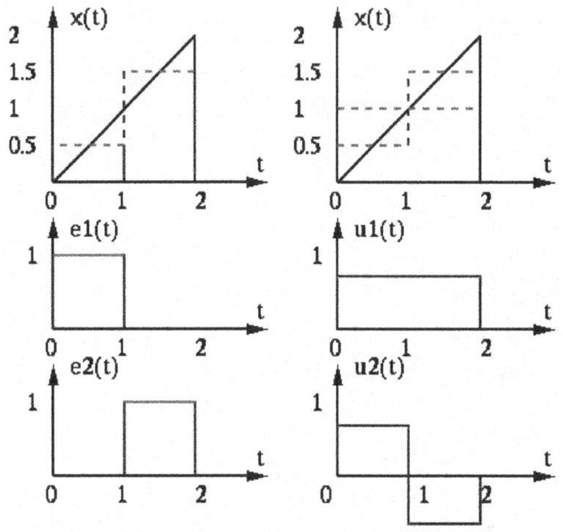

Figure 2.11 Approximation of a function by two different bases $\{e_1(t),\ e_2(t)\}$ (left) and $\{u_1(t),\ u_2(t)\}$ (right).

We can make the following observations:

- The first basis $\{e_1(t), e_2(t)\}$ is the standard basis; the two coefficients $c[1]$ and $c[2]$ represent the average values of the signal during two consecutive time segments.
- The second basis $\{u_1(t), u_2(t)\}$ represents the signal $x(t)$ in a totally different way. The first coefficient $d[1]$ represents the average of the signal (zero frequency), while the second coefficient $d[2]$ represents the variation of the signal in terms of the difference between the first half and the second. (In fact they correspond to the first two frequency components in several orthogonal transforms, including the discrete Fourier transform, discrete cosine transform, Walsh-Hadamard transform, etc.)
- The second basis $\{u_1(t), u_2(t)\}$ is a rotated version of the first basis $\{e_1(t), e_2(t)\}$, as shown in Fig. 2.12, and naturally they produce the same approximation $\hat{x}(t)$. Consequently, the two sets of coefficients $\{c[1], c[2]\}$ and $\{d[1], d[2]\}$ are related by an orthogonal matrix representing the rotation by an angle $\theta = -45°$:

$$\begin{bmatrix} d[2] \\ d[1] \end{bmatrix} = \begin{bmatrix} \cos\theta & \sin\theta \\ -\sin\theta & \cos\theta \end{bmatrix} \begin{bmatrix} c[2] \\ c[1] \end{bmatrix} = \begin{bmatrix} \sqrt{2}/2 & -\sqrt{2}/2 \\ \sqrt{2}/2 & \sqrt{2}/2 \end{bmatrix} \begin{bmatrix} 1/2 \\ 3/2 \end{bmatrix} = \begin{bmatrix} -1/\sqrt{2} \\ \sqrt{2} \end{bmatrix}. \tag{2.257}$$

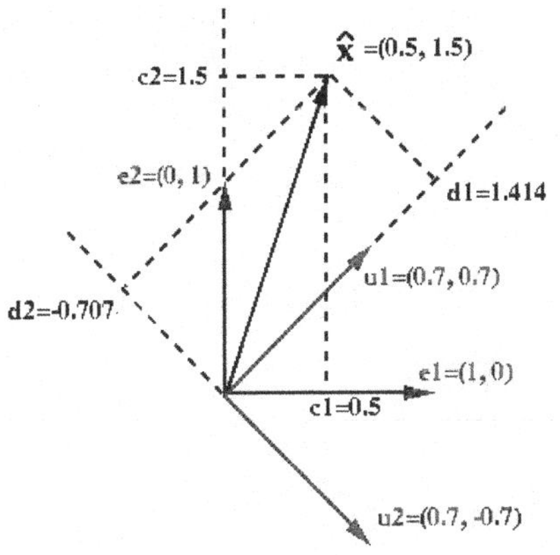

Figure 2.12 Representation of a signal vector under two different bases.

Example 2.8: The temperature is measured every 3 hours in a day to obtain eight samples, as shown below:

Time (hours)	0	3	6	9	12	15	18	21
Temperature (F)	65	60	65	70	75	80	75	70

These time samples can be considered as a vector $\boldsymbol{x} = [x[1], \ldots, x[8]]^{\mathrm{T}} = [65, 60, 65, 70, 75, 80, 75, 70]^{\mathrm{T}}$ in $\mathbb{R}^8$ space under the standard basis implicitly used; i.e., the nth element $x[n]$ is the coefficient for the nth standard basis vector $\boldsymbol{e}_k = [0, \ldots, 0, 1, 0, \ldots, 0]^{\mathrm{T}}$ (all elements are zero except the nth one); i.e.,

$$\boldsymbol{x} = \sum_{k=1}^{8} x[k]\boldsymbol{e}_k. \tag{2.258}$$

This 8-D signal vector $\boldsymbol{x}$ is approximated in an M-D subspace ($M < 8$) as shown below for different M values:

- $M = 1$: $\boldsymbol{x}$ is approximated as $\hat{\boldsymbol{x}} = c[1]\boldsymbol{b}_1$ in a 1-D subspace spanned by $\boldsymbol{b}_1 = [1, 1, 1, 1, 1, 1, 1, 1]^{\mathrm{T}}$. Here, the coefficient can be obtained as

$$c[1] = \frac{\langle \boldsymbol{x}, \boldsymbol{b}_1 \rangle}{\langle \boldsymbol{b}_1, \boldsymbol{b}_1 \rangle} = \frac{560}{8} = 70, \tag{2.259}$$

which represents the average or DC component of the daily temperature. The approximation is

$$\hat{\boldsymbol{x}} = c[1]\boldsymbol{b}_1 = [70, 70, 70, 70, 70, 70, 70, 70]^{\mathrm{T}}. \tag{2.260}$$

The error vector is $\tilde{\boldsymbol{x}} = \boldsymbol{x} - \hat{\boldsymbol{x}} = [-5, -10, -5, 0, 5, 10, 5, 0]^{\mathrm{T}}$ and the error is $||\tilde{\boldsymbol{x}}||^2 = 300$.

- $M = 2$: $\boldsymbol{x}$ can be better approximated in a 2-D subspace spanned by the same $\boldsymbol{b}_1$ and a second basis vector $\boldsymbol{b}_2 = [1, 1, 1, 1, -1, -1, -1, -1]^{\mathrm{T}}$. As $\boldsymbol{b}_2$ is orthogonal to $\boldsymbol{b}_1$, its coefficient $c[2]$ can be found independently:

$$c[2] = \frac{\langle \boldsymbol{x}, \boldsymbol{b}_2 \rangle}{\langle \boldsymbol{b}_2, \boldsymbol{b}_2 \rangle} = \frac{-40}{8} = -5, \tag{2.261}$$

which represents the temperature difference between morning and afternoon. The approximation is

$$\hat{\boldsymbol{x}} = c[1]\boldsymbol{b}_1 + c[2]\boldsymbol{b}_2 = [65.65, 65, 65, 75, 75, 75, 75]^{\mathrm{T}}. \tag{2.262}$$

The error vector is $\tilde{\boldsymbol{x}} = \boldsymbol{x} - \hat{\boldsymbol{x}} = [0, -5, 0, 5, 0, 5, 0, -5]^{\mathrm{T}}$ and the error is $||\tilde{\boldsymbol{x}}||^2 = 100$.

- $M = 3$: The approximation can be further improved if a third basis vector $\boldsymbol{b}_3 = [1, 1, -1, -1, -1, -1, 1, 1]^{\mathrm{T}}$ is added. As all three basis vectors are orthogonal to each other, the coefficient $c[3]$ can also be independently obtained:

$$c[3] = \frac{\langle \boldsymbol{x}, \boldsymbol{b}_3 \rangle}{\langle \boldsymbol{b}_3, \boldsymbol{b}_3 \rangle} = \frac{-20}{8} = -2.5, \tag{2.263}$$

which represents the temperature difference between daytime and nighttime. The approximation can be expressed as

$$\hat{\boldsymbol{x}} = c[1]\boldsymbol{b}_1 + c[2]\boldsymbol{b}_2 + c[3]\boldsymbol{b}_3 = [62.5, 62, 5, 67.5, 67.5, 77.5, 77.5, 72.5, 72.5]^{\mathrm{T}}. \tag{2.264}$$

The error vector is $\tilde{\boldsymbol{x}} = \boldsymbol{x} - \hat{\boldsymbol{x}} = [2.5, -2.5, -2.5, 2.5, -2.5, 2.5, 2.5, -2.5]^{\mathrm{T}}$ and the error is $||\tilde{\boldsymbol{x}}||^2 = 50$.

We can now make the following observations:

- The original 8-D signal vector $\boldsymbol{x}$ can be approximated by $M < 8$ basis vectors spanning an M-D subspace. As more basis vectors are included in the approximation, the error becomes progressively smaller.
- A typical signal contains both slow-varying or low-frequency components and fast-varying or high-frequency components, and the former are likely to contain more energy than the latter. In order to reduce error when approximating the signal, basis functions representing lower frequencies should be used first.
- When progressively more basis functions representing more details or subtle variations in the signal are added in the signal approximation, their coefficients are likely to have lower values than those for the slow-varying basis functions, and they are more likely to be affected by noise such as some random fluctuation; therefore, they are less significant and could be neglected without losing much essential information.
- The three basis vectors $\boldsymbol{b}_1$, $\boldsymbol{b}_2$, and $\boldsymbol{b}_3$ used above are actually the first three basis vectors of the sequency-ordered Hadamard transform to be discussed in Chapter 8.

2.4 Frames and biorthogonal bases

2.4.1 Frames

Previously we considered the representation of a signal vector $\boldsymbol{x} \in H$ as some linear combination of an orthogonal basis $\{\boldsymbol{u}_k\}$ that spans the space (Theorem. 2.3)

$$\boldsymbol{x} = \sum_k c[k]\boldsymbol{u}_k = \sum_k \langle \boldsymbol{x}, \boldsymbol{u}_k \rangle \boldsymbol{u}_k, \tag{2.265}$$

and Parseval's identity $||\boldsymbol{x}||^2 = ||\boldsymbol{c}||^2$ indicates that $\boldsymbol{x}$ is equivalently represented by the coefficients $\boldsymbol{c}$ without any redundancy. However, sometimes it may not be easy or even possible to identify a set of linearly independent and orthogonal basis vectors in the space. In such cases we could still consider representing a signal vector $\boldsymbol{x}$ by a set of vectors $\{\boldsymbol{f}_k\}$ which may not be linearly independent and, therefore, do not form a basis of the space. A main issue is the redundancy that exists among such a set of non-independent vectors. As it is now possible to find a set of coefficients $d[k]$ so that $\sum_k d[k]\boldsymbol{f}_k = 0$, an immediate consequence is that the representation is no longer unique:

$$\boldsymbol{x} = \sum_k c[k]\boldsymbol{f}_k = \sum_k c[k]\boldsymbol{f}_k + \sum_k d[k]\boldsymbol{f}_k = \sum_k (c[k] + d[k])\boldsymbol{f}_k. \tag{2.266}$$

One consequence of the redundancy is that Parseval's identity no longer holds. The energy contained in the coefficients $||\boldsymbol{c}||^2$ may be either higher or lower than the actual energy $||\boldsymbol{x}||^2$ in the signal. Therefore, we need to develop some theory to address this issue when using non-independent vectors for signal representation.

First, in order for the expansion $\boldsymbol{x} = \sum_k c[k]\boldsymbol{f}_k$ to be a precise representation of the signal vector $\boldsymbol{x}$ in terms of a set of coefficients $c[k] = \langle \boldsymbol{x}, \boldsymbol{f}_k \rangle$, we need to guarantee that, for any vectors $\boldsymbol{x}, \boldsymbol{y} \in H$, the following always holds:

$$\langle \boldsymbol{x}, \boldsymbol{f}_k \rangle = \langle \boldsymbol{y}, \boldsymbol{f}_k \rangle \quad \text{iff} \quad \boldsymbol{x} = \boldsymbol{y}. \tag{2.267}$$

Moreover, these representations also need to be stable in the following two aspects.

- **Stable representation**
  If the difference between two vectors is small, the difference between their corresponding coefficients should also be small:

$$\text{if} \quad ||\boldsymbol{x} - \boldsymbol{y}||^2 \to 0, \quad \text{then} \quad \sum_k |\langle \boldsymbol{x}, \boldsymbol{f}_k \rangle - \langle \boldsymbol{y}, \boldsymbol{f}_k \rangle|^2 \to 0; \tag{2.268}$$

  i.e.,

$$\sum_k |\langle \boldsymbol{x}, \boldsymbol{f}_k \rangle - \langle \boldsymbol{y}, \boldsymbol{f}_k \rangle|^2 \le B||\boldsymbol{x} - \boldsymbol{y}||^2, \tag{2.269}$$

where $0 < B < \infty$ is a positive real constant. In particular, if $\boldsymbol{y} = \boldsymbol{0}$ and therefore $\langle \boldsymbol{y}, \boldsymbol{f}_k \rangle = 0$, we have

$$\sum_k |\langle \boldsymbol{x}, \boldsymbol{f}_k \rangle|^2 \leq B||\boldsymbol{x}||^2. \tag{2.270}$$

- **Stable reconstruction**
  If the difference between two sets of coefficients is small, the difference between the reconstructed vectors should also be small:

$$\text{if} \quad \sum_k |\langle \boldsymbol{x}, \boldsymbol{f}_k \rangle - \langle \boldsymbol{y}, \boldsymbol{f}_k \rangle|^2 \to 0, \quad \text{then} \quad ||\boldsymbol{x} - \boldsymbol{y}||^2 \to 0; \tag{2.271}$$

  i.e.,

$$A||\boldsymbol{x} - \boldsymbol{y}||^2 \leq \sum_k |\langle \boldsymbol{x}, \boldsymbol{f}_k \rangle - \langle \boldsymbol{y}, \boldsymbol{f}_k \rangle|^2, \tag{2.272}$$

  where $0 < A < \infty$ is also a positive real constant. Again, if $\boldsymbol{y} = \boldsymbol{0}$ and $\langle \boldsymbol{y}, \boldsymbol{f}_k \rangle = 0$, we have

$$A||\boldsymbol{x}||^2 \leq \sum_k |\langle \boldsymbol{x}, \boldsymbol{f}_k \rangle|^2. \tag{2.273}$$

Combining Eqs. (2.270) and (2.273), we have the following definition:

Definition 2.24. *A family of finite or infinite vectors $\{\boldsymbol{f}_k\}$ in Hilbert space H is a frame if there exist two real constants $0 < A \leq B < \infty$, called the lower and upper bounds of the frame, such that for any $\boldsymbol{x} \in H$, the following holds:*

$$A||\boldsymbol{x}||^2 \leq \sum_k |\langle \boldsymbol{x}, \boldsymbol{f}_k \rangle|^2 \leq B||\boldsymbol{x}||^2. \tag{2.274}$$

In particular, if $A = B$; i.e.,

$$A||\boldsymbol{x}||^2 = \sum_k |\langle \boldsymbol{x}, \boldsymbol{f}_k \rangle|^2, \tag{2.275}$$

then the frame is tight.

2.4.2 Signal expansion by frames and Riesz bases

Our purpose here is to represent a given signal vector $\boldsymbol{x} \in H$ as a linear combination $\boldsymbol{x} = \sum_k c[k]\boldsymbol{f}_k$ of a set of frame vectors $\{\boldsymbol{f}_k\}$. The process of finding the coefficients $c[k]$ needed in the combination can be considered as a *frame transformation*, denoted by F^*, that maps the given $\boldsymbol{x}$ to a coefficient vector $\boldsymbol{c}$:

$$\boldsymbol{c} = F^*\boldsymbol{x} = [\ldots, c[k], \ldots]^{\mathrm{T}} = [\ldots, \langle \boldsymbol{x}, \boldsymbol{f}_k \rangle, \ldots]^{\mathrm{T}}, \tag{2.276}$$

where we have defined $c[k] = \langle \boldsymbol{x}, \boldsymbol{f}_k \rangle$, following the example of the unitary transformation in Eq. (2.204). Here, F^* is the adjoint of another transformation F, which can be found from the following inner product in the definition of a unitary transformation (Eq. (2.141)):

$$\begin{aligned}\langle \boldsymbol{c}', F^*\boldsymbol{x}\rangle &= \sum_k c'[k]\, \overline{\langle \boldsymbol{x}, \boldsymbol{f}_k\rangle} \\ &= \sum_k c'[k]\, \langle \boldsymbol{f}_k, \boldsymbol{x}\rangle = \langle \sum_k c'[k]\boldsymbol{f}_k, \boldsymbol{x}\rangle = \langle F\boldsymbol{c}', \boldsymbol{x}\rangle. \end{aligned} \tag{2.277}$$

We see that F is a transformation that constructs a vector as a linear combination of the frame $\{\boldsymbol{f}_k\}$ based on a given set of coefficients $\boldsymbol{c}'$:

$$\boldsymbol{x}' = F\boldsymbol{c}' = \sum_k c'[k]\boldsymbol{f}_k. \tag{2.278}$$

Based on F and F^* we can further define an operator FF^*:

$$FF^*\boldsymbol{x} = F(F^*\boldsymbol{x}) = F\boldsymbol{c} = \sum_k \langle \boldsymbol{x}, \boldsymbol{f}_k\rangle \boldsymbol{f}_k. \tag{2.279}$$

Different from the unitary transformation satisfying $UU^* = UU^{-1} = I$, here $FF^* \neq I$ is, in general, not an identity operator. Applying its inverse $(FF^*)^{-1}$ to both sides of the equation above, we get

$$\begin{aligned}\boldsymbol{x} &= (FF^*)^{-1}F\boldsymbol{c} = (FF^*)^{-1}[\sum_k \langle \boldsymbol{x}, \boldsymbol{f}_k\rangle \boldsymbol{f}_k] = \sum_k \langle \boldsymbol{x}, \boldsymbol{f}_k\rangle (FF^*)^{-1}\boldsymbol{f}_k \\ &= \sum_k \langle \boldsymbol{x}, \boldsymbol{f}_k\rangle \tilde{\boldsymbol{f}}_k = \sum_k c[k]\tilde{\boldsymbol{f}}_k, \end{aligned} \tag{2.280}$$

where we have defined $\tilde{\boldsymbol{f}}_k$, called the *dual vector* of $\boldsymbol{f}_k$, as

$$\tilde{\boldsymbol{f}}_k = (FF^*)^{-1}\boldsymbol{f}_k; \quad \text{i.e.} \quad \boldsymbol{f}_k = (FF^*)\tilde{\boldsymbol{f}}_k. \tag{2.281}$$

Note that $(FF^*)^{-1}F = (F^*)^-$ above is actually the pseudo-inverse of F^* satisfying (Appendix A):

$$(F^*)^- F^* = (FF^*)^{-1}FF^* = I. \tag{2.282}$$

We can define $(F^*)^-$ as another transformation:

$$\tilde{F} = (FF^*)^{-1}F = (F^*)^-, \tag{2.283}$$

and then rewrite Eq. (2.280) as

$$\boldsymbol{x} = \tilde{F}\boldsymbol{c} = \tilde{F}[\ldots, c[k], \ldots]^{\mathrm{T}} = \sum_k \langle \boldsymbol{x}, \boldsymbol{f}_k\rangle \tilde{\boldsymbol{f}}_k = \sum_k c[k]\tilde{\boldsymbol{f}}_k. \tag{2.284}$$

This is the inverse frame transformation which reconstructs the vector $\boldsymbol{x}$ based on the coefficients $\boldsymbol{c}$ obtained by the forward frame transformation in Eq. (2.276). Equations (2.284) and (2.276) form a frame transformation pair, similar to the unitary transformation pair in Eq. (2.204).

We can find the adjoint of $\tilde{F}$ from the following inner product (by reversing the steps in Eq. (2.277)):

$$\begin{aligned}\langle \tilde{F}\boldsymbol{c}, \boldsymbol{x}\rangle &= \langle \sum_k c[k]\tilde{\boldsymbol{f}}_k, \boldsymbol{x}\rangle = \sum_k c[k]\, \langle \tilde{\boldsymbol{f}}_k, \boldsymbol{x}\rangle \\ &= \sum_k c[k]\, \overline{\langle \boldsymbol{x}, \tilde{\boldsymbol{f}}_k\rangle} = \langle \boldsymbol{c}, \tilde{F}^*\boldsymbol{x}\rangle. \end{aligned} \tag{2.285}$$

Here, $\tilde{F}^*$ is the adjoint of $\tilde{F}$:

$$\tilde{F}^*\boldsymbol{x} = [\ldots, \langle \boldsymbol{x}, \tilde{\boldsymbol{f}}_k\rangle, \ldots]^{\mathrm{T}} = [\ldots, d[k], \ldots]^{\mathrm{T}} = \boldsymbol{d}, \tag{2.286}$$

where we have defined $d[k] = \langle \boldsymbol{x}, \tilde{\boldsymbol{f}}_k\rangle$. Replacing F by F^* in Eq. (2.283), we get

$$\tilde{F}^* = (F^*F)^{-1}F^* = F^-, \tag{2.287}$$

which is the pseudo-inverse of F satisfying:

$$\tilde{F}^*F = (F^*F)^{-1}F^*F = F^-F = I. \tag{2.288}$$

Theorem 2.8. *A vector $\boldsymbol{x} \in H$ can be equivalently represented by either of the two dual frames $\{\boldsymbol{f}_k\}$ or $\{\tilde{\boldsymbol{f}}_k\}$:*

$$\boldsymbol{x} = \sum_k \langle \boldsymbol{x}, \tilde{\boldsymbol{f}}_k\rangle \boldsymbol{f}_k = \sum_k \langle \boldsymbol{x}, \boldsymbol{f}_k\rangle \tilde{\boldsymbol{f}}_k. \tag{2.289}$$

Proof: Consider the inner product $\langle \boldsymbol{x}, \boldsymbol{x}\rangle$, with the first $\boldsymbol{x}$ replaced by the expression in Eq. (2.280):

$$\begin{aligned}\langle \boldsymbol{x}, \boldsymbol{x}\rangle &= \left\langle \sum_k \langle \boldsymbol{x}, \boldsymbol{f}_k\rangle \tilde{\boldsymbol{f}}_k, \boldsymbol{x} \right\rangle = \sum_k \langle \boldsymbol{x}, \boldsymbol{f}_k\rangle \langle \tilde{\boldsymbol{f}}_k, \boldsymbol{x}\rangle \\ &= \left\langle \boldsymbol{x}, \sum_k \overline{\langle \tilde{\boldsymbol{f}}_k, \boldsymbol{x}\rangle} \boldsymbol{f}_k \right\rangle = \left\langle \boldsymbol{x}, \sum_k \langle \boldsymbol{x}, \tilde{\boldsymbol{f}}_k\rangle \boldsymbol{f}_k \right\rangle. \end{aligned} \tag{2.290}$$

Comparing the two sides of the equation, we get

$$\boldsymbol{x} = \sum_k \langle \boldsymbol{x}, \tilde{\boldsymbol{f}}_k\rangle \boldsymbol{f}_k = \sum_k d[k]\boldsymbol{f}_k. \tag{2.291}$$

Combining this result with Eq. (2.280), we get Eq. (2.289). Q.E.D.

Note that, according to Eq. (2.278), Eq. (2.291) can also be written as

$$\boldsymbol{x} = \sum_k \langle \boldsymbol{x}, \tilde{\boldsymbol{f}}_k\rangle \boldsymbol{f}_k = \sum_k d[k]\boldsymbol{f}_k = F\boldsymbol{d}. \tag{2.292}$$

We can now combine Eqs. (2.276) and (2.286) together with Eq. (2.289) to form two alternative frame transformation pairs based on either frame $\{\boldsymbol{f}_k\}$ or its dual $\{\tilde{\boldsymbol{f}}_k\}$:

$$\begin{cases} c[k] = \langle \boldsymbol{x}, \boldsymbol{f}_k \rangle \\ \boldsymbol{x} = \sum_k c[k]\tilde{\boldsymbol{f}}_k = \sum_k \langle \boldsymbol{x}, \boldsymbol{f}_k \rangle \tilde{\boldsymbol{f}}_k \end{cases},$$
$$\begin{cases} d[k] = \langle \boldsymbol{x}, \tilde{\boldsymbol{f}}_k \rangle \\ \boldsymbol{x} = \sum_k d[k]\boldsymbol{f}_k = \sum_k \langle \boldsymbol{x}, \tilde{\boldsymbol{f}}_k \rangle \boldsymbol{f}_k \end{cases}. \tag{2.293}$$

These equations are respectively the forward and inverse frame transformation of $\boldsymbol{x}$ based on the frame $\{\mathbf{f}_k\}$ and its dual $\{\tilde{\mathbf{f}}_k\}$, which can also be expressed (due to Eqs. (2.284) and (2.292)) more concisely as

$$\begin{cases} \boldsymbol{c} = F^*\boldsymbol{x} \\ \boldsymbol{x} = \tilde{F}\boldsymbol{c} = (F^*)^-\boldsymbol{c} \end{cases} \qquad \begin{cases} \boldsymbol{d} = \tilde{F}^*\boldsymbol{x} = F^-\boldsymbol{x} \\ \boldsymbol{x} = F\boldsymbol{d} \end{cases}. \tag{2.294}$$

These frame transformation pairs can be considered as the generalization of the unitary transformation pair given in Eq. (2.204), which is carried out by U and its inverse U^{-1}, while the frame transformation pairs in Eq. (2.294) are carried out by F^* (or F) and its pseudo-inverse $\tilde{F} = (F^*)^-$ (or $\tilde{F}^* = F^-$). We also see from Eq. (2.294) that

$$\tilde{F}F^*\boldsymbol{x} = F\tilde{F}^*\boldsymbol{x} = \boldsymbol{x}; \tag{2.295}$$

i.e., $\tilde{F}F^* = (F^*)^-F^* = I$ and $F\tilde{F}^* = F^-F = I$, similar to $U^{-1}U = U^*U = I$.

Also, similar to the unitary transformation, the signal energy is conserved by the frame transformation:

$$\begin{aligned} ||\boldsymbol{x}||^2 &= \langle \boldsymbol{x}, \boldsymbol{x} \rangle = \langle \tilde{F}\boldsymbol{c}, \boldsymbol{x} \rangle = \langle \boldsymbol{c}, \tilde{F}^*\boldsymbol{x} \rangle = \langle \boldsymbol{c}, \boldsymbol{d} \rangle \\ &= \langle F\boldsymbol{d}, \boldsymbol{x} \rangle = \langle \boldsymbol{d}, F^*\boldsymbol{x} \rangle = \langle \boldsymbol{d}, \boldsymbol{c} \rangle. \end{aligned} \tag{2.296}$$

This relationship can be considered as the generalized version of Parseval's identity. However, we note that:

$$\begin{aligned} ||\boldsymbol{c}||^2 &= \langle \boldsymbol{c}, \boldsymbol{c} \rangle = \langle F^*\boldsymbol{x}, F^*\boldsymbol{x} \rangle = \langle FF^*\boldsymbol{x}, \boldsymbol{x} \rangle \neq \langle \boldsymbol{x}, \boldsymbol{x} \rangle = ||\boldsymbol{x}||^2, \\ ||\boldsymbol{d}||^2 \langle \boldsymbol{d}, \boldsymbol{d} \rangle &= \langle \tilde{F}^*\boldsymbol{x}, \tilde{F}^*\boldsymbol{x} \rangle = \langle \tilde{F}\tilde{F}^*\boldsymbol{x}, \boldsymbol{x} \rangle \neq \langle \boldsymbol{x}, \boldsymbol{x} \rangle = ||\boldsymbol{x}||^2. \end{aligned} \tag{2.297}$$

To find out how the signal energy is related to the energy contained in either of the two sets of coefficients, we need to study further the operator FF^*. Consider the inner product of Eq. (2.279) with some vector $\boldsymbol{y}$:

$$\begin{aligned} \langle FF^*\boldsymbol{x}, \boldsymbol{y} \rangle &= \sum_k \langle \boldsymbol{x}, \boldsymbol{f}_k \rangle \langle \boldsymbol{f}_k, \boldsymbol{y} \rangle = \langle \boldsymbol{x}, \sum_k \overline{\langle \boldsymbol{f}_k, \boldsymbol{y} \rangle} \boldsymbol{f}_k \rangle \\ &= \langle \boldsymbol{x}, \sum_k \langle \boldsymbol{y}, \boldsymbol{f}_k \rangle \boldsymbol{f}_k \rangle = \langle \boldsymbol{x}, FF^*\boldsymbol{y} \rangle. \end{aligned} \tag{2.298}$$

We see that FF^* is a self-adjoint operator, and according to Theorem 2.4, its eigenvalues are real and its eigenvectors are orthogonal; i.e., if

$$FF^*\boldsymbol{\phi}_k = \lambda_k \boldsymbol{\phi}_k, \qquad \text{(for all } k\text{)}, \tag{2.299}$$

then $\overline{\lambda}_k = \lambda_k$ and $\langle \boldsymbol{\phi}_k, \boldsymbol{\phi}_l \rangle = \delta[k-l]$. Now $\boldsymbol{x}$ can also be expanded in terms of these eigenvectors as

$$\boldsymbol{x} = \sum_k \langle \boldsymbol{x}, \boldsymbol{\phi}_k \rangle \boldsymbol{\phi}_k, \tag{2.300}$$

and the energy contained in $\boldsymbol{x}$ is

$$||\boldsymbol{x}||^2 = \langle \boldsymbol{x}, \boldsymbol{x} \rangle = \langle \sum_k \langle \boldsymbol{x}, \boldsymbol{\phi}_k \rangle \boldsymbol{\phi}_k, \sum_l \langle \boldsymbol{x}, \boldsymbol{\phi}_l \rangle \boldsymbol{\phi}_l \rangle$$
$$= \sum_k \sum_l \langle \boldsymbol{x}, \boldsymbol{\phi}_k \rangle \overline{\langle \boldsymbol{x}, \boldsymbol{\phi}_l \rangle} \langle \boldsymbol{\phi}_k, \boldsymbol{\phi}_l \rangle = \sum_k |\langle \boldsymbol{x}, \boldsymbol{\phi}_k \rangle|^2. \tag{2.301}$$

Another operator can be similarly formed by the dual frame transformation $\tilde{F}$:

$$\tilde{F}\tilde{F}^* = [(FF^*)^{-1}F]\,[(FF^*)^{-1}F]^* = (FF^*)^{-1}FF^*(FF^*)^{-1} = (FF^*)^{-1}, \tag{2.302}$$

which is also a self-adjoint operator whose eigenvalues and eigenvectors are respectively $\{1/\lambda_k\}$ and $\{\boldsymbol{\phi}_k\}$; i.e.,

$$\tilde{F}\tilde{F}^* \boldsymbol{\phi}_k = (FF^*)^{-1} \boldsymbol{\phi}_k = \frac{1}{\lambda_k} \boldsymbol{\phi}_k \qquad \text{(for all } k\text{)}. \tag{2.303}$$

Theorem 2.9. *The frame transformation coefficients* $c[k] = \langle \boldsymbol{x}, \boldsymbol{f}_k \rangle$ *and* $d[k] = \langle \boldsymbol{x}, \tilde{\boldsymbol{f}}_k \rangle$ *satisfy respectively the following inequalities:*

$$\lambda_{min} ||\boldsymbol{x}||^2 \leq \sum_k |\langle \boldsymbol{x}, \boldsymbol{f}_k \rangle|^2 = ||\boldsymbol{c}||^2 = ||F^*\boldsymbol{x}||^2 \leq \lambda_{max} ||\boldsymbol{x}||^2, \tag{2.304}$$

$$\frac{1}{\lambda_{max}} ||\boldsymbol{x}||^2 \leq \sum_k |\langle \boldsymbol{x}, \tilde{\boldsymbol{f}}_k \rangle|^2 = ||\boldsymbol{d}||^2 = ||\tilde{F}^*\boldsymbol{x}||^2 \leq \frac{1}{\lambda_{min}} ||\boldsymbol{x}||^2, \tag{2.305}$$

where λ_{min} *and* λ_{max} *are respectively the smallest and largest eigenvalues of the self-adjoint operator* FF^*. *When all eigenvalues are the same, then* $\lambda_{max} = \lambda_{min} = \lambda$ *and the frame is tight:*

$$\sum_k |\langle \boldsymbol{x}, \boldsymbol{f}_k \rangle|^2 = \lambda ||\boldsymbol{x}||^2, \qquad \sum_k |\langle \boldsymbol{x}, \tilde{\boldsymbol{f}}_k \rangle|^2 = \frac{1}{\lambda} ||\boldsymbol{x}||^2. \tag{2.306}$$

Proof: Applying $(FF^*)^{-1}$ to both sides of Eq. (2.292) we get

$$(FF^*)^{-1}\boldsymbol{x} = \sum_k \langle \boldsymbol{x}, \tilde{\boldsymbol{f}}_k \rangle (FF^*)^{-1} \boldsymbol{f}_k = \sum_k \langle \boldsymbol{x}, \tilde{\boldsymbol{f}}_k \rangle \tilde{\boldsymbol{f}}_k. \tag{2.307}$$

This result and Eq. (2.279) form a symmetric pair:

$$(FF^*)\boldsymbol{x} = \sum_k \langle \boldsymbol{x}, \boldsymbol{f}_k \rangle \boldsymbol{f}_k, \tag{2.308}$$

$$(FF^*)^{-1}\boldsymbol{x} = \sum_k \langle \boldsymbol{x}, \tilde{\boldsymbol{f}}_k \rangle \tilde{\boldsymbol{f}}_k. \tag{2.309}$$

Taking the inner product of each of these equations with $\boldsymbol{x}$, we get

$$\begin{aligned}\langle (FF^*)\boldsymbol{x}, \boldsymbol{x}\rangle &= \sum_k \langle \boldsymbol{x}, \boldsymbol{f}_k\rangle\langle \boldsymbol{f}_k, \boldsymbol{x}\rangle = \sum_k |\langle \boldsymbol{x}, \boldsymbol{f}_k\rangle|^2 \\ &= \sum_k |c[k]|^2 = ||\mathbf{c}||^2, \end{aligned} \tag{2.310}$$

$$\begin{aligned}\langle (FF^*)^{-1}\boldsymbol{x}, \boldsymbol{x}\rangle &= \sum_k \langle \boldsymbol{x}, \tilde{\boldsymbol{f}}_k\rangle\langle \tilde{\boldsymbol{f}}_k, \boldsymbol{x}\rangle = \sum_k |\langle \boldsymbol{x}, \tilde{\boldsymbol{f}}_k\rangle|^2 \\ &= \sum_k |d[k]|^2 = ||\boldsymbol{d}||^2. \end{aligned} \tag{2.311}$$

These two expressions represent the energy contained in each of the two sets of coefficients $c[k] = \langle \boldsymbol{x}, \boldsymbol{f}_k\rangle$ and $d[k] = \langle \boldsymbol{x}, \tilde{\boldsymbol{f}}_k\rangle$.

We will now carry out the following two parallel steps. First, we apply FF^* to both sides of Eq. (2.300):

$$\begin{aligned}FF^*\boldsymbol{x} &= FF^*(\sum_k \langle \boldsymbol{x}, \boldsymbol{\phi}_k\rangle\boldsymbol{\phi}_k) = \sum_k \langle \boldsymbol{x}, \boldsymbol{\phi}_k\rangle FF^*\boldsymbol{\phi}_k \\ &= \sum_k \langle \boldsymbol{x}, \boldsymbol{\phi}_k\rangle\lambda_k\boldsymbol{\phi}_k, \end{aligned} \tag{2.312}$$

and take inner product with $\boldsymbol{x}$ on both sides:

$$\begin{aligned}\langle FF^*\boldsymbol{x}, \boldsymbol{x}\rangle &= \langle \sum_k \langle \boldsymbol{x}, \boldsymbol{\phi}_k\rangle\lambda_k\boldsymbol{\phi}_k, \boldsymbol{x}\rangle = \sum_k \langle \boldsymbol{x}, \boldsymbol{\phi}_k\rangle\lambda_k\langle \boldsymbol{\phi}_k, \boldsymbol{x}\rangle \\ &= \sum_k \lambda_k|\langle \boldsymbol{x}, \boldsymbol{\phi}_k\rangle|^2. \end{aligned} \tag{2.313}$$

Replacing the left-hand side by Eq. (2.310), we get

$$\sum_k |\langle \boldsymbol{x}, \boldsymbol{f}_k\rangle|^2 = \sum_k \lambda_k|\langle \boldsymbol{x}, \boldsymbol{\phi}_k\rangle|^2. \tag{2.314}$$

Applying Eq. (2.301) to the right-hand side we get

$$\lambda_{\min}||\boldsymbol{x}||^2 \leq \sum_k |\langle \boldsymbol{x}, \boldsymbol{f}_k\rangle|^2 \leq \lambda_{\max}||\boldsymbol{x}||^2. \tag{2.315}$$

Next, we apply $(FF^*)^{-1}$ to both sides of Eq. (2.300):

$$(FF^*)^{-1}\boldsymbol{x} = \sum_k \langle \boldsymbol{x}, \boldsymbol{\phi}_k\rangle(FF^*)^{-1}\boldsymbol{\phi}_k = \sum_k \langle \boldsymbol{x}, \boldsymbol{\phi}_k\rangle\frac{1}{\lambda_k}\boldsymbol{\phi}_k, \tag{2.316}$$

and take inner product with $\boldsymbol{x}$ on both sides:

$$\langle (FF^*)^{-1}\boldsymbol{x}, \boldsymbol{x}\rangle = \sum_k \langle \boldsymbol{x}, \boldsymbol{\phi}_k\rangle\frac{1}{\lambda_k}\langle \boldsymbol{\phi}_k, \boldsymbol{x}\rangle = \sum_k \frac{1}{\lambda_k}|\langle \boldsymbol{x}, \boldsymbol{\phi}_k\rangle|^2. \tag{2.317}$$

Replacing the left-hand side by Eq. (2.311), we get

$$\sum_k |\langle \boldsymbol{x}, \tilde{\boldsymbol{f}}_k\rangle|^2 = \sum_k \frac{1}{\lambda_k}|\langle \boldsymbol{x}, \boldsymbol{\phi}_k\rangle|^2. \tag{2.318}$$

Applying Eq. (2.301) to the right-hand side we get

$$\frac{1}{\lambda_{\max}}||\boldsymbol{x}||^2 \le \sum_k |\langle \boldsymbol{x}, \tilde{\boldsymbol{f}}_k \rangle|^2 \le \frac{1}{\lambda_{\min}}||\boldsymbol{x}||^2. \tag{2.319}$$

Q.E.D.

This theorem indicates that the frame transformation associated with either F or $\tilde{F}$ does not conserve signal energy, due obviously to the redundancy of the non-independent frame vectors. However, as shown in Eq. (2.296), the energy is conserved when both sets of coefficients are involved.

Theorem 2.10. *Let λ_k and $\boldsymbol{\phi}_k$ be the kth eigenvalue and the corresponding eigenvector of operator FF^*: $FF^*\boldsymbol{\phi}_k = \lambda_k \boldsymbol{\phi}_k$ for all k, Then*

$$\sum_k \lambda_k = \sum_k ||\boldsymbol{f}_k||^2, \qquad \sum_k \frac{1}{\lambda_k} = \sum_k ||\tilde{\boldsymbol{f}}_k||^2. \tag{2.320}$$

Proof: As FF^* is self-adjoint, its eigenvalues λ_k are real and its eigenfunctions are orthogonal $\langle \boldsymbol{\phi}_k, \boldsymbol{\phi}_l \rangle = \delta[k-l]$, therefore we have

$$\begin{aligned}\sum_k \lambda_k &= \sum_k \lambda_k \langle \boldsymbol{\phi}_k, \boldsymbol{\phi}_k \rangle = \sum_k \langle FF^* \boldsymbol{\phi}_k, \boldsymbol{\phi}_k \rangle \\ &= \sum_k \left\langle \sum_k \langle \boldsymbol{\phi}_k, \boldsymbol{f}_k \rangle \boldsymbol{f}_k, \boldsymbol{\phi}_k \right\rangle = \sum_k \sum_k |\langle \boldsymbol{f}_k, \boldsymbol{\phi}_k \rangle|^2. \end{aligned} \tag{2.321}$$

On the other hand,

$$\begin{aligned}||\boldsymbol{f}_k||^2 = \langle \boldsymbol{f}_k, \boldsymbol{f}_k \rangle &= \left\langle \sum_k \langle \boldsymbol{f}_k, \boldsymbol{\phi}_k \rangle \boldsymbol{\phi}_k, \sum_l \langle \boldsymbol{f}_k, \boldsymbol{\phi}_l \rangle \boldsymbol{\phi}_l \right\rangle \\ &= \sum_k \sum_l \langle \boldsymbol{f}_k, \boldsymbol{\phi}_k \rangle \overline{\langle \boldsymbol{f}_k, \boldsymbol{\phi}_l \rangle} \langle \boldsymbol{\phi}_k, \boldsymbol{\phi}_l \rangle = \sum_k |\langle \boldsymbol{f}_k, \boldsymbol{\phi}_k \rangle|^2. \end{aligned} \tag{2.322}$$

Therefore, we get

$$\sum_k ||\boldsymbol{f}_k||^2 = \sum_k \sum_k |\langle \boldsymbol{f}_k, \boldsymbol{\phi}_k \rangle|^2 = \sum_k \lambda_k. \tag{2.323}$$

The second equation in the theorem can be similarly proved. Q.E.D.

Definition 2.25. *If the vectors in a frame are linearly independent, the frame is called a Riesz basis.*

Theorem 2.11. *(Biorthogonality of Riesz basis) A Riesz basis $\{\boldsymbol{f}_k\}$ and its dual $\{\tilde{\boldsymbol{f}}_k\}$ form a pair of biorthogonal bases satisfying*

$$\langle \boldsymbol{f}_k, \tilde{\boldsymbol{f}}_l \rangle = \delta[k-l] \qquad k, l \in \mathbb{Z}. \tag{2.324}$$

Proof: We let $\boldsymbol{x} = \boldsymbol{f}_l$ in Eq. (2.289) and get:

$$\boldsymbol{f}_l = \sum_k \langle \boldsymbol{f}_l, \tilde{\boldsymbol{f}}_k \rangle \boldsymbol{f}_k. \tag{2.325}$$

Since these vectors are linearly independent; i.e., $\boldsymbol{f}_l$ cannot be expressed as a linear combination of the rest of the frame vectors, the equation above has only one interpretation: all coefficients $\langle \boldsymbol{f}_l, \tilde{\boldsymbol{f}}_k \rangle = 0$ for all $k \neq l$ except when $k = l$ and $\langle \boldsymbol{f}_k, \tilde{\boldsymbol{f}}_k \rangle = 1$; i.e., Eq. (2.324) holds. Q.E.D.

If the dual frames $\boldsymbol{f}$ and $\tilde{\boldsymbol{f}}$ in Theorem 2.8 are a pair of biorthogonal bases, then Eq. (2.289) is a *biorthogonal transformation*:

$$\boldsymbol{x} = \sum_k \langle \boldsymbol{x}, \tilde{\boldsymbol{f}}_k \rangle \boldsymbol{f}_k = \sum_k \langle \boldsymbol{x}, \boldsymbol{f}_k \rangle \tilde{\boldsymbol{f}}_k. \tag{2.326}$$

In summary, we see that signal representation by a set of linearly independent and orthogonal basis vectors $\boldsymbol{x} = \sum_k c[k]\boldsymbol{u}_k = \sum_k \langle \boldsymbol{x}, \boldsymbol{u}_k \rangle \boldsymbol{u}_k$ in Eq. (2.87) is now generalized so that the signal is represented by a set of frame vectors, which are in general neither linearly independent nor orthogonal. The representation can be in either of the two dual frames, and the frame transformation and its inverse are pseudo-inverse of each other. Moreover, the signal energy is no longer conserved by the transformation, as Parseval's identity is invalid due to the redundancy in the frame. Instead, the signal energy and the energy in the coefficients are related by Eqs. (2.304), (2.305), and (2.296).

As a special case, when $F = U$ satisfying $UU^* = U^*U = I$, a frame transformation becomes a unitary transformation with the following degenerations:

- The pseudo-inverse in Eq. (2.287) becomes a regular inverse

$$U^- = (U^*U)^{-1}U^* = U^* = U^{-1}. \tag{2.327}$$

- The dual transformation in Eq. (2.283) becomes the same as the transformation itself:

$$\tilde{U} = (UU^*)^{-1}U = U. \tag{2.328}$$

- The biorthogonality in Eq. (2.324) becomes usual orthogonality:

$$\langle \boldsymbol{u}_k, \boldsymbol{u}_l \rangle = \delta[k-l] \qquad k, l \in \mathbb{Z}. \tag{2.329}$$

- The two dual transformation pairs in Eq. (2.293) (or Eq. (2.294)) become identical with $\boldsymbol{c} = \boldsymbol{d}$, the same as the unitary transformation pair in Eq. (2.204):

$$\begin{cases} \boldsymbol{c} = U^*\boldsymbol{x} \\ \boldsymbol{x} = U\boldsymbol{c} \end{cases}. \tag{2.330}$$

- The eigenequations of operators FF^* and $\tilde{F}\tilde{F}^*$ in Eqs. (2.299) and (2.303) become a trivial case:

$$UU^*\boldsymbol{\phi}_k = U^*U\boldsymbol{\phi}_k = I\boldsymbol{\phi}_k = \lambda_k\boldsymbol{\phi}_k = \boldsymbol{\phi}_k. \tag{2.331}$$

with $\lambda_{\max} = \lambda_{\min} = \lambda_k = 1$ (for all k).

- Both Eqs. (2.304) and (2.305), as well as Eqs.2.296, become Parseval's identity in Eq. (2.191):

$$\langle \boldsymbol{x}, \boldsymbol{x} \rangle = ||\boldsymbol{x}||^2 = \langle U\boldsymbol{x}, U\boldsymbol{x} \rangle = \langle \boldsymbol{c}, \boldsymbol{c} \rangle = ||\boldsymbol{c}||^2. \tag{2.332}$$

2.4.3 Frames in finite-dimensional space

Here we consider the frame transformation in $\mathbb{C}^N$. Let $\boldsymbol{F} = [\boldsymbol{f}_1, \dots, \boldsymbol{f}_M]$ be an N by M matrix composed of a set of M frame vectors as its columns. We assume $M > N$, and the M frame vectors are obviously not independent. The dual frame is also an N by M matrix composed of M dual vectors as its columns $\tilde{\boldsymbol{F}} = [\tilde{\boldsymbol{f}}_1, \dots, \tilde{\boldsymbol{f}}_M]$. Any given vector $\boldsymbol{x} \in \mathbb{C}^N$ can now be represented by either the frame $\boldsymbol{F}$ (second transformation pair in Eq. (2.294)) or its dual $\tilde{\boldsymbol{F}}$ (first transformation pair in Eq. (2.294)), in the form of a matrix multiplication (e.g., the generic operator F becomes a matrix $\boldsymbol{F}$):

$$\begin{cases} \boldsymbol{c} = \boldsymbol{F}^*\boldsymbol{x} \\ \boldsymbol{x} = \tilde{\boldsymbol{F}}\boldsymbol{c} = (\boldsymbol{F}^*)^-\boldsymbol{c} \end{cases}, \qquad \begin{cases} \boldsymbol{d} = \tilde{\boldsymbol{F}}^*\boldsymbol{x} = \boldsymbol{F}^-\boldsymbol{x} \\ \boldsymbol{x} = \boldsymbol{F}\boldsymbol{d} \end{cases}. \tag{2.333}$$

Here, $\boldsymbol{F}^*$ and $\tilde{\boldsymbol{F}}^*$ are M by N matrices. These frame transformations are in the same form as the unitary transformations in Eq. (2.204). However, different from matrices $\boldsymbol{U}$ and $\boldsymbol{U}^* = \boldsymbol{U}^{-1}$ there, here $\boldsymbol{F}$ and $\tilde{\boldsymbol{F}}$ in Eq. (2.333) are not square matrices and therefore not invertible. Consequently, the matrices used in the forward and inverse frame transformations are pseudo-inverse of each other:

$$\boldsymbol{F}^- = (\boldsymbol{F}^*\boldsymbol{F})^{-1}\boldsymbol{F}^* = \tilde{\boldsymbol{F}}^*, \qquad (\boldsymbol{F}^*)^- = (\boldsymbol{F}\boldsymbol{F}^*)^{-1}\boldsymbol{F} = \tilde{\boldsymbol{F}}. \tag{2.334}$$

Let us consider the representation of $\boldsymbol{x}$ by the coefficients $\boldsymbol{c}$ or $\boldsymbol{d}$ in the frame transformation domain. First, we represent $\boldsymbol{x}$ in terms of frame $\boldsymbol{F}$ based on the second transformation in Eq. (2.333):

$$\boldsymbol{d} = \tilde{\boldsymbol{F}}^*\boldsymbol{x} = \begin{bmatrix} \tilde{\boldsymbol{f}}_1^* \\ \vdots \\ \tilde{\boldsymbol{f}}_M^* \end{bmatrix} \boldsymbol{x} = \begin{bmatrix} \langle \boldsymbol{x}, \tilde{\boldsymbol{f}}_1 \rangle \\ \vdots \\ \langle \boldsymbol{x}, \tilde{\boldsymbol{f}}_M \rangle \end{bmatrix}, \tag{2.335}$$

and $\boldsymbol{x}$ is reconstructed by the inverse transformation:

$$\boldsymbol{x} = \boldsymbol{F}\boldsymbol{d} = [\boldsymbol{f}_1, \dots, \boldsymbol{f}_M] \begin{bmatrix} \langle \boldsymbol{x}, \tilde{\boldsymbol{f}}_1 \rangle \\ \vdots \\ \langle \boldsymbol{x}, \tilde{\boldsymbol{f}}_M \rangle \end{bmatrix} = \sum_{k=1}^{M} \langle \boldsymbol{x}, \tilde{\boldsymbol{f}}_k \rangle \boldsymbol{f}_k. \tag{2.336}$$

Alternatively, we can also represent $\boldsymbol{x}$ in terms of the dual frame $\tilde{\boldsymbol{F}}$ based on the first transformation in Eq. (2.333):

$$\boldsymbol{c} = \boldsymbol{F}^*\boldsymbol{x} = \begin{bmatrix} \boldsymbol{f}_1^* \\ \vdots \\ \boldsymbol{f}_M^* \end{bmatrix} \boldsymbol{x} = \begin{bmatrix} \langle \boldsymbol{x}, \boldsymbol{f}_1 \rangle \\ \vdots \\ \langle \boldsymbol{x}, \boldsymbol{f}_M \rangle \end{bmatrix}, \tag{2.337}$$

and $\boldsymbol{x}$ is reconstructed by the inverse transformation:

$$\boldsymbol{x} = \tilde{\boldsymbol{F}}\boldsymbol{c} = [\tilde{\boldsymbol{f}}_1, \ldots, \tilde{\boldsymbol{f}}_M] \begin{bmatrix} \langle \boldsymbol{x}, \boldsymbol{f}_1 \rangle \\ \vdots \\ \langle \boldsymbol{x}, \boldsymbol{f}_M \rangle \end{bmatrix} = \sum_{k=1}^{M} \langle \boldsymbol{x}, \boldsymbol{f}_k \rangle \tilde{\boldsymbol{f}}_k. \tag{2.338}$$

Theorem 2.12. *If a frame $\boldsymbol{F} = [\boldsymbol{f}_1, \ldots, \boldsymbol{f}_M]$ in $\mathbb{C}^N$ is tight; i.e., all eigenvalues $\lambda_k = \lambda$ of $\boldsymbol{F}\boldsymbol{F}^*$ are the same, and all frame vectors are normalized $||\boldsymbol{f}_k|| = 1$, then the frame bound is M/N.*

Proof: As $\boldsymbol{F}\boldsymbol{F}^*$ is an N by N matrix, it has N eigenvalues $\lambda_k = \lambda$ $(k = 1, \ldots, N)$. Then Theorem 2.10 becomes

$$\sum_{k=1}^{N} \lambda_k = N\lambda = \sum_{k=1}^{M} ||\boldsymbol{f}_k||^2 = M; \tag{2.339}$$

i.e., $\lambda = M/N$. Q.E.D.

In particular, if $M = N$ linearly independent frame vectors are used, then they form a Riesz basis in $\mathbb{C}^N$, and $\boldsymbol{F} = [\boldsymbol{f}_1, \ldots, \boldsymbol{f}_N]$ becomes an N by N invertible matrix, and its pseudo-inverse is just a regular inverse, and the second equation in Eq. (2.334) becomes $(\boldsymbol{F}^*)^{-1} = \tilde{\boldsymbol{F}}$; i.e.,

$$\boldsymbol{F}^*\tilde{\boldsymbol{F}} = \begin{bmatrix} \boldsymbol{f}_1^* \\ \vdots \\ \boldsymbol{f}_N^* \end{bmatrix} [\tilde{\boldsymbol{f}}_1, \ldots, \tilde{\boldsymbol{f}}_N] = \boldsymbol{I}, \tag{2.340}$$

which indicates that these Riesz vectors are indeed biorthogonal:

$$\langle \boldsymbol{f}_k, \tilde{\boldsymbol{f}}_l \rangle = \delta[k - l], \qquad (k, l = 1, \ldots, N). \tag{2.341}$$

Moreover, if these N vectors are also orthogonal; i.e., $\langle \boldsymbol{f}_k, \boldsymbol{f}_l \rangle = \delta[k - l]$, then $\boldsymbol{F} = \boldsymbol{U}$ becomes a unitary matrix satisfying $\boldsymbol{U}^* = \boldsymbol{U}^{-1}$, and $\tilde{\boldsymbol{U}} = (\boldsymbol{U}^*)^{-1} = \boldsymbol{U}$, i.e., the vectors and their duals become the same, and they form an orthonormal basis of $\mathbb{C}^N$. Now the frame transformation becomes a unitary transformation $\boldsymbol{U}^*\boldsymbol{x} = \boldsymbol{c}$ and the inverse is simply $\boldsymbol{U}\boldsymbol{c} = \boldsymbol{x}$. Also, the eigenvalues of $\boldsymbol{U}\boldsymbol{U}^* = \boldsymbol{I}$ are all $\lambda_k = 1$, and $||\boldsymbol{u}_k||^2 = 1$, so Theorem 2.10 holds trivially.

Example 2.9: $M = 3$ vectors in an $N = 2$ dimensional space $\mathbb{R}^2$ form a frame:

$$\boldsymbol{F} = [\boldsymbol{f}_1, \boldsymbol{f}_2, \boldsymbol{f}_3] = \begin{bmatrix} -1 & 1/2 & 1/2 \\ 0 & \sqrt{3}/2 & -\sqrt{3}/2 \end{bmatrix}. \tag{2.342}$$

Note that these frame vectors are all normalized $||\boldsymbol{f}_k|| = 1$. We also have

$$\boldsymbol{F}\boldsymbol{F}^{\mathrm{T}} = \frac{3}{2}\begin{bmatrix} 1 & 0 \\ 0 & 1 \end{bmatrix}, \quad (\boldsymbol{F}\boldsymbol{F}^{\mathrm{T}})^{-1} = \frac{2}{3}\begin{bmatrix} 1 & 0 \\ 0 & 1 \end{bmatrix}. \tag{2.343}$$

The eigenvalues of these two matrices are obviously $\lambda_1 = \lambda_2 = 3/2$ and $1/\lambda_1 = 1/\lambda_2 = 2/3$, respectively, indicating this is a tight frame with $A = B$. The dual frame $\tilde{\boldsymbol{F}}$ can be found as the pseudo-inverse of $\boldsymbol{F}^{\mathrm{T}}$:

$$\tilde{\boldsymbol{F}} = [\tilde{\boldsymbol{f}}_1, \tilde{\boldsymbol{f}}_2, \tilde{\boldsymbol{f}}_3] = (\boldsymbol{F}\boldsymbol{F}^{\mathrm{T}})^{-1}\boldsymbol{F} = \frac{2}{3}\boldsymbol{F} = \begin{bmatrix} -2/3 & 1/3 & 1/3 \\ 0 & \sqrt{3}/3 & -\sqrt{3}/3 \end{bmatrix}. \tag{2.344}$$

Any given $\boldsymbol{x} = [x[1], x[2]]^{\mathrm{T}}$ can be expanded in terms of either of the two frames:

$$\boldsymbol{x} = \sum_{k=1}^{3} c[k]\tilde{\boldsymbol{f}}_k = \sum_{k=1}^{3} \langle \boldsymbol{x}, \boldsymbol{f}_k \rangle \tilde{\boldsymbol{f}}_k = \sum_{k=1}^{3} d[k]\boldsymbol{f}_k = \sum_{k=1}^{3} \langle \boldsymbol{x}, \tilde{\boldsymbol{f}}_k \rangle \boldsymbol{f}_k, \tag{2.345}$$

where $\boldsymbol{c} = \boldsymbol{F}^*\boldsymbol{x}$ or

$$c[1] = -x[1], \quad c[2] = \frac{1}{2}[x[1] + \sqrt{3}x[2]], \quad c[3] = \frac{1}{2}[x[1] - \sqrt{3}x[2]], \tag{2.346}$$

and $\boldsymbol{d} = \tilde{\boldsymbol{F}}^*\boldsymbol{x}$ or

$$d[1] = -\frac{2}{3}x[1], \quad d[2] = \frac{1}{3}[x[1] + \sqrt{3}x[2]], \quad d[3] = \frac{1}{3}[x[1] - \sqrt{3}x[2]]. \tag{2.347}$$

The energy contained in the coefficients $\boldsymbol{c}$ and $\boldsymbol{d}$ is respectively:

$$||\boldsymbol{c}||^2 = \sum_{k=1}^{3} |\langle \boldsymbol{x}, \boldsymbol{f}_k \rangle|^2 = \frac{3}{2}||\boldsymbol{x}||^2 = \lambda ||\boldsymbol{x}||^2, \tag{2.348}$$

and

$$||\boldsymbol{d}||^2 = \sum_{k=1}^{3} |\langle \boldsymbol{x}, \tilde{\boldsymbol{f}}_k \rangle|^2 = \frac{2}{3}||\boldsymbol{x}||^2 = \frac{1}{\lambda}||\boldsymbol{x}||^2. \tag{2.349}$$

Specifically, if we let $\boldsymbol{x} = [1, 2]^{\mathrm{T}}$, then

$$\boldsymbol{c} = \boldsymbol{F}^{\mathrm{T}}\boldsymbol{x} = \begin{bmatrix} \boldsymbol{f}_1^{\mathrm{T}} \\ \boldsymbol{f}_2^{\mathrm{T}} \\ \boldsymbol{f}_3^{\mathrm{T}} \end{bmatrix} \boldsymbol{x} = \begin{bmatrix} \langle \boldsymbol{x}, \boldsymbol{f}_1 \rangle \\ \langle \boldsymbol{x}, \boldsymbol{f}_2 \rangle \\ \langle \boldsymbol{x}, \boldsymbol{f}_3 \rangle \end{bmatrix} = \begin{bmatrix} -1 \\ 1+\sqrt{3} \\ 1-\sqrt{3} \end{bmatrix}, \tag{2.350}$$

and

$$\boldsymbol{d} = \tilde{\boldsymbol{F}}^{\mathrm{T}}\boldsymbol{x} = \begin{bmatrix} \tilde{\boldsymbol{f}}_1^{\mathrm{T}} \\ \tilde{\boldsymbol{f}}_2^{\mathrm{T}} \\ \tilde{\boldsymbol{f}}_3^{\mathrm{T}} \end{bmatrix} \boldsymbol{x} = \begin{bmatrix} \langle \boldsymbol{x}, \tilde{\boldsymbol{f}}_1 \rangle \\ \langle \boldsymbol{x}, \tilde{\boldsymbol{f}}_2 \rangle \\ \langle \boldsymbol{x}, \tilde{\boldsymbol{f}}_3 \rangle \end{bmatrix} = \frac{2}{3}\begin{bmatrix} -1 \\ 1+\sqrt{3} \\ 1-\sqrt{3} \end{bmatrix}. \tag{2.351}$$

Example 2.10: Vectors $\boldsymbol{f}_1$ and $\boldsymbol{f}_2$ form a basis that spans the 2-D space:

$$\boldsymbol{f}_1 = \begin{bmatrix} 1 \\ 0 \end{bmatrix}, \quad \boldsymbol{f}_2 = \begin{bmatrix} 1 \\ 1 \end{bmatrix}, \quad \boldsymbol{F} = [\boldsymbol{f}_1, \boldsymbol{f}_2] = \begin{bmatrix} 1 & 1 \\ 0 & 1 \end{bmatrix}, \tag{2.352}$$

$$\boldsymbol{F}\boldsymbol{F}^{\mathrm{T}} = \begin{bmatrix} 2 & 1 \\ 1 & 1 \end{bmatrix}, \quad (\boldsymbol{F}\boldsymbol{F}^{\mathrm{T}})^{-1} = \begin{bmatrix} 1 & -1 \\ -1 & 2 \end{bmatrix}. \tag{2.353}$$

The dual frame can be found to be

$$\tilde{\boldsymbol{F}} = (\boldsymbol{F}\boldsymbol{F}^{\mathrm{T}})^{-1}\boldsymbol{F} = \begin{bmatrix} 1 & 0 \\ -1 & 1 \end{bmatrix}; \quad \text{i.e.} \quad \tilde{\boldsymbol{f}}_1 = \begin{bmatrix} 1 \\ -1 \end{bmatrix}, \quad \tilde{\boldsymbol{f}}_2 = \begin{bmatrix} 0 \\ 1 \end{bmatrix}. \tag{2.354}$$

Obviously, the biorthogonality condition in Eq. (2.324) is satisfied by these two sets of bases. Next, to represent a vector $\boldsymbol{x} = [0, 2]^{\mathrm{T}}$ by each of the two bases, we find the coefficients as

$$\begin{aligned} c[1] &= \langle \boldsymbol{x}, \tilde{\boldsymbol{f}}_1 \rangle = 2, \qquad & c[1] &= \langle \boldsymbol{x}, \tilde{\boldsymbol{f}}_2 \rangle = -2, \\ d[1] &= \langle \boldsymbol{x}, \boldsymbol{f}_1 \rangle = 0 & d[2] &= \langle \boldsymbol{x}, \boldsymbol{f}_2 \rangle = -2. \end{aligned}$$

Now we have

$$\boldsymbol{x} = c[1]\boldsymbol{f}_1 + c[2]\boldsymbol{f}_2 = 2\begin{bmatrix} 1 \\ 0 \end{bmatrix} - 2\begin{bmatrix} 1 \\ 1 \end{bmatrix} = \begin{bmatrix} 0 \\ -2 \end{bmatrix}, \tag{2.355}$$

or

$$\boldsymbol{x} = d[1]\tilde{\boldsymbol{f}}_1 + d[2]\tilde{\boldsymbol{f}}_2 = -2\begin{bmatrix} 0 \\ 1 \end{bmatrix} = \begin{bmatrix} 0 \\ -2 \end{bmatrix}. \tag{2.356}$$

2.5 Kernel function and Mercer's theorem

Definition 2.26. *A kernel is a function that maps two continuous variable t, τ to a complex value $K(t, \tau) \in \mathbb{C}$. If the two variables are truncated and sampled to become discrete t_m, t_n $(m, n = 1, \ldots, N)$, then the kernel can be represented by an N by N matrix $\boldsymbol{K}$ with the mnth element being $K(t_m, t_n) = K[m, n]$:*

$$\boldsymbol{K} = \begin{bmatrix} K[1,1] & K[1,2] & \cdots & K[1,N] \\ K[2,1] & K[2,2] & \cdots & K[2,N] \\ \vdots & \vdots & \ddots & \vdots \\ K[N,1] & K[N,2] & \cdots & K[N,N] \end{bmatrix}. \tag{2.357}$$

If $K(t, \tau) = \overline{K}(\tau, t)$ or $K[m, n] = \overline{K}[n, m]$ (i.e., $\boldsymbol{K} = \boldsymbol{K}^*$), the kernel is Hermitian (self-adjoint).

Definition 2.27. *A continuous kernel $K(t, \tau)$ is positive definite if the following holds for any function $x(t)$ defined over $[a, b]$:*

$$\int_a^b \int_a^b x(t) K(t, \tau) \overline{x}(\tau) \, d\tau \, dt > 0. \tag{2.358}$$

A discrete kernel $K[m,n]$ *is positive definite if the following holds for any vector* $\boldsymbol{x} = [x[1], \ldots, x[N]]$*:*

$$\boldsymbol{x}^* \boldsymbol{K} \boldsymbol{x} = \sum_{m=1}^{N} \sum_{n=1}^{N} x[m] K[m,n] \overline{x}[n] > 0. \tag{2.359}$$

Definition 2.28. *An operator* T_K *associated with a continuous kernel* $K(t,\tau)$ *defined below can be applied to a function* $x(t)$ *to generate another function* $y(t)$*:*

$$T_K x(t) = \int_a^b K(t,\tau) x(\tau) \, d\tau = y(t). \tag{2.360}$$

An operator T_K *associated with a discrete kernel* $K[m,n]$ *is simply the matrix* $T_K = \boldsymbol{K}$*, which, when applied to a vector* $\boldsymbol{x}$*, generates another vector* $\boldsymbol{y} = T_K \boldsymbol{x} = \boldsymbol{K}\boldsymbol{x}$*, or in component form:*

$$\sum_{m=1}^{N} K[m,n] x[m] = y[n] \qquad n = 1, \ldots, N. \tag{2.361}$$

Theorem 2.13. *The operator* T_K *associated with a Hermitian kernel is Hermitian (self-adjoint):*

$$\langle T_k x(t), y(t) \rangle = \langle x(t), T_K y(t) \rangle. \tag{2.362}$$

Proof: For operator T_K associated with a continuous kernel, we have

$$\langle T_K x(t), y(t) \rangle = \int_a^b T_K x(t) \, \overline{y}(t) \, dt = \int_a^b \left[\int_a^b K(t,\tau) x(\tau) \, d\tau \right] \overline{y}(t) \, dt$$
$$= \int_a^b \left[\int_a^b \overline{K}(\tau,t) \overline{y}(t) \, dt \right] x(\tau) \, d\tau = \int_a^b x(\tau) \, \overline{T_K y(\tau)} \, d\tau = \langle x(t), T_K y(t) \rangle. \tag{2.363}$$

For operator $T_K = \boldsymbol{K}$ associated with a discrete kernel, we have

$$\langle \boldsymbol{K}\boldsymbol{x}, \boldsymbol{y} \rangle = \sum_{n=1}^{N} \left(\sum_{m=1}^{N} K[m,n] x[m] \right) \overline{y}[n]$$
$$= \sum_{m=1}^{N} x[m] \left(\sum_{n=1}^{N} \overline{K}[m,n] \overline{y}[n] \right) = \langle \boldsymbol{x}, \boldsymbol{K}\boldsymbol{y} \rangle. \tag{2.364}$$

Q.E.D.

A self-adjoint operator T_K has all the properties stated in Theorem 2.4. Specifically, let λ_k be the kth eigenvalue of a self-adjoint operator T_K and $\phi_k(t)$ or $\boldsymbol{\phi}_k$ be the corresponding eigenfunction or eigenvector:

$$\int_a^b K(t,\tau) \phi_k(\tau) \, d\tau = \lambda_k \phi_k(t), \qquad \text{or} \qquad T_K \boldsymbol{\phi}_k = \boldsymbol{K} \boldsymbol{\phi}_k = \lambda_k \boldsymbol{\phi}_k, \tag{2.365}$$

then we have

1. All eigenvalues λ_k are real.
2. All eigenfunctions/eigenvectors are mutually orthogonal:

$$\langle \phi_k(t), \phi_l(t) \rangle = \langle \boldsymbol{\phi}_k, \boldsymbol{\phi}_l \rangle = \delta[k-l]. \tag{2.366}$$

3. All eigenfunctions/eigenvectors form a complete orthogonal system; i.e., they form a basis that spans the function/vector space.

Theorem 2.14. *(Mercer's theorem) Let λ_k $(k = 1, 2, \ldots)$ be the kth eigenvalue of the operator T_K associated with a positive definite Hermitian kernel $K(t, \tau)$, and $\phi_k(t)$ the corresponding eigenfunction, then the kernel can be expanded as*

$$K(t, \tau) = \sum_{k=1}^{\infty} \lambda_k \phi_k(t) \overline{\phi}_k(\tau). \tag{2.367}$$

Let λ_k $(k = 1, 2, \ldots)$ be the kth eigenvalue of the operator $\boldsymbol{K}$ associated with a positive definite Hermitian kernel $K[m, n]$, and $\boldsymbol{\phi}_k$ the corresponding eigenvector, then the kernel can be expanded as

$$K[m, n] = \sum_{k=1}^{N} \lambda_k \phi[m, k] \overline{\phi}[n, k] \qquad m, n = 1, \ldots, N, \tag{2.368}$$

where $\phi[m, k]$ is the mth element of the kth eigenvector $\boldsymbol{\phi}_k = [\phi[1, k], \ldots, \phi[N, k]]^T$.

The general proof of Mercer's theorem in Hilbert space is beyond the scope of this book and therefore omitted, but the discrete version in $\mathbb{C}^N$ given in Eq. (2.368) is simply the element form of Eq. (2.163) for any Hermitian matrix:

$$\boldsymbol{K} = \sum_{k=1}^{N} \lambda_k \boldsymbol{\phi}_k \boldsymbol{\phi}_k^*. \tag{2.369}$$

Note that given Eq. (2.367) in Mercer's theorem, Eq. (2.365) can be easily derived:

$$\int_a^b K(t, \tau) \phi_l(\tau) \, d\tau = \int_a^b \left[\sum_{k=1}^{\infty} \lambda_k \phi_k(t) \overline{\phi}_k(\tau) \right] \phi_l(\tau) \, d\tau$$
$$= \sum_{k=1}^{\infty} \lambda_k \phi_k(t) \int_a^b \overline{\phi}_k(\tau) \phi_l(\tau) \, d\tau = \sum_{k=1}^{\infty} \lambda_k \phi_k(t) \delta[k-l] = \lambda_l \phi_l(t). \tag{2.370}$$

As an example, the covariance of a centered stochastic process $x(t)$ with $\mu_x(t) = 0$ is a Hermitian kernel $K(t, \tau) = \sigma_x^2(t, \tau)$ that maps two variables t and τ to a complex value:

$$Cov(x(t), x(\tau)) = \sigma_x^2(t, \tau) = E[x(t)\overline{x}(\tau)] = \overline{E[x(\tau)\overline{x}(t)]} = \overline{\sigma}_x^2(\tau, t). \tag{2.371}$$

Moreover, we can show that it is also positive definite. For any function $f(t)$ we have

$$\int_a^b \int_a^b f(t)\,\sigma_x^2(t,\tau)\,\overline{f}(\tau)\,dt\,d\tau = \int_a^b \int_a^b E[f(t)x(t)\,\overline{f}(\tau)\overline{x}(\tau)]\,dt\,d\tau$$
$$= E\left[\int_a^b f(t)x(t)\,dt \int_a^b \overline{f}(\tau)\overline{x}(\tau)\,d\tau\right] = E\left|\int_a^b f(t)x(t)\,dt\right|^2 > 0. \quad (2.372)$$

Let T_K be the Hermitian integral operator associated with this kernel $\sigma_x^2(t,\tau) = \overline{\sigma}_x^2(\tau,t)$. Its eigenequation is

$$T_k\phi_k(t) = \int_a^b \sigma_x^2(t,\tau)\phi_k(\tau)\,d\tau = \lambda_k\phi_k(t) \qquad k = 1,2,\dots, \quad (2.373)$$

where all eigenvalues $\lambda_k > 0$ are real and positive, and the eigenfunctions $\phi_k(t)$ are orthogonal:

$$\langle\phi_k(t),\phi_l(t)\rangle = \int_a^b \phi_k(t)\overline{\phi}_l(t)\,dt = \delta[k-l], \quad (2.374)$$

and they form a complete orthogonal basis that spans the function space.

When the stochastic process $x(t)$ is truncated and sampled, it become a random vector $\boldsymbol{x} = [x[1],\dots,x[N]]^{\mathrm{T}}$. The covariance between any two components $x[m]$ and $x[n]$ is a discrete Hermitian kernel that maps two variables m and n to a complex value:

$$Cov(x[m],x[n]) = \sigma_{mn}^2 = E[x[m]\overline{x}[n]] = \overline{E[x[n]\overline{x}[m]]} = \overline{\sigma}_{nm}^2 \qquad m,n = 1,\dots,N. \quad (2.375)$$

The associated operator is an N by N covariance matrix:

$$\boldsymbol{\Sigma}_x = E(\boldsymbol{x}\boldsymbol{x}^*) = \begin{bmatrix} \sigma_{11}^2 & \sigma_{12}^2 & \cdots & \sigma_{1N}^2 \\ \sigma_{21}^2 & \sigma_{22}^2 & \cdots & \sigma_{2N}^2 \\ \vdots & \vdots & \ddots & \vdots \\ \sigma_{N1}^2 & \sigma_{N2}^2 & \cdots & \sigma_{NN}^2 \end{bmatrix}. \quad (2.376)$$

The eigenequation of this matrix operator is

$$\boldsymbol{\Sigma}_x\boldsymbol{\phi}_k = \lambda_k\boldsymbol{\phi}_k \qquad k = 1,\dots,N. \quad (2.377)$$

As $\boldsymbol{\Sigma}^* = \boldsymbol{\Sigma}$ is Hermitian (symmetric if $\boldsymbol{x}$ is real) and positive definite, its eigenvalues λ_k are all real positive, and the eigenvectors are orthogonal:

$$\langle\boldsymbol{\phi}_k,\boldsymbol{\phi}_l\rangle = \boldsymbol{\phi}_k^{\mathrm{T}}\overline{\boldsymbol{\phi}}_l\boldsymbol{\phi}_l^*\boldsymbol{\phi}_k = \delta[k-l] \qquad k,l = 1,\dots,N, \quad (2.378)$$

and they form a unitary matrix $\boldsymbol{\Phi} = [\boldsymbol{\phi}_1,\dots,\boldsymbol{\phi}_N]$ satisfying $\boldsymbol{\Phi}^{-1} = \boldsymbol{\Phi}^*$ i.e., $\boldsymbol{\Phi}^*\boldsymbol{\Phi} = \boldsymbol{I}$. Eq. (2.377) can also be written in the following forms:

$$\boldsymbol{\Sigma}_x\boldsymbol{\Phi} = \boldsymbol{\Phi}\boldsymbol{\Lambda}, \qquad \boldsymbol{\Phi}^*\boldsymbol{\Sigma}_x\boldsymbol{\Phi} = \boldsymbol{\Lambda}, \qquad \boldsymbol{\Sigma}_x = \boldsymbol{\Phi}\boldsymbol{\Lambda}\boldsymbol{\Phi}^* = \sum_{k=1}^N \lambda_k\boldsymbol{\phi}_k\boldsymbol{\phi}_k^*. \quad (2.379)$$

Theorem 2.15. *(Karhunen-Loève Theorem, continuous) Let* $\sigma_x^2(t,\tau) = Cov(x(t), x(\tau))$ *be the covariance of a centered stochastic process* $x(t)$ *with* $\mu_x = E(x(t)) = 0$, *and* λ_k *and* $\phi_k(t)$ *be respectively the kth eigenvalue and the corresponding eigenfunction of the integral operator associated with* $\sigma_x^2(t,\tau)$ *as a kernel:*

$$T_K \phi_k(t) = \int_a^b \sigma_x^2(t,\tau)\phi_k(t)\,dt = \lambda_k \phi_k(t) \qquad \text{for all } k, \tag{2.380}$$

then $x(t)$ *can be series expanded as*

$$x(t) = \sum_k c[k]\phi_k(t). \tag{2.381}$$

Here, $c[k]$ *is the kth random coefficient given by*

$$c[k] = \langle x(t), \phi_k(t)\rangle = \int_a^b x(t)\overline{\phi}_k(t)\,dt \qquad \text{for all } k, \tag{2.382}$$

which are centered (zero mean) $E(c[k]) = 0$ *and uncorrelated:*

$$Cov(c[k], c[l]) = \lambda_k\ \delta[k-l]. \tag{2.383}$$

Proof: As $\sigma_x^2(t,\tau)$ is a self-adjoint kernel, the eigenfunctions $\phi_k(t)$ of the associated operator T_K form a complete orthogonal basis, and any given stochastic process $x(t)$ can be represented as a linear combination of $\phi_k(t)$; i.e., Eq. (2.381) holds.

Taking an inner product with $\phi_l(t)$ on both sides of Eq. (2.381), we get Eq. (2.382)

$$\langle x(t), \phi_l(t)\rangle = \int_a^b x(t)\overline{\phi}_l(t)\,dt = \sum_{k=1}^{\infty} c[k]\langle \phi_k(t), \phi_l(t)\rangle$$
$$= \sum_{k=1}^{\infty} c[k]\delta[k-l] = c[l]. \tag{2.384}$$

The expectation of this equation is indeed zero:

$$E[c[k]] = E\left[\int_a^b x(t)\overline{\phi}_k(t)\,dt\right] = \int_a^b E[x(t)]\,\overline{\phi}_k(t)\,dt = 0. \tag{2.385}$$

Finally, we show Eq. (2.383) holds:

$$Cov(c[k], c[l]) = E(c[k]\overline{c}[l]) = E\left[\int_a^b x(t)\overline{\phi}_k(t)\,dt \int_a^b \overline{x}(\tau)\phi_l(\tau)\,d\tau\right]$$
$$= \int_a^b \left[\int_a^b \phi_l(\tau)E[x(t)\overline{x}(\tau)]\,d\tau\right]\overline{\phi}_k(t)\,dt = \int_a^b \left[\int_a^b \phi_l(\tau)\sigma_x^2(t,\tau)\,d\tau\right]\overline{\phi}_k(t)\,dt$$
$$= \int_a^b \lambda_l\phi_l(t)\overline{\phi}_k(t)\,dt = \lambda_l \int_a^b \phi_l(t)\overline{\phi}_k(t)\,dt = \lambda_l\ \delta[k-l] = \lambda_k\ \delta[k-l]. \tag{2.386}$$

Q.E.D.

When the centered stochastic process $x(t)$ is truncated and sampled to become a finite random vector $\boldsymbol{x} = [x[1], \ldots, x[N]]^{\mathrm{T}}$ with $E(\boldsymbol{x}) = \boldsymbol{\mu}_x = 0$, the Karhunen-Loève theorem takes the following discrete form.

Theorem 2.16. *(Karhunen-Loève Theorem, discrete) Let $\boldsymbol{x}$ be a centered random vector with $\boldsymbol{\mu}_x = E(\boldsymbol{x}) = \mathbf{0}$, and $\boldsymbol{\Sigma}_x$ the covariance matrix with the mn-th component $\sigma_x^2[m,n] = Cov(x[m], x[n])$. Also let λ_k and $\boldsymbol{\phi}_k$ be respectively the kth eigenvalue and the corresponding eigenvector of $\boldsymbol{\Sigma}_x$:*

$$\boldsymbol{\Sigma}_x \boldsymbol{\phi}_k = \lambda_k \boldsymbol{\phi}_k \qquad k = 1, \ldots, N, \tag{2.387}$$

then $\boldsymbol{x}$ can be series expanded as

$$\boldsymbol{x} = \sum_{k=1}^{N} c[k] \boldsymbol{\phi}_k. \tag{2.388}$$

Here $c[k]$ are the kth random coefficients given by

$$c[k] = \langle \boldsymbol{x}, \boldsymbol{\phi}_k \rangle = \boldsymbol{\phi}^* \boldsymbol{x} \qquad k = 1, \ldots, N, \tag{2.389}$$

which are centered (zero mean) $E(c[k]) = 0$ and uncorrelated:

$$Cov(c[k], c[l]) = \lambda_k \ \delta[k - l]. \tag{2.390}$$

Proof: As the covariance matrix $\boldsymbol{\Sigma}_x$ is Hermitian and positive definite, its eigenvalues λ_k are all real positive and eigenvectors $\boldsymbol{\phi}_k$ form a complete orthogonal system by which any $\boldsymbol{x}$ can be series expanded as

$$\boldsymbol{x} = \sum_{k=1}^{N} c[k] \boldsymbol{\phi}_k = \boldsymbol{\Phi} \boldsymbol{c}, \tag{2.391}$$

where $\boldsymbol{c} = [c[1], \ldots, c[N]]^{\mathrm{T}}$ is a random vector formed by the N coefficients, and $\boldsymbol{\Phi} = [\boldsymbol{\phi}_1, \ldots, \boldsymbol{\phi}_N]^{\mathrm{T}}$; i.e., Eq. (2.391) holds.

The random coefficients can be found by pre-multiplying both sides by $\boldsymbol{\Phi}^{-1} = \boldsymbol{\Phi}^*$:

$$\boldsymbol{\Phi}^* \boldsymbol{x} = \boldsymbol{\Phi}^* \boldsymbol{\Phi} \boldsymbol{c} = \boldsymbol{c}; \quad \text{i.e.} \quad c[k] = \langle \boldsymbol{x}, \boldsymbol{\phi}_k \rangle = \boldsymbol{\phi}_k^* \boldsymbol{x} \qquad k = 1, \ldots, N. \tag{2.392}$$

This is Eq. (2.389). The mean vector of $\boldsymbol{c}$ is indeed zero:

$$\boldsymbol{\mu}_c = E(\boldsymbol{c}) = E(\boldsymbol{\Phi}^* \boldsymbol{x}) = \boldsymbol{\Phi}^* E(\boldsymbol{x}) = \mathbf{0}, \tag{2.393}$$

and the covariance matrix of $\boldsymbol{c}$ is

$$\begin{aligned} \boldsymbol{\Sigma}_c &= E(\boldsymbol{c}\boldsymbol{c}^*) = E[(\boldsymbol{\Phi}^* \boldsymbol{x})(\boldsymbol{\Phi}^* \boldsymbol{x})^*] = E[\boldsymbol{\Phi}^* \boldsymbol{x}\boldsymbol{x}^* \boldsymbol{\Phi}] \\ &= \boldsymbol{\Phi}^* E(\boldsymbol{x}\boldsymbol{x}^*) \boldsymbol{\Phi} = \boldsymbol{\Phi}^* \boldsymbol{\Sigma}_x \boldsymbol{\Phi} = \boldsymbol{\Lambda}. \end{aligned} \tag{2.394}$$

Finally, as the covariance matrix $\boldsymbol{\Sigma}_c = \boldsymbol{\Lambda}$ is diagonalized, we get Eq. (2.390):

$$Cov(c[k], c[l]) = \sigma_{kl}^2 = \lambda_k \delta[k - l] \qquad k, l = 1, \ldots, N. \tag{2.395}$$

Q.E.D.

As $Cov(c[k], c[l]) = 0$ for all $k \neq l$; i.e., the N coefficients are not correlated, we see that the random signal $\boldsymbol{x}$ is decorrelated by the transformation $\boldsymbol{c} = \boldsymbol{\Phi}^* \boldsymbol{x}$ in Eq. (2.392). But when $k = l$, we have $Cov(c[k], c[k]) = Var(c[k]) = \sigma_k^2 = \lambda_k$; i.e., the variance of the kth coefficient $c[k]$ is the kth eigenvalue λ_k.

It is interesting to compare the Karhunen-Loève expansion considered above with the generalized Fourier expansion previously considered previously in Subsection 2.1.4. First, for a continuous signal $x(t)$, Eqs. (2.381) and (2.382) can be compared with Eqs. (2.106) and (2.108); then, for a discrete signal $\boldsymbol{x} = [x[1], \ldots, x[N]]^{\mathrm{T}}$, Eqs. (2.391) and (2.392) can be compared with Eqs. (2.96) and (2.98). We see that the generalized Fourier expansion and the Karhunen-Loève expansion are identical in form, for both continuous and discrete signals. However, we need to realize these two types of expansions are of essential difference. The generalized Fourier expansion represents a deterministic signal in terms of a set of predetermined basis functions $u_k(t)$ or vector $\boldsymbol{u}_k$, weighted by deterministic coefficients $c[k]$; whereas the Karhunen-Loève expansion represents a stochastic signal function $x(t)$ in terms of a set of basis functions $\phi_k(t)$, the eigenfunctions of the integral operator associated with the covariance function of the signal, or a random signal vector $\boldsymbol{x}$ in terms of a set of basis vector $\boldsymbol{\phi}_k$, the eigenvectors of the covariance matrix $\boldsymbol{\Sigma}_x$, weighted by random coefficients $c[k]$. These basis functions and vectors cannot be predetermined as they are both dependent on the statistical properties of the specific signal being considered. The Karhunen-Loève theorem and the associated series expansion will be considered in Chapter 9.

2.6 Summary

We summarize below the essential points discussed in this chapter, based on which the various orthogonal transform methods to be specifically considered in the following chapters will all be looked at from a unified point of view.

- A time signal can be considered as a vector $\boldsymbol{x} \in H$ in a Hilbert space, the specific type of which depends on the nature of the signal. For example, a continuous signal $x(t)$ over time interval $a < t < b$ is a vector $\boldsymbol{x} = x(t)$ in $\mathcal{L}^2$ space; and its discrete samples form a vector $\boldsymbol{x} = [\ldots, x[n], \ldots]^{\mathrm{T}}$ in l^2 space. When the signal is truncated to become a set of N samples, then $\boldsymbol{x} = [x[1], \ldots, x[N]]^{\mathrm{T}}$ is a vector in $\mathbb{C}^N$ space.
- A signal vector $\boldsymbol{x}$ given in the default form, either as a time function or a sequence of time samples, can be considered as a linear combination of a set of weighted and shifted time impulses (Eqs. (1.3) and (1.9)):

$$x(t) = \int x(\tau)\delta(t - \tau)\, d\tau, \quad \text{(for all } t\text{)}, \tag{2.396}$$

or

$$x[n] = \sum_m x[m]\delta[m-n], \quad \text{for all } n. \tag{2.397}$$

- Here, $\delta(t-\tau)$ and $\delta[m-n]$ can be treated respectively as the standard basis that spans the corresponding signal space. In other words, the default form of a signal $x[n]$ or $x(t)$ is actually a set of coefficients (countable) or weighting function (uncountable) of the standard basis, which is always implicitly used in the default representation of a time signal.
- The signal vector $\boldsymbol{x}$ can be alternatively and equivalently represented by any of the bases that also span the vector space, such as an orthogonal basis obtained by applying a unitary transformation, a rotation, to the standard basis. Such a basis is composed of a set of either countable vectors $\boldsymbol{b}_k$ or uncountably infinite vectors $\boldsymbol{b}(f)$. If the basis is orthonormal, we represent it as $\boldsymbol{u}_k$ and get:

$$\begin{aligned} \boldsymbol{x} &= \sum_k c[k]\boldsymbol{u}_k = \sum_k \langle \boldsymbol{x}, \boldsymbol{u}_k \rangle \boldsymbol{u}_k, \\ c[k] &= \langle \boldsymbol{x}, \boldsymbol{b}_k \rangle, \end{aligned} \tag{2.398}$$

or

$$\begin{aligned} \boldsymbol{x} &= \int c(f)\boldsymbol{u}(f)\, df = \int \langle \boldsymbol{x}, \boldsymbol{u}(f) \rangle \boldsymbol{u}(f)\, df, \\ c(f) &= \langle \boldsymbol{x}, \boldsymbol{b}(f) \rangle. \end{aligned} \tag{2.399}$$

  The second equation for the weighting coefficient $c[k]$ or function $c(f)$, referred to as the forward orthogonal transform in future chapters, represents the *analysis* of the signal by which the signal is decomposed into a set of components $c[k]\boldsymbol{b}_k$ or $c(f)\boldsymbol{b}(t)$. The summation or integration in the first equation, referred to as the inverse transform, is the *synthesis* of the signal by which the signal is reconstructed from its components.
- The representations of the signal under different orthogonal bases are equivalent, in the sense that the total amount of energy or information contained in the signal, represented by its norm of the vector, is conserved by the unitary transformation (a rotation) relating the two orthogonal bases before and after the transformation due to Parseval's identity.
- In addition to the orthogonal transforms based on a set of orthogonal basis vectors, each of which carries some independent information of the signal, we will also consider the wavelet transforms based on a set of frame vectors that may be non-orthogonal or even non-independent. These frame vectors may be correlated and there may exist certain redundancy in terms of the signal information they each carry. There are both pros and cons in such signal representations with redundancy.

In the rest of the book we will study various orthogonal transforms, each representing a given signal vector as a set of weighting coefficients or weighting function of a chosen orthogonal basis, obtained by applying a unitary transformation to the standard basis of the space. In the discussion we will also address the issues such as why such a unitary transformation is desirable and how to find the optimal transformation according to certain quantifiable criteria.

2.7 Homework problems

1. Approximate a given 3-D vector $\boldsymbol{x} = [1, 2, 3]^{\mathrm{T}}$ in an 2-D subspace spanned by the two standard basis vectors $\boldsymbol{e}_1 = [1, 0, 0]^{\mathrm{T}}$ and $\boldsymbol{e}_2 = [0, 1, 0]^{\mathrm{T}}$. Obtain the error vector $\tilde{\boldsymbol{x}}$ and verify that it is orthogonal to both $\boldsymbol{e}_1$ and $\boldsymbol{e}_2$.
2. Repeat the problem above but now approximate the same 3-D vector $\boldsymbol{x} = [1, 2, 3]^{\mathrm{T}}$ above but now in a different 2-D subspace spanned by two basis vectors $\boldsymbol{a}_1 = [1, 0, -1]^{\mathrm{T}}$ and $\boldsymbol{a}_2 = [-1, 2, 0]^{\mathrm{T}}$. Find a vector in this 2-D subspace $\hat{\boldsymbol{x}} = c[1]\boldsymbol{a}_1 + c[2]\boldsymbol{a}_2$ so that the error $||\boldsymbol{x} - \hat{\boldsymbol{x}}||$ is minimized.
3. Given two vectors $\boldsymbol{u}_1 = [2, 1]^{\mathrm{T}} / \sqrt{5}$ and $\boldsymbol{u}_2 = [-1, 2]^{\mathrm{T}} / \sqrt{5}$ in $\mathbb{R}^2$, do the following:
   (a) Verify that they are orthogonal;
   (b) Normalize them;
   (c) Use them as an orthonormal basis to represent a vector $\boldsymbol{x} = [1, 2]^{\mathrm{T}}$.
4. Use the Gram-Schmidt orthogonalization process to construct two new orthonormal basis vectors $\boldsymbol{b}_1$ and $\boldsymbol{b}_2$ from the two vectors $\boldsymbol{a}_1$ and $\boldsymbol{a}_2$ used in the previous problem, so that they span the same 2-D space, and then approximate the vector $\boldsymbol{x} = [1, 2, 3]^{\mathrm{T}}$ above. Note that as the off-diagonal elements of the 2 by 2 matrix are zero, and both elements on the main diagonal are one, the coefficients $c[1]$ and $c[2]$ can be easily found without solving a linear equation system.
5. Approximate a function $x(t) = t^2$ defined over an interval $[0, 1]$ in a 2-D space spanned by two basis functions $a_1(t)$ and $a_1(t)$:

$$a_1(t) = 1, \qquad a_2(t) = \begin{cases} 0 & 0 \le t < 1/2 \\ 1 & 1/2 \le t < 1 \end{cases}. \tag{2.400}$$

6. Repeat the problem above with the same $\boldsymbol{a}_1$ but a different $\boldsymbol{a}_2$ defined as

$$a_2(t) = \begin{cases} -1 & 0 \le t < 1/2 \\ 1 & 1/2 \le t < 1 \end{cases}. \tag{2.401}$$

   Note that $\boldsymbol{a}_1$ and $\boldsymbol{a}_2$ are orthogonal $\langle a_1(t), a_2(t) \rangle = 0$ (they are actually the first two basis functions of an orthogonal Walsh-Hadamard transform to be discussed in detail later).
7. Repeat the problem above, but now with an additional basis function $\mathbf{a}_3$ defined as

$$a_3(t) = \begin{cases} 1 & 0 \le t < 1/4 \\ -1 & 1/4 \le t < 3/4 \\ 1 & 3/4 \le t < 1 \end{cases}, \tag{2.402}$$

so that the 2-D space is expanded to a 3-D space spanned by $a_1(t)$, $a_2(t)$, and $a_3(t)$ (they are actually the first three basis functions of the Walsh-Hadamard transform).

8. Approximate the same function $x(t) = t^2$ above in a 3-D space spanned by three basis functions $a_0(t) = 1$, $a_1(t) = \sqrt{2}\cos(\pi t)$, and $a_2(t) = \sqrt{2}\cos(2\pi t)$, defined over the same time period. These happen to be the first three basis functions of the cosine transform.
   Hint: The following integral may be needed:
$$\int x^2 \cos(ax)dx = \frac{2x\cos(ax)}{a^2} + \frac{a^2x^2 - 2}{a^3}\sin(ax) + C. \tag{2.403}$$
9. Consider a 2-D space spanned by two orthonormal basis vectors:
$$\boldsymbol{a}_1 = \frac{1}{2}\begin{bmatrix}\sqrt{3}\\1\end{bmatrix}, \qquad \boldsymbol{a}_2 = \frac{1}{2}\begin{bmatrix}-1\\\sqrt{3}\end{bmatrix}. \tag{2.404}$$
   (a) Represent vector $\boldsymbol{x} = [1,\ 2]^{\mathrm{T}}$ under this basis as $\boldsymbol{x} = c[1]\boldsymbol{a}_1 + c[2]\boldsymbol{a}_2$. Find $c[1]$ and $c[2]$.
   (b) Represent a counterclockwise rotation of $\theta = 30°$ by a 2 by 2 matrix $\boldsymbol{R}$.
   (c) Rotate vector $\boldsymbol{x}$ to get $\boldsymbol{y} = \boldsymbol{R}\boldsymbol{x}$.
   (d) Represent $\boldsymbol{y}$ above under basis $\{\boldsymbol{a}_1, \boldsymbol{a}_2\}$ by $\boldsymbol{y} = d[1]\boldsymbol{a}_1 = d[2]\boldsymbol{a}_2$. Find the two coefficients $d[1]$ and $d[2]$.
   (e) Rotate the basis $\{\boldsymbol{a}_1, \boldsymbol{a}_2\}$ in the opposite direction $-\theta = -30°$ represented by $\boldsymbol{R}^{-1} = \boldsymbol{R}^{\mathrm{T}}$ to get $\boldsymbol{b}_1 = \boldsymbol{R}\boldsymbol{a}_1$ and $\boldsymbol{b}_2 = \boldsymbol{R}\boldsymbol{a}_2$.
   (f) Represent $\boldsymbol{x}$ under this new basis $\{\boldsymbol{b}_1, \boldsymbol{b}_2\}$ (which happens to be the standard basis).
   (g) Verify that $d'[1] = d[1]$ and $d'[2] = d[2]$; in other words, the representation $\{d[1], d[2]\}$ of the rotated vector $\boldsymbol{y}$ under the original basis $\{\boldsymbol{a}_1, \boldsymbol{a}_2\}$ is equivalent to the representation $\{d'[1], d'[2]\}$ of the original vector $\boldsymbol{x}$ under the inversely rotated basis $\{\boldsymbol{b}_1, \boldsymbol{b}_2\}$.
10. In Example 2.8 we approximated the temperature signal, an 8-D vector $\boldsymbol{x} = [65, 60, 65, 70, 75, 80, 75, 70]^{\mathrm{T}}$, in a 3-D subspace spanned by three orthogonal basis vectors. This process can be continued by increasing the dimensionality from three to eight, so that the approximation error will be progressively reduced to reach zero, when eventually the signal vector is represented in the entire 8-D vector space. Consider the eight orthogonal basis vectors shown below as the row vectors in this matrix (Walsh-Hadamard transform matrix):
$$\boldsymbol{H}_w = \frac{1}{\sqrt{8}}\begin{bmatrix} 1 & 1 & 1 & 1 & 1 & 1 & 1 & 1\\ 1 & 1 & 1 & 1 & -1 & -1 & -1 & -1\\ 1 & 1 & -1 & -1 & -1 & -1 & 1 & 1\\ 1 & 1 & -1 & -1 & 1 & 1 & -1 & -1\\ 1 & -1 & -1 & 1 & 1 & -1 & -1 & 1\\ 1 & -1 & -1 & 1 & -1 & 1 & 1 & -1\\ 1 & -1 & 1 & -1 & -1 & 1 & -1 & 1\\ 1 & -1 & 1 & -1 & 1 & -1 & 1 & -1 \end{bmatrix}. \tag{2.405}$$

Note that the first three rows are used in the example. Now approximate the same signal by using one to all eight rows as the basis vectors. Plot the original signal and the approximation in k-D subspaces for $k = 1, 2, \dots, 8$, adding one dimension at a time for more detailed variations in the signal. Find the coefficients $c[k]$ and the error in each case. Consider using some software tool such as Matlab.

11. The same temperature signal in Example 2.8, $\boldsymbol{x} = [65, 60, 65, 70, 75, 80, 75, 70]^{\mathrm{T}}$, can also be approximated using a set of different basis vectors obtained by sampling the following cosine functions:

$$a_0(t) = 1, \quad a_1(t) = \sqrt{2}\cos(\pi t), \quad a_2(t) = \sqrt{2}\cos(2\pi t), \tag{2.406}$$

at eight equally spaced points $n_k = 1/16 + n/8 = 0.0625 + n \times 0.125$, $(n = 1, 2, \dots, 8)$. The resulting vectors are actually used in the discrete cosine transform to be discussed later. Find the coefficients $c[k]$ and error for each approximation in a k-D subspace ($k = 1, 2, \dots, 8$), and plot the original signal together with the approximation for each case. Use a software tool such as Matlab.

12. Consider a frame in $\mathbb{R}^2$ containing three vectors that form a frame matrix:

$$\boldsymbol{F} = [\boldsymbol{f}_1, \boldsymbol{f}_2, \boldsymbol{f}_3] = \begin{bmatrix} 1 & -1 & 0 \\ 0 & 1 & 1 \end{bmatrix}. \tag{2.407}$$

- Find the eigenvalues of $\boldsymbol{F}\boldsymbol{F}^{\mathrm{T}}$ and its inverse $(\boldsymbol{F}\boldsymbol{F}^{\mathrm{T}})^{-1}$.
- Find the dual frame $\tilde{\boldsymbol{F}} = [\tilde{\boldsymbol{f}}_1, \tilde{\boldsymbol{f}}_2, \tilde{\boldsymbol{f}}_2]$.
- Find the coefficient vectors $\boldsymbol{c} = [c[1], c[2], c[3]]$ and $\boldsymbol{d} = [d[1], d[2], d[3]]$ for representing $\boldsymbol{x} = [1,\ 2]^{\mathrm{T}}$ so that

$$\boldsymbol{x} = \sum_k c[k]\tilde{\boldsymbol{f}}_k = \sum_k d[k]\boldsymbol{f}_k. \tag{2.408}$$

  Verify that $\boldsymbol{x}$ can be indeed perfectly reconstructed.
- Verify Eqs. (2.304) and (2.305).

13. Consider a frame in $\mathbb{R}^2$ containing two vectors that form a frame matrix:

$$\boldsymbol{F} = [\boldsymbol{f}_1, \boldsymbol{f}_2] = \begin{bmatrix} 2 & -1 \\ 1 & -2 \end{bmatrix}. \tag{2.409}$$

As $\boldsymbol{f}_1$ and $\boldsymbol{f}_2$ are linearly independent, they form a Riesz basis.
- Find the dual frame and verify they are biorthonormal as shown in Eq. (2.324).
- Given $\boldsymbol{x} = [2,\ 3]^{\mathrm{T}}$, find the coefficient vectors $\boldsymbol{c}$ and $\boldsymbol{d}$

$$\boldsymbol{x} = \sum_k c[k]\tilde{\boldsymbol{f}}_k = \sum_k d[k]\boldsymbol{f}_k. \tag{2.410}$$

  Verify that $\boldsymbol{x}$ can be indeed perfectly reconstructed.
- Verify Eqs. (2.304) and (2.305).

14. Consider the following basis in $\mathbb{R}^3$:

$$\boldsymbol{f}_1 = \begin{bmatrix} 1 \\ 0 \\ 0 \end{bmatrix}, \quad \boldsymbol{f}_2 = \begin{bmatrix} 1 \\ 1 \\ 0 \end{bmatrix}, \quad \boldsymbol{f}_3 = \begin{bmatrix} 1 \\ 1 \\ 1 \end{bmatrix}. \tag{2.411}$$

Find its biorthogonal dual $\tilde{\boldsymbol{f}}_1$, $\tilde{\boldsymbol{f}}_2$, $\tilde{\boldsymbol{f}}_3$, and two sets of coefficients $c[k]$ and $d[k]$ $(k = 1, 2, 3)$ to represent a vector $\boldsymbol{x} = [1, 2, 3]^{\mathrm{T}}$.

3 Continuous-time Fourier transform

3.1 The Fourier series expansion of periodic signals

3.1.1 Formulation of the Fourier expansion

As considered in section 2.2.3, the second-order differential operator D^2 over the interval $[0, T]$ is a self-adjoint operator, and its eigenfunctions $\phi_k(t) = e^{j2k\pi t/T}/\sqrt{T}$ $(k = 0, \pm 1, \pm 2, \ldots)$ are orthonormal (Eq. (2.182) i.e. Eq. (2.129)):

$$\langle \phi_k(t), \phi_l(t) \rangle = \frac{1}{T} \int_T e^{j2k\pi t/T} e^{-jln\pi t/T} \, dt = \frac{1}{T} \int_T e^{j2(f-l)\pi t/T} \, dt = \delta[k-l], \tag{3.1}$$

and they form a complete orthogonal system that spans a function space over interval $[0, T]$. Any periodic signal $x_T(t) = x_T(t + T)$ in the space can be expressed as a linear combination of these basis functions:

$$x_T(t) = \sum_{k=-\infty}^{\infty} X[k]\phi_k(t) = \frac{1}{\sqrt{T}} \sum_{k=-\infty}^{\infty} X[k] e^{j2k\pi t/T}. \tag{3.2}$$

Here, a periodic signal is denoted by $x_T(t)$ with a subscript T for its period. However, this subscript may be dropped for simplicity when no confusion will be caused. Note that at the two endpoints $t = 0$ and $t = T$ the summation of the Fourier expansion in Eq. (3.2) is $\sum_{k=-\infty}^{\infty} X[k]/\sqrt{T}$; i.e., the condition in Eq. (2.166) is always satisfied. In fact it can be shown that at these endpoints $t = 0$ and $t = T$ the Fourier expansion is the average of the end values of the original signal $(x_T(0) + x_T(T))/2$.

Owing to the orthogonality of these basis functions, the lth coefficient $X[l]$ can be found by taking an inner product with $\phi_l(t) = e^{j2l\pi t/T}/\sqrt{T}$ on both sides of Eq. (3.2):

$$\begin{aligned} \langle x_T(t), \phi_l(t) \rangle &= \langle x_T(t), e^{j2l\pi t/T}/\sqrt{T} \rangle = \frac{1}{T} \sum_{k=0}^{\infty} X[k] \langle e^{j2k\pi t/T}, e^{j2l\pi t/T} \rangle \\ &= \sum_{k=-\infty}^{\infty} X[k]\delta[k-n] = X[l], \end{aligned} \tag{3.3}$$

we therefore get

$$X[k] = \langle x_T(t), \phi_k(t) \rangle = \frac{1}{\sqrt{T}} \int_T x_T(t) e^{-j2k\pi t/T} \, dt. \tag{3.4}$$

Equations (3.2) and (3.4) form the Fourier series expansion pair:

$$\begin{aligned} X[k] &= \mathcal{F}[x_T(t)] = \frac{1}{\sqrt{T}} \int_T x_T(t) e^{-j2k\pi t/T} \, dt = \langle x_T(t), e^{j2k\pi t/T}/\sqrt{T} \rangle, \\ & \qquad k = 0, \pm 1, \pm 2, \ldots \\ x_T(t) &= \mathcal{F}^{-1}[X[k]] = \frac{1}{\sqrt{T}} \sum_{k=-\infty}^{\infty} X[k] e^{j2k\pi t/T} \\ &= \frac{1}{\sqrt{T}} \sum_{k=-\infty}^{\infty} \langle x_T(t), e^{j2k\pi t/T}/\sqrt{T} \rangle e^{j2k\pi t/T}. \end{aligned} \tag{3.5}$$

This is Eqs. (2.131) and (2.130). As the signal and the basis functions are both periodic, the integral above can be over any interval of T, such as $[0,\ T]$ and $[-T/2,\ t < T/2]$.

As defined in Eq. (2.174), we have $1/T = f_0$ and $2\pi/T = 2\pi f_0 = \omega_0$, where f_0 is the frequency gap between two consecutive eigenfunctions (Eq. (2.178)), and the basis function can also be written as

$$\phi_k(t) = e^{j2k\pi f_0 t}/\sqrt{T} = e^{jk\omega_0 t}/\sqrt{T}. \tag{3.6}$$

We will use any of these equivalent expressions interchangeably, whichever is most convenient in the specific discussion. Moreover, in practice, the constant scaling factor $1/\sqrt{T}$ in the equations above has little significance, we can rescale the two equations in the Fourier series expansion pair and express them in some alternative forms such as

$$\begin{aligned} x_T(t) &= \sum_{k=-\infty}^{\infty} X[k] e^{j2k\pi f_0 t} = \sum_{k=-\infty}^{\infty} X[k] e^{jk\omega_0 t}, \\ X[k] &= \frac{1}{T} \int_T x_T(t) e^{-j2k\pi f_0 t} \, dt = \frac{1}{T} \int_T x_T(t) e^{-jk\omega_0 t} \, dt. \end{aligned} \tag{3.7}$$

In this form, $X[0] = \int_T x_T(t) \, dt/T$ has a clear interpretation, it is the average, offset, or the DC (direct current) component of the signal.

The relationship between the signal period T and the gap f_0 between two consecutive frequencies in the frequency domain is illustrated in Fig. 3.1.

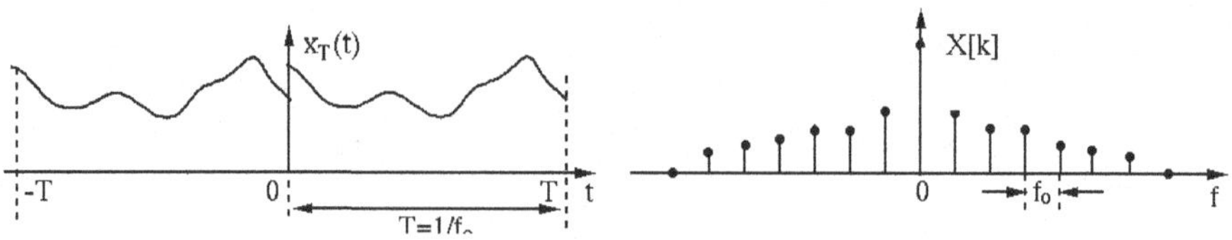

Figure 3.1 Fourier series expansion of periodic signals.

The Fourier series expansion is a unitary transformation that converts a function $x_T(t)$ in the vector space of all periodic time functions into a vector $[\dots, X[-1], X[0], X[1], \dots]^{\mathrm{T}}$ in another space of all vectors of infinite dimensions. Also, the inner product of any two functions $x_T(t)$ and $y_T(t)$ remains the same before and after the unitary transformation:

$$\begin{aligned}\langle x_T(t), y_T(t)\rangle &= \int_T x_T(t)\overline{y}_T(t)\,dt \\ &= \frac{1}{T}\int_T \sum_{k=-\infty}^{\infty} X[k]e^{j2k\pi f_0 t} \sum_{l=-\infty}^{\infty} \overline{Y}[l]e^{-j2n\pi f_0 t}\,dt \\ &= \frac{1}{T}\sum_{k=-\infty}^{\infty}\sum_{l=-\infty}^{\infty} X[k]\overline{Y}[l]\int_T e^{j2(k-l)\pi f_0 t}\,dt \\ &= \sum_{k=-\infty}^{\infty}\sum_{l=-\infty}^{\infty} X[k]\overline{Y}[l]\delta[k-l] = \sum_{k=-\infty}^{\infty} X[k]\overline{Y}[k] = \langle \boldsymbol{X}, \boldsymbol{Y}\rangle. \end{aligned} \tag{3.8}$$

In particular, if $y_T(t) = x_T(t)$, the above becomes Parseval's identity

$$||x_T(t)||^2 = \langle x_T(t), x_T(t)\rangle = \langle \boldsymbol{X}, \boldsymbol{X}\rangle = ||\boldsymbol{X}||^2, \tag{3.9}$$

indicating that the total energy or information contained in the signal is conserved by the Fourier series expansion; therefore; the signal can be equivalently represented in either the time or frequency domain.

3.1.2 Physical interpretation

The Fourier series expansion of a periodic signal $x_T(t)$ can also be expressed in terms of sine and cosine functions of different frequencies:

$$\begin{aligned} x_T(t) &= \sum_{k=-\infty}^{\infty} X[k]e^{jk\omega_0 t} = X[0] + \sum_{k=1}^{\infty}[X[-k]e^{-jk\omega_0 t} + X[k]e^{jk\omega_0 t}] \\ &= X[0] + \sum_{k=1}^{\infty}[X[-k](\cos k\omega_0 t - j\sin k\omega_0 t) + X[k](\cos k\omega_0 t + j\sin k\omega_0 t)] \\ &= X[0] + \sum_{k=1}^{\infty}[(X[k] + X[-k])\cos k\omega_0 t + j(X[k] - X[-k])\sin k\omega_0 t] \\ &= X[0] + 2\sum_{k=1}^{\infty}(a_k \cos k\omega_0 t + b_k \sin k\omega_0 t). \end{aligned} \tag{3.10}$$

Here, we have defined $a_k = (X[k] + X[-k])/2$ and $b_k = (X[k] - X[-k])/2$, which can also be expressed as (Eq. (3.7)):

$$\begin{aligned} a_k &= \frac{1}{2T}\int_T x_T(t)[e^{-jk\omega_0 t} + e^{jk\omega_0 t}]\,dt = \frac{1}{T}\int_T x_T(t)\cos k\omega_0 t\,dt, \\ b_k &= \frac{j}{2T}\int_T x_T(t)[e^{-jk\omega_0 t} - e^{jk\omega_0 t}]\,dt = \frac{1}{T}\int_T x_T(t)\sin k\omega_0 t\,dt. \\ & k = 1, 2, \dots \end{aligned} \tag{3.11}$$

Equations (3.11) and (3.10) are the alternative forms of the Fourier series expansion of $x_T(t)$.

If specially $x_T(t)$ is real, we have

$$X[-k] = \frac{1}{T}\int_T x_T(t) e^{j2k\pi f_0 t}\, dt = \overline{X}[k], \tag{3.12}$$

which means

$$\text{Re}[X[-k]] = \text{Re}[X[k]], \qquad \text{Im}[X[-k]] = -\,\text{Im}[X[k]]; \tag{3.13}$$

i.e., the real part of $X[k]$ is even and the imaginary part is odd. Now we have

$$\begin{aligned} a_k &= \frac{X[k] + X[-k]}{2} = \frac{X[k] + \overline{X}[k]}{2} = \text{Re}[X[k]] \\ b_k &= \frac{j(X[k] - X[-k])}{2} = \frac{j(X[k] - \overline{X}[k])}{2} = -\,\text{Im}[X[k]]; \end{aligned} \tag{3.14}$$

i.e.,

$$\begin{cases} |X[k]| = \sqrt{a_k^2 + b_k^2} \\ \angle X[k] = -\tan^{-1}(b_k/a_k) \end{cases} \qquad \begin{cases} a_k = |X[k]| \cos \angle X[k] \\ b_k = -|X[k]| \sin \angle X[k] \end{cases}. \tag{3.15}$$

The Fourier series expansion of a real signal $x_T(t)$ (Eq. (3.10)) can be rewritten as

$$\begin{aligned} x_T(t) &= X[0] + 2\sum_{k=1}^{\infty}(a_k \cos k\omega_0 t + b_k \sin k\omega_0 t) \\ &= X[0] + 2\sum_{k=1}^{\infty} |X[k]|(\cos \angle X[k] \cos k\omega_0 t - \sin \angle X[k] \sin k\omega_0 t) \\ &= X[0] + 2\sum_{k=1}^{\infty} |X[k]| \cos(k\omega_0 t + \angle X[k]). \end{aligned} \tag{3.16}$$

This is yet another form of the Fourier expansion, which indicates that a real periodic signal $x_T(t)$ can be constructed as a superposition of infinite sinusoids of (a) different frequencies $k\omega_0$, (b) different amplitudes $|X[k]|$, and (c) different phases $\angle X[k]$. In particular, consider the following values for k:

- $k = 0$, the coefficient $X[0] = \int_T x_T(t)\, dt/T$ is the average or DC component of the signal $x_T(t)$;
- $k = 1$, the sinusoid $\cos(\omega_0 t + \angle X[1])$ has the same period T as the signal $x_T(t)$ and its frequency $f_0 = 1/T$ is called the *fundamental frequency* of the signal;
- $k > 1$, the frequency kf_0 of the sinusoidal function $\cos(k\omega_0 t + \angle X[k])$ is k times the frequency f_0 of the fundamental and is called the kth *harmonic* of the signal.

3.1.3 Properties of the Fourier series expansion

Here is a set of properties of the Fourier series expansion:

- **Linearity**

$$\mathcal{F}[a\,x(t) + b\,y(t)] = a\,\mathcal{F}[x(t)] + b\,\mathcal{F}[y(t)]. \tag{3.17}$$

As an integral operator which is by definition linear, the Fourier expansion is obviously linear.

- **Time scaling**
When $x_T(t)$ is scaled in time by a factor of $a > 0$ to become $x(at)$, its period becomes T/a and its fundamental frequency becomes $a/T = af_0$. If $a > 1$, the signal is compressed by a factor a and the frequencies of its fundamental and harmonics become a times higher; if $a < 1$, the signal is expanded and the frequencies of its fundamental and harmonics are a times lower. In either case, the coefficients $X[k]$ remain the same:

$$x(at) = \sum_{k=-\infty}^{\infty} X[k]e^{j2ka\pi f_0 t} = \sum_{k=-\infty}^{\infty} X[k]e^{jka\omega_0 t}. \tag{3.18}$$

- **Time shift**
A time signal $x(t)$ shifted in time by τ becomes $y(t) = x(t-\tau)$. Defining $t' = t - \tau$ we can get its Fourier coefficient as

$$\begin{aligned} Y[k] &= \frac{1}{T}\int_T x(t-\tau)e^{-jk\omega_0 t}\,dt = \frac{1}{T}\int_T x(t')e^{-jk\omega_0 (t'+\tau)}\,dt' \\ &= X[k]e^{-jk\omega_0\tau} = X[k]e^{-j2k\pi f_0\tau}. \end{aligned} \tag{3.19}$$

We see that $|Y[k]| = |X[k]|$, but $\angle Y[k] = \angle X[k] - k\omega_0\tau$, i.e., the phase of $Y[k]$ is that of $X[k]$ shifted by $-k\omega_0\tau$.

- **Differentiation**
Fourier coefficients of the time derivative $y(t) = dx(t)/\,dt$ can be found to be

$$\begin{aligned} Y[k] &= \frac{1}{T}\int_T \left[\frac{d}{dt}x(t)\right] e^{-jk\omega_0 t}\,dt \\ &= \frac{1}{T}\left[e^{-jk\omega_0 t}x(t)\Big|_0^T + jk\omega_0\int_T x(t)e^{-jk\omega_0 t}\,dt \right] = jk\omega_0 X[k] = jk\frac{2\pi}{T}X[k]. \end{aligned} \tag{3.20}$$

- **Integration**
The time integration of $x(t)$ is

$$y(t) = \int_{-\infty}^{t} x(\tau)\,d\tau. \tag{3.21}$$

Note that $y(t)$ is periodic only if the DC component (average) of $x(t)$ is zero; i.e., $X[0] = 0$ (otherwise it would accumulate over time by the integration to result in a ramp).
As $x(t) = dy(t)/\,dt$, according to the differentiation property above, we have

$$X[k] = jk\frac{2\pi}{T}Y[k]; \quad \text{i.e.} \quad Y[k] = \frac{T}{j2k\pi}X[k]. \tag{3.22}$$

Note that $Y[0]$ cannot be obtained from this formula as when $k = 0$, both the numerator and the denominator of $Y[k]$ are zero. However, as the DC component of $y(t)$, $Y[0]$ can be found by the definition:

$$Y[0] = \frac{1}{T}\int_T y(t)\,dt. \tag{3.23}$$

- **Plancherel's identity and Parseval's identity**

$$\frac{1}{T}\int_T x_T(t)\overline{y}_T(t)\,dt = \sum_{k=-\infty}^{\infty} X[k]\overline{Y}[k]. \tag{3.24}$$

The proof of this property is left as a homework problem. In particular, when $y_T(t) = x_T(t)$ in Eq. (3.24), we get

$$\frac{1}{T}\int_T |x_T(t)|^2\,dt = \sum_{k=-\infty}^{\infty} |X[k]|^2. \tag{3.25}$$

This is also given in Eq. (3.9). The left-hand side of the equation represents the average power in $x_T(t)$. The average power of the kth frequency component $X[k]e^{j2\pi kf_0t}$ in the expansion (first equation in Eq. (3.7)) is

$$\frac{1}{T}\int_T |X[k]e^{j2\pi kf_0t}|^2\,dt = \frac{1}{T}\int_T |X[k]|^2\,dt = |X[k]|^2. \tag{3.26}$$

We see that Eq. (3.25) states that the average power of the signal $x_T(t)$ in period T is the sum of the average power of all frequency components; i.e., the power in the signal is conserved in either the time or frequency domain.

- **Multiplication**

The Fourier expansion coefficients of the product of two functions $z_T(t) = x_T(t)\,y_T(t)$ are $X[k] * Y[k]$.

$$\begin{aligned}
Z[k] &= \frac{1}{T}\int_T [x_T(t)\,y_T(t)]\,e^{-jk\omega_0 t}\,dt \\
&= \frac{1}{T}\int_T \left[\sum_{l=-\infty}^{\infty} X[l]e^{jl\omega_0 t}\right]\left[\sum_{l'=-\infty}^{\infty} Y[l']e^{jl'\omega_0 t}\right] e^{-jk\omega_0 t}\,dt \\
&= \sum_{l=-\infty}^{\infty} X[l] \sum_{l'=-\infty}^{\infty} Y[l']\frac{1}{T}\int_T [e^{-jk\omega_0 t}e^{jl\omega_0 t}e^{jl'\omega_0 t}]\,dt \\
&= \sum_{l=-\infty}^{\infty} X[l] \sum_{l'=-\infty}^{\infty} Y[l']\delta[k-l-l']
\end{aligned}$$

$$= \sum_{l=-\infty}^{\infty} X[l]Y[k-l] = X[k] * Y[k], \tag{3.27}$$

where we have used Eq. (1.33) and $X[k] * Y[k]$ is the discrete convolution of two sequences $X[k]$ and $Y[k]$ defined in Eq. (1.107).

- **Circular convolution**
  The *circular convolution* of two periodic functions $x_T(t)$ and $y_T(t)$ is defined as

$$z_T(t) = x_T(t) * y_T(t) = \frac{1}{T}\int_T x_T(\tau)y_T(t-\tau)\,d\tau. \tag{3.28}$$

  Its Fourier expansion coefficients can be found to be

$$Z[k] = X[k]Y[k]. \tag{3.29}$$

  The proof of this property is left as a homework problem.

3.1.4 The Fourier expansion of typical functions

Here we consider the Fourier expansion of a set of typical periodic signals.

- **Constant**
  A constant $x(t) = 1$ can be expressed as a complex exponential $x(t) = e^{j0t}$ with arbitrary period T. The Fourier coefficient for this zero-frequency is $X[0] = 1$, while $X[k] = 0$ for all other coefficients for non-zero ($k \neq 0$ frequencies. Alternatively, we get the same result by following the definition (Eq. (1.33)):

$$X[k] = \frac{1}{T}\int_T e^{-jk\omega_0 t}\,dt = \delta[k]. \tag{3.30}$$

- **Complex exponential**
  A complex exponential $x(t) = e^{j2\pi f_0 t} = e^{j\omega_0 t}$(of period $T = 1/f_0 = 2\pi/\omega_0$) is a special series of complex exponentials containing only one term $k = 1$ with coefficient $X[1] = 1$ and all other $X[k] = 0$ when $k \neq 1$. Alternatively, we can also find $X[k]$ by the definition in Eq. (3.7):

$$X[k] = \frac{1}{T}\int_T x_T(t)e^{-jk\omega_0 t}\,dt = \frac{1}{T}\int_T e^{j\omega_0(1-k)t}\,dt = \delta[k-1]. \tag{3.31}$$

- **Sinusoids**
  The cosine function $x(t) = \cos(2\pi f_0 t) = (e^{j2\pi f_0 t} + e^{-j2\pi f_0 t})/2$ of frequency f_0 is periodic with $T = 1/f_0$, and its Fourier coefficients are

$$\begin{aligned} X[k] &= \frac{1}{T}\int_T \cos(2\pi f_0 t)e^{-j2\pi k f_0 t}\,dt \\ &= \frac{1}{2}\left[\frac{1}{T}\int_T e^{-j2\pi(k-1)f_0 t}\,dt + \frac{1}{T}\int_T e^{-j2\pi(k+1)f_0 t}\,dt\right] \\ &= \frac{1}{2}(\delta[k-1] + \delta[k+1]). \end{aligned} \tag{3.32}$$

In particular, when $f_0 = 0$, $x(t) = 1$ and $X[k] = \delta[k]$, an impulse at zero, representing the constant (zero-frequency) value. Similarly, the Fourier coefficient of $x(t) = \sin(2\pi f_0 t)$ is

$$\begin{aligned} X[k] &= \frac{1}{T}\int_T \sin(2\pi f_0 t) e^{-j2\pi k f_0 t}\,dt \\ &= \frac{1}{2j}\left[\frac{1}{T}\int_T e^{-j2\pi(k-1)f_0 t}\,dt - \frac{1}{T}\int_T e^{-j2\pi(k+1)f_0 t}\,dt\right] \\ &= \frac{1}{2j}(\delta[k-1] - \delta[k+1]). \end{aligned} \tag{3.33}$$

An alternative way to find the Fourier coefficients of $x(t) = \cos(2\pi f_0 t)$ is to express it in terms of complex exponentials and equate it with the desired Fourier expansion:

$$\cos(2\pi f_0 t) = \frac{1}{2}[e^{j2\pi f_0 t} + e^{-j2\pi f_0 t}] = \sum_{k=-\infty}^{\infty} X[k] e^{j2\pi f_0 t}. \tag{3.34}$$

Comparing the two sides of the second equal sign, we see that $X[k] = 0$ for all K except when $k = \pm 1$ we have $X[1] = X[-1] = 1/2$; i.e., $X[k] = (\delta[k - 1] + \delta[k+1])/2$. Similarly, comparing the two sides of the Fourier expansion of the sine function

$$\sin(2\pi f_0 t) = \frac{1}{2j}[e^{j2\pi f_0 t} - e^{-j2\pi f_0 t}] = \sum_{k=-\infty}^{\infty} X[k] e^{j2\pi f_0 t}, \tag{3.35}$$

we see that $X[k] = 0$ for all k except when $k = 1$ with $X[1] = 1/2j$ and $k = -1$ with $X[-1] = -1/2j$; i.e., $X[k] = (\delta[k-1] - \delta[k+1])/2j$.
In general, this method can be used to find the Fourier coefficients of any function that can be expressed as the summation of a set of complex exponentials. By comparing this summation with the desired expression of the Fourier expansion in Eq. (3.7), the coefficients $X[k]$ can be determined.

- **Square wave**
  A square function can be defined as

$$x(t) = \begin{cases} 1 & 0 < t < \tau \\ 0 & \tau < t < T \end{cases}, \tag{3.36}$$

and its Fourier coefficients can be found to be

$$\begin{aligned} X[k] &= \frac{1}{T}\int_0^{\mathrm{T}} x(t) e^{-j2k\pi f_0 t}\,dt = \frac{1}{T}\int_0^{\tau} e^{-j2k\pi f_0 t}\,dt = \frac{1}{j2k\pi}\,(1 - e^{-j2k\pi f_0 \tau}) \\ &= \frac{e^{-jk\pi f_0 \tau}}{k\pi}\,\frac{(e^{jk\pi f_0 \tau} - e^{-j2k\pi f_0 \tau})}{2j} = \frac{e^{-jk\pi f_0 \tau}}{k\pi}\,\sin(k\pi f_0 \tau). \end{aligned} \tag{3.37}$$

A sinc *function* is commonly defined as

$$\mathrm{sinc}(x) = \frac{\sin(\pi x)}{\pi x} \qquad \lim_{x\to 0} \mathrm{sinc}(x) = 1, \tag{3.38}$$

and the expression above for $X[k]$ can be further written as

$$X[k] = f_0\tau \,\frac{\sin(k\pi f_0\tau)}{k\pi f_0\tau} e^{-jk\pi\tau f_0} = \frac{\tau}{T}\,\mathrm{sinc}(kf_0\tau)e^{-jk\pi f_0\tau}. \tag{3.39}$$

The DC component is $X[0] = \tau/T$.
In particular, if $\tau = T/2 = 1/2f_0$, $X[0] = 1/2$ and $X[k]$ above becomes

$$X[k] = \frac{1}{j2k\pi}\,(1 - e^{-jk\pi}) = \frac{e^{-jk\pi/2}}{k\pi}\sin(k\pi/2). \tag{3.40}$$

Moreover, since $e^{\pm j2k\pi} = 1$ and $e^{\pm j(2k-1)\pi} = -1$, all even terms $X[\pm 2k] = 0$ become zero and the odd terms become

$$X[\pm(2k-1)] = \pm 1/j\pi(2k-1) \qquad k = 1, 2, \ldots, \tag{3.41}$$

and the Fourier series expansion of the square wave becomes a linear combination of sinusoids:

$$\begin{aligned} x(t) &= \sum_{k=-\infty}^{\infty} X[k]e^{j2k\pi f_0 t} \\ &= X[0] + \sum_{k=1}^{\infty}\left[\frac{1}{j\pi(2k-1)}e^{j(2k-1)\omega_0 t} + \frac{1}{-j\pi(2k-1)}e^{-j(2k-1)\omega_0 t}\right] \\ &= \frac{1}{2} + \frac{2}{\pi}\sum_{k=1}^{\infty}\frac{\sin((2k-1)\omega_0 t)}{2k-1} \\ &= \frac{1}{2} + \frac{2}{\pi}\left[\frac{\sin(\omega_0 t)}{1} + \frac{\sin(3\omega_0 t)}{3} + \frac{\sin(5\omega_0 t)}{5} + \cdots\right]. \end{aligned} \tag{3.42}$$

As the function $x(t)$ is odd (except the DC), it is composed of only odd sine functions.
A square wave can be alternatively defined as an even function

$$x(t) = \begin{cases} 1 & |t| < T/4 \\ 0 & T/4 < |t| < T/2 \end{cases}. \tag{3.43}$$

We can show that it is composed of only even cosine functions. This is left as a homework problem.

- **Triangle wave**
  A triangle wave is defined as an even function

$$x(t) = 2|t|/T, \qquad (|t| \le T/2). \tag{3.44}$$

First, the DC offset $X[0]$ can be found from the definition:

$$X[k] = \frac{1}{T}\int_T x(t)\,dt = \frac{T}{2}. \tag{3.45}$$

For $k \neq 0$, we realize that this triangle wave can be obtained as an integral of the square wave defined in Eq. (3.36) with these modifications: (a) $\tau = T/2$, (b) DC offset is zero $X[0] = 0$, and (c) vertically scaled by $4/T$. Now according to the integration property, the Fourier coefficients can be easily obtained from Eq. (3.40) as

$$X[k] = \frac{4}{T}\,\frac{T}{j2k\pi}\,\frac{e^{-jk\pi/2}}{k\pi}\sin(k\pi/2) = \frac{2}{j}\frac{\sin(k\pi/2)}{(k\pi)^2}e^{-jk\pi/2}$$
$$= \frac{2\sin(k\pi/2)}{(k\pi)^2}(-j)^{k+1} \qquad k = \pm 1, \pm 2, \ldots. \tag{3.46}$$

It can be shown (homework) that $X[k] = X[-k]$ is real and even with respect to k.

According to the time shift property, the complex exponential $e^{jk\pi/2}$ corresponds to a right shift by $T/4$. If we left shift the signal by $T/4$, the triangle wave $x(t)$ becomes odd, and the complex exponential term in the expression of $X[k]$ disappears:

$$X[k] = \frac{2}{j}\frac{\sin(k\pi/2)}{(k\pi)^2} \qquad k = \pm 1, \pm 2, \ldots. \tag{3.47}$$

This is an imaginary and odd function with respect to k.

The Fourier series expansion of such an odd triangle wave can be written as below. As the function $x(t)$ is odd (except the DC), it is composed of only odd sine functions:

$$\begin{aligned} x(t) &= \sum_{k=-\infty}^{\infty} X[k]e^{j2k\pi f_0 t} = \frac{1}{2} + \sum_{k=1}^{\infty}[X[k]e^{j2k\pi f_0 t} + X[-k]e^{-j2k\pi f_0 t}] \\ &= \frac{1}{2} + \sum_{k=1}^{\infty}\left(\frac{2}{j}\frac{\sin(k\pi/2)}{(k\pi)^2}e^{j2k\pi f_0 t} - \frac{2}{j}\frac{\sin(k\pi/2)}{(k\pi)^2}e^{-j2k\pi f_0 t}\right) \\ &= \frac{1}{2} + \frac{4}{\pi^2}\sum_{k=1}^{\infty}\frac{\sin(k\pi/2)}{k^2}\sin(2k\pi f_0 t) \\ &= \frac{1}{2} + \frac{4}{\pi^2}\left[\sin(2\pi f_0 t) - \frac{1}{9}\sin(6\pi f_0 t) + \frac{1}{25}\sin(10\pi f_0 t) - \cdots\right]. \end{aligned} \tag{3.48}$$

- **Sawtooth**

  A sawtooth function is defined as

$$x(t) = t/T \qquad (0 < t < T). \tag{3.49}$$

We first find $X[0]$, the average or DC component

$$X[0] = \frac{1}{T}\int_T \frac{t}{T}e^{-j0\omega_0 t}\,dt = \frac{1}{2}, \tag{3.50}$$

and then find all remaining coefficients $X[k]$ $(k \neq 0)$

$$X[k] = \frac{1}{T} \int_T \frac{t}{T} e^{-jk\omega_0 t}\, dt. \tag{3.51}$$

In general, these types of integral can be found using integration by parts:

$$\int t e^{at}\, dt = \frac{1}{a^2}(at - 1)e^{at} + C. \tag{3.52}$$

Here, $a = -jk\omega_0 = -j2k\pi/T \neq 0$ and we get

$$X[k] = \frac{1}{T^2(jk\omega_0)^2}[(-jk\omega_0 t - 1)e^{-jk\omega_0 t}]\Big|_0^{\mathrm{T}} = \frac{j}{2k\pi}. \tag{3.53}$$

The Fourier series expansion of the function is

$$x(t) = \frac{1}{2} + \sum_{k=1}^{\infty}\left[\frac{j}{2k\pi}e^{j\omega_0 t} - \frac{j}{2k\pi}e^{-j\omega_0 t}\right] = \frac{1}{2} - \frac{1}{\pi}\sum_{k=1}^{\infty}\frac{1}{k}\sin(k\omega_0 t). \tag{3.54}$$

The sawtooth wave is an odd function and, therefore, it is composed of only odd sine functions. Also note that at the two end points $t = 0$ and $t = T$, the series expansion above takes the same value of $x(0) = x(T) = 1/2$, which is the average of the values of the actual function at the end points, 0 at $t = 0$ but 1 at $t = T$.

Some different versions of the square, triangle and sawtooth waveforms are shown in Fig. 3.2. The corresponding Fourier series expansions of these waveforms are illustrated in Fig. 3.3. The first 10 basis functions for the DC component, fundamental frequency, and progressively higher harmonics are shown on the left, and the reconstructions by inverse transform of the square, triangle, and sawtooth waveforms are shown in the remaining three columns. As we can see, the accuracy of the reconstruction of a waveform improves continuously as more basis functions of higher frequencies are included in the reconstruction so that finer details (corresponding to rapid changes in time) can be better represented.

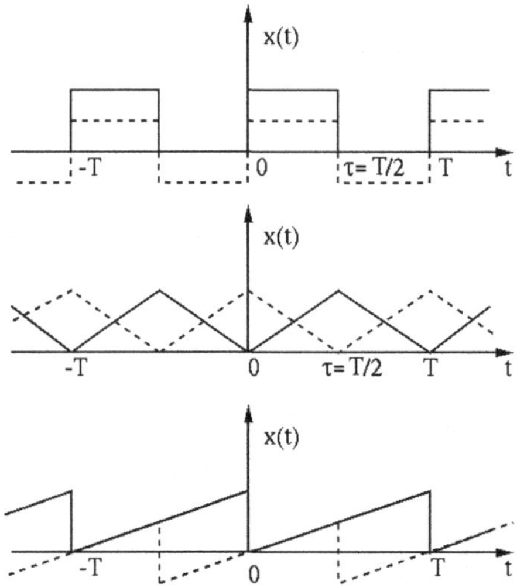

Figure 3.2 Square wave (top), triangle wave (middle), and sawtooth wave (bottom).

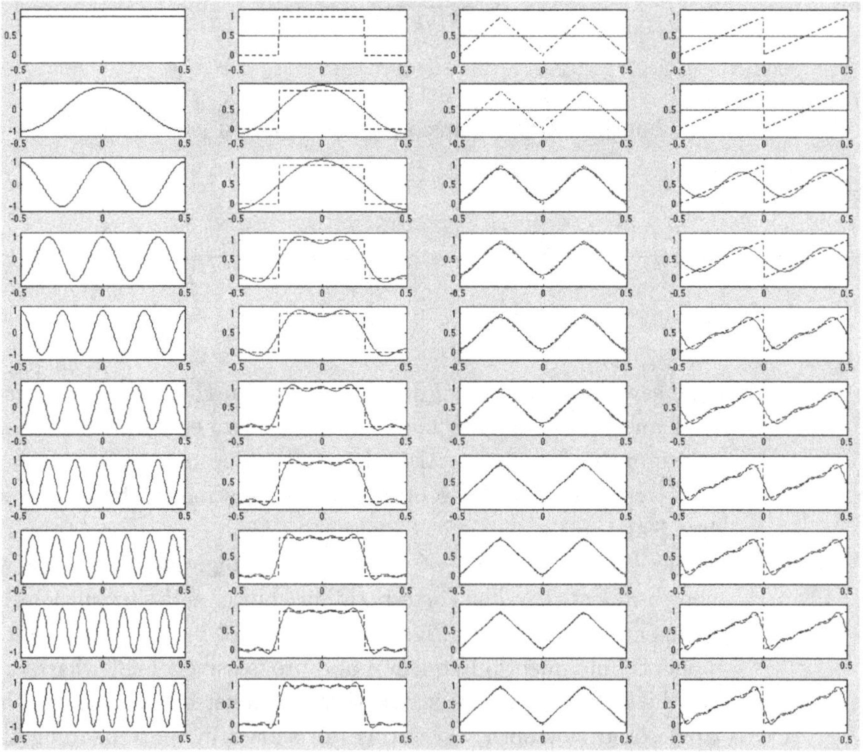

Figure 3.3 Fourier reconstructions of square, triangle, and sawtooth waveforms (second, third, and fourth columns) with progressively more higher harmonics (first column) included.

- **Impulse train**
  An impulse train, also called a Dirac comb function or sampling function, is a sequence of unit impulses separated by a time interval T:

$$x(t) = \text{comb}(t) = \sum_{n=-\infty}^{\infty} \delta(t - nT). \tag{3.55}$$

  As a function with period T, this impulse train can be Fourier expanded:

$$x(t) = \text{comb}(t) = \sum_{k=-\infty}^{\infty} X[k]\, e^{j2k\pi t/T}. \tag{3.56}$$

  with coefficients

$$\begin{aligned} X[k] &= \frac{1}{T}\int_{-T/2}^{T/2} x(t)e^{-j2k\pi t/T}\,dt = \frac{1}{T}\int_{-T/2}^{T/2} \sum_{n=-\infty}^{\infty} \delta(t - nT)e^{-j2k\pi t/T}\,dt \\ &= \frac{1}{T}\int_{-T/2}^{T/2} \delta(t)e^{-j2k\pi t/T}\,dt = \frac{1}{T} \qquad k = 0, \pm 1, \pm 2, \ldots. \end{aligned} \tag{3.57}$$

  The last equation is due to Eq. (1.9). Substituting $X[k] = 1/T$ back into the Fourier series expansion of comb(t), we can also express the impulse train as

$$\text{comb}(t) = \sum_{n=-\infty}^{\infty} \delta(t - nT) = \frac{1}{T}\sum_{k=-\infty}^{\infty} e^{j2k\pi t/T}. \tag{3.58}$$

  This is actually the same as Eq. (1.35).

Fig. 3.4 shows a set of periodic signals (left) and their corresponding Fourier coefficients (right).

To carry out the Fourier series expansion of a given signal function $x(t)$, it is necessary to first determine its period T or equivalently its fundamental frequency $f_0 = 1/T$, which may not be always explicitly available. If $x(t)$ is composed of a set of K terms each of frequency f_k $(k = 1, \ldots, K)$, then the fundamental frequency f_0 is the *greatest common divisor (GCD)* of these frequencies. Or, equivalently, the period $T = 1/f_0$ is the *least common multiple (LCM)* of the periods $T_k = 1/f_k$ of the individual components.

Example 3.1: Find the Fourier coefficients of this signal

$$x(t) = \cos(8\pi t) + \cos(12\pi t) = \cos(2\pi 4t) + \cos(2\pi 6t). \tag{3.59}$$

containing two sinusoids of frequencies $f_1 = 4$ and $f_2 = 6$ or periods $T_1 = 1/f_1 = 1/4$ and $T_2 = 1/f_2 = 1/6$, respectively. The fundamental frequency f_0 of the sum of these component sinusoids is the GCD of the individual frequency components:

$$f_0 = \text{GCD}(f_1, f_2) = \text{GCD}(4, 6) = 2. \tag{3.60}$$

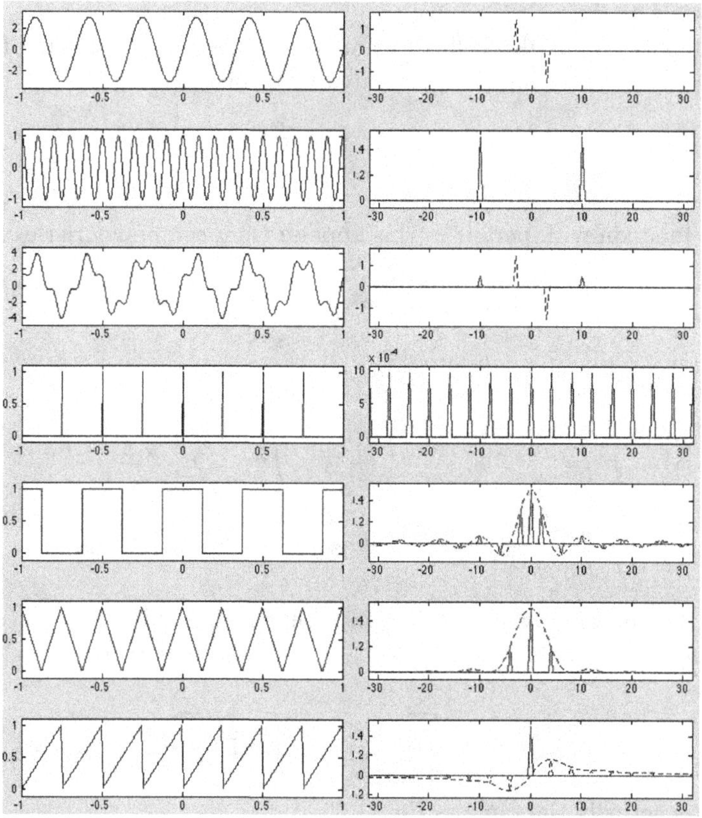

Figure 3.4 Examples of Fourier series expansions. A set of periodic signals (left) and their Fourier expansion coefficients (right) as a function of frequency f (real and imaginary parts are shown in solid and dashed lines, respectively). The first three rows show two sinusoids $x_1(t) = \sin(2\pi 3t)$ and $x_2(t) = \cos(2\pi 10t)$, and their weighted sum $x_1(t) + x_2(t)/5$. The following four rows are for the impulse train, square wave, triangle wave, and sawtooth wave, respectively.

Equivalently, the period T of the sum is the LCM of the periods of the individual components:

$$T = \text{LCM}(T_1, T_2) = \text{LCM}(1/4, 1/6) = 1/2. \tag{3.61}$$

Now the signal can be expressed in terms of its fundamental frequency as $x(t) = \cos(2\pi 2 f_0 t) + \cos(2\pi 3 f_0 t)$ and its Fourier series coefficients can be found to be $X[k] = (\delta(k-2) + \delta[k+2] + \delta[k-3] + \delta[k+3])/2$.

3.2 The Fourier transform of non-periodic signals

3.2.1 Formulation of the CTFT

The Fourier series expansion does not apply to non-periodic signals. To process and analyze such signals in the frequency domain, the concept of the Fourier series expansion needs to be generalized. To do so, we first make some minor modification of the Fourier series expansion pair in Eq. (3.7) by moving the factor $1/T$ from the second equation to the first one, so that the Fourier expansion is redefined as

$$x_T(t) = \sum_{k=-\infty}^{\infty} \frac{1}{T} X[k] e^{jk\omega_0 t} = \sum_{k=-\infty}^{\infty} \frac{1}{T} X[k] e^{j2k\pi f_0 t},$$
$$X[k] = \int_T x_T(t) e^{-jk\omega_0 t}\, dt = \int_T x_T(t) e^{-j2k\pi f_0 t}\, dt. \tag{3.62}$$

Here, the value of $X[k]$ is scaled by T, and its dimension becomes that of the signal $x_T(t)$ multiplied by time, or divided by frequency (the exponential term $\exp(\pm j2\pi f_0 t)$ is dimensionless).

A non-periodic signal $x(t)$ can be considered as a periodic signal $x_T(t)$ with its period increased to approach infinity, as previously discussed in subsection 2.2.3. At this limit $T \to \infty$ the following changes take place:

- The fundamental frequency f_0, the gap between two consecutive frequency components, approaches zero $f_0 = 1/T \to 0$, and the discrete frequencies kf_0 for all integers $-\infty < k < \infty$ is replaced by a continuous variable $-\infty < f < \infty$.
- The discrete and periodic basis functions $\phi_k(t) = e^{j2k\pi f_0 t}$ for all k become uncountable and non-periodic $\phi_f(t) = e^{j2\pi f t}$ for all f, as an orthogonal basis that spans the function space over $(-\infty, \infty)$ (Eq. (1.28)):

$$\langle \phi_f(t), \phi_{f'}(t) \rangle = \int_{-\infty}^{\infty} e^{j2\pi(f-f')t}\, dt = \delta(f - f'). \tag{3.63}$$

  This is Eq. (2.182).
- The coefficients $X[k]$ for the kth basis function, the kth frequency component, $\phi_k(t) = e^{j2k\pi f_0 t}$ for all k are replaced by a continuous weight function $X(f)$ for the continuous and uncountable basis function $\phi_f(t) = e^{j2\pi f t}$ for all f.
- Let $\triangle f = f_0 = 1/T$, then $1/T = \triangle f \to df$ when $T \to \infty$, and the summation in the first equation in Eq. (3.62) becomes an integral.

Owing to the changes above, at the limit, $T \to \infty$, the two equations in Eq. (3.62) become:

$$x(t) = \lim_{T\to\infty} \left[\frac{1}{T} \sum_{k=-\infty}^{\infty} X[k] e^{j2k\pi f_0 t} \right] = \int_{-\infty}^{\infty} X(f)^{j2\pi f t}\, df,$$

$$X(f) = \lim_{T\to\infty}\left[\int_T x_T(t)e^{-j2k\pi f_0 t}\,dt\right] = \int_{-\infty}^{\infty} x(t)e^{-j2\pi ft}\,dt. \tag{3.64}$$

These two equations can be rewritten as the *continuous-time Fourier transform (CTFT)* pair:

$$X(f) = \mathcal{F}[x(t)] = \int_{-\infty}^{\infty} x(t)e^{-j2\pi ft}\,dt,$$
$$x(t) = \mathcal{F}^{-1}[X(f)] = \int_{-\infty}^{\infty} X(f)e^{j2\pi ft}\,df. \tag{3.65}$$

This is Eqs. (2.134) and (2.133). The first equation for $X(f)$ and second equation for $x(t)$ are the forward and inverse CTFT, respectively, which can be more concisely represented as

$$x(t) \overset{\mathcal{F}}{\longleftrightarrow} X(f). \tag{3.66}$$

The weighting function $X(f)$ in Eq. (3.65) is called the *Fourier spectrum* of $x(t)$, representing how the signal energy is distributed over frequency, in comparison with $x(t)$, which represents how the signal energy is distributed over time. A non-periodic signal and its continuous spectrum are illustrated in Fig. 3.5, in comparison to a periodic signal and its discrete spectrum shown in Fig. 3.1.

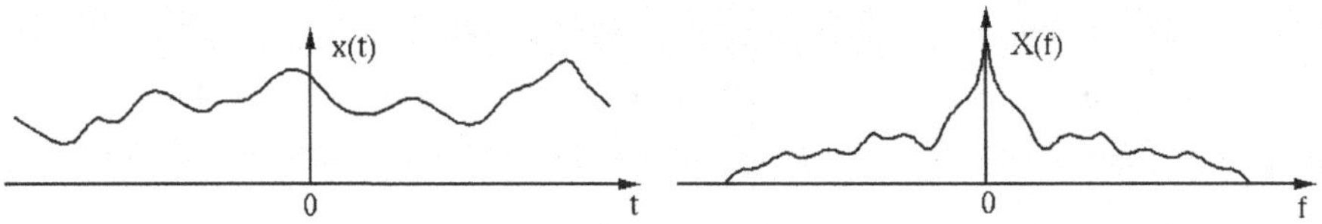

Figure 3.5 Fourier transform of non-periodic and continuous signals. When the time signal is no longer periodic, its discrete spectrum represented by the Fourier series coefficients becomes a continuous function.

Equation (3.65) can be considered as the most generic form of the forward and inverse Fourier transform pair, generally denoted by $\mathcal{F}[\,.\,]$ and $\mathcal{F}^{-1}[\,.\,]$, with different variations depending on the specific nature of the signal $x(t)$, such as whether it is periodic or aperiodic, continuous or discrete (to be considered in the next chapter). For example, the Fourier series expansion in Eq. (3.5) is just a special case of Eq. (3.65), where the Fourier transform is applied to a periodic signal $x_T(t+T) = x_T(t)$, and the Fourier coefficients $X[k]$ are just the discrete spectrum $X(f) = \mathcal{F}[x_T(t)]$ of the periodic signal, as to be shown in the following subsection (Eq. (3.86)).

Comparing Eq. (3.65) with Eqs. (2.133) and (2.134), we see that the CTFT is actually the representation of a signal function $x(t)$ by an uncountably infinite set of orthonormal basis functions (Eq. (2.132)) defined as

$$\phi_f(t) = e^{j2\pi ft}, \qquad -\infty < f < \infty, \tag{3.67}$$

so that the function $x(t)$ can be expressed as a linear combination, an integral, of these basis functions $\phi_f(t)$ over all frequencies f:

$$x(t) = \int_{-\infty}^{\infty} X(f)\phi_f(t)\, df = \int_{-\infty}^{\infty} X(f)e^{j2\pi ft}\, df. \tag{3.68}$$

This is the second equation in Eq. (3.65), and the coefficient function $X(f)$ can be found as the projection of the signal function $x(t)$ onto the basis function $\phi_f(t)$:

$$X(f) = \langle x(t), \phi_f(t) \rangle = \int_{-\infty}^{\infty} x_T(t)e^{-j2\pi kt/T}. \tag{3.69}$$

This is the forward CTFT in Eq. (3.65).

Similar to the Fourier series expansion, the Fourier transform is also a unitary transformation $\mathcal{F}[x(t)] = X(f)$ that conserves the inner product (Theorem 2.6):

$$\begin{aligned}
\langle x(t), y(t) \rangle &= \int_{-\infty}^{\infty} x(t)\overline{y}(t)\, dt \\
&= \int_{-\infty}^{\infty} \left[\int_{-\infty}^{\infty} X(f)e^{j2\pi ft}\, df \right] \left[\int_{-\infty}^{\infty} \overline{Y}(f')e^{-j2\pi f't}\, df' \right] dt \\
&= \int_{-\infty}^{\infty} \int_{-\infty}^{\infty} X(f)\overline{Y}(f') \left[\int_{-\infty}^{\infty} e^{j2\pi(f-f')t}\, dt \right] df\, df' \\
&= \int_{-\infty}^{\infty} X(f) \int_{-\infty}^{\infty} \overline{Y}(f')\delta(f - f')\, df'\, df \\
&= \int_{-\infty}^{\infty} X(f)\overline{Y}(f)\, df = \langle X(f), Y(f) \rangle.
\end{aligned} \tag{3.70}$$

Replacing $y(t)$ by $x(t)$ in Eq. (3.70) above, we get Parseval's identity:

$$||x(t)||^2 = \langle x(t), x(t) \rangle = \langle X(f), X(f) \rangle = ||X(f)||^2. \tag{3.71}$$

As a unitary transformation, the Fourier transform can be considered as a rotation of the basis that spans the function space. Before the transform, any given function $x(t)$ is represented as a linear combination of an uncountably infinite set of standard basis functions $\delta(t - \tau)$ each for a particular moment $t = \tau$, weighted by the coefficient function $x(\tau)$ for the signal amplitude at the moment:

$$x(t) = \int_{-\infty}^{\infty} x(\tau)\delta(t - \tau)\, d\tau. \tag{3.72}$$

After the unitary transformation, the standard basis is rotated to become a different orthonormal basis representing all frequencies f:

$$\mathcal{F}^{-1}[\delta(t - \tau)] = \int_{-\infty}^{\infty} \delta(t - \tau)e^{j2\pi ft}\, dt = e^{j2\pi f\tau} = \cos(2\pi f\tau) + j\sin(2\pi f\tau), \tag{3.73}$$

and the function $x(t)$ can be alternatively represented as a linear combination of these new basis functions weighted by the spectrum $X(f)$:

$$x(t) = \int_{-\infty}^{\infty} X(f) e^{j2\pi ft} \, df. \tag{3.74}$$

The representations of the signal as a function $x(t)$ in the time domain and a spectrum $X(f)$ in the frequency domain are equivalent, in the sense that the total amount of energy or information is conserved due to Parseval's identity. However, how the total energy is distributed through time t or frequency f can be very different, which is an important reason why the Fourier transform is widely used.

The Fourier transform pair in Eq. (3.65) can also be equivalently represented in terms of the angular frequency $\omega = 2\pi f$:

$$\begin{aligned} X(\omega) &= \mathcal{F}[x(t)] = \int_{-\infty}^{\infty} x(t) e^{-j\omega t} \, dt, \\ x(t) &= \mathcal{F}^{-1}[X(\omega)] = \frac{1}{2\pi} \int_{-\infty}^{\infty} X(\omega) e^{j\omega t} \, d\omega. \end{aligned} \tag{3.75}$$

In some of the literature, the CTFT spectrum $X(f)$ or $X(\omega)$ is also denoted by $X(j\omega)$, as it takes this form when treated as a special case of the *Laplace transform*, to be considered in Chapter 6. However, all these different forms are just some notational variations of the same spectrum, a function of frequency f or angular frequency $\omega = 2\pi f$. We will use these notations interchangeably, whichever is most convenient and suitable in the specific discussion, as no confusion should be caused given the context. Moreover, we also note that, when the spectrum is denoted by $X(f)$, the Fourier transform pair in Eq. (3.65) appears symmetric between the time and frequency domains so that the time-frequency duality is more clearly revealed.

For certain signals $x(t)$, the integral in the first equation of Eq. (3.65) may not converge; i.e., their Fourier spectrum $X(f)$ may not exist. Some obvious examples include $x(t) = t$ and $x(t) = t^2$ which grow without bounds as $|t| \to \infty$. The following Dirichlet conditions for $x(t)$ guarantee the convergence of the integral in Eq. (3.65):

1. absolute integrability:

$$\int_{-\infty}^{\infty} |x(t)| \, dt < \infty. \tag{3.76}$$

2. $x(t)$ has a finite number of maxima and minima within any finite interval.
3. $x(t)$ has a finite number of discontinuities within any finite interval.

A more strict condition for the convergence of the integral is that $x(t)$ is an energy signal $x(t) \in L^2(\mathbb{R})$; i.e., it is square-integrable (Eq. 2.29). However, we note that these conditions are sufficient but not necessary, as the Fourier spectra of some signals not satisfying such conditions may still exist. For example, some important and commonly used signals, such $x(t) = 1$ and $x(t) = u(t)$, are neither square-integrable nor absolutely integrable, but their Fourier spectra can still be obtained, due to the introduction of the Dirac delta function, a non-conventional function with a value of infinity. The integrals of these functions can be considered to be marginally convergent.

Example 3.2: Here we consider the Fourier transform of a few special signals

- The unit impulse or Dirac delta:

$$\mathcal{F}[\delta(t)] = \int_{-\infty}^{\infty} \delta(t)e^{-j2\pi ft}\, dt = e^{-j2\pi 0 f} = 1. \tag{3.77}$$

- The constant function:

$$\mathcal{F}[1] = \int_{-\infty}^{\infty} e^{-j2\pi ft}\, dt = \delta(f). \tag{3.78}$$

  This is due to Eq. (1.28).
- The unit step, defined as

$$u(t) = \begin{cases} 0 & t < 0 \\ 1/2 & t = 0 \\ 1 & t > 0 \end{cases}. \tag{3.79}$$

  Its Fourier transform is (Eq. (1.30)):

$$\mathcal{F}[u(t)] = \int_{-\infty}^{\infty} u(t)e^{-j2\pi ft}\, dt = \int_{0}^{\infty} e^{-j2\pi ft}\, dt = \frac{1}{2}\delta(f) + \frac{1}{j2\pi f}. \tag{3.80}$$

  Similarly, we also have (Eq. (1.31))

$$\mathcal{F}[u(-t)] = \int_{-\infty}^{0} e^{-j2\pi ft}\, dt = \frac{1}{2}\delta(f) - \frac{1}{j2\pi f}. \tag{3.81}$$

  Note that the term $\delta(f)/2$ is for the DC component of the unit step. These results can be verified based on the fact that $u(-t) + u(t) = 1$:

$$\mathcal{F}[u(-t)] + \mathcal{F}[u(t)] = \frac{1}{2}\delta(f) - \frac{1}{j2\pi f} + \frac{1}{2}\delta(f) + \frac{1}{j2\pi f} = \delta(f). \tag{3.82}$$

- The sign function $x(t) = \text{sgn}(t)$ defined as

$$\text{sgn}(t) = 2\,u(t) - 1 = \begin{cases} -1 & t < 0 \\ 0 & t = 0 \\ 1 & t > 0 \end{cases}. \tag{3.83}$$

  Owing to linearity of the Fourier transform, its spectrum can be found to be

$$\mathcal{F}[\text{sgn}(t)] = 2\,\mathcal{F}[u(t)] - \mathcal{F}[1] = \delta(f) + \frac{1}{j\pi f} - \delta(f) = \frac{1}{j\pi f}. \tag{3.84}$$

  The term $\delta(f)/2$ disappears as the sign function has zero DC component.

3.2.2 Relation to the Fourier expansion

Now let us consider how the Fourier spectrum of a periodic function is related to its Fourier expansion coefficients. The Fourier expansion of a periodic function $x_T(t)$ is

$$x_T(t) = \sum_{k=-\infty}^{\infty} X[k]e^{j2k\pi t/T} = \sum_{k=-\infty}^{\infty} X[k]e^{j2k\pi f_0 t}. \tag{3.85}$$

where $f_0 = 1/T$ is the fundamental frequency and $X[k]$ the expansion coefficient. The Fourier transform of this periodic function $x_T(t)$ can be found to be

$$\begin{aligned} X(f) &= \int_{-\infty}^{\infty} x_T(t)e^{-j2\pi ft}\, dt = \int_{-\infty}^{\infty} \left[\sum_{k=-\infty}^{\infty} X[k]e^{j2k\pi f_0 t}\right] e^{-j2\pi ft}\, dt \\ &= \sum_{k=-\infty}^{\infty} X[k] \int_{-\infty}^{\infty} e^{-j2\pi(f-kf_0)t}\, dt = \sum_{k=-\infty}^{\infty} X[k]\delta(f-kf_0). \end{aligned} \tag{3.86}$$

Here, we have used the result of Eq. (1.29). It is clear that the spectrum of a periodic function is discrete, in the sense that it is non-zero only at a set of discrete frequencies $f = kf_0$, where $X(f) = X[k]\delta(f - kf_0)$. This result also illustrates an important point: while the dimension of the Fourier coefficient $X[k]$ is the same as that of the signal $x_T(t)$ (the exponential function is dimensionless); i.e., $[X[k]] = [x_T(t)]$, the dimension of the spectrum is

$$[X(f)] = [X[k]][t] = \frac{[X[k]]}{[f]}. \tag{3.87}$$

As the dimension of $X(f)$ is that of the signal $x(t)$ multiplied by time, or divided by frequency, $X(f)$ is actually a *frequency density* function.

In the future we will loosely use the term "spectrum" not only for a continuous function $X(f)$ of frequency f, but also for the discrete transform coefficients $X[k]$, as they can always be associated with a continuous function as in Eq. (3.86).

Next, we consider how the Fourier spectrum $X(t)$ of a signal $x(t)$ can be related to the Fourier series coefficients of its periodic extension defined as

$$x'(t) = \sum_{n=-\infty}^{\infty} x(t+nT) = x'(t+T). \tag{3.88}$$

As $x'(t+T) = x'(t)$ is periodic, it can be Fourier expanded and the kth Fourier coefficient is

$$\begin{aligned} X'[k] &= \frac{1}{T}\int_0^T x'(t)e^{-j2\pi kt/T}\, dt = \frac{1}{T}\int_0^T \left[\sum_{n=-\infty}^{\infty} x(t+nT)\right] e^{-j2\pi kt/T}\, dt \\ &= \frac{1}{T}\sum_{n=-\infty}^{\infty} \int_0^T x(t+nT)]e^{-j2\pi kt/T}\, dt. \end{aligned} \tag{3.89}$$

If we define $\tau = t + nT$; i.e., $t = \tau - nT$, the above becomes

$$X'[k] = \frac{1}{T} \sum_{n=-\infty}^{\infty} \int_{nT}^{(n+1)T} x(\tau) e^{-j2\pi k\tau/T} \, d\tau \; e^{-j2\pi nk}$$
$$= \frac{1}{T} \int_{-\infty}^{\infty} x(\tau) e^{-j2\pi k\tau/T} \, d\tau = \frac{1}{T} X\left(\frac{k}{T}\right). \tag{3.90}$$

($e^{-j2\pi nk} = 1$ as k and n are both integer.) This equation relates the Fourier transform $X(f)$ of a signal $x(t)$ to the Fourier series coefficient $X'[k]$ of the periodic extension $x'(t)$ of the signal. Now the Fourier expansion of $x'(t)$ can be written as

$$x'(t) = \sum_{n=-\infty}^{\infty} x(t + nT) = \sum_{k=-\infty}^{\infty} X'[k] e^{j2\pi kt/T} = \frac{1}{T} \sum_{k=-\infty}^{\infty} X\left(\frac{k}{T}\right) e^{j2\pi kt/T}. \tag{3.91}$$

This equation is called the *Poisson summation formula*. In particular, when $x(t) = \delta(t)$ and $X(f) = \mathcal{F}[\delta(t)] = 1$, the equation above becomes

$$\sum_{n=-\infty}^{\infty} \delta(t + nT) = \frac{1}{T} \sum_{k=-\infty}^{\infty} e^{j2\pi kt/T}. \tag{3.92}$$

This is actually Eq. (1.35) (with f and F replaced by t and T, respectively.)

3.2.3 Properties of the Fourier transform

Here, we consider a set of properties of the Fourier transform, many of which look similar to those of the Fourier series expansion discussed before. This is simply because the Fourier expansion is just a special case (for periodic signals) of the Fourier transform, so it naturally shares all of the properties of the Fourier transform. We assume in the following $x(t)$ and $y(t)$ are two complex functions (real as a special case) and $\mathcal{F}[x(t)] = X(f)$ and $\mathcal{F}[y(t)] = Y(f)$.

- **Linearity**

$$\mathcal{F}[ax(t) + by(t)] = a\mathcal{F}[x(t)] + b\mathcal{F}[y(t)]. \tag{3.93}$$

  The Fourier transform of a function $x(t)$ is simply an inner product of the function with a kernel function $\phi_f(t) = e^{j2\pi ft}$ (Eq. (3.69)). Owing to the linearity of the inner product in the first variable, the Fourier transform is also linear.

- **Time-frequency duality**

$$\text{if } \mathcal{F}[x(t)] = X(f), \quad \text{then} \quad \mathcal{F}[X(t)] = x(-f). \tag{3.94}$$

  Proof:

$$x(t) = \mathcal{F}^{-1}[X(f)] = \int_{-\infty}^{\infty} X(f) e^{j2\pi ft} \, df. \tag{3.95}$$

Defining $t' = -t$, we have

$$x(-t') = \int_{-\infty}^{\infty} X(f)e^{-j2\pi ft'}\,df. \tag{3.96}$$

Interchanging variables t' and f, we get

$$x(-f) = \int_{-\infty}^{\infty} X(t')e^{-j2\pi ft'}\,dt' = \mathcal{F}[X(t)]. \tag{3.97}$$

In particular, if $x(t) = x(-t)$ is even, we have

$$\text{if } \mathcal{F}[x(t)] = X(f), \quad \text{then} \quad \mathcal{F}[X(t)] = x(f). \tag{3.98}$$

This duality is simply the result of the definition of the forward and inverse transforms in Eq. (3.65), which are highly symmetric between time and frequency. Consequently, many of the properties and transforms of typical functions exhibit strong duality between the time and frequency domains.

- **Even and odd signals**
  - If the signal is even, then its spectrum is also even:

$$\text{if} \quad x(t) = x(-t), \quad \text{then} \quad X(f) = X(-f). \tag{3.99}$$

Proof:

$$\begin{aligned} X(f) &= \int_{-\infty}^{\infty} x(t)e^{-j2\pi ft}\,dt = \int_{-\infty}^{\infty} x(-t)e^{-j2\pi ft}\,dt \\ &= \int_{-\infty}^{\infty} x(t')e^{j2\pi ft'}\,dt' = X(-f), \end{aligned} \tag{3.100}$$

where we have assumed $t' = -t$.

  - If the signal is odd, then its spectrum is also odd:

$$\text{if} \quad x(t) = -x(-t), \quad \text{then} \quad X(f) = -X(-f). \tag{3.101}$$

The proof is similar to the above.

- **Time reversal**

$$\mathcal{F}[x(-t)] = X(-f); \tag{3.102}$$

i.e., if the signal $x(t)$ is flipped in time with respect to the origin $t = 0$, its spectrum $X(f)$ is also flipped in frequency with respect to the origin $f = 0$,
Proof:

$$\mathcal{F}[x(-t)] = \int_{-\infty}^{\infty} x(-t)e^{-j2\pi ft}\,dt = \int_{-\infty}^{\infty} x(t')e^{j2\pi ft'}\,dt' = X(-f). \tag{3.103}$$

where we have assumed $-t' = t$. In particular, when $x(t) = \overline{x}(t)$ is real,

$$\mathcal{F}[x(-t)] = X(-f) = \int_{-\infty}^{\infty} x(t)e^{j2\pi ft}\,dt = \overline{\int_{-\infty}^{\infty} x(t)e^{-j2\pi ft}\,dt} = \overline{X(f)}. \tag{3.104}$$

- **Plancherel's identity and Parseval's identity**

$$\langle x(t), y(t)\rangle = \int_{-\infty}^{\infty} x(t)\overline{y}(t)\,dt = \int_{-\infty}^{\infty} X(f)\overline{Y}(f)\,df = \langle X(f), Y(f)\rangle. \quad (3.105)$$

This is Eq. (3.70), indicating that the Fourier transform is a unitary transformation that conserves the inner product. In particular, letting $y(t) = x(t)$, we get Parseval's identity representing signal energy conservation by the Fourier transform:

$$||x(t)||^2 = \int_{-\infty}^{\infty} |x(t)|^2\,dt = \int_{-\infty}^{\infty} |X(f)|^2\,df = \int_{-\infty}^{\infty} S_x(f)\,df = ||X(f)||^2. \quad (3.106)$$

Here, $|x(t)|^2$ is the signal energy distribution over time, and $S_x(f) = |X(f)|^2$, called the *power spectral density (PSD)* of the signal, is the signal energy distribution over frequency. Although in general these two distributions are very different from each other, Parseval's identity indicates that the total signal energy is conserved.

- **Time and frequency scaling**

$$\mathcal{F}[x(at)] = \frac{1}{|a|}X\left(\frac{f}{a}\right). \quad (3.107)$$

Proof: First we assume a positive scaling factor $a > 0$ and get

$$\mathcal{F}[x(at)] = \int_{-\infty}^{\infty} x(at)e^{-j2\pi ft}\,dt = \int_{-\infty}^{\infty} x(u)e^{-j2\pi fu/a}\,d\left(\frac{u}{a}\right) = \frac{1}{a}X\left(\frac{f}{a}\right). \quad (3.108)$$

where we have assumed $u = at$. Applying the time-reversal property to this result we get

$$\mathcal{F}[x(-at)] = \frac{1}{a}X\left(-\frac{f}{a}\right). \quad (3.109)$$

Letting $a' = -a < 0$, we get the following for a negative scaling factor:

$$\mathcal{F}[x(a't)] = \frac{1}{-a'}X\left(\frac{f}{a'}\right). \quad (3.110)$$

Combining the above results for both positive and negative scaling factors, we get Eq. (3.107).

If $|a| < 1$, the signal is stretched and its spectrum is compressed and scaled up. When $|a| \to 0$, $x(at)$ is so stretched that it approaches a constant, and its spectrum is compressed and scaled up to the extent that it approaches an impulse. On the other hand, if $|a| > 1$, then the signal is compressed and its spectrum is stretched and scaled down. When $|a| \to \infty$, we redefine the signal as $a\,x(at)$ with spectrum $X(f/a)$, the signal becomes an impulse and its spectrum $X(f/a)$ becomes a constant.

- **Time and frequency shift**

$$\mathcal{F}[x(t \pm t_0)] = e^{\pm j2\pi f t_0} X(f), \tag{3.111}$$
$$\mathcal{F}^{-1}[X(f \pm f_0)] = e^{\mp j2\pi f_0 t} x(t). \tag{3.112}$$

Proof: We first prove Eq. (3.111):

$$\mathcal{F}[x(t \pm t_0)] = \int_{-\infty}^{\infty} x(t \pm t_0) e^{-j2\pi f t}\, dt. \tag{3.113}$$

Let $t' = t \pm t_0$, then $t = t' \mp t_0$, $dt' = dt$, and the above becomes

$$\mathcal{F}[x(t \pm t_0)] = \int_{-\infty}^{\infty} x(t') e^{-j2\pi f (t' \mp t_0)}\, dt' = e^{\pm j2\pi f t_0} X(f). \tag{3.114}$$

We see that a time shift t_0 of the signal corresponds to a phase shift $2\pi f t_0$ for every frequency component $e^{j2\pi f t}$. This result can be intuitively understood. As the phase shift is proportional to the frequency, a higher frequency component will have a greater phase shift while a lower frequency component will have a smaller phase shift, so that the relative positions of all harmonics remain the same, and the shape of the signal as a superposition of these harmonics remains the same when shifted.
As the spectrum of a shifted signal $y(t) = x(t \pm t_0)$ is $Y(f) = e^{\pm j2\pi f t_0} X(f)$, we see that the magnitude of the spectrum remains the same (shift-invariant), while the phase is shifted by $2\pi f t_0$:

$$|Y(f)| = |X(f)|, \qquad \angle Y(f) = \angle X(f) \pm 2\pi f t_0 \tag{3.115}$$

Applying the time-frequency duality to the time shift property in Eq. (3.111), we get the frequency shift property in Eq. (3.112).

- **Correlation**
The *cross-correlation* between two functions $x(t)$ and $y(t)$ is defined in Eq. (1.45) as

$$r_{xy}(\tau) = x(t) \star y(t) = \int_{-\infty}^{\infty} x(t)\overline{y}(t - \tau)\, dt. \tag{3.116}$$

Its Fourier transform is

$$\mathcal{F}[r_{xy}(\tau)] = X(f)\overline{Y}(f) = S_{xy}(f), \tag{3.117}$$

where $S_{xy}(f) = X(f)\overline{Y}(f)$ is the *cross power spectral density* of the two signals. If both signals $\overline{x}(t) = x(t)$ and $\overline{y}(t) = y(t)$ are real; i.e., $\overline{X}(f) = X(-f)$ and $\overline{Y}(f) = Y(-f)$, then we have

$$\mathcal{F}[r_{xy}(\tau)] = S_{xy}(f) = X(f)Y(-f). \tag{3.118}$$

In particular, when $x(t) = y(t)$, we have

$$\mathcal{F}[r_x(\tau)] = S_x(f) = X(f)\overline{X}(f) = |X(f)|^2, \tag{3.119}$$

where $r_x(\tau) = x(t) \star x(t)$ is the *autocorrelation* and $S_x(f) = |X(f)|^2$ is the PSD of the continuous signal $x(t)$.

Proof:
As $\mathcal{F}[x(t)] = X(f)$ and $\mathcal{F}[y(t-\tau)] = Y(f)e^{-j2\pi f\tau}$, we can easily prove Eq. (3.117) by applying the multiplication theorem:

$$\begin{aligned} r_{xy}(\tau) &= \int_{-\infty}^{\infty} x(t)\overline{y}(t-\tau)\,dt = \int_{-\infty}^{\infty} X(f)\overline{Y}(f)e^{j2\pi f\tau}\,df \\ &= \int_{-\infty}^{\infty} S_{xy}(f)e^{j2\pi f\tau}\,df = \mathcal{F}^{-1}[S_{xy}(f)]. \end{aligned} \tag{3.120}$$

- **Convolution theorem**
As first defined by Eq. (1.86), the convolution of two functions $x(t)$ and $y(t)$ is

$$z(t) = x(t) * y(t) = \int_{-\infty}^{\infty} x(\tau)y(t-\tau)\,d\tau = \int_{-\infty}^{\infty} y(\tau)x(t-\tau)\,d\tau = y(t) * x(t). \tag{3.121}$$

If $y(t) = y(-t)$ is even, then $x(t) * y(t) = x(t) \star y(t)$ is the same as the correlation. The convolution theorem states:

$$\mathcal{F}\left[x(t) * y(t)\right] = X(f)\,Y(f), \tag{3.122}$$

$$\mathcal{F}\left[x(t)y(t)\right] = X(f) * Y(f). \tag{3.123}$$

Proof:

$$\begin{aligned} \mathcal{F}[x(t) * y(t)] &= \int_{-\infty}^{\infty} \left[\int_{-\infty}^{\infty} x(\tau)y(t-\tau)\,d\tau\right] e^{-j2\pi ft}\,dt \\ &= \int_{-\infty}^{\infty} x(\tau)e^{-j2\pi f\tau} \int_{-\infty}^{\infty} y(t-\tau)e^{-j2\pi f(t-\tau)}\,dt\,d\tau \\ &= \int_{-\infty}^{\infty} x(\tau)e^{-j2\pi f\tau}Y(f)\,d\tau = X(f)Y(f). \end{aligned} \tag{3.124}$$

Similarly, we can also prove

$$\mathcal{F}\left[x(t)y(t)\right] = X(f) * Y(f). \tag{3.125}$$

In particular, as shown in Eq. (1.85), the output $y(t)$ of an LTI system can be found as the convolution $y(t) = h(t) * x(t)$ of its impulse response $h(t)$ and the input $x(t)$. Now according to the convolution theorem, the output of the system can be more conveniently obtained in the frequency domain by a multiplication:

$$Y(f) = H(f)X(f). \tag{3.126}$$

where $X(f)$ and $Y(f)$ are respectively the spectra of the input $x(t)$ and the output $y(t)$, and $H(f) = \mathcal{F}[h(t)]$, the Fourier transform of the impulse response function $h(t)$, is the FRF of the system, first defined by Eq. (1.91).

- **Time derivative**

$$\mathcal{F}\left[\frac{d}{dt}x(t)\right] = j2\pi f X(f) = j\omega X(\omega). \tag{3.127}$$

Proof:

$$\frac{d}{dt}x(t) = \frac{d}{dt}\int_{-\infty}^{\infty} X(f)e^{j2\pi ft}\,df = \int_{-\infty}^{\infty} X(f)\frac{d}{dt}e^{j2\pi ft}\,df$$
$$= \int_{-\infty}^{\infty} j2\pi f X(f)e^{j2\pi ft}\,df = \mathcal{F}^{-1}[j2\pi f X(f)]. \quad (3.128)$$

Repeating this process we get

$$\mathcal{F}\left[\frac{d^n}{dt^n}x(t)\right] = (j2\pi f)^n X(f). \quad (3.129)$$

- **Frequency derivative**

$$\mathcal{F}\left[t\, x(t)\right] = j\frac{d}{df}X(f), \quad (3.130)$$
$$\mathcal{F}\left[t^n x(t)\right] = j^n \frac{1}{(2\pi)^n}\frac{d^n}{df^n}X(f). \quad (3.131)$$

The proof is very similar to the above.

- **Time integration**

The Fourier transform of a time integration is

$$\mathcal{F}\left[\int_{-\infty}^{t} x(\tau)\,d\tau\right] = \frac{1}{j2\pi f}\,X(f) + \frac{1}{2}X(0)\delta(f). \quad (3.132)$$

Proof:

The integral of a signal $x(t)$ can be considered as its convolution with $u(t)$:

$$x(t) * u(t) = \int_{-\infty}^{\infty} x(\tau)u(t-\tau)\,d\tau = \int_{-\infty}^{t} x(\tau)\,d\tau. \quad (3.133)$$

Owing to the convolution theorem, we have

$$\mathcal{F}\left[\int_{-\infty}^{t} x(\tau)\,d\tau\right] = \mathcal{F}[x(t) * u(t)] = X(f)\left[\frac{1}{j2\pi f} + \frac{1}{2}\delta(f)\right]$$
$$= \frac{1}{j2\pi f}\,X(f) + \frac{X(0)}{2}\delta(f). \quad (3.134)$$

Comparing Eqs. (3.127) and (3.132), we see that the time derivative and integral are the inverse operations of each other in the frequency domain as well as in the time domain. However, the second term in Eq. (3.132) is necessary for representing the DC component $X(0)$ in signal $x(t)$, while Eq. (3.127) does not have a corresponding term as derivative operation is insensitive to DC component in the signal.

- **Complex conjugate**

$$\mathcal{F}[\overline{x}(t)] = \overline{X}(-f). \quad (3.135)$$

Proof: Taking the complex conjugate of the inverse Fourier transform, we get

$$\overline{x}(t) = \overline{\int_{-\infty}^{\infty} X(f)e^{j2\pi ft}\,df} = \int_{-\infty}^{\infty} \overline{X}(f)e^{-j2\pi ft}\,df$$
$$= \int_{-\infty}^{\infty} \overline{X}(-f')e^{j2\pi f't}\,df' = \mathcal{F}^{-1}[\overline{X}(-f)], \tag{3.136}$$

where we have defined $f' = -f$.

- **Real and imaginary signals**
  - If $x(t)$ is real, then the real part $X_\mathrm{r}(f)$ of its spectrum is even and the imaginary part $X_\mathrm{j}(f)$ is odd,

$$X_\mathrm{r}(f) = X_\mathrm{r}(-f), \quad \text{and} \quad X_\mathrm{j}(f) = -X_\mathrm{j}(-f). \tag{3.137}$$

    Proof: As $\overline{x}(t) = x(t)$ is real; i.e., $\mathcal{F}[\overline{x}(t)] = \mathcal{F}[x(t)]$, from Eq. (3.135) we get

$$X(f) = \overline{X}(-f); \quad \text{i.e.,} \quad X_\mathrm{r}(f) + j\,X_\mathrm{j}(f) = X_\mathrm{r}(-f) - j\,X_\mathrm{j}(-f). \tag{3.138}$$

    Equating the real and imaginary parts on both sides we get Eq. (3.137). Moreover, when the real signal is either even or odd, we have the following results based on Eqs. (3.99) and (3.101):
    * If $x(t) = x(-t)$ is even, then $X(f)$ is also even; i.e. $X_\mathrm{j}(f) = 0$ and $X(f) = X_\mathrm{r}(f) = X_\mathrm{r}(-f)$ is real and even.
    * If $x(t) = -x(-t)$ is odd, then $X(f)$ is also odd; i.e., $X_\mathrm{r}(f) = 0$ and $X(f) = X_\mathrm{j}(f) = -X_\mathrm{j}(-f)$ is imaginary and odd.
  - If $x(t)$ is imaginary, then the real part $X_\mathrm{r}(f)$ of its spectrum is odd and the imaginary part $X_\mathrm{j}(f)$ is even:

$$X_\mathrm{r}(f) = -X_\mathrm{r}(-f), \quad \text{and} \quad X_\mathrm{j}(f) = X_\mathrm{j}(-f). \tag{3.139}$$

    Proof: As $\overline{x}(t) = -x(t)$ is imaginary; i.e., $\mathcal{F}[\overline{x}(t)] = -\mathcal{F}[x(t)]$, from Eq. (3.135) we get

$$-X(f) = \overline{X}(-f); \quad \text{i.e.,} \quad X_\mathrm{r}(f) + j\,X_\mathrm{j}(f) = -X_\mathrm{r}(-f) + j\,X_\mathrm{j}(-f). \tag{3.140}$$

    Equating the real and imaginary parts on both sides we get Eq. (3.139). Moreover, when the imaginary signal is either even or odd, we have the following results based on Eqs. (3.99) and (3.101):
    * If $x(t) = x(-t)$ is even, then $X(f)$ is also even; i.e., $X_\mathrm{r}(f) = 0$ and $X(f) = j\,X_\mathrm{j}(f) = j\,X_\mathrm{j}(-f)$ is imaginary and even.
    * If $x(t) = -x(-t)$ is odd, then $X(f)$ is also odd; i.e., $X_\mathrm{j}(f) = 0$ and $X(f) = X_\mathrm{r}(f) = -X(-f)$ is real and odd.

These results are summarized in Table 3.1.

The complex spectrum $X(f)$ of a time signal $x(t)$ can be expressed in either Cartesian form in terms of the real and imaginary parts $X_\mathrm{r}(f)$ and $X_\mathrm{j}(f)$, or in polar form in terms of the magnitude $|X(f)|$ and phase $\angle X(f)$:

$$X(f) = X_\mathrm{r}(f) + jX_\mathrm{j}(f) = |X(f)|e^{j\angle X(f)}, \tag{3.141}$$

Table 3.1. Symmetry properties of Fourier transform

$x(t) = x_r(t) + jx_j(t)$	$X(f) = X_r(f) + jX_j(f)$
$x(t) = x_r(t)$ real	$X_r(f) = X_r(-f)$ even, $X_j(f) = -X_j(-f)$ odd
$x_r(t) = x_r(-t)$ real, even	$X_r(f) = X_r(-f)$ real, even $X_j(f) = 0$
$x_r(t) = -x_r(-t)$ real, odd	$X_j(f) = -X_j(f)$ imaginary, odd $X_r(f) = 0$
$x(t) = x_j(t)$ imaginary	$X_r(f) = -X_r(-f)$ odd, $X_j(f) = X_j(-f)$ even
$x_j(t) = x_j(-t)$ imaginary, even	$X_j(f) = X_j(-f)$ imaginary, even $X_r(f) = 0$
$x_j(t) = -x_j(-t)$ imaginary, odd	$X_r(f) = -X_r(-f)$ real, odd $X_j(f) = 0$

where

$$\begin{cases} |X(f)| = \sqrt{X_r^2(f) + X_j^2(f)} \\ \angle X(f) = \tan^{-1}[X_j(f)/X_r(f)] \end{cases}, \qquad \begin{cases} X_r(f) = |X(f)| \ \cos \angle X(f) \\ X_j(f) = |X(f)| \ \sin \angle X(f) \end{cases}. \tag{3.142}$$

We see that when the signal is either real or imaginary, $|X(f)|$ is always even and $\angle X(f)$ is always odd.

- **Physical interpretation**

The spectrum of a signal $x(t)$ can be expressed as

$$\begin{aligned} x(t) &= \int_{-\infty}^{\infty} X(f) e^{j2\pi ft} \, df = \int_{-\infty}^{\infty} |X(f)| e^{j2\pi ft + \angle X(f)} \, df \\ &= \int_{-\infty}^{\infty} |X(f)| \cos(2\pi ft + \angle X(f)) \, df + j \int_{-\infty}^{\infty} |X(f)| \sin(2\pi ft + \angle X(f)) \, df. \end{aligned} \tag{3.143}$$

If $x(t)$ is real (as most signals in practice), the second term is zero while the first term (an integral of an even function of f) remains, and we have

$$x(t) = 2 \int_0^{\infty} |X(f)| \cos(2\pi ft + \angle X(f)) \, df. \tag{3.144}$$

We see that the Fourier transform expresses a real time signal as a superposition of infinitely many uncountable frequency components each with a different frequency f, magnitude $|X(f)|$, and phase $\angle X(f)$. Note that Eq. (3.16) for periodical signals is just the discrete version of the equation above.

3.2.4 Fourier spectra of typical functions

- **Unit impulse**

The Fourier transform of the unit impulse function is given in Eq. (3.77) according to the definition of the Fourier transform:

$$\mathcal{F}[\delta(t)] = \int_{-\infty}^{\infty} \delta(t) e^{-j2\pi f t}\, dt = 1. \tag{3.145}$$

- **Sign function**
  The Fourier transform of the sign function $\text{sgn}(t)$ is given in Eq. (3.84):

$$\mathcal{F}[\text{sgn}(t)] = \frac{1}{j\pi f}. \tag{3.146}$$

  Note that $\text{sgn}(t)$ is real and odd, and its spectrum is imaginary and odd. Moreover, based on the time-frequency duality property, we also get

$$\mathcal{F}\left[\frac{1}{t}\right] = -j\pi\ \text{sgn}(f). \tag{3.147}$$

- **Unit step functions**
  As the unit step is the time integral of the unit impulse:

$$u(t) = \int_{-\infty}^{t} \delta(t)\, dt. \tag{3.148}$$

  and $\mathcal{F}[\delta(t)] = 1$, $\mathcal{F}[u(t)]$ can be found according to the time integration property (Eq. (3.132)) to be

$$\mathcal{F}[u(t)] = \frac{1}{j2\pi f} + \frac{1}{2}\delta(f). \tag{3.149}$$

  which is the same as in Eq. (3.80).
  Moreover, owing to the time reversal property $\mathcal{F}[x(-t)] = X(-f)$, we can also get the Fourier transform of a left-sided unit step:

$$\mathcal{F}[u(-t)] = \frac{1}{2}\delta(-f) + \frac{1}{-j2\pi f} = \frac{1}{2}\delta(f) - \frac{1}{j2\pi f} \tag{3.150}$$

  (as $\delta(-f) = \delta(f)$).
- **Constant**
  As a constant time function $x(t) = 1$ is not square-integrable; the integral of its Fourier transform does not converge:

$$\mathcal{F}[1] = \int_{-\infty}^{\infty} e^{-j2\pi f t}\, dt. \tag{3.151}$$

  However, we realize that the constant time function is simply the sum of a right-sided unit step and a left-sided unit step: $x(t) = 1 = u(t) + u(-t)$, and according to the linearity of the Fourier transform, we have

$$\mathcal{F}[1] = \mathcal{F}[u(t)] + \mathcal{F}[u(-t)] = \frac{1}{j2\pi f} + \frac{1}{2}\delta(f) - \frac{1}{j2\pi f} + \frac{1}{2}\delta(f) = \delta(f). \tag{3.152}$$

  Alternatively, the Fourier transform of constant 1 can also be obtained according to the property of time-frequency duality, based on the Fourier transform of the unit impulse:

$$\mathcal{F}[1] = \int_{-\infty}^{\infty} e^{-j2\pi f t}\, dt = \delta(f). \tag{3.153}$$

Owing to the property of time-frequency scaling, if the time function $x(t)$ is scaled by a factor of $1/2\pi$ to become $x(t/2\pi)$, its spectrum $X(f)$ will become $2\pi X(2\pi f) = 2\pi X(\omega)$. Specifically in this case, if we scale the constant 1 as a time function by $1/2\pi$ (still the same), its spectrum $X(f) = \delta(f)$ can be expressed as a function of angular frequency $X(\omega) = 2\pi\delta(\omega)$.

- **Complex exponentials and sinusoids**
  The Fourier transform of a complex exponential $x(t) = e^{j\omega_0 t} = e^{j2\pi f_0 t}$ is

$$\mathcal{F}[e^{j2\pi f_0 t}] = \int_{-\infty}^{\infty} e^{-j2\pi(f-f_0)t}\, dt = \delta(f - f_0), \tag{3.154}$$

and according to Euler's formula, the Fourier transform of cosine function $x(t) = \cos(2\pi f_0 t)$ is

$$\mathcal{F}[\cos(2\pi f_0 t)] = \frac{1}{2}[\delta(f - f_0) + \delta(f + f_0)]. \tag{3.155}$$

Similarly, the Fourier transform of $x(t) = \sin(2\pi f_0 t)$ is

$$\mathcal{F}[\sin(2\pi f_0 t)] = \frac{1}{2j}[\delta(f - f_0) - \delta(f + f_0)]. \tag{3.156}$$

Note that the sine and cosine functions are respectively odd and even, and so are their Fourier spectra. Also none of the step, constant, complex exponential and sinusoidal functions considered above is square or absolutely integrable, and correspondingly their Fourier transform integrals are only marginally convergent, in the sense that their spectra $X(f)$ all contain delta functions (e.g., $\delta(f)$, $\delta(f - f_0)$) with an infinite value at certain frequencies.

- **Exponential decay**
  A right-sided exponential decay function is defined as $e^{-at}u(t)$ $(a > 0)$, and its Fourier transform can be found to be

$$\begin{aligned}\mathcal{F}[e^{-at}u(t)] &= \int_{0}^{\infty} e^{-at} e^{-j2\pi f t}\, dt = \frac{1}{-(a + j2\pi f)} e^{-(a+j2\pi f)t}\bigg|_0^{\infty} \\ &= \frac{1}{a + j2\pi f} = \frac{1}{a + j\omega} = \frac{a - j\omega}{a^2 + \omega^2}.\end{aligned} \tag{3.157}$$

As $\lim_{a\to 0} e^{-at}u(t) \to u(t)$, we have

$$\mathcal{F}[u(t)] = \lim_{a\to 0} \mathcal{F}[e^{-at}u(t)] = \lim_{a\to 0} \frac{1}{a + j2\pi f} = \frac{1}{2}\delta(f) + \frac{1}{j2\pi f}, \tag{3.158}$$

which is the same as in Eq. (3.80). Note that it is tempting to assume at the limit $a = 0$, the second term alone will result, while in fact the first term $\delta(f)/2$ is also necessary. The proof of this result is left to the reader as a homework problem.

Next consider a left-sided exponential decay function $e^{at}u(-t)$, the time-reversal of the right-sided decay function. According to the time reversal property $\mathcal{F}[x(-t)] = X(-f)$, we get

$$\mathcal{F}[e^{at}u(-t)] = \frac{1}{a - j2\pi f} = \frac{1}{a - j\omega}. \tag{3.159}$$

Finally, a two-sided exponential decay $e^{-a|t|}$ is the sum of the right-sided and left-sided decay functions and according to the linearity property, its Fourier transform can be obtained as

$$\begin{aligned} \mathcal{F}[e^{-a|t|}] &= \mathcal{F}[e^{-at}u(t)] + \mathcal{F}[e^{at}u(-t)] = \frac{1}{a + j2\pi f} + \frac{1}{a - j2\pi f} \\ &= \frac{2a}{a^2 + (2\pi f)^2} = \frac{2a}{a^2 + \omega^2}. \end{aligned} \tag{3.160}$$

- **Rectangular function and sinc function**
  A rectangular function, also called a square impulse, of width τ is defined as

$$\text{rect}_\tau(t) = \begin{cases} 1 & |t| < \tau/2 \\ 0 & \text{else} \end{cases}, \tag{3.161}$$

which can be considered as the difference between two unit step functions:

$$\text{rect}(t) = u(t + \tau/2) - u(t - \tau/2). \tag{3.162}$$

Owing to the properties of linearity and time shift, the spectrum of $\text{rect}_\tau(t)$ can be found to be

$$\begin{aligned} \mathcal{F}[\text{rect}(t)] &= \mathcal{F}[u(t + \tau/2)] - \mathcal{F}[u(t - \tau/2)] = \frac{e^{j\pi f\tau}}{j2\pi f} - \frac{e^{-j\pi f\tau}}{j2\pi f} \\ &= \frac{\tau}{\pi f\tau}\sin(\pi f\tau) = \tau\ \text{sinc}(f\tau). \end{aligned} \tag{3.163}$$

This spectrum is zero at $f = k/\tau$ for any integer k. If we let the width $\tau \to \infty$, the rectangular function becomes a constant 1 and its spectrum an impulse function. If we divide both sides of the equation above by τ and let $\tau \to 0$, the time function becomes an impulse and its spectrum a constant.

As both the rectangular function and sinc function are symmetric, the time-frequency duality property applies; i.e., the Fourier spectrum of a sinc function in the time domain is a rectangular function in the frequency domain, called an ideal low-pass (LP) filter:

$$H_{lp}(f) = \begin{cases} 1 & |f| < f_c \\ 0 & |f| > f_c \end{cases}, \tag{3.164}$$

where f_c is called the *cutoff frequency*, then according to time-frequency duality, its time impulse response is

$$h_{lp}(t) = \frac{\sin(2\pi f_c t)}{\pi t} = 2f_c\ \text{sinc}(2f_c t). \tag{3.165}$$

Note that the impulse response $h_{lp}(t)$ is non-zero for $t < 0$, indicating that the ideal LP filter is not causal (response before the input $\delta(0)$ at $t = 0$). In other words, an ideal LP filter is impossible to implement in real time, but it can be trivially realized off-line in the frequency domain.

- **Triangle function**

$$\text{triangle}(t) = \begin{cases} 1 - |t|/\tau & |t| < \tau \\ 0 & |t| \geq \tau \end{cases}. \tag{3.166}$$

Following the definition, the spectrum of the triangle function, as an even function, can be obtained as

$$\begin{aligned} \mathcal{F}[\text{triangle}(t)] &= 2\int_0^\tau (1 - t/\tau)\cos(2\pi f t)\,dt \\ &= 2\left[\int_0^\tau \cos(2\pi f t)\,dt - \frac{1}{\tau}\int_0^\tau t\,\cos(2\pi f t)\,dt\right] \\ &= \frac{1}{\pi f}\left[\sin(2\pi f\tau) - \frac{t}{\tau}\,\sin(2\pi f t)\Big|_0^\tau + \frac{1}{\tau}\int_0^\tau \sin(2\pi f t)\,dt\right] \\ &= \frac{-1}{2\tau(\pi f)^2}\cos(2\pi f t)\Big|_0^\tau = \frac{1}{2\tau(\pi f)^2}(1 - \cos(2\pi f\tau)) \\ &= \tau\frac{\sin^2(\pi f\tau)}{(\pi f\tau)^2} = \tau\,\text{sinc}^2(f\tau). \end{aligned} \tag{3.167}$$

Alternatively, the triangle function (with width 2τ) can be obtained more easily as the convolution of two square functions (with width τ) scaled by $1/\tau$:

$$\text{triangle}(t) = \frac{1}{\tau}\,\text{rect}(t) * \text{rect}(t). \tag{3.168}$$

its Fourier transform can be conveniently obtained based on the convolution theorem:

$$\mathcal{F}[\text{triangle}(t)] = \frac{1}{\tau}\mathcal{F}[\text{rect}(t) * \text{rect}(t)] = \frac{1}{\tau}\,\tau\,\text{sinc}(f)\,\tau\,\text{sinc}(f) = \tau\,\text{sinc}^2(f\tau). \tag{3.169}$$

- **Gaussian function**
Consider the Gaussian function $x(t) = e^{-\pi(t/a)^2}/a$. Note that in particular when $a = \sqrt{2\pi\sigma^2}$, $x(t)$ becomes the normal distribution with variance σ^2 and mean $\mu = 0$. The spectrum of $x(t)$ is

$$\begin{aligned} X(f) &= \mathcal{F}\left[\frac{1}{a}e^{-\pi(t/a)^2}\right] = \frac{1}{a}\int_{-\infty}^{\infty} e^{-\pi(t/a)^2}\,e^{-j2\pi f t}\,dt \\ &= \frac{1}{a}\int_{-\infty}^{\infty} e^{-\pi((t/a)^2 + j2ft)}\,dt = \frac{1}{a}e^{\pi(jaf)^2}\int_{-\infty}^{\infty} e^{-\pi[(t/a)^2 + j2ft + (jaf)^2]}\,dt \\ &= e^{-\pi(af)^2}\int_{-\infty}^{\infty} e^{-\pi(t/a + jaf)^2}\,d(t/a + jaf) = e^{-\pi(af)^2}. \end{aligned} \tag{3.170}$$

The last equation is due to the identity $\int_{-\infty}^{\infty} e^{-\pi x^2} dx = 1$. We see that the Fourier transform of a Gaussian function is another Gaussian function, and the area underneath either $x(t)$ or $X(f)$ is unity. Moreover, if we let $a \to 0$, $x(t)$ will approach $\delta(t)$, while its spectrum $e^{-\pi(af)^2}$ approaches one. On the other hand, if we rewrite the above as

$$X(f) = \mathcal{F}[x(t)] = \mathcal{F}[e^{-\pi(t/a)^2}] = a\, e^{-\pi(af)^2}, \tag{3.171}$$

and let $a \to \infty$, $x(t)$ approaches 1 and $X(f)$ approaches $\delta(f)$.

- **Impulse train**

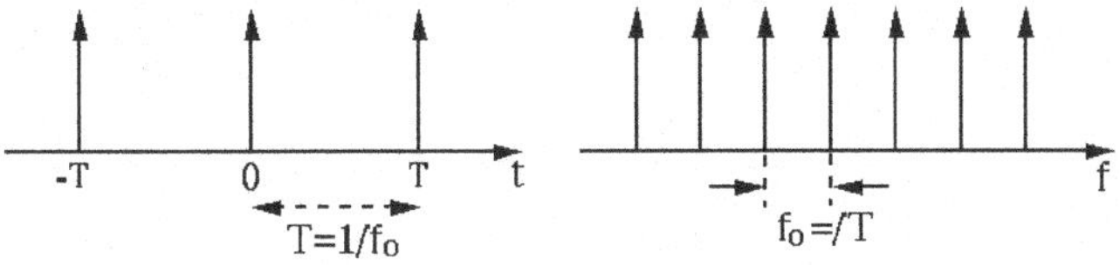

Figure 3.6 Impulse train and its spectrum.

The impulse train is a sequence of infinite unit impulses separated by a constant time interval T:

$$\text{comb}(t) = \sum_{n=-\infty}^{\infty} \delta(t - nT). \tag{3.172}$$

The Fourier transform of this function is

$$\begin{aligned}
\mathcal{F}[\text{comb}(t)] &= \int_{-\infty}^{\infty} \text{comb}(t) e^{-j2\pi ft}\, dt = \int_{-\infty}^{\infty} \left[\sum_{n=-\infty}^{\infty} \delta(t - nT) \right] e^{-j2\pi ft}\, dt \\
&= \sum_{n=-\infty}^{\infty} \int_{-\infty}^{\infty} \delta(t - nT) e^{-j2\pi ft}\, dt = \sum_{n=-\infty}^{\infty} e^{-j2\pi nfT} \\
&= f_0 \sum_{n=-\infty}^{\infty} \delta(f - nf_0) = \frac{1}{T} \sum_{n=-\infty}^{\infty} \delta(f - n/T),
\end{aligned} \tag{3.173}$$

where we have used Eq. (1.35) with F replaced by f_0. We therefore see that the Fourier spectrum of an impulse train is also an impulse train. Also we realize that Eq. (3.173 is a special case of the *Poisson summation formula* given in Eq. (3.92), when $x(t) = \delta(t)$ and $X(f) = \mathcal{F}[\delta(t)] = 1$.

- **Periodic signals**

  As discussed before, a periodic signal $x_T(t + T) = x_T(t)$ can be Fourier expanded into a series with coefficients $X[k]$, as shown in Eq. (3.7). We can also consider this periodic signal as the convolution of a finite signal $x(t)$ defined over the interval $0 < t < T$ with an impulse train with the same interval, as illustrated in Fig. 3.7:

$$x_T(t) = x(t) * \sum_{n=-\infty}^{\infty} \delta(t - nT). \tag{3.174}$$

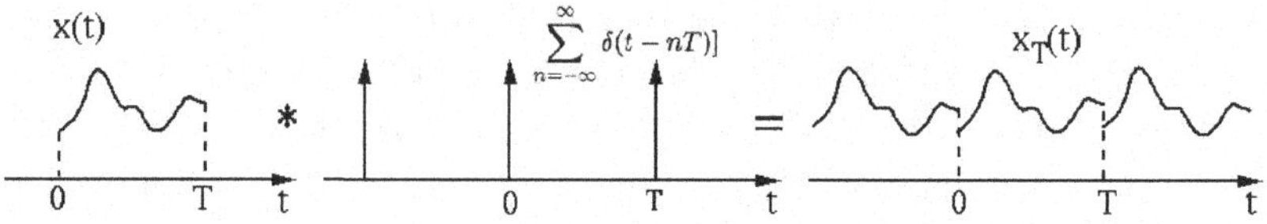

Figure 3.7 Generation of a periodic signal.

According to the convolution theorem, the Fourier transform of this periodic signal can be found to be

$$\mathcal{F}[x_T(t)] = \mathcal{F}[x(t) * \sum_{n=-\infty}^{\infty} \delta(t-nT)] = \mathcal{F}[x(t)]\ \mathcal{F}\left[\sum_{n=-\infty}^{\infty} \delta(t-nT)\right]. \tag{3.175}$$

Here, the two Fourier transforms on the right-hand side above are, respectively:

$$\mathcal{F}[x(t)] = \int_0^{\mathrm{T}} x(t)e^{-j2\pi ft}\,dt, \tag{3.176}$$

and (Eq. (3.173))

$$\mathcal{F}[\sum_{n=-\infty}^{\infty} \delta(t-nT)] = \frac{1}{T}\sum_{k=-\infty}^{\infty} \delta(f-kf_0), \tag{3.177}$$

where $f_0 = 1/T$ is the fundamental frequency. Substituting these into Eq. (3.175) we get

$$\begin{aligned}\mathcal{F}[x_T(t)] &= \left[\int_0^{\mathrm{T}} x(t)e^{-j2\pi ft}\,dt\right]\left[\frac{1}{T}\sum_{k=-\infty}^{\infty} \delta(f-kf_0)\right] \\ &= \sum_{k=-\infty}^{\infty} \frac{1}{T}\int_0^{\mathrm{T}} x(t)e^{-j2\pi kf_0t}\,dt\ \delta(f-kf_0) = \sum_{k=-\infty}^{\infty} X[k]\ \delta(f-kf_0).\end{aligned} \tag{3.178}$$

We realize this is actually Eq. (3.86), indicating that the periodic signal has a discrete spectrum, which can be represented as an impulse train weighted by the Fourier coefficients $X[k]$. As an example, a square wave and its periodic version are shown on the left of Fig. 3.8, and their corresponding spectra are shown on the right. We see that the spectrum of the periodic version is composed of a set of impulses, weighted by the spectrum $X(f) = \mathcal{F}[x(t)]$.

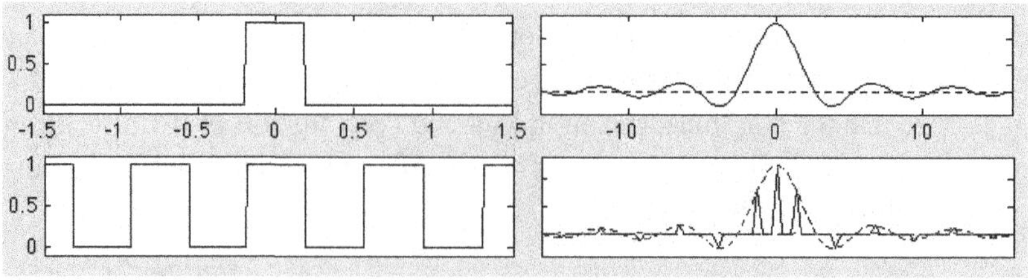

Figure 3.8 A periodic signal and its spectrum.

Fig. 3.9 shows a set of typical signals on the left and their Fourier spectra on the right.

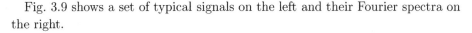

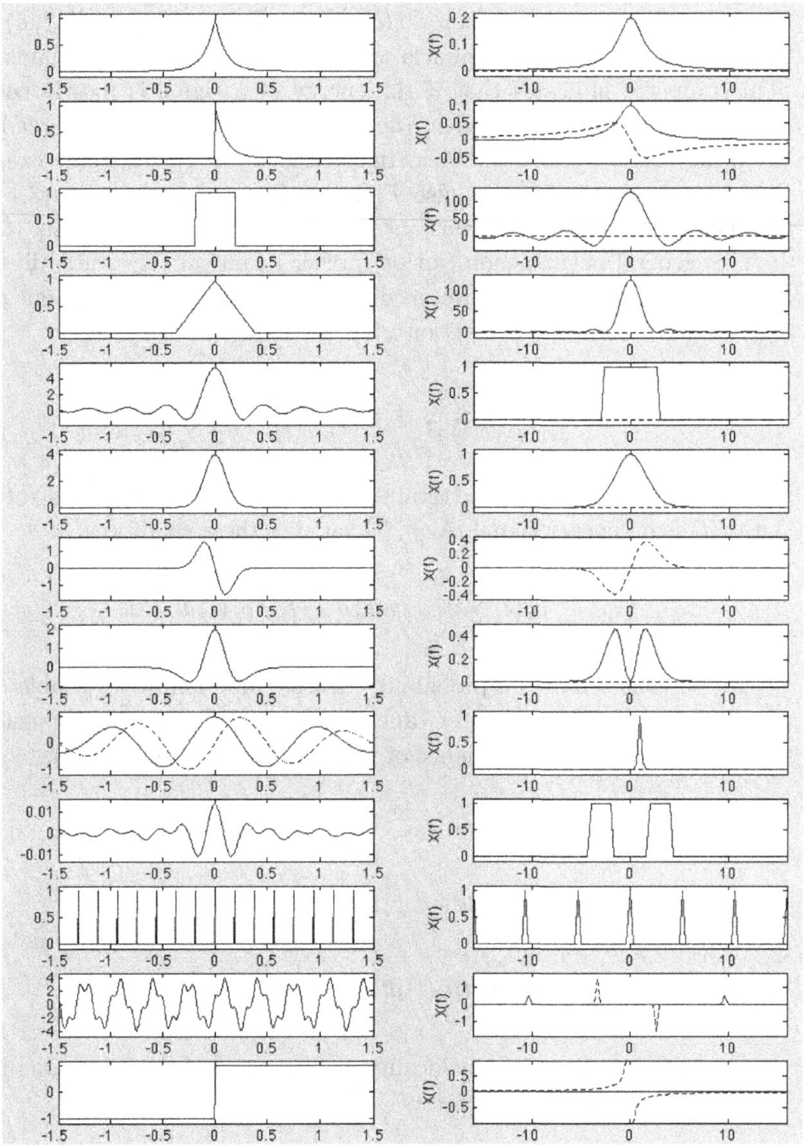

Figure 3.9 Examples of CTFT. A set of signals are shown on the left and their Fourier spectra are shown on the right (real and imaginary parts are shown in solid and dashed lines, respectively).

3.2.5 The uncertainty principle

According to the property of time and frequency scaling (Eq. (3.107)), if a time function is expanded to become $x(at)$ $(a < 1)$, its spectrum $X(f/a)/a$ is compressed. Conversely, if the signal is compressed $(a > 1)$, its spectrum is expanded. This property indicates that if the energy of a signal is mostly concentrated within a short time range, then the energy in its spectrum is spread in a wide frequency range, and vice versa. In particular, as two extreme examples, the Fourier transform of an impulse $\mathcal{F}[\delta(t)] = 1$ is a constant (Eq. (3.145)), while the Fourier transform of a constant $\mathcal{F}[1] = \delta(f)$ is an impulse (Eq. (3.151)).

This general phenomenon can be further quantitatively stated by the *uncertainty principle*. To do so, we need to borrow some concepts from probability theory. First, for a given function $x(t)$, we build another function:

$$p_x(t) = \frac{|x(t)|^2}{||x(t)||^2} = \frac{|x(t)|^2}{\langle x(t), x(t)\rangle} = \frac{|x(t)|^2}{\int_{-\infty}^{\infty} |x(t)|^2\, dt}, \tag{3.179}$$

where the denominator is the total energy of the signal $x(t)$ assumed to be finite; i.e., $x(t)$ is an energy signal. As $p_x(t)$ satisfies these conditions

$$p_x(t) > 0 \qquad \text{and} \qquad \int_{-\infty}^{\infty} p_x(t)\, dt = 1, \tag{3.180}$$

it can be considered as a probability density function over variable t, and how the function $x(t)$ spreads over time; i.e., the locality or the dispersion of $x(t)$, can be measured as the variance of this probability density $p_x(t)$:

$$\sigma_t^2 = \int_{-\infty}^{\infty} (t - \mu_t)^2 p_x(t)\, dt = \frac{1}{||x(t)||^2} \int_{-\infty}^{\infty} (t - \mu_t)^2 |x(t)|^2\, dt, \tag{3.181}$$

where μ_t is the mean of $p_x(t)$:

$$\mu_t = \int_{-\infty}^{\infty} t p_x(t)\, dt = \frac{1}{||x(t)||^2} \int_{-\infty}^{\infty} t|x(t)|^2\, dt. \tag{3.182}$$

In the frequency domain, the locality or dispersion of the spectrum of the signal can also be similarly measured as

$$\sigma_f^2 = \frac{1}{||X(f)||^2} \int_{-\infty}^{\infty} (f - \mu_f)^2 |X(f)|^2\, df = \frac{1}{||x(t)||^2} \int_{-\infty}^{\infty} (f - \mu_f)^2 |X(f)|^2\, df. \tag{3.183}$$

Here, we have used Parseval's identity $||x(t)||^2 = ||X(f)||^2$, and μ_f is defined as

$$\mu_f = \frac{1}{||X(f)||^2} \int_{-\infty}^{\infty} f|X(f)|^2\, df. \tag{3.184}$$

Now the uncertainty principle can be stated as the following theorem.

Theorem 3.1. *Let $X(f) = \mathcal{F}[x(t)]$ be the Fourier spectrum of a given function $x(t)$ and σ_t^2 and σ_f^2 be defined as above. Then*

$$\sigma_t^2 \sigma_f^2 \geq \frac{1}{16\pi^2}. \tag{3.185}$$

Proof:

Without loss of generality, we assume in the proof $\mu_t = \mu_f = 0$ and consider

$$\sigma_t^2 \sigma_f^2 = \frac{1}{||x(t)||^4} \int_{-\infty}^{\infty} |tx(t)|^2 \, dt \int_{-\infty}^{\infty} |fX(f)|^2 \, df. \tag{3.186}$$

Owing to the time derivative property (Eq. (3.127)), we have

$$\frac{1}{j2\pi} \mathcal{F}\left[\frac{d}{dt} x(t)\right] = fX(f). \tag{3.187}$$

also due to Parseval's identity we have

$$\int_{-\infty}^{\infty} |fX(f)|^2 \, df = \frac{1}{4\pi^2} \int_{-\infty}^{\infty} \left| \frac{d}{dt} x(t) \right|^2 \, dt. \tag{3.188}$$

Now Eq. (3.186) becomes

$$\sigma_t^2 \sigma_f^2 = \frac{1}{4\pi^2 ||x(t)||^4} \int_{-\infty}^{\infty} |tx(t)|^2 \, dt \int_{-\infty}^{\infty} \left| \frac{d}{dt} x(t) \right|^2 \, dt. \tag{3.189}$$

Applying the Cauchy-Schwarz inequality (Eq. (2.30)), we get

$$\sigma_t^2 \sigma_f^2 \geq \frac{1}{4\pi^2 ||x(t)||^4} \left| \int_{-\infty}^{\infty} t \, \overline{x}(t) \frac{d}{dt} x(t) \, dt \right|^2. \tag{3.190}$$

But as

$$\begin{aligned} \frac{d}{dt}[\, |x(t)|^2 \,] &= \frac{d}{dt}[x(t)\overline{x}(t)] = \overline{x}(t) \, \frac{d}{dt} x(t) + x(t) \, \frac{d}{dt} \overline{x}(t) \\ &= 2 \text{ Re} \left[\frac{d}{dt} x(t) \, \overline{x}(t) \right] \leq 2 \frac{d}{dt} x(t) \, \overline{x}(t), \end{aligned} \tag{3.191}$$

replacing $\overline{x}(t) \, \frac{d}{dt} x(t)$ in the integrand by $\frac{d}{dt}[|x(t)|^2]/2$ we get

$$\sigma_t^2 \sigma_f^2 \geq \frac{1}{4 \times 4\pi^2 ||x(t)||^4} \left[\int_{-\infty}^{\infty} t \, \frac{d}{dt}[|x(t)|^2] \, dt \right]^2. \tag{3.192}$$

By integration by parts, the integral becomes

$$\int_{-\infty}^{\infty} t \frac{d}{dt}[|x(t)|^2] \, dt = t|x(t)|^2 \Big|_{-\infty}^{\infty} - \int_{-\infty}^{\infty} |x(t)|^2 \, dt = - \int_{-\infty}^{\infty} |x(t)|^2 \, dt. \tag{3.193}$$

Here, we have assumed $\lim_{|t| \to \infty} tx^2(t) = 0$ for the reason that $x(t)$ contains a finite amount of energy. Substituting this back into the inequality, we finally get

$$\sigma_t^2 \sigma_f^2 \geq \frac{1}{4 \cdot 4\pi^2 ||x(t)||^4} \left[\int_{-\infty}^{\infty} |x(t)|^2 \, dt \right]^2 = \frac{1}{16\pi^2}. \tag{3.194}$$

Q.E.D.

This result is also referred to as the *Heisenberg uncertainty*, as it is analogous to the fact in quantum physics that the position and momentum of a particle cannot be accurately measured simultaneously, higher precision in one quantity implies lower precision in the other. Similarly, here the uncertainty principle indicates an important fact: in the Fourier transform, the temporal and frequency localities of a signal cannot be achieved simultaneously.

3.3 Homework problems

1. Prove Plancherel's identity in Eq. (3.24):

$$\frac{1}{T}\int_T x_T(t)\overline{y}_T(t)\,dt = \sum_{k=-\infty}^{\infty} X[k]\overline{Y}[k]. \tag{3.195}$$

   Hint: replace $x_T(t)$ and $y_T(t)$ in the expression by their corresponding Fourier expansions. Use Eq. (1.33).
2. Prove the circular convolution property in Eq. (3.28):

$$z_T(t) = x_T() * y_T(t) = \frac{1}{T}\int_T x_T(\tau)y_T(t-\tau)\,d\tau. \tag{3.196}$$

   Hint: reconstruct function $z_T(t)$ based on $Z[k] = X[k]Y[k]$

$$z_T(t) = \sum_{k=-\infty}^{\infty} [X[k]Y[k]]e^{jk\omega_0 t}. \tag{3.197}$$

   Use Eq. (1.35).
3. Find the Fourier series coefficients of the following signals:
   (a)

$$x_1(t) = \cos\left(\frac{10\pi}{3}t\right) + \cos\left(\frac{5\pi}{4}t\right). \tag{3.198}$$

   (b)

$$x_2(t) = \cos\left(\frac{5\pi}{6}t\right) + \cos\left(\frac{3\pi}{4}t\right) + \sin\left(\frac{\pi}{3}t\right). \tag{3.199}$$

   (c)

$$x_3(t) = \cos\left(\frac{10}{3}t\right) + \cos\left(\frac{5\pi}{4}t\right). \tag{3.200}$$

4. Show that the Fourier coefficients given in Eq. (3.46) for the even triangle function are real and even ($X[k] = X[-k]$), and the Fourier coefficients given in Eq. (3.47) for the odd triangle function are imaginary and odd ($X[k] = -X[-k]$).
5. Show that the Fourier coefficients $X[k]$ of the triangle function given in Eq. (3.46) is a real and even function of k.

6. If the square wave in Eq. (3.36) is shifted to the left by $T/4$, it becomes an even function:

$$x_T(t) = \begin{cases} 1 & |t| < T/4 \\ 0 & T/4 < |t| < T/2 \end{cases}. \tag{3.201}$$

Show that its Fourier series expansion becomes

$$\begin{aligned} x(t) &= \sum_{k=-\infty}^{\infty} X[k] e^{jk\omega_0 t} \\ &= \frac{1}{2} + \frac{2}{\pi}\left[\frac{\cos(\omega_0 t)}{1} - \frac{\cos(3\omega_0 t)}{3} + \frac{\cos(5\omega_0 t)}{5} + \cdots\right], \end{aligned} \tag{3.202}$$

composed of odd harmonics of even cosine functions.

7. Find the Fourier series coefficients of an even triangle wave

$$x(t) = 2|t|/T. \tag{3.203}$$

Express this even triangle wave $x(t)$ in terms of even cosine functions of different frequencies.

8. Given the signal

$$x(t) = 3\cos\left(\frac{\pi(10t-1)}{3}\right) - 2\sin\left(\frac{\pi(5t+2)}{4}\right), \tag{3.204}$$

find its fundamental frequency and period (if it is periodic) and then the Fourier series coefficients.

9. Find the Fourier series coefficients of the following signal:

$$x(t) = 2\cos(12\pi t - \pi/2) - 3\sin(20\pi t + \pi/3). \tag{3.205}$$

10. Find the Fourier spectrum of a truncated sinusoid

$$x(t) = \begin{cases} \cos(2\pi f_0 t) & |t| < \tau/2 \\ 0 & \text{else} \end{cases}. \tag{3.206}$$

Sketch the spectrum.

11. Find the Fourier spectrum of the following signal:

$$x(t) = \begin{cases} -t & -\tau/2 < t < \tau/2 \\ 0 & \text{else} \end{cases}. \tag{3.207}$$

<u>Hint:</u> $x(t)$ can be written as $x(t) = u(t+1) + u(t-1) - s(t)$, where

$$s(t) = \frac{2}{\tau}\int_{-\infty}^{t} r(t)\,dt = \begin{cases} 0 & t < -\tau/2 \\ 2t/\tau + 1 & -\tau/2 < t < \tau/2 \\ 2 & \tau/2 < t \end{cases} \tag{3.208}$$

is the integral of a square impulse with width τ:

$$r(t) = \begin{cases} 1 & |t| < \tau/2 \\ 0 & \text{else} \end{cases}. \tag{3.209}$$

Find the spectrum of each of the three components and then sum them up.

12. Find the Fourier spectrum of the following signal:

$$x(t) = \begin{cases} 1 - t/\tau & 0 < t < \tau \\ 0 & \text{else} \end{cases}. \tag{3.210}$$

13. Show that the Fourier transform of the step function $u(t)$ given in Eq. (3.80) can also be obtained by

$$\mathcal{F}[u(t)] = \lim_{a \to 0} \mathcal{F}[e^{-at}u(t)] = \lim_{a \to 0} \frac{a}{a^2 + \omega^2} + \lim_{a \to 0} \frac{-j\omega}{a^2 + \omega^2}. \tag{3.211}$$

<u>Hint:</u> The first term approaches $\delta(f)/2$; i.e.,

$$\lim_{a \to 0} \frac{a}{a^2 + \omega^2} = \begin{cases} \infty & f = 0 \\ 0 & f \neq 0 \end{cases} \quad \text{and} \quad \int_{-\infty}^{\infty} \frac{a}{a^2 + \omega^2} \, df = \frac{1}{2}, \tag{3.212}$$

This integral may be needed

$$\int \frac{dx}{a^2 + x^2} = \frac{1}{a} \tan^{-1}\left(\frac{x}{a}\right). \tag{3.213}$$

14. Find the Fourier spectra of the following functions:
(a) $e^{-at}u(t) \quad (a > 0)$
(b) $-e^{-at}u(-t) \quad (a < 0)$
(c) $e^{-a|t|} \quad (a > 0)$
(d) $\cos(\omega_0 t)e^{-at}u(t) \quad (a > 0)$
(e) $\sin(\omega_0 t)e^{-at}u(t) \quad (a > 0)$.

15. Find the Fourier spectra of the following functions, and plot the magnitude and phase of each spectrum using any software tool of your choice (e.g., Matlab). (These functions are used as some "mother wavelet functions" in wavelet transforms.)
(a) Shannon wavelet:

$$\psi_1(t) = \frac{1}{\pi t}[\sin(2\pi f_2 t) - \sin(2\pi f_1 t)]. \tag{3.214}$$

(b) Morlet wavelet:

$$\psi_2(t) = \frac{1}{\sqrt{2\pi}} e^{j\omega_0 t} e^{-t^2/2}. \tag{3.215}$$

(c) Marr (Mexican hat) wavelet:

$$\psi_3(t) = \frac{1}{\sqrt{2\pi\sigma^3}} \left(1 - \frac{t^2}{\sigma^2}\right) e^{-t^2/2\sigma^2}. \tag{3.216}$$

16. Find the Fourier spectrum of the following Gaussian modulated sinusoid:

$$x(t) = \cos(2\pi f_0 t)e^{-\pi(t/a)^2}. \tag{3.217}$$

17. The result of the previous problem can be generalized to a sinusoid $\cos(2\pi f_0 t)$ modulated by any signal $s(t)$, the *amplitude modulation (AM)* in radio broad-

casting. Assume $S(f) = \mathcal{F}[s(t)]$ is a triangle function

$$S(f) = 1 - \frac{|f|}{f_{\max}}, \tag{3.218}$$

where $f_{\max}$ is the highest frequency component contained in the signal $s(t)$. Obtain the spectrum $X(f)$ of the AM signal $x(t) = s(t)\cos(2\pi f_0 t)$ and plot $X(f)$ in the frequency domain.

Another signal $y(t) = x(t)\cos(2\pi f_0 t)$ can be generated as the AM version of $x(t)$. Find and plot $Y(f) = \mathcal{F}[y(t)]$.

18. In Eq. (3.174) we considered a convolution of an impulse train and another signal $x(t)$ of finite duration:

$$y(t) = x(t) * \sum_{n=-\infty}^{\infty} \delta(t - nT). \tag{3.219}$$

Here, we assume $x(t)$ is the triangle function given in Eq. (3.166) and its spectrum $X(f) = \tau\,\text{sinc}^2(f\tau)$ given in Eq. (3.167).

Plot (sketch) both the convolution $y(t)$ in the time domain and its spectrum $Y(f) = \mathcal{F}[y(t)]$ in the frequency domain (identify all points at which $X(f) = 0$) in these two situations: (a) $T > 2\tau$ and (b) $T < 2\tau$. What is the essential difference between these two cases in both the time and frequency domains? (Note that as both the triangle function and the impulse train are even functions, time-frequency duality applies; i.e., the time and frequency domains can be interchanged. Then what we observe here is the basis for the sampling theorem to be considered in Chapter 4.)

19. Show that the Fourier series expansion pair in Eq. (3.7) can be treated as a special case of the CTFT in Eq. (3.65). Do this in two steps:
    (a) First apply the forward CTFT (first equation of Eq. (3.65)) to the Fourier series of a periodic signal $x_T(t + T) = x_T(t)$ (first equation of Eq. (3.7)) to obtain its spectrum $X(f)$.
    (b) Next apply the inverse CTFT (second equation of Eq. (3.65)) to the resulting spectrum $X(f)$ obtained above to get the time signal $x(t)$ back. Verify that the result is indeed the Fourier series.

    <u>Hint:</u> Consider using Eq. (1.28).

4 Discrete-time Fourier transform

4.1 Discrete-time Fourier transform

4.1.1 Fourier transform of discrete signals

To process and analyze a continuous time signal $x(t)$ digitally, analog-to-digital conversion (ADC, A/D) is needed to discretize the signal so that it becomes a sequence of time samples $x[n] = x(nt_0) = x(n/F)$ $(n = 0, \pm 1, \pm 2, \ldots)$, where t_0 is the *sampling period*, the time interval between two consecutive samples, and $F = 1/t_0$ is the *sampling rate* or *sampling frequency*, the number of samples per unit time. The sampled signal $x_s(t)$ can be represented mathematically as the product of the signal and the sampling function, an impulse train (also called a Dirac comb):

$$x_s(t) = x(t)\ \text{comb}(t) = x(t) \sum_{n=-\infty}^{\infty} \delta(t - nt_0) = \sum_{n=-\infty}^{\infty} x[n]\delta(t - nt_0), \tag{4.1}$$

where $x[n] = x(nt_0) = x(n/F)$ is the nth sample of the signal $x(t)$ evaluated at $t = nt_0 = n/F$. The Fourier transform of this sampled signal is

$$\begin{aligned} X_F(f) = \mathcal{F}[x_s(t)] &= \int_{-\infty}^{\infty} \left[\sum_{n=-\infty}^{\infty} x[n]\delta(t - nt_0) \right] e^{-j2\pi ft}\, dt \\ &= \sum_{n=-\infty}^{\infty} x[n] \int_{-\infty}^{\infty} \delta(t - nt_0) e^{-j2\pi ft}\, dt = \sum_{n=-\infty}^{\infty} x[n] e^{-j2n\pi ft_0}. \end{aligned} \tag{4.2}$$

This is the Fourier spectrum of the discrete signal $x[n]$, the discrete-time Fourier transform (DTFT), which is periodic with the sampling frequency F as the period:

$$X_F(f + F) = \sum_{n=-\infty}^{\infty} x[n] e^{-j2n\pi(f+F)t_0}\, dt = \sum_{n=-\infty}^{\infty} x[n] e^{-j2n\pi ft_0}\, dt = X(f), \tag{4.3}$$

as $e^{-j2n\pi Ft_0} = e^{-j2n\pi} = 1$. Here, we have used the subscript F to indicate the spectrum has a period of F, to be distinguished from the non-periodic spectrum of the continuous time signal $x(t)$ before discretization, just as we used $x_T(t)$ to denote a periodic time signal with period T, to distinguish it from a non-periodic signal $x(t)$. However, such subscripts may be dropped for simplicity when no confusion will be caused. The relationship between the sampling period t_0 of a discrete signal and the period $F = 1/t_0$ of its Fourier spectrum is illustrated in Fig. 4.1.

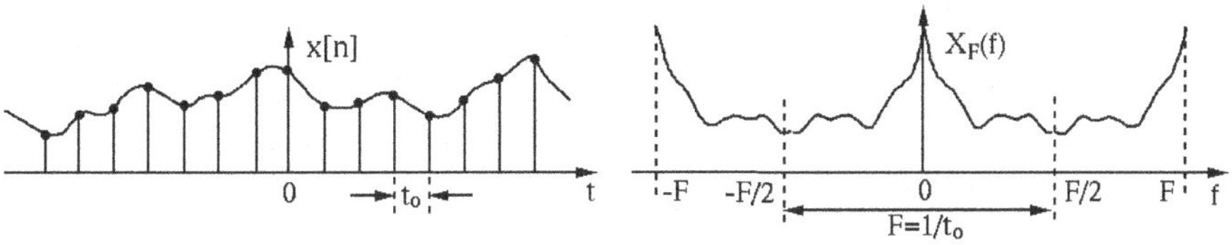

Figure 4.1 Fourier transform of discrete signals.

To get the time samples of the discrete signal back from its spectrum $X_F(f)$, we multiply $e^{j2m\pi f t_0}/F = e^{j2m\pi f/F}/F$ on both sides of Eq. 4.2 and integrate with respect to f over a period F:

$$\frac{1}{F}\int_0^F X_F(f)e^{j2m\pi f t_0}\,df = \frac{1}{F}\sum_{n=-\infty}^{\infty} x[n]\int_0^F e^{-j2(m-n)\pi f t_0}\,df$$
$$= \sum_{n=-\infty}^{\infty} x[n]\delta[n-m] = x[m] \qquad m = 0, \pm 1, \pm 2, \ldots, \tag{4.4}$$

where we have used Eq. (1.33) (with different variables names). This is the inverse DTFT. With a minor modification of the scaling factor $1/F$ of Eqs. (4.2) and (4.4), they can be written as a pair of the DTFT:

$$X_F(f) = \mathcal{F}[x[n]] = \frac{1}{\sqrt{F}}\sum_{n=-\infty}^{\infty} x[n]e^{-j2n\pi t_0 f} = \frac{1}{\sqrt{F}}\sum_{n=-\infty}^{\infty} x[n]e^{-j2n\pi f/F},$$
$$x[n] = \mathcal{F}^{-1}[X_F(f)] = \frac{1}{\sqrt{F}}\int_0^F X_F(f)e^{j2n\pi t_0 f}\,df = \frac{1}{\sqrt{F}}\int_0^F X_F(f)e^{j2n\pi f/F}\,df$$
$$n = 0, \pm 1, \pm 2, \ldots. \tag{4.5}$$

This is Eqs. (2.128) and (2.127). The first equation for $X_F(f)$ and the second equation for $x[n]$ are the forward and inverse DTFT, respectively. Comparing these equations with Eqs. (2.127) and (2.128), we see that the DTFT is actually the representation of a signal vector $\boldsymbol{x} = [\ldots, x[n], \ldots]^{\mathrm{T}}$ by a set of uncountably infinite orthonormal basis vectors:

$$\boldsymbol{\phi}(f) = [\ldots, e^{j2\pi n f/F}, \ldots]^{\mathrm{T}}/\sqrt{F} \qquad 0 < f < F, \tag{4.6}$$

satisfying Eq. (2.125):

$$\langle \boldsymbol{\phi}(f), \boldsymbol{\phi}(f') \rangle = \frac{1}{F}\sum_{n=-\infty}^{\infty} e^{j2\pi n(f-f')/F} = \delta(f-f'). \tag{4.7}$$

Now the vector $\boldsymbol{x} = [\ldots, x[n], \ldots]^{\mathrm{T}}$ in the vector space spanned by the basis can be expressed as a linear combination, an integral, of the basis vectors:

$$\boldsymbol{x} = \int_0^F X_F(f)\boldsymbol{\phi}(f)\, df, \tag{4.8}$$

the element form of which is the inverse DTFT in Eq. (4.5), and the coefficient function $X_F(f)$ can be found as the projection of the vector $\boldsymbol{x}$ onto the basis vector $\boldsymbol{\phi}(f)$:

$$X_F(f) = \langle \boldsymbol{x}, \boldsymbol{\phi}(f) \rangle = \boldsymbol{x}^{\mathrm{T}}\overline{\boldsymbol{\phi}}(f) = \frac{1}{\sqrt{F}} \sum_{n=-\infty}^{\infty} x[n] e^{-j2\pi nf/F}, \tag{4.9}$$

which is the forward DTFT in Eq. (4.5):

As a unitary transform, the DTFT also conserves inner product:

$$\begin{aligned}
\langle \boldsymbol{x}, \boldsymbol{y} \rangle &= \boldsymbol{x}^{\mathrm{T}}\overline{\boldsymbol{y}} = \sum_{n=-\infty}^{\infty} x[n]\overline{y}[n] \\
&= \sum_{n=-\infty}^{\infty} \left[\frac{1}{\sqrt{F}} \int_F X_F(f) e^{j2\pi n t_0 f}\, df \right] \left[\frac{1}{\sqrt{F}} \int_F \overline{Y}_F(f') e^{-j2\pi n t_0 f'}\, df' \right] \\
&= \int_F X_F(f) \int_F \overline{Y}_F(f') \left[\frac{1}{F} \sum_{m=-\infty}^{\infty} e^{-j2\pi m t_0 (f-f')} \right] df'\, df \\
&= \int_F X_F(f) \int_F \overline{Y}_F(f')\, \delta(f-f')\, df'\, df = \int_F X_F(f)\overline{Y}_F(f)\, df \\
&= \langle X_F(f), Y_F(f) \rangle.
\end{aligned} \tag{4.10}$$

This is Plancherel's identity. When $\boldsymbol{x} = \boldsymbol{y}$, we get Parseval's identity:

$$||\boldsymbol{x}||^2 = \langle \boldsymbol{x}, \boldsymbol{x} \rangle = \sum_{n=-\infty}^{\infty} |x[n]|^2 = \int_F |X_F(f)|^2\, df = \langle X_F(f), X_F(f) \rangle = ||X_F(f)||^2, \tag{4.11}$$

indicating that the energy contained in the signal is preserved by the DTFT.

Comparing the pair of equations in Eq. (4.5) with the Fourier series expansion of a periodic signal $x_T(t)$ in Eq. 3.5 we see a duality between the time and frequency domains:

- A continuous and periodic time signal $x_T(t)$ (with period $T = 1/f_0$) is a function in the space spanned by a set of countably infinite periodic functions $\phi_k(t) = e^{j2\pi k f_0 t}/\sqrt{T}$ $(k = 0, \pm 1, \pm 2, \ldots)$. Its spectrum is non-periodic and discrete (with a frequency interval $f_0 = 1/T$ between two consecutive frequency components).
- A non-periodic and discrete time signal $x[n]$ (with a time interval $t_0 = 1/F$ between two consecutive samples) is a vector in the space spanned by a set of uncountably infinite (a continuum of) vectors $\boldsymbol{\phi}(f) = [\ldots, e^{j2\pi n t_0 f}, \ldots]/\sqrt{F}$ $(0 \le f < F)$. Its spectrum $X_F(f)$ is continuous and periodic (with period $F = 1/t_0$).

This duality between time and frequency is obviously due to the symmetry in the most generic definition of the forward and inverse Fourier transforms in Eq. 3.65.

Equations (4.2) and (4.4) can also be expressed in terms of angular frequency $\omega = 2\pi f$ as

$$X_\Omega(\omega) = \sum_{n=-\infty}^{\infty} x[n] e^{-jn\omega t_0},$$

$$x[n] = \frac{1}{\Omega}\int_0^\Omega X_\Omega(\omega) e^{jn\omega t_0}\, d\omega \qquad n = 0, \pm 1, \pm 2, \ldots, \tag{4.12}$$

where $X_\Omega(\omega + \Omega)$ is the spectrum with period $\Omega = 2\pi F$. Moreover, once a continuous signal is sampled to become a sequence of discrete values, the sampling period t_0 may not be of interest anymore during the subsequent digital signal processing, and can be assumed to be $t_0 = 1$, then the sampling frequency also becomes unit $F = 1/t_0 = 1$, and the Fourier transform pair in Eq. (4.5) of the discrete signal can be simply expressed as

$$X(f) = \sum_{n=-\infty}^{\infty} x[n] e^{-j2n\pi f} = \sum_{n=-\infty}^{\infty} x[n] e^{-jn\omega} = X(\omega)$$

$$x[n] = \int_0^1 X(f) e^{j2n\pi f}\, df = \frac{1}{2\pi}\int_0^{2\pi} X(\omega) e^{jn\omega}\, d\omega \qquad n = 0, \pm 1, \pm 2, \ldots \tag{4.13}$$

Now the periodicity of the spectrum becomes $X(f+1) = X(f)$ or $X(\omega + 2\pi) = X(\omega)$.

In some of the literature, the DTFT spectrum $X(f)$ or $X(\omega)$ is also denoted by $X(e^{j\omega})$, because it takes this form when treated as a special case of the *z-transform*, to be discussed in Chapter 6. However, all these different forms are just some notational variations of the same spectrum, a function of frequency f or angular frequency $\omega = 2\pi f$. We will use these notations interchangeably, depending on whichever is most convenient and suitable in the specific discussion, as no confusion should be caused given the context.

Example 4.1: Here, we consider the DTFT of a few special discrete signals:

- The Kronecker delta or a discrete unit impulse $x[n] = \delta[n]$:

$$\mathcal{F}[\delta[n]] = \sum_{n=-\infty}^{\infty} \delta[n] e^{-j2n\pi f} = e^{-j2\pi 0 f} = 1. \tag{4.14}$$

- The constant function, a train of unit impulses, $x[n] = 1$:

$$\mathcal{F}[1] = \sum_{n=-\infty}^{\infty} e^{j2n\pi f} = \sum_{k=-\infty}^{\infty} \delta(f-k) = 2\pi \sum_{k=-\infty}^{\infty} \delta(\omega - 2k\pi). \tag{4.15}$$

Here, we have used Eq. (1.35). The spectrum is also an impulse train in the frequency domain.

- The discrete sign function is defined as

$$\operatorname{sgn}[n] = \begin{cases} -1 & n < 0 \\ 0 & n = 0 \\ 1 & n > 0 \end{cases}. \tag{4.16}$$

Its DTFT spectrum is

$$\mathcal{F}[\operatorname{sgn}[n]] = -\sum_{n=-\infty}^{-1} e^{-jn\omega} + \sum_{n=1}^{\infty} e^{-jn\omega} = -\sum_{m=1}^{\infty} e^{jn\omega} + \sum_{n=1}^{\infty} e^{-jn\omega}. \tag{4.17}$$

Consider the first summation as the following limit when the real parameter $0 < a < 1$ approaches zero:

$$\lim_{a\to 1}\left[-\sum_{n=1}^{\infty}(a\,e^{j\omega})^n\right] = \lim_{a\to 1}\left[a - \sum_{n=0}^{\infty}(a\,e^{j\omega})^n\right] = \lim_{a\to 1}\left[a - \frac{1}{1-a\,e^{j\omega}}\right]. \tag{4.18}$$

Similarly the second summation can be written as

$$\lim_{a\to 1}\left[\sum_{n=1}^{\infty}(a\,e^{-j\omega})^n\right] = \lim_{a\to 1}\left[\sum_{n=0}^{\infty}(a\,e^{-j\omega})^n - a\right] = \lim_{a\to 1}\left[\frac{1}{1-a\,e^{-j\omega}} - a\right]. \tag{4.19}$$

Note that in these limits we cannot simply replace a by one owing to the singularity at $\omega = 2k\pi$ for any integer k. However, we can do so to the sum of the two terms, which is an odd function and is zero at $\omega = 2k\pi$:

$$\mathcal{F}[\operatorname{sgn}[n]] = \lim_{a\to 1}\left[\frac{1}{1-a\,e^{-j\omega}} - \frac{1}{1-a\,e^{j\omega}}\right] = \frac{1+e^{-j\omega}}{1-e^{-j\omega}} = \frac{j\,\sin\omega}{\cos\omega - 1}. \tag{4.20}$$

- The unit step function is defined as

$$u[n] = \begin{cases} 0 & n < 0 \\ 1 & n \geq 0 \end{cases}. \tag{4.21}$$

Note that $u[0] = 1$, unlike $u(0) = 1/2$ in the continuous case. Following the DTFT definition above its spectrum can be directly obtained from Eq. (1.37).

$$\mathcal{F}[u[n]] = \sum_{n=0}^{\infty} e^{-j2\pi n f} = \frac{1}{1-e^{-j2\pi f}} + \frac{1}{2}\sum_{k=-\infty}^{\infty} \delta(f-k). \tag{4.22}$$

Alternatively, we can write $u[n]$ as

$$u[n] = \frac{1}{2}\left[1 + \delta[n] + \operatorname{sgn}[n]\right] = \begin{cases} 1 & n \geq 0 \\ 0 & n < 0 \end{cases}, \tag{4.23}$$

and carry out the Fourier transform on each of the three terms to get:

$$\mathcal{F}[u[n]] = \frac{1}{2}\mathcal{F}\left[1 + \delta[n] + \text{sgn}[n]\ \right]$$
$$= \frac{1}{2}\left[\sum_{k=-\infty}^{\infty} \delta(f-k) + 1 + \frac{1+e^{-j\omega}}{1-e^{-j\omega}}\right] = \frac{1}{1-e^{-j2\pi f}} + \frac{1}{2}\sum_{k=-\infty}^{\infty} \delta(f-k). \tag{4.24}$$

4.1.2 Properties of the DTFT

As one of the variations of the generic Fourier transform in Eq. (3.65), the DTFT shares all of the properties considered in the previous chapter, but in different forms. Here, we assume $X(f) = \mathcal{F}[x[n]]$ and $Y(f) = \mathcal{F}[y[n]]$. Proofs of many of these properties are not given as they can be considered as the special cases of the corresponding CTFT properties discussed previously, and they can also be easily derived from the definition. The reader is encouraged to prove them as homework problems.

- **Linearity**

$$\mathcal{F}[ax[n] + by[n]] = aX(f) + bY(f). \tag{4.25}$$

- **Periodicity**

$$X(f+k) = X(f), \tag{4.26}$$

where k is any integer.

- **Plancherel's identity and Parseval's identity**

$$\langle \boldsymbol{x}, \boldsymbol{y} \rangle = \sum_{n=-\infty}^{\infty} x[n]\overline{y}[n] = \int_0^1 X(f)\overline{Y}(f)\,df = \langle X(f), Y(f) \rangle. \tag{4.27}$$

This was given in Eq. (4.11). In particular, when $\boldsymbol{y} = \boldsymbol{x}$, the equation above becomes Parseval's identity:

$$||\boldsymbol{x}||^2 = \langle \boldsymbol{x}, \boldsymbol{x} \rangle = \sum_{n=-\infty}^{\infty} |x[n]|^2$$
$$= \int_0^1 |X(f)|^2\,df = \langle X(f), X(f) \rangle = ||X(f)||^2. \tag{4.28}$$

This is Eq. (4.11).

- **Complex conjugate**

$$\mathcal{F}[\overline{x}[n]] = \overline{X}(-f). \tag{4.29}$$

- **Time reversal**

$$\mathcal{F}[x[-n]] = X(-f). \tag{4.30}$$

Combining the above with the previous property, we also have

$$\mathcal{F}[\overline{x}[-n]] = \overline{X}(f). \tag{4.31}$$

In particular, if $\overline{x}[n] = x[n]$ is real, then

$$\mathcal{F}[x[-n]] = X(-f) = \overline{X}(f). \tag{4.32}$$

- **Time and frequency shift**

$$\mathcal{F}[x[n \pm n_0]] = e^{\pm j2\pi f n_0} X(f), \tag{4.33}$$
$$\mathcal{F}[e^{\mp j2\pi f_0 n} x[n]] = X(f \pm f_0). \tag{4.34}$$

- **Correlation**
The *cross-correlation* between two discrete signals $x[n]$ and $y[n]$ is defined in Eq. (1.46) as

$$r_{xy}[m] = x[n] \star y[n] = \sum_n x[n]\overline{y}[n-m]. \tag{4.35}$$

Its DTFT is

$$\mathcal{F}[r_{xy}[m]] = X(f)\overline{Y}(f) = S_{xy}(f), \tag{4.36}$$

where $S_{xy}(f) = X(f)\overline{Y}(f)$ is the *cross power spectral density* of the two signals. If both signals $\overline{x}[n] = x[n]$ and $\overline{y}[n] = y[n]$ are real, then we have

$$\mathcal{F}[r_{xy}[m]] = S_{xy}(f) = X(f)Y(-f). \tag{4.37}$$

In particular, when $x[n] = y[n]$, we have

$$\mathcal{F}[r_x[m]] = S_x(f) = X(f)\overline{X}(f) = |X(f)|^2, \tag{4.38}$$

where $r_x[m] = x[n] \star x[n]$ is the *autocorrelation* and $S_x(f) = |X(f)|^2$ is the PDS of the discrete signal $x[n]$.
- **Time and frequency convolution theorems**

$$\mathcal{F}\left[x[n] * y[n]\right] = X(f)\,Y(f), \tag{4.39}$$
$$\mathcal{F}\left[x[n]\,y[n]\right] = X(f) * Y(f). \tag{4.40}$$

Note that both $X(f+1) = X(f)$ and $Y(f+1) = Y(f)$ are periodic, and their convolution is called a circular or periodic convolution.
- **Time differencing**
Corresponding to the first-order derivative of a continuous signal $dx(t)/dt = \lim_{\triangle\to}[x(t+\triangle) - x(t)]/\triangle$, the first-order difference of a discrete signal is simply defined as $x[n] - x[n-1]$. Based on the time shift property, we have

$$\mathcal{F}\left[x[n] - x[n-1]\right] = (1 - e^{-j2\pi f})X(f). \tag{4.41}$$

- **Time accumulation**
Corresponding to the integral of a continuous signal, the accumulation of a discrete signal is a summation of all its samples $x[n]$ from $n = -\infty$ up to $n = m$, and its Fourier transform is

$$\mathcal{F}\left[\sum_{m=-\infty}^{n} x[m]\right] = \frac{1}{1-e^{-j2\pi f}}X(f) + \frac{X(0)}{2}\sum_{k=-\infty}^{\infty}\delta(f-k). \tag{4.42}$$

The accumulation can be expressed as the convolution

$$\sum_{m=-\infty}^{n} x[m] = \sum_{m=-\infty}^{\infty} u[n-m]x[m] = u[n] * x[n], \tag{4.43}$$

where $u[n-m] = 0$ if $m > n$.
The DTFT of this convolution can be easily found according to the time convolution property:

$$\mathcal{F}\left[\sum_{m=-\infty}^{n} x[m]\right] = \mathcal{F}\left[u[n] * x[n]\right]. \tag{4.44}$$

Comparing Eqs. (4.41) and (4.42), we see that differencing and accumulation are the inverse operations of each other, just like the continuous time derivative and integral which are also the inverse operations of each other (Eqs. (3.127) and (3.132)). The second term of the right-hand side in Eq. (4.42) represents the DC component in the signal $x[n]$, which is not needed in Eq. (4.41) as the differencing operation is insensitive to the DC component.
- **Frequency differentiation**

$$\mathcal{F}[n\,x[n]] = \frac{j}{2\pi}\frac{d}{df}X(f). \tag{4.45}$$

- **Modulation**
Here, modulation means every odd sample of the signal $x[n]$ is negated.

$$\mathcal{F}\left[(-1)^n x[n]\right] = X\left(f + \frac{1}{2}\right) = X\left(f - \frac{1}{2}\right). \tag{4.46}$$

Proof: If we let $f_0 = 1/2$ in Eq. (4.34) for the frequency shift property, and note $e^{j2n\pi f_0} = e^{jn\pi} = (-1)^n$, we get Eq. (4.46).
- **Down-sampling**

$$\mathcal{F}\left[x_{(2)}[n]\right] = \mathcal{F}\left[x[2n]\right] = \frac{1}{2}\left[X\left(\frac{f}{2}\right) + X\left(\frac{f+1}{2}\right)\right]. \tag{4.47}$$

Here, the down-sampled version $x_{(2)}[n]$ of a signal $x[n]$ is composed of all the even terms of the signal with all odd terms dropped; i.e., $x_{(2)}[n] = x[2n]$.

Down-sampling of a discrete signal corresponds to the compression of a continuous signal (Eq. (3.107) with $a = 2$):

$$\mathcal{F}[x(2t)] = \frac{1}{2}X\left(\frac{f}{2}\right). \tag{4.48}$$

Proof:

$$\begin{aligned}\mathcal{F}[x_{(2)}[n]] &= \sum_{n=-\infty}^{\infty} x[2n]e^{-j2\pi nf} = \sum_{m=\cdots,-2,0,2,\cdots} x[m]e^{-j\pi mf} \\ &= \frac{1}{2}[\sum_{m=-\infty}^{\infty} x[m]e^{-j\pi mf} + \sum_{m=-\infty}^{\infty} (-1)^m x[n]e^{-j\pi mf}] \\ &= \frac{1}{2}[\sum_{m=-\infty}^{\infty} x[m]e^{-j\pi mf} + \sum_{m=-\infty}^{\infty} x[m]e^{-j\pi m(f+1)}] \\ &= \frac{1}{2}\left[X\left(\frac{f}{2}\right) + X\left(\frac{f+1}{2}\right)\right].\end{aligned} \tag{4.49}$$

Conceptually, the down-sampling of a given discrete signal $x[n]$ can be realized in the following three steps:

- Obtain its modulation $x[n](-1)^n = x[n]e^{jn\pi}$. Owing to the frequency shift property, this corresponds to the spectrum shifted by 1/2:

$$\mathcal{F}\left[(-1)^n x[n]\right] = \mathcal{F}\left[e^{jn\pi}x[n]\right] = X(f+1/2). \tag{4.50}$$

- Obtain the average of the signal and its modulation in both the time and frequency domains:

$$\mathcal{F}\left[\frac{1}{2}[x[n] + x[n](-1)^n]\right] = \frac{1}{2}\left[X(f) + X\left(f + \frac{1}{2}\right)\right]. \tag{4.51}$$

- Remove odd samples of the average to get $x_{(2)}[n]$. In the frequency domain, this corresponds to replacing f by $f/2$:

$$\mathcal{F}\left[x_{(2)}[n]\right] = \frac{1}{2}\left[X(\frac{f}{2}) + X\left(\frac{f+1}{2}\right)\right]. \tag{4.52}$$

- **Up-sampling (time expansion)**

$$\mathcal{F}[x^{(k)}[n]] = X(kf). \tag{4.53}$$

Here, $x^{(k)}[n]$ is defined as

$$x^{(k)}[n] = \begin{cases} x[n/k] & \text{if } n \text{ is a multiple of } k \\ 0 & \text{else} \end{cases}; \tag{4.54}$$

i.e. $x^{(k)}[n]$ is obtained by inserting $k-1$ zeros between every two consecutive samples of $x[n]$. Correspondingly its spectrum $X(kf)$ in the frequency domain is compressed k times with the same magnitude. Note that up-sampling is similar but different from the time scaling of a continuous signal in Eq. (3.107) with $a = 1/k$: $\mathcal{F}[x(t/k)] = kX(kf)$, in which case the signal $x(t)$ is expanded by k, and consequently its Fourier spectrum $X(f)$ is compressed by k, while its magnitude is also scaled up by k.

Proof:

$$\mathcal{F}\left[x^{(k)}[n]\right] = \sum_{n=-\infty}^{\infty} x[n/k]e^{-j2n\pi f} = \sum_{m=-\infty}^{\infty} x[m]e^{-j2km\pi f/k} = X(kf). \quad (4.55)$$

Here, we have assumed $m = n/k$ and this change of the summation index has no effect as the terms skipped are all zeros.

Combining both down- and up-samplings above, we see that if a signal $x[n]$ with $X(f) = \mathcal{F}[x[n]]$ is first down-sampled and then up-sampled, its DTFT transform is

$$\mathcal{F}[(x_{(2)})^{(2)}[n]] = \frac{1}{2}\left[X(f) + X\left(f + \frac{1}{2}\right)\right]. \quad (4.56)$$

Example 4.2: According to the convolution theorem, the convolution $y[n] = h[n] * x[n]$ of the two sequences $\boldsymbol{x} = [\ldots, 0, 1, 2, 3, 4, 5, 6, 7, 8, 0, \ldots]^{\mathrm{T}}$ and $\boldsymbol{h} = [\ldots, 0, 1, 2, 3, 0, \ldots]^{\mathrm{T}}$ in Example 1.139 can also be carried out in the frequency domain as the product of their DTFT spectra $Y(f) = H(f)X(f)$, where $X(f) = \mathcal{F}[x[n]]$ and $H(f) = \mathcal{F}[h[n]]$, and then the convolution in the time domain can be obtained by the inverse DTFT $y[n] = \mathcal{F}^{-1}[Y(f)]$. This process is shown in Fig. 4.2, where $x[n]$, $h[n]$, and $y[n] = x[n] * h[n]$ are shown on the left while their spectra $X(f)$, $H(f)$, and $Y(f)$ are shown on the right. After the inverse DTFT of $Y(f)$, we get $\boldsymbol{y} = [\ldots, 0, 1, 4, 10, 16, 22, 28, 34, 40, 37, 24, 0, \ldots]^{\mathrm{T}}$.

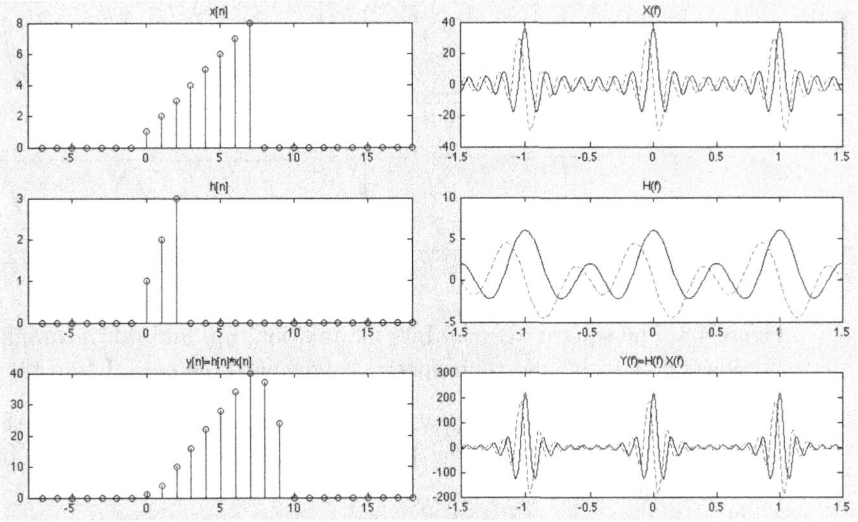

Figure 4.2 Convolution in time and multiplication in frequency. $x[n] * h[n] = \mathcal{F}^{-1}[\mathcal{F}[[x[n]]\ \mathcal{F}[h[n]]]]$. The real and imaginary parts of the spectra are shown in solid and dashed curves respectively.

Example 4.3: Here, we consider the up-sampling, modulation, and down-sampling of a discrete signal of square wave $x[n]$ with seven non-zero samples, as shown in Fig. 4.3.

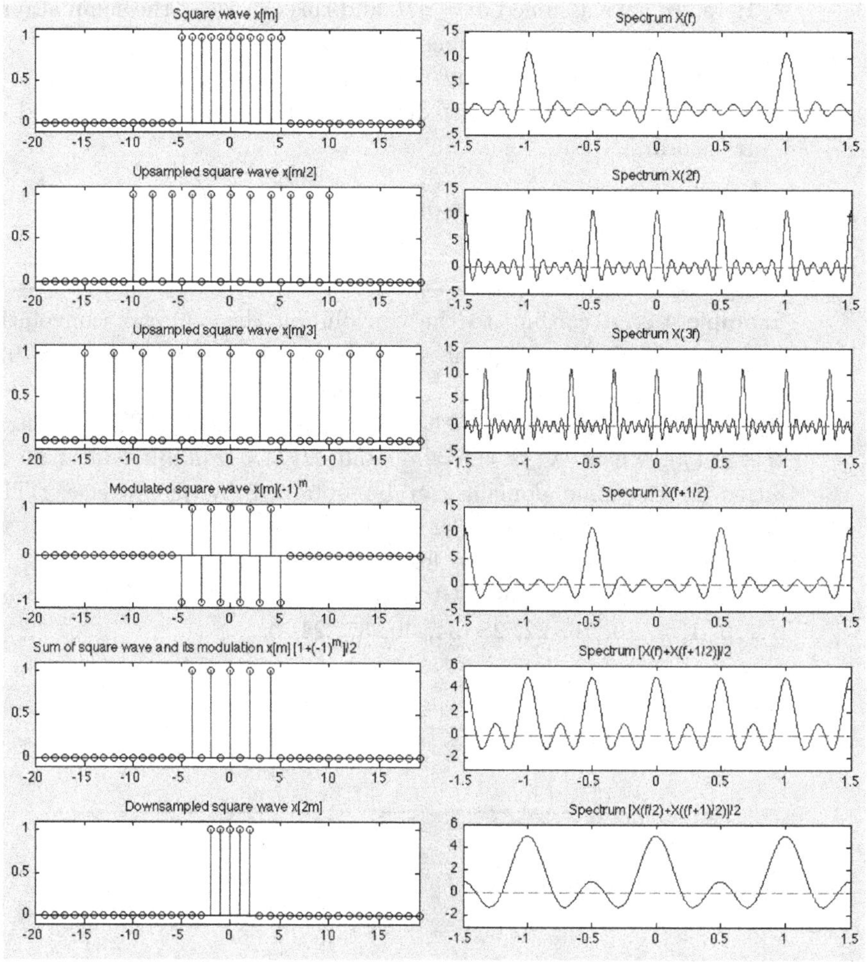

Figure 4.3 The square wave and its modulation, up- and down-sampling versions on the left, and their spectra (showing three periods) on the right.

- The square wave and its spectrum, a sinc function, are shown in the first row of the figure. Note that the DC component is 11, the number of non-zero samples in the signal $x[n]$.
- The up-sampled version $x^{(2)}[n]$ of the signal in both the time and frequency domains are shown in the second row. Note that, unlike time expansion of continuous signals, here the magnitude of the spectrum is not scaled by up-sampling.

- The up-sampled version $x^{(3)}[n]$ of the signal in both the time and frequency domains are shown in the third row.
- The modulation of the signal is shown in the fourth row. Note that all odd-numbered samples are negated and correspondingly the spectrum is shifted by 1/2 (compared with the first plot). Also its DC component is -1 (five positive samples and six negative samples in the time domain).
- The average of the signal (first row) and its modulation (fourth row) are shown in the fifth row. Note that the odd-numbered sampled becomes zero. In the frequency domain, the spectrum is also the average of the corresponding spectra owing to linearity.
- Finally, as shown in the last row, the time signal is compressed by a factor of 2 with all odd-numbered samples (all of value zero) dropped. Correspondingly, the spectrum is expanded by the same factor of 2.

4.1.3 DTFT of typical functions

- **Constant**
  If $x[n] = 1$ in Eq. (4.1), we get an impulse train in the time domain:

$$x_s(t) = \text{comb}(t) = \sum_{n=-\infty}^{\infty} \delta(t - nt_0) = \sum_{n=-\infty}^{\infty} \delta(t - n/F); \tag{4.57}$$

  i.e., $x[n] = 1$ and its discrete-time Fourier transform is also an impulse train in the frequency domain:

$$\mathcal{F}[x[n]] = \mathcal{F}[1] = \sum_{n=-\infty}^{\infty} e^{j2n\pi f} = \sum_{k=-\infty}^{\infty} \delta(f - k). \tag{4.58}$$

  The last equal sign is due to Eq. (1.35).
- **Complex exponential**
  Applying the frequency shift property to the previous result we get

$$\mathcal{F}[e^{j2n\pi f_0}] = \sum_{k=-\infty}^{\infty} \delta(f - f_0 - k). \tag{4.59}$$

  Letting $f_0 = 0$, we get the same result as in Eq. (4.58).
- **Sinusoids**

$$\begin{aligned} \mathcal{F}[\cos(2n\pi f_0)] &= \frac{1}{2}[\mathcal{F}[e^{j2n\pi f_0}] + \mathcal{F}[e^{-j2n\pi f_0}]] \\ &= \frac{1}{2}\left[\sum_{k=-\infty}^{\infty} \delta(f - f_0 - k) + \sum_{k=-\infty}^{\infty} \delta(f + f_0 - k)\right]. \end{aligned} \tag{4.60}$$

Similarly, we have

$$\mathcal{F}[\sin(2n\pi f_0)] = \frac{1}{2j}[\mathcal{F}[e^{j2n\pi f_0}] - \mathcal{F}[e^{-j2n\pi f_0}]]$$
$$= \frac{1}{2j}\left[\sum_{k=-\infty}^{\infty}\delta(f - f_0 - k) - \sum_{k=-\infty}^{\infty}\delta(f + f_0 - k)\right]. \quad (4.61)$$

- **Kronecker delta**

$$\mathcal{F}\left[\delta[n]\right] = \sum_{n=-\infty}^{\infty}\delta[n]e^{j2n\pi f} = e^{j0} = 1. \quad (4.62)$$

- **Sign function**

$$\mathcal{F}\left[\mathrm{sgn}[n]\right] = \frac{-e^{j2\pi f}}{1 - e^{j2\pi f}} + \frac{e^{-j2\pi f}}{1 - e^{-j2\pi f}} = \frac{1 + e^{-j2\pi f}}{1 - e^{-j2\pi f}} = \frac{j\ \sin\omega}{\cos\omega - 1}. \quad (4.63)$$

This is given in Eq. (4.20).

- **Unit step function**

$$\mathcal{F}\left[u[n]\right] = \frac{1}{1 - e^{-j2\pi f}} + \frac{1}{2}\sum_{k=-\infty}^{\infty}\delta(f - k). \quad (4.64)$$

This is given in Eq. (4.24).

- **Exponential decay**

First, consider a right-sided exponential function

$$x[n] = a^n u[n] \qquad (|a| < 1), \quad (4.65)$$

$$\mathcal{F}\left[a^n u[n]\right] = \sum_{n=0}^{\infty}(ae^{-j2\pi f})^n = \frac{1}{1 - a\,e^{-j2\pi f}}. \quad (4.66)$$

Next, consider the two-sided version

$$x[n] = a^{|n|} = a^n u[n] + a^{-n}u[-n-1] \qquad (|a| < 1). \quad (4.67)$$

The transform of the first term is the same as before, while the transform of the second term is

$$\mathcal{F}\left[a^{-n}u[-n-1]\right] = \sum_{m=-\infty}^{-1} a^{-n}e^{-j2n\pi f}$$
$$= \sum_{n=0}^{\infty}(ae^{j2\pi f})^n - 1 = \frac{a\,e^{j2\pi f}}{1 - a\,e^{j2\pi f}}. \quad (4.68)$$

The overall transform is

$$\mathcal{F}\left[a^{|n|}\right] = \frac{1}{1 - a\,e^{-j2\pi f}} + \frac{a\,e^{j2\pi f}}{1 - a\,e^{j2\pi f}} = \frac{1 - a^2}{1 + a^2 - 2a\cos(2\pi f)}. \quad (4.69)$$

- **Square wave**

$$x[n] = \begin{cases} 1 & |n| \le N \\ 0 & |n| > N \end{cases}. \tag{4.70}$$

The Fourier transform of this square wave of width $2N+1$ is found to be

$$\begin{aligned} \mathcal{F}[x[n]] &= \sum_{n=-N}^{N} e^{-jn\omega} = \sum_{n=-N}^{0} e^{-jn\omega} + \sum_{n=0}^{N} e^{-jn\omega} - 1 \\ &= \frac{1-e^{j(N+1)\omega}}{1-e^{j\omega}} + \frac{1-e^{-j(N+1)\omega}}{1-e^{-j\omega}} - 1 = \frac{e^{j(N+1)\omega} - e^{-jN\omega}}{e^{j\omega}-1} \frac{e^{-j\omega/2}}{e^{-j\omega/2}} \\ &= \frac{e^{j(2N+1)\omega/2} - e^{-j(2N+1)\omega/2}}{e^{j\omega/2} - e^{-j\omega/2}} = \frac{\sin((2N+1)\omega/2)}{\sin(\omega/2)}. \end{aligned} \tag{4.71}$$

- **Triangle wave**

$$x[n] = \begin{cases} 1 - |n|/N & |n| \le N \\ 0 & |n| > N \end{cases}. \tag{4.72}$$

This triangle wave function with width $2N+1$ can be constructed as the convolution of two square wave functions of width N, scaled down by N; therefore, its transform can be found by convolution property to be

$$\mathcal{F}[x[n]] = \frac{1}{N} \left[\frac{\sin(N\omega/2)}{\sin(\omega/2)} \right]^2. \tag{4.73}$$

- **Sinc function**

$$x[n] = \frac{\sin(2n\pi f_0)}{n\pi} = \frac{\sin(n\omega_0)}{n\pi}. \tag{4.74}$$

First, consider a square function in frequency:

$$X(\omega) = \begin{cases} 1 & |\omega| \le \omega_0 \\ 0 & |\omega| > \omega_0 \end{cases}. \tag{4.75}$$

The inverse transform of $X(\omega)$ is

$$\mathcal{F}^{-1}[X(\omega)] = \frac{1}{2\pi} \int_{-\omega_0}^{\omega_0} e^{jm\omega}\, d\omega = \frac{1}{2\pi} \frac{1}{jn} [e^{jn\omega_0} - e^{-jn\omega_0}] = \frac{\sin(n\omega_0)}{n\pi}; \tag{4.76}$$

i.e.,

$$\mathcal{F}\left[\frac{\sin(n\omega_0)}{n\pi} \right] = \begin{cases} 1 & |\omega| < \omega_0 \\ 0 & |\omega| > \omega_0 \end{cases}. \tag{4.77}$$

Fig. 4.4 shows a set of typical discrete signals and their DTFT spectra.

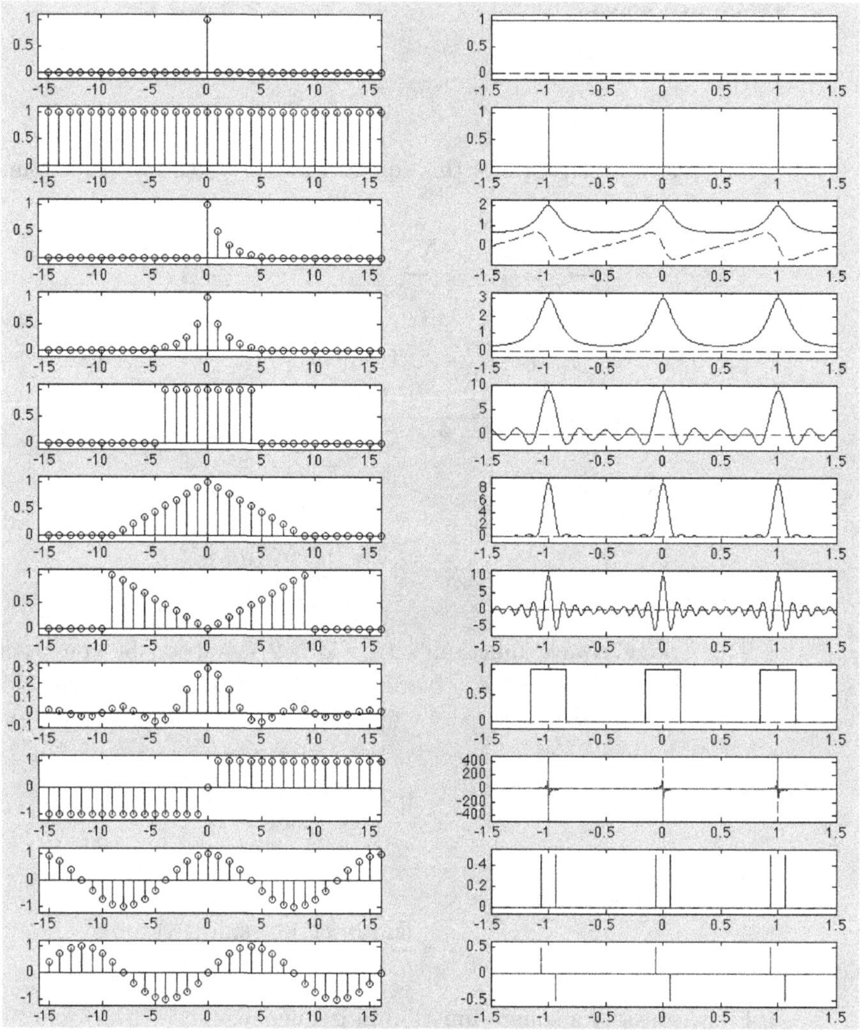

Figure 4.4 A set of discrete signals are shown on the left and their DTFT spectra (three periods shown) are shown on the right (real and imaginary parts are shown in solid and dashed lines, respectively).

4.1.4 The sampling theorem

An important issue in the sampling process in Eq. (4.1) is the determination of the sampling frequency $F = 1/t_0$. On the one hand, it is desirable to minimize the sampling frequency to reduce the data size for lower computational complexity in the subsequent digital signal processing and less space and time are needed for data storage and transmission. On the other hand, the sampling frequency cannot be too low as this may cause certain information contained in the signal to be lost.

This issue can be more conveniently addressed in the frequency domain. We know a time signal $x(t)$ can be perfectly reconstructed from its Fourier spectrum $X(f) = \mathcal{F}[x(t)]$ as its information is equivalently contained in either the time or frequency domain (Parseval's identity). However, after sampling by which $x(t)$ is represented by a sequence of samples $x[n]$ $(n = 0, \pm 1, \pm 2, \ldots)$, can $x(t)$ still be perfectly reconstructed from the spectrum $X_F(f) = \mathcal{F}[x[n]]$?

To answer this question, we consider how the spectrum $X_F(f)$ of the sampled signal $x_s(t)$ is related to the spectrum $X(f)$ of the original signal $x(t)$. Owing to the convolution theorem, the spectrum of the sampled signal $x_s(t) = x(t)\ \text{comb}(t)$ in Eq. (4.1) is the following convolution in the frequency domain:

$$\begin{aligned} X_F(f) &= \mathcal{F}[x(t)\ \text{comb}(t)] = X(f) * \text{Comb}(f) = X(f) * F \sum_{k=-\infty}^{\infty} \delta(f - kF) \\ &= \int_{-\infty}^{\infty} X(f - f') F \sum_{k=-\infty}^{\infty} \delta(f' - kF)\, df' = F \sum_{k=-\infty}^{\infty} X(f - kF), \quad (4.78) \end{aligned}$$

where $\text{Comb}(f) = F \sum_{k=-\infty}^{\infty} \delta(f - kF)$ is the spectrum of the comb function (Eq. 3.173). We see that the spectrum $X_F(f)$ of the sampled signal is a superposition of infinitely many shifted (by kF) and scaled (by F) replicas of the spectrum $X(f)$ of $x(t)$. Obviously if $X(f)$ can be recovered from $X_F(f)$, then $x((t)$ can be reconstructed from $X(f)$.

Consider the following two cases, also illustrated in Fig. 4.5, where the signal $x(t)$ is *band-limited*; i.e., the highest frequency component contained in the signal is f_{max}; i.e., $X(f) = \mathcal{F}[x(t)] = 0$ for any $|f| > f_{\text{max}}$.

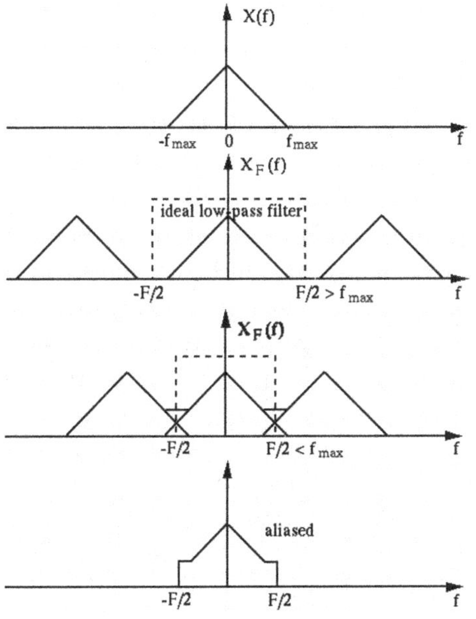

Figure 4.5 Reconstruction of time signal in the frequency domain.

- If $F/2 > f_{\max}$, the neighboring replicas in $X_F(f)$ are separated (second plot) and the original spectrum $X(f)$ (first plot) can be perfectly recovered by a filtering process:

$$X(f) = H_{\text{lp}} X_F(f), \tag{4.79}$$

where $H_{\text{lp}}(f)$ is an ideal LP filter defined as

$$H_{\text{lp}}(f) = \begin{cases} 1/F & |f| < f_c = F/2 \\ 0 & \text{else} \end{cases}. \tag{4.80}$$

This filter scales all frequencies lower than the cutoff frequency $f_c = F/2$ by a factor $1/F = t_0$ but suppresses to zero any frequency higher than $f_c = F/2$.

- If $F/2 < f_{\max}$, then the neighboring replicas in $X_F(f)$ overlap with each other and can no longer be separated. It is impossible to recover $X(f)$ by the LP filtering process, as the output of the ideal filter (last plot) is distorted owing to the overlapping replicas in $X_F(f)$ (third plot). For example, the highest frequency $f_{\max}$ in the signal now appears as a lower frequency $F - f_{\max}$, as if it is folded around $f = F/2$. This phenomenon is called *aliasing* or *folding*, to be further discussed later.

The above result leads to the well-known *sampling theorem*, also called the *Nyquist-Shannon theorem*:

Theorem 4.1. *A signal can be completely reconstructed from its samples taken at a sampling frequency F, if it contains no frequencies higher than $F/2$, referred to as the Nyquist frequency:*

$$f_{max} < f_{Nyquist} = F/2; \qquad i.e. \qquad F > 2 f_{max}. \tag{4.81}$$

This equation is referred to as the Nyquist condition for perfect signal reconstruction.

Now we can answer the original question regarding the proper sampling frequency. The lowest sampling frequency F at which the signal can be sampled without losing any information must be higher than twice the maximum frequency contained in the signal; i.e., $F > 2f_{\max}$, otherwise aliasing or folding will occur and the original signal cannot be perfectly reconstructed. In practice, it is often the case that the signal to be sampled does contain frequency components higher than the Nyquist frequency. To avoid aliasing in such cases, an anti-aliasing LP filtering can be carried out to remove all frequencies higher than the Nyquist frequency before sampling. However, certain signal information contained in the filtered out frequency components is lost in this process.

To fully understand the sampling theorem, we consider the following examples that serve to illustrate the various effects of the sampling process when the Nyquist condition is either satisfied or not.

Example 4.4: Consider a sinusoidal signal

$$x(t) = \sin(2\pi f_0 t) = \frac{1}{2j}[e^{j2\pi f_0 t} - e^{-j2\pi f_0 t}]. \quad (4.82)$$

There are two frequency components in its Fourier spectrum $X(f)$, one at $f = f_0 > 0$ to the right of the center $f = 0$ and another at $f = -f_0 < 0$ to the left of the center. When this signal is sampled at a rate of $F = 4$ samples per second, it becomes a sequence of numbers $x[n]$ separated by the sampling period of $t_0 = 1/F = 1/4$:

$$x[n] = x(t)\big|_{t=nt_0} = x(nt_0) = x(n/F) = x(n/4) = \frac{1}{2j}[e^{j2n\pi f_0/4} - e^{-j2n\pi f_0/4}]. \quad (4.83)$$

Correspondingly the spectrum of this sampled signal becomes periodic $X_F(f + F) = X_F(f)$ containing infinite replicas of the two components at $f = \pm f_0 \pm kF$. This process can also be modeled by the observation of an object rotating counterclockwise at f_0 cycles per second when illuminated only by a strobe light at a fixed rate of $F = 4$ flashes per second (similar to a wagon wheel in a movie), as illustrated in Fig. 4.6.

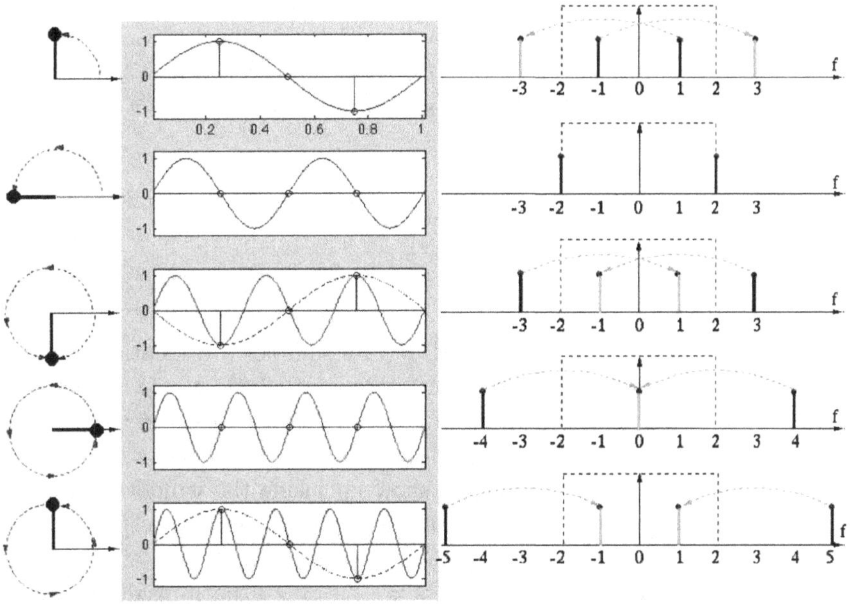

Figure 4.6 Aliasing and folding in time and the frequency domains. Model of rotating object illuminated by a strobe light (left, only the first flash is shown), sampling of the vertical displacement (middle), and the aliased frequency (perceived rotation) (right).

We consider the following five cases of different signal frequencies f_0, and observe where the two frequency components $\pm f_0$ and their replicas $\pm f \pm kF$ appear in the frequency domain. Note that only those frequency components inside the central period, a window determined by the Nyquist frequency $-F/2 \leq f \leq F/2 = 2$ (where $F = 4$), are observed (perceivable).

- $f_0 = 1 < F/2 = 2$:

$$x[n] = x(n/4) = \frac{1}{2j}(e^{j2n\pi/4} - e^{-j2n\pi/4}) = \sin(2n\pi/4). \tag{4.84}$$

  The two frequency components $f = \pm f_0 = \pm 1$ on both right and left are inside the central window $-2 < f < 2$. The replicas of these two frequency components also appear at $f = \pm 1 + 4k$ for any integer k. In our model the object is rotating at a rate of $f_0 = 1$ cycles per second or $90°$ per flash counterclockwise, as shown in the first row of Fig. 4.6.
- $f_0 = 2 = F/2 = 2$:

$$x[n] = x(n/4) = \frac{1}{2j}(e^{j2n\pi 2/4} - e^{-j2n\pi 2/4}) = \frac{1}{2j}[e^{jn\pi} - e^{-jn\pi}] = 0. \tag{4.85}$$

  The signal is sampled twice per period and both samples happen to be zero in this case, as if the samples were taken from a zero signal $x(t) = 0$. In the frequency domain the right and left components are at $f = \pm 2 = \pm F/2$, on the edges of the window. In our model, the object is rotating at a rate of $180°$ per flash, when the vertical displacement of the object happens to be zero, as if it is not rotating. This is shown in the second row of Fig. 4.6.
- $f_0 = 3 > F/2 = 2$:

$$\begin{aligned} x[n] = x(n/4) &= \frac{1}{2j}(e^{j2n\pi 3/4} - e^{-j2n\pi 3/4}) \\ &= \frac{1}{2j}(e^{-j2n\pi/4} - e^{j2n\pi 1/4}) = -\sin(2n\pi/4). \end{aligned} \tag{4.86}$$

  The signal is undersampled and its samples are identical to those obtained from a different signal $-\sin(2\pi t) = \sin(-2\pi t)$ at a frequency $f_0 = 1$. In the frequency domain, the two frequency components at $f = \pm 3$ are both outside the window $-2 < f < 2$, but some of the components of the replicas at $f = \pm f_0 \pm kF = \pm 3 \pm 4k$ could show up inside the window. Specifically the left component for $-f_0 = -3$ of the replica centered at $f = F = 4$ appears at $f = 4 - 3 = 1$, and the right component for $f_0 = 3$ of the replica centered at $f = -F = -4$ appears at $f = -4 + 3 = 1$, both inside the window $-F/2 < f < F/2$, as if they are respectively the right and left components of a sinusoid of frequency $f_0 = 1$ but with opposite polarity. This effect is called *folding*. In the model, the object is rotating at a rate of $270°$ per flash, but it appears to be rotating at a lower rate of $90°$ per flash in the opposite clockwise direction, as shown in the third row of Fig. 4.6.

- $f_0 = 4 = F$:

$$x[n] = x(n/4) = \frac{1}{2j}(e^{j2n\pi 4/4} - e^{-j2n\pi 4/4}) = \frac{1}{2j}[e^{j2n\pi} - e^{-j2n\pi}] = 0. \quad (4.87)$$

  The signal is sampled once per period, the samples are necessarily constant, which is zero in this case. In the frequency domain, the two components are at $f = \pm f_0 = \pm 4$, both outside the window, but their neighboring replicas both appear at the origin at $f = 0$. In the model, the rotating object happens to be at the same position when illuminated, with zero vertical displacement in this case, as if it is standing still, as shown in the fourth row of Fig. 4.6.

- $f_0 = 5 > F/2 = 2$:

$$\begin{aligned} x[n] = x(n/4) &= \frac{1}{2j}(e^{j2n\pi 5/4} - e^{-j2n\pi 5/4}) \\ &= \frac{1}{2j}(e^{j2n\pi/4} - e^{-j2n\pi/4}) = \sin(2n\pi/4). \end{aligned} \quad (4.88)$$

  The samples are identical to those taken from a different signal $\sin(2\pi t)$ with frequency $f_0 = 1$. In the frequency domain, the two components at $f = \pm f_0 = \pm 5$ are both outside the window $-2 < f < 2$, but some components of their replicas could show up inside the window. Specifically the left component for $-f_0 = -5$ of the replica centered at $F = 4$ appears at $f = 4 - 5 = -1$, and the right component for $f_0 = 5$ of the replica centered at $f = -F = -4$ appears at $f = -4 + 5 = 1$, both inside the central $-F/2 < f < F/2$, as if they are respectively the left and right components of a sinusoid of frequency $f_0 = 1$ with the same polarity. This effect is called *aliasing*. In the model, the object rotating at a rate of 450° per flash appears to rotate 90° per flash in the same counterclockwise direction, as shown in the last row of Fig. 4.6.

Note that, in all these cases, the observed frequency f is always the replicas of the lowest frequency inside the window $-F/2 < f < F/2$ in the spectrum, which is the same as the true signal frequency $f = f_0$ only when $f_0 < F/2$. Otherwise, aliasing or folding occurs and the apparent frequency is always lower than the true frequency. In the model of a rotating object, even if we know the object could have rotated an angle of $\phi \pm 2k\pi$ per flash, the perceived frequency by our visual system is always either ϕ or $\phi - 2\pi = -(2\pi - \phi)$ per flash, depending on which has a lower absolute value. In the latter case, as the polarity is changed, not only does the frequency appear to be lower, but also the direction is reversed.

In the marginal case where the signal frequency $f_0 = F/2$ is equal to the Nyquist frequency, the sampled signal may appear zero, as shown above, but this is not necessarily the case in general. Consider the same signal as above with a phase shift $x(t) = \sin(2\pi f_0 t + \phi)$. When it is sampled at exactly the rate of $F = 2f_0$, the values of its samples depend on the phase ϕ:

$$x[n] = x(n/F) = \sin(2n\pi f_0/F + \phi) = \sin(n\pi + \phi). \quad (4.89)$$

This is indeed zero when $\phi = 0$, as shown before. However, when $\phi \neq 0$, we have

$$x[n] = \sin(n\pi + \phi) = \begin{cases} \sin\phi & n \text{ is even} \\ -\sin\phi & n \text{ is odd} \end{cases}. \tag{4.90}$$

In other words, in the marginal case when $f_0 = F/2$, so long as $\phi \neq 0$ and $\phi \neq \pi$, the sign of $x[n]$ alternates and the frequency f_0 of $x(t)$ can be accurately represented, but its amplitude is scaled by $\sin\phi$, and its phase ϕ is not reflected, as shown in Fig. 4.7. In particular, when $\phi = \pi/2$, $x[n] = 1$ if n is even and $x[n] = -1$ if n is odd; i.e., the amplitude of the signal is accurately represented by its samples.

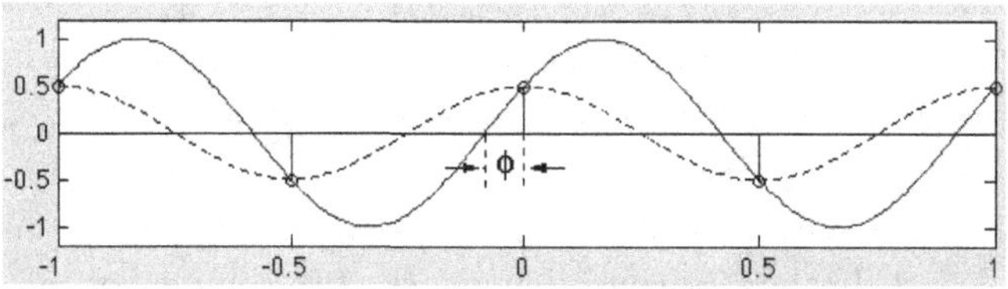

Figure 4.7 Marginal sampling: signal frequency equals Nyquist frequency $f_0 = F/2$.

Example 4.5: This example further illustrates the effect of sampling and aliasing/folding. Consider a continuous signal

$$x(t) = \cos(2\pi f t + \phi) = \frac{1}{2}[e^{j(2\pi f t+\phi)} + e^{-j(2\pi f t+\phi)}] = c_1 e^{j2\pi f t} + c_{-1} e^{-j2\pi f t}, \tag{4.91}$$

where $c_1 = e^{j\phi}/2$ and $c_{-1} = e^{-j\phi}/2$ are respectively the two non-zero coefficients for the frequency components $e^{j2\pi f t}$ and $e^{-j2\pi f t}$. When this signal is sampled at a rate of $F = 1/t_0$, it becomes a discrete signal:

$$x[n] = \cos(2\pi f n t_0 + \phi) = \cos(2\pi f n/F + \phi) = \frac{e^{j\phi}}{2} e^{j2\pi f n/F} + \frac{e^{-j\phi}}{2} e^{-j2\pi f n/F}$$
$$= c_1 e^{-j2\pi f n/F} + c_{-1} e^{-j2\pi f n/F}.$$

Fig. 4.8 shows the signal being sampled at $F = 6$ samples per second, while its frequency f increases from 1 to 12 with increment of 1. In the time domain (left), the original signal (solid line) and the reconstructed one (dashed line) are both plotted. In the frequency domain (right), the spectrum of the sampled version of the signal is periodic with period $F = 6$, and three periods are shown, including two neighboring periods on both the positive and negative sides as well as the middle one. However, note that the signal reconstruction by inverse Fourier transform, and also by human eye, is only based on the information in the middle period.

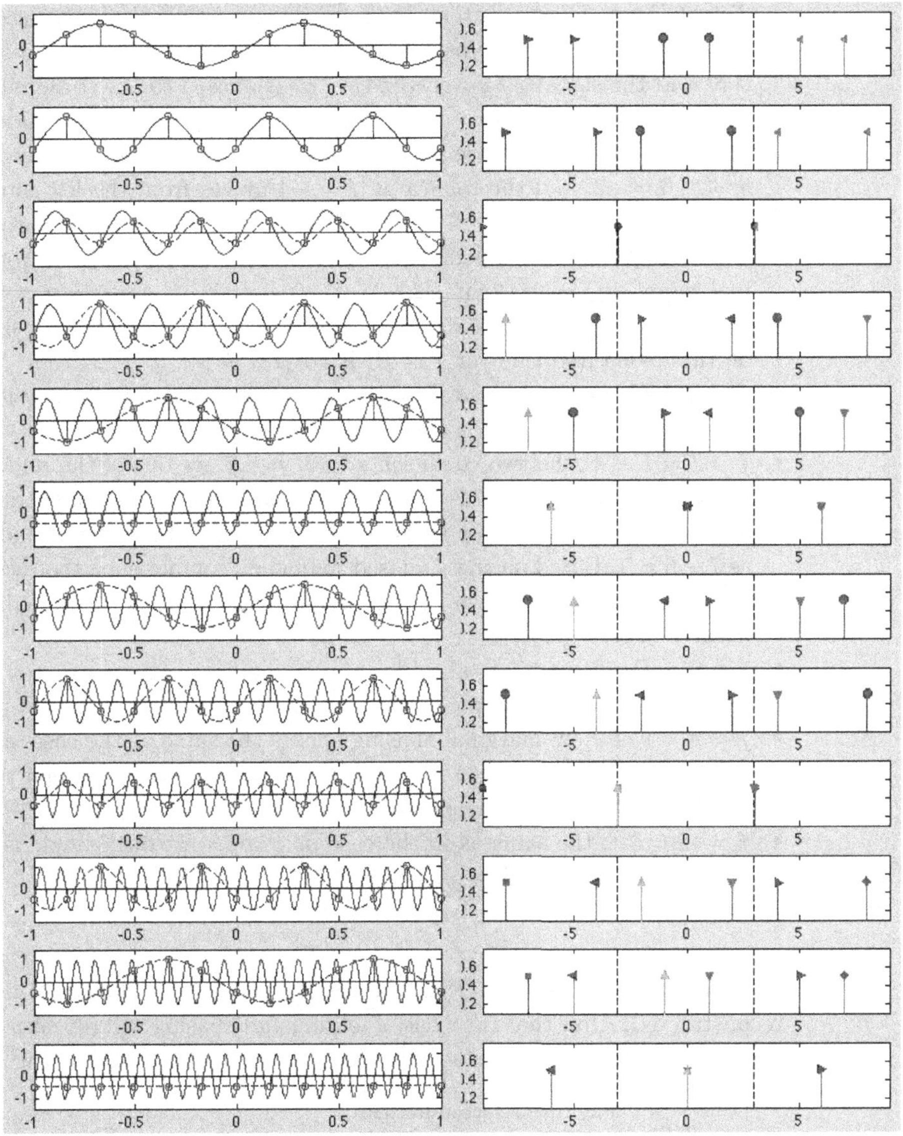

Figure 4.8 Aliasing in the time and frequency domains.

- $f = 1 < F/2 = 3$: the two non-zero frequency components $e^{\pm j2\pi ft}$ are both inside the middle period $-3 < f < 3$ of the spectrum, based on which the signal can be perfectly reconstructed.
- $f = 2 < F/2 = 3$, frequency components $e^{\pm j2\pi ft}$ move outward to a higher frequency of ± 2, which are still inside the middle period, so no aliasing or folding occurs.
- $f = 3 = F/3$: the signal is marginally aliased. Depending on the relative phase difference between the signal and the sampling function, the signal may be

distorted to a different extent. In the worst case, when the two samples happen to be taken at the zero crossings of the signal ($\phi = 0$ or $\phi = \pi$), they are both zero and the signal $x(t) = \cos(2\pi 4f + \phi)$ is aliased to a zero signal $x(t) = 0$.

- $f = 4 > F/2 = 3$: the two coefficients $e^{\pm j2\pi ft}$ are outside the middle period, but the replica at $f = 4$ moves from the right into the middle period to appear as $4 - 6 = -2$, and the replica at $f = -4$ moves from the left into the middle period to appear as $-4 + 6 = 2$. The reconstructed signal based on these folded frequency components is $\cos(2\pi 2t - \phi)$, which is different from the original signal $x(t) = \cos(2\pi 4t + \phi)$.
- $f = 5 >= F/2 = 3$: similar folding occurs and the reconstructed signal based on the folded frequency at $f = \pm 1$ is $\cos(2\pi t - \phi)$.
- $f = 6 = F$: one sample is taken per period, the aliased frequency is zero, and the reconstructed signal is $\cos(\phi)$
- $f = 7 = F + 1$: the two coefficients for $f = \pm 7$ are out of the middle period, but the replica at $f = -7$ is aliased to appear inside the middle period as $-7 + 6 = -1$, and the replica of $f = 7$ is aliased to appear inside the middle period as $7 - 6 = 1$. Based on these aliased frequency components, the reconstructed signal is $\cos(2\pi ft + \phi)$, which appears to be the same as the non-aliased cases when $f = 1$.
- $f = 8 = F + 2$: similar aliasing occurs and the reconstructed signal is $\cos(2\pi 2t + \phi)$, which appears the same as the non-aliased case of $f = 2$.
- $f = 9 = F + F/2$: marginal aliasing occurs the same as the case of $f = 3$.
- When $f = 10 = F + 4$ and $f = 11 = F + 5$, folding occurs similar to the cases when $f = 4$ and $f = 5$, respectively.
- $f = 12 = 2F$: the same as in the case of $f = 6 = F$; one sample is taken per period and the aliased frequency is zero.

We see that only when $f < F/2$ (the first two cases) can the signal be perfectly reconstructed. After that the cycle of folding and aliasing will repeat as the signal frequency f increases continuously. This pattern is illustrated in Fig. 4.9.

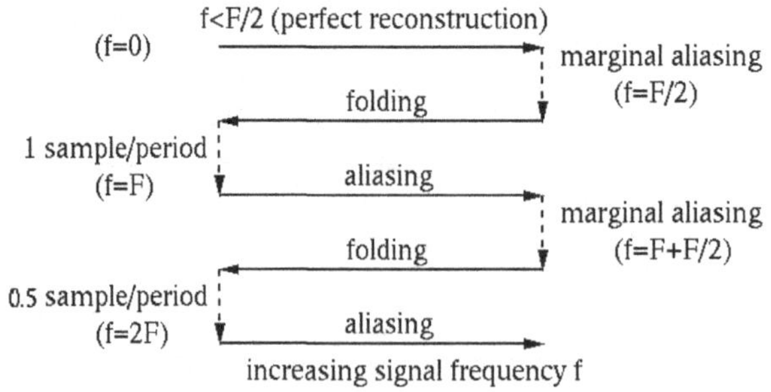

Figure 4.9 Aliasing-folding cycle as signal frequency increases.

Example 4.6: Consider the following three continuous signals that are first sampled at a sampling rate of $F = 10$ samples/second, and then reconstructed based on the resulting samples:

1. $x_1(t) = 2\cos(2\pi 7t) + \cos(2\pi 2t)$
2. $x_2(t) = 2\cos(2\pi 8t) + \cos(2\pi 2t)$
3. $x_3(t) = 2\cos(2\pi 8t) - 2\cos(2\pi 2t)$

As the sampling rate is not higher than twice the highest frequency component in the signal, aliasing/folding happens in all three cases, causing various forms of signal distortion, as shown in Fig. 4.10 that compares the original signals (solid curves) with the reconstructions (dashed curves). It can be seen that the reconstructed signal is distorted in the first case, it becomes a single sinusoid in the second case, and it becomes zero due to the fact that the original signal happens to be sampled at zero crossings.

The mathematical derivation of these results is left for the reader as a homework problem. The Matlab function `guidemo_sampling` used for creating these plots is provided.

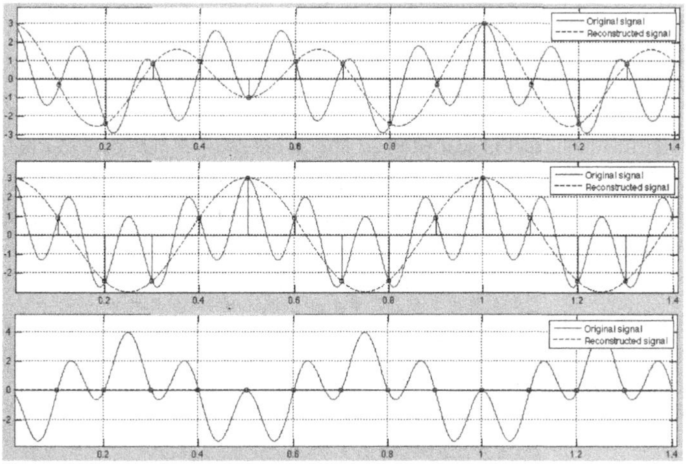

Figure 4.10 Different cases of aliasing/folding.

The sampling theorem is derived based on the assumption that the signal spectrum occupies the entire frequency range $|f| < f_{\max}$, and therefore the signal can be perfectly reconstructed by an ideal LP filter if $F > 2f_{\max}$. However, if the energy of the signal is concentrated within a certain frequency band $f_{\min} < |f| < f_{\max}$, it is possible to reconstruct the signal by a band-pass (BP) filter if $F < 2f_{\max}$. As shown in Fig. 4.11, the signal spectrum which is totally within the frequency range $f_{\min} < |f| < f_{\max} = 3$ can be perfectly reconstructed if the sampling rate F is high enough to satisfy Nyquist condition $F > 2f_{\max} = 6$ (top), but it can also be perfectly reconstructed even though F is much lower $F = 3.5 < 2f_{\max} = 6$ (bottom), so long as in the periodic spectrum after sampling the original spectrum (dark gray) is not distorted by its replica on either the right or left. Then the original spectrum can be recovered by an ideal BP filter that removes all replicas outside the passing band.

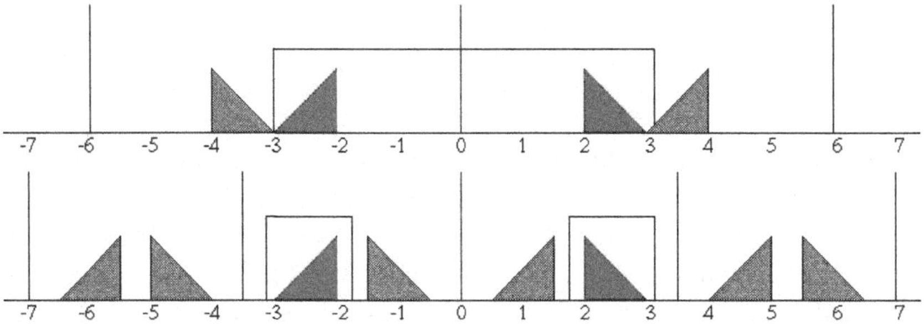

Figure 4.11 LP and BP filtering for signal reconstruction. When the signal frequency is limited to $(f_{\min} = 2) < |f| < (f_{\max} = 3)$, it can be sampled with $F \geq 2f_{\max} = 6$ and reconstructed by an ideal LP filter (top), or sampled with $F = 3.5 < 6$ and reconstructed by an ideal BP filter (bottom).

4.1.5 Reconstruction by interpolation

Once a continuous signal is sampled in the process of A/D conversion, it becomes a discrete signal that can be digitally processed/filtered by some digital signal processing (DSP) system (or just a computer). Often the processed signal needs to be converted back into analog form by a digital-to-analog converter (DAC, D/A), which reconstructs the signal from its samples.

As shown above, the reconstruction of a continuous signal $x(t)$ from its sampled version $x_s(t)$ is a low-pass (LP) filtering process in the frequency domain:

$$\hat{X}(f) = H_{\text{lp}}(f) X_F(f), \tag{4.92}$$

where $H_{\text{lp}}(f)$ is an ideal LP filter defined in Eq. (4.80). If the Nyquist condition is satisfied, the output of the filter is $\hat{X}(f) = X(f)$, from which the signal $x(t) = \mathcal{F}^{-1}[X(f)]$ can be perfectly reconstructed. In practice, as the ideal low-pass filter is hard to implement, sometimes a non-ideal low-pass filter could also be used to approximately reconstruct the signal. On the other hand, if the Nyquist condition is not satisfied, any signal component with frequency $f > F/2$ is outside the central period but one of its aliased or folded versions will appear inside the period, consequently the reconstructed signal is a distorted version of the original signal.

In the time domain, the reconstruction of a signal $x(t)$ from its sampled version $x_s(t)$ is an interpolation process by which the gap between two consecutive samples is filled. The interpolation can be considered as a convolution of the impulses in $x_s(t)$ with a certain function $h(t)$:

$$\hat{x}(t) = h(t) * x_s(t) = h(t) * \sum_{n=-\infty}^{\infty} x[n]\delta(t - nt_0)$$
$$= \sum_{n=-\infty}^{\infty} x[n]h(t) * \delta(t - nt_0) = \sum_{n=-\infty}^{\infty} x[n]h(t - nt_0). \quad (4.93)$$

We consider the following reconstructions based on three different interpretation functions $h_0(t)$, $h_1(t)$, and $h_{\rm lp}(t)$. The time domain interpolation based on these functions and the corresponding LP filtering are illustrated in Fig. 4.12.

- **Zero-order hold**
  The impulse response of a *zero-order hold* filter is

$$h_0(t) = \begin{cases} 1 & 0 \le t < t_0 = 1/F \\ 0 & \text{else} \end{cases}. \quad (4.94)$$

  This is the rectangular function discussed before (Eq. 3.161) with width t_0 and shifted by $t_0/2$. Based on $h_0(t)$, a continuous signal $\hat{x}_0(t)$ can be generated by

$$\hat{x}_0(t) = h_0(t) * x_s(t) = \sum_{n=-\infty}^{\infty} x[n]h_0(t - nt_0). \quad (4.95)$$

  This is a series of square impulses with their heights modulated by $x[n]$. The interpolation corresponds to an LP filtering in the frequency domain (Eq. 3.163 with an exponential factor corresponding to the time shift of $t_0/2$):

$$H_0(f) = \mathcal{F}[h_0(t)] = \frac{1}{\pi f} \sin(\pi f t_0) e^{-j2\pi f t_0/2}. \quad (4.96)$$

- **First-order hold**
  The impulse response of a *first-order hold* filter is

$$h_1(t) = \begin{cases} 1 - |t|/t_0 & |t| < t_0 \\ 0 & \text{else} \end{cases}, \quad (4.97)$$

  which is the triangle function previously discussed (Eq. (3.166)) with $\tau = t_0$. A continuous signal $\hat{x}_1(t)$ can be generated by:

$$\hat{x}_1(t) = h_1(t) * x_s(t) = \sum_{n=-\infty}^{\infty} x[n]h_1(t - nt_0), \quad (4.98)$$

  which is the linear interpolation of the sample train $x[n]$ (a straight line segment connecting every two consecutive samples). This interpolation corresponds to an LP filtering in the frequency domain by the following (Eq. 3.167)

$$H_1(f) = \mathcal{F}[h_1(t)] = \frac{1}{(\pi f)^2 t_0} \sin^2(\pi f t_0) = t_0 \ \text{sinc}^2(f t_0). \quad (4.99)$$

- **Ideal reconstruction**
  The reconstructed signals $\hat{x}_0(t)$ and $\hat{x}_1(t)$ are only approximations of the actual signal $x(t)$, as these interpolations correspond to non-ideal LP filtering in the frequency domain. The interpolation function for a perfect reconstruction is obviously associated with the ideal LP filtering method given in Eq. (4.80). The impulse response of this filter is (Eq. 3.165):

$$h_2(t) = h_{\text{lp}}(t) = \mathcal{F}^{-1}[H_{\text{lp}}(f)] = t_0 \, \frac{\sin(2\pi f_c t)}{\pi t}. \tag{4.100}$$

where $f_c = F/2$. The LP filtering corresponds to the following convolution in the time domain:

$$\begin{aligned} \hat{x}_2(t) &= h_2(t) * x_s(t) = t_0 \, \frac{\sin(2\pi f_c t)}{\pi t} * \sum_{n=-\infty}^{\infty} x[n]\delta(t - nt_0) \\ &= \frac{t_0}{\pi} \sum_{n=-\infty}^{\infty} x[n] \frac{\sin(2\pi f_c (t - nt_0))}{t - nt_0}. \end{aligned} \tag{4.101}$$

This signal generated by the ideal LP filter is the perfect reconstruction of the signal $\hat{x}_2(t) = x(t)$ without any distortion.

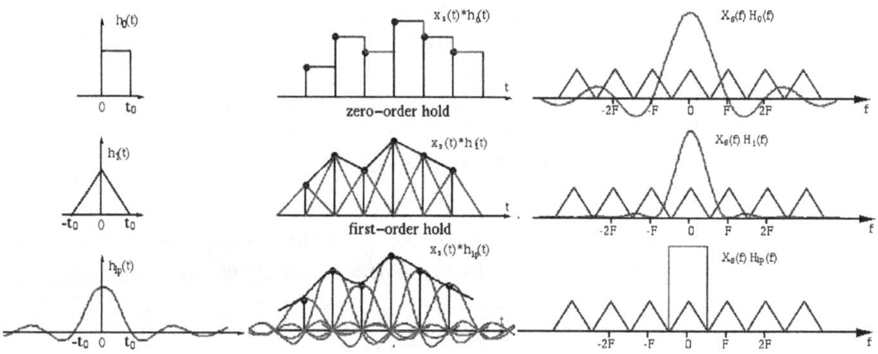

Figure 4.12 Zeroth-, first-, and second-order hold reconstructions. The impulse response of the filter (left), the interpolation in the time domain (middle), and the corresponding LP filtering in the frequency domain (right).

Having considered both signal sampling (A/D conversion) and reconstruction (D/A conversion), we can now put them together with some digital signal processing system to form a pipeline as shown in Fig. 4.13. The discrete signal $x[n]$ obtained by sampling is then processed/filtered to become $y[n] = h[n] * x[n]$, based on which a continuous version $y(t)$ can be reconstructed and used. For example, an analog audio signal can be sampled, digitally processed (e.g., LP filtered to remove some high-frequency noise) and then converted back to analog form to drive the speaker.

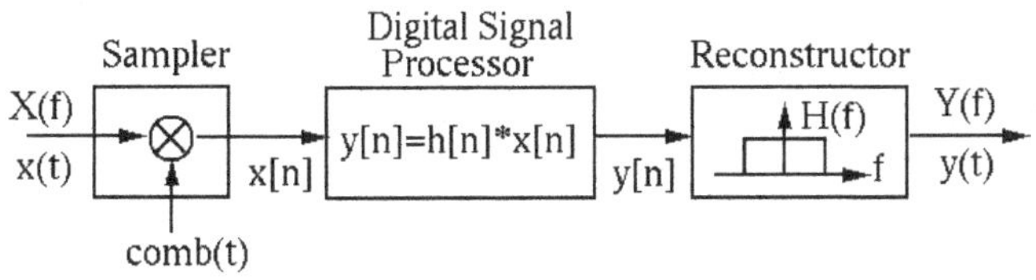

Figure 4.13 Signal sampling, digital processing, and reconstruction. The sampling rate of the comb function $\text{comb}(t) = \sum_k \delta(t - k/F)$ is F, the cutoff frequency of the ideal LP filter $H(f)$ is $F/2$.

4.2 Discrete Fourier transform

4.2.1 Formulation of the DFT

In practice, it is impossible to describe a physical signal, typically continuous and non-periodic, as a time function $x(t)$, as the analytical expression of the function is in general not available. In order to process and analyze such a signal in the frequency domain as well as in the time domain by a digital computer, the signal needs to be digitized in the following two steps.

- First, the signal needs to be truncated so that it has a finite duration from 0 to T, outside which the signal is not defined. However, for certain mathematical convenience we could further assume that the signal repeats itself outside the interval $0 < t < T$; i.e., it is a periodic signal with period T. Correspondingly in the frequency domain, the Fourier spectrum of such a periodic signal becomes discrete, composed of a set of impulses weighted by the Fourier expansion coefficients.
- Second, the signal needs to be discretized by sampling with a sampling rate F so that it can be processed by a digital computer. Correspondingly in the frequency domain, the spectrum of the signal becomes periodic.

Of course, the order of these two steps can be reversed so that the continuous signal is first sampled and then truncated. In either case, when the signal is both finite (periodic) and discrete, its spectrum, also discrete and finite periodic), can be obtained by the *discrete Fourier transform (DFT).*

To formulate the DFT, we recall the two different forms of the Fourier transform. First, a periodic signal $x_T(t+T) = x_T(t)$ has a discrete Fourier spectrum, an impulse train weighted by the coefficients $X[k]$ of its Fourier expansion (Eq. 3.86). The interval between two neighboring frequency components is the fundamental frequency $f_0 = 1/T$. Second, a discrete signal $x[n]$ obtained by sampling a continuous signal $x(t)$ at a sampling rate F (or a gap of $t_0 = 1/F$ between two consecutive samples) has a periodic spectrum $X_F(f+F) = X_F(f)$ (Eq. 4.3). It is obvious, therefore, that if a signal is both periodic with period T and discrete with interval t_0 between two consecutive samples, its spectrum will be both discrete with an interval $f_0 = 1/T$ between two frequency components, and periodic with a period of $F = 1/t_0$. In the time domain, the number of samples in a period T is $N = T/t_0$, while in the frequency domain, the number of frequency components in a period F is

$$\frac{F}{f_0} = \frac{1/t_0}{1/T} = \frac{T}{t_0} = N. \tag{4.102}$$

In other words, the number of independent variables, or *degrees of freedom (DOFs)*, in either the time or frequency domain is conserved by the DFT. This fact is also expected from the viewpoint of information conservation of the transform. We also have the following relations that are useful in the future discussion:

$$TF = \frac{T}{t_0} = N, \qquad f_0 t_0 = \frac{t_0}{T} = \frac{1}{N}. \tag{4.103}$$

Consider a continuous signal already truncated with duration T and assumed to be periodic $x_T(t+T) = x_T(t)$. This signal is further sampled when multiplied by the sampling function comb(t):

$$x_T(t)\ \mathrm{comb}(t) = x_T(t) \sum_{n=-\infty}^{\infty} \delta(t - nt_0) = \sum_{n=-\infty}^{\infty} x[n]\delta(t - nt_0), \tag{4.104}$$

where $x[n] = x_T(nt_0)$ is the nth sample of the signal. Note that $x[n]$ is periodic with period N:

$$x[n+N] = x_T((n+N)t_0) = x_T(nt_0 + T) = x_T(nt_0) = x[n]. \tag{4.105}$$

The Fourier expansion coefficient of this sampled version of the periodic and sampled signal can be found as

$$\begin{aligned} X[k] &= \frac{1}{T}\int_0^{\mathrm{T}} \left[\sum_{n=-\infty}^{\infty} x[n]\delta(t - nt_0)\right] e^{-j2\pi k f_0 t}\, dt \\ &= \frac{1}{T}\sum_{n=0}^{N-1} x[n] \int_0^{\mathrm{T}} \delta(t - nt_0) e^{-j2\pi k f_0 t}\, dt = \frac{1}{T}\sum_{n=0}^{N-1} x[n] e^{-j2\pi k f_0 n t_0} \\ &= \frac{1}{T}\sum_{n=0}^{N-1} x[n] e^{-j2\pi nk/N} \qquad k = 0, 1, \ldots, N-1. \end{aligned} \tag{4.106}$$

The number of terms in the summation is reduced from infinity to N for those inside the integral range from 0 to T, as all terms outside the range make no contribution to the integral. Note that $X[k+N] = X[k]$ is also periodic with period N:

$$X[k+N] = \frac{1}{T}\sum_{n=0}^{N-1} x[n] e^{-j2\pi(k+N)n/N} = \frac{1}{T}\sum_{k=0}^{N-1} x[k] e^{-j2\pi nk/N} e^{-j2n\pi} = X[k]. \tag{4.107}$$

The inverse transform can be obtained by multiplying both sides of Eq. (4.106) by $e^{j2\pi\nu k/N}/F$, and taking summation with respect to n from 0 to $N-1$:

$$\frac{1}{F}\sum_{k=0}^{N-1} X[k]e^{j2\pi\nu k/N} = \frac{1}{F}\sum_{k=0}^{N-1}\left[\frac{1}{T}\sum_{n=0}^{N-1} x[n]e^{-j2\pi nk/N}\right]e^{j2\pi\nu k/N}$$
$$= \sum_{n=0}^{N-1} x[n]\frac{1}{N}\sum_{k=0}^{N-1} e^{j2\pi n[\nu-k]/N} = \sum_{n=0}^{N-1} x[n]\delta[\nu-n] = x[\nu]. \tag{4.108}$$

Here, we have used Eq. (1.40). Now we put Eqs. (4.106) and (4.108) together to form the DFT pair:

$$X[k] = \mathcal{F}[x[n]] = \frac{1}{T}\sum_{n=0}^{N-1} x[n]e^{-j2\pi nk/N} \qquad k = 0, 1, \ldots, N-1,$$
$$x[n] = \mathcal{F}^{-1}[X[k]] = \frac{1}{F}\sum_{k=0}^{N-1} X[k]e^{j2\pi nk/N} \qquad n = 0, 1, \ldots, N-1. \tag{4.109}$$

The first equation is the forward DFT while the second one is the inverse DFT. The inverse DFT represents a periodic discrete signal $x[n] = x[n+N]$ as a linear combination of N complex exponentials $e^{j2\pi nk/N} = e^{j2\pi f_k n}$ $(k = 0, \ldots, N-1)$ each of frequency $f_k = k/N$ cycles per N points. In particular, when $k = 0$, $e^0 = 1$ represents the DC component of the signal, when $k = 1$, $e^{j2\pi n/N} = e^{j2\pi f_0 n}$ is the lowest frequency component, of fundamental frequency $f_0 = 1/N$ cycle per N points, and when $k > 1$, $e^{j2\pi nk/N} = e^{j2\pi k f_0 n}$ is the kth harmonic of frequency $f_k = kf_0$. The coefficients $X[k]$ for the N coefficients of the linear combination are given in the DFT in the first equation. As both $x[n]$ and $X[k]$ are periodic with period N, the summation in either the forward or inverse transform can be over any consecutive N points, such as from $-N/2$ to $N/2-1$. The relationship between the sampling period t_0 of the discrete signal, gap between $x[n]$ and $x[n+1]$, and the period T of its spectrum, and the relationship between the period T of the signal and the gap f_0 between two consecutive frequency components are illustrated in Fig. 4.14.

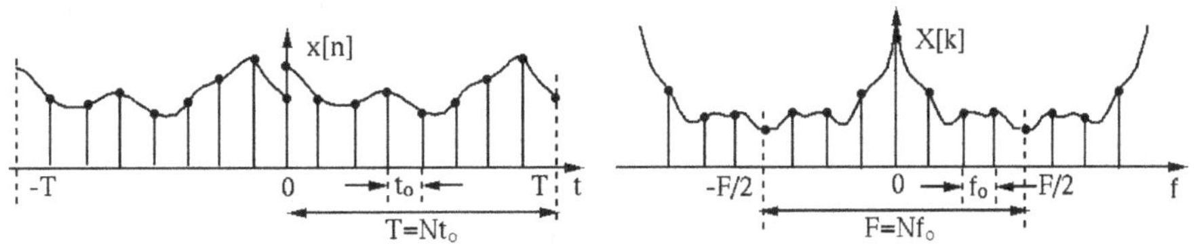

Figure 4.14 From continuous Fourier transform to DFT.

We can modify the scaling factors $1/T$ and $1/F$ for the forward and inverse transforms in Eq. (4.109) by redistributing the total scaling factor of $1/FT = 1/N$ differently between the two transforms. For example, we can scale either of the two by $1/N$, or alternatively, we can also evenly distribute it on both sides:

$$X[k] = \mathcal{F}[x[n]] = \frac{1}{\sqrt{N}} \sum_{n=0}^{N-1} x[n] e^{-j2\pi nk/N} \qquad k = 0, 1, \ldots, N-1,$$
$$x[n] = \mathcal{F}^{-1}[X[k]] = \frac{1}{\sqrt{N}} \sum_{k=0}^{N-1} X[k] e^{j2\pi nk/N} \qquad n = 0, 1, \ldots, N-1. \tag{4.110}$$

This is Eqs. (2.124) and (2.123). The advantage of this notation is that the signal can be represented as a vector $\boldsymbol{x} = [x[0], \ldots, x[N-1]]^{\mathrm{T}}$ in an N-D vector space $\mathbb{C}^N$ spanned by a set of N orthonormal basis vectors:

$$\boldsymbol{w}_k = \frac{1}{\sqrt{N}} [e^{j2\pi 0k/N}, \ldots, e^{j2\pi(N-1)k/N}]^{\mathrm{T}} \qquad k = 0, \ldots, N-1, \tag{4.111}$$

satisfying (Eq. (1.40))

$$\langle \boldsymbol{w}_k, \boldsymbol{w}_l \rangle = \frac{1}{N} \sum_{n=0}^{N-1} e^{j2\pi n(k-l)/N} = \delta[k-l]. \tag{4.112}$$

Under this basis, the given signal vector $\boldsymbol{x}$ can be represented as

$$\boldsymbol{x} = \sum_{k=0}^{N-1} X[k] \boldsymbol{w}_k, \qquad \text{where} \qquad X[k] = \langle \boldsymbol{x}, \boldsymbol{w}_k \rangle = \boldsymbol{x}^{\mathrm{T}} \overline{\boldsymbol{w}}_k. \tag{4.113}$$

The first equation above is the vector form of the first equation in Eq. (4.110) for the inverse DFT, the second equation is the same as the second equation in Eq. (4.110) for the forward DFT. In this case, Parseval's identity holds; i.e., $||\boldsymbol{x}||^2 = ||\boldsymbol{X}||^2$.

The discrete spectrum $X[k]$ of the samples $x[n]$ of a signal $x(t)$ is obviously related to, but certainly not equal to, the spectrum $X(f) = \mathcal{F}[x(t)]$, as the signal has been significantly modified by the truncation and sampling process before the DFT can be carried out. First, owing to the truncation and the assumed periodicity, the signal may no longer be continuous and smooth. Discontinuity will occur at the end point between two consecutive periods if $x(0) \neq x(T)$, as shown on the left of Fig. 4.14. Second, owing to the sampling process, aliasing or folding may occur if the Nyquist condition is not satisfied. Consequently, the spectrum may be contaminated by various artifacts, most likely some faulty high-frequency components corresponding to the discontinuities, together with some faulty low frequencies owing to aliasing or folding. Therefore, special attention needs to be paid to the truncation and sampling process in order to minimize such artifacts. For example, certain windowing methods can be used to smooth the truncated signal, and some anti-aliasing LP filtering can be used to reduce the high-frequency components before sampling to avoid aliasing. Only then can the DFT generate meaningful data representative of the actual signal of interest.

Example 4.7: Consider a discrete sinusoid of $N = 5$ samples with frequency $f = 1/N = 1/5$ (one cycle per $N = 5$ points):

$$x[n] = \cos(n\frac{2\pi}{5}) = \frac{1}{2}[e^{j2\pi n/5} + e^{-j2\pi n/5}] \qquad n = 0, \dots, N-1 = 4. \quad (4.114)$$

Comparing this expression with the DFT expansion

$$x[n] = \sum_{k=0}^{4} X[k]e^{j2\pi nk/5}, \quad (4.115)$$

we see that $X[1] = 1/2$ and $X[4] = X[-1] = 1/2$. Alternatively, following the DFT definition we can also get the kth Fourier coefficient as

$$\begin{aligned} X[k] &= \frac{1}{N}\sum_{n=0}^{N-1} x[n]e^{-j2\pi nk/N} = \frac{1}{10}\sum_{n=0}^{4}\left[e^{-j2\pi n(n-1)/5} + e^{-j2\pi n(n+1)/5}\right] \\ &= \frac{1}{2}[\delta[n+1] + \delta[n-1]]. \end{aligned} \quad (4.116)$$

This result is shown in Fig. 4.15.

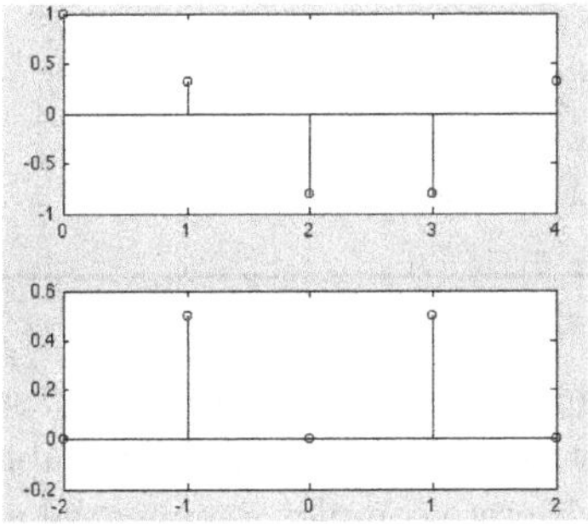

Figure 4.15 Discrete cosine and its DFT spectrum.

Example 4.8: Consider a symmetric square wave with a period of N and width $2M < N$:

$$x[n] = \begin{cases} 1 & |n| \le M \\ 0 & M < |n| \le N/2 \end{cases}. \quad (4.117)$$

For convenience, we choose the limits of the Fourier transform summation from $-N/2$ to $N/2 - 1$ (instead of from 0 to $N - 1$) and get:

$$X[k] = \sum_{n=-N/2}^{N/2-1} x[n]e^{-j2\pi nk/N} = \sum_{n=-M}^{M} e^{-j2\pi nk/N}. \quad (4.118)$$

Let $n' = n + M$, we have $n = n' - M$ and

$$X[k] = \sum_{n'=0}^{2M} e^{-j2\pi n'k/N} e^{j2\pi Mn/N} = e^{j2\pi Mk/N} \frac{1 - e^{-j2\pi(2M+1)k/N}}{1 - e^{-j2\pi k/N}}$$

$$= e^{j2\pi Mk/N} \frac{e^{-j\pi(2M+1)k/N}(e^{j\pi(2M+1)k/N} + e^{-j\pi(2M+1)k/N})}{e^{-j\pi k/N}(e^{j\pi k/N} - e^{-j\pi k/N})}$$

$$= \frac{\sin((2M+1)k\pi/N)}{\sin(k\pi/N)}. \tag{4.119}$$

The signal and its DFT spectrum are shown in Fig. 4.16 ($N = 64$, $M = 8$).

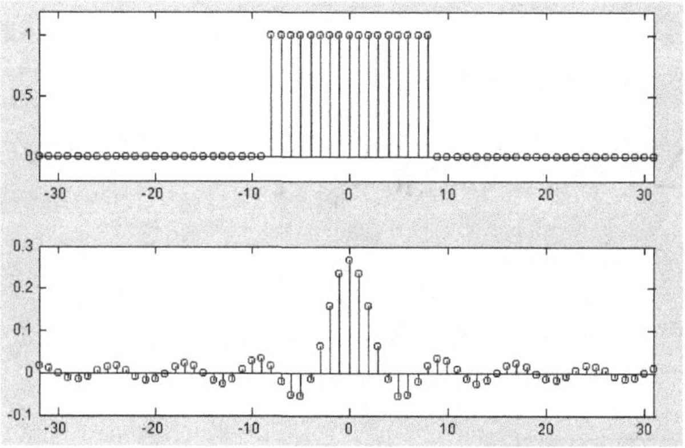

Figure 4.16 Square impulse and its DFT spectrum.

To carry out the DFT of a given signal function $x[n]$, it is necessary to first determine its period N, which may not be always explicitly available. We first consider how to find the period N of a discrete complex exponential given in the form of $e^{j2\pi fn}$. For it to be of period N, it has to satisfy

$$e^{j2\pi f(n+N)} = e^{j2\pi fn} e^{j2\pi N)} = e^{j2\pi fn}; \quad \text{i.e.} \quad e^{j2\pi fN} = e^{j2\pi k} = 1. \tag{4.120}$$

In other words, $fN = k$ has to be an integer or $f = k/N$ has to be a rational number (a ratio of two integers). And in order for $N = k/f$ to be the period, k has to be the smallest integer for $N = k/f$ to be an integer. The complex exponential can now be written as $e^{j2\pi kn/N}$ with N being its period.

If a discrete signal $x[n]$ is composed of a set of K complex exponential terms each of period N_k ($k = 1, \ldots, K$), then the period of $x[n]$ can be found as the LCM of these individual periods.

Example 4.9: Find the DFT of a given signal $x[n] = 2\cos(5\pi n/6) + \sin(3\pi n/4)$. We first find period of the signal by writing it in the standard form:

$$x[n] = 2\cos(2\pi 5n/12) + \cos(2\pi 3n/8). \tag{4.121}$$

The periods of these two terms are $N_1 = 12$ and $N_2 = 8$ and their LCM is

$$N = \text{LCM}(N_1,\ N_2) = \text{LCM}(12, 8) = 24. \tag{4.122}$$

Now the signal can be written as

$$\begin{aligned} x[n] &= 2\cos(2\pi 10n/N) + \sin(2\pi 9n/N) \\ &= 2\cos(2\pi 10 f_0 n) + \sin(2\pi 9 f_0 n), \end{aligned} \tag{4.123}$$

with period $N = 24$ and fundamental frequency $f_0 = 1/N = 1/24$, composed of the 9th and 10th harmonics. Its DFT coefficients are

$$X[k] = \delta[k+10] + \delta[k-10] + [\delta[k+9] + \delta[k-9]]/2j. \tag{4.124}$$

In Matlab the forward and inverse DFTs can be carried out by functions `fft` and `ifft`, respectively. However, these functions are scaled differently. For Parseval's identity to hold, the Matlab forward transform function needs to be rescaled: `X=fft(x)/sqrt(length(x))`.

4.2.2 Array representation

Similar to the kth basis function for the Fourier series expansion $\phi_k(t) = e^{j2\pi f_k t} = \cos(2\pi f_k t) + j\sin(2\pi f_k t)$ representing a continuous sinusoid of frequency $f_k = kf_0 = k/T$ (k cycles per period of T), here the N samples $w_k[n] = e^{j2\pi nk/N} = \cos(2\pi nk/N) + j\sin(2\pi nk/N)$ $(n = 0, \ldots, N-1)$ of the kth basis vector $\boldsymbol{w}_k$ in Eq. (4.111) represent a sinusoid of frequency $f_k = k/N$ (k cycles per period of N samples). However, we also note that kf_0 is the frequency for a continuous sinusoid that grows without limit as k increases; k/N is for the samples of a continuous sinusoid that does not grow without limit. For example, when $k = N-1$, $k/N = (N-1)/N$ actually represents a frequency of 1 (instead of $N-1$) cycles per period of N samples, as any frequency higher than $N/2$ is undersampled and appears as a frequency lower than $N/2$ cycles per N samples caused by aliasing.

Figure 4.17 (first and second columns) shows the first $N = 8$ basis functions $\phi_k(t) = e^{j2\pi kt/T}$ $(k = 0, \ldots, 7)$ for the Fourier series expansion (continuous curves), together with the discrete samples $w_k[n] = e^{j2\pi nk/N}$ for each of the basis vectors of the corresponding eight-point DFT (the circles). We see that while the frequency $kf_0 = k/T$ for the continuous sinusoid $e^{j2\pi kt/T}$ increases with k, the frequency of $e^{j2\pi nk/N}$ does not increase with k monotonically. Actually, its frequency k/N is proportional to k only when $k < N//2 = 4$, but it becomes $(N-k)/N$ when $k > N/2 = 4$, due obviously to aliasing. We also note that the zeroth basis vector $\boldsymbol{w}_0$ represents the DC component of the signal, and the fourth $(N/2)$ basis vector $\boldsymbol{w}_{N/2}$ is the highest representable frequency of $N/2 = 4$ cycles per period T. The third and fourth columns of Fig. 4.17 are for an example to be considered later.

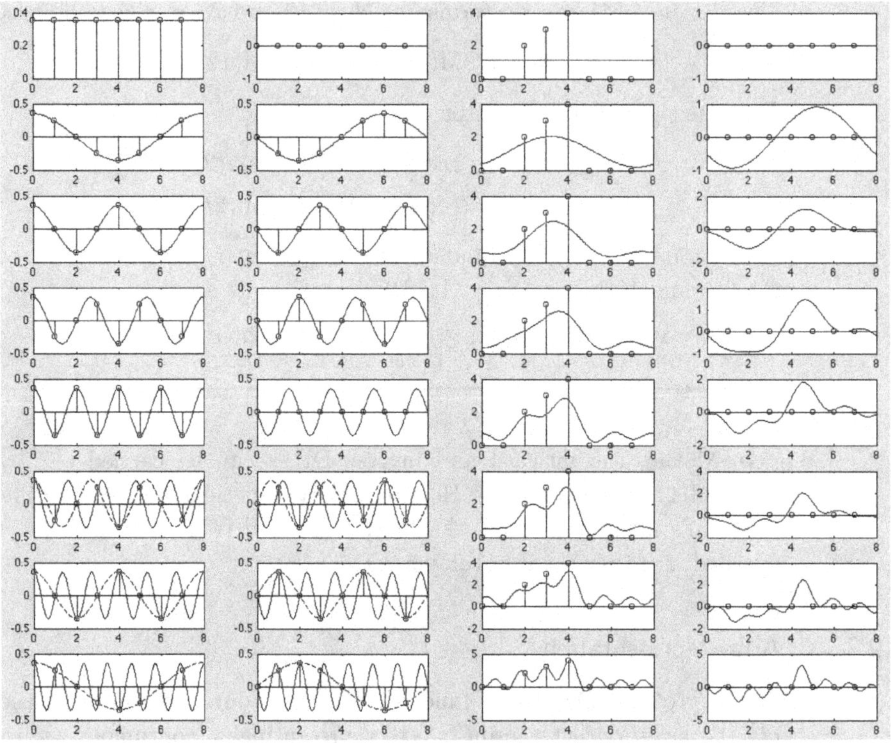

Figure 4.17 Basis functions and vectors of eight-point DFT. The $N = 8$ real and imaginary parts of the DFT basis vectors and the associated basis functions are shown respectively in the first and second columns; the real and imaginary parts of the reconstructions by the inverse DFT of a discrete signal (see Example 4.10) with progressively more components are shown respectively in the third and fourth columns.

In general, the N by N matrix of an N-point DFT is composed of the N basis vectors $\boldsymbol{w}_k$ $(k = 0, \dots, N-1)$ as the N column vectors:

$$\boldsymbol{W} = [\boldsymbol{w}_0, \dots, \boldsymbol{w}_{N-1}] = \frac{1}{\sqrt{N}} \begin{bmatrix} e^{j2\pi 00/N} & e^{j2\pi 01/N} & \dots & e^{j2\pi 0(N-1)/N} \\ e^{j2\pi 10/N} & e^{j2\pi 11/N} & \dots & e^{j2\pi 1(N-1)/N} \\ \vdots & \vdots & \ddots & \vdots \\ e^{j2\pi (N-1)0/N} & e^{j2\pi (N-1)1/N} & \dots & e^{j2\pi (N-1)(N-1)/N} \end{bmatrix}. \tag{4.125}$$

As $\boldsymbol{w}_k$ are orthogonal, $\boldsymbol{W}$ is unitary $\boldsymbol{W}^*\boldsymbol{W} = \boldsymbol{I}$ or $\boldsymbol{W}^* = \boldsymbol{W}^{-1}$. Also, as $w[l,k] = e^{j2\pi kl/N} = w[k,l]$, $\boldsymbol{W} = \boldsymbol{W}^{\mathrm{T}}$ is a symmetric. Therefore we have

$$\boldsymbol{W}^{-1} = \overline{\boldsymbol{W}}; \quad \text{i.e.} \quad \boldsymbol{W}\overline{\boldsymbol{W}} = \boldsymbol{I}. \tag{4.126}$$

Now the DFT of a signal vector $\boldsymbol{x}$ can be expressed in the following matrix forms:

$$\boldsymbol{x} = \boldsymbol{W}\boldsymbol{X} = [\boldsymbol{w}_0, \ldots, \boldsymbol{w}_{N-1}] \begin{bmatrix} X[0] \\ \vdots \\ X[N-1] \end{bmatrix} = \sum_{k=0}^{N-1} X[k]\boldsymbol{w}_k. \tag{4.127}$$

Left multiplying $\boldsymbol{W}^{-1} = \overline{\boldsymbol{W}}$ on both sides, we get

$$\overline{\boldsymbol{W}}\boldsymbol{x} = \overline{\boldsymbol{W}}\boldsymbol{W}\boldsymbol{X} = \boldsymbol{X}; \tag{4.128}$$

i.e.,

$$\boldsymbol{X} = \begin{bmatrix} X[0] \\ \vdots \\ X[N-1] \end{bmatrix} = \overline{\boldsymbol{W}}\boldsymbol{x} = \begin{bmatrix} \overline{\boldsymbol{w}}_0^{\mathrm{T}} \\ \vdots \\ \overline{\boldsymbol{w}}_{N-1}^{\mathrm{T}} \end{bmatrix} \boldsymbol{x}, \tag{4.129}$$

where the kth coefficient is the projection of the signal vector $\boldsymbol{x}$ onto the kth basis vector $\boldsymbol{w}_k$:

$$X[k] = \langle \boldsymbol{x}, \boldsymbol{w}_k \rangle = \overline{\boldsymbol{w}}_k^{\mathrm{T}} \boldsymbol{x} = \boldsymbol{x}^{\mathrm{T}} \overline{\boldsymbol{w}}_k. \tag{4.130}$$

Equations 4.127 and 4.128 form the DFT pair in matrix form (while Eq. 4.110 is the component form):

$$\begin{cases} \boldsymbol{X} = \overline{\boldsymbol{W}}\boldsymbol{x} & \text{(forward)} \\ \boldsymbol{x} = \boldsymbol{W}\boldsymbol{X} & \text{(inverse)} \end{cases}. \tag{4.131}$$

As a unitary operation, the DFT is actually a rotation in $\mathbb{C}^N$, represented by the unitary matrix $\boldsymbol{W}$. Any signal vector $\boldsymbol{x} = [x[0], \ldots, x[N-1]]^{\mathrm{T}}$ given under the standard basis $\boldsymbol{e}_n$ $(n = 0, \ldots, N-1)$ can also be expressed in terms of a different set of basis vectors $\boldsymbol{w}_k$ $(k = 0, \ldots, N-1)$:

$$\begin{aligned} \boldsymbol{x} = \boldsymbol{I}\boldsymbol{x} = [\boldsymbol{e}_0, \ldots, \boldsymbol{e}_{N-1}]\boldsymbol{x} &= \sum_{n=0}^{N-1} x[n]\boldsymbol{e}_n \\ = \boldsymbol{W}\boldsymbol{X} = [\boldsymbol{w}_0, \ldots, \boldsymbol{w}_{N-1}]\boldsymbol{X} &= \sum_{k=0}^{N-1} X[k]\boldsymbol{w}_k, \end{aligned} \tag{4.132}$$

where $\boldsymbol{w}_k = \boldsymbol{W}\boldsymbol{e}_k$ is obtained by rotating the standard basis vector $\boldsymbol{e}_k$ $(k = 0, \ldots, N-1)$. Equivalently, the signal vector is rotated in the opposite direction to become $\boldsymbol{X} = \boldsymbol{W}^{-1}\boldsymbol{x} = \overline{\boldsymbol{W}}\boldsymbol{x}$. As rotation does not change the vector norm (Parseval's identity), the signal energy is conserved $||\boldsymbol{x}|| = ||\boldsymbol{X}||$; i.e., either the original signal $\boldsymbol{x}$ in the time domain or its Fourier coefficients X in the frequency domain contains the same amount of energy or information.

We now consider specifically the following three examples for $N = 2$, 4 and 8.

- $N = 2$, the element of the lth row and kth column $(l, k = 0, 1)$ of the two-point DFT matrix is

$$w[l, k] = \frac{1}{\sqrt{2}}(e^{j2\pi/N})^{kl} = \frac{1}{\sqrt{2}}(e^{j\pi})^{kl} = \frac{1}{\sqrt{2}}(-1)^{kl}, \tag{4.133}$$

and the DFT matrix is

$$\boldsymbol{W}_{2\times 2} = \frac{1}{\sqrt{2}} \begin{bmatrix} 1 & 1 \\ 1 & -1 \end{bmatrix}. \tag{4.134}$$

The DFT of a two-point signal $\boldsymbol{x} = [x[0], x[1]]^{\mathrm{T}}$ can be trivially found as

$$\boldsymbol{X} = \begin{bmatrix} X[0] \\ X[1] \end{bmatrix} = \overline{\boldsymbol{W}}\boldsymbol{x} = \frac{1}{\sqrt{2}} \begin{bmatrix} 1 & 1 \\ 1 & -1 \end{bmatrix} \begin{bmatrix} x[0] \\ x[1] \end{bmatrix} = \frac{1}{\sqrt{2}} \begin{bmatrix} x[0] + x[1] \\ x[0] - x[1] \end{bmatrix}. \tag{4.135}$$

We see that the first component $X[0]$ is proportional to the sum of the two signal samples representing the average or DC component of the signal, and the second $X[1]$ is proportional to the difference between the two samples representing the variations (details) in the signal.

- $N = 4$, the element of the lth row and kth column $(l, k = 0, \ldots, 3)$ of the four-point DFT matrix is

$$w[l,k] = \frac{1}{\sqrt{N}}(e^{j2\pi/N})^{kl} = \frac{1}{2}(e^{j\pi/2})^{kl} = j^{kl}. \tag{4.136}$$

The 4 by 4 DFT matrix is

$$\boldsymbol{W}_{4\times 4} = \frac{1}{2} \begin{bmatrix} 1 & 1 & 1 & 1 \\ 1 & j & -1 & -j \\ 1 & -1 & 1 & -1 \\ 1 & -j & -1 & j \end{bmatrix} = \frac{1}{2} \begin{bmatrix} 1 & 1 & 1 & 1 \\ 1 & 0 & -1 & 0 \\ 1 & -1 & 1 & -1 \\ 1 & 0 & -1 & 0 \end{bmatrix} + \frac{j}{2} \begin{bmatrix} 0 & 0 & 0 & 0 \\ 0 & 1 & 0 & -1 \\ 0 & 0 & 0 & 0 \\ 0 & -1 & 0 & 1 \end{bmatrix}. \tag{4.137}$$

- $N = 8$, we have

$$w[l,k] = \frac{1}{\sqrt{N}}(e^{j2\pi/N})^{kl} = \frac{1}{\sqrt{8}}(e^{j\pi/4})^{kl} = \frac{1}{\sqrt{8}}(0.707 + j\,0.707)^{kl}. \tag{4.138}$$

The real and imaginary parts of the DFT matrix $\boldsymbol{W} = \boldsymbol{W}_{\mathrm{r}} + j\boldsymbol{W}_{\mathrm{j}}$ are respectively:

$$\boldsymbol{W}_{\mathrm{r}} = \frac{1}{\sqrt{8}} \begin{bmatrix} 1.0 & 1.0 & 1.0 & 1.0 & 1.0 & 1.0 & 1.0 & 1.0 \\ 1.0 & 0.7 & 0.0 & -0.7 & -1.0 & -0.7 & 0.0 & 0.7 \\ 1.0 & 0.0 & -1.0 & 0.0 & 1.0 & 0.0 & -1.0 & -0.0 \\ 1.0 & -0.7 & 0.1 & 0.7 & -1.0 & 0.7 & 0.0 & -0.7 \\ 1.0 & -1.0 & 1.0 & -1.0 & 1.0 & -1.0 & 1.0 & -1.0 \\ 1.0 & -0.7 & 0.0 & 0.7 & -1.0 & 0.7 & 0.0 & -0.7 \\ 1.0 & 0.0 & -1.0 & 0.0 & 1.0 & 0.0 & -1.0 & -0.0 \\ 1.0 & 0.7 & 0.0 & -0.7 & -1.0 & -0.7 & 0.0 & 0.7 \end{bmatrix}, \tag{4.139}$$

and

$$\boldsymbol{W}_{\rm j} = \frac{1}{\sqrt{8}} \begin{bmatrix} 0.0 & 0.0 & 0.0 & 0.0 & 0.0 & 0.0 & 0.0 & 0.0 \\ 0.0 & -0.7 & -1.0 & -0.7 & 0.0 & 0.7 & 1.0 & 0.7 \\ 0.0 & -1.0 & 0.0 & 1.0 & 0.0 & -1.0 & 0.0 & 1.0 \\ 0.0 & -0.7 & 1.0 & -0.7 & 0.0 & 0.7 & -1.0 & 0.7 \\ 0.0 & 0.0 & 0.0 & 0.0 & 0.0 & 0.0 & 0.0 & 0.0 \\ 0.0 & 0.7 & -1.0 & 0.7 & 0.0 & -0.7 & 1.0 & -0.7 \\ 0.0 & 1.0 & 0.0 & -1.0 & 0.0 & 1.0 & 0.0 & -1.0 \\ 0.0 & 0.7 & 1.0 & 0.7 & 0.0 & -0.7 & -1.0 & -0.7 \end{bmatrix}. \tag{4.140}$$

The values of $\boldsymbol{W}_{\rm r}$ and $\boldsymbol{W}_{\rm j}$ are also plotted in the first two columns of Fig. 4.17.

4.2.3 Properties of the DFT

As one of the variations of the generic continuous-time Fourier transform (CTFT), the DFT shares all the properties of the CTFT discussed previously, but in different forms. Here, we consider only a set of selected properties and leave out some of the proofs which should be very similar to those for the corresponding CTFT properties.

- **Time and frequency shift**

$$\mathcal{F}[x[n \pm n_0]] = X[k]e^{\pm j2\pi n_0 k/N}, \tag{4.141}$$

$$\mathcal{F}\left[x[n]e^{\mp j2\pi n k_0/N}\right] = X[k \pm k_0]. \tag{4.142}$$

As the spectrum of a shifted signal $y[n] = x[n \pm n_0]$ is $Y[k] = e^{\pm j2\pi n_0 k/N}X[k]$, we see that the magnitude of the spectrum remains the same (shift-invariant), while the phase is shifted by $2\pi n_0 k/N$ (a multiple of $2\pi n_0/N$):

$$|Y[k]| = |X[k]|, \qquad \angle Y[k] = \angle X[k] \pm 2\pi n_0 k/N \tag{4.143}$$

- **Plancherel's identity and Parseval's identity**

$$\langle \boldsymbol{x}, \boldsymbol{y}\rangle = \sum_{n=0}^{N-1} x[n]\overline{y}[n] = \sum_{k=0}^{N-1} X[k]\overline{Y}[k] = \langle \boldsymbol{X}, \boldsymbol{Y}\rangle. \tag{4.144}$$

In particular, when $\boldsymbol{y} = \boldsymbol{x}$, this equation becomes

$$||\boldsymbol{x}||^2 = \langle \boldsymbol{x}, \boldsymbol{x}\rangle = \sum_{n=0}^{N-1} |x[n]|^2 = \sum_{k=0}^{N-1} |X[k]|^2 = \langle \boldsymbol{X}, \boldsymbol{X}\rangle = ||\boldsymbol{X}||^2. \tag{4.145}$$

- **DC and highest frequency representable**

$X_{\rm r}[0]$ represents the DC offset of the signal (zero frequency):

$$X_{\rm r}[0] = \sum_{n=0}^{N-1} x_{\rm r}[n]\cos(\frac{2\pi n 0}{N}) = \sum_{n=0}^{N-1} x_{\rm r}[n], \tag{4.146}$$

and $X_r[N/2]$ represents the highest frequency component:

$$X_r[N/2] = \sum_{n=0}^{N-1} x_r[n] \cos\left(\frac{2\pi n N/2}{N}\right) = \sum_{n=0}^{N-1} x_r[n](-1)^n. \tag{4.147}$$

When $k = 0$ and $n = N/2$, the imaginary parts $X_j[0] = X_j[N/2] = 0$ are zero as $\sin(0) = \sin(n\pi) = 0$.

- **Symmetry**
  The DFT is a complex transform which can be separated into real and imaginary parts:

$$X[k] = \sum_{n=0}^{N-1} x[n]e^{-j2\pi nk/N}$$
$$= \sum_{n=0}^{N-1} [x_r[n] + jx_j[n]] \left[\cos\left(\frac{2\pi nk}{N}\right) - j\,\sin\left(\frac{2\pi nk}{N}\right)\right] = X_r[k] + jX_j[k], \tag{4.148}$$

where

$$X_r[k] = \sum_{n=0}^{N-1} x_r[n] \cos\left(\frac{2\pi nk}{N}\right) + \sum_{n=0}^{N-1} x_j[n] \sin\left(\frac{2\pi nk}{N}\right),$$
$$X_j[k] = \sum_{n=0}^{N-1} x_j[n] \cos\left(\frac{2\pi nk}{N}\right) - \sum_{n=0}^{N-1} x_r[n] \sin\left(\frac{2\pi nk}{N}\right). \tag{4.149}$$

In particular, if $x[n] = x_r[n]$ is real ($x_j[n] = 0$), then $X_r[k]$ is even

$$X_r[k] = \sum_{n=0}^{N-1} x_r[n] \cos\left(\frac{2\pi nk}{N}\right) = X_r[-k], \tag{4.150}$$

and $X_j[k]$ is odd

$$X_j[k] = -\sum_{n=0}^{N-1} x_r[n] \sin\left(\frac{2\pi nk}{N}\right) = -X_j[-k]. \tag{4.151}$$

- **Convolution theorem** The convolution of two finite and discrete $x[n]$ and $h[n]$ $(n = 0, \ldots, N-1)$ is defined as

$$y[n] = h[n] * x[n] = \sum_{m=0}^{N-1} x[m]h[n-m] \qquad n = 0, \ldots, N-1. \tag{4.152}$$

As both $x[n+N] = x[n]$ and $h[n+N] = h[n]$ are assumed to be periodic with period N, it is obvious that the result $y[n]$ of the convolution is also periodic: $y[n+N] = y[n]$. The convolution is therefore also referred to as a *circular convolution*.
Let $X[k] = \mathcal{F}[x[n]]$ and $H[k] = \mathcal{F}[h[n]]$, then the convolution theorem states:

$$\mathcal{F}[h[n] * x[n]] = H[k]X[k], \qquad \mathcal{F}[h[n]x[n]] = H[k] * X[k]. \tag{4.153}$$

We now prove the first part of Eq. (4.153):

$$\mathcal{F}[x[n] * h[n]] = \sum_{n=0}^{N-1} \left[\sum_{m=0}^{N-1} x[m]h[n-m] \right] e^{-j2\pi nk/N}$$
$$= \sum_{m=0}^{N-1} x[m] \left[\sum_{n=0}^{N-1} h[n-m]e^{-j2\pi(n-m)k/N} \right] e^{-j2\pi mk/N}$$
$$= H[k] \sum_{m=0}^{N-1} x[m]e^{-j2\pi km/N} = H[k]X[k]. \tag{4.154}$$

Note that, owing to the assumed periodicity, the upper and lower limits of the summation are not important so long as they cover all N terms in the period. The second part of Eq. (4.153) can be similarly proved.

- **Diagonalization of circulant matrix**

An N by N matrix $\boldsymbol{H}$ can be constructed based on $h[n]$ of the convolution above, with its element in the mth row and nth column defined as $h[m,n] = h[m-n]$, so that the circular convolution in Eq. (4.152) can be expressed as a matrix multiplication $\boldsymbol{y} = \boldsymbol{H}\boldsymbol{x}$:

$$\begin{bmatrix} y[0] \\ y[1] \\ \vdots \\ y[N-2] \\ y[N-1] \end{bmatrix} = \begin{bmatrix} h[0] & h[N-1] & \cdots & h[2] & h[1] \\ h[1] & h[0] & \cdots & h[3] & h[2] \\ \vdots & \vdots & \ddots & \vdots & \vdots \\ h[N-2] & h[N-3] & \cdots & h[0] & h[N-1] \\ h[N-1] & h[N-2] & \cdots & h[1] & h[0] \end{bmatrix} \begin{bmatrix} x[0] \\ x[1] \\ \vdots \\ x[N-2] \\ x[N-1] \end{bmatrix}. \tag{4.155}$$

This matrix $\boldsymbol{H}$ is a *circulant matrix*, each row of which is a circularly right-rotated version of the row above. Let $H[k] = \mathcal{F}[h[n]]$ be the DFT of $h[n]$, and $\boldsymbol{w}_k = [w^{j2\pi 0k/N}, \ldots, w^{j2\pi(N-1)k/N}]^{\mathrm{T}}$ be the kth column vector of the DFT matrix $\boldsymbol{W}$, then we can show that they are respectively the eigenvalue and eigenvector of the matrix $\boldsymbol{H}$:

$$\boldsymbol{H}\boldsymbol{w}_k = H[k]\boldsymbol{w}_k, \qquad k = 0, \ldots, N-1. \tag{4.156}$$

To show this, we first consider the mth element of the left-hand side:

$$\sum_{n=0}^{N-1} h[m,n]w^{j2\pi nk/N} = \sum_{n=0}^{N-1} h[m-n]w^{j2\pi nk/N}$$
$$= \sum_{l=0}^{N-1} h[l]w^{-j2\pi lk/N} w^{j2\pi mk/N} = H[k]w^{j2\pi mk/N}$$
$$m = 0, \ldots, N-1, \tag{4.157}$$

where we have assumed $m - n = l$. This result happens be the mth element of the right-hand side of Eq. (4.156); i.e., Eq. (4.156) holds. If we further define $\boldsymbol{D} = \mathrm{diag}(H[0], \ldots, H[N-1])$ as a diagonal matrix composed of all N DFT coefficients along the main diagonal, then Eq. (4.156) can be written in matrix form as

$$\boldsymbol{H}\boldsymbol{W} = \boldsymbol{W}\boldsymbol{D}; \qquad \text{i.e.,} \qquad \boldsymbol{W}^{-1}\boldsymbol{H}\boldsymbol{W} = \overline{\boldsymbol{W}}\boldsymbol{H}\boldsymbol{W} = \boldsymbol{D}. \tag{4.158}$$

We see that the circulant matrix $\boldsymbol{H}$ is diagonalized by the DFT matrix $\boldsymbol{W} = [\boldsymbol{w}_0, \ldots, \boldsymbol{w}_{N-1}]$. Now by taking the DFT on both sides of $\boldsymbol{y} = \boldsymbol{H}\boldsymbol{x}$ in Eq. (4.155) (by pre-multiplying $\overline{\boldsymbol{W}}$), we get

$$\boldsymbol{Y} = \mathcal{F}[\boldsymbol{y}] = \overline{\boldsymbol{W}}\boldsymbol{y} = \overline{\boldsymbol{W}}\boldsymbol{H}\boldsymbol{x} = \overline{\boldsymbol{W}}\boldsymbol{H}\boldsymbol{W}\,\overline{\boldsymbol{W}}\boldsymbol{x} = \boldsymbol{D}\boldsymbol{X}, \tag{4.159}$$

or in component form as:

$$\begin{bmatrix} Y[0] \\ Y[1] \\ \vdots \\ Y[N-1] \end{bmatrix} = \begin{bmatrix} H[0] & 0 & \cdots & 0 \\ 0 & H[1] & \cdots & 0 \\ \vdots & \vdots & \ddots & \vdots \\ 0 & 0 & \cdots & H[N-1] \end{bmatrix} \begin{bmatrix} X[0] \\ X[1] \\ \vdots \\ X[N-1] \end{bmatrix}. \tag{4.160}$$

The kth element of this vector equation is

$$Y[k] = H[k]\,X[k]. \tag{4.161}$$

Of course, we realize this is the matrix form of the discrete convolution theorem.

We further consider two issues regarding the N DFT coefficients. First, we consider the interpretation of the DFT coefficients $X[k]$ $(k = 0, \ldots, N-1)$ in order to know how they can be properly modified for various desired data processing purposes such as filtering (e.g., low-, band- or high-pass/stop). Here, we assume the time signal $\boldsymbol{x} = [x[0], \ldots, x[N-1]]^{\mathrm{T}}$ is real and N is even, therefore the real part of its spectrum $X_{\mathrm{r}}[k] = X_{\mathrm{r}}[-k]$ is even and the imaginary part $X_{\mathrm{j}}[k] = -X_{\mathrm{j}}[-k]$ is odd, and the inverse DFT can be written as

$$x[n] = \mathrm{Re}\left[\sum_{k=0}^{N-1} X[k] e^{j2\pi nk/N}\right] \tag{4.162}$$

$$= \sum_{k=0}^{N-1} [X_{\mathrm{r}}[k]\cos(2\pi nk/N) - X_{\mathrm{j}}[k]\sin(2\pi nk/N)]$$

$$= \sum_{k=0}^{N-1} |X[k]| \cos(2\pi nk/N + \angle X[k]), \tag{4.163}$$

where

$$\begin{cases} |X[k]| = \sqrt{X_{\mathrm{r}}^2[k] + X_{\mathrm{j}}^2[k]} \\ \angle X[k] = \tan^{-1}(X_{\mathrm{j}}[k]/X_{\mathrm{r}}[k]) \end{cases} \qquad \begin{cases} X_{\mathrm{r}}[k] = |X[k]|\cos\angle X[k] \\ X_{\mathrm{j}}[k] = |X[k]|\sin\angle X[k] \end{cases}. \tag{4.164}$$

The forward DFT is

$$X[k] = \frac{1}{N}\sum_{n=0}^{N-1} x[n]e^{-j2\pi nk/N} = \frac{1}{N}\sum_{n=0}^{N-1} x[n][\cos(2\pi nk/N) - j\sin 2\pi nk/N]. \tag{4.165}$$

These N DFT coefficients and the frequency components they represent are illustrated in Fig. 4.18.

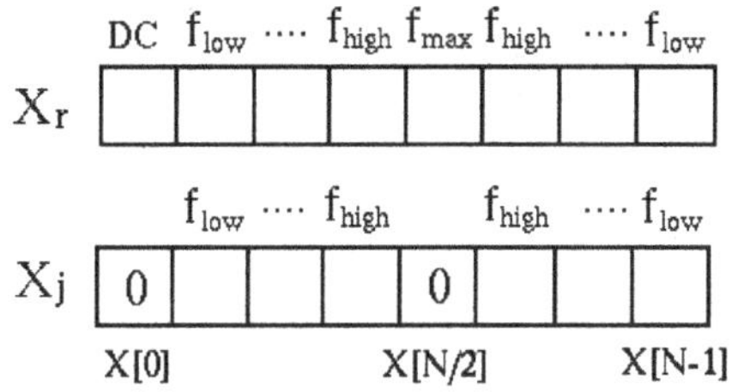

Figure 4.18 DFT coefficients for different frequency components.

Consider specifically the following terms in the summation in Eq. (4.163).

- $k = 0$:

$$X[0] = \frac{1}{N}\sum_{n=0}^{N-1} x[n]. \tag{4.166}$$

This is the DC component, which is real with zero phase.

- $k = N/2$:

$$X[N/2] = \frac{1}{N}\sum_{n=0}^{N-1} x[n]\cos(n\pi) = \frac{1}{N}\sum_{n=0}^{N-1} x[n](-1)^n$$
$$= \frac{1}{N}\sum_{n=0,2,\cdots,N-2} [x[n] - x[n+1]]. \tag{4.167}$$

This is the coefficient for the highest frequency component $\cos(n\pi) = (-1)^n$ with frequency $f_{\max} = 1/2$ with period $1/f_{\max} = 2$. Same as $X[0]$, $X[N/2]$ is also real with zero phase shift.

- $k = 1, \dots, N/2 - 1$:
These terms represent $(N-2)/2$ sinusoids $|X[k]|\cos(2\pi nk/N + \angle X[k])$ with frequency k/N, amplitude $|X[k]|$ and phase shift $\angle X[k]$.
- $k = N/2 + 1, \dots, N-1$:
Owing to the periodicity $X[k-N] = X[k]$, these terms are the same as those in the range $k = -1, \dots, -(N/2-1)$ or $-k = 1, \dots, N/2 - 1$ and we have

$$|X[-k]|\cos(-2\pi nk/N + \angle X[-k]) = |X[k]|\cos(2\pi nk/N + \angle X[k]). \tag{4.168}$$

(Note that $\angle X[k]$ is odd and the cosine function is even.) These are the same sinusoids as those in the previous range.

Combining all the terms above together we can rewrite Eq. (4.163) as

$$x[n] = X[0] + X[N/2]\cos(n\pi) + 2\sum_{k=1}^{N/2-1} |X[k]|\cos(2\pi nk/N + \angle X[k]). \tag{4.169}$$

This is the discrete version of Eq. 3.144 in the case of the continuous Fourier transform.

Example 4.10: The $N = 8$ samples of a real signal $x[n]$ are given as a complex vector with zero imaginary part:

$$\boldsymbol{x} = [(0,0),(0,0),(2,0),(3,0),(4,0),(0,0),(0,0),(0,0)]^{\mathrm{T}}. \tag{4.170}$$

The real and imaginary parts of the eight-point DFT matrix $\boldsymbol{W}$ are given in Eqs. (4.139) and (4.140) respectively. The DFT of the signal can be carried out by matrix multiplication:

$$\boldsymbol{X} = \overline{\boldsymbol{W}}\boldsymbol{x}, \tag{4.171}$$

where $\boldsymbol{X} = \boldsymbol{X}_{\mathrm{r}} + j\boldsymbol{X}_{\mathrm{j}}$ are the $N = 8$ DFT coefficients:

$$\begin{aligned}\boldsymbol{X}_{\mathrm{r}} &= \left[3.18,\ -2.16,\ 0.71,\ -0.66,\ 1.06,\ -0.66,\ 0.71,\ -2.16\right]^{\mathrm{T}},\\ \boldsymbol{X}_{\mathrm{j}} &= \left[0.0,\ -1.46,\ 1.06,\ -0.04,\ 0.0,\ 0.04,\ -1.06,\ 1.46\right]^{\mathrm{T}}.\end{aligned} \tag{4.172}$$

Note that the real and imaginary parts are even and odd, respectively. The kth complex DFT coefficient $X[k] = X_r[k] + j\,X_j[k]$ can also be expressed as the magnitude and phase $|X[k]| = \sqrt{X_r^2[k] + X_j^2[k]}$ and $\angle X[k] = \tan^{-1}[X_j[k]/X_r[k]]$ $(k = 0, \cdots, N-1)$. In vector form we have:

$$\begin{aligned}|\boldsymbol{X}| &= \left[3.18,\ 2.61,\ 1.27,\ 0.67,\ 1.06,\ 0.67,\ 1.27,\ 2.61\right]^{\mathrm{T}},\\ \angle\boldsymbol{X} &= \left[0.0^\circ,\ -146.05^\circ,\ 56.31^\circ,\ -176.30^\circ,\ 0.0^\circ,\ 176.30^\circ,\ -56.31^\circ,\ 146.05^\circ\right]^{\mathrm{T}}.\end{aligned} \tag{4.173}$$

Note that the magnitude and phase are even and odd, respectively.

The signal $x[n]$ can be reconstructed by the inverse DFT from its DFT coefficients $X[k]$:

$$\boldsymbol{x} = \begin{bmatrix} x[0] \\ \vdots \\ x[7] \end{bmatrix} = \boldsymbol{W}\boldsymbol{X} = [\boldsymbol{w}_0, \ldots, \boldsymbol{w}_7] \begin{bmatrix} X[0] \\ \vdots \\ X[7] \end{bmatrix} = \sum_{k=0}^{7} X[k]\boldsymbol{w}_k. \tag{4.174}$$

The reconstruction of this eight-point discrete signal as a linear combination of its frequency components is illustrated in columns 3 (real) and 4 (imaginary) of Fig. 4.17, as the discrete version of the corresponding Fourier series expansion of a continuous signal. Here, progressively more and higher frequency components are included in the reconstruction for better approximation of the signal, from the DC component alone (top row) until all N frequency components are used for a perfect reconstruction (last row).

Consider further the shifted version of the 8-point signal $y[n] = x[n - n_0] = x[n-1]$ (with shift amount $n_0 = 1$):

$$\boldsymbol{y} = [(0,0),(0,0),(0,0),(2,0),(3,0),(4,0),(0,0),(0,0)]^{\mathrm{T}}. \tag{4.175}$$

Its complex DFT coefficients are

$$\begin{aligned}\boldsymbol{Y}_{\mathrm{r}} &= \left[3.18,\ -2.56,\ 1.06,\ 0.44,\ -1.06,\ 0.44,\ 1.06,\ -2.56\right]^{\mathrm{T}},\\ \boldsymbol{Y}_{\mathrm{j}} &= \left[0.0,\ 0.50,\ -0.71,\ 0.50,\ 0.0,\ -0.50,\ 0.71,\ -0.50\right]^{\mathrm{T}}.\end{aligned} \tag{4.176}$$

or in the form of magnitude and phase:

$$|\boldsymbol{Y}| = \left[3.18,\ 2.61,\ 1.27,\ 0.67,\ 1.06,\ 0.67,\ 1.27,\ 2.61\right]^{\mathrm{T}},$$
$$\angle\boldsymbol{Y} = \left[0.0^\circ,\ 168.95^\circ,\ -33.69^\circ,\ 48.69^\circ,\ -180.0^\circ,\ -48.69^\circ,\ 33.69^\circ,\ -168.95^\circ\right]^{\mathrm{T}}. \tag{4.177}$$

The time shift property of the DFT (first equation of Eq. (4.142)) can be verified by Comparing Eqs. (4.173) and (4.177). Specifically, the DFT of the time shifted signal has the same magnitude as that of the original signal, but its phase is shifted by $-2\pi n_0 k/N = -k\pi/4$ ($k = 0, \ldots, 7$); i.e., a multiple of $-\pi/4$ or -45°.

We consider each of the $N = 8$ complex coefficients $X[k] = X_{\mathrm{r}}[k] + jX_{\mathrm{j}}[k]$ given in Eq. (4.172) in Example 4.10.

- $X_{\mathrm{r}}[0] = 3.18/\sqrt{8}$ is proportional to the sum of all signal samples $x[n]$; therefore, it represents the average of the signal. As $X_{\mathrm{j}}[0] = 0$, $\angle X[0] = 0$.
- $X_{\mathrm{r}}[4] = 1.06/\sqrt{8}$ is the amplitude of the highest frequency component with $f_4 = 4/8$. As $X_{\mathrm{j}}[4] = 0$, $\angle X[4] = 0$.
- The remaining $(N-2)/2 = 3$ pairs of terms corresponding to $k = 1, 7$, $k = 2, 6$ and $k = 3, 5$ represent three sinusoids with frequency $f_k = k/N$, amplitude $|X[k]| = \sqrt{X_{\mathrm{r}}^2[k] + X_{\mathrm{j}}^2[k]}$, and phase $\angle X[k] = \tan^{-1}(X_{\mathrm{j}}[k]/X_{\mathrm{r}}[k])$.
  - $k = 1, 7$:
    $f_1 = 1/8,\ \omega_1 = 0.79,\ |X[1]| = 2.61/\sqrt{8},\ \angle X[1] = -2.55$ rad/s.
  - $k = 2, 6$:
    $f_2 = 2/8,\ \omega_2 = 1.57,\ |X[2]| = 1.28/\sqrt{8},\ \angle X[2] = 0.98$ rad/s.
  - $k = 3, 5$:
    $f_3 = 3/8,\ \omega_3 = 2.36,\ |X[3]| = 0.67/\sqrt{8},\ \angle X[3] = -3.08$ rad/s.

Now the signal can be expanded as (Eq. 4.169)

$$\begin{aligned} x[n] &= \frac{1}{\sqrt{N}}\left[X[0] + 2\sum_{n=1}^{3}|X[k]|\cos\left(\frac{2\pi nk}{N} + \angle X[k]\right) + X[4]\cos(m\pi)\right] \\ &= \frac{1}{\sqrt{8}}\{3.18 + 2[2.61\cos(0.79n - 2.55) + 1.28\cos(1.57n + 0.98) \\ &\quad +0.67\cos(2.36n - 3.08)] + 1.06\cos(3.14n)\} \qquad n = 0, \ldots, 7. \end{aligned} \tag{4.178}$$

Next, we consider the centralization of the DFT spectrum. In all previous discussions regarding the Fourier spectrum, the DC component of zero frequency at the origin is always conceptually assumed to be in the middle of the frequency axis, while the higher frequencies (both positive and negative) are farther away from the middle point on both sides of the origin. However, on the other hand, the N DFT coefficients $X[k]$ in the vector $\boldsymbol{X} = \overline{\boldsymbol{W}}\boldsymbol{x}$ generated by the DFT algorithm are indexed in such a way that the DC component $X[0]$ of zero frequency is the first (leftmost) element of $\boldsymbol{X}$ while the highest frequency component $X[N/2]$ is in the middle. Therefore, it is sometimes desirable to rearrange the DFT spectrum

$\boldsymbol{X}$ so that it is consistent with the conceptual form of the spectrum. Specifically, this centralization process can be carried out by right shifting all components $X[k]$ in $\boldsymbol{X}$ by $N/2$, so that in the resulting vector $X'[k+N/2] = X[k]$ the DC component ($k=0$) appears in the middle at $N/2$, while the elements in the first half ($k < N/2$) for the positive frequencies are shifted to the second half to the right of the DC component, and those originally in the second half ($k > N/2$) for the negative frequencies are shifted to the first half to the left of the DC, owing to the periodicity $X'[k+N/2] = X'[k+N/2-N] = X'[k-N/2]$ (i.e., right shift by $N/2$ is equivalent to left shift by $N/2$). This process is illustrated in Fig. 4.19.

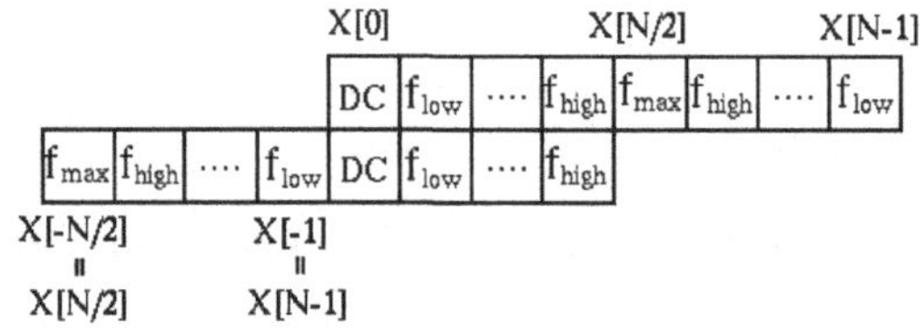

Figure 4.19 Centralization of the DFT spectrum. Coefficient indexing of DFT algorithm (top) and conceptual DFT spectrum (bottom). Note that $X[-N/2] = X[N/2], \ldots, X[-1] = X[N-1]$.

Computationally, according to the frequency shift property of the DFT, the centralization process can also be realized in the time domain before the DFT, by multiplying the time sample $x[n]$ by $e^{jn\pi} = (-1)^n$:

$$\mathcal{F}[x[n]e^{jn\pi}] = \mathcal{F}[x[n](-1)^n] = X[k-N/2]. \tag{4.179}$$

In other words, the DFT spectrum of a discrete signal is centralized if all of its odd-indexed samples are negated so that it becomes $x[0], -x[1], x[2], -x[3], \ldots$.

As an example, the real signal $\boldsymbol{x} = [0,0,2,3,4,0,0,0]^{\mathrm{T}}$ and its DFT coefficients in Example 4.10 are plotted in the top two panels of Fig. 4.20, respectively. If we negate all odd-indexed elements of the signal, the spectrum becomes centralized, as plotted in the third panel of the figure. Note that as the time signal is real, the real part of the spectrum is even: $X_{\mathrm{r}}[1] = X_{\mathrm{r}}[7]$, $X_{\mathrm{r}}[2] = X_{\mathrm{r}}[6]$, $X_{\mathrm{r}}[3] = X_{\mathrm{r}}[5]$; and the imaginary part is odd: $X_{\mathrm{j}}[1] = -X_{\mathrm{j}}[7]$, $X_{\mathrm{j}}[2] = -X_{\mathrm{j}}[6]$, $X_{\mathrm{j}}[3] = -X_{\mathrm{j}}[5]$. Also note that $X_{\mathrm{r}}[0] \neq X_{\mathrm{r}}[4]$ and $X_{\mathrm{j}}[0] = X_{\mathrm{j}}[4] = 0$ are always zero.

$$\begin{aligned} \boldsymbol{X}_{\mathrm{r}} &= \left[1.06,\ -0.66,\ 0.71,\ -2.16\ 3.18,\ -2.16,\ 0.71,\ -0.66\right]^{\mathrm{T}}, \\ \boldsymbol{X}_{\mathrm{j}} &= \left[0.0,\ 0.04,\ -1.06,\ 1.46\ 0.0,\ -1.46,\ 1.06,\ -0.04\right]^{\mathrm{T}}. \end{aligned} \tag{4.180}$$

Example 4.11: We reconsider the discrete convolution in Example 1.139 of two sequences $\boldsymbol{h} = [1\ 2\ 3]^{\mathrm{T}}$ and $\boldsymbol{x} = [1\ 2\ 3\ 4\ 5\ 6\ 7\ 8]^{\mathrm{T}}$ of three and eight elements respectively. As their convolution contains $N = 8+3-1 = 10$ elements, we

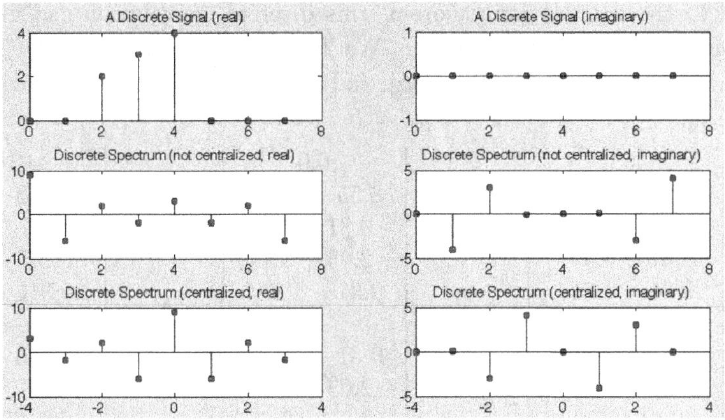

Figure 4.20 A discrete signal (top) and its DFT spectrum before (middle) and after (bottom) centralization. The real and imaginary parts of these complex coefficients are shown respectively in the middle and bottom panels.

augment both sequences so that they become N-D vectors:

$$\boldsymbol{x} = [1\,2\,3\,4\,5\,6\,7\,8\,0\,0]^{\mathrm{T}}, \qquad \boldsymbol{h} = [1\,2\,3\,0\,0\,0\,0\,0\,0\,0]^{\mathrm{T}}. \tag{4.181}$$

and assumed to be periodic $x[n+N] = x[n]$ and $h[n+N] = h[n]$.

Now their convolution $y[n] = h[n] * x[n]$ becomes a *circular convolution*:

$$y[n] = h[n] * x[n] = \sum_{m=0}^{N-1} h[n-m]x[m] \qquad n = 0, \ldots, N-1, \tag{4.182}$$

which is obviously also periodic $y[n+N] = y[n]$. This circular convolution is shown in the table in Eq. (4.183).

m	$\cdots$	-4	-3	-2	-1	0	1	2	3	4	5	6	7	8	9	10	11	12	$\cdots$
$x[m]$	$\cdots$	7	8	0	0	1	2	3	4	5	6	7	8	0	0	1	2	3	$\cdots$
$h[-2-m]$	$\cdots$	3	2	1															$\cdots$
$h[-1-m]$	$\cdots$		3	2	1														$\cdots$
$h[0-m]$	$\cdots$			3	2	1													$\cdots$
$h[1-m]$	$\cdots$				3	2	1												$\cdots$
$h[2-m]$	$\cdots$					3	2	1											$\cdots$
$h[3-m]$	$\cdots$						3	2	1										$\cdots$
$h[4-m]$	$\cdots$							3	2	1									$\cdots$
$h[5-m]$	$\cdots$								3	2	1								$\cdots$
$h[6-m]$	$\cdots$									3	2	1							$\cdots$
$h[7-m]$	$\cdots$										3	2	1						$\cdots$
$h[8-m]$	$\cdots$											3	2	1					$\cdots$
$h[9-m]$	$\cdots$												3	2	1				$\cdots$
$h[10-m]$	$\cdots$													3	2	1			$\cdots$
$h[11-m]$	$\cdots$														3	2	1		$\cdots$
$h[12-m]$	$\cdots$															3	2	1	$\cdots$
$y[n]$	$\cdots$	34	40	37	24	1	4	10	16	22	28	34	40	37	24	1	4	10	$\cdots$

(4.183)

Owing to the convolution theorem, this discrete convolution can also be carried out in the Fourier domain. We first find the 10-point DFT of both sequences $\boldsymbol{X} = \mathcal{F}[\boldsymbol{x}]$ and $\boldsymbol{H} = \mathcal{F}[\boldsymbol{h}]$, and then obtain their element-wise product $\boldsymbol{Y} = [Y[0], \cdots, Y[9]]^{\mathrm{T}}$, where $Y[k] = H[k]X[k]$ $(k = 0, \cdots, N-1 = 9)$:

$$\boldsymbol{X} = \begin{bmatrix} 36.0 \\ -15.87 - 0.95j \\ -0.81 + 7.92j \\ 5.37 + 0.59j \\ 0.31 - 4.31j \\ -4.0 \\ 0.31 + 4.31j \\ 5.37 - 0.59j \\ -0.81 - 7.92j \\ -15.87 + 0.95j \end{bmatrix}, \quad \boldsymbol{H} = \begin{bmatrix} 6.0 \\ 3.55 + 4.03j \\ -0.81 + 3.67j \\ -2.05 + 0.14j \\ 0.31 - 1.68j \\ 2.0 \\ 0.31 + 1.68j \\ -2.05 - 0.14j \\ -0.81 - 3.67j \\ 3.55 - 4.03j \end{bmatrix}, \quad \boldsymbol{Y} = \begin{bmatrix} 216.0 \\ -52.43 - 67.31j \\ 28.37 - 9.37j \\ 11.07 - 0.46j \\ -7.13 - 1.85j \\ -8.0 \\ -7.13 + 1.85j \\ -11.07 + 0.46j \\ -28.37 + 9.37j \\ -52.43 + 67.31j \end{bmatrix}. \tag{4.184}$$

The convolution $y[n] = h[n] * x[n]$ can be obtained by inverse DFT to be

$$\boldsymbol{y} = \mathcal{F}^{-1}[\boldsymbol{Y}] = [1\ 4\ 10\ 16\ 22\ 28\ 34\ 40\ 37\ 24]^{\mathrm{T}}, \tag{4.185}$$

which is the same as the result obtained by the circular convolution.

4.2.4 Four different forms of the Fourier transform

The various forms of the Fourier transform for different types of signals (periodic or non-periodic, continuous or discrete) discussed in the current and previous chapters can be considered as the following four different variations of the most generic Fourier transform as shown below.

- **I. Non-periodic continuous signal, continuous, non-periodic spectrum**
  This is the most generic form of the Fourier transform for any continuous and non-periodic signal $x(t)$, considered as a function in a function space spanned by a set of uncountably infinite basis functions $\phi_f(t) = e^{j2\pi ft}$ $(-\infty < f < \infty)$ that are orthonormal according to Eq. (1.28):

$$\langle \phi_f(t), \phi_{f'}(t) \rangle = \int_{-\infty}^{\infty} e^{j2\pi(f-f')t}\, dt = \delta(f - f'). \tag{4.186}$$

  The signal $x(t)$ can therefore be expressed as a linear combination (integral) of these uncountable basis functions as

$$x(t) = \int_{-\infty}^{\infty} X(f)\phi_f(t)\, df = \int_{-\infty}^{\infty} X(f)e^{j2\pi ft}\, df. \tag{4.187}$$

  This is the inverse transform and the coefficient function $X(f)$ can be obtained as the projection of the signal onto each of the basis functions:

$$X(f) = \langle x(t), \phi_f(t) \rangle = \langle x(t), e^{j2\pi ft} \rangle = \int_{-\infty}^{\infty} x(t)e^{-j2\pi ft}\, dt. \tag{4.188}$$

  This is the forward transform.

- **II. Periodic continuous signal, discrete non-periodic spectrum**
This is the Fourier series expansion of a continuous and periodic signal $x_T(t + T) = x_T(t)$, considered as a vector in the space of periodic functions spanned by a set of countable basis functions $\phi_k(t) = e^{j2\pi kt/T}/\sqrt{T}$ (for all integer k) that are orthonormal according to Eq. (1.33):

$$\langle \phi_k(t), \phi_l(t) \rangle = \int_T e^{j2\pi(k-l)t/T} \, dt = \delta[k - l]. \tag{4.189}$$

The signal $x_T(t)$ can therefore be expressed as a linear combination (summation) of these basis functions as

$$x_T(t) = \sum_{k=-\infty}^{\infty} X[k] \phi_k(t) = \frac{1}{\sqrt{T}} \sum_{k=-\infty}^{\infty} X[k] e^{j2\pi kt/T}. \tag{4.190}$$

This is the inverse transform and the coefficient $X[k]$ can be obtained as the projection of the signal onto the kth basis function:

$$X[k] = \langle x_T(t), \phi_k(t) \rangle = \left\langle x_T(t), \frac{1}{\sqrt{T}} e^{j2\pi kt/T} \right\rangle = \frac{1}{\sqrt{T}} \int_T x_T(t) e^{-j2\pi kt/T} \, dt. \tag{4.191}$$

This is the forward transform. These Fourier expansion coefficients for a periodic signal can be considered as the samples of a continuous spectrum:

$$X(f) = \sum_{k=-\infty}^{\infty} X[k] \delta(f - k f_0), \tag{4.192}$$

where any two consecutive frequency components are separated by $f_0 = 1/T$.

- **III. Non-periodic discrete signal, continuous periodic spectrum**
This is the discrete-time Fourier transform of a discrete and non-periodic signal

$$x(t) = \sum_{n=-\infty}^{\infty} x[n] \delta(t - n t_0). \tag{4.193}$$

The sequence of signal samples $x[n]$ (for all integer n) form an infinite dimensional vector $\boldsymbol{x} = [\ldots, x[n], \ldots]^{\mathrm{T}}$ in the vector space of all such vectors spanned by an uncountably infinite set of basis vectors $\boldsymbol{\phi}_f = [\ldots, e^{j2\pi nf/F}/\sqrt{F}, \ldots]^{\mathrm{T}}$ $(0 < f < F)$ that are orthonormal according to Eq. (1.35):

$$\langle \boldsymbol{\phi}_f, \boldsymbol{\phi}_{f'} \rangle = \frac{1}{F} \sum_{n=-\infty}^{\infty} e^{j2\pi n(f-f')} = \sum_{k=-\infty}^{\infty} \delta(f - f' - kF). \tag{4.194}$$

The signal $\boldsymbol{x}$ can therefore be expressed as a linear combination (integral) of these uncountable basis vectors as

$$\boldsymbol{x} = \int_F X(f) \boldsymbol{\phi}_f \, df, \tag{4.195}$$

or in component form:

$$x[n] = \frac{1}{\sqrt{F}} \int_F X(f) e^{j2\pi nf/F} \, df. \tag{4.196}$$

This is the inverse transform, and the coefficient function $X(f)$ can be obtained as the projection of the signal onto each basis function:

$$X(f) = \langle \boldsymbol{x}, \boldsymbol{\phi}_f \rangle = \frac{1}{\sqrt{F}} \sum_{n=-\infty}^{\infty} x[n] e^{-j2\pi n f/F}. \tag{4.197}$$

This is the forward transform. Here, $X(f+F) = X(f)$ is periodic.

- **IV. Periodic discrete signal, discrete periodic spectrum**
This is the DFT of a discrete and periodic signal $\boldsymbol{x} = [x[0], \ldots, x[N-1]]^{\mathrm{T}}$, which is an N-D vector in an N-D unitary space spanned by a set of N N-D vectors $\boldsymbol{\phi}_k = [e^{j2\pi 0k/N}, \ldots, e^{j2\pi(N-1)k/N}]^{\mathrm{T}}/\sqrt{N}$ that are orthonormal according to Eq. (1.40):

$$\langle \boldsymbol{\phi}_k, \boldsymbol{\phi}_l \rangle = \frac{1}{N} \sum_{n=0}^{N-1} e^{j2\pi n(k-l)/N} = \sum_{n=-\infty}^{\infty} \delta[k-l-nN]. \tag{4.198}$$

The signal vector can therefore be expressed as a linear combination (summation) of the N basis vectors:

$$\boldsymbol{x} = \sum_{k=0}^{N-1} X[k] \boldsymbol{\phi}_k; \tag{4.199}$$

or in component form:

$$x[n] = \frac{1}{\sqrt{N}} \sum_{k=0}^{N-1} X[k] e^{j2\pi nk/N} \qquad n = 0, 1, \ldots, N-1. \tag{4.200}$$

This is the inverse transform, and the weighting coefficient $X[k]$ can be obtained as the projection of the signal onto each basis function:

$$X[k] = \langle \boldsymbol{x}, \boldsymbol{\phi}_k \rangle = \frac{1}{\sqrt{N}} \sum_{n=0}^{N-1} x[n] e^{-j2\pi nk/N} \qquad k = 0, 1, \ldots, N-1. \tag{4.201}$$

Here, the discrete signal $x[n]$ are the samples of a continuous function:

$$x_T(t) = \sum_{n=0}^{N-1} x[n] \delta(t - nt_0); \tag{4.202}$$

and similarly, the frequency coefficients can be considered as the samples of a continuous spectrum:

$$X_F(f) = \sum_{k=0}^{N-1} X[k] \delta(f - kf_0). \tag{4.203}$$

The four forms of Fourier transform can be summarized in Table 4.1 (where $T = 1/f_0$, $F = 1/t_0$, $T/t_0 = F/f_0 = N$).

Table 4.1. Four different forms of the Fourier transform.

	Signal $x(t)$	Spectrum $X(f)$
I	Continuous, Non-periodic $x(t) = \int_{-\infty}^{\infty} X(f) e^{j2\pi ft}\, df$	Non-periodic, Continuous $X(f) = \int_{-\infty}^{\infty} x(t) e^{-j2\pi ft}\, dt$
II	Continuous, Periodic (T) $x_T(t) = \sum_{k=-\infty}^{\infty} X[k] e^{j2\pi k f_0 t}$	Non-periodic, Discrete (f) $X[k] = \int_T x_T(t) e^{-j2\pi k f_0 t}\, dt/T$ $X(f) = \sum_{k=-\infty}^{\infty} X[k] \delta(f - kf_0)$
III	Discrete (t_0), Non-periodic $x(t) = \sum_{n=-\infty}^{\infty} x[n] \delta(t - nt_0)$ $x[n] = \int_F X_F(f) e^{j2\pi fnt_0}\, df/F$	Periodic (F), Continuous $X_F(f) = \sum_{n=-\infty}^{\infty} x[n] e^{-j2\pi fnt_0}$
IV	Discrete (t_0), Periodic (T) $x[n] = \sum_{k=0}^{N-1} X[k] e^{j2\pi nk/N} / \sqrt{N}$ $x(t) = \sum_{n=0}^{N-1} x[n] \delta(t - nt_0)$ $T/t_0 = N$	Periodic (F), Discrete (f_0) $X[k] = \sum_{n=0}^{N-1} x[n] e^{-j2\pi nk/N} / \sqrt{N}$ $X(f) = \sum_{k=0}^{N-1} X[k] \delta(f - kf_0)$ $F/f_0 = T/t_0 = N$

These four forms of Fourier transform are also illustrated graphically in Fig. 4.21.

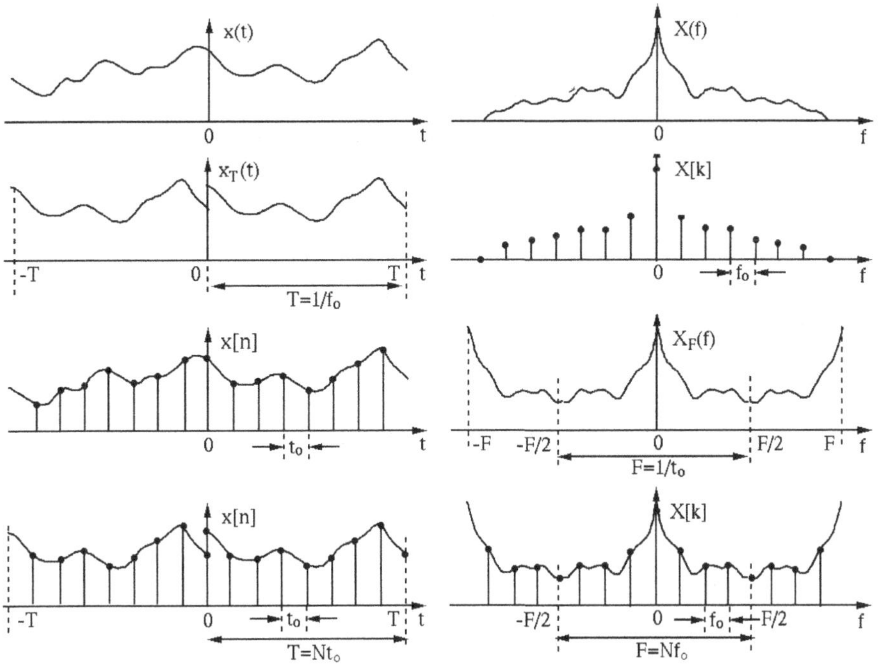

Figure 4.21 Four different forms of the Fourier transform. Various types of time signals (continuous/discrete, periodic/non-periodic) on the left and their spectra on the right.

Note in particular the relationship between the time and frequency domains: the continuity and discreteness in one domain correspond to, respectively, the non-periodicity and periodicity in the other.

All four forms of the Fourier transform share the same set of properties, discussed most thoroughly for the continuous and non-periodic case, although they take different forms for each of the four cases.

4.2.5 DFT computation and fast Fourier transform

The Fourier transform of a signal can be carried out numerically by a computer only if the signal is both discrete and finite; i.e., out of all four different forms of the Fourier transform discussed above, only the DFT can actually be carried out. Based on Eq. (4.110, the forward and inverse DFT can be implemented by the following Matlab code:

```
X=[exp(-j*2*pi*[0:N-1]'*[0:N-1]/N)/sqrt(N)]*x;   // forward DFT

x=[exp(j*2*pi*[0:N-1]'*[0:N-1]/N)/sqrt(N)]*X;    // inverse DFT
```

where the signal `x` and its spectrum `X` are both assumed to be N-D column vectors, same as in the text. Matlab also has a built-in function for the fast Fourier transform (FFT) to be discussed below.

The C code for DFT and inverse DFT based on array multiplications in Eq. (4.131) is listed below. The function `dft` takes both the real part `xr` and imaginary part `xi` of a complex signal vector as the input, and returns the complex DFT coefficients. This is an in-place algorithm; i.e., the input vector `xr[n]+j xi[n]` ($n = 0, \ldots, N-1$) for the time signal will be overwritten by the output, its DFT coefficients. The same function is also used for the inverse DFT, in which case the input is the DFT coefficients while the output is the reconstructed signal vector in the time domain. The function carries out forward DFT when the parameter `inv=0`, or inverse DFT when `inv=1`.

```
void dft(xr,xi,N,inv)
     float *xr, *xi;        // real and imaginary parts of data
     int N;                 // size of data
     int inverse;           // inv=0 forward DFT, inv=1 inverse DFT
{ int k,m,n;
  float arg,s,c,*yr,*yi;
  yr=(float *) malloc(N*sizeof(float));
  yi=(float *) malloc(N*sizeof(float));
  for (k=0; k<N; k++) {    // for all N frequency components
    yr[k]=yi[k]=0;
    for (n=0; n<N; n++) { // for all N data samples
      arg=-2*Pi*n*k/N;
```

```
        if (inv) arg=-arg; // minus sign not needed for inverse DFT
        c=cos(arg); s=sin(arg);
        yr[k]+=xr[n]*c-xi[n]*s;
        yi[k]+=xi[n]*c+xr[n]*s;
      }
    }
    arg=1.0/sqrt((float)N);
    for (k=0; k<N; k++)
      { xr[k]=arg*yr[k]; xi[k]=arg*yi[k]; }
    free(yr); free(yi);
}
```

The computational complexity of this algorithm is $O(N^2)$, due obviously to the two nested loops each of size N; i.e., it takes $O(N)$ operations to obtain each of the N coefficients $X[k]$. Owing to such a high computational complexity, the actual application of the Fourier transform was quite limited in practice before the fast algorithm was available.

To speed up the computation, a revolutionary *fast Fourier transform (FFT)* algorithm was developed in the 1960s by which the complexity of a DFT is reduced from $O(N^2)$ to $O(N \log_2 N)$. For example, if the signal size is $N = 10^3 \approx 2^{10}$, then $O(N^2) = 10^6$ but $O(N \log_2 N) \approx 10^4$; the complexity is reduced by 100-fold. Owing to this significant improvement in computational efficiency, the Fourier transform became highly valuable not only theoretically but also practically.

The FFT algorithm is based on the following properties of the elements of the matrix $\boldsymbol{W}$. We first define $w_N = e^{-j2\pi/N}$ with the following properties:

$$w_N^{kN} = e^{-j2k\pi N/N} = e^{-j2k\pi} = 1, \tag{4.204}$$

$$w_{2N}^{N} = e^{-j2N\pi/2N} = e^{-j\pi} = -1. \tag{4.205}$$

$$w_{2N}^{2k} = e^{-j2k2\pi/2N} = e^{-jk2\pi/N} = w_N^k, \tag{4.206}$$

We let $N = 2M$ and write an N-point DFT as

$$\begin{aligned} X[k] &= \sum_{n=0}^{N-1} x[n] e^{j2\pi nk/N} = \sum_{n=0}^{N-1} x[n] w_N^{nk} \\ &= \sum_{n=0}^{M-1} x[2n] w_{2M}^{2nk} + \sum_{n=0}^{M-1} x[2n+1] w_{2M}^{(2n+1)k} \\ &= \sum_{n=0}^{M-1} x[2n] w_M^{nk} + \sum_{n=0}^{M-1} x[2n+1] w_M^{nk} w_{2M}^{k} \\ &= X_{\rm e}[k] + X_{\rm o}[k] w_{2M}^{k}, \end{aligned} \tag{4.207}$$

where we have used Eq. (4.206) and defined

$$X_{\rm e}[k] = \sum_{n=0}^{N-1} x[2n] w_M^{nk}, \qquad \text{and} \quad X_{\rm o}[k] = \sum_{n=0}^{N-1} x[2n+1] w_M^{nk}. \tag{4.208}$$

These are two $N/2$-point DFTs for the even- and odd- indexed signal samples, respectively. In other words, an N-point DFT is now converted into two $N/2$-point DFTs. Also note that this is only for the first half of the N coefficients $X[k]$ for $k = 0, \ldots, M-1$. The coefficients in the second half can be obtained by replacing k in Eq. 4.207 by $k + M$:

$$X[k+M] = X_e[k+M] + X_o[k+M]w_{2M}^{k+M}. \tag{4.209}$$

Owing to Eq. (4.204), we have

$$X_e[k+M] = \sum_{n=0}^{M-1} x[2n]w_M^{n(k+M)} = \sum_{n=0}^{M-1} x[2n]w_M^{nk} = X_e[k], \tag{4.210}$$

and similarly $X_o[k+M] = X_o[k]$. Also, owing to Eq. (4.205), we have

$$w_{2M}^{k+M} = w_{2M}^k w_{2M}^M = -w_{2M}^k, \tag{4.211}$$

then Eq. (4.209) can be written as

$$X[k+M] = X_e[k] - X_o[k]w_{2M}^k. \tag{4.212}$$

The N-point DFT can now be obtained from Eqs. (4.207) and (4.212) with complexity of $O(N)$, once $X_e[k]$ and $X_o[k]$ are obtained by the two $N/2$-point DFTs in Eq. (4.208), each of which can be carried out in exactly the same way. In other words, this process of reducing the data size by half can be carried out recursively $\log_2 N$ times until eventually the size is unity and the DFT coefficient is simply the same as the signal sample. This recursion is illustrated in Fig. 4.22. We see that the N-point DFT is carried out in $\log_2 N$ stages, each with $O(N)$ complexity, with total complexity of $O(N \log_2 N)$.

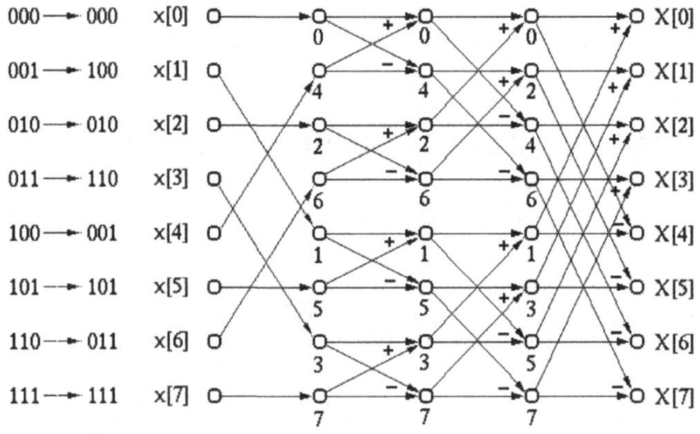

Figure 4.22 The FFT algorithm.

The C code for the FFT algorithm is given below. The function `fft` takes as input two vectors `xr` and `xi` of N elements each for the real and imaginary parts of the complex time signal, and returns its complex DFT coefficients as the outputs. Here, the total number of vector elements N is assumed to be a power of 2, so that the FFT algorithm can be conveniently implemented. This is an in-place algorithm; i.e., the input vector `xr[n]+j xi[n]` ($n = 0, \ldots, N-1$) for the time signal will be overwritten by its DFT coefficients. The same function is also used for the inverse DFT, in which case the input will be the DFT coefficients while the output is the reconstructed signal vector in the time domain. The function carries out the forward DFT when the argument `inv=0` or inverse DFT when `inv=1`. The main body of the function is composed of an outer loop of size $log_2 N$, the total number of stages, and an inner loop of size N for the computation for each stage. The computational complexity is therefore $O(N \log_2 N)$.

```
void fft(xr,xi,N,inv)
     float *xr,*xi;             // real and imaginary parts of data
     int N;                     // size of data
     int inv;                   // inv=0 for FFT, inv=1 for IFFT
{ int i,i1,j,k,l,ln,n,m;
  float arg,s,c,w,tmpr,tmpi;
  ln=log2f((float)N);
  for (i=0; i<N; ++i) {         // for all N elements of data
    j=0;
    for (k=0; k<ln; ++k)
      j=(j<<1) | (1&(i>>k)); // bit reversal
    if (j < i) {                // swap x[i] and x[j]
      w=xr[i]; xr[i]=xr[j]; xr[j]=w;
      w=xi[i]; xi[i]=xi[j]; xi[j]=w;
    }
  }
  for (i=0; i<ln; i++) {        // for log2(N) stages
    n=pow(2.0,(float)i);        // section size in current stage
    w=-Pi/n;
    if (inv) w=-w;              // no minus sign for inverse DFT
    k=0;
    while (k<N-1) {             // for N elements in a stage
      for (j=0; j<n; j++) {  // for all points in each section
        l=k+j;
        c=cos(j*w); s=sin(j*w);
        tmpr=xr[l+n]*c-xi[l+n]*s;
        tmpi=xi[l+n]*c+xr[l+n]*s;
        xr[l+n]=xr[l]-tmpr;
        xi[l+n]=xi[l]-tmpi;
        xr[l]=xr[l]+tmpr;
        xi[l]=xi[l]+tmpi;
      }
      k=k+2*n;                  // move on to next section
```

```
        }
    }
    arg=1.0/sqrt((float)N);
    for (i=0; i<N; i++)
      { xr[i]*=arg; xi[i]*=arg; }
}
```

The computational complexity for DFT can be further reduced if the signal is real, in which case the imaginary part of the signal is zero $x_{\mathrm{r}}[n]=0$ in the time domain and the real part of the spectrum is even $X_{\mathrm{r}}[-k]=X_{\mathrm{r}}[k]$ while the imaginary part is odd $X_{\mathrm{j}}[-k]=-X_{\mathrm{j}}[k]$ in the frequency domain. This 50% redundancy in either the time or frequency domain can be avoided to reduce the complexity by half.

Also, two real signal vectors $x_1[n]$ and $x_2[n]$ $(n=0,\ldots,N-1)$ can be transformed by one DFT by the following steps.

1. Construct a complex vector composed of $x_1[n]$ as its real part and $x_2[n]$ as its imaginary part:

$$x[n]=x_1[n]+j\,x_2[n] \qquad n=0,\ldots,N-1. \tag{4.213}$$

2. Obtain the DFT of $x[n]$:

$$X[k]=\mathcal{F}[x[n]]=X_{\mathrm{r}}[k]+jX_{\mathrm{j}}[k] \qquad k=0,\ldots,N-1. \tag{4.214}$$

3. Obtain $\mathcal{F}[x_1[n]]=X_1[k]=X_{1r}[k]+jX_{1j}[k]$.
   As $x_1[n]$ is real, the real part of its spectrum $X_{1r}[k]$ is even and the imaginary part $X_{1j}[k]$ is odd; i.e.,

$$X_1[k]=X_{1r}[k]+jX_{1j}[k]=\frac{X_{\mathrm{r}}[k]+X_{\mathrm{r}}[-k]}{2}+j\frac{X_{\mathrm{j}}[k]-X_{\mathrm{j}}[-k]}{2}. \tag{4.215}$$

   The two fractions extract respectively the even component of $X_{\mathrm{r}}[k]$ and the odd component of $X_{\mathrm{j}}[k]$.
4. Obtain $\mathcal{F}[x_2[n]]=X_2[k]=X_{2r}[k]+jX_{2j}[k]$.
   As $jx_2[n]$ is imaginary, the real part of its spectrum $jX_{2r}[k]$ is odd and the imaginary part $jX_{2j}[k]$ is even; i.e.,

$$jX_2[k]=jX_{2r}[k]+j(jX_{2j}[k])=\frac{X_{\mathrm{r}}[k]-X_{\mathrm{r}}[-k]}{2}+j\frac{X_{\mathrm{j}}[k]+X_{\mathrm{j}}[-k]}{2}. \tag{4.216}$$

   The two fractions extract respectively the odd component of $X_{\mathrm{r}}[k]$ and the even component of $X_{\mathrm{j}}[k]$. Dividing both sides by j, we get the spectrum $X_2[k]$ of real signal $x_2[n]$:

$$X_2[k]=X_{2r}[k]+jX_{2j}[k]=\frac{X_{\mathrm{j}}[k]+X_{\mathrm{j}}[-k]}{2}-j\frac{X_{\mathrm{r}}[k]-X_{\mathrm{r}}[-k]}{2}. \tag{4.217}$$

As we can now obtain the spectra of two signal vectors with the computation of only one, the complexity can be reduced by half.

4.3 Two-dimensional Fourier transform

4.3.1 Two-dimensional signals and their spectra

All signals considered so far are assumed to be 1-D time functions. However, a signal could also be a function over a 1-D space, with the spatial frequency defined as the number of cycles in unit length (distance), instead of in unit time. Moreover, the concept of frequency analysis can be extended to various signals in 2- or 3-D spaces. For example, an image can be considered as a 2-D signal, and computer image processing has been a very active field of study for several decades with a wide variety of applications. Like in the 1-D case, the Fourier transform is also a powerful tool in two-, or higher-dimensional signals processing and analysis. We will consider the Fourier transform of some generic 2-D continuous signal denoted by $f(x, y)$, with x and y for the two spatial dimensions.

The Fourier transform of a 2-D signal $f(x, y)$ is defined as

$$F(u, v) = \int_{-\infty}^{\infty} \int_{-\infty}^{\infty} f(x, y) e^{-j2\pi(ux+vy)} \, dx \, dy. \tag{4.218}$$

This is the forward transform, where u and v represent two spatial frequencies (cycles per unit distance) along two perpendicular directions of x and y in the 2-D space respectively. The signal can be reconstructed by the inverse transform:

$$f(x, y) = \int_{-\infty}^{\infty} \int_{-\infty}^{\infty} F(u, v) e^{j2\pi(ux+vy)} \, du \, dv, \tag{4.219}$$

by which the signal is expressed as a linear combination of infinite set of uncountable 2-D orthogonal basis functions $\phi_{u,v}(x, y) = e^{j2\pi(ux+vy)}$, weighted by the Fourier coefficient function $F(u, v)$, the 2-D spectrum of the signal.

In the following discussion, we will always assume $f(x, y) = \overline{f}(x, y)$ is a real 2-D signal. The integrand in the 2-D Fourier transform is the product of two functions: the kernel function $\phi_{u,v}(x, y) = e^{j2\pi(ux+vy)}$ of the integral transform, the orthogonal basis functions, and the spectrum $F(u, v)$, the weighting function for the basis. Below, we consider each of them separately.

- First, we consider the basis function $e^{j2\pi(ux+vy)}$.
  We define two vectors, one in the spatial domain, another in the spatial frequency domain:
  - $\boldsymbol{r}$ is a vector associated with each point (x, y) in a 2-D spatial domain:
  $$\boldsymbol{r} = [x, y]^{\mathrm{T}}. \tag{4.220}$$
  - $\boldsymbol{w}$ is a vector associated with each point (u, v) in a 2-D frequency domain:
  $$\boldsymbol{w} = [u, v]^{\mathrm{T}} = w[u/w, v/w]^{\mathrm{T}} = w\boldsymbol{n}, \tag{4.221}$$
  where $w = \sqrt{u^2 + v^2}$ is the magnitude and $\boldsymbol{n} = [u/w, v/w]^{\mathrm{T}}$ is the unit vector ($||\boldsymbol{n}|| = 1$) along the direction of $\boldsymbol{w}$.

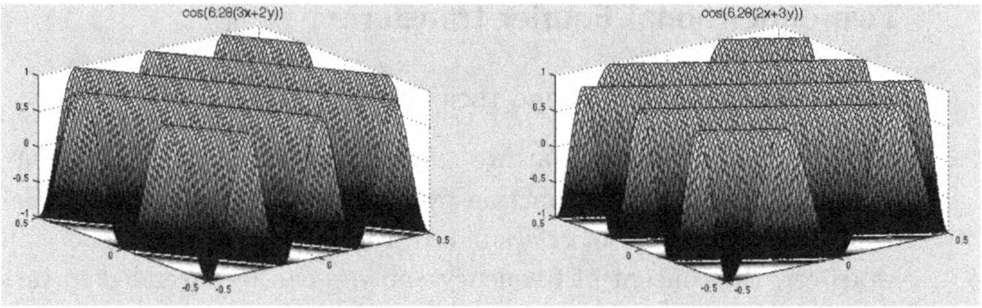

Figure 4.23 Different propagation directions of 2-D sinusoid $\cos(2\pi(ux + vy))$. The left plot is for $\cos[2\pi(3x + 2y)]$ with $u = 3$ and $v = 2$ cycles per unit length along x and y directions (the right and left sides of plot), respectively. The right plot is for $\cos[2\pi(2x + 3y)]$ with $u = 2$ and $v = 3$ cycles per unit length along x and y.

The inner product $\langle \boldsymbol{r}, \boldsymbol{n} \rangle = \boldsymbol{r}^{\mathrm{T}} \boldsymbol{n} = xu + yv$ is the projection of a vector $\boldsymbol{r}$ onto the direction of $\boldsymbol{n}$, and the 2-D basis function $\phi_{u,v}(x, y)$ can be written as

$$\begin{aligned}\phi_{u,v}(x, y) &= e^{j2\pi(xu+yv)} = e^{j2\pi w\langle \boldsymbol{r}, \boldsymbol{n}\rangle} \\ &= \cos(2\pi w\langle \boldsymbol{r}, \boldsymbol{n}\rangle) + j\,\sin(2\pi w\langle \boldsymbol{r}, \boldsymbol{n}\rangle). \end{aligned} \tag{4.222}$$

As all spatial points $\boldsymbol{r} = (x, y)$ along a straight line perpendicular to the direction $\boldsymbol{n}$ have the same projection $\langle \boldsymbol{r}, \boldsymbol{n}\rangle$, the function $\cos(2\pi w\langle \boldsymbol{r}, \boldsymbol{n}\rangle)$ takes the same value along the straight line; i.e., it is a planar sinusoid with *frequency* $w = \sqrt{u^2 + v^2}$ along the *direction* $\boldsymbol{n}$, at an angle $\theta = \tan^{-1}(v/u)$ from the positive direction of u. The same is true for the sine function of the imaginary part $\sin(2\pi w(\boldsymbol{r}^{\mathrm{T}}\boldsymbol{n}))$. For example, two 2-D sinusoidal functions $\cos(2\pi(3x + 2y))$ and $\cos(2\pi(2x + 3y))$ are shown in Fig. 4.23.

- Second, we consider the weighting function $F(u, v)$.
  As the signal $f(x, y)$ is assumed real, its Fourier coefficient $F(u, v)$ can be written as below in terms of the real and imaginary parts:

$$\begin{aligned}F(u,v) &= \int_{-\infty}^{\infty}\int_{-\infty}^{\infty} f(x,y)e^{-j2\pi(xu+yv)}\,dx\,dy \\ &= \int_{-\infty}^{\infty}\int_{-\infty}^{\infty} f(x,y)\ \cos(2\pi(xu+yv))\,dx\,dy \\ &\quad - j\int_{-\infty}^{\infty}\int_{-\infty}^{\infty} f(x,y)\ \sin(2\pi(xu+yv))\,dx\,dy \\ &= F_{\mathrm{r}}(u,v) + jF_{\mathrm{j}}(u,v) = |F(u,v)|e^{j\angle F(u,v)}, \end{aligned} \tag{4.223}$$

where $F_{\mathrm{r}}(u, v)$ and $F_{\mathrm{j}}(u, v)$ are respectively the real and imaginary parts:

$$\begin{cases} F_{\mathrm{r}}(u,v) = \int_{-\infty}^{\infty}\int_{-\infty}^{\infty} f(x,y)\cos(2\pi(xu+yv))\,dx\,dy \\ F_{\mathrm{j}}(u,v) = -\int_{-\infty}^{\infty}\int_{-\infty}^{\infty} f(x,y)\sin(2\pi(xu+yv))\,dx\,dy \end{cases}, \tag{4.224}$$

and $|F(u,v)|$ and $\angle F(u,v)$ are respectively the amplitude and phase of $F(u,v)$:

$$\begin{cases} |F(u,v)| = \sqrt{F_r^2(u,v) + F_j^2(u,v)} \\ \angle F(u,v) = \tan^{-1}[F_j(u,v)/F_r(u,v)] \end{cases}, \quad \begin{cases} F_r(u,v) = |F(u,v)| \cos \angle F(u,v) \\ F_j(u,v) = |F(u,v)| \sin \angle F(u,v) \end{cases}. \tag{4.225}$$

Note that $F_r(u,v)$ is even and $F_j(u,v)$ is odd:

$$\begin{cases} F_r(-u,-v) = F_r(u,v) \\ F_r(u,-v) = F_r(-u,v) \end{cases}, \quad \begin{cases} F_j(-u,-v) = -F_j(u,v) \\ F_j(u,-v) = -F_j(-u,v) \end{cases}, \tag{4.226}$$

and $|F(u,v)|$ is even and $\angle F(u,v)$ is odd:

$$\begin{cases} |F(-u,-v)| = |F(u,v)| \\ |F(u,-v)| = |F(-u,v)| \end{cases}, \quad \begin{cases} \angle F(-u,-v) = -\angle F(u,v) \\ \angle F(u,-v) = -\angle F(-u,v) \end{cases}. \tag{4.227}$$

Combining the two aspects considered above, we can rewrite the inverse 2-D Fourier transform as

$$\begin{aligned} f(x,y) &= \int_{-\infty}^{\infty}\int_{-\infty}^{\infty} |F(u,v)| e^{j\angle F(u,v)} e^{j2\pi(xu+yv)}\, du\, dv \\ &= \int_{-\infty}^{\infty}\int_{-\infty}^{\infty} |F(u,v)| \cos[2\pi(ux+vy) + \angle F(u,v)]\, du\, dv \\ &\quad + j\int_{-\infty}^{\infty}\int_{-\infty}^{\infty} |F(u,v)| \sin[2\pi(ux+vy) + \angle F(u,v)]\, du\, dv \\ &= \int_{-\infty}^{\infty}\int_{-\infty}^{\infty} |F(u,v)| \cos[2\pi w\langle \boldsymbol{r}, \boldsymbol{n}\rangle + \angle F(u,v)]\, du\, dv. \end{aligned} \tag{4.228}$$

Note that the imaginary part is dropped as $f(x,y)$ is real. We now see that $f(x,y)$ is a superposition of uncountably infinite 2-D spatial sinusoids $|F(u,v)| \cos[2\pi w\langle \boldsymbol{r}, \boldsymbol{n}\rangle + \angle F(u,v)]$ with

- **frequency** $w = \sqrt{u^2+v^2}$,
- **direction** $\boldsymbol{n}$ (with angle $\theta = \tan^{-1}(v/u)$ from the positive direction of u in the 2-D spatial frequency domain),
- **amplitude** $|F(u,v)| = \sqrt{F_r^2(u,v) + F_j^2(u,v)}$,
- **phase** $\angle F(u,v) = \tan^{-1}[F_j(u,v)/F_r(u,v)]$.

The frequency w and directional angle θ are determined by position (u, v) on the 2-D spatial frequency domain, and the amplitude $|F(u,v)|$ and phase $\angle F(u,v)|$ are determined by 2-D spectrum $F(u,v)$. Moreover, as $|H(u,v)|$ is even and $\angle H(u,v)$ is odd, Eq. 4.228 can be further rewritten as

$$\begin{aligned} f(x,y) &= 2\int_{0}^{\infty}\int_{0}^{\infty} |F(u,v)| \cos(2\pi(ux+vy) + \angle F(u,v))\, du\, dv \\ &\quad + 2\int_{-\infty}^{0}\int_{0}^{\infty} |F(u,v)| \cos(2\pi(ux-vy) + \angle F(u,-v))\, du\, dv \\ &= 2\int_{0}^{\infty}\int_{0}^{\infty} |F(u,v)| \cos(2\pi w\langle \boldsymbol{r}, \boldsymbol{n}\rangle + \angle F(u,v))\, du\, dv \\ &\quad + 2\int_{-\infty}^{0}\int_{0}^{\infty} |F(u,v)| \cos(2\pi w\langle \boldsymbol{r}, \boldsymbol{n}'\rangle + \angle F(u,-v))\, du\, dv, \end{aligned} \tag{4.229}$$

where $\boldsymbol{n}'$ is the unit vector in the direction determined by the angle $\tan^{-1}(-v/u) = -\tan^{-1}(v/u) = -\theta$. This equation is the 2-D version of Eq. (3.144). The first integral represents superposition of sinusoids in the directions $0° < \theta < 90°$ (NE to SW), while the second integral represents a superposition of sinusoids in the directions $-90° > \theta > 0°$ (NW to SE).

4.3.2 Fourier transform of typical 2-D functions

- **Planar sinusoidal wave**

$$f(x,y) = \cos(2\pi(3x-2y)) = \frac{1}{2}[e^{j2\pi(3x-2y)} + e^{-j2\pi(3x-2y)}]. \tag{4.230}$$

This is a planar sinusoid of spatial frequency $\sqrt{3^2+2^2} = \sqrt{13}$ in the direction of $\theta = \tan^{-1}(-2/3)$ with unit amplitude and zero phase. Its 2-D Fourier spectrum is

$$\begin{aligned} F(u,v) &= \int\int_{-\infty}^{\infty} f(x,y)e^{-j2\pi(ux+vy)}\,dx\,dy \\ &= \frac{1}{2}\int\int_{-\infty}^{\infty} [e^{j2\pi(3x-2y)} + e^{-j2\pi(3x-2y)}]e^{-j2\pi(ux+vy)}\,dx\,dy \\ &= \frac{1}{2}\int\int_{-\infty}^{\infty} e^{-j2\pi((u-3)x+(v+2)y)}\,dx\,dy + \frac{1}{2}\int\int_{-\infty}^{\infty} e^{-j2\pi((u+3)x+(v-2)y)}\,dx\,dy \\ &= \frac{1}{2}\int_{-\infty}^{\infty} e^{-j2\pi(u-3)x}\,dx \int_{-\infty}^{\infty} e^{-j2\pi(v+2)y}\,dy \\ &\quad + \frac{1}{2}\int_{-\infty}^{\infty} e^{-j2\pi(u+3)x}\,dx \int_{-\infty}^{\infty} e^{-j2\pi(v-2)y}\,dy \\ &= \frac{1}{2}[\delta(u-3)\delta(v+2) + \delta(u+3)\delta(v-2)]. \end{aligned} \tag{4.231}$$

This transform pair is shown in Fig. 4.24(a).

- **Superposition of three planar sinusoidal waves**

$$f(x,y) = 3\,\cos(2\pi 2x) + 2\,\cos(2\pi 3y) + \cos(2\pi 5(x-y)). \tag{4.232}$$

Its 2-D Fourier spectrum is

$$\begin{aligned} F(u,v) &= \int\int_{-\infty}^{\infty} f(x,y)e^{-j2\pi(ux+vy)}\,dx\,dy \\ &= \frac{3}{2}[\delta(u-2) + \delta(u+2)]\delta(v) + \delta(u)[\delta(v-3) + \delta(v+3)] \\ &\quad + \frac{1}{2}[\delta(u-5)\delta(v+5) + \delta(u+5)\delta(v-5)]. \end{aligned} \tag{4.233}$$

This transform pair is shown in Fig. 4.24(b).

- **Rectangular impulse in 2-D space**

$$f(x,y) = \begin{cases} 1 & -\frac{a}{2} < x < \frac{a}{2},\ -\frac{b}{2} < y < \frac{b}{2} \\ 0 & \text{else} \end{cases}. \tag{4.234}$$

This 2-D function is separable as it can be written as the product of two 1-D functions $f(x, y) = f_x(x) f_y(y)$, where $f_x(x)$ and $f_y(y)$ are each a 1-D square impulse function. The spectrum is the product of the spectra $F_x(u) = \mathcal{F}[f_x(x)]$ and $F_y(v) = \mathcal{F}[f_y(y)]$, a 2-D sinc function.

$$
\begin{aligned}
F(u, v) &= \int\int_{-\infty}^{\infty} f(x, y) e^{-j2\pi(ux+vy)}\, dx\, dy \\
&= \int_{-\infty}^{\infty} f_x(x) e^{-j2\pi ux}\, dx \int_{-\infty}^{\infty} f_y(y) e^{-j2\pi vy}\, dy \\
&= \int_{-a/2}^{a/2} e^{-j2\pi ux}\, dx \int_{-b/2}^{b/2} e^{-j2\pi vy}\, dy = \frac{\sin(\pi u a)}{\pi u} \frac{\sin(\pi v b)}{\pi v}.
\end{aligned} \tag{4.235}
$$

This transform pair is shown in Fig. 4.24(c).

- **Cylindrical impulse**

$$
f(x, y) = \begin{cases} 1 & x^2 + y^2 < R^2 \\ 0 & \text{else} \end{cases}. \tag{4.236}
$$

As $f(x, y)$ is not separable but central symmetric, it is more convenient to use a polar coordinate system in both the spatial and frequency domains. We let

$$
\begin{cases} x = r \cos\theta, & r = \sqrt{x^2 + y^2} \\ y = r \sin\theta & \theta = \tan^{-1}(y/x) \end{cases}, \tag{4.237}
$$

$$
dx\, dy = r\, dr\, d\theta, \tag{4.238}
$$

and

$$
\begin{cases} u = \rho \cos\phi, & \rho = \sqrt{u^2 + v^2} \\ v = \rho \sin\phi, & \phi = \tan^{-1}(v/u) \end{cases}, \tag{4.239}
$$

$$
du\, dv = \rho\, d\rho\, d\phi, \tag{4.240}
$$

then we have

$$
\begin{aligned}
F(u, v) &= \int\int_{-\infty}^{\infty} f(x, y) e^{-j2\pi(ux+vy)}\, dx\, dy \\
&= \int_0^R \left[\int_0^{2\pi} e^{-j2\pi r\rho(\cos\theta\cos\phi + \sin\theta\sin\phi)}\, d\theta \right] r\, dr \\
&= \int_0^R \left[\int_0^{2\pi} e^{-j2\pi r\rho\cos(\theta-\phi)}\, d\theta \right] r\, dr = \int_0^R \left[\int_0^{2\pi} e^{-j2\pi r\rho\cos\theta}\, d\theta \right] r\, dr.
\end{aligned} \tag{4.241}
$$

To continue, we need to use the zeroth-order Bessel function $J_0(x)$, defined as

$$
J_0(x) = \frac{1}{2\pi} \int_0^{2\pi} e^{-jx\cos\theta}\, d\theta, \tag{4.242}
$$

which is related to the first-order Bessel function $J_1(x)$ by

$$\frac{d}{dx}(x\,J_1(x)) = x\,J_0(x); \tag{4.243}$$

i.e.,

$$\int_0^x x\,J_0(x)\,dx = x\,J_1(x). \tag{4.244}$$

Substituting $2\pi r\rho$ for x, we have

$$F(u,v) = F(\rho,\phi) = \int_0^R 2\pi r\,J_0(2\pi r\rho)\,dr = \frac{1}{\rho}R\,J_1(2\pi\rho R). \tag{4.245}$$

We see that the spectrum $F(u,v) = F(\rho,\phi)$ is independent of angle ϕ and, therefore, is a central symmetric sinc-like function.

- **Ideal LP filter**

$$F(u,v) = \begin{cases} 1 & u^2+v^2 < R^2 \\ 0 & \text{else} \end{cases}. \tag{4.246}$$

This cylindrical impulse in the frequency domain is called an ideal LP filter. When the spectrum of any given 2-D signal is multiplied by the ideal filter, all of its low-frequency components inside the radius R are kept, while all higher frequency components outside the circle are suppressed to zero.
Owing to the symmetry property of the Fourier transform, the inverse transform of this ideal LP filter is the same 2-D sinc-like function shown in Eq. (4.245) in the spatial domain, as shown in Fig. 4.24(d).

- **Gaussian function in 2-D space**

$$f(x,y) = \frac{1}{a^2}e^{-\pi(x^2+y^2)/a^2} = \frac{1}{a}e^{-\pi(x/a)^2}\frac{1}{a}e^{-\pi(y/a)^2}. \tag{4.247}$$

The spectrum of this function can be found as

$$\begin{aligned} F(u,v) &= \int\int_{-\infty}^{\infty} f(x,y)e^{-j2\pi(ux+vy)}\,dx\,dy \\ &= \frac{1}{a}\int_{-\infty}^{\infty} e^{-\pi(x/a)^2}e^{-j2\pi ux}\,dx\;\frac{1}{a}\int_{-\infty}^{\infty} e^{-\pi(y/a)^2}e^{-j2\pi vy}\,dy \\ &= e^{-\pi(au)^2}e^{-\pi(av)^2}. \end{aligned} \tag{4.248}$$

The last equation is owing to Eq. (3.171). Now we see that the Fourier transform of a 2-D Gaussian function is also a Gaussian, the product of two 1-D Gaussian functions along directions of u and v, respectively, as shown in Fig. 4.24(e).

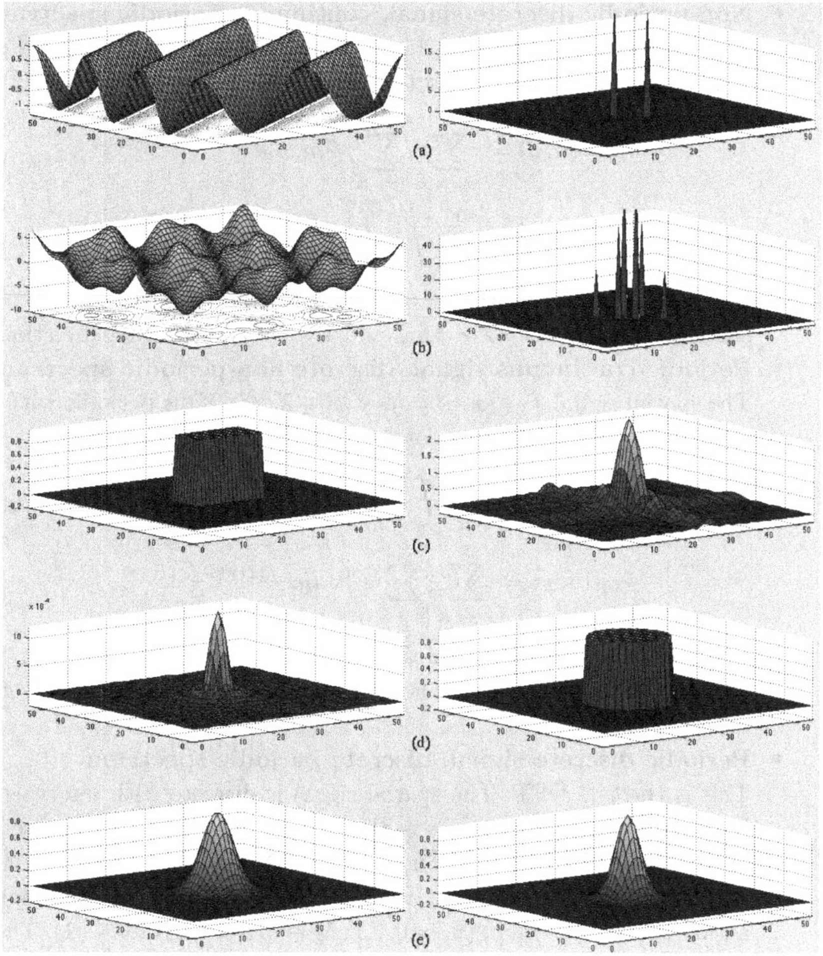

Figure 4.24 Some 2-D signals (left) and their spectra (right).

4.3.3 Four forms of 2-D Fourier transform

As with the 1-D case, there also exist four different forms of 2-D Fourier transform, depending on whether the given 2-D signal $f(x,y)$ is periodic or non-periodic, and whether it is discrete or continuous.

- **Non-periodic continuous signal, continuous non-periodic spectrum**

$$F(u,v) = \int_{-\infty}^{\infty}\int_{-\infty}^{\infty} f(x,y)e^{-j2\pi(ux+vy)}\,dx\,dy, \tag{4.249}$$

$$f(x,y) = \int_{-\infty}^{\infty}\int_{-\infty}^{\infty} F(u,v)e^{j2\pi(ux+vy)}\,du\,dv. \tag{4.250}$$

This is the most generic 2-D Fourier transform pair from Eqs. (4.218) and (4.218).

- **Non-periodic discrete signal, continuous periodic spectrum**
  The spatial signal $f[m,n]$ is discrete with spatial intervals x_0 and y_0 between consecutive signal samples in the x and y directions, respectively:

$$F_{UV}(u,v) = \sum_{m=-\infty}^{\infty} \sum_{n=-\infty}^{\infty} f[m,n] e^{-j2\pi(umx_0+vny_0)}, \tag{4.251}$$

$$f[m,n] = \frac{1}{UV} \int_0^U \int_0^V F(u,v) e^{j2\pi(umx_0+vny_0)}\, du\, dv. \tag{4.252}$$

  The 2-D spectrum $F_{UV}(u,v) = F(u+U, v+V)$ is periodic with periods (the sampling frequencies) $U = 1/x_0$ and $V = 1/y_0$ in the two directions.
- **Periodic continuous signal, discrete non-periodic spectrum**
  The spatial signal $f_{XY}(x,y) = f_{XY}(x+X, y+Y)$ is periodic with periods X and Y in x and y directions of the 2-D space, respectively:

$$F[k,l] = \frac{1}{XY} \int_0^X \int_0^Y f_{XY}(x,y) e^{j2\pi(kxu_0+lyv_0)}\, dx\, dy, \tag{4.253}$$

$$f_{XY}(x,y) = \sum_{k=-\infty}^{\infty} \sum_{l=-\infty}^{\infty} F[k,l] e^{-j2\pi(xku_0+ylv_0)}. \tag{4.254}$$

  The 2-D spectrum is discrete with intervals $u_0 = 1/X$ and $v_0 = 1/Y$ between consecutive frequency components $F[k,l]$ in spatial frequency directions u and v, respectively.
- **Periodic discrete signal, discrete periodic spectrum**
  This is the 2-D DFT. The spatial signal is discrete with intervals x_0 and y_0 between consecutive samples in the x and y directions, respectively, and it is also periodic with period X and Y. The 2-D signal has $X/x_0 = M$ and $Y/y_0 = N$ samples along each of the two spatial directions and can be represented as an $M \times N$ array $x[m,n]$ $(m = 0, \ldots, M-1, n = 0, \ldots, M-1)$. The 2-D DFT pair is

$$F[k,l] = \frac{1}{\sqrt{MN}} \sum_{n=0}^{N-1} \sum_{m=0}^{M-1} f[m,n] e^{-j2\pi(\frac{mk}{M}+\frac{nl}{N})}, \tag{4.255}$$

$$f[m,n] = \frac{1}{\sqrt{MN}} \sum_{l=0}^{N-1} \sum_{k=0}^{M-1} F[k,l] e^{j2\pi(\frac{mk}{M}+\frac{nl}{N})}. \tag{4.256}$$

$$0 \le m,k \le M-1, \quad 0 \le n,l \le N-1$$

  The spectrum is both discrete and periodic with periods (sampling rates) $U = 1/x_0$ and $V = 1/y_0$ and intervals $u_0 = 1/X$ and $v_0 = 1/Y$ between consecutive frequency components $F[k,l]$ along u and v, respectively. The signal is periodic $f[m+M, n+N] = x[m,n]$, and so is its spectrum $F[k+M, l+N] = F[k,l]$.

Note that the kernel function of the 2-D Fourier transform is separable in the sense that it can be expressed as a product of two 1-D kernel functions in each of the two dimensions:

$$\phi_{u,v}(x,y) = e^{j2\pi(ux+vy)} = e^{j2\pi ux} e^{j2\pi vy} = \phi_u(x)\phi_v(y). \tag{4.257}$$

The 2-D transform can be carried out as

$$\begin{aligned} F(u,v) &= \int_{-\infty}^{\infty}\int_{-\infty}^{\infty} f(x,y)e^{-j2\pi ux}e^{-j2\pi vy}\,dx\,dy \\ &= \int_{-\infty}^{\infty}\left[\int_{-\infty}^{\infty} f(x,y)e^{-j2\pi ux}\,dx\right]e^{-j2\pi vy}\,dy \\ &= \int_{-\infty}^{\infty} F'(u,y)e^{-j2\pi vy}\,dy, \end{aligned} \tag{4.258}$$

where $F'(u,y)$ is an intermediate result obtained by a 1-D transform in the dimension of x:

$$F'(u,y) = \int_{-\infty}^{\infty} f(x,y)e^{-j2\pi ux}\,dx, \tag{4.259}$$

and the 2-D spectrum $F(u,v)$ can be obtained by another 1-D transform in the dimension of y. In other words, the 2-D transform can be carried out in two steps, each for one of the two dimensions. Obviously, the order of the two steps can be reversed.

As in the case of the 1-D Fourier transform, among all four forms of 2-D Fourier transform, only the 2-D DFT with finite and discrete signal samples and frequency components can be carried out numerically. Also, how the total scaling factor $1/MN$ is distributed between the forward and inverse transforms is of little significance.

4.3.4 Computation of the 2-D DFT

We represent a 2-D discrete signal $x[m,n]$ $(m = 0,\dots,M-1,\ n = 0,\dots,N-1)$ by an M by N matrix $\boldsymbol{x}_{M\times N} = [\boldsymbol{x}_0,\dots,\boldsymbol{x}_{N-1}]$ consisting of N M-D column vectors $\boldsymbol{x}_n$ $(n = 0,\dots,N-1)$ (or M N-D row vectors). As the kernel function is separable, the 2-D DFT of $x[m,n]$ can be carried as

$$\begin{aligned} X[k,l] &= \frac{1}{\sqrt{MN}}\sum_{n=0}^{N-1}\left[\sum_{m=0}^{M-1} x[m,n]e^{-j2\pi\frac{mk}{M}}\right]e^{-j2\pi\frac{nl}{N}} \\ &= \frac{1}{\sqrt{N}}\sum_{n=0}^{N-1} X'[k,n]e^{-j2\pi\frac{nl}{N}} \qquad k = 0,\dots,M-1,\ \ l = 0,\dots,N-1, \end{aligned} \tag{4.260}$$

where $X'[k,n]$ is the intermediate result of the DFT of the nth column of $\boldsymbol{x}$:

$$X'[k,n] = \frac{1}{\sqrt{M}}\sum_{m=0}^{M-1} x[m,n]e^{-j2\pi\frac{mk}{M}} \qquad k = 0,\dots,M-1 \tag{4.261}$$

We see that the 2-D DFT can be carried out in the following two steps (or in reverse order).

- **Column transforms**
  Carry out the 1-D DFT in Eq. (4.261) to each of the N column vectors of $\boldsymbol{x}$ (the column index n is treated as a parameter):

$$\boldsymbol{X'}_n = \overline{\boldsymbol{W}}_M \boldsymbol{x}_n \qquad n = 0, \ldots, N-1, \tag{4.262}$$

  where $\boldsymbol{X'}_n = [X[0,n], \ldots, X[M-1,n]]^{\mathrm{T}}$ is an M-D vector and $\boldsymbol{W}_M$ is an $M \times M$ DFT matrix (Eq. (4.125)). Putting all N such columns together we get

$$[\boldsymbol{X'}_0, \ldots, \boldsymbol{X'}_{N-1}] = \boldsymbol{X'}_{M\times N} = \overline{\boldsymbol{W}}_M\, [\boldsymbol{x}_0, \ldots, \boldsymbol{x}_{N-1}] = \overline{\boldsymbol{W}}_M\, \boldsymbol{x}_{M\times N}, \tag{4.263}$$

  where we have defined $\boldsymbol{X'}_{M\times N} = [\boldsymbol{X'}_0, \ldots, \boldsymbol{X'}_{N-1}]$.
- **Row transforms**
  Rewrite $\boldsymbol{X'}_{M\times N}$ in terms of its M row vectors

$$\boldsymbol{X'}_{M\times N} = \begin{bmatrix} \boldsymbol{X}'^{\mathrm{T}}_0 \\ \vdots \\ \boldsymbol{X}'^{\mathrm{T}}_{M-1} \end{bmatrix}_{M\times N}, \tag{4.264}$$

  where $\boldsymbol{X}'^{\mathrm{T}}_m$ is the mth row vector of $\boldsymbol{X'}$. Then carry out the 1-D DFT in the second equation of Eq. (4.260) to each of the M row vectors (the row index k is treated as a parameter) to get $\overline{\boldsymbol{W}}_N \boldsymbol{X'}_m$ for the mth row vector of the desired 2-D DFT $\boldsymbol{X}$:

$$\boldsymbol{X}^{\mathrm{T}}_m = (\overline{\boldsymbol{W}}_N \boldsymbol{X'}_m)^{\mathrm{T}} = \boldsymbol{X}'^{\mathrm{T}}_m \overline{\boldsymbol{W}}_N \qquad m = 0, \ldots, M-1, \tag{4.265}$$

  where $\boldsymbol{W}_N = \boldsymbol{W}_N^{\mathrm{T}}$ is an $N \times N$ DFT matrix. Putting all M such rows together we get the 2-D DFT in matrix form:

$$\boldsymbol{X}_{M\times N} = \begin{bmatrix} \boldsymbol{X}^{\mathrm{T}}_0 \\ \vdots \\ \boldsymbol{X}^{\mathrm{T}}_{M-1} \end{bmatrix} = \begin{bmatrix} \boldsymbol{X}'^{T}_0 \\ \vdots \\ \boldsymbol{X}'^{T}_{M-1} \end{bmatrix} \overline{\boldsymbol{W}}_N = \boldsymbol{X}'_{M\times N} \overline{\boldsymbol{W}}_N. \tag{4.266}$$

Substituting Eq. (4.263) into this equation we get

$$\boldsymbol{X}_{M\times N} = \boldsymbol{X}'_{M\times N} \overline{\boldsymbol{W}}_N = \overline{\boldsymbol{W}}_M\, \boldsymbol{x}_{M\times N} \overline{\boldsymbol{W}}_N, \tag{4.267}$$

Pre-multiplying $\boldsymbol{W}_M$ and post-multiplying $\boldsymbol{W}_N$ on both sides, we get the inverse 2-D DFT:

$$\boldsymbol{x}_{M\times N} = \boldsymbol{W}_M\, \boldsymbol{X}_{M\times N}\, \boldsymbol{W}_N. \tag{4.268}$$

We can now rewrite these two equations as a 2-D DFT pair:

$$\begin{cases} \boldsymbol{X} = \overline{\boldsymbol{W}}\, \boldsymbol{x}\, \overline{\boldsymbol{W}} & \text{(forward)} \\ \boldsymbol{x} = \boldsymbol{W}\, \boldsymbol{X}\, \boldsymbol{W} & \text{(inverse)} \end{cases}. \tag{4.269}$$

Here, the subscripts of all matrices are dropped.

The DFT matrix $\boldsymbol{W}$ can be expressed in terms of its rows as well as its columns, and the matrix form of the inverse transform can be expanded to become

$$\begin{aligned}
\boldsymbol{x} &= [\boldsymbol{w}_0, \ldots, \boldsymbol{w}_{M-1}] \begin{bmatrix} X[0,0] & \cdots & X[0,N-1] \\ \vdots & \ddots & \vdots \\ X[M-1,0] & \cdots & X[M-1,N-1] \end{bmatrix} \begin{bmatrix} \boldsymbol{w}_0^{\mathrm{T}} \\ \vdots \\ \boldsymbol{w}_{N-1}^{\mathrm{T}} \end{bmatrix} \\
&= [\boldsymbol{w}_0, \ldots, \boldsymbol{w}_{M-1}] \begin{bmatrix} \sum_{l=0}^{N-1} X[0,l]\boldsymbol{w}_l^{\mathrm{T}} \\ \vdots \\ \sum_{l=0}^{N-1} X[M-1,l]\boldsymbol{w}_l^{\mathrm{T}} \end{bmatrix} \\
&= \sum_{k=0}^{M-1} \boldsymbol{w}_k \sum_{l=0}^{N-1} X[k,l]\boldsymbol{w}_l^{\mathrm{T}} = \sum_{k=0}^{M-1}\sum_{l=0}^{N-1} X[k,l]\boldsymbol{w}_k\boldsymbol{w}_l^{\mathrm{T}} = \sum_{k=0}^{M-1}\sum_{l=0}^{N-1} X[k,l]\boldsymbol{B}_{kl},
\end{aligned} \tag{4.270}$$

where $\boldsymbol{B}_{kl} = \boldsymbol{w}_k\boldsymbol{w}_l^{\mathrm{T}}$ $(k = 0, \ldots, M-1,\ l = 0, \ldots, N-1)$ is an M by N matrix with the mnth element being $e^{j2\pi(\frac{mk}{M}+\frac{nl}{N})}/\sqrt{MN}$. This result indicates that the 2-D signal $\boldsymbol{x}$ can be expressed as a linear combination of a set of MN 2-D basis functions $\boldsymbol{B}_{kl}$, each weighted by the coefficient $X[k,l]$, which is given in the first equation of Eq. (4.269) for the forward 2-D DFT:

$$\boldsymbol{X} = \begin{bmatrix} \overline{\boldsymbol{w}}_0^{\mathrm{T}} \\ \vdots \\ \overline{\boldsymbol{w}}_{M-1}^{\mathrm{T}} \end{bmatrix} \boldsymbol{x}[\overline{\boldsymbol{w}}_0, \ldots, \overline{\boldsymbol{w}}_{N-1}], \tag{4.271}$$

as the klth coefficient:

$$\begin{aligned}
X[k,l] &= \overline{\boldsymbol{w}}_k^{\mathrm{T}} \begin{bmatrix} x[0,0] & \cdots & x[0,N-1] \\ \vdots & \ddots & \vdots \\ x[M-1,0] & \cdots & x[M-1,N-1] \end{bmatrix} \overline{\boldsymbol{w}}_l \\
&= \sum_{m=0}^{M-1}\sum_{n=0}^{N-1} x[m,n]\overline{B}_{kl}[m,n] = \langle \boldsymbol{x}, \boldsymbol{B}_{kl} \rangle.
\end{aligned} \tag{4.272}$$

This inner product of two 2-D matrices $\boldsymbol{x}$ and $\boldsymbol{B}_{kl}$ (Eq. (2.16)) can be considered as the projection of the signal $\boldsymbol{x}$ onto the klth 2-D DFT basis function $\boldsymbol{B}_{kl}$, which can be found by letting all coefficients in the summation in Eq. (4.270) be zero except $X[k,l] = 1$. For example, when $M = N = 8$, the $M \times N = 64$ such 2-D basis functions are shown in Fig. 4.25.

The C code of a function `fft2d` for both the forward and inverse 2-D DFT is listed below, where `xxr` and `xxi` are the real and imaginary parts of a 2-D `M` by `N` array, to be replaced by the real and imaginary parts of the 2-D array as the result of the transform. As with the 1-D FFT function, the function carries out either the forward transform if `inv=0` or the inverse transform if `inv=1`.

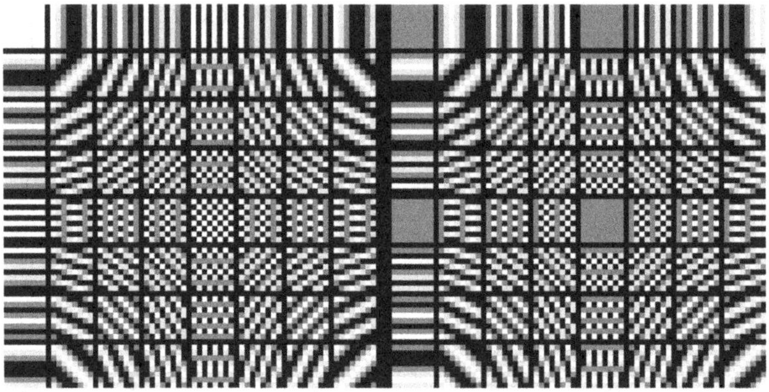

Figure 4.25 $8 \times 8 = 64$ 2-D DFT basis functions $\boldsymbol{B}_{kl}$, $(k, l = 0, \ldots, 7)$. The left half of the image shows the real part of the 8 by 8 2-D basis functions, while the right half shows the imaginary part. The DC component is at the top-left corner of the real part, and the highest frequency component in both the horizontal and vertical directions is in the middle of the real part.

```
fft2d(xxr,xxi,M,N,inv)
    float **xxr, **xxi;
    int M,N,inv;
{ float *xr, *xi;
  int m,n,k;
  k=M; if (N>N) k=N;
  xr = (float *) malloc(k*sizeof(float));
  xi = (float *) malloc(k*sizeof(float));
  for (n=0; n<N; n++) {      // for N column xforms
    for (m=0; m<M; m++)
      { xr[m]=xxr[m][n]; xi[m]=xxi[m][n]; }
    fft(xr,xi,M,inv);
    for (m=0; m<M; m++)
      { xxr[m][n]=xr[m]; xxi[m][n]=xi[m]; }
  }
  for (m=0; m<M; m++) {      // for M column xforms
    for (n=0; n<N; n++)
      { xr[n]=xxr[m][n]; xi[n]=xxi[m][n]; }
    fft(xr,xi,N,inv);
    for (n=0; n<N; n++)
      { xxr[m][n]=xr[n]; xxi[m][n]=xi[n]; }
  }
  free(xr);  free(xi);
}
```

Example 4.12: Consider the 2-D DFT of a real 8 × 8 2-D signal (imaginary part is zero):

$$\boldsymbol{x} = \begin{bmatrix} 0.0 & 0.0 & 0.0 & 0.0 & 0.0 & 0.0 & 0.0 & 0.0 \\ 0.0 & 0.0 & 70.0 & 80.0 & 90.0 & 0.0 & 0.0 & 0.0 \\ 0.0 & 0.0 & 90.0 & 100.0 & 110.0 & 0.0 & 0.0 & 0.0 \\ 0.0 & 0.0 & 110.0 & 120.0 & 130.0 & 0.0 & 0.0 & 0.0 \\ 0.0 & 0.0 & 130.0 & 140.0 & 150.0 & 0.0 & 0.0 & 0.0 \\ 0.0 & 0.0 & 0.0 & 0.0 & 0.0 & 0.0 & 0.0 & 0.0 \\ 0.0 & 0.0 & 0.0 & 0.0 & 0.0 & 0.0 & 0.0 & 0.0 \\ 0.0 & 0.0 & 0.0 & 0.0 & 0.0 & 0.0 & 0.0 & 0.0 \end{bmatrix}. \tag{4.273}$$

The eight-point DFT matrix $\boldsymbol{W}_8$ is the same as the one shown in Eqs. (4.139) and (4.140). The real and imaginary parts of the 2-D DFT of this signal $\boldsymbol{x}$ are $\boldsymbol{X} = \overline{\boldsymbol{W}}_8\, \boldsymbol{x}\, \overline{\boldsymbol{W}}_8$, as below:

$$\boldsymbol{X}_{\rm r} = \left[\begin{array}{r|rrr|r|rrr} 165.0 & -98.9 & 10.0 & -21.1 & 55.0 & -21.1 & 10.0 & -98.9 \\ \hline -63.1 & -11.3 & 27.7 & 13.2 & -21.0 & 1.6 & -32.7 & 85.7 \\ 15.0 & 0.0 & -5.0 & -2.9 & 5.0 & 0.0 & 5.0 & 17.1 \\ -41.9 & 16.8 & 2.7 & 6.3 & -14.0 & 4.3 & -7.7 & 33.4 \\ \hline 15.0 & -8.5 & 0.0 & -1.5 & 5.0 & -1.5 & 0.0 & -8.5 \\ \hline -41.9 & 33.4 & -7.7 & 4.3 & -14.0 & 6.3 & 2.7 & 16.8 \\ 15.0 & -17.1 & 5.0 & 0.0 & 5.0 & -2.9 & -5.0 & 0.0 \\ -63.1 & 85.7 & -32.7 & 1.6 & -21.0 & 13.2 & 27.7 & -11.3 \end{array}\right], \tag{4.274}$$

and

$$\boldsymbol{X}_{\rm j} = \left[\begin{array}{r|rrr|r|rrr} 0.0 & -88.9 & 55.0 & 11.1 & 0.0 & -11.1 & -55.0 & 88.9 \\ \hline -90.5 & 89.2 & -27.1 & 6.9 & -30.2 & 16.8 & 15.0 & 19.9 \\ 15.0 & -17.1 & 5.0 & 0.0 & 5.0 & -2.9 & -5.0 & 0.0 \\ -15.5 & 31.9 & -15.0 & -0.8 & -5.2 & 4.9 & 12.9 & -13.2 \\ \hline 0.0 & -8.5 & 5.0 & 1.5 & 0.0 & -1.5 & -5.0 & -8.5 \\ \hline 15.5 & 13.2 & -12.9 & -4.9 & 5.2 & 0.8 & 15.0 & -31.9 \\ -15.0 & 0.0 & 5.0 & 2.9 & -5.0 & 0.0 & -5.0 & 17.1 \\ 90.5 & -19.9 & -15.0 & -16.8 & 30.2 & -6.9 & 27.1 & -89.2 \end{array}\right]. \tag{4.275}$$

These $8 \times 8 = 64$ elements $X[k,l] = X_{\rm r}[k,l] + j\, X_{\rm j}[k,l]$ of $\boldsymbol{X}$ are the complex coefficients for the amplitudes $|X[k,l]|$ and phase $\angle X[k,l]$ of the 64 2-D frequency components shown in Fig. 4.25. Note that as the signal $x[m,n]$ is real, the real part of its spectrum is even: $X_{\rm r}[k,l] = X_{\rm r}[M-k, N-l]$, $X_{\rm r}[k, N-l] = X_{\rm r}[M-k, l]$, while the imaginary part is odd: $X_{\rm j}[k,l] = -X_{\rm j}[M-k, N-l]$, $X_{\rm j}[k, N-l] = -X_{\rm j}[M-k, l]$. Consider specifically the following coefficients:

- $X[0,0]$ is the amplitude of the DC offset (average) of the signal.

- $X[0, N/2]$ is the amplitude of the highest frequency component $(-1)^n$ in the horizontal direction.
- $X[M/2, 0]$ is the amplitude of the highest frequency component $(-1)^m$ in the vertical direction.
- $X[M/2, N/2]$ is the amplitude of the highest frequency component $(-1)^{m+n}$ in both directions. The above four coefficients are real with zero phase.
- $X[0, l]$ pairs up with $X[0, N-l]$ $(l = 1, \ldots, N/2-1)$ to represent the amplitude and phase of a planar sinusoid $|X[0, l]| \cos[2\pi(nl/N) + \angle X[0, l]]$ in horizontal direction.
- $X[k, 0]$ pairs up with $X[M-k, 0]$ $(k = 1, \ldots, M/2-1)$ to represent the amplitude and phase of a planar sinusoid $|X[k, 0]| \cos[2\pi(mk/M) + \angle X[k, 0]]$ in the vertical direction.

The coefficients in the rest of the array $X[k, l]$ can be divided into four quadrants with the top-left paired up with the low-right to represent sinusoids in the NW-SE directions, while the top-right paired up with the low-right to represent sinusoids in the NE-SW directions.

As shown in the example above, when the signal in the spatial domain is real with its imaginary part $x_{\rm j}[m, n] = 0$, half of the data points are redundant; correspondingly, in the spatial frequency domain, both the real and imaginary parts of $X[k, l]$ are symmetric (even or odd). More specifically, we note that the real part $X_{\rm r}[k, l]$ has $MN/2 + 2$ independent variables, and the imaginary part $X_{\rm j}[k, l]$ has $MN/2 - 2$ independent variables. Taking advantage of the symmetry property, an algorithm can be designed to cut by half the computation for the 2-D DFT of real signals.

In the 2-D spectrum matrix, the DC component $X[0, 0]$ at zero frequency is at the upper left corner, the low-frequency components are around the edges, and the high-frequency components are in the area around the center $(M/2, N/2)$, as shown in the example above. However, sometimes it is preferable to centralize the spectrum so that the DC component $X[0, 0]$ is in the middle, and the high-frequency components are farther away from the center around the corners and edges, so that the 2-D spectrum is consistent with the convention that the DC component at the origin is always in the center of the 2-D coordinate system of the frequency domain. Similar to the case of 1-D DFT discussed before, the centralization of the 2-D spectrum can be simply realized by shifting the 2-D spectrum in both dimensions by half of the corresponding length. Alternatively, based on the frequency shift property, the centralization can be equivalently realized in the spatial domain by negating every other spatial sample, similar to the 1-D case in Eq. (4.179):

$$\begin{aligned} \mathcal{F}^{-1}[X[k - M/2, l - N/2]] &= x[m, n] e^{j2\pi\left(\frac{m\,M/2}{M} + \frac{n\,N/2}{N}\right)} \\ &= x[m, n] e^{j\pi(m+n)} = x[m, n](-1)^{m+n}. \end{aligned} \quad (4.276)$$

If we negate the sign of any spatial sample $x[m, n]$ when $m + n$ is odd; i.e.

$$\begin{bmatrix} x[0,0] & -x[0,1] & x[0,2] & \cdots \\ -x[1,0] & x[1,1] & -x[1,2] & \cdots \\ x[2,0] & -x[2,1] & x[2,2] & \cdots \\ \vdots & \vdots & \vdots & \ddots \end{bmatrix}, \tag{4.277}$$

then the resulting 2-D Fourier spectrum will be centralized. For the example above, the centralized spectrum becomes

$$\boldsymbol{X}_{\rm r} = \left[\begin{array}{r|rrr|r|rrr} 5.0 & -1.5 & 0.0 & -8.5 & 15.0 & -8.5 & 0.0 & -1.5 \\ \hline -14.0 & 6.3 & 2.7 & 16.8 & -41.9 & 33.4 & -7.7 & 4.3 \\ 5.0 & -2.9 & -5.0 & 0.0 & 5.0 & -17.1 & 5.0 & 0.0 \\ -21.0 & 13.2 & 27.7 & -11.3 & -63.1 & 85.7 & -32.7 & 1.6 \\ \hline 55.0 & -21.1 & 10.0 & -98.9 & 165.0 & -98.9 & 10.0 & -21.1 \\ \hline -21.0 & 1.6 & -32.7 & 85.7 & -63.1 & -11.3 & 27.7 & 13.2 \\ 5.0 & 0.0 & 5.0 & 17.1 & 15.0 & 0.0 & -5.0 & -2.9 \\ -14.0 & 4.3 & -7.7 & 33.4 & -41.9 & 16.8 & 2.7 & 6.3 \end{array}\right], \tag{4.278}$$

and

$$\boldsymbol{X}_{\rm j} = \left[\begin{array}{r|rrr|r|rrr} 0.0 & -1.5 & -5.0 & -8.5 & 0.0 & -8.5 & 5.0 & 1.5 \\ \hline 5.2 & 0.8 & 15.0 & -31.9 & 15.5 & 13.2 & -12.9 & -4.9 \\ -5.0 & 0.0 & -5.0 & 17.1 & -15.0 & 0.0 & 5.0 & 2.9 \\ 30.2 & -6.9 & 27.1 & -89.2 & 90.5 & -19.9 & -15.0 & -16.8 \\ \hline 0.0 & -11.1 & -55.0 & 88.9 & 0.0 & -88.9 & 55.0 & 11.1 \\ \hline -30.2 & 16.8 & 15.0 & 19.9 & -90.5 & 89.2 & -27.1 & 6.9 \\ 5.0 & -2.9 & -5.0 & 0.0 & 15.0 & -17.1 & 5.0 & 0.0 \\ -5.2 & 4.9 & 12.9 & -13.2 & -15.5 & 31.9 & -15.0 & -0.8 \end{array}\right]. \tag{4.279}$$

4.4 Homework problems

1. Show that the DTFT pair in Eq. (4.5) can be treated as a special case of the CTFT in Eq. (3.65). Do this in two steps:
   (a) First apply the inverse CTFT (second equation of Eq. (3.65)) to the DTFT (first equation of Eq. (4.5)) to obtain the time signal $x(t)$.
   (b) Next apply the forward CTFT (first equation of Eq. (3.65)) to the resulting signal $x(t)$ obtained above to get its spectrum. Verify that the result is indeed the DFTF.
2. The DTFT in Eq. (4.2) is a special case of the CTFT in Eq. (3.65), when the discrete time signal is expressed as an impulses train in Eq. (4.1). Now show that by applying the inverse CTFT (second equation of Eq. (3.65)) to the spectrum $X_F(f)$ in Eq. (4.2), the time signal in Eq. (4.1) can be obtained. <u>Hint:</u> Consider using Eq. (1.28).

3. Prove the following DTFT properties:
   (a) DTFT of time and frequency shift (Eqs. (4.33) and (4.32))
   (b) DTFT of correlation (Eq. (4.36))
   (c) DTFT of time convolutions (Eq. (4.39))
   (d) DTFT of frequency convolutions (Eq. (4.40))
   (e) DTFT of accumulation (Eq. (4.42))
   Hint: Note that $X(f+n) = X(f)$ is periodic with period 1 and $X(n) = X(0)$.
   (f) DTFT of frequency differentiation (Eq. (4.45))
   (g) DTFT of modulation (Eq. (4.46))
   Hint: Note that $(-1)^n = (e^{-j\pi})^n$. Alternatively, we could also let $f_0 = 1/2$ in Eq. (4.34) for the frequency shift property.
4. Find the Fourier transform of each of the following signals:
   (a)
$$x[n] = \begin{cases} 1 & |n| \le l \\ 0 & |n| > l \end{cases} \tag{4.280}$$
   (b)
$$x[n] = n(1/2)^n u[n]. \tag{4.281}$$
   (c)
$$x[n] = (1/2)^{|n|} \sin((n-1)\pi/4). \tag{4.282}$$
   (d)
$$x[n] = (n-1)(1/2)^{|n|}. \tag{4.283}$$
   (e)
$$x[n] = \cos(n\pi/3)\frac{\sin(n\pi/4)}{n\pi}. \tag{4.284}$$
5. Given the input $x[n]$ and the corresponding output $y[n]$ of an LTI system:
$$x[n] = (1/2)^n u[n], \qquad y[n] = (1/3)^n u[n]. \tag{4.285}$$
   (a) find its frequency response function $H(f)$ and impulse response function $h[n]$;
   (b) carry out the convolution $h[n] * x[n] = y[n]$ to verify your result.
6. A signal $x[n] = 2\cos(n\pi/8) + 2\cos(n\pi/3)$ is taken as the input to each of the following LTI systems.
   (a)
$$h_1[n] = \frac{\sin(n\pi/6)}{n\pi}. \tag{4.286}$$
   (b)
$$h_2[n] = \frac{\sin(n\pi/2)}{n\pi} + \frac{\sin(n\pi/6)}{n\pi}. \tag{4.287}$$

(c)

$$h_3[n] = \frac{\sin(n\pi/2)}{n\pi} - \frac{\sin(n\pi/6)}{n\pi}. \tag{4.288}$$

(d)

$$h_4[n] = \frac{\sin(n\pi/6)}{n\pi}\frac{\sin(n\pi/2)}{n\pi}\frac{3}{\pi}. \tag{4.289}$$

Find the corresponding output $y_i[n]$ $(i = 1, 2, 3, 4)$.

7. Two signals $x_1(t)$ and $x_2(t)$ are both band-limited; i.e., $X_1(f) = 0$ for $|f| > f_{\max 1}$ and $X_2(f) = 0$ for $|f| > f_{\max 2}$. Find the minimum sampling frequency for sampling each of the following signals without aliasing or folding:
   (a) $x_1(t) + x_2(t - \tau)$
   (b) $x_1(t)x_2(t)$
   (c) $x_1(t) * x_2(t)$
   (d) $x_1(t)\cos(2\pi f_0 t)$
   (e) $dx_1(t)/dt$
   (f) $x_1(at)$.
8. The following signal is sampled with sampling frequency $F = 1/t_0 = 8$ Hz ($t_0 = 1/F = 1/8$ is the sampling period):

$$x(t) = \sin(2\pi f_0 t) = \frac{1}{2j}(e^{j2\pi f_0 t} - e^{-j2\pi f_0 t}). \tag{4.290}$$

   The resulting discrete samples can be represented as

$$x[n] = x(t)\big|_{t=nt_0} = x(nt_0) = x(n/F) = x(n/8). \tag{4.291}$$

   For each of the following possible frequencies f_0 of $x(t)$,
   (a) give the expression $x[n]$ for the sampled signal and indicate whether the signal is sufficiently sampled, aliased, or folded.
   (b) plot the sampling process in the time domain to show how the continuous signal is sampled
   (c) show the spectrum of the sampled signal in the frequency domains, thereby explaining whether and why aliasing/folding happens.
   - $f_0 = 3 < F/2 = 4$ Hz
   - $f_0 = 5 > F/2 = 4$ Hz
   - $f_0 = 9 > F/2 = 4$ Hz.
9. Assume the signal energy is concentrated within a frequency band $f_{\min} < |f| < f_{\max}$ where $f_{\min} = 2$ kHz and $f_{\min} = 3$ kHz, as shown in Fig. 4.11 in the text. What is the lowest sampling frequency F with which a perfect reconstruction is possible. Which of the possible sampling frequencies 2, 2.5, 3, 3.5, 4, 4.5, 5, 5.5, 6 kHz will allow a perfect reconstruction of the signal from its samples?
   Hint: Consider using some graphic tool to visualize the periodic spectrum after sampling with different rates F.

10. Figure 4.26 shows the combination of a sampler and the corresponding reconstructor by which the sampling frequency could be significantly reduced, based on the assumption that the signal $x(t)$ is real and its energy is concentrated within a frequency band $f_{\min} < |f| < f_{\max}$. In the sampler, before $x(t)$ is sampled, it is first multiplied by $e^{j\omega_0 t} = e^{j2\pi f_0 t}$ where $f_0 = (f_{\min} + f_{\max})/2$ and then filtered by an ideal filter $H(f)$. In the reconstructor, after the same ideal filter $H(f)$, the signal is further multiplied by $e^{-j\omega_0 t}$ and then its real part is taken as the output.
Assume the energy of signal $x(t)$ is again concentrated within a frequency band $f_{\min} < |f| < f_{\max}$ as shown in Fig. 4.11.
   (a) Sketch the spectra $X_i(f)$ $(i = 1, 2, 3, 4, 5)$ of the signals along the path in both the sampler and reconstructor.
   (b) Determine the minimum cutoff frequency f_c of the two ideal LP filters.
   (c) Determine the lowest sampling frequency F of the comb function $\text{comb}(t) = \sum_m \delta(t - m/F)$ for a perfect reconstruction. (Without using this method, the lowest sampling frequency for perfect reconstruction is $F > 2f_{\max}$.)
   (d) Show that the reconstructed signal $y(t)$ or its spectrum $Y(f)$ is the same as the input $x(t)$ or its spectrum $X(f)$ (up to a scaling factor which is neglected).

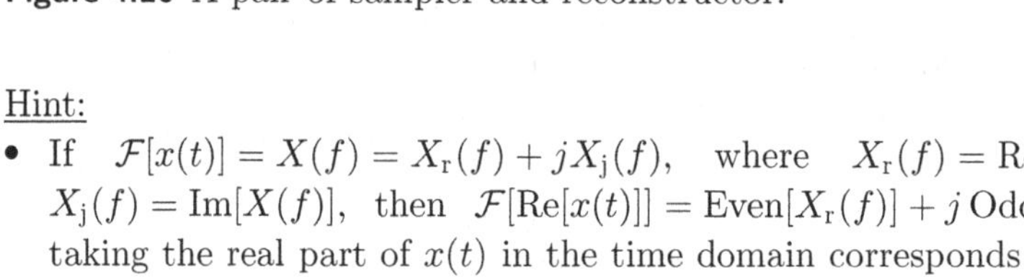

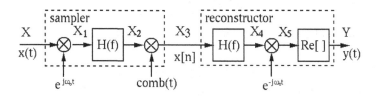

Figure 4.26 A pair of sampler and reconstructor.

Hint:
- If $\mathcal{F}[x(t)] = X(f) = X_r(f) + jX_j(f)$, where $X_r(f) = \text{Re}[X(f)]$ and $X_j(f) = \text{Im}[X(f)]$, then $\mathcal{F}[\text{Re}[x(t)]] = \text{Even}[X_r(f)] + j\,\text{Odd}[X_j(f)]$; i.e., taking the real part of $x(t)$ in the time domain corresponds to taking the even and odd parts of the real and imaginary parts of $X(f)$ in frequency, respectively.
- $X_e(f) = [X(f) + X(-f)]/2$, and $X_o(f) = [X(f) - X(-f)]/2$.

11. Provided in the website for the book, is a Matlab function `guidemo_sampling` which allows the user to specify the parameters (frequency, amplitude, and phase) of two sinusoids, as well as the sampling rate, and displays the combination of the two sinusoids and the discrete samples in both the time and frequency domains. Use this function to explore different combinations of the two signals, as well as the sampling rate, and inspect the possible aliasing and folding in both the time and frequency domain.

12. Find the discrete signal $x[n]$ obtained by sampling each of the following continuous signals at sampling rate $F = 10$ samples/second:
(a) $x_1(t) = 2\cos(2\pi 7t) + \cos(2\pi 2t)$
(b) $x_2(t) = 2\cos(2\pi 8t) + \cos(2\pi 2t)$
(c) $x_3(t) = 2\cos(2\pi 8t) - 2\cos(2\pi 2t)$.
Confirm the three cases of different folding shown in the three panels of Fig. 4.10. Use the Matlab function `guidemo_sampling` provided to reproduce these cases and explore other possible combinations of different signal frequencies and sampling rates.
13. Find the DFT of the following discrete signal

$$x[n] = 3\sin(5\pi n/6) + 4\cos(3\pi n/4) + 5\sin(\pi n/3). \tag{4.292}$$

14. Let $x[n] = n$ be a discrete signal with period $N = 4$. Find its DFT by matrix multiplication $\boldsymbol{X} = \overline{\boldsymbol{W}}\boldsymbol{x}$, where $\boldsymbol{W}$ is given in Eq. (4.137). Then carry out the inverse DFT by $\boldsymbol{x} = \boldsymbol{W}\boldsymbol{X}$ to confirm the signal is perfectly reconstructed.
15. Let $\boldsymbol{x} = [1, 1, -1, -1, 1, 1, -1, -1]^{\mathrm{T}}$ be the input to an LTI system with impulse response $\boldsymbol{h} = [1, 2, 3]^{\mathrm{T}}$. Find the output $y[n] = h[n] * x[n]$ in two different ways: (1) time-domain convolution, and (2) frequency-domain multiplication. Write a Matlab program to confirm your result.
Note that, given any two of three variables in the frequency domain, the frequency response function $H[k] = \mathcal{F}[h[n]]$, the input $X[k] = \mathcal{F}[x[n]]$, and the corresponding output $Y[k] = \mathcal{F}[y[n]]$, we can always easily find the third based on the simple relationship $Y[k] = H[k]X[k]$. This is not possible in the time domain. Now verify your solution $y[n]$ in two ways: (1) find $x[n]$ given $h[n]$ and $y[n]$, and (2) find $h[n]$ given $x[n]$ and $y[n]$.
16. The impulse response of a discrete LTI system is $h[n] = a^n u[n]$ with $|a| < 1$ and the input is $x[n] = \cos(2\pi n f_0)$. Find the corresponding output $y[n] = h[n] * x[n]$ in both the time and frequency domains.

5 Applications of the Fourier transforms

As a general mathematical tool, the Fourier transform finds a wide variety of applications in both science and engineering. Essentially, any field that deals with signals, either sinusoidal waves or any combination thereof, may benefit from the Fourier transform method for data processing and analysis. In this chapter we only consider a small set of some typical applications.

5.1 LTI systems in time and frequency domains

Previously, we considered mostly the Fourier transform of a given signal $x(t)$ and the resulting spectrum $X(f) = \mathcal{F}[x(t)]$ representing the frequency contents of the signal. However, the Fourier transform can also be used to characterize a linear, time-invariant (LTI) system. Recall that the output of an LTI system can be found as the convolution of the input and the impulse response function $h(t)$ of the system (Eq. (1.85)):

$$y(t) = \mathcal{O}[x(t)] = h(t) * x(t) = \int_{-\infty}^{\infty} h(\tau)x(t-\tau)\,d\tau. \tag{5.1}$$

Carrying out the Fourier transform on both sides, we get the output in the frequency domain as a product (Eq. (3.126))

$$Y(f) = H(f)X(f), \tag{5.2}$$

where $X(f)$ and $Y(f)$ are respectively the spectra of the input $x(t)$ and output $y(t)$ in time domain, and $H(f)$ is the Fourier transform of the impulse response $h(t)$, the *frequency response function (FRF)* of the system (Eq. (1.91)):

$$H(f) = \mathcal{F}[h(t)] = \int_{-\infty}^{\infty} h(t)e^{-j2\pi ft}\,dt = \int_{-\infty}^{\infty} h(t)e^{-j\omega t}\,dt = H(\omega). \tag{5.3}$$

As the Fourier transform of the impulse response function $h(t)$, the FRF is a function of frequency f or angular frequency $\omega = 2\pi f$ and therefore can be denoted by either $H(f)$ or $H(\omega)$, whichever is more convenient and suitable in the context.

As the FRF $H(f)$ of an LTI system is in general complex, it can be represented in terms of either its real and imaginary parts, or its magnitude and phase:

$$H(f) = \text{Re}[H(f)] + j\,\text{Im}[H(f)] = |H(f)|\angle H(f) = H(f)e^{j\angle H(f)}, \tag{5.4}$$

where

$$\begin{cases} |H(f)| = \sqrt{\text{Re}[H(f)]^2 + \text{Im}[H(f)]^2} \\ \angle H(f) = \tan^{-1}(\text{Im}[H(f)]/\,\text{Re}[H(f)]) \end{cases} \qquad \begin{cases} \text{Re}[H(f)] = |H(f)|\cos\angle H(f) \\ \text{Im}[H(f)] = |H(f)|\sin\angle H(f) \end{cases}. \tag{5.5}$$

The magnitude $|H(f)|$ and phase angle $\angle H(f)$ are called the *gain* and *phase shift* of the system, respectively. The FRF $H(f)$ of a given LTI system can be plotted in several different ways.

- The real part $\text{Re}[H(f)]$ and imaginary part $\text{Im}[H(f)]$ can be plotted individually as a real function of frequency f or ω.
- The gain $|H(f)|$ and phase shift $\angle H(f)$ can be plotted individually as a function of frequency f or ω.
- In a *Bode plot* of $H(f)$, the gain $|H(f)|$ and phase shift $\angle H(f)$ are plotted as two functions of the frequency in base-10 logarithmic scale, so that the frequency range can be extended to cover several orders of magnitude (or decades). Moreover, the gain $|H(f)|$ is also plotted on a logarithmic scale, called *log-magnitude*, defined as

$$\text{Lm}\,H(f) = 20\,\log_{10}|H(f)|. \tag{5.6}$$

  The unit of the log-magnitude is *decibel*, denoted by dB. For example, if the gain of an LTI system is $|H(f)| = 100$, then the log-magnitude of the gain is $\text{Lm}\,H(f) = 20\,\log_{10}100 = 40$ dB. If $|H(f)| = 0.01$ (an attenuation system), then the log-magnitude of the gain is -40 dB.
  The main convenience of the log-magnitude used in the Bode plot is that the FRF of an LTI system composed of multiple cascading components can be easily obtained as the algebraic sum of the individual FRFs of these components. For example, if an LTI system is composed of four subsystems: $H_1(f) = A(f)$, $H_2(f) = B(f)$, $H_3(f) = 1/C(f)$, and $H_4(f) = 1/D(f)$, then the FRF of the system $H(f) = H_1(f)H_2(f)H_3(f)H_4(f)$ can be found to be

$$\text{Lm}\,H(f) = \text{Lm}\left[\frac{A(f)B(f)}{C(f)D(f)}\right] = \text{Lm}\,A(f) + \text{Lm}\,B(f) - \text{Lm}\,C(f) - \text{Lm}\,D(f), \tag{5.7}$$

  with the same operations as for the phase plot:

$$\angle H(f) = \angle\left[\frac{A(f)B(f)}{C(f)D(f)}\right] = \angle A(f) + \angle B(f) - \angle C(f) - \angle D(f). \tag{5.8}$$

- In a *Nyquist diagram*, the value of $H(f)$ at any frequency f is plotted in the 2-D complex plane, either as a point in terms of $\text{Re}[H(f)]$ and $\text{Im}[H(0)]$ as its horizontal and vertical coordinates in a Cartesian coordinate system, or, equivalently, as a vector in terms of $|H(f)|$ and $\angle H(f)$ as its length and angle in a polar coordinate system. The Nyquist diagram of $H(f)$ is the locus of all such points $H(f)$ while f varies over the entire frequency range from $-\infty$ to ∞. The Nyquist diagram is typically used to determine the stability of an LTI system with a feedback loop.

Example 5.1: The FRF of a first-order LTI system (to be considered in detail in Example 5.2) is given as:

$$H(f) = \frac{1}{j2\pi f\tau + 1} = \frac{1}{jf/f_c + 1} \tag{5.9}$$

where $\tau = 1/2\pi f_c$ is the *time constant* of the system and $f_c = 1000$ is the *cutoff frequency.* Note that $|H(f_c)| = 1/\sqrt{2} = 0.707$ and $\angle H(f_c) = 45°$. This FRF is plotted in each of the four ways in Fig. 5.1.

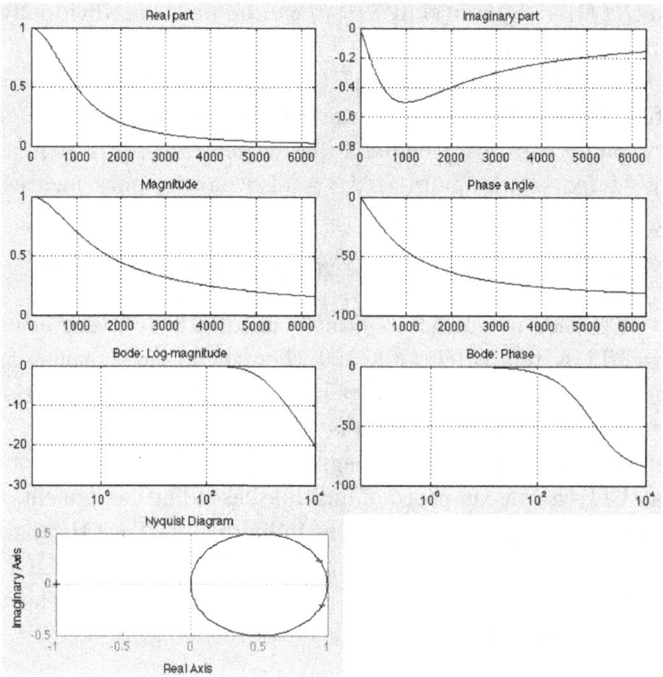

Figure 5.1 Different plots of the FRF of a first-order system. From the top down: the real and imaginary parts of the FRF, the magnitude and phase of the FRF, the Bode plots, and the Nyquist diagram.

Note that the Nyquist diagram of this first-order system is a circle. When the order of the FRF becomes higher, the Nyquist diagram may take some more complicated shapes. Fig. 5.2 shows the Nyquist diagram of the FRF of a third-order system:

$$H(\omega) = \frac{(j\omega)^2 + 5j\omega + 2}{3(j\omega)^2 + 2j\omega + 1} \tag{5.10}$$

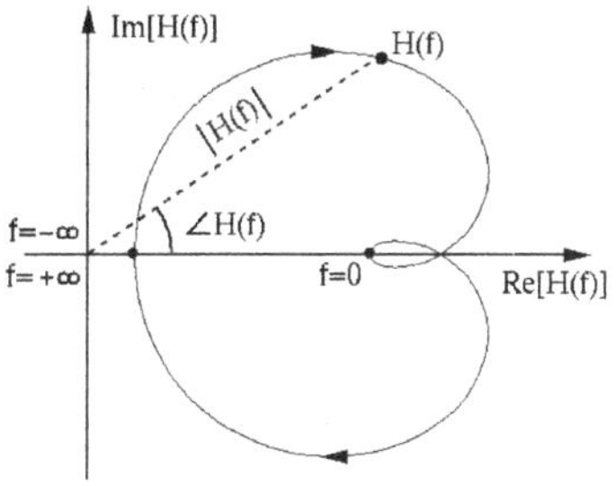

Figure 5.2 The Nyquist diagram of the FRF of a third-order system. points corresponding to specific frequencies at $f = -\infty$, $f = 0$, and $f = \infty$ are indicated.

The process of a signal passing through an LTI system can be described in both the time and frequency domains. Consider, in particular, when the input is a complex exponential $x(t) = e^{j2\pi ft}$, the corresponding output is

$$\begin{aligned} y(t) &= \mathcal{O}[e^{j2\pi ft}] = h(t) * e^{j2\pi ft} = \int_{-\infty}^{\infty} h(\tau) e^{j2\pi f(t-\tau)}\, d\tau \\ &= e^{j2\pi ft} \int_{-\infty}^{\infty} h(\tau) e^{-j2\pi f\tau}\, d\tau = H(f)\, e^{j2\pi ft} \\ &= |H(f)| e^{j\angle H(f)} e^{j2\pi ft} = |H(f)| e^{j[2\pi ft + \angle H(f)]}. \end{aligned} \tag{5.11}$$

This is the eigenequation of the LTI system indicating that when its input is a complex exponential, its output is the same exponential scaled by its FRF $H(f) = |H(f)|e^{j\angle H(f)}$.

If the system is real with $h(t) = \overline{h}(t)$, then taking the real part on both sides of the equation above we get

$$\begin{aligned} \mathcal{O}[\mathrm{Re}[e^{j2\pi ft}]] &= \mathcal{O}[\cos 2\pi ft] = \mathrm{Re}[\ |H(f)| e^{j(2\pi ft + \angle H(f))}\ ] \\ &= |H(f)|\ \cos(2\pi ft + \angle H(f)). \end{aligned} \tag{5.12}$$

Of course, we can also take the imaginary part of Eq. (5.11) to get

$$\mathcal{O}[\sin 2\pi ft] = |H(f)| \sin(2\pi ft + \angle H(f)). \tag{5.13}$$

We see that the response of any real LTI system to a sinusoidal input is the same sinusoid with its amplitude scaled by the magnitude of the FRF, and its phase shifted by the phase angle of the FRF.

The result in Eq. (5.11) can be generalized to cover any input that can be expressed as a linear combination of a set of sinusoids (inverse Fourier transform in Eq. (3.65)):

$$x(t) = \int_{-\infty}^{\infty} X(f) e^{j\omega t}\, df. \tag{5.14}$$

The corresponding output of the LTI system is $y(t) = h(t) * x(t)$. However, owing to the linearity of the system, we can also get the output as

$$y(t) = \mathcal{O}[x(t)] = \mathcal{O}[\int_{-\infty}^{\infty} X(f)e^{j2\pi ft}\,df] = \int_{-\infty}^{\infty} X(f)\mathcal{O}[e^{j2\pi ft}]\,df$$
$$= \int_{-\infty}^{\infty} X(f)H(f)e^{j2\pi ft}\,df = \mathcal{F}^{-1}[X(f)H(f)] = \mathcal{F}^{-1}[Y(f)], \quad (5.15)$$

where $Y(f) = H(f)X(f)$ (Eq. (5.2)). We see that the output $y(t)$ happens to be the inverse Fourier transform of $Y(f) = H(f)X(f)$. In other words, while in the time domain the output is the convolution of the input and the impulse response function $y(t) = h(t) * x(t)$, in the frequency domain the output is the product $Y(f) = H(f)X(f)$ of the input and the frequency response function.

All results derived above for continuous signals can be extended to discrete signals. If the discrete input to an LTI system is a complex exponential $x[n] = e^{j2\pi fn} = e^{j\omega n}$, the corresponding output is

$$y[n] = \mathcal{O}[x[n]] = h[n] * x[n] = \sum_{\nu=-\infty}^{\infty} h[\nu]e^{j2\pi f(n-\nu)}$$
$$= e^{j2\pi fn} \sum_{\nu=-\infty}^{\infty} h[\nu]e^{-j2\pi f\nu} = e^{j2\pi fn}\, H(f) = e^{j\omega n}\, H(\omega), \quad (5.16)$$

where

$$H(f) = \mathcal{F}[h[n]] = \sum_{n=-\infty}^{\infty} h[n]e^{-j2\pi fn} \quad (5.17)$$

is the Fourier transform of the impulse response $h[n]$ (Eq. (4.13), also the frequency response function of the system, first given in Eq. (1.112). Also, similar to the continuous case, we have

$$\mathcal{O}[\cos(2\pi nf)] = \mathrm{Re}[\,|H(f)|e^{j(2\pi nf+\angle H(f))}] = |H(f)|\cos[2\pi nf + \angle H(f)]. \quad (5.18)$$

As in general a discrete input $x[n]$ to the LTI system can be expressed as (Eq. (4.13))

$$x[n] = \int_0^1 X(f)e^{j2n\pi f}\,df, \quad (5.19)$$

the corresponding output can be found to be

$$y[n] = \mathcal{O}[x[n]] = \mathcal{O}[\int_0^1 X(f)e^{j2n\pi f}\,df] = \int_0^1 X(f)\mathcal{O}[e^{j2n\pi f}]\,df$$
$$= \int_0^1 X(f)H(f)e^{j2n\pi f}\,df = \mathcal{F}^{-1}[Y(f)], \quad (5.20)$$

which is the inverse DTFT of $Y(f) = X(f)H(f)$.

The results above for both continuous and discrete cases are of course the same as the convolution theorems given in Eqs. (3.123) and (4.39), which is illustrated in Fig. 5.3.

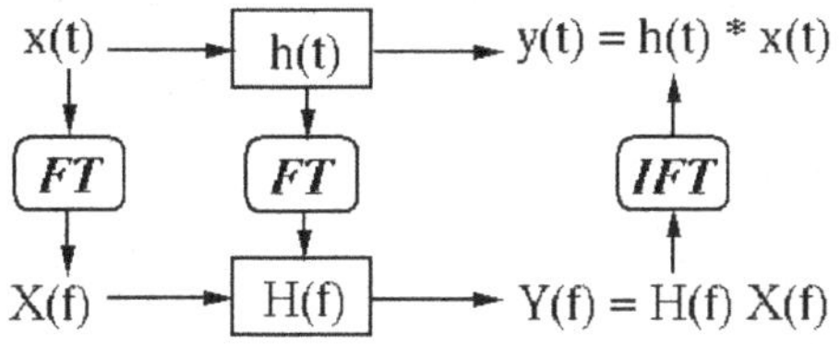

Figure 5.3 Signal through system in the time and frequency domains.

We see that an LTI system can be described by its impulse response function $h(t)$ in the time domain, or by its frequency response function $H(f) = \mathcal{F}[h(t)]$ in the frequency domain. Correspondingly, the response of the system to a given input $x(t)$ can be obtained as a convolution $y(t) = h(t) * x(t)$ in the time domain, or as a product $Y(f) = H(f)X(f)$ in the frequency domain. Although both the forward and inverse Fourier transforms are needed for the frequency domain method, we can gain some benefits not possible in the time domain. Most obviously, the response of an LTI system to an input $x(t)$ can be conveniently obtained in the frequency domain by a multiplication, instead of the corresponding convolution in the time domain.

Moreover, as the output of an LTI system can be expressed as a product $Y(f) = H(f)X(f)$, given any two of the three variables $X(f)$, $H(f)$, and $Y(f)$, we can always conveniently find the third, as shown in the following three cases.

1. Prediction of system output:
   Given input $X(f)$ to the FRF $H(f)$, we can find the output $Y(f)$. This operation can also be carried out equivalently as a convolution in the time domain.
2. System identification/filter design:
   Given input $X(f)$ and the observed output $Y(f)$, we can determine the FRF $H(f) = Y(f)/X(f)$ of an unknown system. This process is also useful in design of a system, called a filter in signal processing, given the input and desired output. Correspondingly, in the time domain, it is difficult to find $h(t)$ given $x(t)$ and $y(t)$.
3. Signal restoration:
   Based on the observed output $Y(f)$ from a measuring system with known FRF $H(f)$, we can find the input $X(f)$ without the distortion caused by the system. In the time domain, it is difficult to find $x(t)$ given $y(t)$ and $h(t)$.

5.2 Solving differential and difference equations

An important type of LTI systems can be described by a *linear constant-coefficient differential equation (LCCDE)* that relates its output $y(t)$ to its input

$x(t)$

$$\sum_{k=0}^{N} a_k \frac{d^k}{dt^k} y(t) = \sum_{k=0}^{M} b_k \frac{d^k}{dt^k} x(t). \tag{5.21}$$

If the input is a complex exponential $x(t) = e^{j\omega t}$, then, according to Eq. (5.11), the output is also a complex exponential $y(t) = H(\omega)e^{j\omega t}$ with a complex coefficient $H(\omega)$, the FRF of the system. Note that this output here is the *steady state response* of the system to the complex exponential input. (Initial conditions and transient response of the system will be considered later). Substituting such $x(t)$ and $y(t)$ into the LCCDE above and applying the time differentiation property (Eq. (3.127)), we get

$$H(\omega) \sum_{k=0}^{N} a_k (j\omega)^k e^{j\omega t} = \sum_{k=0}^{M} b_k (j\omega)^k e^{j\omega t}. \tag{5.22}$$

Solving this we get the FRF of the system:

$$H(\omega) = \frac{\sum_{k=0}^{M} b_k (j\omega)^k}{\sum_{k=0}^{N} a_k (j\omega)^k} = \frac{N(\omega)}{D(\omega)}, \tag{5.23}$$

where $N(\omega) = \sum_{k=0}^{M} b_k (j\omega)^k$ and $D(\omega) = \sum_{k=0}^{N} a_k (j\omega)^k$ are the numerator and denominator of $H(\omega)$, respectively.

More generally, consider an input $x(t) = X(\omega)e^{j\omega t}$ with a complex coefficient $X(\omega) = |X(\omega)|e^{j\angle X(\omega)}$ called the *phasor* of $x(t)$. The corresponding output can be assumed to be also a complex exponential $y(t) = Y(\omega)e^{j\omega t}$ with a phasor coefficient $Y(\omega)$. Substituting such $x(t)$ and $y(t)$ into the differential equation, we get

$$Y(\omega) \sum_{k=0}^{N} a_k (j\omega)^k e^{j\omega t} = X(\omega) \sum_{k=0}^{M} b_k \frac{d^k}{dt^k} x(t). \tag{5.24}$$

This result can also be directly obtained by taking the Fourier transform on both sides of the LCCDE in Eq. (5.21). We see that the FRF of the LTI system can also be found as the ratio of the output phasor $Y(\omega)$ and input phasor $X(\omega)$:

$$H(\omega) = \frac{Y(\omega)}{X(\omega)} = \frac{\sum_{k=0}^{M} b_k (j\omega)^k}{\sum_{k=0}^{N} a_k (j\omega)^k} = \frac{N(\omega)}{D(\omega)}. \tag{5.25}$$

This is also the definition of the FRF of a continuous LTI system described by the LCCDE Eq. (5.21). In this case of a continuous LTI system described by a LCCDE, the frequency $\omega = 2\pi f$ only appears in the form of $j\omega$ in all functions in the frequency domain, including $H(\omega)$, $X(\omega)$, $Y(\omega)$, $N(\omega)$, and $D(\omega)$. For this reason, these functions could also be denoted as functions of $j\omega$, such as $H(j\omega)$.

Moreover, owing to the linearity of the system, if the input is a linear combination of complex exponentials,

$$x(t) = \int_{-\infty}^{\infty} X(\omega)e^{j\omega t}\, df, \tag{5.26}$$

we can get $X(\omega) = \mathcal{F}[x(t)]$. Given the FRF $H(\omega)$ of the system, we can find the output:

$$y(t) = \mathcal{F}^{-1}[Y(\omega) = \mathcal{F}^{-1}[H(\omega)X(\omega)] = \int_{-\infty}^{\infty} H(\omega)X(\omega)e^{j\omega t}\, df. \tag{5.27}$$

In parallel with the continuous LTI systems described by the LCCDE in Eq. (5.21), one particular type of discrete LTI systems can be described by a *linear constant-coefficient difference equation (LCCDE)* that relates the output $y[n]$ to the input $x[n]$:

$$\sum_{k=0}^{N} a_k y[n-k] = \sum_{k=0}^{M} b_k x[n-k]. \tag{5.28}$$

If the input is a complex exponential $x[n] = e^{j\omega n}$, then according to Eq. (5.16) the output is also a complex exponential $y[n] = H(\omega)e^{j\omega n}$. Substituting such $x[n]$ and $y[n]$ into the equation above, we get

$$H(\omega)\sum_{k=0}^{N} a_k e^{-j\omega k} = \sum_{k=0}^{M} b_k e^{-j\omega k}. \tag{5.29}$$

Solving for $H(\omega)$, we get

$$H(\omega) = \frac{\sum_{k=0}^{M} b_k e^{-j\omega k}}{\sum_{k=0}^{N} a_k e^{-j\omega k}} = \frac{N(\omega)}{D(\omega)}, \tag{5.30}$$

where $N(\omega) = \sum_{k=0}^{M} b_k e^{-j\omega k}$ and $D(\omega) = \sum_{k=0}^{N} a_k e^{-j\omega k}$. Alternatively, taking the DTFT on both sides of Eq. (5.28) and applying the time shift property (Eq. (4.33)), we get

$$Y(\omega)\sum_{k=0}^{N} a_k e^{-j\omega k} = X(\omega)\sum_{k=0}^{M} b_k e^{-j\omega k}, \tag{5.31}$$

and we also get the FRF:

$$H(\omega) = \frac{Y(\omega)}{X(\omega)} = \frac{\sum_{k=0}^{M} b_k e^{-j\omega k}}{\sum_{k=0}^{N} a_k e^{-j\omega k}} = \frac{N(\omega)}{D(\omega)}. \tag{5.32}$$

This is also the definition of the FRF of a discrete LTI system described by the LCCDE Eq. (5.28). In this case of a discrete LTI system described by an LCCDE, the frequency $\omega = 2\pi f$ only appears in the form of $e^{j\omega}$ in all functions in the frequency domain including $H(\omega)$, $X(\omega)$, $Y(\omega)$, $N(\omega)$, and $D(\omega)$. For this reason, these functions could also be denoted as functions of $e^{j\omega}$, such as $H(e^{j\omega})$.

Given the input $X(\omega) = \mathcal{F}[x[n]]$ and the FRF $H(\omega)$ of the system, we can find the output in the time domain:

$$y[n] = \mathcal{F}^{-1}[Y(\omega)] = \mathcal{F}^{-1}[H(\omega)X(\omega)]. \tag{5.33}$$

In summary, we can solve an LCCDE system, either continuous or discrete, by following these steps:

- find the FRF of the system $H(\omega) = N(\omega)/D(\omega)$;
- carry out the CTFT of a continuous input $x(t)$ to find $X(\omega) = \mathcal{F}[x(t)]$, or the DTFT of a discrete input $x[n]$ to find $X(\omega) = \mathcal{F}[x[n]]$;
- obtain the response in the frequency domain $Y(\omega) = H(\omega)X(\omega)$;
- carry out the inverse DTFT or CTFT on $Y(\omega)$ to get $y(t)$ or $y[n]$.

Example 5.2: In a circuit composed of a resistor R and a capacitor C as shown in Fig. 5.4, the input $x(t) = v_{\text{in}}(t)$ is the voltage across both R and C in series, and the the output $y(t) = v_C(t)$ is the voltage across C. Find both the step and impulse response of the system, and the FRF $H(\omega)$ of the system.

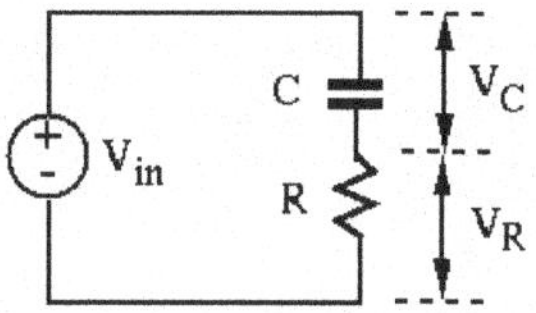

Figure 5.4 An RC circuit.

- Set up the differential equation
  The current through both C and R is $i(t) = C\, dv_C(t)/dt = C\dot{y}(t)$, and by Ohm's law, the voltage across R is $v_R(t) = Ri(t) = RC\,\dot{y}(t)$. The input voltage $x(t)$ is the sum of $v_R(t)$ and $v_C(t)$:

$$v_R(t) + v_C(t) = RC\,\dot{y}(t) + y(t) = \tau\dot{y}(t) + y(t) = x(t) = v_{\text{in}}(t), \tag{5.34}$$

  where $\tau = RC$ is the *time constant* of the system. Dividing both sides by τ we get

$$\dot{y}(t) + \frac{1}{\tau}y(t) = \frac{1}{\tau}x(t). \tag{5.35}$$

- Find the step response
  - Find homogeneous solution $y_{\text{h}}(t)$ when $x(t) = 0$:
    Assume $y_{\text{h}}(t) = A\,e^{st}$ and we have $\dot{y}_{\text{h}}(t) = sAe^{st}$, now the homogeneous differential equation becomes

$$(s\tau + 1)A\,e^{st} = 0; \quad \text{i.e.,} \quad s\tau + 1 = 0. \tag{5.36}$$

    We therefore get $s = -1/\tau$ and $y_{\text{h}}(t) = A\,e^{-t/\tau}$.
  - Find the particular solution $y_{\text{p}}(t)$ when $x(t) = u(t)$:
    As the right-hand side is a constant $1/\tau$ for $t > 0$, we assume the corresponding output is also a constant $y_{\text{p}}(t) = C$ and $\dot{y}_{\text{p}}(t) = 0$. Substituting these into the equation we get $y_{\text{p}}(t) = 1$.
  - Find the complete response to unit step:

$$y(t) = y_{\text{h}}(t) + y_{\text{p}}(t) = (A\,e^{-t/\tau} + 1)u(t). \tag{5.37}$$

Given the initial condition $y(t)\big|_{t<0} = y_0$ (initial voltage across C), we get $A = y_0 - 1$, and the complete response to $x(t) = u(t)$ is

$$y(t) = [(y_0 - 1)e^{-t/\tau} + 1]u(t) = [(1 - e^{-t/\tau}) + y_0 e^{-t/\tau}]u(t). \tag{5.38}$$

Physically, the first term is for the charging of the capacitor owing to the step input while the second term is for the discharge of the capacitor with a non-zero initial voltage. In particular, when $y_0 = 0$, we have

$$y(t) = (1 - e^{-t/\tau})u(t). \tag{5.39}$$

- Find the impulse response $h(t)$
  Owing to the fact that if $\mathcal{O}[x(t)] = y(t)$, then $\mathcal{O}[\dot{x}(t)] = \dot{y}(t)$ (Eq. (1.75)) (valid for a DE under zero initial condition), we can get the impulse response $h(t)$ to $\delta(t) = du(t)/dt$ by taking the derivative of the unit response to $u(t)$ obtained above:

$$\begin{aligned} h(t) = \dot{y}(t) &= \frac{d}{dt}[(1 - e^{-t/\tau})u(t)] \\ &= \frac{1}{\tau}e^{-t/\tau}u(t) + (1 - e^{-t/\tau})\delta(t) = \frac{1}{\tau}e^{-t/\tau}u(t). \end{aligned}$$

- A different method to find $h(t)$
  As the system is causal, $h(t) = 0$ for all $t < 0$ when the input is zero, we can assume

$$h(t) = f(t)u(t) = \begin{cases} f(t) & t > 0 \\ 0 & t < 0 \end{cases}, \tag{5.40}$$

where $f(t)$ is a function to be determined, and have

$$\dot{h}(t) = \dot{f}(t)u(t) + f(t)\dot{u}(t) = \dot{f}(t)u(t) + f(0)\delta(t). \tag{5.41}$$

Now Eq. (5.35) becomes

$$\tau\dot{f}(t)u(t) + \tau f(0)\delta(t) + f(t)u(t) = \delta(t). \tag{5.42}$$

Separating terms containing $u(t)$ and $\delta(t)$ respectively, we get two equations:

$$\begin{cases} \tau\dot{f}(t) + f(t) = 0 \\ f(0) = 1/\tau \end{cases}. \tag{5.43}$$

This homogeneous equation with an initial condition can be solved to get

$$f(t) = \frac{1}{\tau}e^{-t/\tau}, \tag{5.44}$$

and the impulse response same as above:

$$h(t) = f(t)u(t) = \frac{1}{\tau}e^{-t/\tau}u(t). \tag{5.45}$$

- Find the impulse responses and FRF

Taking the CTFT on both sides of Eq. (5.35), we get

$$Y(\omega)\left(j\omega+\frac{1}{\tau}\right)=X(\omega)\frac{1}{\tau}, \tag{5.46}$$

and the FRF of the system is

$$H(\omega)=\frac{Y(\omega)}{X(\omega)}=\frac{1/\tau}{j\omega+1/\tau}=\frac{1}{j\omega\tau+1}, \tag{5.47}$$

which is plotted in various ways in Example 5.1. Taking the inverse CTFT of $H(\omega)$ we get the impulse response:

$$h(t)=\mathcal{F}^{-1}[H(\omega)]=\mathcal{F}^{-1}\left[\frac{1/\tau}{j\omega+1/\tau}\right]=\frac{1}{\tau}e^{-t/\tau}u(t). \tag{5.48}$$

- Find the step response
  In time domain, the step response can be found by convolution (see Example 1.4):

$$h(t)*u(t)=\frac{1}{\tau}\int_0^t e^{-(t-t')/\tau}\,dt'=\frac{1}{\tau}e^{-t/\tau}\tau(e^{t/\tau}-1)u(t)=(1-e^{-t/\tau})u(t), \tag{5.49}$$

where $u(t)$ is included to reflect the fact that this result is valid only if $t>0$. Alternatively, in the frequency domain, the response to a unit step $U(\omega)=\mathcal{F}[u(t)]$ is

$$\begin{aligned}Y(\omega)=H(\omega)U(\omega)&=\frac{1}{j\omega\tau+1}\left[\frac{1}{2}\delta(f)+\frac{1}{j\omega}\right]\\&=\frac{1}{2}\frac{\delta(f)}{j\omega\tau+1}+\frac{1}{j\omega\tau+1}\frac{1}{j\omega}=\frac{1}{2}\delta(f)+\frac{1}{j\omega}-\frac{\tau}{j\omega\tau+1}.\end{aligned} \tag{5.50}$$

Note that $\delta(f)x(f)=\delta(f)x(0)$. Taking the inverse CTFT of the above we get the step response in the time domain:

$$y(t)=\mathcal{F}^{-1}[Y(\omega)]=\mathcal{F}^{-1}\left[\frac{1}{2}\delta(f)+\frac{1}{j\omega}\right]-\mathcal{F}^{-1}\left[\frac{\tau}{j\omega\tau+1}\right]=(1-e^{-t/\tau})u(t). \tag{5.51}$$

This result is the same as that in Eq. (5.39) obtained by solving the DE in Eq. (5.35).

Example 5.3: Consider an LTI system described by a first-order difference equation:

$$y[n]-a\,y[n-1]=x[n]\quad\text{or}\quad y[n]=x[n]+a\,y[n-1]. \tag{5.52}$$

This system is a *recursive filter* as the current output $y[n]$ depends on the past output $y[n-1]$ as well as the current input $x[n]$. We assume the system is causal; i.e., $h[n]=0$ for $n<0$.

- Find the impulse response by solving the difference equation
  If the input is $x[n] = \delta[n]$, then the output is $y[n] = h[n]$ and Eq. (5.52) becomes $h[n] - a\,h[n-1] = \delta[n]$. $h[n]$ can be found recursively:

$$\begin{cases} n=0 & h[0]-ah[-1]=h[0]=\delta[0]=1; \quad \text{i.e. } h[0]=1 \\ n=1 & h[1]-ah[0]=h[1]-a=\delta[1]=0; \quad \text{i.e. } h[1]=a \\ n=2 & h[2]-ah[1]=h[2]-a^2=\delta[2]=0; \quad \text{i.e. } h[2]=a^2 \\ \cdots \end{cases}. \tag{5.53}$$

  Summarizing the above, we get $h[n] = a^n u[n]$.
  Alternatively, we can also assume a general solution $h[n] = A\,e^{jn\omega}$ and Eq. (5.52) becomes

$$A\,e^{jn\omega} - a\,A\,e^{j(n-1)\omega} = \delta[n] = 0 \quad (n > 0), \tag{5.54}$$

  from which we get $e^{j\omega} = a$ and $h[n] = Aa^n$. Using the initial condition $h[0] = 1$ obtained above, we get $A = 1$ and, therefore, we also get $h[n] = a^n$. Note that the system is stable only if $|a| < 1$.
- Find the step response by convolution
  The response to a unit step $x[n] = u[n]$ can be found by convolution

$$y[n] = h[n] * u[n] = \sum_{m=-\infty}^{\infty} a^m u[m]u[n-m] = \sum_{m=0}^{n} a^m = \frac{1-a^{n+1}}{1-a}u[n]. \tag{5.55}$$

- Find the impulse response by the DTFT
  Taking the DTFT on both sides on Eq. (5.52) we get

$$Y(\omega)(1 - a\,e^{-j\omega}) = X(\omega), \tag{5.56}$$

  and the FRF of the system is

$$H(\omega) = \frac{Y(\omega)}{X(\omega)} = \frac{1}{1 - a\,e^{-j\omega}}. \tag{5.57}$$

  Taking inverse DTFT of $H(f)$, we get the impulse response:

$$h[n] = \mathcal{F}^{-1}[H(\omega)] = a^n u[n]. \tag{5.58}$$

- Find the step response
  In the frequency domain, the response to a unit step $U(\omega) = \mathcal{F}[u[n]]$ is

$$\begin{aligned} Y(\omega) = H(\omega)U(\omega) &= \frac{1}{1-a\,e^{j\omega}}\left[\frac{1}{1-e^{-j\omega}} + \frac{1}{2}\sum_{n=-\infty}^{\infty}\delta(f-n)\right] \\ &= \frac{1}{1-a\,e^{j\omega}}\,\frac{1}{1-e^{-j\omega}} + \frac{1}{2}\sum_{n=-\infty}^{\infty}\frac{\delta(f-n)}{1-a\,e^{j\omega}} \\ &= \frac{1}{1-a}\left[\frac{1}{1-e^{-j\omega}} - \frac{a}{1-a\,e^{-j\omega}}\right] + \frac{1}{2}\sum_{n=-\infty}^{\infty}\frac{\delta(f-n)}{1-a} \end{aligned}$$

$$= \frac{1}{1-a}\left[\frac{1}{1-e^{-j\omega}} - \frac{a}{1-a\,e^{-j\omega}} + \frac{1}{2}\sum_{n=-\infty}^{\infty}\delta(f-n)\right]. \qquad (5.59)$$

Taking the inverse DTFT of the above we get the step response in the time domain (Eqs. (4.64) and (4.66))

$$y[n] = \mathcal{F}^{-1}[Y(\omega)] = \mathcal{F}^{-1} = \frac{1-a^{n+1}}{1-a}u[n]. \qquad (5.60)$$

5.3 Magnitude and phase filtering

In the context of signal processing, an LTI system can be treated as a filter, and the process of a signal $x(t)$ going through the system $h(t)$ becomes a filtering process, which is either as a convolution in the time domain or, equivalently, a multiplication in the frequency domain (Fig. 5.3):

$$y(t) = h(t) * x(t) \quad \text{or} \quad Y(f) = H(f)X(f). \qquad (5.61)$$

Note that sometimes a filter may not be causal as its impulse response function $h(t)$ may not be zero for $t < 0$. Obviously, a non-causal filter is not actually implementable in real time unless a certain delay is allowed; i.e., the filtering is not truly real-time. However, non-causal filter can be readily implemented off-line, as it can be applied to pre-recorded data containing all signal samples at the same time.

The filtering process in the frequency domain has the benefit that the signal can be easily manipulated by various filters based on the frequency contents of the signal. We could modify and manipulate the phase as well as the magnitude of the frequency components of the signal. The complex multiplication $Y(f) = H(f)\,X(f)$ in Eq. (5.61 can be written in terms of both magnitude and phase:

$$Y(f) = |Y(f)|e^{j\angle Y(f)} = H(f)X(f) = |H(f)|e^{j\angle H(f)}\,|X(f)|e^{j\angle X(f)}; \qquad (5.62)$$

i.e.,

$$\begin{cases} |Y(f)| = |H(f)|\,|X(f)| \\ \angle Y(f) = \angle H(f) + \angle X(f) \end{cases}. \qquad (5.63)$$

We now consider both aspects of the filtering process.

- **Magnitude filtering**
  Various filtering schemes can be implemented based on the gain $|H(f)|$ of the filter. Typically, depending on which part of the signal spectrum is enhanced or attenuated, a filter can be classified as one of these different types: low-pass (LP), high-pass (HP), band-pass (BP), and band-stop (BS) filters, as illustrated in Fig. 5.5. Moreover, if the gain $|H(f)| = 1$ is unity (or a constant

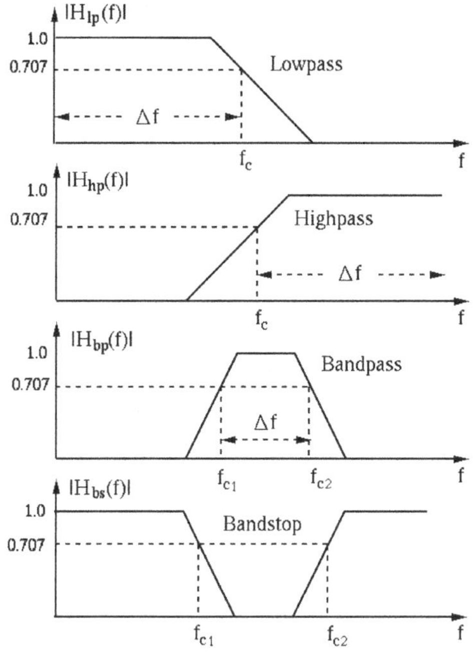

Figure 5.5 Illustration of four different types of filters (LP, HP, BP, and BS).

independent of frequency f), then $H(f) = |H(f)|e^{j\angle H(f)} = e^{j\angle H(f)}$ is said to be an all-pass (AP) filter.

In general, a filter $H(f)$ can be characterized by two parameters:

- The *cutoff frequency* f_c of a filter is the frequency at which $|H(f)|$ is reduced to $1/\sqrt{2} = 0.707$ of the maximum magnitude (gain) $H_{max} = |H(f_0)| \geq |H(f)|$ at some peak frequency f_0; i.e.,

$$|H(f_c)| = \frac{1}{\sqrt{2}} H_{max} \quad \text{or} \quad |H(f_c)|^2 = \frac{1}{2} H_{max}^2. \tag{5.64}$$

As the power of the filtered signal (proportional to its magnitude squared) at the cutoff frequency f_c is half of the maximum power at the peak frequency f_0, the cutoff frequency is also called the *half-power frequency.* This attenuation of $|H(f)|$ at $f = f_c$ is also commonly represented in terms of the log-magnitude in decibel:

$$20 \log_{10} \left(\frac{|H(f_c)|}{H_{max}} \right) = 20 \log_{10} \frac{1}{\sqrt{2}} = 20 \log_{10} 0.707 = -3.01 \text{ dB} \approx -3 \text{ dB}. \tag{5.65}$$

We see that $\text{Lm}\, H(f_c)$ at the cutoff or half-power frequency is 3 dB lower than $\text{Lm}\, H_{max} = \text{Lm}\, H(f_0)$ at the peak frequency.

- The *bandwidth* $\triangle f$ of a BP filter is the interval between two cutoff frequencies on either side of the peak frequency:

$$\triangle f = f_{c2} - f_{c1}. \tag{5.66}$$

The *quality factor* Q of a BP filter is defined as the ratio of its bandwidth $\triangle f$ and the peak frequency f_0 at which $|H(f_0)| = H_{\max}$:

$$Q = \frac{\triangle f}{f_0}. \tag{5.67}$$

Note that the higher the value of Q, the narrower the BP filter is.

For an LP filter, we define the lower cutoff frequency to be zero $f_{c1} = 0$, and the bandwidth is the same as the cutoff frequency $\triangle f = f_c$. The RC circuit considered in Example 5.2 is actually an LP filter. The magnitude of the FRF given in Eq. (5.47) is:

$$|H(\omega)| = \frac{1}{|j\omega\tau + 1|} = \frac{1}{\sqrt{(\omega\tau)^2 + 1}} = \begin{cases} 1 & \omega = 0 \\ 1/\sqrt{2} & \omega = 1/\tau \\ 0 & \omega = \infty \end{cases}. \tag{5.68}$$

We see that the cutoff frequency is $\omega_c = 1/\tau$ or $f_c = 1/2\pi\tau$. An HP filter is solely described by its lower cutoff frequency f_{c1}, as the higher cutoff frequency f_{c2} is undefined (or $f_{c2} = \infty$).

- **Phase filtering**

The filtering process affects the phase angles of the frequency components in a signal as well as their magnitudes owing to the phase shift $\angle H(f)$ of the filter, which is non-zero in general.

- *Linear-phase filtering:*

We first note that if a sinusoidal time function $\cos(\omega t) = \cos(2\pi f t)$ of frequency f or period $T = 1/f$ is phase-shifted by $-\phi$:

$$\cos(\omega t - \phi) = \cos(\omega(t - \phi/\omega)) = \cos(\omega(t - \tau)), \tag{5.69}$$

it is time-delayed by

$$\tau = \frac{\phi}{\omega} = \frac{\phi}{2\pi f} = \phi\frac{T}{2\pi}, \quad \text{i.e.,} \quad \frac{\phi}{2\pi} = \frac{\tau}{T}. \tag{5.70}$$

If $\cos(\omega t)$ is filtered by $H(f)$ to become $|H(f)|\cos(\omega t + \angle H(f))$, it is phase-shifted by $\angle H(f)$, or time-delayed by

$$\tau = -\frac{\angle H(f)}{2\pi f} = -\angle H(f)\frac{T}{2\pi}. \tag{5.71}$$

Moreover, when a signal $x(t)$ is filtered by an AP linear-phase filter $H(f)$ with $|H(f)| = 1$ and $\angle H(f) = -\tau 2\pi f$, it becomes:

$$Y(f) = H(f)X(f) = |H(f)|e^{j\angle H(f)}X(f) = X(f)e^{-j2\pi f\tau}. \tag{5.72}$$

Integrating over frequency, we get the output signal in the time domain:

$$y(t) = \int_{-\infty}^{\infty} Y(f) e^{j2\pi ft} df = \int_{-\infty}^{\infty} X(f) e^{j2\pi f(t-\tau)} df = x(t-\tau). \quad (5.73)$$

Note that this is actually the time-shift property of the Fourier transform, and the shape of the signal remain the same except it is delayed by τ.
In general, a linear phase filter $H(f)$ (not necessarily AP) with $\angle H(f) = -\tau 2\pi f$ will delay all frequency components of an input signal by the same amount:

$$\tau_\phi = -\frac{\angle H(f)}{2\pi f} = -\angle H(f) \frac{T}{2\pi}, \quad (5.74)$$

which is called the *phase delay* of the linear-phase filter. The relative positions of these frequency components remain the same, only their magnitudes are modified by $|H(f)|$. Note, however, $\angle H(f) = \theta - \tau 2\pi f$ is not a linear phase shift due to the constant θ.

Example 5.4: Consider a signal $x(t) = \cos(2\pi f_1 t) + \cos(2\pi f_2 t)$ composed of two sinusoidal components with frequencies $f_1 = 2$ and $f_2 = 4$, respectively (top of Fig. 5.6). When filtered by an AP filter $H(f) = |H(f)| e^{j\angle H(f)}$ with a unity gain $|H(f)| = 1$ and a linear phase shift $\angle H(f) = -\tau\, 2\pi f = -\tau\, 2\pi/T$ where $\tau = 0.2$, the two components of the signal are phase-shifted by $\phi_1 = -\tau 2\pi f_1$ and $\phi_2 = -\tau 2\pi f_2$, respectively, and the signal becomes:

$$\begin{aligned} y(t) &= \cos(2\pi f_1 t - \phi_1) + \cos(2\pi f_2 t - \phi_2) \\ &= \cos(2\pi f_1 (t-\tau)) + \cos(2\pi f_2 (t-\tau)). \end{aligned} \quad (5.75)$$

As the two sinusoids are time-shifted by the same amount, their relative positions remain the same and so does the waveform of the signal, except that it is delayed by τ (middle of Fig. 5.6).

- *Nonlinear-phase filtering:* If the phase shift of a filter $H(f)$ is not a linear function of frequency, relative temporal positions of the various frequency components contained in the input signal will not be maintained during the filtering process. Consequently, the waveform of the output will not be the same as that of the input; i.e., the signal will be distorted by the filter, even though it is an AP filter with a constant gain. For example, while filtering the same signal $x(t) = \cos(2\pi f_1 t) + \cos(2\pi f_2 t)$ above, if the phase shift of the filter is $\phi_1 = \phi_2 = 6\pi\tau$ for both components, the output becomes

$$\begin{aligned} y(t) &= \cos(2\pi f_1 t - 6\pi\tau) + \cos(2\pi f_2 t - 6\pi\tau) \\ &= \cos(2\pi f_1 (t - 3\tau/f_1)) + \cos(2\pi f_2 (t - 3\tau/f_2)). \end{aligned} \quad (5.76)$$

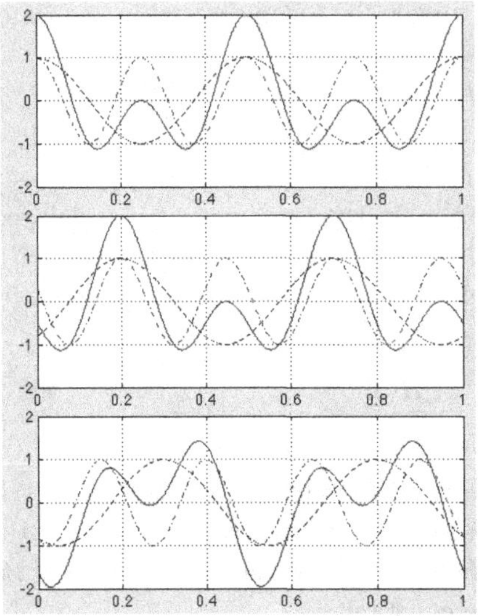

Figure 5.6 Filtering with linear and non-linear phase shift. The original signal (top) containing two frequency components of $f_1 = 2$ and $f_2 = 4$ is filtered by an AP filter with linear phase (middle) and non-linear (constant) phase (bottom). The signals are plotted in solid lines while the two frequency components are plotted in dashed lines.

When $\tau = 0.2$, the sinusoid with $f_1 = 2$ is shifted by $\tau_1 = 0.3$ and the sinusoid with $f_2 = 4$ is shifted by $\tau_2 = 0.15$ (bottom of Fig. 5.6). As the relative positions of the two components no longer remain the same after filtering, the waveform of the signal is different from the original. Another example is shown in Fig. 5.7, where a signal, a square impulse, is filtered, first by a linear phase AP filter (top), which causes a pure time delay without any distortion, and then by a constant-phase (non-linear) AP filter (bottom), by which the signal is distorted.

For a non-linear phase filter we can define the *group delay* as

$$\tau_g(f) = -\frac{d\angle H(f)}{2\pi\, df} = -\frac{d\angle H(\omega)}{d\omega}. \tag{5.77}$$

the significance of which can be understood by considering the filtering of a signal containing two sinusoids:

$$x(t) = \cos(\omega_1 t) + \cos(\omega_2 t) = 2\cos\left(\frac{(\omega_1 - \omega_2)t}{2}\right)\cos\left(\frac{(\omega_1 + \omega_2)t}{2}\right). \tag{5.78}$$

This is a sinusoid of a high frequency $(\omega_1 + \omega_2)/2$ with its amplitude modulated by a sinusoid of a low frequency $(\omega_1 - \omega_2)/2$ as the envelope. When filtered by an AP filter with phase-shift $\angle H(\omega_1) = -\phi_1$ and $\angle H(\omega_2) = -\phi_2$,

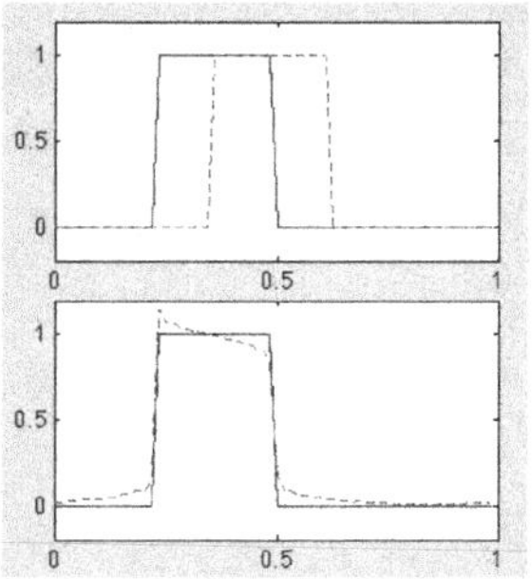

Figure 5.7 Filtering with linear and constant phase shift. The square wave (solid line) is filtered first by a linear-phase filter without distortion (dashed line in top panel) and then by a constant (non-linear) filter with distortion (dashed line in bottom panel).

the signal becomes:

$$\begin{aligned} y(t) &= \cos(\omega_1 t - \phi_1) + \cos(\omega_2 t - \phi_2) \\ &= 2\cos\left(\frac{(\omega_1 - \omega_2)t - (\phi_1 - \phi_2)}{2}\right)\cos\left(\frac{(\omega_1 + \omega_2)t - (\phi_1 + \phi_2)}{2}\right) \\ &= 2\cos\left(\frac{\omega_1 - \omega_2}{2}(t - \tau_d)\right)\cos\left(\frac{\omega_1 + \omega_2}{2}(t - \tau_s)\right), \end{aligned} \tag{5.79}$$

where

$$\tau_d = \frac{\phi_1 - \phi_2}{\omega_1 - \omega_2}, \qquad \tau_s = \frac{\phi_1 + \phi_2}{\omega_1 + \omega_2}. \tag{5.80}$$

are respectively the delays of the envelope and the amplitude-modulated sinusoid. If the filter has linear phase; i.e., $\phi_1/\omega_1 = \phi_2/\omega_2 = \tau$, then $\tau_d = \tau_s = \tau$ and the output $y(t) = x(t - \tau)$ is simply a delayed version of the input. Otherwise, $\tau_d \neq \tau_s$; i.e., the envelope is delayed by $\tau_d = \Delta\phi/\Delta\omega$, different from the delay of the amplitude-modulated sinusoid, and the waveform of the signal is no longer maintained. This example can be generalized to a continuum of frequencies within a narrow band, in which case the group delay is defined as $\tau_g(\omega) = d\phi/d\omega = -d\angle H(\omega)/d\omega$ (Eq. (5.77)).

Example 5.5: Fig. 5.8 shows the non-linear filtering of signals. First, when a signal composed of two sinusoids with $f_1 = 7$ and $f_2 = 8$ (top panel) is filtered by a non-linear phase AP filter with $\angle H(f_1) = -\phi_1 = \pi$ and $\angle H(f_2) = -\phi_2 = \pi/2$, its envelope is delayed by $\tau_g = (\phi_1 - \phi_2)/(\omega_1 - \omega_2) = 0.25$ (second panel). Also, when a signal composed of four sinusoids with $f_1 = 7$, $f_2 = 8$, $f_3 = 9$, and $f_4 = 10$ (third panel) is filtered by a non-linear phase AP filter with $\angle H(\omega) = \theta - \omega\tau$ with $\tau = 0.6$ and $\theta = 2$, its envelope is shifted by $\tau_g = -d\angle H(\omega)/d\omega = \tau = 0.6$ (bottom panel).

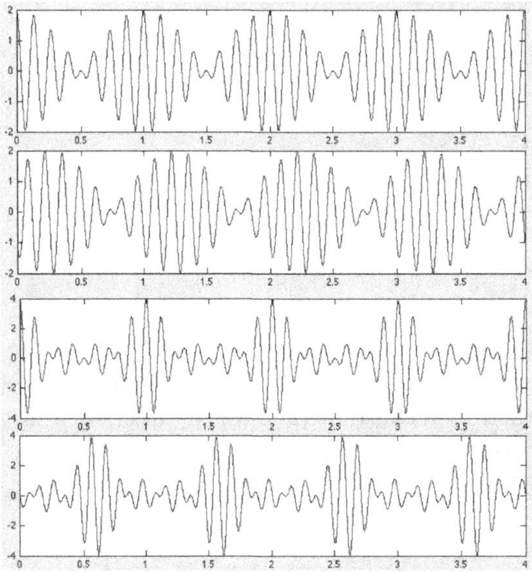

Figure 5.8 Non-linear phase filtering of two signals to show the envelope is delayed by the group delay τ_g.

5.4 Implementation of 1-D filtering

Here, we consider how the filtering process is carried out computationally, using four different types of LP filters as examples. We will discuss their implementation and filtering effects when applied to a square impulse train shown in the top row of Fig. 5.9.

- The *moving-average LP filtering* is carried out in the time domain by replacing each sample in the discrete signal by the average of a sequence of neighboring samples. This operation of moving average is actually a convolution of the signal with a square window covering the neighborhood of the sample in question:

$$y(t) = h(t) * x(t), \qquad \text{where} \quad h(t) = \begin{cases} 1 & -w/2 < t < w/2 \\ 0 & \text{else} \end{cases}, \tag{5.81}$$

  and w is the width of the square window. Correspondingly in the frequency domain, the moving-average filtering is a multiplication of the signal spectrum $X(f)$ by the FRF $H(f) = \mathcal{F}[h(t)]$, a sinc function first shown in Eq. (3.163). As shown in the second and third rows of Fig. 5.9, we see that the sinc FRF $H(f)$ has a lot of leakage; i.e., many high-frequency components can still leak through the filter to appear in the output.

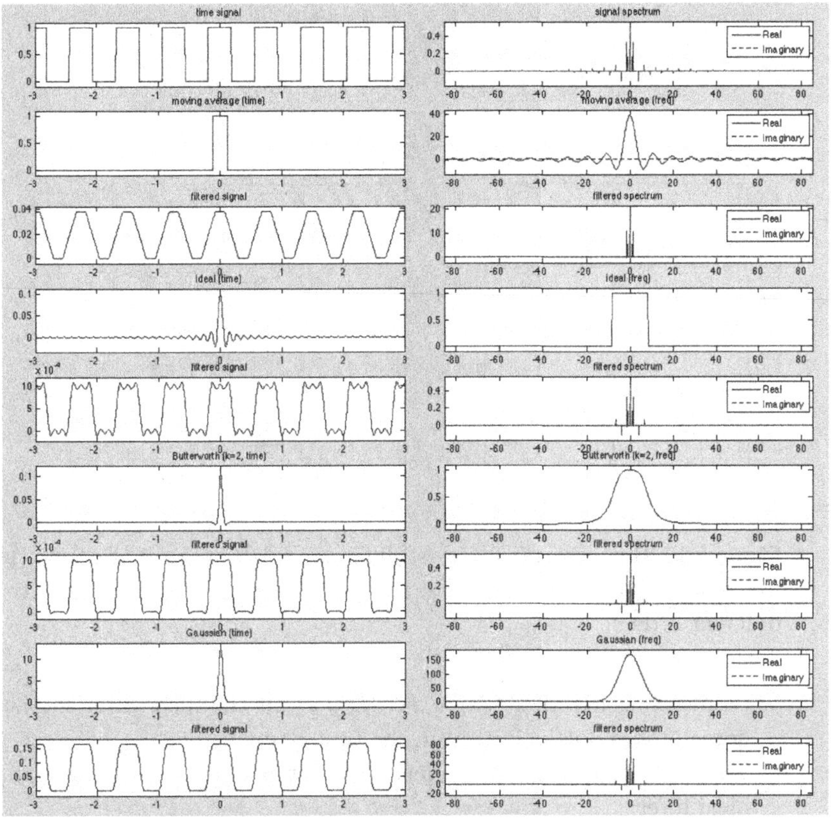

Figure 5.9 1-D LP filters in both the time (left) and frequency (right) domains. A square impulse train $x(t)$ and its spectrum $X(f)$ are shown in the top row. The following eight rows show $h(t)$ and $H(f)$ of four filters (moving average, ideal, Butterworth and Gaussian), and the corresponding output $y(t) = h(t) * x(t)$ and its spectrum $Y(f) = H(f)X(f)$.

- The *ideal LP filter (rectangular)* is defined in the frequency domain as

$$H(f) = \begin{cases} 1 & |f| < f_c \\ 0 & |f| > f_c \end{cases}, \tag{5.82}$$

where f_c is the cutoff frequency. As shown in Eq. (3.165), in the time domain, the impulse response of the ideal filter is a sinc function:

$$h(t) = \frac{\sin(2\pi f_c t)}{\pi t} = 2f_c \ \mathrm{sinc}(2f_c t). \tag{5.83}$$

After filtering, all frequency components outside the passing band are totally removed while those within remain unchanged. As shown in the fourth and fifth rows of Fig. 5.9, the ideal LP filter causes some severe ringing artifacts in the filtered signal, due obviously to the convolution of the signal with the

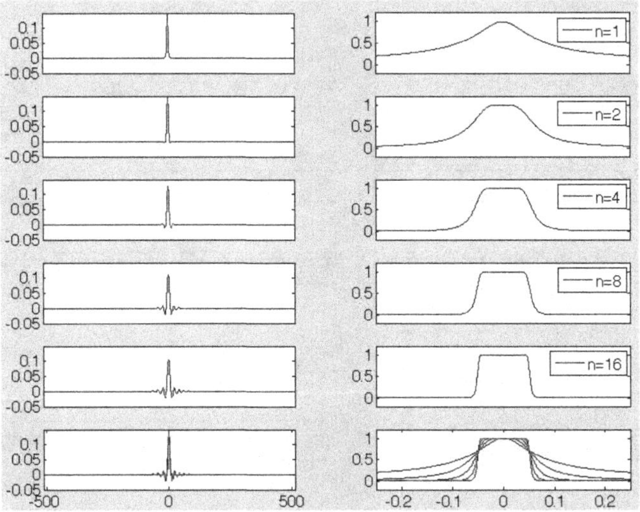

Figure 5.10 Butterworth filters of different orders in both the time (left) and frequency (right) domains. The plot in the last row compares all five filters of different orders.

ringing sinc function $h(t) = \mathcal{F}^{-1}[H(f)]$. So the ideal filter in the frequency domain does not look ideal in the time domain.

- The *Butterworth LP filter* defined below avoids the ringing artifacts of the ideal filter:

$$H(f) = \frac{1}{\sqrt{1 + (f/f_c)^{2n}}} = \begin{cases} 1 & f = 0 \\ 1/\sqrt{2} & f = f_c \\ 0 & f = \infty \end{cases}, \tag{5.84}$$

where f_c is the cutoff frequency at which $H(f) = H(f_c) = 1/\sqrt{2}$, and n is a positive integer for the order of the filter. By adjusting n one can control the shape of the filter and thereby make a proper tradeoff between the ringing effects and how accurately the passing band is specified. As shown in Fig. 5.10, when n is small, the shape of the filter is smooth (low frequency accuracy) with little ringing; when n is large, the filter becomes sharper (higher frequency accuracy) but with stronger ringing effect. When $n \to \infty$, the Butterworth filter becomes an ideal filter. The Butterworth filter with $n = 4$ and its effect are shown respectively in the sixth and seventh rows of Fig. 5.9.

- The *Gaussian filter* can be defined in either the frequency or time domain as (3.170):

$$H(f) = e^{-a(f/f_c)^2} \quad \text{or} \quad h(t) = f_c\sqrt{\pi/a}\, e^{-(\pi f_c t)^2/a}, \tag{5.85}$$

where $a = \ln 2/2 = 0.347$ so that at the cutoff frequency we have $H(f_c) = H(0)/\sqrt{2} = 1/\sqrt{2}$. Obviously, the Gaussian filter is smooth in both the time and frequency domains without any ringing effect. The Gaussian filter and its effect are shown respectively in the eighth and ninth rows of Fig. 5.9.

Inspecting the filtered signal in Fig. 5.9, we see that the sharp corners of the ideal filter corresponding to some high-frequency components are smoothed out by all other types of LP filter, consequently, the undesirable ringing artifact is much reduced. However, the tradeoff is the sacrifice of the accuracy in defining the passing band in the frequency domain, as a smooth LP filtering window means necessarily certain high-frequency leakage. Some other smooth filters also exist based on the cosine function, such as Hann, Hamming, and cosine windows.

As all of these filters $H(f)$ are real with zero phase $\angle H(f) = 0$ (i.e., they are special linear phase filters with zero delay $\tau_\phi = -\angle H(f)/2\pi f = 0$) only the magnitude of the signal spectrum is modified by the filtering process, while the phase remains the same:

$$|Y(f)| = |H(f)|\,|X(f)| \neq |X(f)|,$$
$$\angle Y(f) = \angle H(f) + \angle X(f) = \angle X(f). \tag{5.86}$$

Consequently, the relative positions of the frequency components remain the same, and the waveform of the signal is modified only owing to the magnitude of the filter FRF $H(f)$.

Other types of HP, BP, and BS filters can be easily derived from the LP filters considered above. Specifically, let $H_{\rm lp}(f)$ be an LP filter with $H_{\rm lp}(0) = 1$, then an HP filter can be easily obtained as

$$H_{\rm hp}(f) = 1 - H_{\rm lp}(f). \tag{5.87}$$

Also, a BP filter can be obtained as the difference between two LP filters $H_{\rm lp\,1}(f)$ and $H_{\rm lp\,2}(f)$ with their corresponding cutoff frequencies satisfying $f_1 > f_2$:

$$H_{\rm bp}(f) = H_{\rm lp\,1}(f) - H_{\rm lp\,2}(f), \tag{5.88}$$

and a BS filter is obtained simply as

$$H_{\rm bs}(f) = 1 - H_{\rm bp}(f). \tag{5.89}$$

Fig. 5.11 shows such filters based on a fourth-order Butterworth LP filter.

The discussion above is based on the assumption that the DC component corresponding to origin at zero frequency $f = 0$ is in the middle of the spectrum, while the higher frequency components are farther away on each side. However, in computational implementation of these filters, the signal and the filter are discrete and finite in both the time and frequency domains, and the DC component $X[0]$ is the first element, the leftmost element of the N-D array $[X[0], \ldots, X[N-1]]$ for the discrete spectrum, while the high-frequency components are around the middle point $n = N/2$. In other words, in order to use the filters given above as LP filters, all spectra need to be centralized as discussed in in Chapter 4 (Eq. (4.179)). Alternatively, without centralizing, an LP filter can still be used as an HP filter and vice versa.

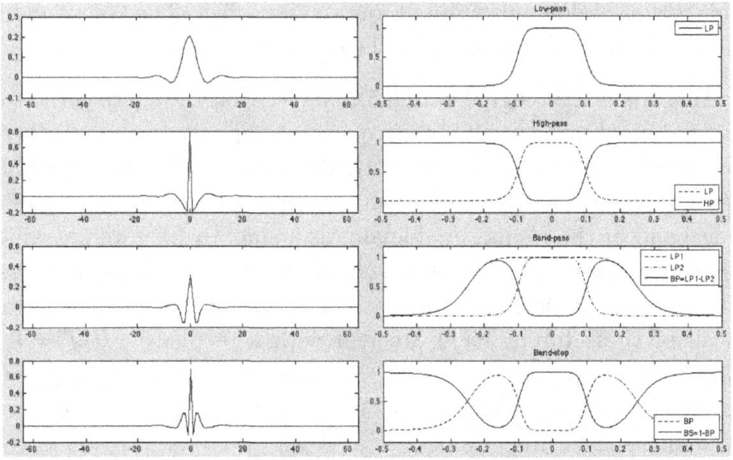

Figure 5.11 The FRFs of LP, HP, BP, and BS filters in the frequency domain (right) and their corresponding impulse response functions $h(t)$ in the time domain (left).

Note that all of these filters are even functions $H(f) = H(-f)$; i.e., both the positive and negative frequencies in $X(f)$ are modified identically by the filter. This is an important requirement of any filter in order to maintain the symmetry property of any real signal being filtered ($X_\mathrm{r}(f) = X_\mathrm{r}(-f)$, $X_\mathrm{j}(f) = -X_\mathrm{j}(f)$), so that the output signal obtained by the inverse Fourier transform of the filtered spectrum remains real. Any non-even filter will necessarily change the symmetry of the signal spectrum and thereby cause the output to be complex, which makes little sense in general.

Also note that as all filters discussed above are non-causal, as their impulse response $h(t)$ is non-zero for $t < 0$, they cannot be implemented in real time. This non-causality can be avoided if the impulse response $h(t)$ has a finite duration, or if it can be truncated without major distortion, so that when it is right shifted in time by certain amount τ it becomes causal $h(t-\tau) = 0$ for $t < 0$. Correspondingly, the filtered signal is delayed by τ. For example, if we delay the moving-average filter by $\tau = w/2$; i.e.,

$$h(t - w/2) = \begin{cases} 1 & 0 < t < w \\ 0 & \text{else} \end{cases}. \tag{5.90}$$

This delayed version of the average filter is causal and realizable in real time. Of course these non-causal filtering can all be implemented off-line when all data samples are available and can, therefore, be arbitrarily manipulated.

Example 5.6: An audio signal (Handel's "Hallelujah Chorus") and its Fourier spectrum are shown on the right and left in the top row of Fig. 5.12. Then the signal is filtered in the frequency domain by an ideal LP filter and four subsequent ideal BP filters with progressively higher passing frequency bands, as shown in

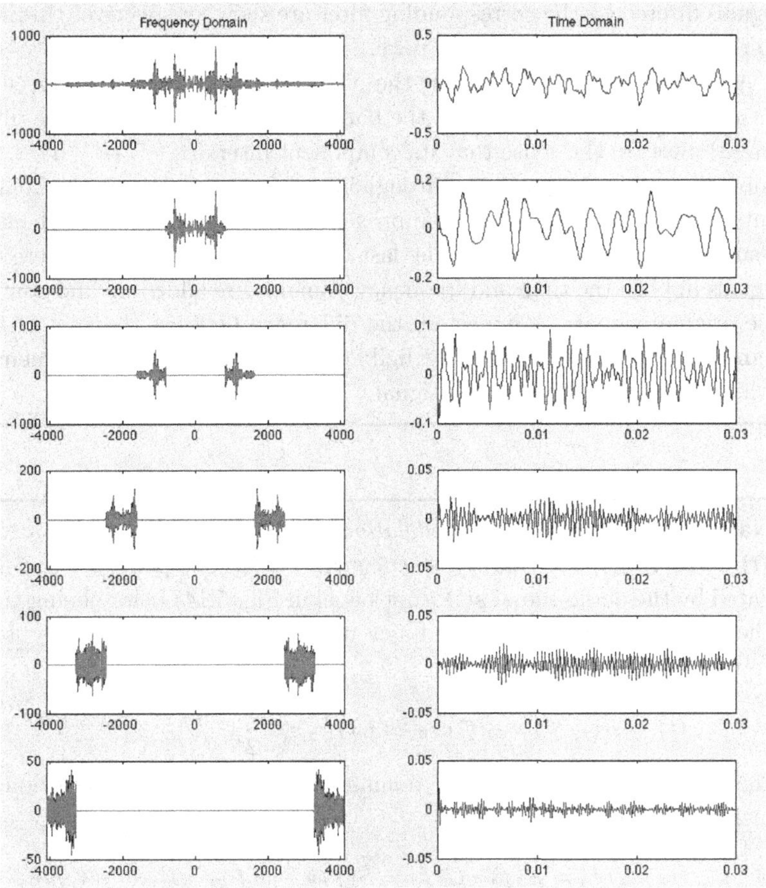

Figure 5.12 The filtering of an audio signal in the frequency domain (left) and time domain (right). From top down: original signal, LP filtered, and BP filtered by a sequence of five filters with progressively higher passing bands.

the panels on the left in Fig. 5.12, with a short piece of the corresponding filtered signal shown on the right. This filtering process can be easily implemented in Matlab while the filtered signal can be played so that the actual filtering effect in terms of the sound quality can be heard.

Example 5.7: The annual precipitation in Los Angeles area in the $N = 126$ years from 1878 to 2003, treated as a discrete time signal $x[n]$, and the DFT spectrum $X[k]$ are shown in the top row of Fig. 5.13. Here, the average of the data is removed (i.e., the DC component in the middle of the spectrum is zero), so that other frequency components with much smaller magnitudes can be better seen. Four Butterworth filters, including an LP filter and three BP filters with different passing bands, are shown in the second, fourth, sixth, and eighth rows, while the signals filtered by the corresponding filter are shown respectively in the following third, fifth, seventh, and ninth rows.

A *filter bank* can be formed by these four filters. Owing to the specific arrangement of the passing bands and the bandwidths of these filters, the filter bank is an AP filter, in the sense that its component filters $H_k(f)$ $(k = 1, \ldots, 4)$ add up approximately to a constant throughout all frequencies; i.e., the combined outputs of the filter bank contain approximately all information in the signal. This result is further confirmed by the last (tenth) row in Fig. 5.13 where the filtered signals in both the time and frequency domains are added up and compared with the original signal. As expected, the difference between the sum of the filtered signal and the original one is negligible; i.e., the filtered signals, when combined, contain all information in the signal.

Example 5.8: In *amplitude modulation (AM)* radio broadcasting, a *carrier wave* $c(t) = \cos(2\pi f_c t)$ with *radio frequency (RF)* f_c (second panel in Fig. 5.14) is modulated by the audio signal $s(t)$ (first panel in Fig. 5.14) before being transmitted. The modulation is implemented as a multiplication carried out by a modulator (mixer):

$$x(t) = s(t)\; c(t) = s(t)\cos(2\pi f_c t) = s(t)\frac{1}{2}[e^{j2\pi f_c t} + e^{-j2\pi f_c t}]. \tag{5.91}$$

This multiplication in the time domain corresponds to a convolution in the frequency domain:

$$\begin{aligned} X(f) &= S(f) * C(f) = S(f) * \frac{1}{2}[\delta(f - f_c) + \delta(f + f_c)] \\ &= \frac{1}{2}[S(f - f_c) + S(f + f_c)]. \end{aligned} \tag{5.92}$$

This AM signal in the frequency domain is shown in the third panel in Fig. 5.14. Let $f_m \ll f_c$ be the highest frequency contained in the signal; i.e., $S(f) = 0$ for $|f| > f_m$, then the bandwidth occupied by the AM signal is $\triangle f = 2f_m$ ($f_c \pm f_m$ and $-f_c \pm f_m$). The AM signal is transmitted and then received by a radio receiver, where the audio signal is separated from the carrier wave by a *demodulation* process, which is essentially implemented by another multiplication (fourth panel of Fig. 5.14):

$$y(t) = x(t)\; \cos(2\pi f_c t) = s(t)\cos^2(2\pi f_c t) = \frac{s(t)}{2} + \frac{s(t)\cos(4\pi f_c t)}{2}. \tag{5.93}$$

To obtain the audio signal $s(t)$, an LP filter is used to remove the higher frequency components centered around $\pm 2f_c$, while the audio signal centered around the origin $f = 0$ is further amplified and then sent to the speaker.

This process of both modulation and demodulation in the frequency domain is illustrated in Fig. 5.14 for an artificial signal with a triangular spectrum and also in Fig. 5.16 for a real music signal.

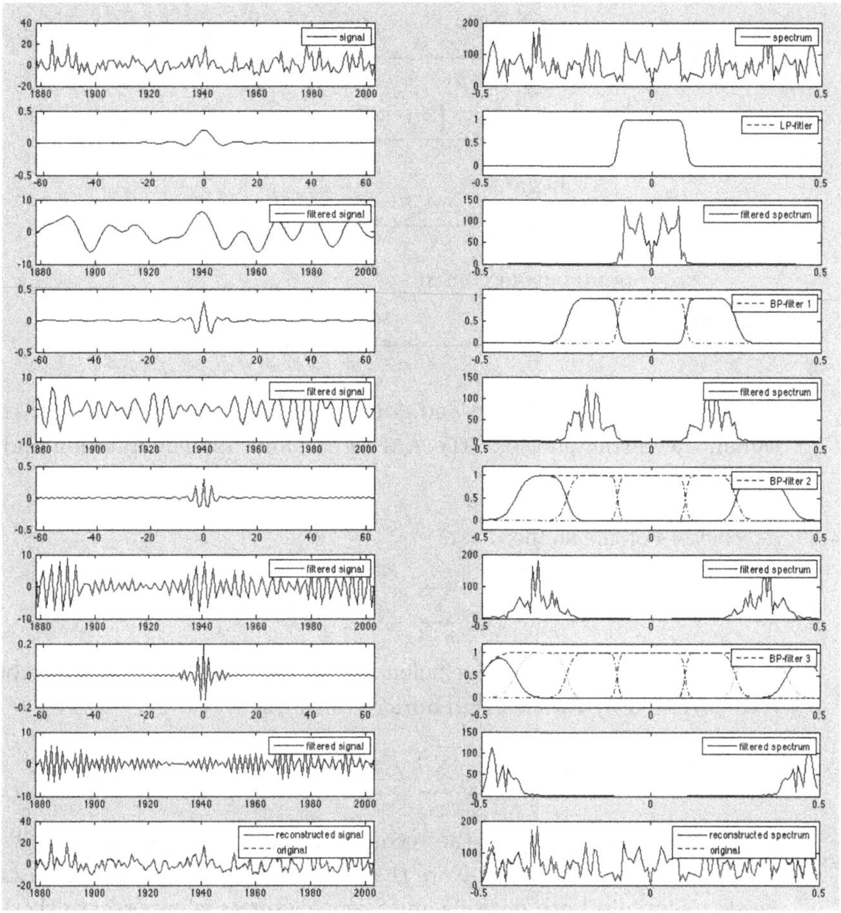

Figure 5.13 Annual precipitation from 1878 to 2003 (left) and its spectrum (right). Here, only the magnitude of each spectrum is shown while the phase is neglected. For each filter, the impulse response function $h(f)$ and the filtered time signal $y(t) = h(t) * x(t)$ are shown on the left, while the frequency response function $H(f)$ and the filtered signal spectrum $Y(f) = H(f)X(f)$ are shown on the right. The dashed curves in the plots on the right show all previous filters and their partial sum.

Example 5.9: A two-dimensional shape in an image can be described by all the pixels along its boundary, in terms of their coordinates $(x[n], y[n])$, $(n = 1, \dots, N)$, where N is the total number of pixels along the boundary. The coordinates $x[n]$ and $y[n]$ can be treated, respectively, as the real and imaginary components of a complex number $z[n] = x[n] + j\,y[n]$, and the Fourier transform can be carried out to obtain the Fourier coefficients, called the *Fourier*

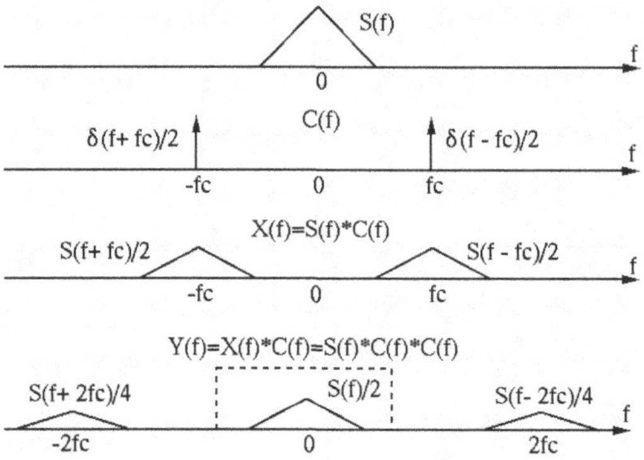

Figure 5.14 AM modulation and demodulation. In top-down order: the audio signal, the carrier sinusoid, the AM signal, and its demodulation and lowpass filtering.

descriptors of the shape:

$$Z[k] = \frac{1}{\sqrt{N}} \sum_{n=1}^{N} z[n] e^{-j2\pi nk/N} \qquad k = 1, \ldots, N. \tag{5.94}$$

Based on all N of these coefficients $Z[k]$, the original shape can be perfectly reconstructed by the inverse Fourier transform:

$$z[n] = \frac{1}{\sqrt{N}} \sum_{k=1}^{N} Z[k] e^{j2\pi nk/N} \qquad n = 1, \ldots, N. \tag{5.95}$$

It is interesting to observe the reconstructed shape using only the first $M < N$ low-frequency components. Note that the inverse transform with M components needs to contain both positive and negative terms symmetric to the DC component in the middle:

$$\hat{z}[n] = \sum_{k=-M/2}^{M/2} Z[k] e^{j2\pi nk/N} \qquad n = 1, \ldots, N. \tag{5.96}$$

As an example, the shape of Gumby in Fig. 5.15 is represented by a chain of $N = 1157$ pixels along the boundary, in terms of their coordinates $\{x[n], y[n]\}$ ($n = 0, 1, \ldots, N-1$), which are then Fourier transformed to get the same number of Fourier coefficients as the Fourier descriptors of the figure.

Figure 5.15 Gumby (left) and its boundary pixels (right).

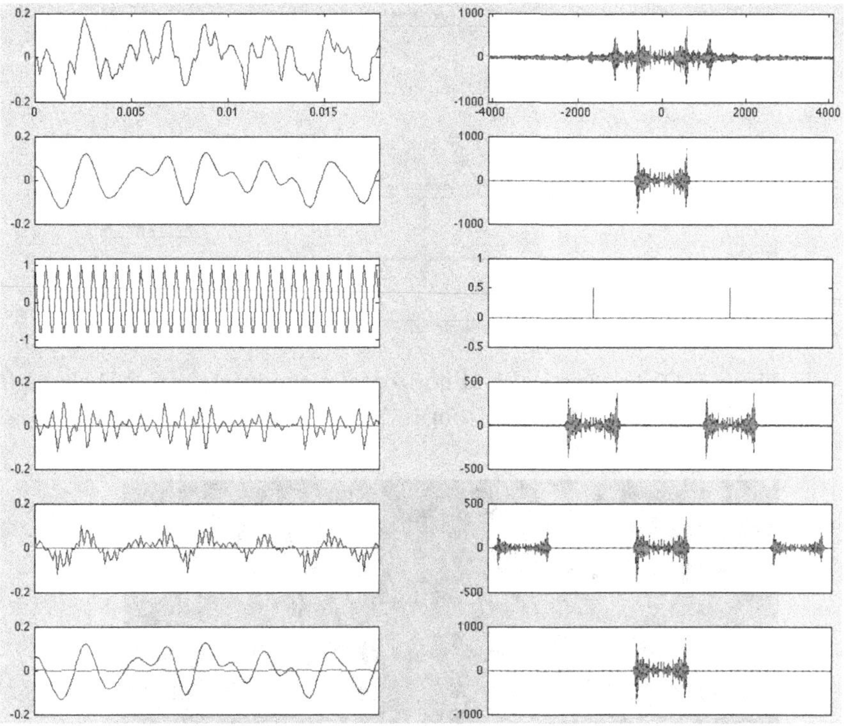

Figure 5.16 AM modulation and demodulation of an audio signal. The signals are shown on the left and their spectra are on the right. From the top down: the audio signal, the LP filtered audio signal, the carrier wave, the AM modulated signal, the demodulated signal, and LP filtered signal.

The two different representations of the shape, $z[n]$ in the spatial domain and $Z[k]$ in the frequency domain, are plotted in Fig. 5.17. Note that as the magnitudes of a small number of complex coefficients for DC and some low-frequency components are much larger than the rest of the coefficients, a mapping $y = x^{0.5}$ is applied to the magnitudes of all DFT coefficients, so that those coefficients with small magnitudes do not appear to be zero in the plots. The reconstructed shapes corresponding to different M values are shown in Fig. 5.18. We see that the original shape can be almost perfectly reconstructed using only the first few tens of the frequency components. For example, the second to last figure in the bottom row reconstructed based on the first $M = 30$ components looks almost identical to the last figure based on all $N = 1257$ components, except the latter may have some very minor details of the shape, such as the sharper corners corresponding to very high-frequency components. This result shows that the remaining $N - M = 1257 - 30 = 1127$ frequency components contain little information, and can, therefore, be neglected (treated as zero) in the inverse DFT with little effect in terms of the quality of the reconstruction. Moreover, it

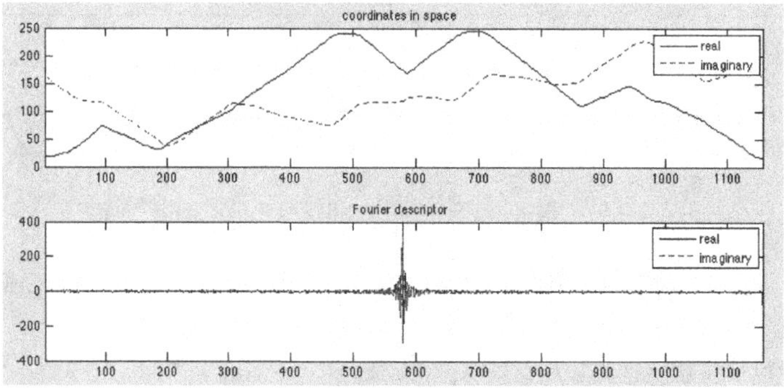

Figure 5.17 The vertical and horizontal components of a 2-D shape (top) and its Fourier descriptors (bottom).

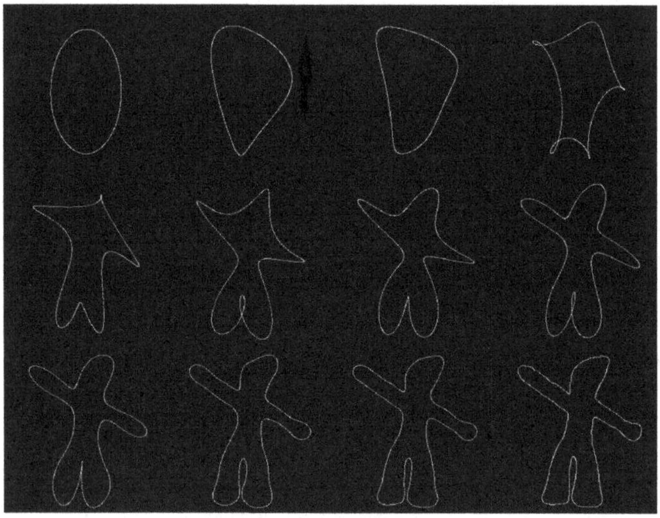

Figure 5.18 Reconstructions of Gumby based on the first M frequency components. Top: $M = 1$, 2, 3, and 4; middle: $M = 5$, 6, 7, and 8; bottom: $M = 10$, 20, 30, and $M = N = 1,257$.

may be beneficial to remove the higher frequency components anyway as they are likely to be caused by some random noise instead of the signal of interest.

Some observations can be made based on this example.

- A few coefficients corresponding to mostly low-frequency components have significantly higher magnitudes than the rest, indicating that most of the signal energy is concentrated around the low-frequency region of the spectrum. This phenomenon is common in general, owing to the fact that in most physical signals relatively slow changes over time or space are more significant than rapid and sudden changes; i.e., they tend to be continuous and smooth owing to their physical nature.

- The plots of the x- and y-coordinates in space are much smoother than the real and imaginary parts of the Fourier coefficients. Given a signal value $x[n]$ at position n, one can estimate the value $x[n+1]$ at the next position with reasonable confidence. However, this is not the case in the spatial frequency domain. The magnitudes of the DFT coefficients seem random. Given $X[k]$, one has little idea about the next value $X[k+1]$. In other words, the signal is highly correlated in the spatial domain but significantly decorrelated in the frequency domain after the Fourier transform.
- As most of the signal energy is concentrated in a small number of low-frequency components, little error will result if only $M \ll N$ of the coefficients corresponding to low frequencies are used in the inverse DFT for the reconstruction of the figure in space. Such an LP filtering may also have the effect of removing unwanted high-frequency noise.

This example illustrates some general applications of the Fourier transform; namely, information extraction and data compression. Useful features contained in a signal, such as the basic shape of a figure in an image, may be extracted by keeping a small number of the Fourier coefficients with most others ignored. It is possible to process, store, and transmit only a small portion of the data without losing much information. Moreover, the observations made here for the Fourier transform are also valid in general for all other orthogonal transforms, as we will see in later chapters.

5.5 Implementation of 2-D filtering

The filtering of a 2-D spatial signal $f(x, y)$ (e.g., an image) can be carried out in the frequency domain by multiplying its spectrum $F(u, v)$ by the FRF $H(u, v)$ of a filter:

$$G(u, v) = H(u, v)\, F(u, v). \tag{5.97}$$

The filtered spectrum can then be inverse transformed back to the spatial domain to get the filtered signal:

$$g(x, y) = \mathcal{F}^{-1}[G(u, v)]. \tag{5.98}$$

We consider below a few 2-D filters which are 2-D extensions of the 1-D filters discussed above. These filters are centrally symmetric, and all of them keep the frequency components around the central area unchanged and suppress the frequency components farther away from the center around the corners and edges of the 2-D discrete spectrum. They are LP filters if the 2-D spectrum is centralized (Eq. (4.276)) so that the DC component $F(0, 0)$ at the origin $u = v = 0$ is in the middle of the spectrum, and the distance of any frequency component $F(u, v)$ to the origin is simply $\sqrt{u^2 + v^2}$.

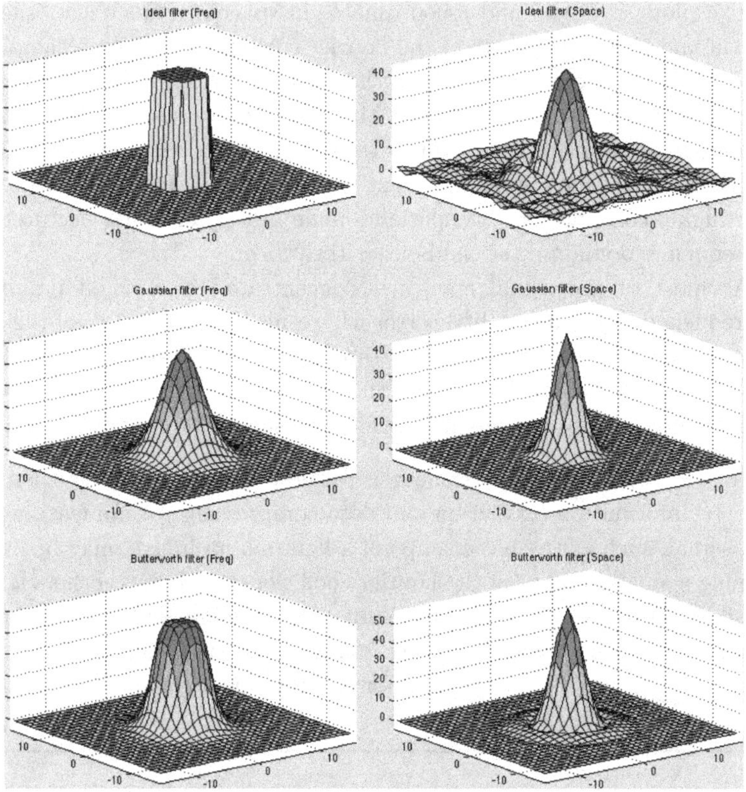

Figure 5.19 2-D filters in both the frequency (left) and spatial (right) domains. From the top down: ideal, Gaussian, and Butterworth LP filters

- **Ideal filter**

$$H_{\text{ideal}}(u,v) = \begin{cases} 1 & \sqrt{u^2+v^2} < w_c \\ 0 & \text{else} \end{cases}, \tag{5.99}$$

where w_c is the cutoff frequency. The ideal filter completely removes any frequency components outside the circle determined by the cutoff frequency. Similar to the 1-D case, some severe ringing artifacts will be caused in 2-D ideal LP filtering.

- **Gaussian filter**

$$H_{\text{Gaussian}}(u,v) = \exp[-a\,(u^2+v^2)/w_c^2], \tag{5.100}$$

where $a = \ln 2/2$ so that at the cutoff frequency $u^2+v^2 = w_c^2$, $H_{\text{Gaussian}}(u,v) = H_{\text{Gaussian}}(0,0)/\sqrt{2} = 1/\sqrt{2}$.

- **Butterworth filter**

$$H_{\text{Butterworth}}(u,v) = \frac{1}{\sqrt{1+((u^2+v^2)/w_c^2)^n}}, \tag{5.101}$$

where w_c is the cutoff frequency at which $|H(u,v)| = 1/\sqrt{2}$ (when $u^2 + v^2 = w_c^2$). When the order n of the Butterworth filter is low it is smooth, but when $n \to \infty$, the Butterworth filter approaches an ideal filter.

These filters are shown in Fig. 5.19, in terms of their impulse response functions (right) in the spatial domain and the frequency response function (left) in the frequency domains.

These filters can be readily used for HP filtering in either of two ways. First, if the spectrum is not centralized, then the high-frequency components around the middle area of the spectrum will be mostly kept unchanged while the low-frequency components farther away from the center are reduced by these filters. Alternatively, corresponding to each LP filter $H_{\text{lp}}(u,v)$ above, an HP filter can be easily obtained as $H_{\text{hp}}(u,v) = 1 - H_{\text{lp}}(u,v)$ for a centralized spectrum.

We also note that if the 2-D signal is real, the real and imaginary parts of its spectrum are respectively even and odd, and when it is filtered by any of the central symmetric filters above, the even/odd symmetry of the spectrum is to be maintained, and the filtered signal obtained by the inverse transform remains real. Any filter that fails to maintain the even/odd symmetry of the spectrum of a real signal will necessarily cause the output to be complex.

Example 5.10: Consider the Fourier transform of a 2-D signal shown in the left panel of Fig. 5.20. Here, the image of a panda is treated as the real part of a 2-D complex signal, while the imaginary part is set to zero. The real (even) and imaginary (odd) parts of the spectrum are also shown in image forms in the middle and right panel in the figure respectively.

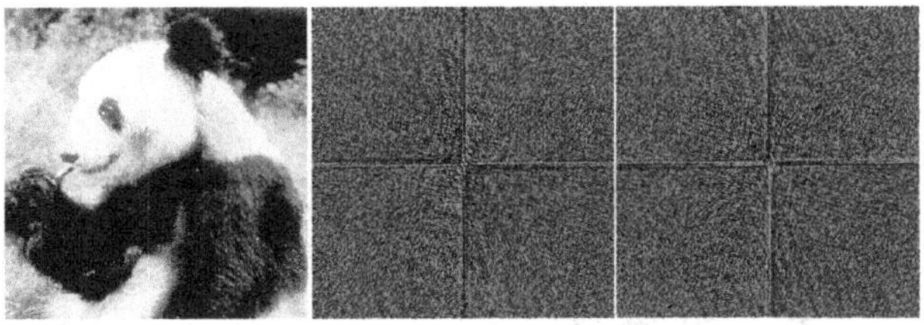

Figure 5.20 An image (left) and the real (middle) and imaginary (right) parts of its Fourier spectrum.

As the signal energy is mostly concentrated in a small number of low-frequency components around the DC component (typical for most 2-D signals), they show up as a bright spot in the middle of the centralized spectrum, while the rest of the image corresponding to higher frequency components containing little energy appears dark. In order for all frequency components to be visible, a non-linear mapping $y = x^\alpha$ ($\alpha = 0.3$ in this case) is applied to all pixel values of the image, so that the low pixel values representing frequency components of low magnitudes are relatively enhanced and become visible in the image. The spectrum can also be represented alternatively in terms of its magnitude and phase components, as shown in Fig. 5.21, where two images, one of a panda and the other a cat, together with the magnitude and phase of their corresponding spectra are shown (the first three panels of both rows).

Obviously, the real and imaginary parts of the spectrum are equally important in terms of the amount of information they each carry to represent the image signal. But are the magnitude and phase components of the spectrum also equally important in this regard? To answer this question, two images, a panda and a cat, are reconstructed based on the magnitude of the spectrum of one image but the phase of the other, as shown in the two panels on the right in Fig. 5.21, where the top image is based on the phase of the panda, and the bottom one is based on the phase of the cat. As an image so reconstructed always looks similar to the image whose phase is used in the reconstruction, it is obvious that the phase component plays a more significant and dominant role than the magnitude components. This result can be easily understood in light of the previous discussion regarding linear phase filtering. Specifically, if the relative positions of all frequency components of a signal remain unchanged by a linear phase filter, then the waveform of a signal remains the same (although they may all be delayed by the same amount of time), otherwise distortion will result if the signal is filtered by a non-linear phase filter. In other words, the phases of the frequency components are more essential in terms of maintaining the waveform of a signal, in comparison with their magnitudes.

For this reason, the real and imaginary parts $\text{Re}[X]$ and $\text{Im}[X]$ of the spectrum should always be filtered identically so that the phase angle $\angle X = \tan^{-1} \text{Im}[X]/\text{Re}[X]$ of each frequency component remains the same, and so do the relative positions of the different frequency components; thereby, the waveform of the signal is only modified by the magnitude of the filter as desired.

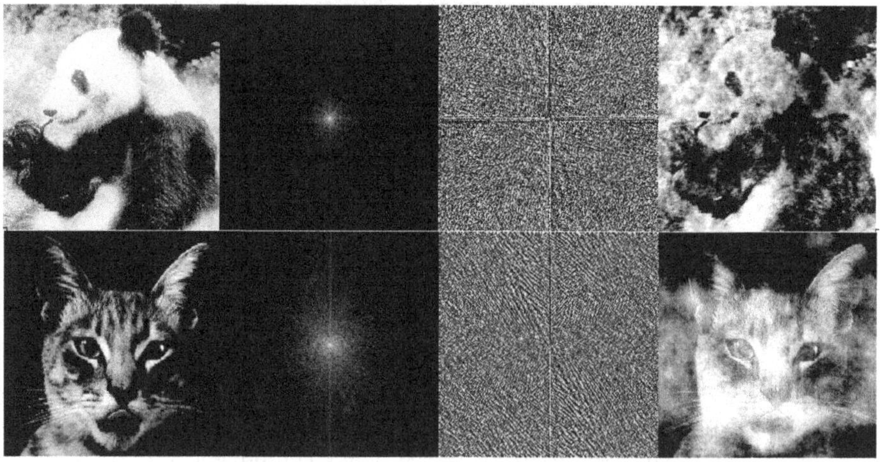

Figure 5.21 The images of panda and cat and the magnitudes and phases of their spectra. The top-right image is constructed based on the phase of the panda but magnitude of the cat, while the bottom-right image is constructed based on the phase of the cat but the magnitude of the panda.

Next we illustrate the effects of different types of filtering of the image in Fig. 5.20. First, the effects of ideal filtering are shown in Fig. 5.22. Corresponding to such filtering in the frequency domain shown in the top row, the original image in the spatial domain is convolved with a 2-D sinc function, the inverse DFT of the ideal LP filter (Eq. (4.245)), as shown in the bottom row. Note that in both LP and HP cases the filtered images have some obvious ringing artifacts caused by the convolution with the ringing sinc function. If the Butterworth filter without sharp edges is used instead, the filtered images no longer suffer from the ringing artifacts, as shown in Fig. 5.23.

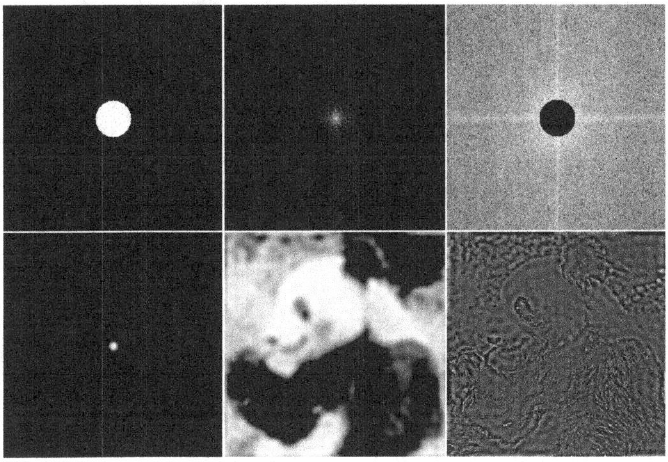

Figure 5.22 Ideal filtering of an image. This figure shows an ideal filter (left) and the LP (middle) and HP (right) filtered images. The top row shows the spectra of the filter and the filtered images in the frequency domain, while the bottom row shows the corresponding images in the spatial domain.

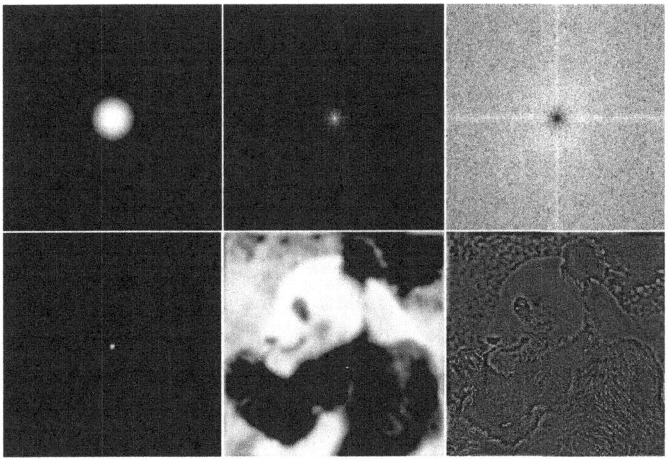

Figure 5.23 Butterworth filtering (from left to right, ideal filter, LP, and HP).

Moreover, in 2-D filtering we can also modify the coefficients for different frequency components in terms of their spatial directions and their spatial frequencies. In Fig. 5.24, the 2-D spectrum of the image of the panda is LP filtered in four different directions: N-S, NW-SE, E-W, and NE-SW (top). In the corresponding images reconstructed by the inverse transform of each directionally LP filtered spectrum (bottom), the image features in the orientation favored by the directional filtering are emphasized. Note that all of these four directional filters maintain the even/odd symmetry of the spectrum of the real image.

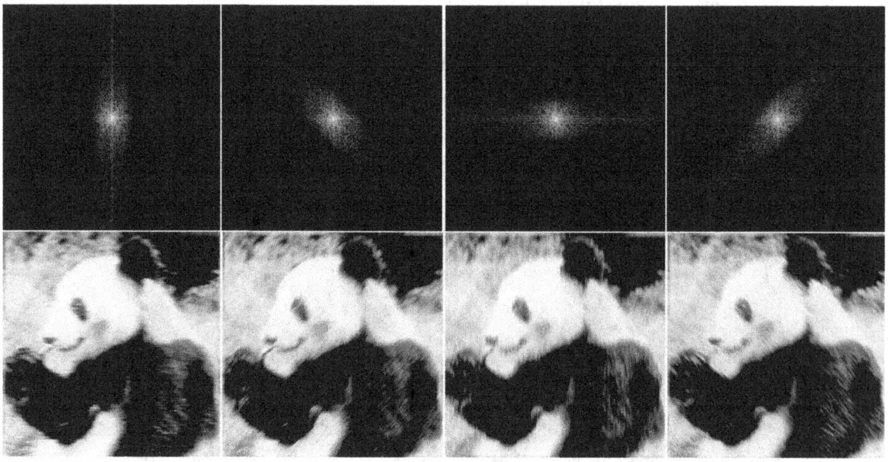

Figure 5.24 Directional LP filtering.

Finally, we show that the Fourier transform can be used for data compression as shown in Fig. 5.25. After 80% of the DFT coefficients with magnitudes less than a certain threshold value (corresponding mostly to high-frequency components) are suppressed to zero (upper right panel), the image is reconstructed based on the remaining 20% of the coefficients still containing over 99% of the signal energy (lower right panel). We see that the reconstructed image looks very much the same as the original one except that some very fine details (e.g., the fur on the left arm) corresponding to those high-frequency components are suppressed.

Why can we throw away 80% of the coefficients but still keep over 99% of the energy in the frequency domain, while it is highly unlikely we can do so in the spatial domain? This is obviously owing to the two general properties of all orthogonal transforms: (a) decorrelation of signal components and (b) compaction of signal energy. Of course, this is an oversimplified example only to illustrate the basic ideas of transform-based data compression. In practice, there are some other aspects in a compression process, such as the quantization and encoding of frequency components. The interested reader can do some further reading about image compression standards, such as the JPEG (Joint Photographic Experts Group) image compression standard.

Figure 5.25 Image compression based on DFT. An image (lower left) and its 2-D DFT spectrum (upper left), together with the reconstructed image (lower right) based on 20% of its DFT coefficients containing 99% of the total energy (upper right).

Example 5.11: In image recognition a set of 2-D image patterns are to be identified or recognized. For example, in optical character recognition (OCR), the ten digits (from 0 to 9) or the 26 English alphabets (from A to Z) are to be automatically recognized by some pattern recognition (PR) algorithm. These patterns can be represented in the original image form, or, to gain certain advantages in the DFT domain, they can also be represented by their 2-D DFT coefficients. For example, as shown in the previous example, image data can be significantly compressed in the DFT domain, and the computational complexity of the subsequent pattern recognition can be much reduced. Shown in Fig. 5.26 are the ten digits and their DFT spectra, both before and after the compression based on LP filtering. We see that the digits can be approximately represented by a small fraction of their DFT coefficients. Also, translational invariance, a desired feature in image recognition, can be conveniently achieved in the DFT domain. Specifically, owing to the shift property of the Fourier transform; i.e., only the phase of the spectrum of a shifted signal will be changed while its magnitude remains the same, a 2-D pattern represented in the DFT domain can be identified independent of its spatial position in the image. For example, comparing the spectra of two images of the same digit 5 of different locations in the image as shown in Fig. 5.27, we note that the real and imaginary parts of the two spectra are different from each other, but their magnitudes are the same, independent of the spatial translation of the image pattern.

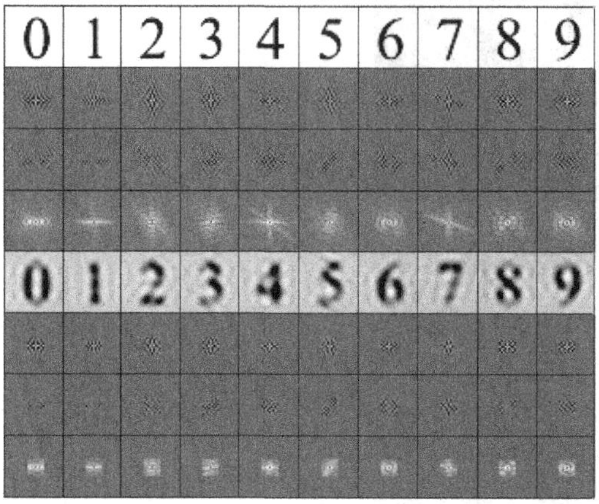

Figure 5.26 The ten digits and their DFT (top) and their compressed versions (8% compression rate) based on ideal LP-filtering (bottom). The real and imaginary parts of the DFT spectra are shown in the second and third rows, together with their magnitudes shown in the bottom row.

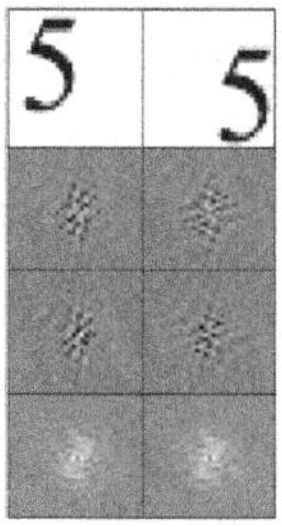

Figure 5.27 Translational invariance in DFT domain. The real and imaginary parts of the 2-D DFT of the same digit 5 but located in different positions are different, but their magnitudes (bottom) are the same.

5.6 Hilbert transform and analytic signals

The *Hilbert transform* of a time function $x(t)$ is another time function, denoted by $\hat{x}(t)$, defined as the following convolution with $1/\pi t$:

$$\mathcal{H}[x(t)] = \hat{x}(t) = x(t) * \frac{1}{\pi t} = \frac{1}{\pi}\int_{-\infty}^{\infty} \frac{x(\tau)}{t-\tau}\,d\tau = \frac{1}{\pi}\int_{-\infty}^{\infty} \frac{x(t-\tau)}{\tau}\,d\tau. \quad (5.102)$$

As the integrand is not integrable owing to its pole at $\tau = 0$, the integral of the Hilbert transform is defined in the sense of the *Cauchy principal value* of the integral as

$$\mathcal{H}[x(t)] = \frac{1}{\pi} \lim_{\epsilon \to 0} \left[\int_{-\infty}^{-\epsilon} \frac{x(t-\tau)}{\tau} d\tau + \int_{\epsilon}^{\infty} \frac{x(t-\tau)}{\tau} d\tau \right]. \tag{5.103}$$

In particular, if $x(t) = c$ is a constant, the integrand becomes an odd function c/τ, and the two integrals are negations of each other; i.e., the Hilbert transform of a constant is zero.

The Hilbert transform can be more conveniently studied in the frequency domain as a multiplication corresponding to the time convolution in Eq. (5.102). First, to find the spectrum of $1/\pi t$, we apply the property of time-frequency duality to the Fourier transform of the sign function sgn(t) (Eq. (3.84)) and get

$$\mathcal{F}\left(\frac{1}{\pi t}\right) = -j \ \mathrm{sgn}(f) = -j \begin{cases} -1 & f < 0 \\ 0 & f = 0 \\ 1 & f > 0 \end{cases} = \begin{cases} j & f < 0 \\ 0 & f = 0 \\ -j & f > 0 \end{cases}. \tag{5.104}$$

Now the Hilbert transform $\hat{x}(t) = x(t) * 1/\pi t$ can be expressed in the frequency domain as a multiplication:

$$\hat{X}(f) = \mathcal{F}[\hat{x}(t)] = [-j \ \mathrm{sgn}(f)] \ X(f) = \begin{cases} jX(f) & f < 0 \\ 0 & f = 0 \\ -jX(f) & f > 0 \end{cases}. \tag{5.105}$$

The effect of the Hilbert transform applied to a signal $x(t)$ becomes clear: it multiplies the negative part of the signal spectrum $X(f)$ by $j = e^{j\pi/2}$ (a rotation by an angle of $\pi/2$ in complex plane) and the positive part by $-j = e^{-j\pi/2}$ (a rotation by an angle of $-\pi/2$). Therefore the Hilbert transform is also called a *quadrature filter.*

As the Hilbert transform of a time function is still a time function, it can be applied to a signal $x(t)$ multiple times, and the result is most conveniently obtained in the frequency domain:

$$\mathcal{F}[\mathcal{H}^n[x(t)]] = [-j \ \mathrm{sgn}(f)]^n X(f). \tag{5.106}$$

In particular, as $\mathrm{sgn}^2(f) = 1$, we have

$$[-j \ \mathrm{sgn}(f)]^2 = -1, \qquad [-j \ \mathrm{sgn}(f)]^3 = j \ \mathrm{sgn}(f), \qquad [-j \ \mathrm{sgn}(f)]^4 = 1. \tag{5.107}$$

Correspondingly in the time domain, we have

$$\mathcal{H}[x(t)] = \hat{x}(t), \quad \mathcal{H}^2[x(t)] = -x(t), \quad \mathcal{H}^3[x(t)] = -\hat{x}(t), \quad \mathcal{H}^4[x(t)] = x(t). \tag{5.108}$$

We see that applying the Hilbert transform to $x(t)$ once we get $\mathcal{H}[x(t)] = \hat{x}(t)$, and applying the transform three more times we get the original signal back, which is actually the inverse Hilbert transform:

$$\begin{cases} \mathcal{H}[x(t)] = x(t) * 1/\pi t = \hat{x}(t) \\ \mathcal{H}^{-1}[\hat{x}(t)] = \mathcal{H}^3[\hat{x}(t)] = -\mathcal{H}[\hat{x}(t)] = x(t) \end{cases}. \tag{5.109}$$

Example 5.12: When the Hilbert transform is applied to this simple sinusoid

$$\cos(2\pi f_0 t) = \frac{1}{2} e^{j2\pi f_0 t} + \frac{1}{2} e^{-j2\pi f_0 t}. \tag{5.110}$$

the coefficient $1/2$ for $f < 0$ is rotated by 90° to become $e^{j\pi/2}/2 = -1/2j$ while the other coefficient $1/2$ for $f > 0$ is rotated by -90° to become $e^{-j\pi/2}/2 = 1/2j$, and the transformed signal becomes

$$\mathcal{H}[\cos(2\pi f t)] = \frac{1}{2j} e^{j2\pi f_0 t} - \frac{1}{2j} e^{-j2\pi f_0 t} = \sin(2\pi f t). \tag{5.111}$$

Similarly, we have $\mathcal{H}[\sin(2\pi f t)] = -\cos(2\pi f t)$, $\mathcal{H}[-\cos(2\pi f t)] = -\sin(2\pi f t)$ and $\mathcal{H}[-\sin(2\pi f t)] = \cos(2\pi f t)$.

Next we consider the concept of analytic signals. A real-valued signal $x_{\rm a}(f)$ is said to be *analytic* if its Fourier spectrum $X_{\rm a}(f) = \mathcal{F}[x_{\rm a}(t)]$ is zero when $f < 0$. Any signal $x(t)$ can be turned into an analytic signal by multiplying its spectrum $X(f) = \mathcal{F}[x(t)]$ with a step function $2u(f)$ in the frequency domain:

$$X_{\rm a}(f) = X(f) 2u(f) = \begin{cases} 0 & f < 0 \\ X(0) & f = 0 \\ 2X(f) & f > 0 \end{cases}. \tag{5.112}$$

Applying the time-frequency duality to the Fourier transform of the unit step in Eq. (3.80) we get the inverse Fourier transform of the unit step spectrum $u(f)$:

$$\mathcal{F}^{-1}[u(f)] = \frac{1}{-j2\pi t} + \frac{1}{2}\delta(-t) = \frac{j}{2\pi t} + \frac{1}{2}\delta(t). \tag{5.113}$$

The analytic signal can then be obtained by taking the inverse Fourier transform on both sides of Eq. (5.112):

$$\begin{aligned} x_{\rm a}(t) &= \mathcal{F}^{-1}[X_{\rm a}(f)] = \mathcal{F}^{-1}[X(f)] * \mathcal{F}^{-1}[2u(f)] = x(t) * \left[\delta(t) + \frac{j}{\pi t}\right] \\ &= x(t) + j\, x(t) * \frac{1}{\pi t} = x(t) + j\, \hat{x}(t). \end{aligned} \tag{5.114}$$

Alternatively, an analytic signal can also be initially defined in the time domain by Eq. (5.114). Taking the Fourier transform on both sides, we get

$$X_{\rm a}(f) = X(f) + j\,\hat{X}(f) = X(f) + j \begin{cases} jX(f) & f < 0 \\ 0 & f = 0 \\ -jX(f) & f > 0 \end{cases} = \begin{cases} 0 & f < 0 \\ X(0) & f = 0 \\ 2X(f) & f > 0 \end{cases}, \tag{5.115}$$

where $\hat{X}(f) = \mathcal{F}[\hat{x}(t)]$.

If the signal $x(t)$ is real, its spectrum satisfies $X(f) = \overline{X}(-f)$, indicating the corresponding analytic signal $x_a(t) = x(t) + j\,\hat{x}(t)$ contains the complete information in $x(t)$, even though the negative half of its spectrum is suppressed to zero. In fact, the original spectrum $X(f)$ can also be reconstructed from $X_a(f)$. When $f > 0$, obviously we get $X(f) = X_a(f)/2$, when $f < 0$, we have

$$X(f) = \overline{X}(-f) = \overline{X}(|f|) = \frac{1}{2}\overline{X}_a(|f|). \tag{5.116}$$

Combining these two cases, we have

$$X(f) = \frac{1}{2}\begin{cases} X_a(f) & f > 0 \\ \overline{X}_a(|f|) & f < 0 \end{cases} = \frac{X_a(f) + \overline{X}_a(-f)}{2}, \tag{5.117}$$

the second equality is due to the fact that $\overline{X}_a(-f) = 0$ when $f > 0$ and $X_a(f) = 0$ when $f < 0$.

Example 5.13: In Example 5.8 concerning the AM modulation and demodulation, the bandwidth $\triangle f = 2f_m$ is twice the highest frequency f_m contained in the signal, one sideband of f_m on each side of the carrier frequency f_c (double sideband). In order to efficiently use the broadcast spectrum as a limited resource, it is desirable to minimize the bandwidth needed for each broadcast transmission. The *single-sideband (SSB) modulation* is such a method by which the bandwidth is reduced by half (from $2f_m$ to f_m). One implementation of the SSB modulation is based on the Hilbert transform and analytic signals, taking advantage of the fact that the negative half of the spectrum of an analytic signal is always zero and, therefore, does not need to be transmitted. Specifically, an analytic signal is first constructed based on the signal $s(t)$ to be transmitted:

$$s_a(t) = s(t) + j\,\hat{s}(t), \tag{5.118}$$

where $\hat{s}(t) = \mathcal{H}[x(t)]$ is the Hilbert transform of $s(t)$. Then $s_a(t)$ is used to modulate a carrier frequency represented as a complex exponential $e^{j2\pi f_c}$. The real part of the resulting AM signal $s_a(t)e^{j2\pi f_c t}$ is then transmitted:

$$\begin{aligned} x(t) &= \text{Re}[s_a(t)e^{j2\pi f_c t}] = \text{Re}[(s(t) + j\,\hat{s}(t))\,(\cos(2\pi f_c t) + j\sin(2\pi f_c t))] \\ &= s(t)\cos(2\pi f_c t) - \hat{s}(t)\sin(2\pi f_c t) = x_0(t) - x_1(t), \end{aligned} \tag{5.119}$$

where $x_0(t) = s(t)\cos(2\pi f_c t)$ and $x_1(t) = \hat{s}(t)\sin(2\pi f_c t)$ are two modulated RF signals with 90° phase difference. The block diagram of the SSB modulation is illustrated in Fig. 5.28. In the frequency domain Eq. (5.119) becomes

$$\begin{aligned} X(f) &= X_0(f) - X_1(f) \\ &= S(f) * \frac{1}{2}[\delta(f - f_c) + \delta(f + f_c)] - \hat{S}(f) * \frac{1}{2j}[\delta(f - f_c) - \delta(f + f_c)] \\ &= \frac{1}{2}[S(f - f_c) + S(f + f_c) + j\hat{S}(f - f_c) - j\hat{S}(f + f_c)]. \end{aligned} \tag{5.120}$$

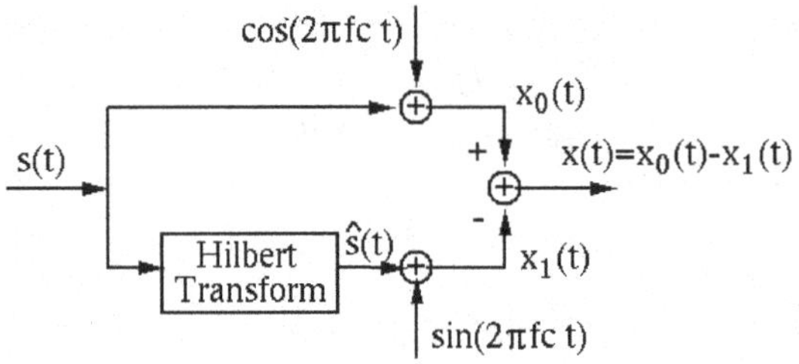

Figure 5.28 SSB modulation using Hilbert transform.

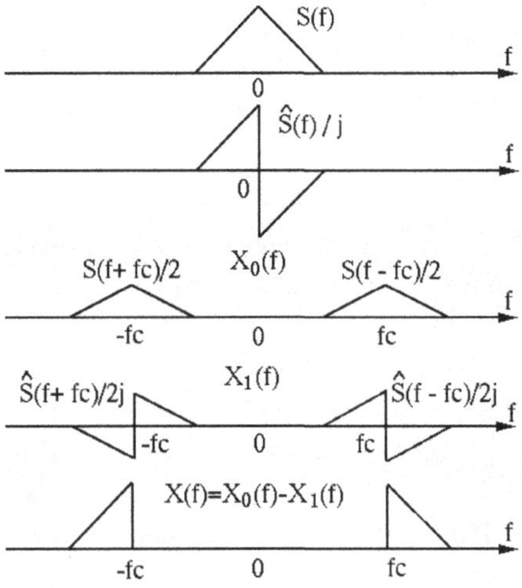

Figure 5.29 The SSB modulation in the frequency domain.

Note that $\hat{S}(f - f_c)$ and $\hat{S}(f + f_c)$ are related to $S(f - f_c)$ and $S(f + f_c)$ by Eq. (5.105), and in two of the following four cases, they cancel each other:

$$\begin{array}{l} f + f_c < 0 : \hat{S}(f + f_c) = jS(f + f_c), \;\; X(f + f_c) = 2S(f + f_c), \\ f + f_c > 0 : \hat{S}(f + f_c) = -jS(f + f_c), \;\; X(f + f_c) = 0, \\ f - f_c < 0 : \hat{S}(f - f_c) = jS(f - f_c), \;\; X(f - f_c) = 0, \\ f - f_c > 0 : \hat{S}(f - f_c) = -jS(f - f_c), \;\; X(f - f_c) = 2S(f - f_c). \end{array} \tag{5.121}$$

The spectra of the signals in the process are shown in Fig. 5.29, from which we see that the bandwidth of this modulated signal $x(t)$ is indeed reduced by half. The SSB modulation is carried out on a real music signal as shown in Fig. 5.30, where the signal and its spectrum at various stages of the process are shown on the left and right respectively.

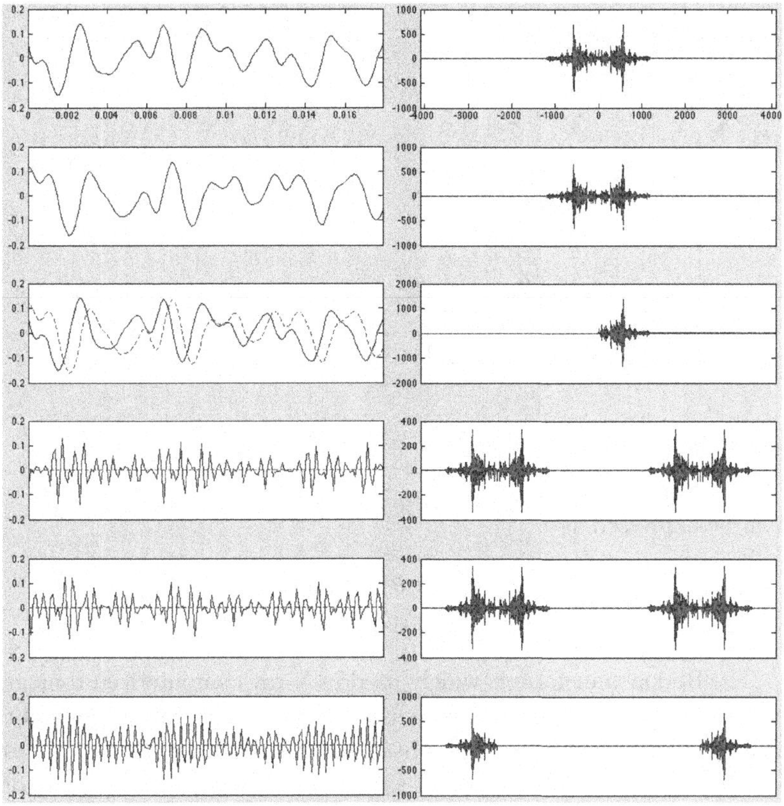

Figure 5.30 The SSB modulation of a music signal. In top-down order: the original signal $s(t)$, its Hilbert transform $\hat{s}(t)$, the corresponding analytic signal $s_a(t) = s(t) + j\hat{s}(t)$ (whose spectrum $S_a(f) = 0$ for $f < 0$), AM modulation of $x_0(t)$ and $x_1(t)$, SSB-modulated $x(t) = x_0(t) - x_1(t)$.

5.7 Radon transform and image restoration from projections

Like the Fourier transform, the Radon transform is also an integral transform, as illustrated in Fig. 5.31, that integrates a 2-D function $f(x, y)$ along a straight line $L(\theta)$ specified by an angle θ (measured from the positive direction of x). The resulting 1-D function $g_\theta(s)$ of s, the distance between the origin and line $L(\theta)$, is in fact the projection of $f(x, y)$ onto a straight line in the direction of s. In particular, if the direction is along either x or y (corresponding to $\theta = 0$ or $\theta = \pi/2$), we get

$$g(y) = \int_{-\infty}^{\infty} f(x, y)\, dx \quad \text{or} \quad g(x) = \int_{-\infty}^{\infty} f(x, y)\, dy. \tag{5.122}$$

The projections along all different directions θ can be considered as a 2-D function $g(s, \theta)$, from which the original 2-D function $f(x, y)$ can be reconstructed by the inverse Radon transform. This forward and inverse Radon transform pair

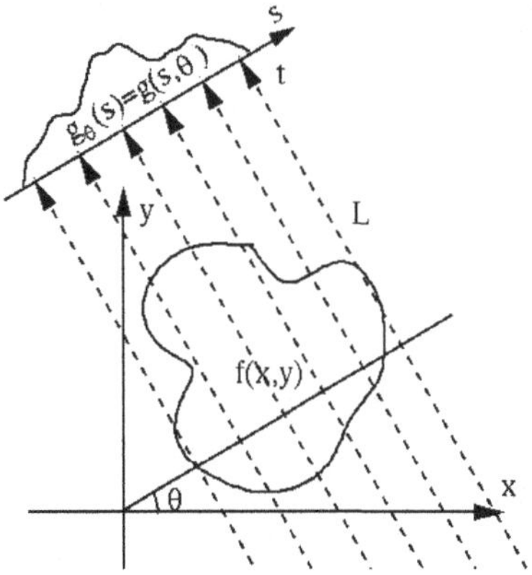

Figure 5.31 Radon transform.

can be expressed as

$$\begin{cases} g(s,\theta) = \mathcal{R}[f(x,y)] \\ f(x,y) = \mathcal{R}^{-1}[g(s,\theta)] \end{cases}. \tag{5.123}$$

The Radon transform is widely used in X-ray computerized tomography (CT) to get the image of a cross-section, a slice, of a certain part of the body. Moreover, a 3-D volume of data can be obtained as a sequence of such slices along the direction perpendicular to cross-sections. Let I_0 denote the intensity of the source X-ray and $f(x,y)$ denote the absorption coefficient of the tissue at position (x,y). The detected signal intensity I can be obtained according to this simple model:

$$I = I_0\, e^{-\int_{L(\theta)} f(x,y)\,dt}. \tag{5.124}$$

Here, t is the integral variable along the pathway $L(\theta)$ of the X-ray through the tissue. The exponent, the absorption coefficient integrated along $L(\theta)$, is just the Radon transform $g(s,\theta)$ of $f(x,y)$, which can be obtained given the detected I:

$$g(s,\theta) = \int_{L(\theta)} f(x,y)\,dt = \ln\,(I_0/I), \tag{5.125}$$

and the cross-section $f(x,y)$ representing the tissue absorption coefficient can then be obtained by the inverse Radon transform.

Now let us further formulate the Radon transform. The straight line $L(\theta)$ along which the projection of a 2-D function is obtained can be specified by the following equation (i.e., any point (x,y) on $L(\theta)$ satisfies the equation):

$$x\cos\theta + y\sin\theta - s = 0, \tag{5.126}$$

with two parameters s and θ, as shown in Fig. 5.32 (left). Now the 1-D integral along $L(\theta)$ of the Radon transform in Eq. (5.125) can be written as the following

2-D integral:

$$g(s,\theta) = \mathcal{R}[f(x,y)] = \int\int_{-\infty}^{\infty} f(x,y)\delta(x\cos\theta + y\sin\theta - s)\, dx\, dy$$
$$(-\infty < s < \infty,\quad 0 \le \theta < 2\pi), \tag{5.127}$$

which converts the 2-D spatial function $f(x,y)$ into a function $g(s,\theta)$ in a 2-D parameter space.

Next we define a new coordinate system (s,t) in the 2-D space by rotating the (x,y) coordinate system by an angle θ:

$$\begin{cases} s = x\cos\theta + y\sin\theta \\ t = -x\sin\theta + y\cos\theta \end{cases} \quad \text{or} \quad \begin{cases} x = s\cos\theta - t\sin\theta \\ y = s\sin\theta + t\cos\theta \end{cases}, \tag{5.128}$$

where t is the coordinate along the direction of the projection line $L(\theta)$, perpendicular to the direction of s. Note that this rotation is a unitary transformation which conserves vector norm; i.e., $x^2 + y^2 = s^2 + t^2$. In the new (s,t) coordinate system, the Radon transform can be expressed as a 1-D integral along the direction of t:

$$g(s,\theta) = \mathcal{R}[f(x,y)] = \int_{-\infty}^{\infty} f(s\cos\theta - t\sin\theta, s\sin\theta + t\cos\theta)\, dt. \tag{5.129}$$

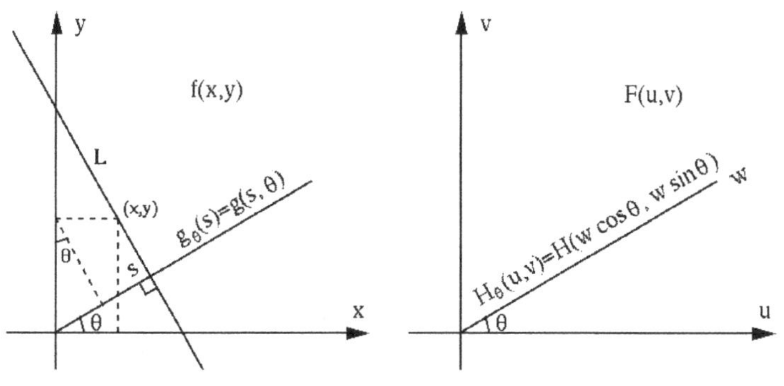

Figure 5.32 Radon transform and projection-slice theorem.

Example 5.14: First consider the Radon transform of a 2-D Gaussian function $f(x,y) = e^{-(x^2+y^2)} = e^{-(s^2+t^2)}$:

$$g(s,\theta) = \int_{-\infty}^{\infty} e^{-(s^2+t^2)}\, dt = e^{-s^2}\int_{-\infty}^{\infty} e^{-t^2}\, dt = \sqrt{\pi}\, e^{-s^2}. \tag{5.130}$$

We see that $g(s,\theta)$ is a 1-D Gaussian function of s, independent of θ, as a 2-D Gaussian function is central symmetric.

Next consider the Radon transform of a plane wave

$$f(x,y) = \cos[2\pi(2x+3y)] = \frac{1}{2}[e^{j2\pi(2x+3y)} + e^{-j2\pi(2x+3y)}], \tag{5.131}$$

which propagates along the direction of $\phi = \tan^{-1}(3/2)$ (with respect to the horizontal direction). As the Radon transform is obviously linear, we can find the transforms of $e^{j2\pi(2x+3y)}$ and $e^{-j2\pi(2x+3y)}$ separately. The first term can be expressed in terms of the rotated coordinate system (s,t)

$$\begin{aligned} e^{j2\pi(2x+3y)} &= e^{j2\pi 2x} e^{j2\pi 3y} = e^{j2\pi[2(s\cos\theta - t\sin\theta)]} e^{j2\pi[3(s\sin\theta + t\cos\theta)]} \\ &= e^{j2\pi 2(2\cos\theta+3\sin\theta)} e^{j2\pi t(-2\sin\theta+3\cos\theta)}. \end{aligned}$$

Its Radon transform is

$$\begin{aligned} \mathcal{G}[e^{j2\pi(2x+3y)}] &= e^{j2\pi s(2\cos\theta+3\sin\theta)} \int_{-\infty}^{\infty} e^{j2\pi t(-2\sin\theta+3\cos\theta)}\, dt \\ &= e^{j2\pi s(2\cos\theta+3\sin\theta)} \delta(-2\sin\theta + 3\cos\theta). \end{aligned} \tag{5.132}$$

Similarly, we can get

$$\mathcal{G}[e^{-j2\pi(2x+3y)}] = e^{-j2\pi s(2\cos\theta+3\sin\theta)} \delta(2\sin\theta - 3\cos\theta). \tag{5.133}$$

Adding these two results we get

$$\mathcal{G}[\cos[2\pi(2x+3y)]] = \cos[2\pi s(2\cos\theta + 3\sin\theta)]\delta(2\sin\theta - 3\cos\theta). \tag{5.134}$$

We see that this Radon transform is zero except when $2\sin\theta = 3\cos\theta$ or $\theta = \tan^{-1}(3/2) = \phi$; i.e., the straight line $L(\theta)$ for the Radon transform is perpendicular to the propagation direction of the plane wave. In this case the Radon transform is a delta function (due to the infinite integral of a constant along the direction of $L(\theta)$), weighted by a sinusoidal function of s along the direction of propagation. When $\theta \neq \phi$, the integrand in Eq. (5.132) along $L(\theta)$ is a sinusoid with frequency $3\cos\theta - 2\sin\theta$, and the infinite integral is always zero.

Projection-slice theorem: The 1-D Fourier transform of the Radon transform $g(s,\theta) = \mathcal{R}[f(x,y)]$ with respect to s (with θ treated as a parameter) is equal to the slice of the 2-D Fourier transform $F(u,v) = \mathcal{F}[f(x,y)]$ through the origin along the direction θ:

$$G(w,\theta) = \mathcal{F}[g(s,\theta)] = F_\theta(u,v), \tag{5.135}$$

where $F_\theta(u,v)$ denotes a slice of $F(u,v)$ through the origin along direction θ.

Proof: First find the 1-D Fourier transform of the Radon transform $g(s,\theta) = \mathcal{R}[f(x,y)]$ with respect to s:

$$G(w,\theta) = \mathcal{F}[g(s,\theta)] = \int_{-\infty}^{\infty} g(s,\theta) e^{-j2\pi ws}\, ds, \tag{5.136}$$

where w is the spatial frequency of $f(x, y)$ along the direction of s. Substituting the expression of $g(s, \theta)$ in Eq. (5.127) into the above equation, we get

$$\begin{aligned} G(w, \theta) &= \int_{-\infty}^{\infty} \left[\int\int_{-\infty}^{\infty} f(x, y)\delta(x\cos\theta + y\sin\theta - s)\, dx\, dy \right] e^{-j2\pi ws}\, ds \\ &= \int\int_{-\infty}^{\infty} f(x, y) \left[\int_{-\infty}^{\infty} \delta(x\cos\theta + y\sin\theta - s)\, e^{-j2\pi ws}\, ds \right] dx\, dy \\ &= \int\int_{-\infty}^{\infty} f(x, y)\, e^{-j2\pi w(x\cos\theta + y\sin\theta)}\, dx\, dy \\ &= F(w\cos\theta,\ w\sin\theta) = F_\theta(u, v), \end{aligned}$$

where

$$\begin{cases} u = w\cos\theta \\ v = w\sin\theta \end{cases} \quad \text{or} \quad \begin{cases} w = \sqrt{u^2 + v^2} \\ \theta = \tan^{-1}(v/u) \end{cases}, \tag{5.137}$$

and $F(w\cos\theta,\ w\sin\theta) = F_\theta(u, v)$ is the 2-D Fourier transform $F(u, v)$ of the signal $f(x, y)$ evaluated at $u = w\cos\theta$ and $v = w\sin(\theta)$, along the direction of θ.

Inverse Radon theorem: Given its Radon transform $g(s, \theta)$, the original 2-D signal $f(x, y)$ can be reconstructed by

$$f(x, y) = \mathcal{R}^{-1}[g(s, \theta)] = \frac{1}{2\pi^2} \int_0^{\pi} \int_{-\infty}^{\infty} \left[\frac{\partial}{\partial s} g(s, \theta) \right] \frac{1}{x\cos\theta + y\sin\theta - s}\, ds\, d\theta, \tag{5.138}$$

or in polar form:

$$f(r, \phi) = \frac{1}{2\pi^2} \int_0^{\pi} \int_{-\infty}^{\infty} \left[\frac{\partial}{\partial s} g(s, \theta) \right] \frac{1}{r\cos(\phi - \theta) - s}\, ds\, d\theta, \tag{5.139}$$

where

$$\begin{cases} x = r\cos\phi \\ y = r\sin\phi \end{cases}, \qquad \begin{cases} r = \sqrt{x^2 + y^2} \\ \phi = \tan^{-1}(y/x) \end{cases}. \tag{5.140}$$

Proof: Based on Eq. (5.137), the Fourier spectrum $F(u, v)$ can be written in polar form as $F(w, \theta)$ and the inverse transform $f(x, y) = \mathcal{F}^{-1}[F(u, v)]$ becomes

$$\begin{aligned} f(x, y) &= \int_{-\infty}^{\infty} \int_{-\infty}^{\infty} F(u, v) e^{j2\pi(ux + vy)}\, du\, dv \\ &= \int_0^{2\pi} \int_0^{\infty} F(w, \theta) e^{j2\pi w(x\cos\theta + y\sin\theta)}\, w\, dw\, d\theta \\ &= \int_0^{\pi} \int_{-\infty}^{\infty} F(w, \theta) e^{j2\pi w(x\cos\theta + y\sin\theta)}\, |w| dw\, d\theta. \end{aligned}$$

Here, $F(w, \theta)$ is a slice of $F(u, v)$ along the direction θ, which, according to the projection-slice theorem Eq. (5.135), is equal to the Fourier transform of the Radon transform of $f(x, y)$; i.e., $F(w, \theta) = G(w, \theta) = \mathcal{F}[g(s, \theta)]$, the equation above then becomes

$$f(x,y) = \int_0^{\pi} \left[\int_{-\infty}^{\infty} |w| G(w,\theta) e^{j2\pi w(x\cos\theta + y\sin\theta)}\, dw \right] d\theta$$
$$= \int_0^{\pi} g'(x\cos\theta + y\sin\theta, \theta)\, d\theta, \tag{5.141}$$

where $g'(s,\theta)$ is defined as the inverse Fourier transform of $|w|G(w,\theta)$:

$$g'(s,\theta) = g'(x\cos\theta + y\sin\theta, \theta)$$
$$= \int_{-\infty}^{\infty} |w| G(w,\theta) e^{j2\pi w(x\cos\theta + y\sin\theta)}\, dw = \mathcal{F}^{-1}[|w|G(w,\theta)].$$

We can consider $|w|\, G(w,\theta)$ as a filtering process of $g(s,\theta)$ by a filter $|w|$ in the frequency domain w; i.e., $g'(s,\theta)$ is the filtered version of $g(s,\theta)$ in the spatial domain s. As $|w|$ can be written as a product $|w| = w\ \mathrm{sgn}(w)$ (an HP filter), the inverse Fourier transform above for $|w|G(w,\theta) = wG(w,\theta)\,\mathrm{sgn}(w)$ becomes

$$g'(s,\theta) = \mathcal{F}^{-1}[wG(w,\theta)\ \mathrm{sgn}(w)] = \mathcal{F}^{-1}[wG(w,\theta)] * \mathcal{F}^{-1}[\mathrm{sgn}(w)]$$
$$= \left[\frac{1}{j2\pi}\frac{\partial}{\partial s} g(s,\theta)\right] * \left[\frac{1}{-j\pi s}\right] = \frac{1}{2\pi^2} \int_{-\infty}^{\infty} \left[\frac{\partial}{\partial t} g(t,\theta)\right] \frac{1}{s-t}\, dt. \tag{5.142}$$

Here, we have used the convolution theorem and also Eqs. (3.127) and (3.147) for the two inverse transforms. Comparing this expression with the definition of the Hilbert transform in Eq. (5.102), we see that $g'(s,\theta)$ is also the Hilbert transform of $\partial g(s,\theta)/\partial s/2\pi$:

$$g'(s,\theta) = \mathcal{H}\left[\frac{1}{2\pi}\frac{\partial}{\partial s} g(s,\theta)\right]. \tag{5.143}$$

Substituting Eq. (5.142) back into Eq. (5.141) for $f(x,y)$, we get

$$f(x,y) = \frac{1}{2\pi^2} \int_0^{\pi} \int_{-\infty}^{\infty} \left[\frac{\partial}{\partial t} g(t,\theta)\right] \frac{1}{s-t}\, dt\, d\theta. \tag{5.144}$$

Replacing s by $x\cos\theta + y\sin\theta$, we get Eq. (5.138). Q.E.D.

In practice, the inverse Radon transform can be carried out based on Eq. (5.141), instead of Eq. (5.138) or Eq. (5.139), in the following steps:

1. Fourier transform of $g(s,\theta)$ with respect to s for all directions θ:

$$G(w,\theta) = \mathcal{F}[g(s,\theta)]. \tag{5.145}$$

2. Filtering in the frequency domain by $|w|$:

$$G'(w,\theta) = |w|G(w,\theta). \tag{5.146}$$

3. Inverse Fourier transform:

$$g'(s,\theta) = \mathcal{F}^{-1}[G'(w,\theta)]. \tag{5.147}$$

4. Summation of $g'(x\cos\theta + y\sin\theta, \theta)$ over all directions θ (called "back projection"):

$$f(x,y) = \int_0^\pi g'(s,\theta)\,d\theta = \int_0^\pi g'(x\ \cos\theta + y\ \sin\theta, \theta)\,d\theta. \tag{5.148}$$

As the higher frequency components of most signals contain little energy and are more susceptible to noise (lower signal-to-noise ratio), the HP filter $|w|$ that is likely to amplify noise in the signal is typically modified so that its magnitude is reduced in the high-frequency range.

Example 5.15: Consider the Radon transform, both the forward transform for projection and the inverse transform for reconstruction, of two 2-D signals, a shape in a black-and-white image and a gray-scale image, as shown on the left in Fig. 5.33. In each of the two cases, we obtain the projections $g(s,\theta)$ (second from left) of all 180 angles, 1 apart, of the image $f(x,y)$, and then reconstruct the image, first without filtering to produce a blurred reconstruction (third from left), and then with HP filtering by $|w|$ to produce an almost perfect reconstruction (right).

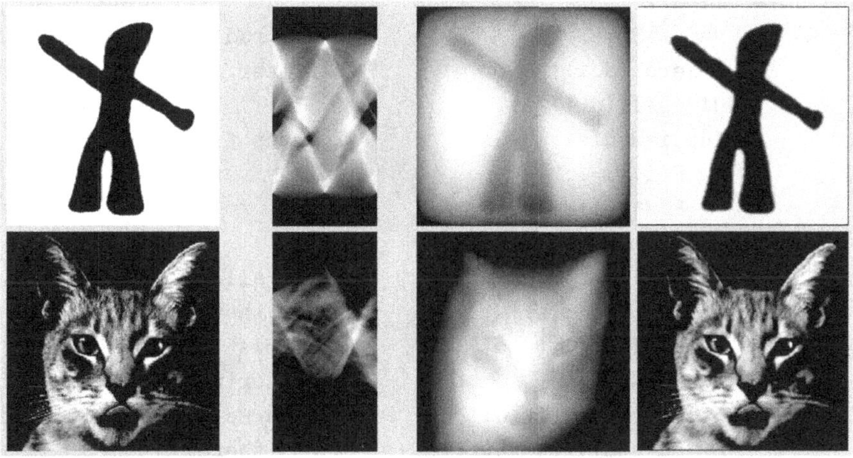

Figure 5.33 The forward and inverse Radon transform. From left to right: original image $f(x,y)$, Radon projections $g(s,\theta)$, back projection without filtering, back projection with filtering.

The Matlab code for both the forward and inverse Radon transforms is listed below. The projection directions are given in vector theta in degrees.

```
function proj = Radon(im,theta) % forward Radon transform
    K=length(theta);              % number of projection directions
```

```
    [m,n]=size(im);                  % size of image
    d=fix(sqrt(2)*max(m,n));         % diagonal of image
    tmp=zeros(d);                    % size of projection, d=1.414*n
    i=(d-m)/2;
    j=(d-n)/2;
    tmp(i:i+m-1,j:j+n-1)=im;         % copy input image to tmp
    proj=zeros(d,K);                 % K projections of length d
    for k=1:K                        % for all directions
        a=theta(k);                  % rotation angle
        proj(:,k)=sum(imrotate(tmp,a,'bilinear','crop'));
                                     % image rotation and projection
    end
end

function im=iRadon(proj,theta)   % inverse Radon transform
    [d,K]=size(proj);                % diagonal of image
    n=ceil(d/sqrt(2));               % size of image
    im=zeros(n);
    n2=n/2;
    d2=d/2;
    v=pi/180;                        % for radian/degree conversion
    F=zeros(d,1);                    % filter in the frequency domain
    d1=ceil((d-1)/2);
    for i=2:d1+1;                    % setup filter
       F(i)=i-1;
       F(d+2-i)=i-1;
    end
    for k=1:K                        % for all directions
        g=proj(:,k);                 % g(s,theta)
        G=fft(g);                    % Fourier transform of g
        G=G.*F;                      % filtering in the frequency domain
        g=real(ifft(G));             % inverse Fourier transform
        c=cos(v*theta(k));           % cos(theta)
        s=sin(v*theta(k));           % sin(theta)
        for i=1:n
           for j=1:n                 % for all pixels in image
                y=i-n2;
                x=j-n2;              % image center is at origin
                t=fix(x*c+y*s)+d2;
                im(i,j)=im(i,j)+g(t); % back projection
           end
        end
    end
end
```

5.8 Orthogonal frequency-division modulation (OFDM)

A digital communication system transmits and receives messages consisting of a finite number of symbols representing various types of information. Consider the transmission of a block of N_c complex numbers (symbols) $\{d_1, d_2, \ldots, d_{N_c}\}$, where $d_k = a_k + jb_k$, during a time interval of T seconds, which can be carried out in either of the following two ways:

- **Serial transmission**
  Represent each symbol by a unique waveform over a time interval of $T_s = T/N_c$, called *symbol time*, and transmit the waveforms sequentially.
- **Parallel transmission**
  Represent each symbol by a unique waveform over the entire time interval $T_s = T$ as the symbol time and sum the N_c waveforms representing the group of N_c symbols for parallel transmission. The individual waveforms must then be separated at the receiver to recover all N_c symbols.

Note that for either method the transmission rate is $R = N_c/T$ symbols per second.

Many of today's wireless communication systems operate in an environment where signals are reflected from a variety of objects, such as buildings and walls, on their way from transmitter to receiver. This means that the signal at the receiver is the sum of a number of copies of the transmitted signal with various delays and attenuations (referred to as multi-path). As the maximum delay increases from a small fraction of the symbol interval, inter-symbol interference increases with a consequent increase in the probability of error at the receiver. Thus, the parallel transmission of a group of symbols is advantageous owing to the longer symbol time $T_s = T$.

We now consider implementation of the parallel transmission by the *orthogonal frequency division multiplexing (OFDM)* method. Specifically, we use each of the N_C orthogonal sinusoids $e^{j2\pi kt/T} = \cos(2\pi kt/T) + j\sin(2\pi kt/T)$ of different frequencies $kf_0 = k/T$ $(k = 1, \ldots, N_c)$ to represent one of the N_c complex values d_k, so that the signal to be transmitted is a linear combination of these sinusoids weighted by d_k:

$$x(t) = \text{Re}\left[\sum_{k=1}^{N_c} d_k e^{j2\pi kt/T}\right] = \sum_{k=1}^{N_c}[a_k \cos(2\pi kt/T) - b_k \sin(2\pi kt/T)]$$
$$0 \le t < T = 1/f_0. \tag{5.149}$$

This continuous signal is then discretized by sampling at an interval T/N or sampling rate $F_s = N/T$ to become

$$x[n] = x(nT/N) = \sum_{k=1}^{N_c}[a_k \cos(2\pi kn/N) - b_k \sin(2\pi kn/N)]$$
$$n = 0, \ldots, N-1. \tag{5.150}$$

Note that if the sampling rate $F_s = N/T$ is higher than twice the maximum frequency component $N_c f_0 = N_c/T$ in the signal; i.e., $F_s = N/T > 2N_c f_0 = 2N_c/T$; i.e., $N > 2N_c$, then the Nyquist condition is satisfied and the continuous signal $x(t)$ can be reconstructed by a D/A converter from the N samples $x[0], x[1], \ldots, x[N-1]$.

The transmission of the N_c symbols can be carried out in the following steps:

1. Generate N samples $x[n]$ ($n = 0, \ldots, N-1$) based on the N_c complex values d_k ($k = 1, \ldots, N_c$) as in Eq. (5.150).
2. Transmit $x[n]$ through the digital communication channel.
3. Reconstruct $x(t)$ from the received $x[n]$ by D/A conversion.
4. Separate $x(t)$ to recover the N_c symbols d_k.

Some analog circuits are necessary to generate the signal $x[n]$ in step 1 above. However, we now show that such a hardware requirement can be avoided, as $x[n]$ can be completely generated by the following digital signal processing approach. First we construct the following vector of $N = 2(N_c + 1)$ elements:

$$\begin{aligned}&[Y[0], Y[1], \ldots, Y[N_c], Y[N_c+1], Y[N_c+2], \ldots, Y[2N_c+1]] \\ &= [0, d_1, \ldots, d_{N_c}, 0, \overline{d}_{N_c}, \ldots, \overline{d}_1],\end{aligned} \tag{5.151}$$

and then carry out the inverse DFT to get

$$\begin{aligned}y[n] &= \mathcal{F}^{-1}[Y[k]] = \sum_{k=0}^{N-1} Y[k] e^{j2\pi nk/N} \\ &= \sum_{k=1}^{N_c} d_k e^{j2\pi kn/N} + \sum_{k=N_c+2}^{2N_c+1} \overline{d}_{2N_c+2-k} e^{j2\pi kn/N} \\ &\qquad n = 0, \ldots, N-1 = 2N_c + 1.\end{aligned} \tag{5.152}$$

We let $m = N - k = 2N_c + 2 - k$ (i.e., $k = N - m = 2N_c + 2 - m$) so that the second summation becomes

$$\sum_{m=N_c}^{1} \overline{d}_m e^{-j2\pi mn/N} \underbrace{e^{jN(2\pi/N)n}}_{1} = \sum_{m=1}^{N_c} \overline{d}_m e^{-j2\pi mn/N}. \tag{5.153}$$

Now $y[n]$ above can be further written as

$$\begin{aligned}y[n] &= \sum_{k=1}^{N_c} [d_k e^{j2\pi kn/N} + \overline{d}_k e^{-j2\pi kn/N}] \\ &= \sum_{k=1}^{N_c} [(a_k + jb_k) e^{j2\pi kn/N} + (a_k - jb_k) e^{-j2\pi kn/N}] \\ &= 2\sum_{k=1}^{N_c} \left[a_k \left(\frac{e^{j2\pi kn/N} + e^{-j2\pi kn/N}}{2} \right) - b_k \left(\frac{e^{j2\pi kn/N} - e^{-j2\pi kn/N}}{2j} \right) \right]\end{aligned}$$

$$= 2\sum_{k=1}^{N_c}[a_k \cos(2\pi kn/N) - b_k \sin(2\pi kn/N)] = 2x[n] \qquad n = 0, \ldots, N-1, \tag{5.154}$$

which happens to be the signal we need to generate in step 1 above. After this signal is transmitted and then received, we can carry out the DFT to get

$$Y[k] = \frac{1}{N}\sum_{n=0}^{N-1} y[n]e^{-j2\pi kn/N} \qquad k = 0, \ldots, N-1. \tag{5.155}$$

The N_c original symbols d_k carried in the signal $x[n]$ can then be easily recovered as $d_k = Y[k]$ $(k = 1, \ldots, N_c)$ according to Eq. 5.151.

5.9 Homework problems

Some of the problems below can be carried out in Matlab (or any other programming language of choice).

1. Assume a real LTI system $h(t) = \overline{h}(t)$; re-derive Eqs. (5.12) and (5.13) by applying Eq. (5.11) to the following:

$$\mathcal{O}[\cos(2\pi ft)] = \frac{1}{2}\mathcal{O}[e^{j2\pi ft}] + \mathcal{O}[e^{-j2\pi ft}], \tag{5.156}$$

   and

$$\mathcal{O}[\sin(2\pi ft)] = \frac{1}{2j}\mathcal{O}[e^{j2\pi ft}] - \mathcal{O}[e^{-j2\pi ft}]. \tag{5.157}$$

2. Find and sketch the response $y(t)$ of the system in Example 5.2 to each of the following inputs $x(t)$.
   - The input $x(t)$ is a square impulse:

$$x(t) = \begin{cases} 1 & 0 \le t < \triangle \\ 0 & \triangle \le t; \end{cases} . \tag{5.158}$$

     Find $y(t)$ for $0 \le t < \triangle$, $t = \triangle$, and $\triangle < t$. Assume zero initial condition $y(0) = 0$.
   - The input $x(t + T) = x(t)$ has a period $T > \triangle$ for all $-\infty < t < \infty$. Find $y(0)$, $y(\triangle)$, $y(T)$, and $y(t)$ for $0 \le t < \triangle$ and $\triangle < t < T$.
     Hint: The output $y(t + T) = y(t)$ is also periodic, in particular, $y(0) = y(T)$.
3. Find and sketch the Bode plots, including both the log-magnitude (Lm) plot of $\text{Lm}\, H(\omega)$ and phase plot $\angle H(\omega)$ versus $log_{10}\omega$, of the following frequency response functions of some typical LTI systems.
   (a) Constant gain $H(\omega) = k$ (consider both cases $k > 0$ and $k < 0$).
   (b) Derivative factor $H(\omega) = j\omega\tau$. What is the slope of the Lm plot in terms of dB/dec (decibel per 10-fold frequency increase).

(c) Integral factor $H(\omega) = 1/j\omega\tau$. What is the slope of the Lm plot?
(d) First-order factor in numerator $H(\omega) = 1 + j\omega\tau$. First give the general expressions of $\text{Lm}\, H(\omega)$ and $\angle H(\omega)$. Then consider the following three special cases:
  * $\omega\tau = 1$; i.e., $\omega = 1/\tau$.
  * $\omega\tau \ll 1$, find the asymptote of both Lm and phase plots.
  * $\omega\tau \gg 1$, find the asymptote of both Lm and phase plots. What is the slope of the Lm plot?

  Sketch the complete plots by combining the three cases.
(e) First-order factor in denominator $H(\omega) = 1/(1 + j\omega\tau)$.

4. Consider the same RC circuit in Example 5.2 (Fig. 5.4), with an input voltage $x(t) = v_{\text{in}}(t)$ across the two components in series, but the output $y(t) = v_R(t)$ is the voltage across the resistor R, instead of across the capacitor C.
The impulse response of this system can be most easily obtained based on the result of Example 5.2 and Kirchhoff's voltage law stating $v_{\text{in}}(t) = v_C(t) + v_R(t)$; i.e.,

$$v_R(t) = v_{\text{in}}(t) - v_C(t) = \delta(t) - \frac{1}{\tau} e^{-t/\tau} u(t). \tag{5.159}$$

However, let us not use the previous result; instead, we solve this system independently by the following steps.
  - Set up the differential equation of the system.
  - Find the impulse response function $h(t)$ in two methods when $x(t) = \delta(t)$:
    (a) $v_R(t) = v_{\text{in}}(t) - v_C(t)$. When $v_{\text{in}}(t) = \delta$, $v_R(t) = h(t)$ and $v_C(t)$ is obtained in Example 5.2.
    (b) Solve the differential equation for $y'(t) = f(t)u(t)$ when $x'(t) = u(t)$. Then find $h(t) = y(t) = \dot{y}'(t)$ corresponding to $x(t) = \dot{x}(t) = \delta(t)$.
  - Find the frequency response function $H(f)$ by assuming $x(t) = e^{j\omega t}$.
  - Verify that $H(\omega) = \mathcal{F}[h(t)]$.

5. Which of the following three FRFs is a linear filter?
(a) A first-order system is $H_1(\omega) = 1/(1 + j\omega\tau)$ where $\tau = 0.01$.
(b) $H_2(\omega) = |H_1(\omega)| e^{-j\omega\tau\pi/4}$.
(c) $H_2(\omega) = |H_1(\omega)| e^{-j\pi/4}$.
Plot the phase $\angle H_i(\omega)$ $(i = 1, 2, 3)$ of each of these FRFs as a function of ω. For each of the systems $H_i(\omega)$, find its response $y_i(t)$ to an input $x(t) = \cos(50t) + \cos(200t)$, and then plot both $y(t)$ and $x(t)$.

6. The FRF $H(\omega)$ of a given LTI system can be plotted in different ways:
  - The plot of the real and imaginary parts $\text{Re}[H(\omega)]$ and $\text{Im}[H(\omega)]$ as a function of frequency ω.
  - The plot of the magnitude $|H(\omega)|$ and phase $\angle H(\omega)$ as a function of frequency ω.
  - The Bode plot of both the log-magnitude of the gain $\text{Lm}\, H(\omega) = 20 \log_{10} |H(\omega)|$ and the phase $\angle H(\omega)$ over the logarithmic scale of ω.

- The Nyquist diagram of the real and imaginary parts in the complex plane (or the gain and phase as a polar plot).

Write a Matlab function that generates all four plots of any given FRF. Then plot the FRFs for the following systems.

(a) Plot the FRF obtained in Example 5.2 (Eq. (5.47)) with $\tau = 1/2\pi 1000$. In the Bode plot, identify the cutoff frequency ω_c and find the values of $\text{Lm}\, H(\omega_c)$ and $\angle H(\omega_c)$. In the Nyquist diagram, identify the points corresponding to $\omega = -\infty$, $\omega = 0$, and $\omega = \infty$, as well as $\omega = \omega_c$, and find the values of the gain $|H(\omega)|$ and phase $\angle H(\omega)$ at each of these frequencies. Repeat the above for the FRF of the RC circuit in the previous problem with the same τ.

(b) Plot the three second-order FRFs in the following form:

$$H(\omega) = \frac{N(\omega)}{(j\omega)^2 + 2\zeta\omega_n j\omega + \omega_n^2} \tag{5.160}$$

where the numerator $N(\omega)$ takes the form of each one of the three terms of the denominator $D(\omega)$, and $\zeta = 0.05$, $\omega_n = 2\pi 1000$ are two system parameters.

(c) Plot the following FRF:

$$H(\omega) = \frac{(j\omega)^2 + 5j\omega + 2}{3(j\omega)^2 + 2j\omega + 1} \tag{5.161}$$

Note that Matlab has the built-in functions `bode` for the Bode plot and `nyquist` for the Nyquist diagram. However it is still worth the effort to write your own functions for these plots, in order to thoroughly understand how they are realized. Verify your Bode plots and Nyquist diagrams with those generated by the Matlab functions.

Hint: use the frequency range $0 \le \omega \le \omega_{\max}$ for your Bode plot and $-\omega_{\max} \le \omega \le \omega_{\max}$ for your Nyquist diagram, where $\omega_{\max} = 2\pi 10^4$ approximates $\omega = \infty$. You could use frequency increment of $\triangle\omega = 0.1$.

7. Implement various filtering schemes to filter an input sound signal.

(a) Load a sound file such as a piece of music. For example, `load Handel` in Matlab will load the first 9 seconds of "Hallelujah Chorus" by Handel into a vector `y` with sampling rate in variable `Fs`.

(b) Implement a set of five BP filters, both ideal and Butterworth, that divide the entire frequency range into five frequency bands, so that the first filter (LP) passes only low-frequency components including DC, the next BP filter passes the frequency components in the next higher frequency band, etc., until the last filter (HP) passes all high-frequency components in the last band including the highest frequency component contained in the signal. Note that your filters need to cover negative frequencies as well as positive ones.

(c) Listen to the output from each of these filters to experience the different filtering effects. Compare the filtering effects of the ideal and Butterworth filters of different orders. (In Matlab, to listen to a signal in vector y, do `play(audioplayer(y,Fs))`.)

(d) Repeat the above with BP filters replaced by BS filters.

(e) If the gains of the BP filters can be individually adjusted, they are called *equalization (EQ) filters* and used to compensate for the unequal (uneven) frequency response of the signal processing system to reduce the signal distortion and improve the sound quality. Experiment with different gains for each of the five filters in part (b) to experience different sound effects (e.g., high or low frequency component enhancement/reduction).

8. Construct an analytic signal based on (a) $x(t) = \cos(\omega_0 t)$ and (b) $y(t) = \sin(\omega_0 t)$. Verify that the negative half of the spectrum of the constructed analytic signal is zero.

9. Implement AM modulation and demodulation as discussed in Example 5.8.
   (a) Create a triangular spectrum as shown in the top panel of Fig. 5.14 with the highest frequency $f_{\rm m}$ and obtain the time signal by the inverse Fourier transform.
   (b) Carry out AM modulation (Eq. (5.91)) of the signal with a carrier frequency $f_{\rm c} > 2f_{\rm m}$; display the spectrum of the resulting signal in both the time and frequency domains.
   (c) Carry out AM demodulation (Eq. (5.93)); display the spectrum of the resulting signal in both the time and frequency domains.
   (d) Carry out an ideal LP filtering to remove all frequencies higher than $f_{\rm m}$. Display the spectrum of the resulting signal in both the time and frequency domains.
   (e) Replace the artificial signal above by a real sound signal, and repeat the steps above. You may need to LP filter the signal to make sure $f_{\rm m} < f_{\rm c}/2$. Listen to the original signal and the reconstructed signal to convince yourself it is perfectly reconstructed.

10. Implement SSB modulation discussed in Example 5.13 by following the diagram shown in Fig. 5.28.
    (a) Use the same artificial signal $x(t)$ with a triangular spectrum and obtain its Hilbert transform $\hat{x}(t)$. (In Matlab, the analytic version of a given signal vector **x** can be obtained by function `hilbert(x)`, whose imaginary part `imag(hilbert(x))` is the Hilbert transform of x.) Display both $x(t)$ and $\hat{x}(t)$ and their spectra.
    (b) Use $x(t)$ and $\hat{x}(t)$ to amplitude modulate respectively $\cos(2\pi f_{\rm c} t)$ and $\sin(2\pi f_{\rm c} t)$ with $f_{\rm c} > 2f_{\rm m}$. Display the resulting signals and their spectra.
    (c) Find the difference as given in Eq. (5.119), display the signal and its spectrum to verify that it has only one sideband.
    (d) Repeat the steps above with the artificial signal replaced by a real sound signal. You may need to LP filter the signal first to make sure $f_{\rm m} < f_{\rm c}/2$.

11. Implement image filtering and compression as discussed in Examples 5.10.

(a) Carry out a 2-D DFT of an image of your choice and display its spectrum first in terms of the real X_r and imaginary X_j parts, and then its magnitude and phase. The spectral information can be displayed as 3-D plots as well as 2-D images. Note that a non-linear mapping (such as $y = x^{0.3}$) may be needed in order to see most of the frequency components.
(b) Carry out ideal LP and HP filtering of the image and display the filters as well as the image after filtering in both the spatial and spatial frequency domains.
(c) Repeat the step above with the ideal filters replaced by the corresponding Butterworth filters.
(d) Carry out directional filtering as shown in Fig. 5.24.
(e) Carry out image compression as shown in Fig. 5.25 by suppressing to zero all frequency components lower than a certain threshold. Obtain the percentage of such suppressed frequency components, and the percentage of lost energy (in terms of signal value squared). (Note that this exercise only serves to illustrate the basic idea of image compression, but it is not how image compression is practically done, where those components suppressed need to be recorded as well.)

12. In OCR, a set of 2-D image patterns, such as the ten digits or the 26 English alphabets, are to be recognized. Here we show that these image patterns can be alternatively represented by their 2-D DFT coefficients. Moreover, representing these image patterns in DFT domain has certain advantages.
(a) Carry out the DFT of each of the 26 images for digits 0 to 9, display the real, imaginary parts of each DFT spectrum, as well as its magnitude.
(b) Use an ideal LP-filter (rectangular) to keep about 5% of the 2-D DFT coefficients corresponding to the lowest frequency components. Then carry out inverse DFT based on these frequency components and display the resulting 2-D image patterns of the ten digits. Convince yourself that 5% of the data is enough to approximately represent the original information; so that all subsequent recognition can be carried in a much reduced data set.
(c) A desired feature in image recognition is translational invariance; i.e., a pattern should be identified independent of its spatial position in the image. The time shift property of the Fourier transform, the magnitude of the spectrum is independent of translation of signal, can be used to achieve translational invariance. Compare the spectra of two images of the same pattern with different locations. Note that the real and imaginary parts of the the two spectra are different from each other, while their magnitudes are the same, independent of the spatial translation of the image pattern.

13. The m-file QPSK_OFDMTxRx.m simulates a baseband OFDM system for transmitting an ASCII file such as a text message using quadrature phase shift keying (QPSK) to represent the bit stream.

(a) Create an ASCII text file and using QPSK_OFDMTxRx.m experiment with different FFT and cyclic prefix lengths. Also try slightly mismatching the channel impulse responses at the transmitter and receiver.
(b) Write your own Matlab function x=d2x(d) that uses the inverse DFT to produce the sequence $\bar{x}$ given the sequence $\bar{d}$.
(c) Write your own Matlab function d=x2d(x) that uses the DFT to produce the sequence $\bar{d}$ given the sequence $\bar{x}$.
(d) Test your functions using the sequence $\bar{d} = (1, 2, 3, 4, 5, 6, 7, 8)$.

6 The Laplace and z-transforms

The Laplace and z-transforms are respectively the natural generalization of the CTFT and DTFT, and both find a wide variety of applications in many fields of science and engineering in general, and in signal processing and system analysis/design in particular. Owing to some of its most favorable properties, such as the conversion of ordinary differential and difference equations into easily solvable algebraic equations, a problem presented in the time domain can be much more conveniently tackled in the transform domain.

6.1 The Laplace transform

6.1.1 From Fourier transform to Laplace transform

The Laplace transform of a signal $x(t)$ can be considered as the generalization of the CTFT of the signal:

$$\mathcal{F}[x(t)] = \int_{-\infty}^{\infty} x(t)e^{-j\omega t}\,dt = X(j\omega). \tag{6.1}$$

Here, we adopt the notation $X(j\omega)$ for the CTFT spectrum, instead of $X(f)$ or $X(\omega)$ used previously, the reason for which will become clear later. The transform above is based on the underlying assumption that the signal $x(t)$ is square-integrable so that the integral converges and the spectrum $X(j\omega)$ exists. However, this assumption is not valid for signals such as $x(t) = t$, $x(t) = x^2$, and $x(t) = e^{at}$, all of which are not square-integrable, as they grow without a bound when $|t| \rightarrow \infty$. In such cases, we could still consider the Fourier transform of a modified version of the signal $x'(t) = x(t)e^{-\sigma t}$, where $e^{-\sigma t}$ is an exponential factor with a real parameter σ, which can force the given signal $x(t)$ to decay exponentially for a properly chosen value of σ (either positive or negative). For example, $x(t) = e^{at}u(t)$ $(a > 0)$ does not converge when $t \rightarrow \infty$; therefore, its Fourier spectrum does not exist. However, if we choose $\sigma > a$, the modified version $x'(t) = x(t)e^{-\sigma t} = e^{-(\sigma - a)t}u(t)$ will converge as $t \rightarrow \infty$.

In general, the Fourier transform of the modified signal is

$$\mathcal{F}[x'(t)] = \mathcal{F}[x(t)e^{-\sigma t}] = \int_{-\infty}^{\infty} x(t)e^{-(\sigma + j\omega)t}\,dt = \int_{-\infty}^{\infty} x(t)e^{-st}\,dt, \tag{6.2}$$

where we have defined a complex variable $s = \sigma + j\omega$. If the integral above converges, it results in a complex function $X(s)$, called the *bilateral Laplace transform* of $x(t)$, formally defined as

$$X(s) = \mathcal{L}[x(t)] = \mathcal{F}[x(t)e^{-\sigma t}] = \int_{-\infty}^{\infty} x(t)\phi(t,s)\,dt = \int_{-\infty}^{\infty} x(t)e^{-st}\,dt. \quad (6.3)$$

As with the CTFT, the Laplace transform can also be considered as an integral transform with a kernel function:

$$\phi(t,s) = e^{-st} = e^{-(\sigma+j\omega)t} = e^{-\sigma t}e^{-j\omega t}, \quad (6.4)$$

which is a modified version of the kernel function $\phi(t,f) = e^{j2\pi ft}$ for the Fourier transform. However, different from the parameter f for frequency in the Fourier kernel function, the parameter $s = \sigma + j\omega$ in the Laplace kernel is complex with both real and imaginary parts $\text{Re}[s] = \sigma$ and $\text{Im}[s] = \omega$, and the transform $X(s)$, a complex function, is defined in a 2-D complex plane, called the s-plane, with Cartesian coordinates of σ for the real (horizontal) axis and $j\omega$ for the imaginary (vertical) axis.

The Laplace transform $X(s)$ exists only inside a certain region of the s-plane, called the *region of convergence (ROC)*, composed of all s values that guarantee the convergence of the integral in Eq. 6.3. Owing to the introduction of the exponential decay factor $e^{-\sigma t}$, we can properly choose the parameter σ so that the Laplace transform can be applied to a broader class of signals than the Fourier transform.

If the imaginary axis $s = j\omega$ (when $\text{Re}[s] = \sigma = 0$) is inside the ROC, we can evaluate the 2-D function $X(s)$ along the imaginary axis from $\omega = -\infty$ to $\omega = \infty$ to obtain the Fourier transform $X(j\omega)$ of $x(t)$. In other words, the 1-D Fourier spectrum of the signal is the cross-section of the 2-D function $X(s) = X(\sigma + j\omega)$ along the imaginary axis $s = j\omega$, if it is inside the ROC; i.e., the CTFT is just a special case of the Laplace transform when $\sigma = 0$ and $s = j\omega$:

$$\mathcal{F}[x(t)] = \int_{-\infty}^{\infty} x(t)e^{-j\omega t}dt = \int_{-\infty}^{\infty} x(t)e^{-st}dt\Big|_{s=j\omega} = X(s)\Big|_{s=j\omega} = X(j\omega) \quad (6.5)$$

This is why the CTFT spectrum can also be denoted by $X(j\omega)$.

Given the Laplace transform $X(s) = \mathcal{L}[x(t)]$, the time signal $x(t)$ can be obtained by the inverse Laplace transform, which can be derived from the corresponding Fourier transform:

$$\mathcal{L}[x(t)] = X(s) = X(\sigma + j\omega) = \mathcal{F}[x(t)e^{-\sigma t}]. \quad (6.6)$$

Taking the inverse Fourier transform of the above, we get

$$x(t)e^{-\sigma t} = \mathcal{F}^{-1}[X(\sigma + j\omega)] = \frac{1}{2\pi}\int_{-\infty}^{\infty} X(\sigma + j\omega)e^{j\omega t}\,d\omega. \quad (6.7)$$

Multiplying both sides by $e^{\sigma t}$, we get

$$x(t) = \frac{1}{2\pi}\int_{-\infty}^{\infty} X(\sigma + j\omega)e^{(\sigma+j\omega)t}\,d\omega. \quad (6.8)$$

To further represent this inverse transform in terms of s (instead of ω), we note

$$ds = d(\sigma + j\omega) = j\,d\omega, \quad \text{i.e.,} \quad d\omega = ds/j. \tag{6.9}$$

The integral over $-\infty < \omega < \infty$ with respect to ω corresponds to the integral with respect to s over $\sigma - j\infty < s < \sigma + j\infty$:

$$x(t) = \mathcal{L}^{-1}[X(s)] = \frac{1}{j2\pi}\int_{\sigma-j\infty}^{\sigma+j\infty} X(s)e^{st}\,ds. \tag{6.10}$$

Now we get the forward and inverse Laplace transform pair:

$$X(s) = \mathcal{L}[x(t)] = \int_{-\infty}^{\infty} x(t)e^{-st}\,dt$$

$$x(t) = \mathcal{L}^{-1}[X(s)] = \frac{1}{j2\pi}\int_{\sigma-j\infty}^{\sigma+j\infty} X(s)e^{st}\,ds, \tag{6.11}$$

which can also be more concisely represented as

$$x(t) \overset{\mathcal{L}}{\longleftrightarrow} X(s). \tag{6.12}$$

In practice, we hardly need to carry out the integral in the inverse transform with respective to the complex variable s, as the Laplace transform pairs of most signals of interest can be obtained in some other ways and made available in table form.

In many applications the Laplace transform is a rational function as a ratio of two polynomials:

$$X(s) = \frac{N(s)}{D(s)} = \frac{\sum_{k=0}^{M} b_k s^k}{\sum_{k=0}^{N} a_k s^k} = \frac{b_M}{a_N}\,\frac{\prod_{k=1}^{M}(s - z_k)}{\prod_{k=1}^{N}(s - p_k)}. \tag{6.13}$$

The last equal sign in Eq. (6.13) is due to the fundamental theorem of algebra, stating that an Nth-order polynomial has N roots (some of which may be repeated with multiplicity greater than 1). Here, the roots $z_k\,(k = 1, 2, \ldots, m)$ of the numerator polynomial of order M are called the *zeros* of $X(s)$, and the roots $p_k\,(k = 1, 2, \ldots, n)$ of the denominator polynomial of order N are called the *poles* of $X(s)$; i.e.,

$$X(z_k) = 0 \quad \text{and} \quad X(p_k) = \infty. \tag{6.14}$$

The locations of the zeros and poles of $X(s)$ in the s-plane are of great importance, as they characterize some most essential properties of a signal $x(t)$, such as whether it is right- or left-sided, whether it grows or decays over time, as to be discussed later.

Moreover, if $N > M$, then $X(\infty) = 0$; i.e., $s = \infty$ is a zero. On the other hand, if $M > N$, then $X(\infty) = \infty$; i.e., $s = \infty$ is a pole. In general, we always assume $M < N$, as otherwise we can carry out a long division to expand $X(s)$ into multiple terms so that $M < N$ is true for each fraction. For example,

$$X(s) = \frac{s^2 - 3s + 1}{s - 2} = s - 1 - \frac{1}{s - 2}. \tag{6.15}$$

Now Eq. 6.13 can be converted into a sum of N terms by the method of partial fraction expansion:

$$X(s) = \frac{b_M \prod_{k=1}^{M}(s - z_k)}{a_N \prod_{k=1}^{N}(s - p_k)} = \frac{b_M}{a_N} \sum_{k=1}^{N} \frac{c_k}{s - p_k}. \tag{6.16}$$

6.1.2 The region of convergence

A Laplace transform $X(s) = \mathcal{L}[x(t)]$ always needs to be associated with the corresponding ROC, without which the inverse transform $x(t) = \mathcal{L}^{-1}[X(s)]$ cannot be meaningfully carried out. This point can be best illustrated in the following example.

Example 6.1:

1. A right-sided signal $x(t) = e^{-at}u(t)$ (a is a real constant):

$$X(s) = \int_0^{\infty} e^{-at} e^{-st}\,dt = \int_0^{\infty} e^{-at} e^{-(\sigma+j\omega)t}\,dt = \int_0^{\infty} e^{-(a+\sigma)t} e^{-j\omega t}\,dt. \tag{6.17}$$

For this integral to converge, it is necessary to have $a + \sigma > 0$; i.e. the ROC is $\text{Re}[s] = \sigma > -a$, inside which the above becomes

$$X(s) = \frac{1}{-(a+\sigma+j\omega)} e^{-(a+\sigma+j\omega)t}\Big|_0^{\infty} = \frac{1}{(\sigma+a)+j\omega} = \frac{1}{s+a}. \tag{6.18}$$

In particular, if $a = 0$, $x(t) = u(t)$ and we have

$$U(s) = \mathcal{L}[u(t)] = \frac{1}{s}, \qquad \sigma > 0. \tag{6.19}$$

If we let $\sigma \to 0$, then $U(s)$ is evaluated along the imaginary axis $s = j\omega$ and becomes $U(j\omega) = 1/j\omega$, which is seemingly the Fourier transform of $u(t)$. However, this result is actually invalid, as $\sigma = 0$ is not inside the ROC $\text{Re}[s] = \sigma > 0$. Comparing this result with the real Fourier transform of $u(t)$ in Eq. (3.80),

$$\mathcal{F}[u(t)] = \frac{1}{2}\delta(f) + \frac{1}{j\omega}, \tag{6.20}$$

we see an extra term $\delta(f)/2$ in the Fourier spectrum, which reflects the fact that the integral is only marginally convergent when $s = j\omega$.

2. A left-sided signal $x(t) = -e^{-at}u(-t)$:

$$X(s) = -\int_{-\infty}^{0} e^{-at} e^{-st}\,dt = -\int_{-\infty}^{0} e^{-(a+\sigma+j\omega)t}\,dt, \tag{6.21}$$

where a is a real constant. For this integral to converge, it is necessary that $a+\sigma<0$; i.e., the ROC is $\text{Re}[s]=\sigma<-a$, inside which the above becomes

$$X(s)=\frac{1}{a+\sigma+j\omega}\,e^{-(a+\sigma+j\omega)t}\Big|_{-\infty}^{0}=\frac{1}{a+\sigma+j\omega}=\frac{1}{s+a},\qquad \sigma<-a. \tag{6.22}$$

When $a=0$, $x(t)=-u(-t)$ we have

$$\mathcal{L}[-u(-t)]=\frac{1}{s},\qquad \sigma<0. \tag{6.23}$$

We see that the Laplace transforms of two different signals $e^{-at}u(t)$ and $-e^{-at}u(-t)$ are identical, but their corresponding ROCs are different.

Based on the examples above, we summarize a set of properties of the ROC:

- If a signal $x(t)$ of finite duration is absolutely integrable then its transform $X(s)$ exists for any s; i.e., its ROC is the entire s-plane.
- The ROC does not contain any poles at which $X(s)=\infty$.
- Two different signals may have identical transform but different ROCs. The inverse transform can be carried out only if an associated ROC is also specified.
- Only the real part $\text{Re}[s]=\sigma$ of s determines the convergence of the integral in the Laplace transform and thereby the ROC. The imaginary part $\text{Im}[s]$ has no effect on the convergence. Consequently the ROC is always bounded by two vertical lines parallel to the imaginary axis $s=j\omega$, corresponding to two poles p_1 and p_2 with $\text{Re}[p_1]<\text{Re}[p_2]$. It is possible that $\text{Re}[p_1]=-\infty$ and/or $\text{Re}[p_2]=\infty$.
- The ROC of a right-sided signal is the right-sided half plane to the right of the rightmost pole; the ROC of the transform of a left-sided signal is a left-sided half plane to the left of the leftmost pole. If a signal is two-sided, its ROC is the intersection of the two ROCs corresponding to its two one-sided parts, which can be either a vertical strip or an empty set.
- The Fourier transform $X(j\omega)$ of a signal $x(t)$ exists if the ROC of the corresponding Laplace transform $X(s)$ contains the imaginary axis $\text{Re}[s]=0$; i.e., $s=j\omega$.

6.1.3 Properties of the Laplace transform

The Laplace transform has a set of properties most of which are in parallel with those of the Fourier transform. The proofs of most of these properties are omitted as they are similar to that of their counterparts in the Fourier transform. However, here, we need to pay special attention to the ROCs. Here, we always assume

$$\mathcal{L}[x(t)]=X(s),\qquad \mathcal{L}[y(t)]=Y(s), \tag{6.24}$$

with ROCs R_x and R_y, respectively.

- **Linearity**

$$\mathcal{L}[ax(t) + by(t)] = aX(s) + bY(s), \quad \text{ROC} \supseteq (R_x \cap R_y). \tag{6.25}$$

It is obvious that the ROC of the linear combination of $x(t)$ and $y(t)$ should be the intersection $R_x \cap R_y$ of their individual ROCs, in which both $X(s)$ and $Y(s)$ exist. However, note that in some cases the ROC of the linear combination may be larger than $R_x \cap R_y$. For example, $\mathcal{L}[u(t)] = 1/s$ and $\mathcal{L}[u(t-\tau)] = e^{-s\tau}/s$ have the same ROC $\text{Re}[s] > 0$, but their difference $u(t) - u(t-\tau)$ has finite duration and the corresponding ROC is the entire s-plane. Also when *zero-pole cancellation* occurs the ROC of the linear combination may also be larger than $R_x \cap R_y$. For example, let

$$X(s) = \mathcal{L}[x(t)] = \frac{1}{s+1}, \quad \text{Re}[s] > -1, \tag{6.26}$$

and

$$Y(s) = \mathcal{L}[y(t)] = \frac{1}{(s+1)(s+2)}, \quad \text{Re}[s] > -1, \tag{6.27}$$

then

$$\mathcal{L}[x(t) - y(t)] = \frac{1}{s+1} - \frac{1}{(s+1)(s+2)} = \frac{s+1}{(s+1)(s+2)} = \frac{1}{s+2}$$
$$\text{Re}[s] > -2. \tag{6.28}$$

- **Time shift**

$$\mathcal{L}[x(t - t_0)] = e^{-t_0 s} X(s), \quad \text{ROC} = R_x. \tag{6.29}$$

- **Time reversal**

$$\mathcal{L}[x(-t)] = X(-s), \quad \text{ROC} = -R_x. \tag{6.30}$$

- **s-domain shift**

$$\mathcal{L}[e^{-s_0 t} x(t)] = X(s + s_0), \quad \text{ROC} = R_x + \text{Re}[s_0]. \tag{6.31}$$

Note that the ROC is shifted by s_0; i.e., it is shifted vertically by $\text{Im}[s_0]$ (with no effect on ROC) and horizontally by $\text{Re}[s_0]$.

- **Time scaling**

$$\mathcal{L}[x(at)] = \frac{1}{|a|} X\left(\frac{s}{a}\right), \quad \text{ROC} = \frac{R_x}{a}. \tag{6.32}$$

Note that the ROC is horizontally scaled by $1/a$, which could be either positive ($a > 0$) or negative ($a < 0$), in which case both the function $x(t)$ and the ROC of its Laplace transform are horizontally flipped.

- **Conjugation**

$$\mathcal{L}[x^*(t)] = X^*(s^*), \quad \text{ROC} = R_x. \tag{6.33}$$

- **Convolution**

$$\mathcal{L}[x(t) * y(t)] = X(s)Y(s), \quad \text{ROC} \supseteq (R_x \cap R_y). \tag{6.34}$$

Note that the ROC of the convolution could be larger than the intersection of R_x and R_y, due to the possible pole-zero cancellation caused by the convolution, similar to the linearity property. For example, assume

$$X(s) = \mathcal{L}[x(t)] = \frac{s+1}{s+2}, \quad \text{Re}[s] > -2, \tag{6.35}$$

$$Y(s) = \mathcal{L}[y(t)] = \frac{s+2}{s+1}, \quad \text{Re}[s] > -1, \tag{6.36}$$

then

$$\mathcal{L}[x(t) * y(t)] = X(s)Y(s) = 1, \tag{6.37}$$

with an ROC of the entire s-plane.

- **Differentiation in the time domain**

$$\mathcal{L}[\frac{d}{dt}x(t)] = sX(s), \quad \text{ROC} \supseteq R_x. \tag{6.38}$$

This is an important property based on which the Laplace transform finds a lot of applications in systems analysis and design. This property can be proven by differentiating the inverse Laplace transform:

$$\frac{d}{dt}x(t) = \frac{1}{j2\pi}\int_{\sigma-j\infty}^{\sigma+j\infty} X(s)\frac{d}{dt}e^{st}\,ds = \frac{1}{j2\pi}\int_{\sigma-j\infty}^{\sigma+j\infty} sX(s)e^{st}\,ds. \tag{6.39}$$

Again, multiplying $X(s)$ by s may cause pole-zero cancellation and therefore the resulting ROC may be larger than R_x. For example, let $x(t) = u(t)$ and $X(s) = \mathcal{L}[u(t)] = 1/s$ with ROC Re$[s] > 0$, then we have $\mathcal{L}[dx(t)/dt] = \mathcal{L}[\delta(t)] = sX(s) = 1$, but its ROC is the entire s-plane. Repeating this property we get

$$\mathcal{L}[\frac{d^n}{dt^n}x(t)] = s^n X(s). \tag{6.40}$$

In particular, when $x(t) = \delta(t)$, we have

$$\mathcal{L}[\frac{d^n}{dt^n}\delta(t)] = s^n, \quad \text{ROC} = \text{entire } s\text{-plane}. \tag{6.41}$$

- **Differentiation in the s-domain**

$$\mathcal{L}[tx(t)] = -\frac{d}{ds}X(s), \quad \text{ROC} = R_x. \tag{6.42}$$

This can be proven by differentiating the Laplace transform:

$$\frac{d}{ds}X(s) = \int_{-\infty}^{\infty} x(t)\frac{d}{ds}e^{-st}\,dt = \int_{-\infty}^{\infty} (-t)x(t)e^{-st}\,dt. \tag{6.43}$$

Repeating this process we get

$$\mathcal{L}[t^n x(t)] = (-1)^n \frac{d^n}{ds^n} X(s), \quad \text{ROC} = R_x. \tag{6.44}$$

- **Integration in the time domain**

$$\mathcal{L}\left[\int_{-\infty}^{t} x(\tau)\, d\tau\right] = \frac{X(s)}{s}, \qquad \text{ROC} \supseteq (R_x \cap \{\text{Re}[s] > 0\}). \tag{6.45}$$

This can be proven by realizing that

$$x(t) * u(t) = \int_{-\infty}^{\infty} x(\tau)u(t-\tau)\, d\tau = \int_{-\infty}^{t} x(\tau)\, d\tau, \tag{6.46}$$

and therefore by convolution property we have

$$\mathcal{L}[x(t) * u(t)] = X(s)\frac{1}{s}. \tag{6.47}$$

As the ROC of $\mathcal{L}[u(t)] = 1/s$ is the right half plane $\text{Re}[s] > 0$, the ROC of $X(s)/s$ is the intersection $R_x \cap \{\text{Re}[s] > 0\}$, except when pole-zero cancellation occurs. For example, when $x(t) = d\delta(t)/dt$ with $X(s) = s$, $\mathcal{L}[\int_{-\infty}^{t} x(\tau)\, d\tau] = s/s = 1$ with the ROC being the entire s-plane.

6.1.4 The Laplace transform of typical signals

- $\delta(t)$, $\delta(t-\tau)$

$$\mathcal{L}[\delta(t)] = \int_{-\infty}^{\infty} \delta(t) e^{-st}\, dt = e^0 = 1, \quad \text{ROC: entire } s\text{-plane.} \tag{6.48}$$

Moreover, owing to time-shift property, we have

$$\mathcal{L}[\delta(t-\tau)] = e^{-s\tau}, \quad \text{ROC: entire } s\text{-plane.} \tag{6.49}$$

As the Laplace integration converges for any s, the ROC is the entire s-plane.

- $u(t)$, $t\,u(t)$, $t^n\, u(t)$

Owing to the property of the time domain integration, we get

$$\mathcal{L}[u(t)] = \mathcal{L}\left[\int_{-\infty}^{t} \delta(\tau)\, d\tau\right] = \frac{1}{s} \qquad \text{Re}[s] > 0. \tag{6.50}$$

Applying the s-domain differentiation property to the above, we get

$$\mathcal{L}[tu(t)] = -\frac{d}{ds}\left[\frac{1}{s}\right] = \frac{1}{s^2}, \qquad \text{Re}[s] > 0. \tag{6.51}$$

In general we have

$$\mathcal{L}[t^n\, u(t)] = \frac{n!}{s^{n+1}}, \quad \text{Re}[s] > 0. \tag{6.52}$$

- $e^{-at}\, u(t)$, $t\, e^{-at}\, u(t)$

Applying the s-domain shifting property to

$$\mathcal{L}[u(t)] = \frac{1}{s}, \quad \text{Re}[s] > 0, \tag{6.53}$$

we have

$$\mathcal{L}[e^{-at}u(t)] = \frac{1}{s+a}, \quad \text{Re}[s] > -a. \tag{6.54}$$

Applying the same property to

$$\mathcal{L}[t^n u(t)] = \frac{n!}{s^{n+1}}, \quad \text{Re}[s] > 0, \tag{6.55}$$

we have

$$\mathcal{L}[t^n e^{-at}u(t)] = \frac{n!}{(s+a)^{n+1}}, \quad \text{Re}[s] > -a. \tag{6.56}$$

- $e^{-j\omega_0 t}u(t)$, $\sin(\omega_0 t)u(t)$, $\cos(\omega_0 t)u(t)$
  Letting $a = \pm j\omega_0$ in

$$\mathcal{L}[e^{-at}u(t)] = \frac{1}{s+a}, \qquad \text{Re}[s] > -\text{Re}[a], \tag{6.57}$$

we get

$$\mathcal{L}[e^{\mp j\omega_0 t}u(t)] = \frac{1}{s \pm j\omega_0}, \qquad \text{Re}[s] > 0, \tag{6.58}$$

based on which we further get

$$\mathcal{L}[\cos(\omega_0 t)u(t)] = \frac{1}{2}\mathcal{L}[e^{j\omega_0 t} + e^{-j\omega_0 t}] = \frac{1}{2}\left[\frac{1}{s-j\omega_0} + \frac{1}{s+j\omega_0}\right] = \frac{s}{s^2+\omega_0^2} \tag{6.59}$$

and

$$\mathcal{L}[\sin(\omega_0 t)u(t)] = \frac{1}{2j}\mathcal{L}[e^{j\omega_0 t} - e^{-j\omega_0 t}] = \frac{1}{2j}\left[\frac{1}{s-j\omega_0} - \frac{1}{s+j\omega_0}\right] = \frac{\omega_0}{s^2+\omega_0^2}. \tag{6.60}$$

- $t\,\cos(\omega_0 t)u(t)$, $t\,\sin(\omega_0 t)u(t)$
  Letting $a = \pm j\omega_0$ in

$$\mathcal{L}[te^{-at}u(t)] = \frac{1}{(s+a)^2}, \qquad \text{Re}[s] > -a, \tag{6.61}$$

we get

$$\mathcal{L}[t\,e^{\mp j\omega_0 t}u(t)] = \frac{1}{(s \pm j\omega_0)^2}, \qquad \text{Re}[s] > -a, \tag{6.62}$$

based on which we further get

$$\begin{aligned}\mathcal{L}[t\,\cos(\omega_0 t)u(t)] &= \frac{1}{2}\mathcal{L}[t\,(e^{j\omega_0 t} + e^{-j\omega_0 t})] \\ &= \frac{1}{2}[\frac{1}{(s-j\omega_0)^2} + \frac{1}{(s+j\omega_0)^2}] = \frac{s^2-\omega_0^2}{(s^2+\omega_0^2)^2},\end{aligned} \tag{6.63}$$

and

$$\mathcal{L}[t\ \sin(\omega_0 t)u(t)] = \frac{1}{2j}\mathcal{L}[t\ (e^{j\omega_0 t} - e^{-j\omega_0 t})]$$
$$= \frac{1}{2j}\left[\frac{1}{(s-j\omega_0)^2} - \frac{1}{(s+j\omega_0)^2}\right] = \frac{2s\omega_0}{(s^2+\omega_0^2)^2}. \quad (6.64)$$

- $e^{-at}\cos(\omega_0 t)u(t)$, $e^{-at}\sin(\omega_0 t)u(t)$
  Applying the s-domain shifting property to

$$\mathcal{L}[\cos(\omega_0 t)u(t)] = \frac{s}{s^2+\omega_0^2} \quad \text{and} \quad \mathcal{L}[\sin(\omega_0 t)u(t)] = \frac{\omega_0}{s^2+\omega_0^2}, \quad (6.65)$$

we get, respectively,

$$\mathcal{L}[e^{-at}\cos(\omega_0 t)u(t)] = \frac{s+a}{(s+a)^2+\omega_0^2} \quad (6.66)$$

and

$$\mathcal{L}[e^{-at}\sin(\omega_0 t)u(t)] = \frac{\omega_0}{(s+a)^2+\omega_0^2}. \quad (6.67)$$

6.1.5 Analysis of continuous LTI systems by Laplace transform

The Laplace transform is a convenient tool for the analysis and design of continuous LTI systems $y(t) = \mathcal{O}[x(t)]$ whose output $y(t)$ is the convolution of the input $x(t)$ and its impulse response function $h(t)$:

$$y(t) = \mathcal{O}[x(t)] = h(t) * x(t) = \int_{-\infty}^{\infty} h(\tau)x(t-\tau)\,d\tau. \quad (6.68)$$

In particular, if the input is an impulse $x(t) = \delta(t)$, then the output is the impulse response function $y(t) = \mathcal{O}[\delta(t)] = h(t) * \delta(t) = h(t)$. Also if the input is a complex exponential $x(t) = e^{st} = e^{\sigma + j\omega}$, then the output can be found to be

$$y(t) = \mathcal{O}[e^{st}] = \int_{-\infty}^{\infty} h(\tau)e^{s(t-\tau)}\,d\tau = e^{st}\int_{-\infty}^{\infty} h(\tau)e^{-s\tau}\,d\tau = H(s)e^{st}, \quad (6.69)$$

where $H(s)$ is the *transfer function* of the system, first defined in Eq. (1.89), which is actually the Laplace transform of the impulse response $h(t)$ of the system:

$$H(s) = \mathcal{L}[h(t)] = \int_{-\infty}^{\infty} h(t)e^{-st}\,dt. \quad (6.70)$$

Equation (6.69) is the eigenequation of *any* continuous LTI system, where the transfer function $H(s)$ is the eigenvalue, and the complex exponential input $x(t) = e^{st}$ is the corresponding eigenfunction. In particular, if we let $\sigma = 0$, then $s = j\omega$ and the transfer function $H(s)$ becomes the Fourier transform of the impulse response $h(t)$ of the system:

$$H(s)\Big|_{s=j\omega} = H(j\omega) = \int_{-\infty}^{\infty} h(t)e^{-j\omega t}\,dt = \mathcal{F}[h(t)]. \quad (6.71)$$

This is the frequency response function of the LTI system first defined in Eq. (5.3). Various properties and behaviors such as the stability and filtering effects of a continuous LTI system can be qualitatively characterized based on the locations of the zeros and poles of its transfer function $H(s) = \mathcal{L}[h(t)]$ due to the properties of the ROC of the Laplace transform.

Also, owing to its convolution property of the Laplace transform, the convolution in Eq. (6.68) can be converted to a multiplication in the s-domain

$$y(t) = h(t) * x(t) \xrightarrow{\mathcal{L}} Y(s) = H(s)X(s). \tag{6.72}$$

Based on this relationship, the transfer function $H(s)$ can also be found in the s-domain as the ratio $H(s) = Y(s)/X(s)$ of the output $Y(s)$ and input $X(s)$, which can also be used as the definition of the transfer function of an LTI system. The ROC and poles of the transfer function $H(s)$ of an LTI system dictate the behaviors of system, such as its causality and stability.

- **Stability**
  Also as discussed in Chapter 1, an LTI system is stable if to any bounded input $|x(t)| < B$ its response $y(t)$ is also bounded for all t, and its impulse response function $h(t)$ needs to be absolutely integrable (Eq. (1.101)):

  $$\int_{-\infty}^{\infty} \big|h(\tau)\big| \, d\tau < \infty; \tag{6.73}$$

  i.e., the frequency response function $\mathcal{F}[h(t)] = H(j\omega) = H(s)\big|_{s=j\omega}$ exists. In other words, an LTI system is stable if and only if the ROC of its transfer function $H(s)$ includes the imaginary axis $s = j\omega$.
- **Causality**
  As discussed in Chapter 1, an LTI system is causal if its impulse response $h(t)$ is a consequence of the impulse input $\delta(t)$; i.e., $h(t)$ comes after $\delta(t)$:

  $$h(t) = h(t)u(t) = \begin{cases} h(t) & t \geq 0 \\ 0 & t < 0 \end{cases}. \tag{6.74}$$

  Its output is (Eq. (1.102))

  $$y(t) = \int_{-\infty}^{\infty} h(\tau)x(t-\tau) \, d\tau = \int_{0}^{\infty} h(\tau)x(t-\tau) \, d\tau.. \tag{6.75}$$

  The ROC of $H(s)$ is a right-sided half plane. In particular, when $H(s)$ is rational, the system is causal if and only if its ROC is the right half plane to the right of the rightmost pole, and the order of the numerator is no greater than that of the denominator so that $s = \infty$ is not a pole ($H(\infty)$ exists).

Combining the two properties above, we see that a causal LTI system with a rational transfer function $H(s)$ is stable if and only if all poles of $H(s)$ are in the left half of the s-plane; i.e., the real parts of all poles p_k are negative: $\text{Re}[p_k] < 0$.

One type of continuous LTI system can be characterized by an LCCDE:

$$\sum_{k=0}^{N} a_k \frac{d^k}{dt^k} y(t) = \sum_{k=0}^{M} b_k \frac{d^k}{dt^k} x(t). \tag{6.76}$$

Taking the Laplace transform on both sides of this equation, we get an algebraic equation in the s-domain:

$$Y(s)\left[\sum_{k=0}^{N} a_k s^k\right] = X(s)\left[\sum_{k=0}^{M} b_k s^k\right]. \tag{6.77}$$

The transfer function of such a system is rational:

$$H(s) = \frac{Y(s)}{X(s)} = \frac{\sum_{k=0}^{M} b_k s^k}{\sum_{k=0}^{N} a_k s^k} = \frac{b_M}{a_N} \frac{\prod_{k=0}^{M}(s - z_k)}{\prod_{k=0}^{N}(s - p_k)}, \tag{6.78}$$

where z_k $(k = 1, 2, \ldots, M)$ and p_k $(k = 1, 2, \ldots, N)$ are respectively the zeros and poles of $H(s)$. For simplicity, and without loss of generality, we will assume $N > M$ and $b_M/a_N = 1$ below.

The output $Y(s)$ of the LTI system can be represented as

$$Y(s) = H(s)X(s) = \left(\sum_{k=0}^{M} b_k s^k\right) \frac{1}{\sum_{k=0}^{N} a_k s^k} X(s) = \left(\sum_{k=0}^{M} b_k s^k\right) W(s), \tag{6.79}$$

or in the time domain as

$$y(t) = \sum_{k=0}^{M} b_k \frac{d^k w(t)}{dt^k}, \tag{6.80}$$

where we have defined $W(s) = X(s)/(\sum_{k=0}^{N} a_k s^k)$ as an intermediate variable, or in the time domain:

$$\sum_{k=0}^{N} a_k \frac{d^k w(t)}{dt^k} = x(t), \quad \text{or} \quad a_N \frac{d^N w(t)}{dt^N} = x(t) - \sum_{k=0}^{N-1} a_k w^{(k)}(t). \tag{6.81}$$

Without loss of generality, we assume $a_N = 1$, and the LTI system can now be represented as a *block diagram*, as shown in Fig. 6.1 (for $M = 2$ and $N = 3$).

To find the impulse response $h[n]$ we first convert $H(z)$ to a summation by partial fraction expansion:

$$H(s) = \frac{\sum_{k=0}^{M} b_k s^k}{\sum_{k=0}^{N} a_k s^k} = \sum_{k=1}^{N} \frac{c_k}{s - p_k}, \tag{6.82}$$

(assume no repeated poles) and then carry out the inverse transform (the LTI system in Eq. (6.76) is causal) to get

$$h(t) = \mathcal{L}^{-1}[H(s)] = \sum_{k=1}^{N} \mathcal{L}^{-1}\left[\frac{c_k}{s - p_k}\right] = \sum_{k=1}^{N} c_k e^{p_k t} u(t). \tag{6.83}$$

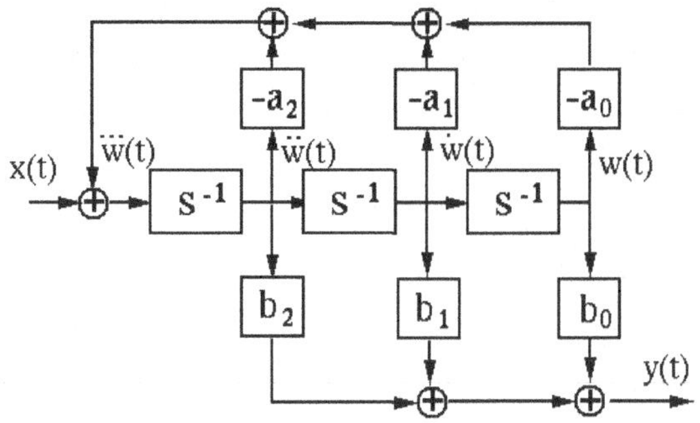

Figure 6.1 Block diagram of a continuous LTI system described by an LCCDE.

The output $y(t)$ of the LTI system can be found by solving the differential equation in Eq. (6.76). Alternatively, it can also be found by the convolution $y(t) = h(t) * x(t)$, or the inverse Laplace transform:

$$y(t) = \mathcal{L}^{-1}[Y(s)] = \mathcal{L}^{-1}[H(s)X(s)]. \tag{6.84}$$

As the LCCDE in Eq. (6.76) is an LTI system, it can also be solved in the following two steps. First, we assume the input on the right-hand side is simply $x(t)$ and find the corresponding output $y(t)$. Then the response to the true input $\sum_k b_k d^k x(t)/dt^k$ can be found to be $\sum_k b_k d^k y(t)/dt^k$.

Note that the output $y(t)$ obtained this way is only the particular solution due to input $x(t)$, but the homogeneous solution due to non-zero initial conditions is not represented by the bilateral Laplace transform. This problem will be addressed by the unilateral Laplace transform to be discussed later, which takes the initial conditions into consideration.

According to the fundamental theorem of algebra, if all coefficients a_k of the denominator polynomial of $H(s)$ are real, then its roots p_k are either real or complex conjugate pairs, corresponding to the following system behaviors in the time domain:

- If at least one pole $\text{Re}[p_k] > 0$ is on the right half s-plane, then the corresponding term $c_k e^{p_k t} u(t)$ grows exponentially without bounds, and the system is unstable.
- If all poles $\text{Re}[p_k] < 0$ $(1 < k < N)$ are on the left half s-plane (i.e., all terms in the summation of $h(t)$ above decay to zero exponentially), then $h(t)$ is absolutely integrable and the system is stable.
- Any pair of complex conjugate poles $p_{1,2} = \sigma \pm j\omega$ corresponds to a sinusoid of frequency ω:

$$e^{p_1 t} + e^{p_2 t} = e^{\sigma}[e^{j\omega t} + e^{-j\omega t}] = \frac{1}{2} e^{\sigma} \cos(\omega t). \tag{6.85}$$

This term either decays, if $\sigma < 0$, or grows, if $\sigma > 0$ exponentially. In the latter case, the system is unstable.

Table 6.1. Pole locations in s-domain and waveform in time domain

	Pole locations in s-plane	Waveforms in the time domain
1	single real pole: $p > 0$	exponential growth: $h(t) = e^{pt}$
2	complex conjugate poles: $p_{1,2} = \sigma \pm j\omega$ $(\sigma > 0)$	exponentially growing sinusoid: $h(t) = \cos(\omega t)e^{\sigma t}$
3	complex conjugate poles: $p_{1,2} = \pm j\omega$	sinusoid: $h(t) = \cos(\omega t)$
4	complex conjugate poles: $p_{1,2} = \sigma \pm j\omega$ $(\sigma < 0)$	exponentially decaying sinusoid: $h(t) = \cos(\omega t)e^{-\|\sigma\|t}$
5	single real pole: $p < 0$	exponential decay: $h(t) = e^{-\|p\|t}$

- If $0 < \text{Re}[p_2] \ll \text{Re}[p_1]$, then $e^{p_1 t}$ grows much more rapidly than $e^{p_2 t}$, i.e., the behavior of an unstable system is dominated by the rightmost pole on the right half s-plane. On the other hand, if $\text{Re}[p_2] \ll \text{Re}[p_1] < 0$, then $e^{p_1 t} = e^{-|p_1|t}$ decays much more slowly then $e^{p_2 t}$; i.e., the behavior of a stable system is also dominated by the rightmost pole on the left half s-plane. Based on this observation, the behavior of a high-order system with a large number of poles can be approximated based only on its most dominant poles.

These different pole locations in s-plane and the corresponding waveforms in the time domain are further illustrated in Fig. 6.2 and summarized in Table 6.1.

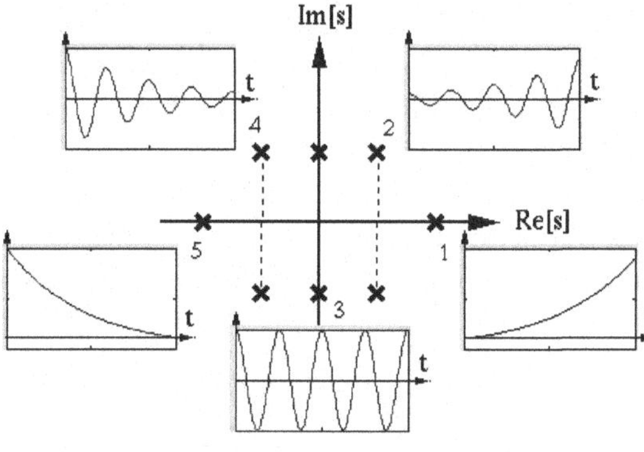

Figure 6.2 Different pole locations of $H(s)$ and the corresponding waveforms of $h(t)$.

An LTI system can be considered as a filter characterized by the magnitude and phase of its frequency response function $H(j\omega) = H(s)\big|_{s=j\omega}$:

$$|H(j\omega)| = \frac{\prod_{k=1}^{M} |j\omega - z_k|}{\prod_{k=1}^{N} |j\omega - p_k|} = \frac{\prod_{k=1}^{M} |\boldsymbol{u}_k|}{\prod_{k=1}^{N} |\boldsymbol{v}_k|},$$
$$\angle H(j\omega) = \frac{\sum_{k=1}^{M} \angle(j\omega - z_k)}{\sum_{k=1}^{N} \angle(j\omega - p_k)} = \frac{\sum_{k=1}^{M} \angle\boldsymbol{u}_k}{\sum_{k=1}^{N} \angle\boldsymbol{v}_k}, \tag{6.86}$$

where each factor $\boldsymbol{u}_k = j\omega - z_k$ or $\boldsymbol{v}_k = j\omega - p_k$ is a vector in the s-plane that connects a point $j\omega$ on the imaginary axis and one of the zeros or poles. The filtering effects of the system are therefore dictated by the zero and pole locations on the s-plane and can be qualitatively determined by observing how $|H(j\omega)|$ and $\angle H(j\omega)$ change when frequency ω increases along the imaginary axis from 0 toward ∞.

Example 6.2: The input to an LTI is

$$x(t) = e^{-3t}u(t), \tag{6.87}$$

and its corresponding output is

$$y(t) = h(t) * x(t) = (e^{-t} - e^{-2t})u(t). \tag{6.88}$$

We want to identify the system by determining $h(t)$ and $H(s)$. In the s-domain, input and output signals are

$$X(s) = \frac{1}{s+3}, \qquad \text{Re}[s] > -3, \tag{6.89}$$

and

$$Y(s) = H(s)X(s) = \frac{1}{s+1} - \frac{1}{s+2} = \frac{1}{(s+1)(s+2)}, \qquad \text{Re}[s] > -1. \tag{6.90}$$

The transfer function can therefore be obtained

$$H(s) = \frac{Y(s)}{X(s)} = \frac{s+3}{(s+1)(s+2)} = \frac{s+3}{s^2+3s+2}. \tag{6.91}$$

This system $H(s)$ has two poles $p_1 = -1$ and $p_2 = -2$ and therefore three possible ROCs: $\text{Re}[s] < -2$, $-2 < \text{Re}[s] < -1$, and $\text{Re}[s] > -1$ corresponding to left-sided (anti-causal), two-sided and right-sided (causal) system, respectively. To determine which of these ROCs the system has, recall that the ROC of a convolution $Y(s) = H(s) * X(s)$ should be no less than the intersection of the ROCs of $H(s)$ and $X(s)$; i.e., the ROC of $H(s)$ must be $\text{Re}[s] > -1$; i.e., the system is causal and stable. The inverse Laplace transform of $Y(s) = H(s)X(s)$ is the LCCDE of the system:

$$\frac{d^2}{dt^2}y(t) + 3\frac{d}{dt}y(t) + 2y(t) = \frac{d}{dt}x(t) + 3x(t). \tag{6.92}$$

In the following, we will consider two specific systems $H(s) = N(s)/D(s)$, where $D(s)$ is either a first-order ($n = 1$) or a second-order ($n = 2$) polynomial.

6.1.6 First-order system

In the transfer function $H(s)$ of a first-order LTI system, the denominator $D(s)$ is a first-order polynomial of order $N = 1$, and $H(s)$ is conventionally written in the following *canonical form*:

$$H(s) = \frac{N(s)}{D(s)} = \frac{1}{s-p} = \frac{1}{s+1/\tau}, \tag{6.93}$$

where τ is the time constant, and $p = -1/\tau$ is the pole of $H(s)$. In practice, $\tau > 0$ is always positive and the pole $p = -1/\tau < 0$ is on the left side of the s-plane; i.e., the system is stable.

Here, we reconsider the RC circuit in Example 5.2 to illustrate the essential properties of the first-order system. The input is the voltage $x(t) = v_{\text{in}}(t)$ applied across R and C in series, and the output can be either the voltage $v_C(t)$ across C or the voltage $v_R(t)$ across R. First, we let the output be $y(t) = v_C(t)$, the system can be described by a differential equation:

$$RC\dot{y}(t) + y(t) = x(t); \qquad \text{i.e.,} \qquad \dot{y}(t) + \frac{1}{\tau}y(t) = \frac{1}{\tau}x(t), \tag{6.94}$$

where $\tau = RC$ is the time constant of the system. Now we solve this LCCDE by taking the Laplace transform on both sides of this equation to get

$$\left[s + \frac{1}{\tau}\right]Y(s) = \frac{1}{\tau}X(s); \qquad \text{i.e.,} \qquad H_C(s) = \frac{Y(s)}{X(s)} = \frac{1/\tau}{s+1/\tau}. \tag{6.95}$$

Given $H_c(s)$, we can also get $H_R(s)$ when $v_R(t)$ is treated as output based on Kirchhoff's voltage law $\delta(t) = h_C(t) + h_R(t)$:

$$H_R(s) = 1 - H_C(s) = 1 - \frac{1/\tau}{s+1/\tau} = \frac{s}{s+1/\tau}. \tag{6.96}$$

We now consider both the impulse and step responses as well as the filtering effects of this first-order system.

1. **Impulse response function:**
   Taking the inverse Laplace transform on both sides of Eqs. (6.95) and (6.96), we get

$$h_C(t) = \mathcal{L}^{-1}[H_C(s)] = \frac{1}{\tau}e^{-t/\tau}u(t), \tag{6.97}$$

   and

$$h_R(t) = \mathcal{L}^{-1}[H_R(s)] = \delta(t) - \frac{1}{\tau}e^{-t/\tau}u(t). \tag{6.98}$$

2. **Step response:**

The step response of this system to input $x(t) = u(t)$ or $X(s) = 1/s$ can also be found in the s-domain as

$$Y(s) = H_C(s)\,X(s) = \frac{1/\tau}{s(s+1/\tau)} = \frac{1}{s} - \frac{1}{s+1/\tau}. \tag{6.99}$$

Taking the inverse transform we get the step response:

$$y(t) = v_C(t) = (1 - e^{-t/\tau})u(t). \tag{6.100}$$

The step response of the system when the voltage $v_R(t)$ across R is treated as output can be obtained based on Kirchhoff's voltage law:

$$v_R(t) = u(t) - v_C(t) = u(t) - (1 - e^{-t/\tau})u(t) = e^{-t/\tau}u(t). \tag{6.101}$$

The impulse and step response functions for the two first-order systems are shown in Fig. 6.3.

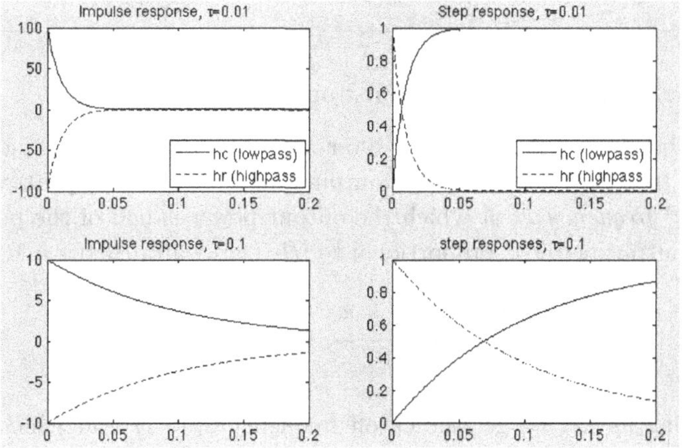

Figure 6.3 Impulse (left) and step (right) responses of first-order systems.

3. **First-order systems as filters:**
The filtering effects of the first-order system are characterized by the magnitudes and phases of their frequency response functions $H(j\omega) = H(s)\big|_{s=j\omega}$:

$$|H_C(j\omega)| = \left|\frac{1/\tau}{j\omega + 1/\tau}\right| = \frac{1}{\sqrt{(\omega\tau)^2+1}},$$
$$\angle H_C(j\omega) = -\angle(j\omega + 1/\tau) = -\tan^{-1}\omega\tau, \tag{6.102}$$

and

$$|H_R(j\omega)| = \left|\frac{j\omega}{j\omega + 1/\tau}\right| = \frac{\omega\tau}{\sqrt{(\omega\tau)^2+1}},$$
$$\angle H_R(j\omega) = \angle(j\omega\tau) - \angle(j\omega\tau + 1) = \frac{\pi}{2} - \tan^{-1}(\omega\tau). \tag{6.103}$$

Both the linear and Bode plots of the two systems are given in Fig. 6.4, where the magnitudes of the two frequency response functions are plotted for

$\tau = 0.01$ (top) and $\tau = 0.1$ (bottom), and in both linear scale (left) and Bode plots for their magnitudes (middle) and phases (right) are also plotted. We see that H_C and H_R attenuate high and low frequencies, and, therefore, are correspondingly LP and HP filters, respectively.

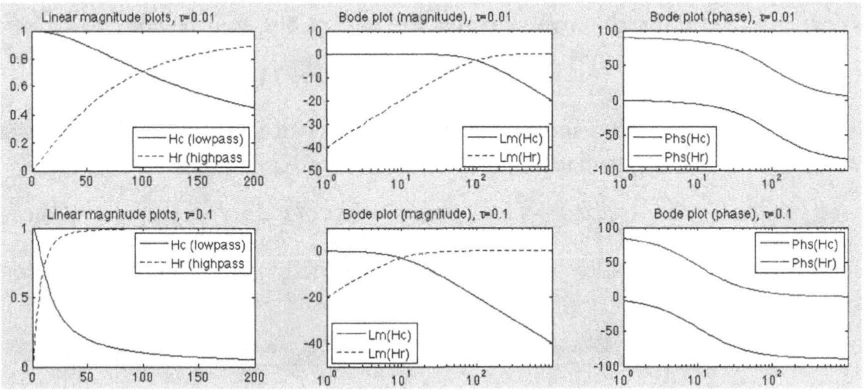

Figure 6.4 Filtering effects of first-order systems.

The bandwidth $\Delta\omega$ of the LP filter $H_C(j\omega)$ is defined as the interval between zero frequency at which the output power reaches its peak value and the *cutoff frequency* ω_c at which the output power is half of the peak power. As the output power is proportional to $|H_C(j\omega)|^2$ and $H_C(0) = 1$, we have

$$\frac{|H_C(j\omega_c)|^2}{|H_C(0)|^2} = \frac{1}{(\omega_c\tau)^2 + 1} = \frac{1}{2}. \tag{6.104}$$

Solving for ω_c, we get the cutoff frequency $\omega_c = 1/\tau$, at which $|H_C(j\omega_c)| = 1/\sqrt{2} = 0.707$ and $Lm\,H(j\omega_c) = 20\log_{10} 0.707 \approx -3$ dB.

The filtering effects of a system can be qualitatively determined based on the locations of the zeros and poles of the transfer function $H(s)$ of the system. For each point $j\omega$ along the imaginary axis representing a frequency, we define two vectors connecting $j\omega$ to the zero $s_z = 0$ of $H_R(s)$ and the common pole of both $H_C(s)$ and $H_R(s)$, respectively, as shown in Fig. 6.5:

$$\boldsymbol{u} = j\omega, \qquad \boldsymbol{v} = j\omega + 1/\tau. \tag{6.105}$$

Now the magnitudes of $H_C(j\omega)$ and $H_R(j\omega)$ can be expressed as

$$|H_C(j\omega)| = 1/\tau|\boldsymbol{v}|, \qquad |H_R(j\omega)| = |\boldsymbol{u}|/|\boldsymbol{v}|. \tag{6.106}$$

Based on the following two extreme cases:

- when $\omega = 0$, $|\boldsymbol{u}| = 0$ and $|\boldsymbol{v}| = 1/\tau$, we have $H_C(0) = 1$ and $H_R(0) = 0$;
- when $\omega = \infty$, $|\boldsymbol{v}| = |\boldsymbol{u}| = \infty$, we have $H_C(j\infty) = 0$ and $H_R(j\infty) = 1$;

we see that indeed $H_C(j\omega)$ and $H_R(j\omega)$ are LP and HP filters, respectively.

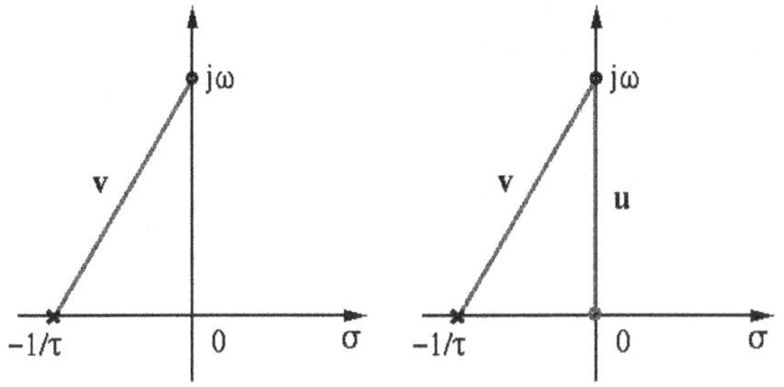

Figure 6.5 Qualitative determination of filtering behavior of first-order systems.

6.1.7 Second-order system

In the transfer function $H(s)$ of a second-order LTI system, the denominator polynomial $D(s)$ is a second-order polynomial of order $N = 2$, and $H(s)$ is conventionally written in the following *canonical form*:

$$H(s) = \frac{N(s)}{D(s)} = \frac{N(s)}{s^2 + 2\zeta\omega_n s + \omega_n^2} = \frac{N(s)}{(s - p_1)(s - p_2)}, \tag{6.107}$$

where ω_n and ζ in $D(s)$ are the two system parameters called the *natural frequency*, which is always positive, and the *damping coefficient* respectively. The two poles p_1 and p_2 of $H(s)$ are the two roots of the denominator quadratic function $D(s) = s^2 + 2\zeta\omega_n s + \omega_n^2$:

$$\begin{cases} p_1 = (-\zeta + \sqrt{\zeta^2 - 1})\omega_n = (-\zeta + j\sqrt{1 - \zeta^2})\omega_n \\ p_2 = (-\zeta - \sqrt{\zeta^2 - 1})\omega_n = (-\zeta - j\sqrt{1 - \zeta^2})\omega_n \end{cases}. \tag{6.108}$$

We also have the following relations:

$$p_1 p_2 = \omega_n^2, \qquad p_1 + p_2 = -2\zeta\omega_n, \qquad p_1 - p_2 = 2j\omega_n\sqrt{1 - \zeta^2} = 2j\omega_d, \tag{6.109}$$

where

$$\omega_d = \omega_n\sqrt{1 - \zeta^2} < \omega_n \tag{6.110}$$

is called the *damped natural frequency.*

If $|\zeta| \geq 1$, both poles are real, otherwise they form a complex conjugate pair located on a circle in the s-plane with radius ω_n:

$$p_{1,2} = (-\zeta \pm j\sqrt{1 - \zeta^2})\omega_n = -\omega_n e^{\mp j\phi}, \tag{6.111}$$

where, as shown in Fig. 6.6

$$\phi = \tan^{-1}\left(\frac{\sqrt{1 - \zeta^2}}{\zeta}\right), \tag{6.112}$$

and

$$\sin\phi = \sqrt{1 - \zeta^2}, \quad \cos\phi = \zeta, \quad \tan\phi = \sqrt{1 - \zeta^2}/\zeta. \tag{6.113}$$

Table 6.2. Root locus of the second-order system

ζ	p_1, p_2	comments on poles
$\zeta = -\infty$	$\infty, 0$	
$-\infty < \zeta < -1$	$(-\zeta \pm \sqrt{\zeta^2 - 1})\omega_n$	real, $0 < p_2 < p_1$
$\zeta = -1$	ω_n	real, repeated, $0 < p_1 = p_2 = \omega_n$
$-1 < \zeta < 0$	$(-\zeta \pm j\sqrt{1 - \zeta^2})\omega_n$	conjugate pair in quadrants I, VI
$\zeta = 0$	$\pm j\omega_n$	imaginary pair
$0 < \zeta < 1$	$(-\zeta \pm j\sqrt{1 - \zeta^2})\omega_n$	conjugate pair in quadrants II, III
$\zeta = 1$	$-\omega_n$	real, repeated $p_1 = p_2 = -\omega_n < 0$
$1 < \zeta < \infty$	$(-\zeta \pm \sqrt{\zeta^2 - 1})\omega_n$	real, $p_2 < p_1 < 0$
$\zeta = \infty$	$0, -\infty$	

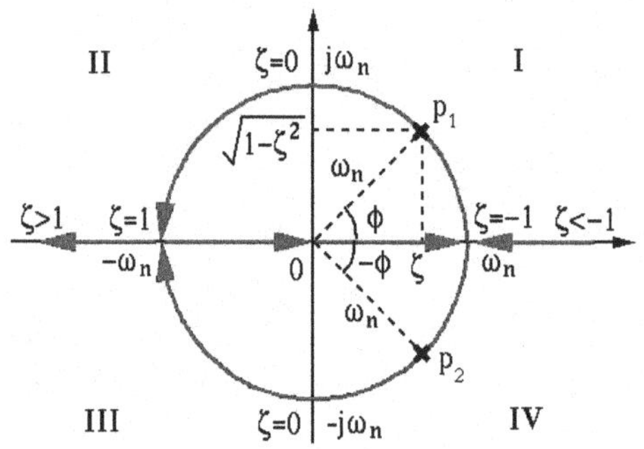

Figure 6.6 Root locus of the poles of a second-order system.

As also shown in Fig. 6.6, the positions of the poles on the circle are determined by the angle ϕ. When the value of ζ increases from $-\infty$ to ∞, the pole locations change along the *root locus* in the s-plane, as shown in Fig. 6.6, from which we see that each of the two poles follows its own root locus when ζ moves from $-\infty$ to ∞:

- Locus of p_1: $\infty \Longrightarrow \omega_n \Longrightarrow j\omega_n \Longrightarrow -\omega_n \Longrightarrow 0$.
- Locus of p_2: $0 \Longrightarrow \omega_n \Longrightarrow -j\omega_n \Longrightarrow -\omega_n \Longrightarrow -\infty$.

The root locus is further summarized in Table 6.2. We see that only when $\zeta > 0$ will the two poles $p_{1,2}$ be in the left half of the s-plane and the system is stable. When $\zeta = 0$, the poles are on the imaginary axis and the system is marginally stable, and when $\zeta < 0$ the poles are on the right half plane and the system is unstable.

The behavior of a second-order system in terms of its impulse response function $h(t)$ is determined by the two system parameters ω_n and ζ, which are directly associated with the locations of the poles of the transfer function $H(s)$. In the following, we show how $h(t)$ can be determined by inverse Laplace transform of $H(s)$, based on the given pole locations in the s-plane. Here, we assume $N(s) = 1$

so that the transfer function is

$$H(s) = \frac{1}{s^2 + 2\zeta\omega_n s + \omega_n^2} = \frac{1}{(s-p_1)(s-p_2)}. \tag{6.114}$$

If $\zeta = \pm 1$, we have

$$H(s) = \frac{1}{s^2 \pm 2\omega_n s + \omega_n^2} = \frac{1}{(s \pm \omega_n)^2} = \frac{1}{(s-p)^2}, \tag{6.115}$$

where $p = \pm\omega_n$ is the repeated pole of $H(s)$; then we have

$$h(t) = \mathcal{L}^{-1}[H(s)] = t\,e^{\pm\omega_n t}u(t). \tag{6.116}$$

If $|\zeta| \neq 1$, then $p_1 \neq p_2$, and $H(s)$ can be written as the following by partial fraction expansion:

$$H(s) = \frac{1}{p_1 - p_2}\left[\frac{1}{s-p_1} - \frac{1}{s-p_2}\right]. \tag{6.117}$$

The impulse response can be found by inverse Laplace transform:

$$h(t) = \mathcal{L}^{-1}[H(s)] = \frac{1}{p_1 - p_2}\left[e^{p_1 t} - e^{p_2 t}\right]u(t) = C\left[e^{p_1 t} - e^{p_2 t}\right]u(t), \tag{6.118}$$

where

$$C = \frac{1}{p_1 - p_2} = \frac{1}{2\omega_n\sqrt{\zeta^2 - 1}} = \frac{1}{2j\omega_n\sqrt{1-\zeta^2}}. \tag{6.119}$$

In the following we consider specifically each of the cases listed in Table 6.2 to see how $h(t)$ given in Eq. (6.118) varies when the value of ζ changes from $-\infty$ to ∞.

- $-\infty < \zeta < -1$, $0 < p_2 < p_1$. Both poles are on the real axis on the right side of the s-plane, and both terms $e^{p_1 t}$ and $e^{p_2 t}$ grow exponentially as $t \to \infty$; so does their difference, i.e., the system is unstable:

$$h(t) = C\left[e^{p_1 t} - e^{p_2 t}\right]u(t), \quad (p_1 > p_2). \tag{6.120}$$

- $\zeta = -1$. $p_1 = p_2 = -\zeta\omega_n = \omega_n$ are repeated poles still on the right side of the s-plane. We have

$$h(t) = t\,e^{\omega_n t}u(t), \tag{6.121}$$

  which grows without bound when $t \to \infty$, the system is unstable.
- $-1 < \zeta < 0$. The two poles form a conjugate pair in quadrants I and IV:

$$p_{1,2} = (-\zeta \pm j\sqrt{1-\zeta^2})\omega_n = -\omega_n\zeta \pm j\omega_d. \tag{6.122}$$

  Now we have

$$h(t) = \frac{1}{2j\omega_n\sqrt{1-\zeta^2}}e^{-\zeta\omega_n t}[e^{j\omega_d t} - e^{-j\omega_d t}]u(t) = \frac{e^{-\zeta\omega_n t}}{\omega_d}\sin(\omega_d t)u(t). \tag{6.123}$$

  As $\zeta < 0$ and therefore $-\zeta\omega_n t > 0$, $h(t)$ is an exponentially growing sinusoid and the system is still unstable.

- $\zeta = 0$. $p_{1,2} = \pm j\omega_{\rm n}$ are on the imaginary axis, and the system is marginally stable:

$$h(t) = \frac{1}{2j\omega_{\rm n}}[e^{j\omega_{\rm n}t} - e^{-j\omega_{\rm n}t}]u(t) = \frac{1}{\omega_{\rm n}}\sin(\omega_{\rm n}t)\,u(t). \tag{6.124}$$

In particular, when the frequency of the input $x(t) = e^{j\omega_{\rm n}t}$ is the same as the system's natural frequency $\omega_{\rm n}$, the output can be found to be (Eq. (6.69)):

$$y(t) = H(s)\big|_{s=j\omega_{\rm n}} e^{j\omega_{\rm n}t} = \frac{1}{s^2+\omega_{\rm n}^2}\bigg|_{s=j\omega_{\rm n}} e^{j\omega_{\rm n}t} = \frac{e^{j\omega_{\rm n}t}}{\omega_{\rm n}^2-\omega_{\rm n}^2} = \infty. \tag{6.125}$$

The response of the system becomes infinity; i.e., *resonance* occurs.

- $0 < \zeta < 1$. The two poles form a complex conjugate pair in quadrants II and III. Similar to the case when $-1 < \zeta < 0$, we have the same expression for $h(t)$:

$$h(t) = \frac{e^{-\zeta\omega_{\rm n}t}}{\omega_{\rm n}\sqrt{1-\zeta^2}}\sin(\omega_{\rm d}t)u(t) = \frac{e^{-\zeta\omega_{\rm n}t}}{\omega_{\rm d}}\sin(\omega_{\rm d}t)u(t). \tag{6.126}$$

As $\zeta > 0$, p_1 and p_2 are on the left half plane, and the impulse response $h(t)$ is an exponentially decaying sinusoid with frequency $\omega_{\rm d}$, the system is *underdamped* and stable.

- $\zeta = 1$. $p_1 = p_2 = -\zeta\omega_{\rm n} = -\omega_{\rm n} < 0$ are two repeated poles on the left side, the system is *critically damped* and stable:

$$h(t) = t\,e^{-\omega_{\rm n}t}u(t). \tag{6.127}$$

- $1 < \zeta < \infty$, $p_2 < p_1 < 0$, both poles are on the real axis on the left of the s-plane; the impulse response is the difference of two exponentially decaying functions:

$$h(t) = C(e^{p_1t} - e^{p_2t})u(t) = C(e^{-|p_1|t} - e^{-|p_2|t})u(t), \qquad |p_1| < |p_2|, \tag{6.128}$$

which decays to zero in time. The system is *overdamped* and stable.

All seven cases considered above are summarized in Table 6.3, as a continuation of Table 6.2. These different impulse response functions $h(t)$ corresponding to different values of ζ are plotted in Fig. 6.7. Consider, in particular, the following two cases:

- $\zeta \ll -1$, we have $0 < p_2 \ll p_1$ and

$$h(t) = C(e^{p_1t} - e^{p_2t})u(t) \approx Ce^{p_1t}; \tag{6.129}$$

i.e., p_1, which is farther away from the origin, dominates the system behavior.

- $\zeta \gg 1$, we have $p_2 \ll p_1 < 0$ and

$$h(t) = C(e^{-|p_1|t} - e^{-|p_2|t})u(t) \approx Ce^{-|p_1|t}; \tag{6.130}$$

i.e., p_1, which is closer to the origin, dominates the system behavior.

Table 6.3. Pole locations corresponding to different ζ values

ζ	$H(s)$	$h(t) = C(e^{p_1 t} - e^{p_2 t})$	Comments
$\zeta < -1$		$C(e^{p_1 t} - e^{p_2 t})u(t)$	exponential growth
$\zeta = -1$	$1/(s - \omega_\text{n})^2$	$t\, e^{\omega_\text{n} t} u(t)$	exponential growth
$-1 < \zeta < 0$		$\frac{e^{-\zeta\omega_\text{n} t}}{\omega_\text{d}} \sin(\omega_\text{d} t)u(t)$	exponentially growing sinusoid
$\zeta = 0$	$1/(s^2 + \omega_\text{n}^2)$	$\frac{1}{\omega_\text{n}} \sin(\omega_\text{n} t)\, u(t)$	sinusoid
$0 < \zeta < 1$		$\frac{e^{-\zeta\omega_\text{n} t}}{\omega_\text{d}} \sin(\omega_\text{d} t)u(t)$	exponentially decaying sinusoid
$\zeta = 1$	$1/(s + \omega_\text{n})^2$	$t\, e^{-\omega_\text{n} t}\, u(t)$	critically damped
$\zeta > 1$		$C(e^{-\lvert p_1 \rvert t} - e^{-\lvert p_2 \rvert t})u(t)$	exponential decay

In either case, when the non-dominant pole can be neglected, the behavior of the second-order system can be approximated by a first-order system with a single pole $p = -1/\tau$.

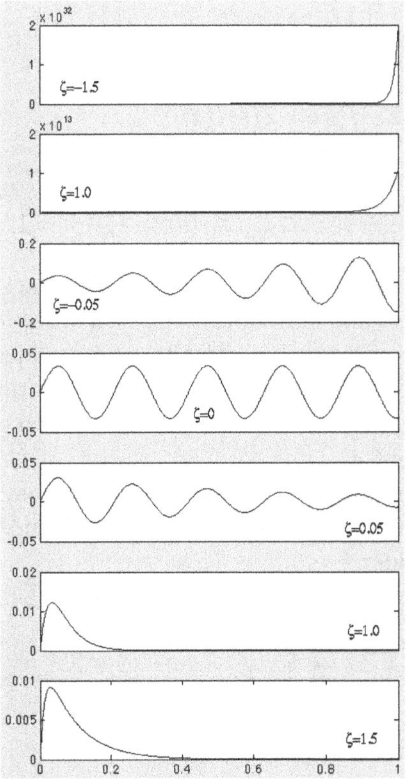

Figure 6.7 Impulse response of second order system for different ζ.

Example 6.3: As a typical example of the second-order system, consider a circuit composed of a resistor R, a capacitor C, and an inductor L connected in series as shown in Fig. 6.8. An input voltage $v_{\text{in}}(t)$ is applied to the series combination of the three elements and the output is $v_L(t)$, $v_R(t)$, or $v_C(t)$, the voltage across one of the three elements. The system is described by the following differential equation in the time domain:

$$v_{\text{in}}(t) = v_L(t) + v_R(t) + v_C(t) = L\frac{d}{dt}\,i(t) + Ri(t) + \frac{1}{C}\int_{-\infty}^{t} i(\tau)\,d\tau. \quad (6.131)$$

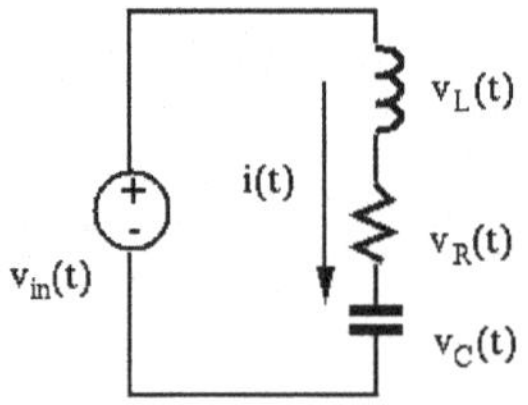

Figure 6.8 A second order RCL series circuit.

Taking the Laplace transform on both sides, we get an algebraic equation in the s-domain:

$$X(s) = V_L(s) + V_R(s) + V_C(s) = \left[sL + R + \frac{1}{sC}\right] I(s)$$
$$= [Z_L + Z_R + Z_C]I(s) = Z(s)I(s),$$

where

$$Z_L(s) = \frac{V_L(s)}{I(s)} = sL, \qquad Z_R = \frac{V_R(s)}{I(s)} = R, \qquad Z_C(s) = \frac{V_C(s)}{I(s)} = 1/sC \quad (6.132)$$

are the *impedances* of the circuit elements L, R, and C, respectively, defined as the ratio between the voltage across and current through each of the components in the s-domain, similar to the resistance $R = v(t)/i(t)$ of a resistor R defined by Ohm's law as the ratio between the voltage and current in the time domain. The relations between the voltage and current associated with each of the three elements are summarized in Table 6.4.

The transfer function $H(s)$, the ratio between the voltage across one of the three elements (V_L, V_R, or V_C) and input voltages $V(s)$, can be found by treating the series circuit as a voltage divider:

- Output is voltage across the capacitor $v_C(t)$

$$H_C(s) = \frac{V_C(s)}{V(s)} = \frac{Z_C(s)}{Z(s)} = \frac{1/sC}{Ls + R + 1/sC} = \frac{\omega_{\text{n}}^2}{s^2 + 2\zeta\omega_{\text{n}} s + \omega_{\text{n}}^2}. \quad (6.133)$$

- Output is voltage across the resistor $v_R(t)$

$$H_R(s) = \frac{V_R(s)}{V(s)} = \frac{Z_R(s)}{Z(s)} = \frac{R}{Ls + R + 1/sC} = \frac{2\zeta\omega_{\text{n}} s}{s^2 + 2\zeta\omega_{\text{n}} s + \omega_{\text{n}}^2}. \quad (6.134)$$

Table 6.4. Impedances of capacitor, resistor, and inductor

	Capacitor C	Resistor R	Inductor L
Time domain	$v_C(t) = \int i(t)\,dt/C$	$v_R(t) = Ri(t)$	$v_L(t) = Li'(t)$
s-domain	$V_C(s) = I(s)/Cs$	$V_R(s) = RI(s)$	$V_L(s) = I(s)sL$
Impedance $Z(s)$	$1/sC$	R	sL

- Output is voltage across the inductor $v_L(t)$

$$H_L(s) = \frac{V_L(s)}{V(s)} = \frac{Z_L(s)}{Z(s)} = \frac{sL}{Ls + R + 1/sC} = \frac{s^2}{s^2 + 2\zeta\omega_{\rm n}s + \omega_{\rm n}^2}. \tag{6.135}$$

Here, we have converted the denominator $D(s)$ into the canonical second-order form:

$$D(s) = s^2 + (R/L)s + (1/LC) = s^2 + 2\zeta\omega_{\rm n}s + \omega_{\rm n}^2 = (s - p_1)(s - p_2), \tag{6.136}$$

where the damping coefficient ζ and natural frequency $\omega_{\rm n}$ are defined as

$$\zeta = \frac{R}{2}\sqrt{\frac{C}{L}} > 0, \qquad \omega_{\rm n} = \frac{1}{\sqrt{LC}} > 0. \tag{6.137}$$

If we assume $0 < \zeta < 1$, then the two poles are:

$$p_{1,2} = (-\zeta \pm j\sqrt{1-\zeta^2})\omega_{\rm n} = -\omega_{\rm n}e^{\mp j\phi}, \tag{6.138}$$

where $\phi = \tan^{-1}(\sqrt{1-\zeta^2}/\zeta)$ as defined in Eq. (6.112).

The total impedance $Z(s)$ of the three elements in series is the sum of the individual impedances:

$$Z(s) = \frac{V(s)}{I(s)} = sL + R + \frac{1}{sC} = Z_L + Z_R + Z_C. \tag{6.139}$$

In the following, we further consider some important characteristics of the second-order systems in both the time and frequency domains.

1. **Impulse response function**
   When voltage $v_C(t)$ across C is treated as the output, the impulse response $h_C(t)$ can be found by inverse transform of Eq. (6.133):

$$h_C(t) = \mathcal{L}^{-1}[H_C(s)] = \mathcal{L}^{-1}\left[\frac{\omega_{\rm n}^2}{(s-p_1)(s-p_2)}\right] = \frac{\omega_{\rm n}e^{-\zeta\omega_{\rm n}t}}{\sqrt{1-\zeta^2}}\sin(\omega_{\rm d}t)u(t). \tag{6.140}$$

   This is based on the assumption $0 < \zeta < 1$ (Eq. (6.126) multiplied by $\omega_{\rm n}^2$). Alternatively, when the voltage across R or L is treated as the output, the corresponding impulse response $h_R(t)$ or $h_L(t)$ can also be found by inverse transform of Eqs. (6.134) or (6.135). The derivations of these impulse responses $h_L(t)$ and $h_R(t)$ are left as homework problems, but their waveforms are plotted together with $h_C(t)$ obtained above in Fig. 6.9 ($\zeta = 0.05$), from which we see that the three responses do add up to the step input $\delta(t) = h_C(t) + h_L(t) + h_R(t)$; i.e., Kirchhoff's voltage law holds.

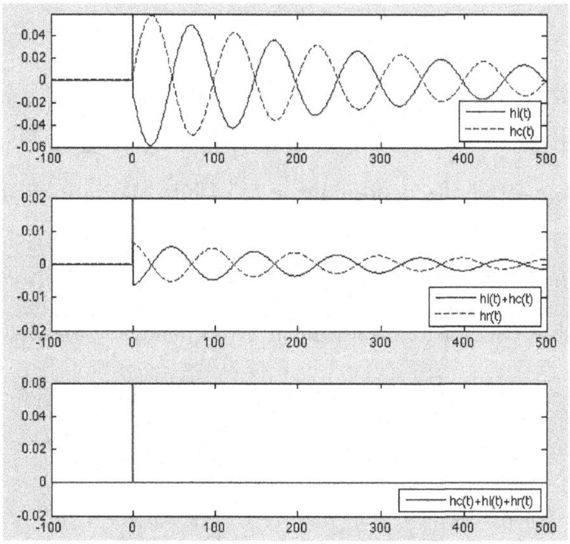

Figure 6.9 Impulse responses by R, L, and C of an RCL system. Top: impulse responses $h_L(t)$ (solid curve) and $h_C(t)$ (dashed curve); middle: their sum $h_L(t) + h_C(t)$ (solid curve) and impulse response $h_R(t)$ (dashed curve); bottom: the sum of all three: $\delta(t) = h_L(t) + h_R(t) + h_C(t)$.

2. **Step response**

When $V_C(s)$ across C is treated as the output, in the s-domain the step response to a step input $U(s) = \mathcal{L}[u(t)] = 1/s$ is

$$Y_C(s) = H_C(s)U(s) = \frac{\omega_\text{n}^2}{s^2 + 2\zeta\omega_\text{n} s + \omega_\text{n}^2}\frac{1}{s}$$
$$= \frac{1}{s} + \frac{p_2}{p_1 - p_2}\frac{1}{s - p_1} - \frac{p_1}{p_1 - p_2}\frac{1}{s - p_2}, \tag{6.141}$$

and the step response in the time domain can be obtained by inverse transform:

$$y_C(t) = \mathcal{L}^{-1}[Y(s)] = \left[1 + \frac{1}{p_1 - p_2}\left(p_2 e^{p_1 t} - p_1 e^{p_2 t}\right)\right] u(t)$$
$$= \left[1 - \frac{\omega_\text{n}}{p_1 - p_2}\left(e^{j\phi} e^{(-\zeta\omega_\text{n} + j\omega_\text{d})t} - e^{-j\phi} e^{(-\zeta\omega_\text{n} - j\omega_\text{d})t}\right)\right] u(t)$$
$$= \left[1 - \frac{e^{-\zeta\omega_\text{n} t}}{\sqrt{1 - \zeta^2}} \sin(\omega_\text{d} t + \phi)\right] u(t). \tag{6.142}$$

This step response function is plotted in Fig. 6.10 for different ζ values.

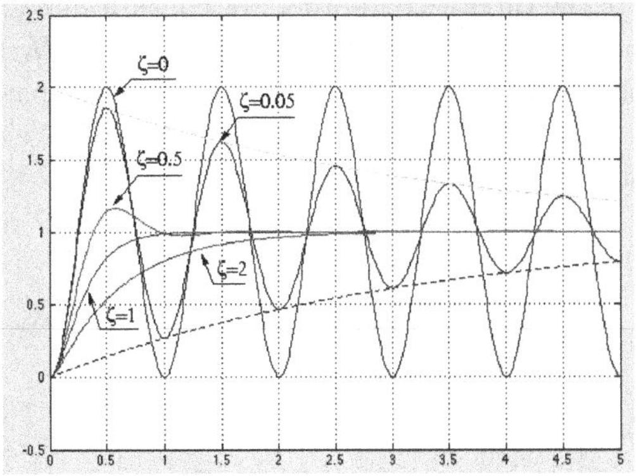

Figure 6.10 Step response of second-order system for different ζ. Step responses corresponding to five different values of ζ: 0, 0.05, 0.5, 1, and 2. The envelope of the step response for $\zeta = 0.05$ is also plotted to show the exponential decay of the sinusoid.

Alternatively, the voltage across R or L can also be treated as the output of the second-order system, and we can find the system's step response in both the s- and time domains for these cases. The derivations of these step responses $y_L(t)$ and $y_R(t)$ are left as homework problems, but their waveforms are plotted together with $y_C(t)$ in Fig. 6.11 ($\zeta = 0.05$), from which we see that the three responses do add up to the step input $u(t) = y_C(t) + y_L(t) + y_R(t)$; i.e., Kirchhoff's voltage law holds.

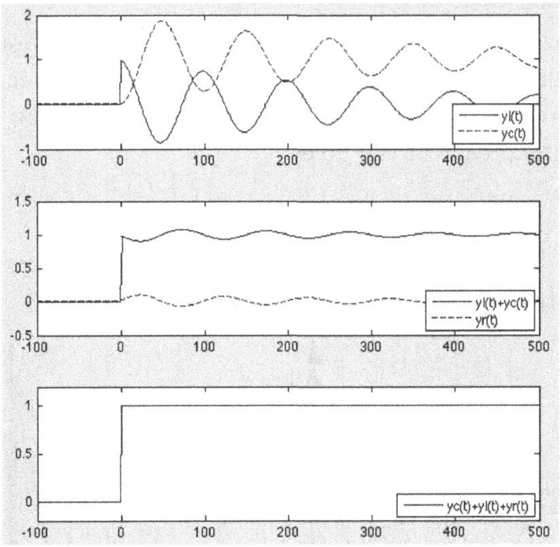

Figure 6.11 Step responses by R, L as well as C of an RCL system. Top: step responses $y_L(t)$ (solid curve) and $y_C(t)$ (dashed curve); middle: their sum $y_L(t) + y_C(t)$ (solid curve) and step responses $y_R(t)$ (dashed curve); bottom: the sum of all three: $y_C(t) + y_L(t) + y_R(t) = u(t)$.

3. **Second-order systems as filters**
The filtering effects of the three second-order systems are characterized by the magnitudes and phases of their frequency response functions $H_C(j\omega)$, $H_R(j\omega)$, and $H_L(j\omega)$, as plotted in Fig. 6.12, based on the assumed parameters $\omega_n = 2\pi 1000$ and $\zeta = 0.1$ (top) and $\zeta = 1/\sqrt{2} = 0.707$ (bottom). We see that when $\zeta = 0.1 < 0.707$, both $H_C(j\omega)$ and $H_L(j\omega)$ behave like a BP filter similar to $H_R(j\omega)$ (top row), but when $\zeta \geq 0.707$, they behave as LP and HP filters without any peak (bottom row), respectively.

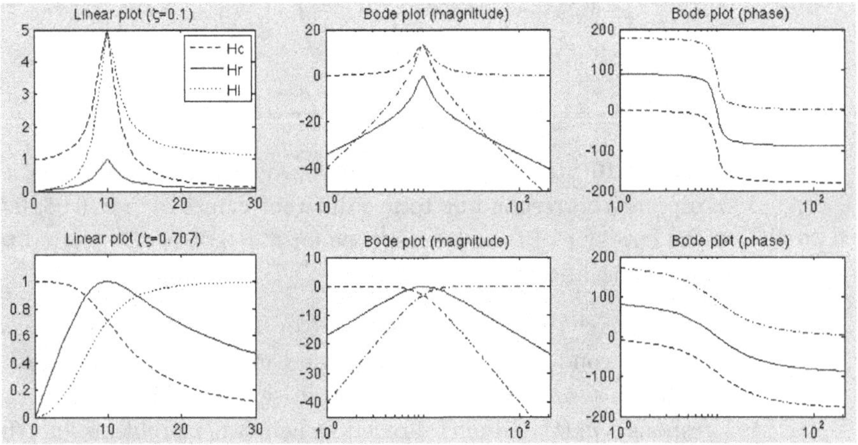

Figure 6.12 Frequency response functions $H_C(j\omega)$, $H_R(j\omega)$, and $H_L(\omega)$. Top: $\zeta = 0.1$; bottom: $\zeta = 0.707$; left: linear magnitude plots; middle and right: bode log-magnitude and phase plots.

The filtering effects of the three systems can be qualitatively estimated based on the location of the zeros and poles of their corresponding transfer functions. We first define three vectors connecting an arbitrary point $j\omega$ on the imaginary axis to the origin and each of the poles:

$$\boldsymbol{u} = j\omega, \qquad \boldsymbol{v}_1 = j\omega - p_1, \qquad \boldsymbol{v}_2 = j\omega - p_2, \tag{6.143}$$

and then observe how each of the three frequency response functions changes when ω increase from 0 toward ∞, as illustrated in Fig. 6.13.

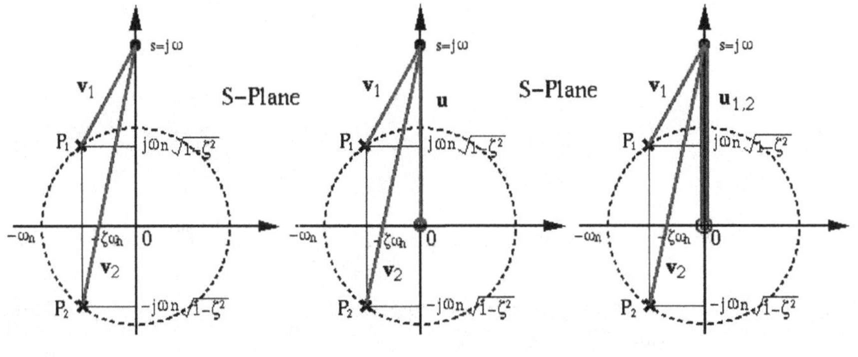

Figure 6.13 Graphic determination of filtering behavior of second-order systems $H_C(s)$ (left), $H_R(s)$ (middle), and $H_L(s)$ (right).

Consider each of the three transfer functions:

- $H_C(s) = \omega_n^2/D(s)$ with two poles but no zero:

$$|H_C(j\omega)| = \frac{\omega_n^2}{|j\omega - p_1||j\omega - p_2|} = \frac{\omega_n^2}{|\boldsymbol{v}_1||\boldsymbol{v}_2|}, \tag{6.144}$$

  which is some constant when $\omega = 0$ but approaches 0 when $\omega \to \infty$ causing both $|\boldsymbol{v}_1| \to \infty$ and $|\boldsymbol{v}_2| \to \infty$; i.e., the system is a LP filter.

- $H_R(s) = 2\zeta\omega_n s/D(s)$ with two poles and one zero:

$$|H_R(j\omega)| = \frac{2\zeta\omega_n\,|j\omega|}{|j\omega - p_1||j\omega - p_2|} = \frac{2\zeta\omega_n\,|\boldsymbol{u}|}{|\boldsymbol{v}_1||\boldsymbol{v}_2|}, \tag{6.145}$$

  which is zero when $\omega = 0$ or $\omega \to \infty$, but greater than 0 when $0 < \omega < \infty$; i.e., the system is a BP filter.

- $H_L(s) = s^2/D(s)$ with two poles and two repeated zeros (corresponding to two vectors $\boldsymbol{u}_1 = \boldsymbol{u}_2 = \boldsymbol{u}$):

$$|H_L(j\omega)| = \frac{|j\omega|^2}{|j\omega - p_1||j\omega - p_2|} = \frac{|\boldsymbol{u}|^2}{|\boldsymbol{v}_1||\boldsymbol{v}_2|}, \tag{6.146}$$

  which is zero when $\omega = 0$, but approaches constant 1 when $\omega \to \infty$; i.e., the system is an HP filter.

4. **Peak frequency of second-order filters**

The *peak frequency* ω_p of a filter $H(j\omega)$ is the frequency at which $|H(j\omega_p)| = |H_{max}|$ is maximized. To simplify the algebra, we first define a variable $u = (\omega/\omega_n)^2$ (frequency ω normalized by ω_n) so that the squared magnitudes of the frequency response functions can be expressed as

$$\begin{aligned}
|H_C(j\omega)|^2 &= \left|\frac{\omega_n^2}{(j\omega)^2 + 2\zeta\omega_n j\omega + \omega_n^2}\right|^2 = \frac{1}{(u-1)^2 + 4\zeta^2 u},\\
|H_R(j\omega)|^2 &= \left|\frac{2\zeta\omega_n j\omega}{(j\omega)^2 + 2\zeta\omega_n j\omega + \omega_n^2}\right|^2 = \frac{4\zeta^2 u}{(u-1)^2 + 4\zeta^2 u},\\
|H_L(j\omega)|^2 &= \left|\frac{(j\omega)^2}{(j\omega)^2 + 2\zeta\omega_n j\omega + \omega_n^2}\right|^2 = \frac{u^2}{(u-1)^2 + 4\zeta^2 u}.
\end{aligned} \tag{6.147}$$

To find the value u_p at which each of these functions is maximized, we take the derivative of each of the functions with respect to u, set the results to zero, and then solve the resulting equations to get

$$\begin{cases} u_{p_C} = 1 - 2\zeta^2 \\ u_{p_R} = 1 \\ u_{p_L} = 1/(1-2\zeta^2) \end{cases}; \quad \text{i.e.} \quad \begin{cases} \omega_{p_C} = \omega_n\sqrt{u_{p_C}} = \omega_n\sqrt{1-2\zeta^2} \\ \omega_{p_R} = \omega_n\sqrt{u_{p_R}} = \omega_n \\ \omega_{p_L} = \omega_n\sqrt{u_{p_L}} = \omega_n/\sqrt{1-2\zeta^2} \end{cases}. \tag{6.148}$$

We see that the three peak frequencies are different:

$$\omega_{\mathrm{p}_C} \le \omega_{\mathrm{p}_R} \le \omega_{\mathrm{p}_L}. \tag{6.149}$$

Substituting these peak frequencies into Eq. (6.147), we get the peak values of the three filters:

$$|H_{\max_R}| = |H(j\omega_{\mathrm{p}_R})| = 1,$$
$$|H_{\max_C}| = |H(j\omega_{\mathrm{p}_C})| = |H_{\max_L}| = |H(j\omega_{\mathrm{p}_L})| = \frac{1}{2\zeta\sqrt{1-\zeta^2}}. \tag{6.150}$$

Also note that for the peak frequencies ω_{p_c} and ω_{p_L} given in Eq. (6.148) to be real, the following has to be satisfied:

$$1 - 2\zeta^2 > 0; \quad \text{i.e.,} \quad \zeta < 1/\sqrt{2} = 0.707, \tag{6.151}$$

Otherwise these peak frequencies do not exist, and $H_C(j\omega)$ becomes an LP filter that reaches its maximum of 1 at $\omega = 0$, and $|H_L(j\omega)|$ becomes a HP filter that reaches its maximum of 1 at $\omega = \infty$, as shown in Fig. 6.12.

5. **Bandwidth of second-order BP filter**

The bandwidth $\Delta\omega = \omega_1 - \omega_2$ of a BP filter $H(j\omega)$ is defined as the interval between two *cutoff frequencies* ω_1 and ω_2 at which the output power is half of that at the peak frequency ω_{p}:

$$|H(j\omega_1)|^2 = |H(j\omega_2)|^2 = \frac{1}{2}|H(j\omega_{\mathrm{p}})|^2 = \frac{1}{2}|H_{\max}|^2. \tag{6.152}$$

Specifically, for the BP filter $H_R(j\omega)$, $H_{\max}(0) = 1$, and at the two cutoff frequencies we have

$$\frac{|H_R(j\omega)|^2}{|H_{\max_R}|^2} = |H_R(j\omega)|^2 = \frac{4\zeta^2 u}{(u-1)^2 + 4\zeta^2 u} = \frac{1}{2}. \tag{6.153}$$

Solving this quadratic equation we get two solutions:

$$u_{1,2} = 1 + 2\zeta^2 \pm 2\zeta\sqrt{1+\zeta^2}, \tag{6.154}$$

and the corresponding cutoff frequencies are

$$\omega_{1,2} = \omega_{\mathrm{n}}\sqrt{1 + 2\zeta^2 \pm 2\zeta\sqrt{1+\zeta^2}}, \tag{6.155}$$

and the bandwidth is

$$\Delta\omega_R = \omega_1 - \omega_2 = 2\zeta\omega_{\mathrm{n}}. \tag{6.156}$$

Based on this result, the denominator of the second-order transfer function can also be written as

$$D(s) = s^2 + \Delta\omega_{\mathrm{n}} s + \omega_{\mathrm{n}}^2. \tag{6.157}$$

6.1.8 The unilateral Laplace transform

When applied to solving LCCDEs, the bilateral Laplace transform considered so far can only find the particular solutions, not the homogeneous solution due to non-zero initial conditions, which are not taken into consideration. This problem can be overcome by the *unilateral* or one-sided Laplace transform, which can solve a given LCCDE to find the homogeneous and the particular solutions.

The unilateral Laplace transform of a given signal $x(t)$ is defined as

$$\mathcal{UL}[x(t)] = \int_{-\infty}^{\infty} x(t)u(t)e^{-st}\,dt = \int_{0}^{\infty} x(t)e^{-st}\,dt. \tag{6.158}$$

When the unilateral Laplace transform is applied to a signal $x(t)$, it is always assumed that the signal starts at time $t = 0$; i.e., $x(t) = 0$ for all $t < 0$. When it is applied to the impulse response function $h(t)$ of an LTI system to find the transfer function $H(s) = \mathcal{UL}[h(t)]$, it is always assumed that its impulse response $h(t) = 0$ for $t < 0$; i.e., the system is causal. In either case, all poles have to be on the left half s-plane; i.e., the ROC is always in the right half s-plane. Obviously, if $x(t) = x(t)u(t)$, its unilateral and bilateral Laplace transforms are identical. Otherwise the two Laplace transforms are different.

The unilateral Laplace transform shares all of the properties of the bilateral Laplace transform, although some may be expressed in different forms. Here, we will not repeat all the properties, except those that are most relevant to solving the LCCDE of an LTI system.

- **Time derivative**

$$\mathcal{UL}\left[\frac{d}{dt}x(t)\right] = sX(s) - x(0). \tag{6.159}$$

Proof:

$$\mathcal{UL}\left[\frac{d}{dt}x(t)\right] = \int_{0}^{\infty}\left[\frac{d}{dt}x(t)\right]e^{-st}\,dt = \int_{0}^{\infty} e^{-st}\,d[x(t)]$$
$$= x(t)e^{-st}\Big|_0^{\infty} - \int_0^{\infty} x(t)\,d(e^{-st}) = -x(0) + s\int_0^{\infty} x(t)e^{-st}\,dt = sX(s) - x(0). \tag{6.160}$$

We can further get the transform of the second derivative of $x(t)$:

$$\mathcal{UL}\left[\frac{d^2}{dt^2}x(t)\right] = s\,\mathcal{UL}\left[\frac{d}{dt}x(t)\right] - \dot{x}(0) = s^2X(s) - sx(0) - \dot{x}(0), \tag{6.161}$$

and in general we have

$$\mathcal{UL}[x^{(n)}(t)] = s^nX(s) - \sum_{k=0}^{n-1} s^k x^{(n-1-k)}(0). \tag{6.162}$$

- **The initial-value theorem**
  If a right-sided signal $x(t)$ contains no impulse or higher order singularities at $t = 0$, its initial value $x(0^+)$ ($t \to 0$ from $t > 0$) can be found to be

$$x(0^+) = \lim_{t\to 0} x(t) = \lim_{s\to\infty} sX(s). \tag{6.163}$$

  Proof: At the limit $s \to 0$, Eq. (6.160) becomes

$$\lim_{s\to 0} \int_0^\infty \frac{d}{dt} x(t) e^{-st}\, dt = \int_0^\infty dx(t) = x(\infty) - x(0) = \lim_{s\to 0} [sX(s) - x(0)]; \tag{6.164}$$

  i.e.,

$$\lim_{s\to 0} sX(s) = x(\infty). \tag{6.165}$$

- **The final-value theorem**
  If a right-sided signal $x(t)$ approaches a finite value $x(\infty)$ as $t \to \infty$, it can be found to be

$$x(\infty) = \lim_{t\to\infty} x(t) = \lim_{s\to 0} sX(s). \tag{6.166}$$

  Proof: At the limit $s \to \infty$, Eq. (6.160) becomes

$$\lim_{s\to\infty} \int_0^\infty \frac{d}{dt} x(t) e^{-st}\, dt = 0 = \lim_{s\to\infty} [sX(s) - x(0)]; \tag{6.167}$$

  i.e.,

$$\lim_{s\to\infty} sX(s) = x(0). \tag{6.168}$$

Owing to these properties, the unilateral Laplace transform is a useful tool for solving LCCDEs with non-zero initial conditions.

Example 6.4: We consider Example 5.2 one more time, where the LCCDE of the first-order system is

$$\tau \dot{y}(t) + y = x \tag{6.169}$$

and $y(0) = y_0$ is the initial condition. Taking the unilateral Laplace transform on both sides, we get

$$\tau[sY(s) - y_0] + Y(s) = X(s); \quad \text{i.e.} \quad Y(s) = \frac{X(s)}{s\tau + 1} + \frac{\tau y_0}{s\tau + 1}. \tag{6.170}$$

Consider the following two inputs:

- When $x(t) = \delta(t)$, $X(s) = 1$ and the output is

$$Y(s) = \frac{1}{s\tau + 1} + \frac{\tau y_0}{s\tau + 1}. \tag{6.171}$$

Taking the inverse transform we get

$$y(t) = \left(\frac{1}{\tau} + y_0\right) e^{-t/\tau} u(t) = \frac{1}{\tau} e^{-t/\tau} u(t) + y_0 e^{-t/\tau} u(t). \qquad (6.172)$$

This first term is the particular solution representing the discharge of the capacitor charge $Q = \frac{1}{RC} \int \delta(t)\, dt = u(t)/\tau$ instantly charged by the input voltage $x(t) = \delta(t)$, and the second term is the homogeneous solution representing the discharge of the initial voltage y_0. Comparing the result above with Eq. (6.97), we see that that the bilateral Laplace transform fails to find the homogeneous solution.

- When $x(t) = u(t)$, $X(s) = 1/s$ and the output is

$$Y(s) = \frac{1}{s} \frac{1}{s\tau + 1} + \frac{\tau y_0}{s\tau + 1} = \frac{1}{s} - \frac{\tau}{s\tau + 1} + \frac{\tau y_0}{s\tau + 1}. \qquad (6.173)$$

Taking the inverse transform we get

$$y(t) = [1 + (y_0 - 1)e^{-t/\tau}]u(t) = (1 - e^{-t/\tau})u(t) + y_0 e^{-t/\tau} u(t). \qquad (6.174)$$

The first term is the particular solution representing the charge of the capacitor C by the input $x(t) = u(t)$, while the second is the homogeneous solution representing the discharge of the initial voltage $y(0) = y_0$. Comparing the result above with Eq. (6.100), we see that that the bilateral Laplace transform fails to find the homogeneous solution.

All these results are consistent with Example 5.2. Note that the bilateral Laplace transform fails to find the homogeneous solutions.

Example 6.5: Solve the following second-order LCCDE:

$$\frac{d^2}{dt^2} y(t) + 3\frac{d}{dt} y(t) + 2y(t) = x(t) = \alpha u(t) \qquad (6.175)$$

with initial conditions

$$y(0) = \beta, \quad \dot{y}(0) = \gamma. \qquad (6.176)$$

Applying the unilateral Laplace transform to the LCCDE we get

$$\begin{aligned} &s^2 Y(s) - \beta s - \gamma + 3sY(s) - 3\beta + 2Y(s) \\ &= (s^2 + 3s + 2)Y(s) - \beta s - \gamma - 3\beta = \alpha/s. \end{aligned} \qquad (6.177)$$

Solving for $Y(s)$ we get

$$Y(s) = \frac{\alpha}{s(s+1)(s+2)} + \frac{\beta(s+3)}{(s+1)(s+2)} + \frac{\gamma}{(s+1)(s+2)} = Y_{\mathrm{p}}(s) + Y_{\mathrm{h}}(s). \qquad (6.178)$$

This is the general solution of the LCCDE which is composed of two parts:

- The homogeneous (zero-input) solution due to the non-zero initial conditions $\beta \neq 0$ and $\gamma \neq 0$ with zero input $\alpha = 0$:

$$Y_h(s) = \frac{\beta(s+3)}{(s+1)(s+2)} + \frac{\gamma}{(s+1)(s+2)}. \tag{6.179}$$

- The particular (zero-state) solution due to the non-zero input $\alpha \neq 0$ but with zero initial conditions $\beta = \gamma = 0$:

$$Y_p(s) = \frac{\alpha}{s(s+1)(s+2)}. \tag{6.180}$$

Given specific values $\alpha = 2$, $\beta = 3$, and $\gamma = -5$ and using the method of partial fraction expansion, we can write $Y(s)$ as

$$\begin{aligned} Y(s) = Y_p(s) + Y_h(s) &= \left(\frac{2}{s(s+1)(s+2)} + \frac{3(s+3)}{(s+1)(s+2)} \right) - \frac{5}{(s+1)(s+2)} \\ &= \left(\frac{1}{s} - \frac{2}{s+1} + \frac{1}{s+2} \right) + \left(\frac{1}{s+1} + \frac{2}{s+2} \right). \end{aligned}$$

Taking the inverse transform on both sides we get the solution in the time domain solution:

$$\begin{aligned} y_p(t) &= \mathcal{U}L^{-1}[Y_p(s)] = \mathcal{U}L^{-1}\left(\frac{1}{s} - \frac{2}{s+1} + \frac{1}{s+2} \right) = [1 - 2e^{-t} + e^{-2t}]u(t), \\ y_h(t) &= \mathcal{U}L^{-1}[Y_h(s)] = \mathcal{U}L^{-1}\left(\frac{1}{s+1} + \frac{2}{s+2} \right) = [e^{-t} + 2e^{-2t}]u(t), \end{aligned}$$

and

$$y(t) = y_h(t) + y_p(t) = [1 - e^{-t} + 3e^{-2t}]u(t). \tag{6.181}$$

If the bilateral Laplace transform is applied to the same LCCDE, we get

$$s^2Y(s) + 3sY(s) + 2Y(s) = (s^2 + 3s + 2)Y(s) = \frac{\alpha}{s} = \frac{2}{s}. \tag{6.182}$$

Solving this for $Y(s)$ and taking the inverse transform, we get

$$Y(s) = \frac{2}{s(s+1)(s+2)}, \quad y(t) = (e^{-t} + 2e^{-2t})\, u(t). \tag{6.183}$$

This is the particular solution above with zero initial conditions. From this we see that the bilateral Laplace transform can only solve an LCCDE system of zero initial conditions. When the initial conditions of the system are not all zero, the unilateral Laplace transform has to be used.

6.2 The z-transform

Similar to the Laplace transform, the z-transform is also a powerful tool widely used in many fields, especially in digital signal processing and discrete system analysis/design. Much of the discussion below is in parallel with that for the Laplace transform, with the only essential difference that all signals and systems considered here are discrete in time.

6.2.1 From Fourier transform to z-transform

The z-transform of a discrete signal $x[n]$ can be considered as the generalization of the DTFT of the signal in Eq. (4.13):

$$\mathcal{F}[x[n]] = \sum_{n=-\infty}^{\infty} x[n]e^{-jn\omega} = X(e^{j\omega}). \tag{6.184}$$

Here, we adopt the notation $X(e^{j\omega})$ for the DTFT spectrum, instead of $X(f)$ or $X(\omega)$ used previously, for a reason which will become clear later. The transform above is based on the underlying assumption that the signal $x[n]$ is square-summable so that the summation converges and $X(e^{j\omega})$ exists. However, this assumption is not true for signals such as $x[n] = n$, $x[n] = n^2$, and $x[n] = e^n$, all of which are not square-summable as they grow without a bound when $|n| \to \infty$. In such cases, we could still consider the Fourier transform of a modified version of the signal $x'[n] = x[n]e^{-\sigma n}$, where $e^{-\sigma n}$ is an exponential factor with a real parameter σ, which can force the given signal $x[n]$ to decay exponentially for some properly chosen value of σ (either positive or negative). For example, $x[n] = e^{an}u[n]$ $(a > 0)$ does not converge when $n \to \infty$; therefore, its Fourier spectrum does not exist. However, if we choose $\sigma > a$, the modified version $x'[n] = x[n]e^{-\sigma} = e^{-(\sigma-a)n}u[n]$ will converge as $m \to \infty$.

In general, the Fourier transform of the modified signal is

$$\mathcal{F}[x'[n]] = \mathcal{F}[x[n]e^{-\sigma n}] = \sum_{n=-\infty}^{\infty} x[n]e^{-n(\sigma+j\omega)} = \sum_{n=-\infty}^{\infty} x[n]z^{-n}, \tag{6.185}$$

where we have defined a complex variable

$$z = e^{s} = e^{\sigma+j\omega} = e^{\sigma}\, e^{j\omega} = |z|\angle z, \tag{6.186}$$

which can be represented most conveniently in polar form in terms of its magnitude $|z| = e^{\sigma}$ and angle $\angle z = \omega$. If the summation above converges, it results in a complex function $X(z)$, which is called the *bilateral z-transform* of $x[n]$, formally defined as

$$X(z) = \mathcal{Z}[x[n]] = \mathcal{F}[x[n]e^{-\sigma n}] = \sum_{n=-\infty}^{\infty} x[n]z^{-n}. \tag{6.187}$$

Here, $X(z)$ is a function defined over a 2-D complex z-plane typically represented in polar coordinates of $|z|$ and $\angle z$. Similar to the Laplace transform, here the z-transform $X(z)$ exists only inside the corresponding ROC in the z-plane, composed of all z values that guarantee the convergence of the summation in Eq. 6.187. Owing to the introduction of the exponential decay factor $e^{-\sigma n}$, we can properly choose the parameter σ so that the z-transform can be applied to a broader class of signals than the Fourier transform.

If the unit circle $|z| = e^{\sigma} = 1$ (when $\sigma = 0$ and $s = j\omega$) is inside the ROC, we can evaluate the 2-D function $X(z)$ along the unit circle with respect to $z = e^{j\omega}$ from $\omega = -\pi$ to $\omega = \pi$ to obtain the Fourier transform of $x[n]$. We see that the 1-D Fourier spectrum $X(e^{j\omega})$ of the discrete signal $x[n]$ is simply the cross-section of the 2D function $X(z) = X(|z|e^{j\omega})$ along the unit circle $z = e^{j\omega}$, which is obviously periodic with period 2π. In other words, the DTFT in Eq. (4.13) is just a special case of the z-transform when $\sigma = 0$ and $z = e^{j\omega}$:

$$\mathcal{F}[x[n]] = \sum_{n=-\infty}^{\infty} x[n]e^{-j\omega n} = \sum_{n=-\infty}^{\infty} x[n]z^{-n}\Big|_{z=e^{j\omega}} = X(z)\Big|_{z=e^{j\omega}} = X(e^{j\omega}). \tag{6.188}$$

This is why the DTFT spectrum can also be denoted by $X(e^{j\omega})$.

Given the z-transform $X(z) = \mathcal{Z}[x[n]]$, the time signal $x[n]$ can be found by the inverse z-transform, which can be derived from the corresponding Fourier transform of discrete signals:

$$\mathcal{Z}[x[n]] = X(z) = X(e^{\sigma+j\omega}) = \mathcal{F}[x[n]e^{-\sigma n}]. \tag{6.189}$$

Taking the inverse Fourier transform of the above, we get

$$x[n]e^{-m\sigma} = \mathcal{F}^{-1}[X(e^{\sigma+j\omega})] = \frac{1}{2\pi}\int_0^{2\pi} X(e^{\sigma+j\omega})e^{jn\omega}\, d\omega. \tag{6.190}$$

Multiplying both sides by $e^{n\sigma}$, we get

$$x[n] = \frac{1}{2\pi}\int_0^{2\pi} X(e^{\sigma+j\omega})e^{(\sigma+j\omega)n}\, d\omega. \tag{6.191}$$

To represent the inverse z-transform in terms of z (instead of ω), we note that

$$dz = d(e^{\sigma+j\omega}) = e^{\sigma} je^{j\omega}\, d\omega = jz\, d\omega; \quad \text{i.e.,} \quad d\omega = z^{-1}\, dz/j, \tag{6.192}$$

and the integral of the inverse transform with respect to ω from 0 to 2π becomes an integral with respect to z along a circle of radius e^{σ}:

$$x[n] = \frac{1}{2\pi}\oint X(z)z^{n}z^{-1}\, dz/j = \frac{1}{2\pi j}\oint X(z)z^{n-1}\, dz. \tag{6.193}$$

Now we get the forward and inverse z-transform pair:

$$X(z) = \mathcal{Z}[x[n]] = \sum_{n=-\infty}^{\infty} x[n]z^{-n},$$
$$x[n] = \mathcal{Z}^{-1}[X(z)] = \frac{1}{2\pi j} \oint X(z) z^{n-1}\, dz, \tag{6.194}$$

which can also be more concisely represented as

$$x[n] \overset{\mathcal{Z}}{\longleftrightarrow} X(z). \tag{6.195}$$

In practice, we rarely need to carry out the integral in the inverse transform with respect to the complex variable z, as the z-transform pairs of most of the signals of interest can be obtained in some alternative methods and they are commonly made available in table form.

As shown in Eq. (6.186), the z-transform is related to the Laplace transform by an analytic function $z = e^s = e^{\sigma} e^{j\omega}$, which maps a complex variable s in the s-plane to another complex variable z in the z-plane and vice versa. This function is called a *conformal mapping*, as it preserves the angle formed by any two intersecting curves in the complex plane.

We consider specifically the following mappings.

- Any vertical line $\text{Re}[s] = \sigma_0$ in the s-plane is mapped repeatedly to a circle of radius $|z| = e^{\sigma_0}$ centered at the origin in the z-plane. This is a many-to-one mapping as an infinite number of points $s_k = \sigma_0 + j(\omega - 2k\pi)$ ($-\pi \le \omega \le \pi$, $-\infty < k < \infty$) are mapped to a single point $z = e^{\sigma_0} e^{j\omega}$. The following three cases are of particular interest:
  - when $\sigma_0 = -\infty$, the vertical line in the s-plane is mapped to the origin $z = e^{-\infty} = 0$ in the z-plane.
  - when $\sigma_0 = 0$, the imaginary axis $s = j\omega$ in the s-plane is mapped to the unit circle $z = e^{j\omega}$ ($|z| = 1$) in the z-plane.
  - when $\sigma_0 = \infty$, the vertical line in the s-plane is mapped to a circle $|z| = e^{\infty}$ of infinite radius in the z-plane.
- Any horizontal line $\text{Im}[s] = j\omega_0$ in the s-plane is mapped to a ray $\angle z = \omega_0$ in the z-plane from the origin along the direction of angle ω_0.
- The right angle formed by any pair of vertical and horizontal lines in the s-plane is mapped to the right angle formed by the corresponding circle and ray in the z-plane; i.e., the right angle is preserved by the mapping $z = e^s$.

Note that the continuous-time Fourier spectrum $X(j\omega) = \mathcal{F}[x(t)]$ is a non-periodic function defined over the entire imaginary axis $s = j\omega$ of the s-plane in the infinite range $-\infty < \omega < \infty$. But when the signal $x(t)$ is sampled to become a discrete signal $x[n]$, the corresponding discrete-time Fourier spectrum $X(e^{j\omega}) = \mathcal{F}[x[n]]$ becomes a periodic function over a finite range $0 \le \omega < 2\pi$ around the unit circle $z = e^{j\omega}$ in the z-plane. These results are of course consistent with those obtained in the previous chapters.

In many applications the z-transform takes the form of a rational function as a ratio of two polynomials:

$$X(z) = \frac{\sum_{k=0}^{M} b_k z^k}{\sum_{k=0}^{N} a_k z^k} = \frac{b_M}{a_N} \frac{\prod_{k=1}^{M}(z - z_k)}{\prod_{k=1}^{N}(z - p_k)}. \tag{6.196}$$

Here, the roots $z_k (k = 1, 2, \ldots, m)$ of the numerator polynomial of order M are the zeros of $X(z)$, and the roots $p_k, (k = 1, 2, \ldots, n)$ of the denominator polynomial of order N are the poles of $X(z)$. Some of these roots may be repeated. Moreover, if $N > M$, then $X(\infty) = 0$; i.e., $z = \infty$ is a zero. On the other hand, if $M > N$, then $X(\infty) = \infty$; i.e., $z = \infty$ is a pole. In general, we always assume $M < N$, as otherwise we can carry out a long division to expand $X(z)$ into multiple terms so that $M < N$ is true for each fraction. The locations of the zeros and poles of $X(z)$ characterize some essential properties of a signal $x[n]$.

6.2.2 Region of convergence

As with the Laplace transform, the ROC plays an important role in the z-transform. Here, we consider the z-transform of a set of signals which are in parallel with those in Example 6.1 of the Laplace transform.

Example 6.6: Consider the z-transform of the following two discrete signals.

1. A right-sided signal $x[n] = a^{-n} u[n]$:

$$X(z) = \sum_{n=-\infty}^{\infty} x[n] z^{-n} = \sum_{n=0}^{\infty} (az)^{-n}, \tag{6.197}$$

where a is a real constant. This summation is a geometric series which does not converge unless $|(az)^{-1}| < 1$; i.e., the ROC can be specified as $|z| > 1/|a|$, which is the entire region outside the circle with radius $|z| = 1/|a|$. Now the z-transform above can be further written as

$$X(z) = \sum_{n=0}^{\infty} (az^{-1})^n = \frac{1}{1 - (az)^{-1}}, \quad \text{if } |z| > 1/|a|. \tag{6.198}$$

Specially when $a = 1$, we have $x[n] = u[n]$ and

$$U(z) = \mathcal{Z}[u[n]] = \frac{1}{1 - z^{-1}}, \quad \text{if } |z| > 1. \tag{6.199}$$

If we let $\text{Re}[s] = \sigma \to 0$; i.e., $|z| = 1$, $U(z)$ will be evaluated along the unit circle $z = e^{j\omega}$ and become $\mathcal{Z}[u[n]] = 1/(1 - e^{-j\omega})$, which is seemingly the Fourier spectrum of $u[n]$. However, this result is actually invalid, as $|z| = 1$ is not inside the ROC $|z| > 1$. Comparing this result with the real Fourier transform of $u[n]$

in Eq. (4.24),

$$\mathcal{F}[u[n]] = \frac{1}{1-e^{-j2\pi f}} + \frac{1}{2}\sum_{k=-\infty}^{\infty}\delta(f-k), \tag{6.200}$$

we see an extra term $\sum_{k=-\infty}^{\infty}\delta(f-k)/2$ in the Fourier spectrum which reflects the fact that the summation is only marginally convergent when $|z|=1$.

2. A left-sided signal $x[n] = -a^{-n}u[-n-1]$:

$$X(z) = -\sum_{n=-\infty}^{\infty} a^{-n}u[-n-1]z^{-n} = -\sum_{n=-\infty}^{-1}(az)^{-m} = 1-\sum_{n=0}^{\infty}(az)^{n}. \tag{6.201}$$

We see that only when $|az|<1$; i.e., z is inside the ROC $|z|<1/|a|$, will this summation converge and $X(z)$ exist:

$$X(z) = 1-\frac{1}{1-az} = \frac{1}{1-(az)^{-1}}, \qquad \text{if } |z|<1/|a|. \tag{6.202}$$

A set of properties of the ROC can be summarized based on these two examples.

- If a signal $x[n]$ of finite duration is absolutely summable then its z-transform $X(z)$ exists for any z; i.e., its ROC is the entire z-plane.
- The ROC does not contain any poles because, by definition, $X(z)$ does not exist at any pole.
- Two different signals may have an identical transform but different ROCs. The inverse transform can be carried out only if an associated ROC is also specified.
- Only the magnitude $|z|=e^{\sigma}$ of z determines the convergence of the summation in the z-transform and thereby the ROC. The angle $\angle z$ has no effect on the convergence. Consequently, the ROC is always bounded by two concentric circles centered at the origin corresponding to two poles p_1 and p_2 with $|p_1|<|p_2|$. It is possible that $|p_1|=0$ and/or $|p_2|=\infty$.
- The ROC of a right-sided signal is outside the outermost pole; The ROC of a left-sided signal is inside the innermost pole. If a signal is two-sided, its ROC is the intersection of the two ROCs corresponding to its two one-sided parts, which can be either a ring between two circles or an empty set.
- The Fourier transform $X(e^{j\omega})$ of a signal $x[n]$ exists if the ROC of the corresponding z-transform $X(z)$ contains the unit circle $|z|=1$; i.e., $z=e^{j\omega}$.

The zeros and poles of $X(z)=\mathcal{Z}[x[n]]$ dictate the ROC and thereby the most essential properties of the corresponding signal $x[n]$, such as whether it is right- or left-sided, whether it grows or decays over time.

Example 6.7: Find the time signal corresponding to the following z-transform:

$$X(z) = \frac{1}{(1-\frac{1}{3}z^{-1})(1-2z^{-1})} = -\frac{1/5}{1-\frac{1}{3}z^{-1}} + \frac{6/5}{1-2z^{-1}}. \tag{6.203}$$

This function has two poles: $p_1 = 1/3$ and $p_2 = 2$. Now consider three possible ROCs corresponding to three different time signals.

- If $|z| > 2$, the ROC is outside the outermost pole $p_2 = 2$, both terms of $X(z)$ correspond to right-sided time functions:

$$x[n] = -\frac{1}{5}\left(\frac{1}{3}\right)^n u[n] + \frac{1}{5}\left(\frac{1}{3}\right)^n u[n]. \tag{6.204}$$

- If $|z| < 1/3$, the ROC is inside the innermost pole $p_1 = 1/3$, both terms of $X(z)$ correspond to left-sided time functions:

$$x[n] = \frac{1}{5}\left(\frac{1}{3}\right)^n u[-n-1] - \frac{1}{5}\left(\frac{1}{3}\right)^n u[-n-1]. \tag{6.205}$$

- If $1/3 < |z| < 2$, the ROC is a ring between the two poles, the two terms correspond to two different types of functions, one right-sided; and the other left-sided:

$$x[n] = -\frac{1}{5}\left(\frac{1}{3}\right)^n u[n] - \frac{1}{5}\left(\frac{1}{3}\right)^n u[-n-1]. \tag{6.206}$$

In particular, note that only the last ROC includes the circle $|z| = 1$ and the corresponding time function $x[n]$ has a DFT. The Fourier transform of the other two functions does not exist.

6.2.3 Properties of the z-transform

The z-transform has a set of properties many of which are in parallel with those of the DTFT. The proofs of such properties are, therefore, omitted, as they are similar to that of their counterparts in the Fourier transform. However, here, we need to pay special attention to the ROCs. In the following, we always assume

$$\mathcal{Z}[x[n]] = X(z), \qquad \mathcal{Z}[y[n]] = Y(z), \tag{6.207}$$

associated with R_x and R_y as their corresponding ROCs. If a property can be easily derived from the definition, the proof is not provided.

- **Linearity**

$$\mathcal{Z}[ax[n] + by[n]] = aX(z) + bY(z), \qquad \text{ROC} \supseteq (R_x \cap R_y). \tag{6.208}$$

Similar to the case of the Laplace transform, the ROC of the linear combination of $x[n]$ and $y[n]$ may be larger than the intersection of their individual ROCs $R_x \cap R_y$, due to reasons such as zero-pole cancellation.

- **Time shift**

$$\mathcal{Z}[x[n-n_0]] = z^{-n_0}X(z), \qquad \text{ROC} = R_x. \tag{6.209}$$

Time delay is a very important and useful operation that delays a signal $x[n]$ by one time unit to become $x[n-1]$. This operation is easily realized in the z-domain by a multiplication with z^{-1}, which can be readily used as a delay unit.

- **Time reversal**

$$\mathcal{Z}[x[-n]] = X(z^{-1}), \qquad \text{ROC} = 1/R_x. \tag{6.210}$$

Proof:

$$\mathcal{Z}[x[-n]] = \sum_{n=-\infty}^{\infty} x[-n]z^{-n} = \sum_{n'=-\infty}^{\infty} x[n'](z^{-1})^{-n'} = X(z^{-1}). \tag{6.211}$$

- **Modulation**

$$\mathcal{Z}[(-1)^n x[n]] = X(-z). \tag{6.212}$$

Here, modulation means every other sample of the signal is negated.
Proof:

$$\mathcal{Z}[(-1)^n x[n]] = \sum_{n=-\infty}^{\infty} x[n](-1)^n z^{-n} = \sum_{n=-\infty}^{\infty} x[n](-z)^{-n} = X(-z). \tag{6.213}$$

- **Down-sampling**

$$\mathcal{Z}[x_{(2)}[n]] = \frac{1}{2}[X(z^{1/2}) + X(-z^{1/2})]. \tag{6.214}$$

Here, the down-sampled version $x_{(2)}[n]$ of a signal $x[n]$ is composed of all the even terms of the signal with all odd terms dropped; i.e., $x_{(2)}[n] = x[2n]$.
Proof:

$$\begin{aligned}\mathcal{Z}[x_{(2)}[n]] &= \sum_{n=-\infty}^{\infty} x[2n]z^{-n} = \sum_{m=\cdots,-2,0,,2,\cdots} x[m](z^{1/2})^{-m} \\ &= \frac{1}{2}[\sum_{m=-\infty}^{\infty} x[m](z^{1/2})^{-m} + \sum_{m=-\infty}^{\infty} x[m](-z^{1/2})^{-m}] \\ &= \frac{1}{2}[X(z^{1/2}) + X(-z^{1/2})],\end{aligned} \tag{6.215}$$

where we have assumed $m = 2n$. The third equal sign is due to the fact that the two terms are the same when m is even but their sum is zero when m is odd.

- **Up-sampling**

$$\mathcal{Z}[x^{(k)}[n]] = X(z^k). \tag{6.216}$$

Here, $x^{(k)}[n]$ is defined as

$$x^{(k)}[n] = \begin{cases} x[n/k] & \text{if } n \text{ is a multiple of } k \\ 0 & \text{else} \end{cases}; \tag{6.217}$$

i.e., $x^{(k)}[n]$ is obtained by inserting $k-1$ zeros between every two consecutive samples of $x[n]$.

Proof:

$$\mathcal{Z}[x^{(k)}[n]] = \sum_{n=-\infty}^{\infty} x[n/k] z^{-n} = \sum_{m=-\infty}^{\infty} x[m] z^{-km} = X(z^k). \tag{6.218}$$

Note that the change of the summation index from n to $m = n/k$ has no effect as the terms skipped are all zeros.

Combining the down- and up-sampling above, we see that if a signal $x[n]$ with $X(z) = \mathcal{Z}[x[n]]$ is first down-sampled and then up-sampled, its z-transform is

$$\mathcal{Z}[(x_{(2)})^{(2)}[n]] = \frac{1}{2}[X(z) + X(-z)]. \tag{6.219}$$

However, also note that if the signal is first up- and then down-sampled, it remains the same: $(x^{(2)})_{(2)}[n] = x[n]$.

- **Convolution**

$$\mathcal{Z}\left[x[n] * y[n]\right] = X(z)Y(z), \qquad \text{ROC} \supseteq (R_x \cap R_y). \tag{6.220}$$

The ROC of the convolution could be larger than the intersection of R_x and R_y, due to the possible pole-zero cancellation caused by the convolution.

- **Autocorrelation**

$$\mathcal{Z}\left[\sum_k x[k]x[k-n]\right] = X(z)X(z^{-1}). \tag{6.221}$$

Proof:

The autocorrelation of a signal $x[n]$ is the convolution of the signal with its time-reversed version. Applying the properties of time reversal and convolution, the above can be proven.

- **Time difference**

$$\mathcal{Z}\left[x[n] - x[n-1]\right] = (1 - z^{-1})X(z), \qquad \text{ROC} = R_x. \tag{6.222}$$

Proof:

$$\mathcal{Z}[x[n] - x[n-1]] = X(z) - z^{-1}X(z) = (1 - z^{-1})X(z). \tag{6.223}$$

Note that, owing to the additional zero $z = 1$ and pole $z = 0$, the resulting ROC is the same as R_x except the possible deletion of $z = 0$ caused by the added pole and/or addition of $z = 1$ caused by the added zero, which may cancel an existing pole.

- **Time accumulation**

$$\mathcal{Z}\left[\sum_{k=-\infty}^{n} x[k]\right] = \frac{1}{1-z^{-1}} X(z). \tag{6.224}$$

Proof: First, we realize that the accumulation of $x[n]$ can be written as its convolution with $u[n]$:

$$u[n] * x[n] = \sum_{k=-\infty}^{\infty} u[n-k]x[k] = \sum_{k=-\infty}^{n} x[k]. \tag{6.225}$$

Applying the convolution property, we get

$$\mathcal{Z}\left[\sum_{k=-\infty}^{n} x[k]\right] = \mathcal{Z}[u[n] * x[n]] = \frac{1}{1-z^{-1}} X(z), \tag{6.226}$$

as $\mathcal{Z}[u[n]] = 1/(1-z^{-1})$.

- **Scaling in the z-domain**

$$\mathcal{Z}\left[z_0^n x[n]\right] = X\left(\frac{z}{z_0}\right), \qquad \text{ROC} = |z_0| R_x. \tag{6.227}$$

Proof:

$$\mathcal{Z}\left[z_0^n x[n]\right] = \sum_{n=-\infty}^{\infty} x[n]\left(\frac{z}{z_0}\right)^{-1} = X\left(\frac{z}{z_0}\right). \tag{6.228}$$

In particular, if $z_0 = e^{j\omega_0}$, the above becomes

$$\mathcal{Z}\left[e^{jn\omega_0} x[n]\right] = X(e^{-j\omega_0} z), \qquad \text{ROC} = R_x. \tag{6.229}$$

The multiplication by $e^{-j\omega_0}$ to z corresponds to a rotation by angle ω_0 in the z-plane; i.e., a frequency shift by ω_0. The rotation is either clockwise ($\omega_0 > 0$) or counter clockwise ($\omega_0 < 0$) corresponding to, respectively, either a left-shift or a right shift in the s-domain. The property is essentially the same as the frequency shifting property of the DTFT.

- **Conjugation**

$$\mathcal{Z}\left[x^*[n]\right] = X^*(z^*), \qquad \text{ROC} = R_x. \tag{6.230}$$

Proof: Complex conjugate of the z-transform of $x[n]$ is

$$X^*(z) = \left[\sum_{n=-\infty}^{\infty} x[n] z^{-n}\right]^* = \sum_{n=-\infty}^{\infty} x^*[n](z^*)^{-n}. \tag{6.231}$$

Replacing z by z^*, we get the desired result.

- **Differentiation in the z-domain**

$$\mathcal{Z}\left[nx[n]\right] = -\frac{d}{dz} X(z), \quad \text{ROC} = R_x. \tag{6.232}$$

Proof:

$$\frac{d}{dz}X(z) = \sum_{n=-\infty}^{\infty} x[n]\frac{d}{dz}(z^{-n}) = \sum_{n=-\infty}^{\infty}(-n)x[n]z^{-n-1}; \tag{6.233}$$

i.e.,

$$\mathcal{Z}\left[n\,x[n]\right] = -z\frac{d}{dz}X(z). \tag{6.234}$$

Example 6.8: Given a signal $x[n]$ and its z-transform $X(f)$, find the z-transform of the signal modified as shown below.

- When the signal is modulated, time-reversed, and shifted to become $(-1)^n x[k-n]$, its z-transform becomes

$$\mathcal{Z}[(-1)^n x[k-n]] = \sum_{n=-\infty}^{\infty}(-1)^n x[k-n]z^{-n} = \sum_{n=-\infty}^{\infty} x[k-n](-z)^{-n}$$
$$= \sum_{m=-\infty}^{\infty} x[m](-z)^{m-k} = (-z)^{-k}\sum_{m=-\infty}^{\infty} x[m](-z^{-1})^{-m} = (-z)^{-k}X(-z^{-1}), \tag{6.235}$$

  where $m = k - n$.
- When the signal is first down-sampled and then up-sampled as shown in Fig. 6.14, its z-transform becomes

$$X'(z) = \frac{1}{2}[X(z) + X(-z)], \tag{6.236}$$

  which can be obtained by applying the properties of down-sampling and up-sampling in Eqs. (6.214) and (6.216). To verify this result, we apply the property of modulation in Eq. (6.212) to the second term and get

$$x'[n] = \mathcal{Z}^{-1}[X'(z)] = \frac{1}{2}[\mathcal{Z}^{-1}[X(z)] + \mathcal{Z}^{-1}[X(-z)]]$$
$$= \frac{1}{2}[x[n] + (-1)^n x[n]] = \begin{cases} x[n] & \text{even } n \\ 0 & \text{odd } n \end{cases}. \tag{6.237}$$

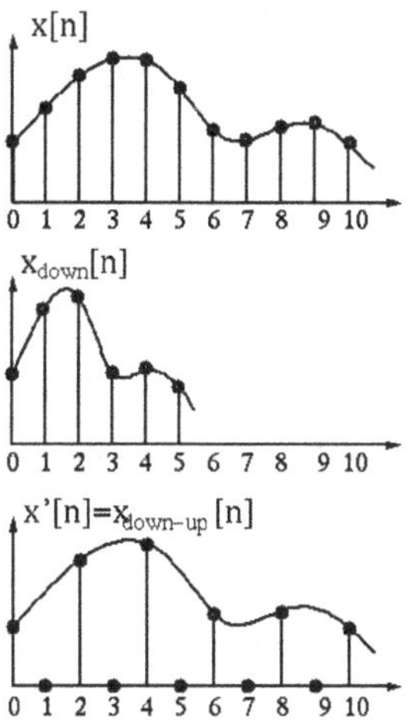

Figure 6.14 Down- and up-sampling.

6.2.4 The z-transform of typical signals

- $\delta[n]$, $\delta[n-m]$

$$\mathcal{Z}[\delta[n]] = \sum_{n=-\infty}^{\infty} \delta[n] z^{-n} = 1, \quad \text{for all } z. \tag{6.238}$$

Owing to the time-shift property, we also have

$$\mathcal{Z}[\delta[n-m]] = z^{-m}, \quad \text{for all } z. \tag{6.239}$$

- $u[n]$, $a^n u[n]$, $n a^n u[n]$

$$\mathcal{Z}[u[n]] = \sum_{n=0}^{\infty} z^{-n} = \frac{1}{1-z^{-1}}, \quad |z| > 1. \tag{6.240}$$

Owing to the scaling in the z-domain property, we have

$$\mathcal{Z}[a^n u[n]] = \frac{1}{1-(z/a)^{-1}} = \frac{1}{1-az^{-1}}, \quad |z| > a. \tag{6.241}$$

Applying the property of differentiation in the z-domain to the above, we have

$$\mathcal{Z}[n\, a^n u[n]] = -z\frac{d}{dz}\left[\frac{1}{1-az^{-1}}\right] = -z\frac{-az^{-2}}{(1-az^{-1})^2} = \frac{az^{-1}}{(1-az^{-1})^2}, \quad |z| > a. \tag{6.242}$$

- $e^{\pm jn\omega_0} u[n]$, $\cos[n\omega_0]u[n]$, $\sin[n\omega_0]u[n]$

Applying the scaling in the z-domain property to $\mathcal{Z}[u[n]] = 1/(1-z^{-1})$, we have

$$\mathcal{Z}[e^{jm\omega_0} u[n]] = \frac{1}{1-(e^{j\omega_0} z)^{-1}} = \frac{1}{1-e^{-j\omega_0} z^{-1}}, \qquad |z| > 1, \tag{6.243}$$

and similarly, we have

$$\mathcal{Z}[e^{-jm\omega_0} u[n]] = \frac{1}{1-e^{j\omega_0} z^{-1}}, \qquad |z| > 1. \tag{6.244}$$

Moreover, we have

$$\mathcal{Z}[\cos(n\omega_0)u[n]] = \mathcal{Z}\left[\frac{e^{jn\omega_0}+e^{-jn\omega_0}}{2}u[n]\right]$$
$$= \frac{1}{2}\left[\frac{1}{1-e^{j\omega_0} z^{-1}} + \frac{1}{1-e^{-j\omega_0} z^{-1}}\right] = \frac{2-(e^{j\omega_0}+e^{-j\omega_0})z^{-1}}{2[1-(e^{j\omega_0}+e^{-j\omega_0})z^{-1}+z^{-2}]}$$
$$= \frac{1-\cos\omega_0 z^{-1}}{1-2\cos\omega_0 z^{-1}+z^{-2}}, \qquad |z| > 1. \tag{6.245}$$

Similarly we have

$$\mathcal{Z}[\sin(n\omega_0)u[n]] = \frac{\sin\omega_0 z^{-1}}{1-2\cos\omega_0 z^{-1}+z^{-2}}, \qquad |z| > 1. \tag{6.246}$$

- $r^n \cos[n\omega_0]u[n]$, $r^n \sin[n\omega_0]u[n]$
  Applying the z-domain scaling property to the above, we have

$$\mathcal{Z}[r^n \cos(n\omega_0)u[n]] = \frac{1-r\cos\omega_0 z^{-1}}{1-2r\, cos\omega_0 z^{-1}+r^2 z^{-2}}, \qquad |z| > r, \tag{6.247}$$

and

$$\mathcal{Z}[r^n \sin(n\omega_0)u[n]] = \frac{r\sin\omega_0 z^{-1}}{1-2r\cos\omega_0 z^{-1}+r^2 z^{-2}}, \qquad |z| > r. \tag{6.248}$$

6.2.5 Analysis of discrete LTI systems by z-transform

The z-transform is a convenient tool for the analysis and design of discrete LTI systems $y[n] = \mathcal{O}[x[n]]$ whose output $y[n]$ is the convolution of the input $x[n]$ and its impulse response function $h[n]$:

$$y[n] = \mathcal{O}[x[n]] = h[n] * x[n] = \sum_{m=-\infty}^{\infty} h[m]x[n-m]. \tag{6.249}$$

In particular, if the input is an impulse $x[n] = \delta[n]$, then the output is the impulse response function $y[n] = \mathcal{O}[\delta[n]] = h[n] * \delta[n] = h[n]$. Also if the input is a complex exponential $x[n] = e^{sn} = z^n$ $(z = e^s)$, then the output is

$$y[n] = \mathcal{O}[z^n] = \sum_{m=-\infty}^{\infty} h[m]z^{n-m} = z^n \sum_{m=-\infty}^{\infty} h[m]z^{-m} = H(z)z^n, \qquad (6.250)$$

where $H(z)$ is the *transfer function* of the discrete system, first defined in Eq. (1.111), which is actually the z-transform of the impulse response $h[n]$ of the system:

$$H(z) = \mathcal{Z}[h[n]] = \sum_{n=-\infty}^{\infty} h[n]z^{-n}. \qquad (6.251)$$

Equation (6.250) is the eigenequation of *any* discrete LTI system, where the transfer function $H(z)$ is the eigenvalue, and the complex exponential input $x[n] = e^{sn} = z^n$ is the corresponding eigenfunction. In particular, if we let $\sigma = 0$; i.e., $z = e^{j\omega}$, then the transfer function $H(z)$ becomes the discrete-time Fourier transform of the impulse response $h[n]$ of the system:

$$H(z)\big|_{s=j\omega} = H(e^{j\omega}) = \sum_{n=-\infty}^{\infty} h[n]e^{-j\omega n} = \mathcal{F}[h[n]]. \qquad (6.252)$$

This is the frequency response function of the discrete LTI system first defined in Eq. (5.17). Various properties and behaviors, such as the stability and filtering effects of a discrete LTI system, can be qualitatively characterized based on the locations of the zeros and poles of its transfer function $H(z) = \mathcal{Z}[h[n]]$ due to the properties of the ROC of the z-transform.

Also, owing to its convolution property of the z-transform, the convolution in Eq. (6.249) can be converted to a multiplication in the z-domain:

$$y[n] = h[n] * x[n] \xrightarrow{\mathcal{Z}} Y(z) = H(z)X(z). \qquad (6.253)$$

Based on this relationship the transfer function $H(z)$ can also be found in the z-domain as the ratio $H(z) = Y(z)/X(z)$ of the output $Y(z)$ and input $X(z)$. The ROC and poles of the transfer function $H(s)$ of an LTI system dictate the behaviors of system, such as its causality and stability.

- **Stability**

  Also as discussed in Chapter 1, a discrete LTI system is stable if to any bounded input $|x[n]| < B$ its response $y[n]$ is also bounded for all n, and its impulse response function $h[n]$ needs to be absolutely summable (Eq. (1.122)):

$$\sum_{n=-\infty}^{\infty} \big|h[n]\big| < \infty; \qquad (6.254)$$

  i.e., the frequency response function $\mathcal{F}[h[n]] = H(e^{j\omega}) = H(z)\big|_{z=e^{j\omega}}$ exists. In other words, an LTI system is stable if and only if the ROC of its transfer function $H(z)$ includes the unit circle $|z| = 1$.

- **Causality**
  A discrete LTI system is causal if its impulse response $h[n]$ is a consequence of the impulse input $\delta[n]$; i.e., $h[n]$ comes after $\delta[n]$:

$$h[n] = h[n]u[n] = \begin{cases} h[n] & n \geq 0 \\ 0 & n < 0 \end{cases}, \tag{6.255}$$

  and its output is (Eq. (1.123)):

$$y[n] = \sum_{m=-\infty}^{\infty} h[m]x[n-m] = \sum_{n=0}^{\infty} h[n]x[m-n]. \tag{6.256}$$

  The ROC of $H(z)$ is the exterior of a circle. In particular, when $H(z)$ is rational, the system is causal if and only if its ROC is the exterior of a circle outside the outermost pole, and the order of numerator is no greater than that of the denominator so that $z = \infty$ is not a pole ($H(\infty)$ exists).

Combining the two properties above, we see that a causal LTI system with a rational transfer function $H(z)$ is stable if and only if all poles of $H(z)$ are inside the unit circle of the z-plane; i.e., the magnitudes of all poles are smaller than 1: $|p_k| < 1$.

One type of discrete LTI system can be characterized by an LCCDE:

$$\sum_{k=0}^{N} a_k y[n-k] = \sum_{k=0}^{M} b_k x[n-k]. \tag{6.257}$$

Taking the z-transform of this equation, we get an algebraic equation in the z-domain:

$$Y(z)\left[\sum_{k=0}^{N} a_k z^{-k}\right] = X(z)\left[\sum_{k=0}^{M} b_k z^{-k}\right]. \tag{6.258}$$

The transfer function of such a system is rational:

$$H(z) = \frac{Y(z)}{X(z)} = \frac{\sum_{k=0}^{M} b_k z^{-k}}{\sum_{k=0}^{N} a_k z^{-k}} = c\,\frac{\prod_{k=1}^{M}(z - z_{0_k})}{\prod_{k=1}^{N}(z - z_{0_k})}, \tag{6.259}$$

where z_k, $(k = 1, 2, \ldots, M)$ and p_k, $(k = 1, 2, \ldots, N)$ are respectively the zeros and poles of $H(z)$. For simplicity, and without loss of generality, we will assume $N > M$ and $c = 1$ below.

The output $Y(z)$ of the LTI system can be represented as

$$Y(z) = H(z)X(z) = \left(\sum_{k=0}^{M} b_k z^{-k}\right) \frac{1}{\sum_{k=0}^{N} a_k z^{-k}} X(z) = \left(\sum_{k=0}^{M} b_k z^{-k}\right) W(z), \tag{6.260}$$

or in the time domain as

$$y[n] = \sum_{k=0}^{M} b_k w[n-k], \tag{6.261}$$

where we have defined $W(z) = X(z)/(\sum_{k=0}^{N} a_k z^{-k})$ as an intermediate variable, or in the time domain:

$$\sum_{k=0}^{N} a_k w[n-k] = x[n], \quad \text{or} \quad a_N w[n-N] = x[n] - \sum_{k=0}^{N-1} a_k w[n-k]. \quad (6.262)$$

Without loss of generality, we assume $a_N = 1$, and the LTI system can now be represented as a *block diagram*, as shown in Fig. 6.15 (for $M = 2$ and $N = 3$).

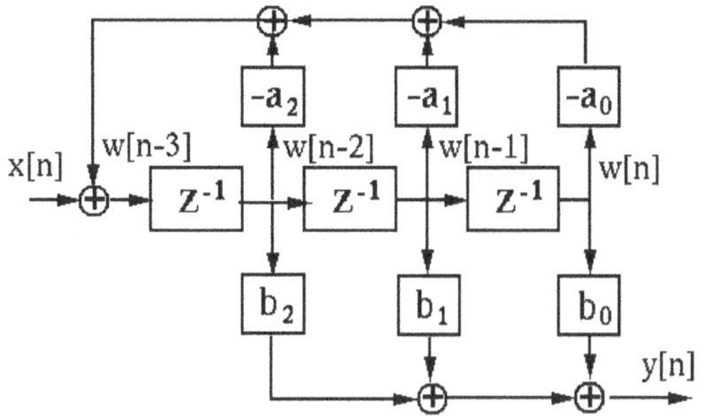

Figure 6.15 Block diagram of a discrete LTI system described by an LCCDE.

To find the impulse response $h[n]$ we first convert $H(z)$ to a summation by partial fraction expansion:

$$H(z) = \frac{\prod_{k=1}^{M}(z - z_{0_k})}{\prod_{k=1}^{N}(z - z_{0_k})} = \sum_{k=1}^{N} \frac{c_k}{1 - p_k z^{-1}}, \quad (6.263)$$

(assume no repeated poles) and then carry out the inverse transform (the LTI system in Eq. (6.257) is causal) to get

$$h[n] = \mathcal{Z}^{-1}[H(z)] = \sum_{k=1}^{N} \mathcal{Z}^{-1}\left[\frac{c_k}{1 - p_k z^{-1}}\right] = \sum_{k=1}^{N} c_k p_k^n u[n]. \quad (6.264)$$

The output $y[n]$ of the LTI system can be found by solving the difference equation in Eq. (6.257). Alternatively, it can also be found by the convolution $y[n] = h[n] * x[n]$, or the inverse z-transform:

$$y[n] = \mathcal{Z}^{-1}[Y(z)] = \mathcal{Z}^{-1}[H(z)X(z)]. \quad (6.265)$$

As the LCCDE in Eq. (6.257) is an LTI system, it can also be solved in the following two steps. First, we assume the input on the right-hand side is simply $x[n]$ and find the corresponding output $y[n]$. Then the response to the true input $\sum_k b_k x[n-k]$ can be found to be $\sum_k b_k y[n-k]$.

Note that the output $y[n]$ obtained this way is only the particular solution due to input $x[n]$, but the homogeneous solution due to any non-zero initial conditions is not represented by the bilateral Laplace transform. This problem will be addressed by the unilateral z-transform to be discussed later, which takes the initial conditions into consideration.

Same as in the case of a continuous LTI system, here, the behavior of a discrete LTI system in terms of stability and oscillation is also dictated by the pole locations in the z-plane. The poles are either real or form complex conjugate pairs, either inside or outside the unit circle, as shown in Fig. 6.16, where the numbered pole locations correspond to those in the s-plane for the continuous case as shown in Fig. 6.2 with similar waveforms in the time domain.

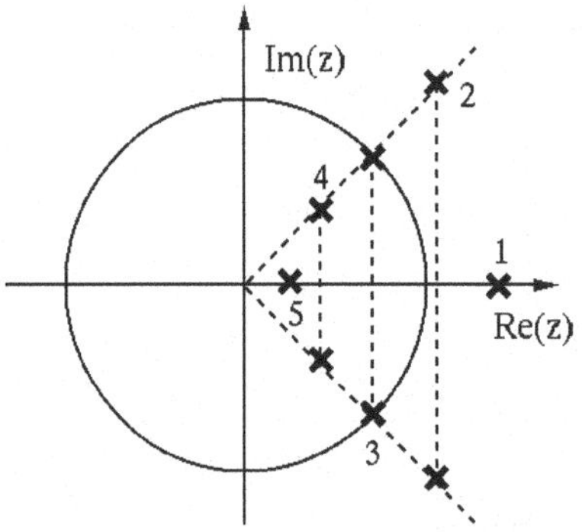

Figure 6.16 Different pole locations of $H(z)$.

A discrete LTI system can be treated as a filter, called a *digital filter*. Depending on the specific form of the LCCDE (Eq. (6.257)) that describes the filter, it belongs to either of the following two types.

- *Finite impulse response (FIR) filters*
  In Eq. (6.257), specially if $a_0 = 1$ and $a_k = 0$ for all $k > 0$, then the impulse response of the system becomes

$$h[n] = \sum_{k=0}^{M} b_k \delta[n-k], \qquad n = 0, \ldots, M. \tag{6.266}$$

  As $h[n]$ has only a finite number of non-zero terms, it is absolutely summable, and the transfer function

$$H(z) = \sum_{n=0}^{M} h[n] = \sum_{n=0}^{M} b_n z^{-n} \tag{6.267}$$

  does not have any poles; i.e., an FIR filter is always stable. In particular, if $b_k = 1/(M+1)$, this system becomes a discrete moving-average filter; i.e., the output $y[n]$ is the average of the last $M+1$ inputs.
- *Infinite impulse response (IIR) filters*
  Any LTI system described by Eq. (6.257) without the special condition $a_k = 0$ $(k > 0)$ is an IIR filter, as there are, in general, an infinite number of terms in its impulse response in Eq. (6.264). As discussed previously, an IIR filter is stable if all of its poles are inside the unit circle. For example, consider this simple LTI system

$$y[n] - ay[n-1] = x[n], \tag{6.268}$$

with impulse response $h[n] = a^n u[n]$. This system is stable only if $|a| < 1$ and its transfer function is $H(z) = 1/(1 - az^{-1})$. As the impulse response $h[n]$ has infinite non-zero terms ($n = 0, 1, \ldots$), this is an IIR filter.

Example 6.9: The input and output of an LTI system are related by

$$y[n] - \frac{1}{2}y[n-1] = x[n] + \frac{1}{3}x[n-1]. \tag{6.269}$$

Note that without further information such as the initial condition, this equation does not uniquely specify $y[n]$ when $x[n]$ is given. Taking the z-transform of this equation and using the time-shift property, we get

$$Y(z) - \frac{1}{2}z^{-1}Y(z) = X(z) + \frac{1}{3}z^{-1}X(z), \tag{6.270}$$

and the transfer function can be obtained

$$H(z) = \frac{Y(z)}{X(z)} = \frac{1 + \frac{1}{3}z^{-1}}{1 - \frac{1}{2}z^{-1}} = \frac{1}{1 - \frac{1}{2}z^{-1}}\left(1 + \frac{1}{3}z^{-1}\right). \tag{6.271}$$

Note that the causality and stability of the system is not provided by this equation, unless the ROC of this $H(z)$ is specified. Consider the following two possible ROCs.

- If ROC is $|z| > 1/2$, it is outside the pole $z_p = 1/2$ and includes the unit circle. The system is causal and stable:

$$h[n] = \left(\frac{1}{2}\right)^n u[n] + \frac{1}{3}\left(\frac{1}{2}\right)^{n-1} u[n-1]. \tag{6.272}$$

- If ROC is $|z| < 1/2$, it is inside the pole $z_p = 1/2$ and does not include the unit circle. The system is anti-causal and unstable:

$$h[n] = -\left(\frac{1}{2}\right)^n u[-n-1] - \frac{1}{3}\left(\frac{1}{2}\right)^{n-1} u[-n]. \tag{6.273}$$

6.2.6 First- and second-order systems

As discussed previously in Example 5.3, a first-order causal system is described by the following difference equation:

$$y[n] - ay[n-1] = x[n]. \tag{6.274}$$

Its impulse response is $h[n] = a^n u[n]$ with $|a| < 1$ (stable system). The transfer function of the system is

$$H(z) = \mathcal{Z}[h[n]] = \sum_{n=0}^{\infty} a^n z^{-n} = \frac{1}{1 - az^{-1}} = \frac{z}{z - a}. \tag{6.275}$$

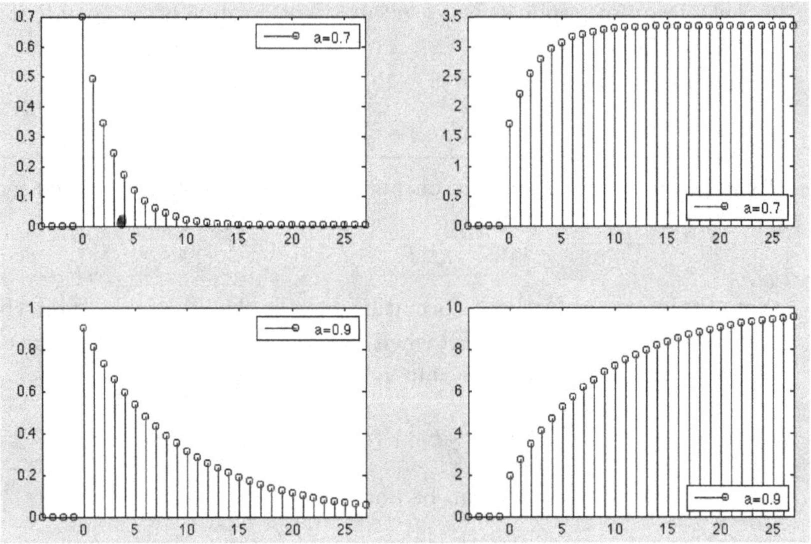

Figure 6.17 Impulse and step responses of first-order system. The impulse responses (left) and step responses (right) for $a = 0.7$ (top) and $a = 0.9$ (bottom).

Like τ for a continuous first-order system, the pole a here is the only parameter needed to characterize a first-order discrete system. Also as shown in Example 5.3, the step response of the first-order system is

$$y[n] = h[n] * u[n] = \frac{1 - a^{n+1}}{1 - a} u[n]. \tag{6.276}$$

The impulse and step responses of the first-order system are shown in Fig. 6.17.

The canonical form of the difference equation for a second-order system is

$$y[n] - 2r\cos\theta \, y[n-1] + r^2 = x[n]. \tag{6.277}$$

Like ζ and ω_n for a continuous second-order system, here r and θ are the two parameters needed to characterize a second-order discrete system. Taking the z-transform on both sides we get

$$(1 - 2r\cos\theta z^{-1} + r^2 z^{-2})Y(z) = X(z), \tag{6.278}$$

and the transfer function is

$$\begin{aligned} H(z) &= \frac{Y(z)}{X(z)} = \frac{1}{1 - 2r\cos\theta z^{-1} + r^2 z^{-2}} = \frac{1}{(1 - p_1 z^{-1})(1 - p_2 z^{-1})} \\ &= \frac{1}{(1 - r\,e^{j\theta} z^{-1})(1 - r\,e^{-j\theta} z^{-1})} = \frac{z^2}{(z - r\,e^{j\theta})(z - r\,e^{-j\theta})}, \end{aligned} \tag{6.279}$$

where p_1 and p_2 are the two poles, the two solutions of the quadratic equation $z^2 - 2r\cos\theta z + r^2 = 0$:

$$p_{1,2} = r\cos\theta \pm jr\sin\theta = r(\cos\theta \pm j\sin\theta) = r\,e^{\pm j\theta}. \tag{6.280}$$

When θ is not zero or π, therefore, $e^{j\theta} \neq e^{-j\theta}$, the two poles are different. We see that, for the system to be stable, we must have $|p_{1,2}| = |r| < 1$ for both poles to be inside the unit circle.

To find the impulse response of the system by the inverse z-transform, we first carry out the partial fraction expansion:

$$H(z) = \frac{1}{(1 - r\,e^{j\theta}z^{-1})(1 - r\,e^{-j\theta}z^{-1})} = \frac{A}{1 - r\,e^{j\theta}z^{-1}} + \frac{B}{1 - r\,e^{-j\theta}z^{-1}} \tag{6.281}$$

to find

$$A = \frac{e^{j\theta}}{2j\sin\theta}, \qquad B = \frac{-e^{-j\theta}}{2j\sin\theta}. \tag{6.282}$$

Now we can get the impulse response:

$$h[n] = \mathcal{Z}^{-1}\left[\frac{A}{1 - r\,e^{j\theta}z^{-1}} + \frac{B}{1 - r\,e^{-j\theta}z^{-1}}\right] = \left[A(r\,e^{j\theta})^m + B(r\,e^{-j\theta})^m\right]u[m]$$
$$= r^n\frac{\sin((n+1)\theta)}{\sin\theta}\,u[n]. \tag{6.283}$$

This is the underdamped impulse response of a discrete second-order system, also plotted in Fig. 6.18 for different system parameters. We see that r and θ dictate the decay rate and oscillation frequency of the response, respectively, corresponding to ζ and ω_n in the impulse response $h(t)$ of a continuous system given in Eq. (6.126).

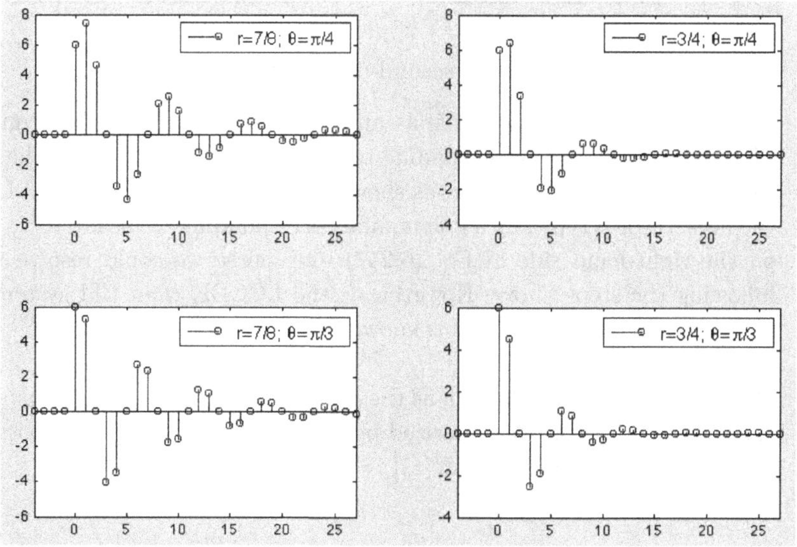

Figure 6.18 Impulse response of second-order system.

The step response of a discrete second-order system can be found as

$$y[n] = h[n] * u[n] = \sum_{m=0}^{n} h[m] = \left[A \sum_{m=0}^{n} (r\, e^{j\theta})^m + B \sum_{m=0}^{n} (r\, e^{-j\theta})^m \right] u[n]$$
$$= \left[A \frac{1 - r\, e^{j(n+1)\theta}}{1 - r\, e^{j\theta}} - B \frac{1 - r\, e^{-j(n+1)\theta}}{1 - r\, e^{-j\theta}} \right] u[n]$$
$$= \frac{\sin\theta - r^{n+1} \sin((n+2)\theta) + r^{n+2} \sin((n+1)\theta)}{\sin\theta (1 - 2r\cos\theta + r^2)} u[n], \qquad (6.284)$$

which is also plotted in Fig. 6.19 for different system parameters.

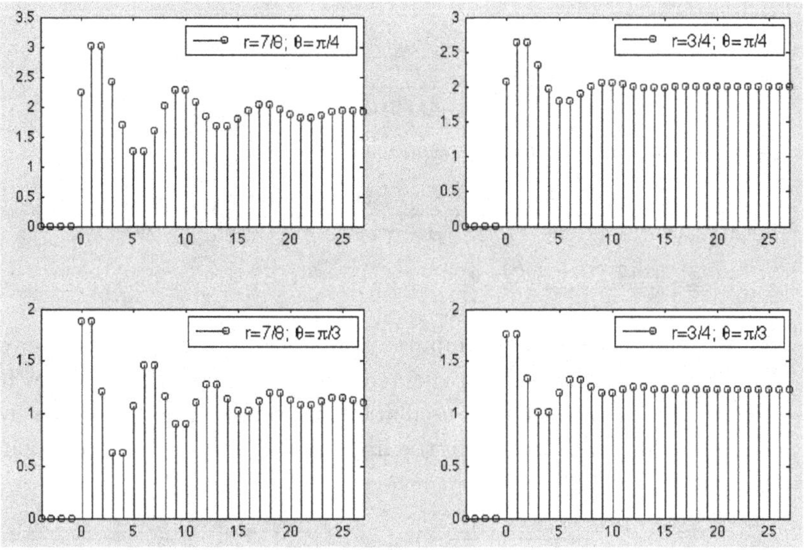

Figure 6.19 Step response of second-order system.

In these examples for the first- and second-order systems, the numerator of $H(z)$ is always unity corresponding to the assumed input $x[n]$ on the right-hand side of the LCCDE. However, as shown in Eq. (6.259) for a general LTI system, the numerator is typically a polynomial corresponding to the input $\sum_k b_k x[n-k]$ on the right-hand side of Eq. (6.257). Of course we could resolve the system following the steps above. However, as the LCCDE is an LTI system, once the response $y[n]$ to an input $x[n]$ is known, we can find its response to $\sum_k b_k x[n-k]$ to be $\sum_k b_k y[n-k]$.

A discrete LTI system, such as the first- and second-order systems, can also be considered as a filter characterized by the magnitude and phase of its frequency response function $H(e^{j\omega}) = H(z)\big|_z = e^{j\omega}$:

$$|H(e^{j\omega})| = \frac{\prod_{k=1}^{M} |e^{j\omega} - z_k|}{\prod_{k=1}^{N} |e^{j\omega} - p_k|} = \frac{\prod_{k=1}^{M} |\boldsymbol{u}_k|}{\prod_{k=1}^{N} |\boldsymbol{v}_k|},$$

$$\angle H(e^{j\omega}) = \frac{\sum_{k=1}^{M} \angle(e^{j\omega} - z_k)}{\sum_{k=1}^{N} \angle(e^{j\omega} - p_k)} = \frac{\sum_{k=1}^{M} \angle \boldsymbol{u}_k}{\sum_{k=1}^{N} \angle \boldsymbol{v}_k}, \tag{6.285}$$

where each factor $\boldsymbol{u}_k = e^{j\omega} - z_k$ or $\boldsymbol{v}_k = e^{j\omega} - p_k$ is a vector in the z-plane that connects the point $e^{j\omega}$ on the unit circle and one of the zeros or poles. The filtering effects of the system are therefore dictated by the zero and pole locations on the z-plane and can be qualitatively determined by observing how $|H(e^{j\omega})|$ and $\angle H(e^{j\omega})$ change as frequency ω varies along the unit circle from $-\pi$ to π.

The frequency response function of the first-order system in Eq. 6.275 is

$$H(e^{j\omega}) = H(z)\big|_{z=e^{j\omega}} = \frac{1}{1 - pe^{-j\omega}} = \frac{e^{j\omega}}{e^{j\omega} - p} = \frac{\boldsymbol{u}}{\boldsymbol{v}}, \tag{6.286}$$

where $p = r\,e^{j\theta}$ is the pole of the system and the zero is at the origin, and $\boldsymbol{u} = e^{j\omega}$ and $\boldsymbol{v} = e^{j\omega} - p = e^{j\omega} - r\,e^{j\theta}$ are the two vectors connecting $e^{j\omega}$ to the zero and pole, respectively, as shown on the left in Fig. 6.20. While the magnitude of $\boldsymbol{u}$ is unity, the magnitude of $\boldsymbol{v}$ varies as ω moves from $-\pi$ to π, and it reaches a minimum when $\omega = 0$ and $e^{j\omega} = 1$. We can qualitatively determine that the system is an LP filter with maximum magnitude at zero frequency, as shown in the top panel of Fig. 6.21.

The frequency response function of the first-order system in Eq. 6.279 is

$$H(e^{j\omega}) = H(z)\big|_{z=e^{j\omega}} = \frac{e^{j2\omega}}{(e^{j\omega} - p_1)(e^{j\omega} - p_2)} = \frac{\boldsymbol{u}_1\boldsymbol{u}_2}{\boldsymbol{v}_1\boldsymbol{v}_2}, \tag{6.287}$$

where $p_{1,2} = r\,e^{\pm j\theta}$ are the two poles and the double zeros are at the origin, and $\boldsymbol{u}_1 = \boldsymbol{u}_2 = e^{j\omega}$ and $\boldsymbol{v}_{1,2} = e^{j\omega} - p_{1,2} = e^{j\omega} - r\,e^{\pm j\theta}$ are the vectors connecting $e^{j\omega}$ to the two zeros and two poles, respectively, as shown on the right in Fig. 6.20. While the magnitude of $\boldsymbol{u}_1 = \boldsymbol{u}_2$ is unity, the magnitudes of $\boldsymbol{v}_1$ and $\boldsymbol{v}_2$ vary as ω moves from $-\pi$ to π, and they reach a minimum when $\omega = \pm\theta$. We can qualitatively determine that the system is a BS filter with center frequency of the passing band around $\omega = \theta$, as shown in the bottom panel of Fig. 6.21.

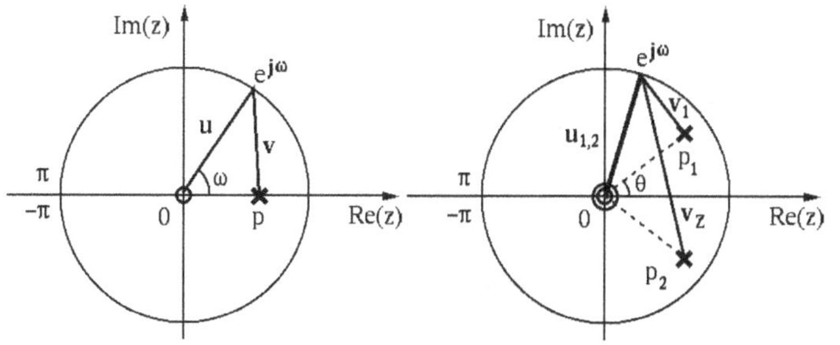

Figure 6.20 First-order (left) and second-order (right) filters in z-plane.

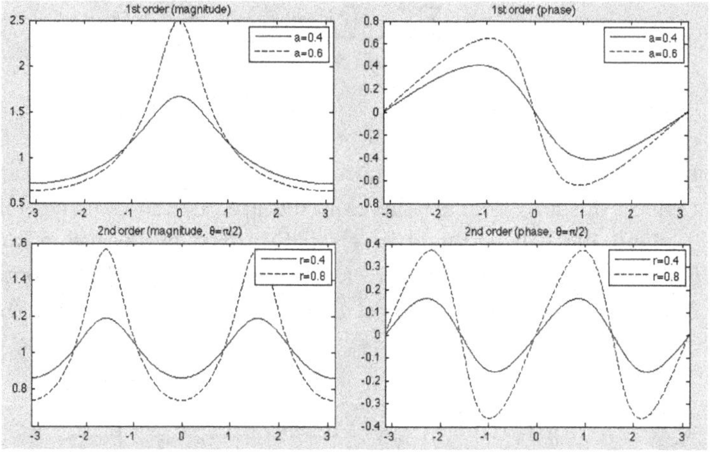

Figure 6.21 Filters of first-order (top) and second-order (bottom). The magnitudes of the filters are shown on the left and the phases are shown on the right.

6.2.7 The unilateral z-transform

As with the bilateral Laplace transform, the bilateral z-transform does not take the initial conditions into consideration while solving difference equations, and this problem can be resolved by the *unilateral* z-transform defined below:

$$\mathcal{UZ}[x[n]] = X(z) = \sum_{n=-\infty}^{\infty} x[n]u[n]z^{-m} = \sum_{n=0}^{\infty} x[n]z^{-n}. \tag{6.288}$$

When the unilateral z-transform is applied to a signal $x[n]$, it is always assumed that the signal starts at time $n = 0$; i.e., $x[n] = 0$ for $n < 0$. When it is applied to the impulse response function of an LTI system to find the transfer function $H(z) = \mathcal{UZ}[h[n]]$, it is always assumed that the system is causal; i.e., $h[n] = 0$ for $n < 0$. In both cases, the ROC is always the exterior of a circle.

By definition, the unilateral z-transform of any signal $x[n] = x[n]u[n]$ is identical to its bilateral z-transform. However, when $x[n] \neq x[n]u[n]$, the two z-transforms are different. Some of the properties of the unilateral z-transform that are different from the bilateral z-transform are listed below.

- **Time advance**

$$\mathcal{UZ}[x[n+1]] = \sum_{n=0}^{\infty} x[n+1]z^{-n} = z\sum_{m=1}^{\infty} x[m]z^{-m}$$
$$= z\left[\sum_{m=0}^{\infty} x[m]z^{-m} - x[0]\right] = zX(z) - zx[0], \tag{6.289}$$

where $m = n + 1$.

- **Time delay**

$$\mathcal{UZ}[x[n-1]] = \sum_{n=0}^{\infty} x[n-1]z^{-n} = z^{-1} \sum_{m=-1}^{\infty} x[m]z^{-m}$$

$$= z^{-1}\left[\sum_{m=0}^{\infty} x[m]z^{-m} + zx[-1]\right] = z^{-1}X(z) + x[-1], \tag{6.290}$$

where $m = n - 1$. Similarly, we have

$$\mathcal{UZ}[x[n-2]] = \sum_{n=0}^{\infty} x[n-2]z^{-n} = z^{-2} \sum_{m=-2}^{\infty} x[m]z^{-m}$$

$$= z^{-2}\left[\sum_{n=0}^{\infty} x[n]z^{-n} + zx[-1] + z^2x[-2]\right] = z^{-2}X(z) + x[-1]z^{-1} + x[-2], \tag{6.291}$$

where $m = n - 2$. In general, we have

$$\mathcal{UZ}[x[n-n_0]] = z^{-n_0}X(z) + \sum_{k=0}^{n_0-1} z^{-k}x[k-n_0]. \tag{6.292}$$

- **Initial-value theorem**
If $x[n] = x[n]\,u[n]$; i.e., $x[n] = 0$ for $n < 0$, then

$$x[0] = \lim_{z\to\infty} X(z). \tag{6.293}$$

Proof:

$$\lim_{z\to\infty} X(z) = \lim_{z\to\infty}\left[\sum_{n=0}^{\infty} x[n]z^{-n}\right] = x[0]. \tag{6.294}$$

This is because all terms with $n > 0$ become zero as $z^{-n} = 1/z^n \to 0$ as $z \to \infty$.

- **Final-value theorem**
If $x[n] = x[n]u[n]$; i.e., $x[n] = 0$ for $n < 0$, then

$$\lim_{n\to\infty} x[n] = \lim_{z\to 1}(1 - z^{-1})X(z). \tag{6.295}$$

Proof:

$$\mathcal{Z}[x[n] - x[n-1]] = \sum_{n=0}^{\infty}[x[n] - x[n-1]]z^{-n} = X(z) - X(z)z^{-1}; \tag{6.296}$$

i.e.,

$$(1 - z^{-1})X(z) = \lim_{m\to\infty}\sum_{n=0}^{m}[x[n] - x[n-1]]z^{-n}. \tag{6.297}$$

Letting $z \to 1$, we get

$$\lim_{z\to 1}(1-z^{-1})X(z) = \lim_{m\to\infty}\sum_{n=0}^{m}[x[n]-x[n-1]]$$
$$= \lim_{m\to\infty}\left[\sum_{n=0}^{m-1}[x[n]-x[n]] + x[m] - x[-1]\right] = \lim_{m\to\infty} x[m].$$

Note that $x[-1] = 0$.

Owing to the initial- and final-value theorems, the unilateral z-transform is a powerful tool for solving LCCDEs with non-zero initial conditions.

Example 6.10: A system is described by the LCCDE

$$y[n] + 3y[n-1] = x[n] = \alpha u[n]. \tag{6.298}$$

Taking the unilateral z-transform on both sides, we get

$$Y(z) + 3Y(z)z^{-1} + 3y[-1] = X(z) = \frac{\alpha}{1-z^{-1}}. \tag{6.299}$$

- **The particular (zero-state) solution**
  If the system is initially at rest; i.e., $y[-1] = 0$, the above equation can be solved for the output $Y(z)$ to get

$$Y(z) = H(z)X(z) = \frac{1}{1+3z^{-1}}\,\frac{\alpha}{1-z^{-1}} = \frac{3\alpha/4}{1+3z^{-1}} + \frac{\alpha/4}{1-z^{-1}}, \tag{6.300}$$

  where $H(z) = 1/(1+3z^{-1})$ is the system's transfer function. In the time domain this is the particular (or zero-state) solution (caused by the input with zero initial condition):

$$y_{\rm p}[n] = \alpha\left[\frac{1}{4} + \frac{3}{4}(-3)^n\right]u[n]. \tag{6.301}$$

- **The homogeneous (zero-input) solution**
  When the initial condition is non-zero,

$$y[-1] = \beta, \tag{6.302}$$

  but the input is zero $x[n] = 0$, the z-transform of the difference equation becomes

$$Y(z) + 3Y(z)z^{-1} + 3\beta = 0. \tag{6.303}$$

  Solving this for $Y(z)$ we get

$$Y(z) = \frac{-3\beta}{1+3z^{-1}} \tag{6.304}$$

in the time domain; this is the homogeneous (or zero-input) solution (caused by the initial condition with zero input):

$$y_h[n] = -3\beta(-3)^n u[n]. \tag{6.305}$$

When neither $y[-1]$ nor $x[n]$ is zero, we have

$$Y(z) + 3Y(z)z^{-1} + 3\beta = X(z) = \frac{\alpha}{1 - z^{-1}}. \tag{6.306}$$

Solving this algebraic equation in the z-domain for $Y(z)$ we get

$$Y(z) = \frac{\alpha}{(1 + 3z^{-1})(1 - z^{-1})} - \frac{3\beta}{1 + 3z^{-1}}. \tag{6.307}$$

The first term is the particular solution caused by the input alone and the second term is the homogeneous solution caused by the initial condition alone. The $Y(z)$ can be further written as

$$Y(z) = \frac{1}{1 + 3z^{-1}}\left(\frac{3}{4}\alpha - 3\beta\right) + \frac{\alpha}{4}\frac{1}{1 - z^{-1}}, \tag{6.308}$$

and in the time domain, we have the general solution

$$y_g[n] = \left[\left(\frac{3}{4}\alpha - 3\beta\right)(-3)^n + \frac{\alpha}{4}\right] u[n] = y_h[n] + y_p[n], \tag{6.309}$$

which is the sum of both the homogeneous and particular solutions.

Note that the bilateral z-transform can also be used to solve LCCDEs. However, as the bilateral z-transform does not take initial conditions into account, it is always implicitly assumed that the system is initially at rest. If this is not the case, the unilateral z-transform has to be used.

6.3 Homework problems

1. Find the Laplace transform and the corresponding ROC of the following signals.
   (a) $x(t) = [e^{-2t} + e^t \cos(3t)\, u(t)$. (Write $X(s) = \mathcal{L}[x(t)]$ in the form of a rational function, a ratio of two polynomials.)
   (b) $x(t) = e^{-a|t|} = e^{-at}u(t) + e^{at}u(-t)$. (Consider both cases: (1) $a > 0$ and (2) $a < 0$.)
   (c) Another two-sided signal $x(t) = e^{-at}u(t) - e^{-bt}u(-t)$.
   (d) $x(t) = u(-1) - u(1)$.
2. Given the following Laplace transform $X(s)$, find the time function $x(t)$ corresponding to each of the possible ROCs. In each case, decide if $x(t)$ is stable or not, if it is left-sided or right-sided.
   (a) $X(s) = \frac{s^2-3}{s+2}$

(b) $X(s) = \frac{1}{(s+1)(s+2)} = \frac{1}{s+1} - \frac{1}{s+2}$

3. Given the transfer functions $H_R(s)$ and $H_L(s)$ in Eqs. (6.134) and (6.135) respectively, find the impulse responses $h_R(t)$ and $h_L(t)$. Assume $0 < \zeta < 1$. Check to confirm Kirchhoff's voltage law:

$$h_C(t) + h_R(t) + h_L(t) = \delta(t), \tag{6.310}$$

where $h_C(t)$ is given in Eq. (6.140). Do this in two different ways.

4. In Eqs. (6.141) and (6.142) we considered only the step response $Y_C(s)$ in the s-domain and $y_C(t)$ in the time domain when the voltage across C in an RCL system is treated as the output.
   - Find the step responses $Y_L(s)$ and $y_L(t)$ when the voltage across L is treated as the output.
   - Find the step responses $Y_R(s)$ and $y_R(t)$ when the voltage across R is treated as the output.
   - Verify your results by Kirchhoff's voltage law:

$$Y_C(s) + Y_R(s) + Y_L(s) = 1/s, \qquad y_C(t) + y_R(t) + y_L(t) = u(t). \tag{6.311}$$

5. Consider an RC circuit with input voltage $v_{\text{in}}(t) = A\cos(\omega t)u(t)$ applied to the series combination of a resistor R and a capacitor C (representing a sinusoidal input and a switch which is closed at $t = 0$). The initial voltage on C for $t \le 0$ is $v_C(0)$. Use the unilateral Laplace transform method to find the voltage $v_C(t)$ across C for $t > 0$.

6. Use the Laplace transform method to find the response of a second-order system to a sinusoidal input $x(t) = \cos(\omega_0 t)u(t)$:

$$\ddot{y}(t) + 2\zeta\omega_{\text{n}}\dot{y}(t) + \omega_{\text{n}}^2 y(t) = x(t) = \cos(\omega_0 t)u(t). \tag{6.312}$$

   Assume zero initial conditions: $y(0) = \dot{y}(0) = 0$.

7. An LTI system is described by the following LCCDE:

$$\frac{d^2y(t)}{dt^2} + 3\frac{dy(t)}{dt} + 2y(t) = x(t). \tag{6.313}$$

   - Find the particular solution $y_{\text{p}}(t)$ (with zero initial conditions) when the input is $x(t) = e^{-3t}u(t)$.
   - Find the homogeneous solution $y_{\text{h}}(t)$ (with zero input $x(t) = 0$) with initial conditions:

$$y(0) = 1, \quad \dot{y}(0) = \left.\frac{dy(t)}{dt}\right|_{t=0} = -1. \tag{6.314}$$

   - Find the complete solution $y(t) = y_{\text{p}}(t) + y_{\text{h}}(t)$.

8. Find the z-transform and the corresponding ROC of the following signals.
   (a) $x[n] = 0$ for all n except $x[-1] = x[0] = x[1] = 1$
   (b) $x[n] = b^{|n|}$. Consider both cases $b > 1$ and $b < 1$.
   (c) Another two-sided signal $x[n] = a^{-n}u[n] - b^{-n}u[-n-1]$

9. Given the following z-transform $X(z)$, find the corresponding discrete signals.

(a)

$$X(z)=\frac{1-\frac{1}{2}z^{-1}}{1+2z^{-1}-3z^{-2}},\qquad |z|>3. \tag{6.315}$$

(b)

$$X(z)=\frac{1-\frac{1}{2}z^{-1}}{1+\frac{1}{2}z^{-1}},\qquad |z|>1/2. \tag{6.316}$$

10. Given a discrete signal $x[n]$ shown below:

n	$\cdots$	-1	0	1	2	3	4	$\cdots$
$x[n]$	$\cdots$	0	1	2	3	4	0	$\cdots$

find the z-transforms of $y[n]=x_{(2)}[n]$ and then $z[n]=y^{(2)}[n]$ using the up- and down-sampling properties, and compare them with $Y(z)$ and $Z(z)$ obtained directly from the definition of the z-transform.

11. Given the input $x[n]$ and the response $y[n]$ below, find the impulse response $h[n]$ of the LTI system, and decide if the system is causal and stable.

$$x[n]=\left(\frac{1}{5}\right)^n u[n],\qquad y[n]=\left[3\left(\frac{1}{2}\right)^n-2\left(\frac{1}{3}\right)^n\right]u[n]. \tag{6.317}$$

12. Find the impulse response $h[n]$ and step response $y[n]$ of the following discrete LTI system:

$$y[n]-2r\cos\theta y[n-1]+r^2=x[n-1]. \tag{6.318}$$

First, take the approach used in the text ($y[n]=\mathcal{Z}^{-1}[H(z)X(z)]$) to find the responses when the right-hand side is $x[n]$, and then confirm your results by time invariance: if $\mathcal{O}[x[n]]=y[n]$ then $\mathcal{O}[x[n-k]]=y[n-k]$.

13. An LTI system is described by the following LCCDE:

$$6y[n]+5y[n-1]+y[n-2]=2x[n-1]-x[n-2]. \tag{6.319}$$

(a) Find the transfer function $H(z)=Y(z)/X(z)$ and impulse response $h[n]$.

(b) Obtain an inverse system $G(z)=1/H(z)$, so that when the two systems are cascaded the output of $G(z)$ is the same as the input to $H(z)$; i.e., $Z(z)=G(x)Y(z)=G(z)H(z)X(z)=X(z)$ or $z[n]=x[n]$.

(c) Find the impulse response $g[n]$ and the corresponding LCCDE of the inverse system in terms of input $y[n]$ and output $z[n]$. Show the inverse system is not causal.

(d) Introduce a unit delay z^{-1} in the system so that the resulting system $G'(z)=G(z)z^{-1}$ is causal; i.e., its output is the same as the input $z[n]=x[n-1]$. Get its input response $g'[n]$ and give the corresponding LCCDE in terms of input $y[n]$ and output $z[n]$.

14. Design the following four types of filters by specifying the zero and pole positions of a rational transfer function $H(s)$. Use minimum number of zeros and poles (no more than two for each). For each of the four cases, determine the

expression of the frequency response function $H(j\omega)$ and sketch the magnitude plots $|H(j\omega)|$ for $-2\pi 100 < \omega < 2\pi 100$ to verify your design.
- LP filter;
- HP filter;
- BP filter with passing band centered around $\pm 50\pi$;
- BS filter with passing band centered around $\pm 50\pi$.

15. Design the following four types of filters by specifying the zero and pole positions of a rational transfer function $H(z)$. Use minimum number of zeros and poles (no more than two for each). For each of the four cases, determine the expression of the frequency response function $H(e^{j\omega})$ and sketch the magnitude plots $|H(e^{j\omega})|$ for $-\pi < \omega < \pi$ to verify your design.
    - LP filter;
    - HP filter;
    - BP filter with passing band centered around $\pi/2$;
    - BS filter with passing band centered around $\pi/2$.
16. Use the provided Matlab function `ZeroPolePlots.m` to explore the following:
    - The filtering effect of a continuous system with frequency response function $H(j\omega)$ (including both magnitude and phase) with different numbers of zeros and poles and various locations. In particular, set the order of the denominator polynomial to $N = 2$, and explore the Bode plots of $H(j\omega)$ for different order of the numerator polynomial $M = 1, 2, 3$.
    - The filtering effect of a discrete system with frequency response functions $H(e^{j\omega})$ (including both magnitude and phase) with different numbers of zeros and poles and various locations.

7 Fourier-related orthogonal transforms

The Fourier transform converts a complex signal into its complex spectrum. If the signal is real, as in most applications, the imaginary part of the signal is zero, and its spectrum is symmetric; i.e., in both the time and frequency domains half of the data is redundant, causing unnecessary computational time and storage space. In this chapter, we will consider three real orthogonal transforms, all closely related to the Fourier transform with similar behaviors, but the problem of data redundancy is avoided. Here we will always assume the signal in question is real.

7.1 The Hartley transform

7.1.1 Continuous Hartley transform

The Hartley transform is an integral transform based on a real kernel function:

$$\begin{aligned}\phi_f(t) &= \text{cas}(2\pi ft) = \cos(2\pi ft) + \sin(2\pi ft) \\ &= \sqrt{2}\,\sin\left(2\pi ft + \frac{\pi}{4}\right) = \sqrt{2}\,\cos\left(2\pi ft - \frac{\pi}{4}\right) = \phi_t(f) \\ &\qquad (-\infty < t, f < \infty).\end{aligned} \tag{7.1}$$

Here, $\text{cas}(2\pi ft)$ is the *cosine-and-sine (CAS) function* defined as

$$\text{cas}(2\pi ft) = \cos(2\pi ft) + \sin(2\pi ft). \tag{7.2}$$

We can show that this is an uncountable set of orthonormal functions satisfying:

$$\langle \phi_f(t), \phi_{f'}(t) \rangle = \delta(f - f'), \qquad \text{and} \qquad \langle \phi_t(f), \phi_{t'}(f) \rangle = \delta(t - t'). \tag{7.3}$$

Proof:

$$\begin{aligned}\langle \phi_f(t), \phi_{f'}(t) \rangle &= \int_{-\infty}^{\infty} \phi_f(t)\phi_{f'}(t)\,dt \\ &= \int_{-\infty}^{\infty} [\cos(2\pi ft) + \sin(2\pi ft)]\,[\cos(2\pi f't) + \sin(2\pi f't)]\,dt \\ &= \int_{-\infty}^{\infty} [\cos(2\pi ft)\cos(2\pi f't) + \sin(2\pi ft)\sin(2\pi f't)]\,dt\end{aligned}$$

$$+\int_{-\infty}^{\infty} [\cos(2\pi ft)\sin(2\pi f't) + \sin(2\pi ft)\cos(2\pi f't)]\,dt$$
$$= \int_{-\infty}^{\infty} \cos(2\pi(f-f')t)\,dt + \int_{-\infty}^{\infty} \sin(2\pi(f+f')t)\,dt = \delta(f-f'). \quad (7.4)$$

Here, the first term is a Dirac delta $\delta(f-f')$ according to Eq. (1.28), while the second term, an integral of an odd function $\sin(2\pi(f+f')t)$ over all t, is zero and therefore dropped. The second equation of Eq. (7.3) follows immediately as $\phi_f(t) = \phi_t(f)$ is symmetric with respect to t and f.

Given the transform kernel $\phi_f(t) = \text{cas}(2\pi ft)$, the Hartley transform is defined as

$$X_{\text{H}}(f) = \mathcal{H}[x(t)] = \langle x(t), \phi_f(t)\rangle = \int_{-\infty}^{\infty} x(t)\,\text{cas}(2\pi ft)\,dt$$
$$= \int_{-\infty}^{\infty} x(t)[\cos(2\pi ft) + \sin(2\pi ft)]\,dt. \quad (7.5)$$

Here, $X_{\text{H}}(f)$ is a function of frequency f and is called the Hartley spectrum of the signal $x(t)$, similar to its Fourier spectrum $X_{\text{F}}(f)$.

The inverse Hartley transform can be obtained by taking an inner product with $\phi_f(t') = \phi_{t'}(f)$ on both sides of the forward transform above:

$$\langle X_{\text{H}}(f), \phi_{t'}(f)\rangle = \int_{-\infty}^{\infty} X_{\text{H}}(f)\phi_{t'}(f)\,df = \int_{-\infty}^{\infty}\left[\int_{-\infty}^{\infty} x(t)\phi_f(t)\,dt\right]\phi_{t'}(f)\,df$$
$$= \int_{-\infty}^{\infty} x(t)\left[\int_{-\infty}^{\infty} \phi_f(t)\phi_{t'}(f)\,df\right]dt$$
$$= \int_{-\infty}^{\infty} x(t)\delta(t-t')\,dt = x(t'). \quad (7.6)$$

Putting both the forward and inverse Hartley transforms together, we get the following pair of equations:

$$X_{\text{H}}(f) = \mathcal{H}[x(t)] = \langle x(t), \phi_f(t)\rangle = \int_{-\infty}^{\infty} x(t)\ \text{cas}(2\pi ft)\,dt,$$
$$x(t) = \mathcal{H}^{-1}[X_{\text{H}}(f)] = \langle X_{\text{H}}(f), \phi_t(f)\rangle = \int_{-\infty}^{\infty} X_{\text{H}}(f)\ \text{cas}(2\pi ft)\,df. \quad (7.7)$$

We see that the inverse transform is identical to the forward transform:

$$x(t) = \mathcal{H}^{-1}[X_{\text{H}}(f)] = \mathcal{H}[X_{\text{H}}(f)] = \mathcal{H}[\,\mathcal{H}[x(t)]\,]. \quad (7.8)$$

7.1.2 Properties of the Hartley transform

- **Relation to Fourier transform**
Here, we assume the signal $x(t) = \overline{x}(t)$ is real. Its Hartley spectrum can be written as

$$\begin{aligned} X_{\rm H}(f) = \mathcal{H}[x(t)] &= \int_{-\infty}^{\infty} x(t)[\cos(2\pi ft) + \sin(2\pi ft)]\,dt \\ &= \int_{-\infty}^{\infty} x(t)\cos(2\pi ft)\,dt + \int_{-\infty}^{\infty} x(t)\sin(2\pi ft)\,dt \\ &= X_{\rm e}(f) + X_{\rm o}(f), \end{aligned} \tag{7.9}$$

where $X_{\rm e}(f)$ and $X_{\rm o}(f)$ are respectively the even and odd components of the Hartley spectrum $X_{\rm H}(f)$:

$$X_{\rm e}(f) = \frac{1}{2}[X_{\rm H}(f) + X_{\rm H}(-f)] = \int_{-\infty}^{\infty} x(t)\cos(2\pi ft)\,dt,$$
$$X_{\rm o}(f) = \frac{1}{2}[X_{\rm H}(f) - X_{\rm H}(-f)] = \int_{-\infty}^{\infty} x(t)\sin(2\pi ft)\,dt.$$

On the other hand, the Fourier spectrum of $x(t)$ is

$$\begin{aligned} X_{\rm F}(f) = \mathcal{F}[x(t)] &= \int_{-\infty}^{\infty} x(t)e^{-j2\pi ft}\,dt = \int_{-\infty}^{\infty} x(t)[\cos(2\pi ft) - j\sin(2\pi ft)]\,dt \\ &= \int_{-\infty}^{\infty} x(t)\cos(2\pi ft)\,dt - j\int_{-\infty}^{\infty} x(t)\sin(2\pi ft)\,dt \\ &= X_{\rm e}(f) - j\,X_{\rm o}(f). \end{aligned} \tag{7.10}$$

We see that both the Hartley and Fourier spectra of a real signal $x(t)$ are composed of the same even and odd components $X_{\rm e}(f)$ and $X_{\rm o}(f)$, which are also the real and imaginary parts (negative version) of the Fourier spectrum $X_{\rm F}(f)$:

$$X_{\rm e}(f) = \mathrm{Re}[X_{\rm F}(f)], \qquad X_{\rm o}(f) = -\,\mathrm{Im}[X_{\rm F}(f)]; \tag{7.11}$$

i.e., the Hartley spectrum can be obtained as a linear combination of the real and imaginary parts of the Fourier spectrum:

$$X_{\rm H}(f) = X_{\rm e}(f) + X_{\rm o}(f) = \mathrm{Re}[X_{\rm F}(f)] - \mathrm{Im}[X_{\rm F}(f)]. \tag{7.12}$$

In particular, we consider the two cases when the real signal $x(t)$ is either even or odd:
  - If $x(t) = x(-t)$ is even, its Fourier spectrum is real and even with $\mathrm{Im}[X_{\rm F}(f)] = 0$ and $X_{\rm H}(f) = X_{\rm F}(f)$; i.e., the Hartley spectrum is identical to the Fourier spectrum.
  - If $x(t) = -x(-t)$ is odd, its Fourier spectrum is imaginary and odd with $\mathrm{Re}[X_{\rm F}(f)] = 0$ and $X_{\rm H}(f) = -X_{\rm F}(f)$; i.e., the Hartley spectrum is the negative version of its Fourier spectrum.

- **Convolution in both time and frequency domain**
  Let $z(t) = x(t) * y(t)$ be the convolution of $x(t)$ and $y(t)$, then the Hartley spectrum $Z_H(f) = \mathcal{H}[z(t)]$ is

$$\begin{aligned} Z_H(f) &= \mathcal{H}[x(t) * y(t)] \\ &= \frac{1}{2}\left[X_H(f)Y_H(f) - X_H(-f)Y_H(-f) + X_H(f)Y_H(-f) + X_H(-f)Y_H(f)\right], \end{aligned} \tag{7.13}$$

  where $X_H(f) = \mathcal{H}[x(t)]$ and $Y_H(f) = \mathcal{H}[y(t)]$ are the Hartley spectra of $x(t)$ and $y(t)$, respectively.
  Proof:
  According to the convolution theorem of the Fourier transform (Eq. (3.122)), the Fourier spectrum $Z_F(f) = \mathcal{F}[z(t)]$ is the product of the spectra $X_F(f) = \mathcal{F}[x(t)]$ and $Y_F(f) = \mathcal{F}[y(t)]$:

$$\begin{aligned} Z_F(t) = X_F(f)\,Y_F(f) &= [X_e(f) - j\,X_o(f)]\,[Y_e(f) - j\,Y_o(f)] \\ &= [X_e(f)Y_e(f) - X_o(f)Y_o(f)] \\ &\quad - j\,[X_o(f)Y_e(f) + X_e(f)Y_o(f)] \\ &= Z_e(f) - j\,Z_o(f), \end{aligned} \tag{7.14}$$

  where $Z_e(f)$ and $Z_o(f)$ are respectively the even and odd components of $Z_H(f)$:

$$\begin{aligned} Z_e(f) &= X_e(f)Y_e(f) - X_o(f)Y_o(f) \\ &= \frac{1}{2}[X_H(f)Y_H(-f) + X_H(-f)Y_H(f)], \\ Z_o(f) &= X_e(f)Y_o(f) + X_o(f)Y_e(f) \\ &= \frac{1}{2}[X_H(f)Y_H(f) - X_H(-f)Y_H(-f)]. \end{aligned} \tag{7.15}$$

  Substituting these into $Z_H(f) = Z_e(f) + Z_o(f)$, we get Eq. (7.13).
  Also, based on Eq. (3.123), we can similarly prove the Hartley spectrum of the product of two functions $z(t) = x(t)\,y(t)$ is

$$\begin{aligned} Z_H(t) &= \mathcal{H}[x(t)\,y(t)] \\ &= \frac{1}{2}[X_H(f) * Y_H(f) - X_H(-f) * Y_H(-f) \\ &\quad + X_H(f) * Y_H(-f) + X_H(-f) * Y_H(f)]. \end{aligned} \tag{7.16}$$

- **Correlation**
  Let $z(t) = x(t) \star y(t)$ be the correlation of $x(t)$ and $y(t)$, then the Hartley spectrum $Z_H(f) = \mathcal{H}[z(t)]$ is

$$\begin{aligned} Z_H(f) &= \mathcal{H}[x(t) \star y(t)] \\ &= \frac{1}{2}\left[X_H(f)Y_H(f) + X_H(-f)Y_H(-f) + X_H(f)Y_H(-f) - X_H(-f)Y_H(f)\right]. \end{aligned} \tag{7.17}$$

In particular, when $x(t) = y(t)$; i.e., $X_{\rm H}(f) = Y_{\rm H}(f)$, then the odd part $Z_{\rm o}(f)$ of its spectrum is zero, and the correlation $x(t) \star y(t) = x(t) \star x(t)$ becomes autocorrelation, the Eq. (7.17) becomes

$$\mathcal{H}[x(t) \star x(t)] = \frac{1}{2}[X_H^2(f) + X_H^2(-f)]. \tag{7.18}$$

Proof:

According to the correlation property of the Fourier transform (Eq. (3.117)), the Fourier spectrum $Z_{\rm F}(f) = \mathcal{F}[z(t)]$ is the product of the spectra $X_{\rm F}(f) = \mathcal{F}[x(t)]$ and $Y_{\rm F}(f) = \mathcal{F}[y(t)]$:

$$\begin{aligned} Z_{\rm F}(t) &= X_{\rm F}(f)\,\overline{Y}_{\rm F}(f) = [X_{\rm e}(f) - j\,X_{\rm o}(f)]\;[Y_{\rm e}(f) + j\,Y_{\rm o}(f)] \\ &= [X_{\rm e}(f)Y_{\rm e}(f) + X_{\rm o}(f)Y_{\rm o}(f)] \\ &- j\,[X_{\rm o}(f)Y_{\rm e}(f) - X_{\rm e}(f)Y_{\rm o}(f)] \\ &= Z_{\rm e}(f) - j\,Z_{\rm o}(f), \end{aligned} \tag{7.19}$$

where $X_{\rm e}(f)$, $X_{\rm e}(f)$ and $Y_{\rm o}(f)$, $Y_{\rm o}(f)$ are the even and odd components of $X_{\rm H}(f)$ and $Y_{\rm H}(f)$, respectively:

$$X_{\rm e}(f) = \frac{1}{2}[X_{\rm H}(f) + X_{\rm H}(-f)], \qquad X_{\rm o}(f) = \frac{1}{2}[X_{\rm H}(f) - X_{\rm H}(-f)]$$
$$Y_{\rm e}(f) = \frac{1}{2}[Y_{\rm H}(f) + Y_{\rm H}(-f)], \qquad Y_{\rm o}(f) = \frac{1}{2}[Y_{\rm H}(f) - Y_{\rm H}(-f)],$$

and $Z_{\rm e}(f)$ and $Z_{\rm o}(f)$ are the even and odd components of $Z_{\rm H}(f)$:

$$\begin{aligned} Z_{\rm e}(f) &= X_{\rm e}(f)Y_{\rm e}(f) + X_{\rm o}(f)Y_{\rm o}(f) \\ &= \frac{1}{2}[X_{\rm H}(f)Y_{\rm H}(f) + X_{\rm H}(-f)Y_{\rm H}(-f)], \\ Z_{\rm o}(f) &= X_{\rm o}(f)Y_{\rm e}(f) - X_{\rm e}(f)Y_{\rm o}(f) \\ &= \frac{1}{2}[X_{\rm H}(f)Y_{\rm H}(-f) - X_{\rm H}(-f)Y_{\rm H}(f)]. \end{aligned}$$

Substituting these into $Z_{\rm H}(f) = Z_{\rm e}(f) + Z_{\rm o}(f)$, we get Eq. (7.17).

7.1.3 Hartley transform of typical signals

As the Hartley transform is closely related to the Fourier transform, the Hartley spectra of many signals are similar to or the same as their Fourier spectra. In particular, if the signal is either real even or real odd, its Hartley spectrum is either identical or the negative version of its Fourier spectrum. Therefore we only consider the following two examples where the real signal is neither even nor odd.

- **Combination of sinusoids**

$$x(t) = \cos(2\pi f_0 t + \theta) = \frac{1}{2}\left[e^{j2\pi f_0 t}e^{j\theta} + e^{-j2\pi f_0 t}e^{-j\theta}\right]. \tag{7.20}$$

The Fourier transform is

$$\begin{aligned} X_F(f) &= \frac{1}{2}[\delta(f-f_0)e^{j\theta} + \delta(f+f_0)e^{-j\theta}] \\ &= \frac{1}{2}[\delta(f-f_0)(\cos\theta + j\sin\theta) + \delta(f+f_0)(\cos\theta - j\sin\theta)] \\ &= \frac{1}{2}[\delta(f-f_0)\cos\theta + \delta(f+f_0)\cos\theta] \\ &\quad + \frac{j}{2}[\delta(f-f_0)\sin\theta - \delta(f+f_0)\sin\theta]. \end{aligned}$$

Its Hartley transform is

$$\begin{aligned} X_H(f) &= \text{Re}[X_F(f)] - \text{Im}[X_F(f)] \\ &= \frac{1}{2}[\delta(f-f_0)(\cos\theta - \sin\theta) + \delta(f+f_0)(\cos\theta + \sin\theta)]. \end{aligned}$$

In particular, if $\theta = 0$, the signal becomes even $x(t) = \cos(2\pi f_0 t)$, and its Hartley spectrum becomes the same as the Fourier spectrum $X_F(f)$:

$$X_H(f) = \mathcal{H}[\cos(2\pi f_0 t)] = \frac{1}{2}[\delta(f-f_0) + \delta(f+f_0)]. \quad (7.21)$$

Also if $\theta = -\pi/2$, we have $x(t) = \cos(2\pi f_0 t - \pi/2) = \sin(2\pi f_0 t)$, and its Hartley spectrum becomes

$$X_H(f) = \mathcal{H}[\sin(2\pi f_0 t)] = \frac{1}{2}[\delta(f-f_0) - \delta(f+f_0)], \quad (7.22)$$

which is the negative version of the imaginary part of the Fourier spectrum

$$X_F(f) = \frac{1}{2j}[\delta(f-f_0) - \delta(f+f_0)] = \frac{j}{2}[-\delta(f-f_0) + \delta(f+f_0)]. \quad (7.23)$$

For a specific example, consider a signal containing four terms:

$$x(t) = 1 + 3\cos(2\pi 16t) + 2\sin(2\pi 64t) + 2\cos(2\pi 128t + \pi/3). \quad (7.24)$$

In Fig. 7.1 this signal, together with its reconstruction (dashed line) from its Hartley spectrum, is plotted (top), and its Hartley and Fourier spectra are plotted in the middle and bottom panels respectively. We see that the DC (first term) and cosine component without phase shift (second term) appear the same in the two spectra, and the sine component (third term) appears in the two spectra as the negative version of each other. Finally, the cosine function with a phase shift of $\pi/3$ (fourth term) shows up in the Hartley spectrum as the difference between the real and imaginary parts of the Fourier spectrum.

- **Exponential decay function**

$$x(t) = e^{-at}u(t). \quad (7.25)$$

This function together with its Hartley and Fourier spectra are shown respectively in top, middle, and bottom panels of Fig. 7.2.

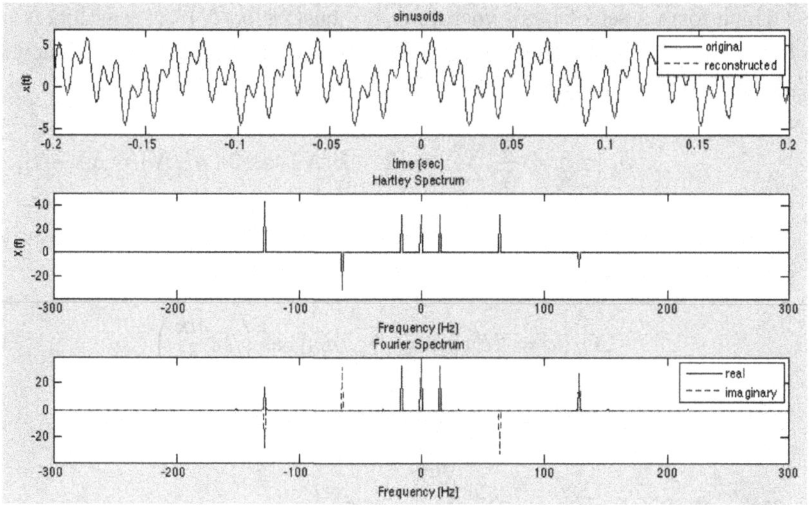

Figure 7.1 The Hartley and Fourier spectra of sinusoidal components of a signal.

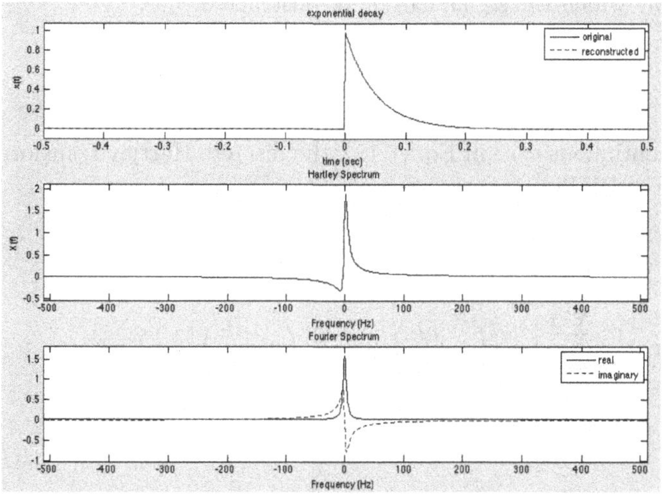

Figure 7.2 The Hartley and Fourier spectra of exponential decay.

7.1.4 Discrete Hartley transform

When a continuous signal $x(t)$ is truncated to have a finite duration $0 < t < T$ and sampled with sampling rate $F = 1/t_0$, it becomes a set of $N = T/t_0$ samples that form a vector $\boldsymbol{x} = [x[0], \dots, x[N-1]]^{\mathrm{T}}$ in the N-D space. Correspondingly, the Hartley transform also becomes a discrete Hartley transform based on a discrete kernel:

$$\phi_k[n] = \frac{1}{\sqrt{N}} \operatorname{cas}\left(2\pi \frac{nk}{N}\right) = \frac{1}{\sqrt{N}} \left[\cos\left(2\pi \frac{nk}{N}\right) + \sin\left(2\pi \frac{nk}{N}\right)\right], \tag{7.26}$$

which form a set of basis vectors $\boldsymbol{\phi}_k = [\mathrm{cas}(2\pi\, 0k/N), \ldots, \mathrm{cas}(2\pi\,(N-1)k/N)]^{\mathrm{T}}$ $(k = 0, \ldots, N-1)$ that span the N-D vector space. We can show that these vectors are orthogonal:

$$\langle \boldsymbol{\phi}_k, \boldsymbol{\phi}_l \rangle = \frac{1}{N} \sum_{n=0}^{N-1} \mathrm{cas}(2\pi\, nk/N)\, \mathrm{cas}(2\pi\, nl/N) = \delta[k-l]. \tag{7.27}$$

The proof is left for the reader as a homework problem. The discrete Hartley transform of a signal vector $\boldsymbol{x}$ is then defined as

$$\begin{aligned} X_H[k] = \mathcal{H}[x[n]] &= \sum_{n=0}^{N-1} x[n]\, \mathrm{cas}\left(2\pi \frac{nk}{N}\right) \\ &= \frac{1}{\sqrt{N}} \sum_{n=0}^{N-1} \left[\cos\left(2\pi \frac{nk}{N}\right) + \sin\left(2\pi \frac{nk}{N}\right)\right]. \end{aligned} \tag{7.28}$$

Here, $X_H[k]$ $(k = 0, \ldots, N-1)$ are N frequency components of the signal, similar to the case of the discrete Fourier transform. Owing to the orthogonality of $\boldsymbol{\phi}_k$ and following the same method used to derive Eq. (7.6), we get the inverse transform by which the signal can be reconstructed:

$$x[n] = \mathcal{H}^{-1}[X_H[k]] = \frac{1}{\sqrt{N}} \sum_{k=0}^{N-1} X[k]\, \mathrm{cas}\left(2\pi \frac{nk}{N}\right). \tag{7.29}$$

As in the continuous case in Eq. (7.12), the discrete Hartley transform is closely related to the DFT:

$$\begin{aligned} X_F[k] = \mathcal{F}[x[n]] &= \frac{1}{\sqrt{N}} \sum_{n=0}^{N-1} x[n] e^{-j2\pi nk/N} \\ = \frac{1}{\sqrt{N}} \sum_{n=0}^{N-1} &\left[\cos\left(2\pi \frac{nk}{N}\right) - j \sin\left(2\pi \frac{nk}{N}\right)\right] = X_{\mathrm{e}}[k] - j\, X_{\mathrm{o}}[k] \\ &(k = 0, \ldots, N-1), \end{aligned} \tag{7.30}$$

where

$$X_{\mathrm{e}}[k] = \mathrm{Re}[X_F[k]] = \frac{1}{\sqrt{N}} \sum_{n=0}^{N-1} x[n] \cos\left(2\pi \frac{nk}{N}\right),$$

$$X_{\mathrm{o}}[k] = -\,\mathrm{Im}[X_F[k]] = \frac{1}{\sqrt{N}} \sum_{n=0}^{N-1} x[n] \sin\left(2\pi \frac{nk}{N}\right).$$

and the discrete Hartley spectrum can also be obtained from the DFT:

$$X_H[k] = \mathcal{H}[x[n]] = X_{\mathrm{e}}[k] + X_{\mathrm{o}}[k] = \mathrm{Re}[X_F[k]] - \mathrm{Im}[X_F[k]]. \tag{7.31}$$

Based on Eq. (7.12), the discrete Hartley transform can be trivially implemented as the difference between the real and imaginary parts of the corresponding DFT.

Correspondingly, the Hartley transform matrix can also be easily obtained as the difference between the real and imaginary parts of the DFT matrix (Eq. (4.125):

$$\boldsymbol{H} = \text{Re}[\overline{\boldsymbol{W}}] - \text{Im}[\overline{\boldsymbol{W}}]. \tag{7.32}$$

In particular, the discrete Hartley transform matrices for $N = 2,\ 4,\ 8$ are listed below:

$$\boldsymbol{H}_{2\times 2} = \frac{1}{\sqrt{2}}\begin{bmatrix} 1 & 1 \\ 1 & -1 \end{bmatrix} = \begin{bmatrix} 0.71 & 0.71 \\ 0.71 & -0.71 \end{bmatrix}, \tag{7.33}$$

$$\boldsymbol{H}_{4\times 4} = \frac{1}{2}\begin{bmatrix} 1 & 1 & 1 & 1 \\ 1 & 1 & -1 & -1 \\ 1 & -1 & 1 & -1 \\ 1 & -1 & -1 & 1 \end{bmatrix}, \tag{7.34}$$

$$\boldsymbol{H}_{8\times 8} = \frac{1}{\sqrt{8}}\begin{bmatrix} 1.0 & 1.00 & 1.0 & 1.00 & 1.0 & 1.00 & 1.0 & 1.00 \\ 1.0 & 1.41 & 1.0 & 0.00 & -1.0 & -1.41 & -1.0 & -0.00 \\ 1.0 & 1.00 & -1.0 & -1.00 & 1.0 & 1.00 & -1.0 & -1.00 \\ 1.0 & 0.00 & -1.0 & 1.41 & -1.0 & -0.00 & 1.0 & -1.41 \\ 1.0 & -1.00 & 1.0 & -1.00 & 1.0 & -1.00 & 1.0 & -1.00 \\ 1.0 & -1.41 & 1.0 & -0.00 & -1.0 & 1.41 & -1.0 & 0.00 \\ 1.0 & -1.00 & -1.0 & 1.00 & 1.0 & -1.00 & -1.0 & 1.00 \\ 1.0 & -0.00 & -1.0 & -1.41 & -1.0 & 0.00 & 1.0 & 1.41 \end{bmatrix}. \tag{7.35}$$

Note that these matrices are real, orthogonal, and symmetric, $\boldsymbol{H}^{-1} = \boldsymbol{H}^{\mathrm{T}} = \boldsymbol{H} = \overline{\boldsymbol{H}}$; i.e., they are used for both forward and inverse transforms. The $N = 8$ elements of each of the N row or column vectors can be considered as N samples of the corresponding continuous Hartley CAS functions $\text{cas}(2\pi ft) = \cos(2\pi ft) + \sin(2\pi ft)$ (third column of Fig. 7.3), as the sum of the corresponding cosine and sine functions (first and second columns of Fig. 7.3).

Example 7.1: As considered before, the DFT of a 8-D signal vector $\boldsymbol{x} = [0, 0, 2, 3, 4, 0, 0, 0]^{\mathrm{T}}$ is (Eq. (4.172)) $\boldsymbol{X} = \boldsymbol{X}_{\mathrm{r}} + j\boldsymbol{X}_{\mathrm{j}}$ where

$$\boldsymbol{X}_{\mathrm{r}} = \left[3.18,\ -2.16,\ 0.71,\ -0.66,\ 1.06,\ -0.66,\ 0.71,\ -2.16\right]^{\mathrm{T}},$$
$$\boldsymbol{X}_{\mathrm{j}} = \left[0.0,\ -1.46,\ 1.06,\ -0.04,\ 0.0,\ 0.04,\ -1.06,\ 1.46\right]^{\mathrm{T}}. \tag{7.36}$$

The discrete Hartley transform of this signal vector is

$$\boldsymbol{X}_H = \boldsymbol{X}_{\mathrm{r}} - \boldsymbol{X}_{\mathrm{j}} = [3.18, -0.71, -0.35, -0.62, 1.06, -0.71, 1.77, -3.62]^{\mathrm{T}}. \tag{7.37}$$

The original signal can be reconstructed by the inverse Hartley transform as a linear combination of the CAS functions with progressively higher different frequencies, as shown in the right column of Fig. 7.3

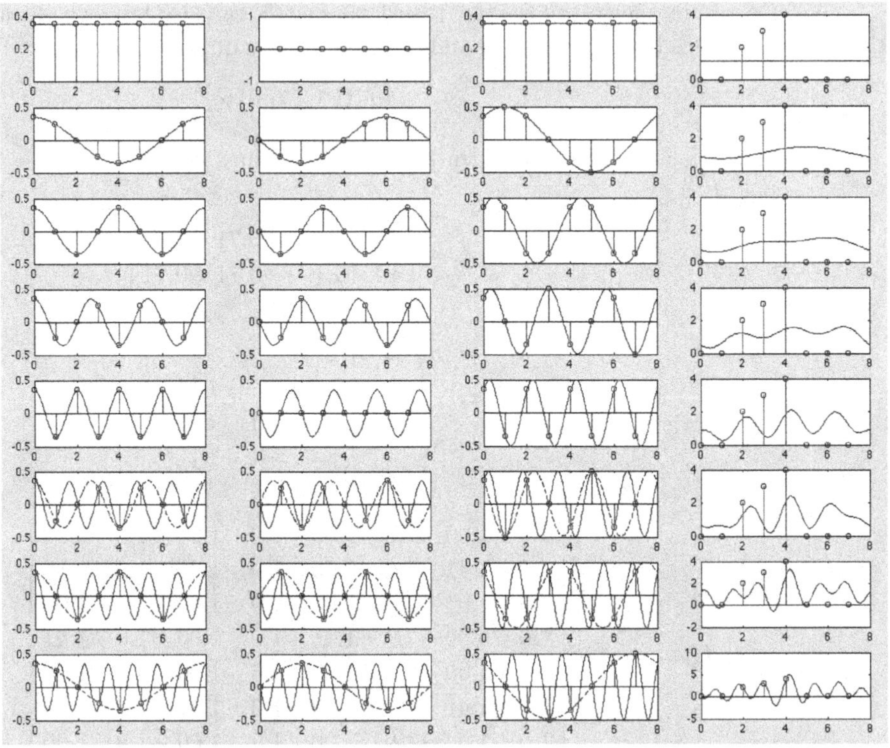

Figure 7.3 Basis functions and vectors of eight-point Hartley transform. The $N = 8$ cosine and sine (real and imaginary) parts of the Fourier basis and their sum, the Hartley CAS transform, are shown in the first, second, and third columns, respectively; the reconstructions of a discrete signal with progressively more and higher frequency components are shown in the fourth columns. The signal is perfectly reconstructed when all N components are included.

7.1.5 The 2-D Hartley transform

Similar to the 2-D Fourier transform, the 2-D Hartley transform of a signal array $x[m, n]$ $(0 \le m \le M-1,\ 0 \le n \le N-1)$ can be defined as

$$X[k, l] = \mathcal{H}[x[m, n]] = \sum_{m=0}^{M-1} \sum_{n=0}^{N-1} x[m, n] \phi_{k,l}[m, n], \tag{7.38}$$

where $\phi_{k,l}[m, n]$ is a discrete 2-D kernel function. Unlike the 2-D Fourier transform with a unique kernel function $\phi_{k,l}[m, n] = e^{j2\pi(\frac{mk}{M} + \frac{nl}{N})} = e^{j2\pi\frac{mk}{M}} e^{j2\pi\frac{nl}{N}}$, there exist two different versions of the 2-D Hartley transform depending on which of the following kernel functions is used:

$$\phi'_{k,l}[m,n] = \text{cas}\left(2\pi\left(\frac{mk}{M} + \frac{nl}{N}\right)\right), \tag{7.39}$$

$$\phi''_{k,l}[m,n] = \text{cas}\left(2\pi\frac{mk}{M}\right)\,\text{cas}\left(2\pi\frac{nl}{N}\right). \tag{7.40}$$

Note that, like the Fourier kernel, the second kernel is separable (i.e., it can be written as a product of two 1-D kernels, one for each of the two dimensions), while the first one is not. As shown below, these two different kernel functions are very similar to, but different from, each other:

$$\begin{aligned}
&\text{cas}\left(2\pi\frac{mk}{M}\right)\,\text{cas}\left(2\pi\frac{nl}{N}\right)\\
&= \left[\cos\left(2\pi\frac{mk}{M}\right) + \sin\left(2\pi\frac{mk}{M}\right)\right]\left[\cos\left(2\pi\frac{nl}{N}\right) + \sin\left(2\pi\frac{nl}{N}\right)\right]\\
&= \left[\cos\left(2\pi\frac{mk}{M}\right)\cos\left(2\pi\frac{nl}{N}\right) + \sin\left(2\pi\frac{mk}{M}\right)\sin\left(2\pi\frac{nl}{N}\right)\right]\\
&\quad \left[\sin\left(2\pi\frac{mk}{M}\right)\cos\left(2\pi\frac{nl}{N}\right) + \cos\left(2\pi\frac{mk}{M}\right)\sin\left(2\pi\frac{nl}{N}\right)\right]\\
&= \cos\left(2\pi\left(\frac{mk}{M} - \frac{nl}{N}\right)\right) + \sin\left(2\pi\left(\frac{mk}{M} + \frac{nl}{N}\right)\right)\\
&\neq \cos\left(2\pi\left(\frac{mk}{M} + \frac{nl}{N}\right)\right) + \sin\left(2\pi\left(\frac{mk}{M} + \frac{nl}{N}\right)\right)\\
&= \text{cas}\left(2\pi\left(\frac{mk}{M} + \frac{nl}{N}\right)\right).
\end{aligned} \tag{7.41}$$

We see that the only difference between the two kernels is the sign of the argument of the cosine function. Both of these kernel functions satisfy the orthogonality

$$\sum_{m=0}^{M-1}\sum_{n=0}^{N-1} \phi_{k,l}[m,n]\phi_{k',l'}[m,n] = \delta[k-k', l-l'], \tag{7.42}$$

either of which can be used for the 2-D Hartley transform.

- Based on the inseparable kernel $\phi'_{k,l}[m,n] = \text{cas}[2\pi(\frac{mk}{M} + \frac{nl}{N})]$, the forward Hartley transform is carried out following the definition:

$$\begin{aligned}
X'_H[k,l] &= \frac{1}{\sqrt{MN}}\sum_{m=0}^{M-1}\sum_{n=0}^{N-1} x[m,n]\,\text{cas}\left(2\pi\left(\frac{mk}{M} + \frac{nl}{N}\right)\right)\\
&= \frac{1}{\sqrt{MN}}\sum_{m=0}^{M-1}\sum_{n=0}^{N-1} x[m,n]\left[\cos\left(2\pi\left(\frac{mk}{M} + \frac{nl}{N}\right)\right) + \sin\left(2\pi\left(\frac{mk}{M} + \frac{nl}{N}\right)\right)\right].
\end{aligned} \tag{7.43}$$

This Hartley transform can be compared with the 2-D Fourier transform:

$$X_F[k,l] = \frac{1}{\sqrt{MN}} \sum_{m=0}^{M-1} \sum_{n=0}^{N-1} x[m,n] e^{-2\pi(\frac{mk}{M}+\frac{nl}{N})}$$
$$= \frac{1}{\sqrt{MN}} \sum_{m=0}^{M-1} \sum_{n=0}^{N-1} x[m,n] \left[\cos\left(2\pi\left(\frac{mk}{M}+\frac{nl}{N}\right)\right) - j\sin\left(2\pi\left(\frac{mk}{M}+\frac{nl}{N}\right)\right)\right]$$
$$= X_e[k,l] - jX_o[k,l] = \mathrm{Re}[X_e[k,l]] + j\,\mathrm{Im}[X_o[k,l]], \tag{7.44}$$

where

$$X_e[k,l] = \mathrm{Re}[X_F[k,l]] = \frac{1}{\sqrt{MN}} \sum_{m=0}^{M-1} \sum_{n=0}^{N-1} x[m,n] \cos\left(2\pi\left(\frac{mk}{M}+\frac{nl}{N}\right)\right) \tag{7.45}$$

and

$$X_o[k,l] = -\,\mathrm{Im}[X_F[k,l]] = \frac{1}{\sqrt{MN}} \sum_{m=0}^{M-1} \sum_{n=0}^{N-1} x[m,n] \sin\left(2\pi\left(\frac{mk}{M}+\frac{nl}{N}\right)\right) \tag{7.46}$$

are respectively the 2-D even and odd components of $X_F[k,l]$. We see the same relationship between the Hartley and Fourier transforms as in 1-D case in Eq. (7.12):

$$X'_H[k,l] = X_e[k,l] + X_o[k,l] = \mathrm{Re}[X_F[k,l]] - \mathrm{Im}[X_F[k,l]]. \tag{7.47}$$

Extending the orthogonality in Eq. (7.27) from 1-D to 2-D, we get

$$\frac{1}{MN} \sum_{m=0}^{M-1} \sum_{n=0}^{N-1} \mathrm{cas}\left(2\pi\left(\frac{mk}{M}+\frac{nl}{N}\right)\right) \mathrm{cas}\left(2\pi\left(\frac{mk'}{M}+\frac{nl'}{N}\right)\right)$$
$$= \delta[k-k', l-l']. \tag{7.48}$$

Based on this orthogonality, and following the same method used to derive Eq. (7.29), we get the inverse transform by which the signal can be reconstructed:

$$x[m,n] = \frac{1}{\sqrt{MN}} \sum_{m=0}^{M-1} \sum_{n=0}^{N-1} X'_H[k,l] \,\mathrm{cas}\left(2\pi\left(\frac{mk}{M}+\frac{nl}{N}\right)\right). \tag{7.49}$$

- Based on the separable kernel $\phi''_{k,l}[m,n] = \mathrm{cas}(2\pi\, mk/M)\,\mathrm{cas}(2\pi\, nl/N)$, the 2-D Hartley transform can also be carried out in two steps of 1-D transforms each for one of the two dimensions, just as the 2-D Fourier kernel $e^{j2\pi(mk/M+nl/N)}$:

$$X''_H[k,l] = \frac{1}{\sqrt{MN}} \sum_{n=0}^{N-1} \sum_{m=0}^{M-1} x[m,n] \,\mathrm{cas}\left(2\pi\,\frac{mk}{M}\right) \mathrm{cas}\left(2\pi\,\frac{nl}{N}\right)$$
$$= \frac{1}{\sqrt{N}} \sum_{n=0}^{N-1} \left[\frac{1}{\sqrt{M}} \sum_{m=0}^{M-1} x[m,n] \,\mathrm{cas}\left(2\pi\,\frac{mk}{M}\right)\right] \mathrm{cas}\left(2\pi\,\frac{nl}{N}\right). \tag{7.50}$$

According to Eq. (7.41), this transform can be further written as

$$X''_H[k,l] = \frac{1}{\sqrt{MN}} \sum_{m=0}^{M-1} \sum_{n=0}^{N-1} x[m,n] \left[\cos\left(2\pi\left(\frac{mk}{M} - \frac{nl}{N}\right)\right) + \sin\left(2\pi\left(\frac{mk}{M} + \frac{nl}{N}\right)\right)\right]$$
$$= X_e[k,-l] + X_o[k,l] = \mathrm{Re}[X_F[k,-l]] - \mathrm{Im}[X_F[k,l]]. \tag{7.51}$$

Similarly, the inverse transform can also be carried out in two stages

$$x[m,n] = \frac{1}{\sqrt{MN}} \sum_{m=0}^{M-1} \sum_{n=0}^{N-1} X''_H[k,l] \operatorname{cas}\left(2\pi\frac{mk}{M}\right) \operatorname{cas}\left(2\pi\frac{nl}{N}\right)$$
$$= \frac{1}{\sqrt{N}} \sum_{m=0}^{M-1} \left[\frac{1}{\sqrt{M}} \sum_{n=0}^{N-1} X''_H[k,l] \operatorname{cas}\left(2\pi\frac{mk}{M}\right)\right] \operatorname{cas}\left(2\pi\frac{nl}{N}\right). \tag{7.52}$$

The inverse 2-D Hartley transforms in Eqs. (7.49) and (7.52) represent the given 2-D signal as a linear combination of a set of MN 2-D basis functions each of size $M \times N$. Such 2-D basis functions can be either inseparable such as $\operatorname{cas}(2\pi(mk/M + nl/N))$ in Eq. (7.49), or inseparable such as $\operatorname{cas}(2\pi mk/M)\operatorname{cas}(2\pi nl/N)$ in Eq. (7.52). The weighting coefficients $X'_H[k,l]$ or $X''_H[k,l]$ are obtained in the forward transform in Eqs. (7.43) or (7.50) as the projection of the signal onto each corresponding basis function. Fig. 7.4 shows two sets of $M \times N = 8 \times 8 = 64$ such 2-D Hartley basis functions that are either separable (left) or inseparable (right).

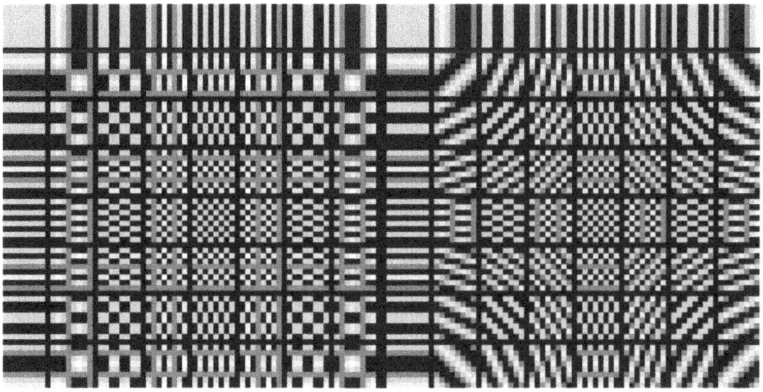

Figure 7.4 The $M \times N = 8 \times 8 = 64$ basis functions for the 2-D Hartley transform. The left half of the image shows the basis functions based on the separable kernel $\phi''_{k,l}[m,n]$, and the right half based on the inseparable kernel $\phi'_{k,l}[m,n]$. The DC component is at the top-left corner, and the highest frequency component in both horizontal and vertical directions is at the middle, which is the same as the 2-D Fourier basis.

The inverse transform in either Eq. (7.49) or Eq. (7.52) is identical to the forward transform. Also, to better compare the two versions of the 2-D Hartley transform, we put Eqs. (7.47) and (7.51) side by side

$$X'_H[k,l] = X_e[k,l] + X_o[k,l] = \mathrm{Re}[X_F[k,l]] - \mathrm{Im}[X_F[k,l]],$$
$$X''_H[k,l] = X_e[k,-l] + X_o[k,l] = \mathrm{Re}[X_F[k,-l]] - \mathrm{Im}[X_F[k,l]],$$

and note that the difference between the two methods is simply the sign of the argument in the even term, it is either $X_e[k,l]$ or $X_e[k,-l] = X_e[-k,l]$ (even). As $X_e[k+M, l+N] = X_e[k,l]$ are periodic, we have $X_e[k,-l] = X_e[k, N-l]$.

Example 7.2: Given the 2-D signal array

$$\boldsymbol{x} = \begin{bmatrix} 0.0 & 0.0 & 0.0 & 0.0 & 0.0 & 0.0 & 0.0 & 0.0 \\ 0.0 & 0.0 & 70.0 & 80.0 & 90.0 & 0.0 & 0.0 & 0.0 \\ 0.0 & 0.0 & 90.0 & 100.0 & 110.0 & 0.0 & 0.0 & 0.0 \\ 0.0 & 0.0 & 110.0 & 120.0 & 130.0 & 0.0 & 0.0 & 0.0 \\ 0.0 & 0.0 & 130.0 & 140.0 & 150.0 & 0.0 & 0.0 & 0.0 \\ 0.0 & 0.0 & 0.0 & 0.0 & 0.0 & 0.0 & 0.0 & 0.0 \\ 0.0 & 0.0 & 0.0 & 0.0 & 0.0 & 0.0 & 0.0 & 0.0 \\ 0.0 & 0.0 & 0.0 & 0.0 & 0.0 & 0.0 & 0.0 & 0.0 \end{bmatrix}, \tag{7.53}$$

the Hartley spectrum corresponding to inseparable kernel $\phi'_{k,l}[m,n]$ is

$$\boldsymbol{X}' = \begin{bmatrix} 165.0 & -10.0 & -45.0 & -32.2 & 55.0 & -10.0 & 65.0 & -187.8 \\ 27.4 & -100.5 & 54.8 & 6.3 & 9.1 & -15.2 & -47.7 & 65.8 \\ 0.0 & 17.1 & -10.0 & -2.9 & 0.0 & 2.9 & 10.0 & -17.1 \\ -26.4 & -15.2 & 17.7 & 7.1 & -8.8 & -0.5 & -20.6 & 46.6 \\ 15.0 & 0.0 & -5.0 & -2.9 & 5.0 & 0.0 & 5.0 & -17.1 \\ -57.4 & 20.2 & 5.3 & 9.2 & -19.1 & 5.5 & -12.3 & 48.7 \\ 30.0 & -17.1 & 0.0 & -2.9 & 10.0 & -2.9 & 0.0 & -17.1 \\ -153.6 & 105.5 & -17.7 & 18.4 & -51.2 & 20.2 & 0.6 & 77.9 \end{bmatrix} \tag{7.54}$$

and the Hartley spectrum corresponding to separable kernel $\phi''_{k,l}[m,n]$ is

$$\boldsymbol{X}'' = \begin{bmatrix} 165.0 & -10.0 & -45.0 & -32.2 & 55.0 & -10.0 & 65.0 & -187.8 \\ 27.4 & -3.5 & -5.6 & -5.4 & 9.1 & -3.5 & 12.7 & -31.2 \\ 0.0 & 0.0 & 0.0 & 0.0 & 0.0 & 0.0 & 0.0 & 0.0 \\ -26.4 & 1.5 & 7.3 & 5.2 & -8.8 & 1.5 & -10.3 & 30.0 \\ 15.0 & 0.0 & -5.0 & -2.9 & 5.0 & 0.0 & 5.0 & -17.1 \\ -57.4 & 3.5 & 15.6 & 11.2 & -19.1 & 3.5 & -22.7 & 65.4 \\ 30.0 & 0.0 & -10.0 & -5.9 & 10.0 & 0.0 & 10.0 & -34.1 \\ -153.6 & 8.5 & 42.7 & 30.0 & -51.2 & 8.5 & -59.8 & 174.9 \end{bmatrix}. \tag{7.55}$$

In either case, the signal is perfectly reconstructed by the inverse transform (identical to the forward transform) corresponding to each of the two kernels.

Example 7.3: An image and both of its Fourier and Hartley spectra are shown in the top row of Fig. 7.5. The real and imaginary parts of the Fourier spectrum are shown respectively in the second and third panels, and the Hartley spectrum is shown in the fourth. These spectra are then LP filtered and then inverse transformed as shown in the bottom row of the figure. The Hartley filtering effect is identical to that of the Fourier filtering, shown in the first panel of the bottom row.

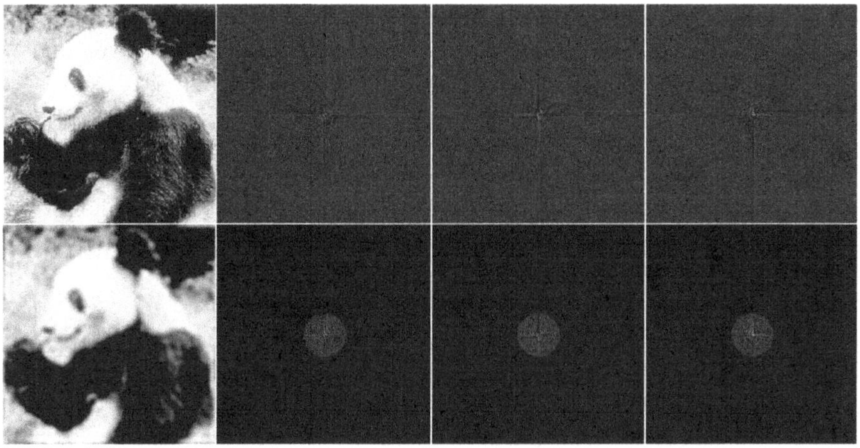

Figure 7.5 The Hartley and Fourier filtering of an image. The image and its Fourier and Hartley spectra before and after an LP filtering are shown in the top and bottom rows respectively. The second and third panel of each row are the real and imaginary parts of the Fourier spectrum, while the fourth panel is for the Hartley spectrum.

7.2 The discrete sine and cosine transforms

Like the Hartley transform, both the sine and cosine transforms, also derived from the Fourier transform, convert a real signal into its real spectrum. Moreover, their discrete versions, the discrete sine transform (DST) and discrete cosine transform (DCT), can also be carried out based on the FFT.

7.2.1 The continuous cosine and sine transforms

We first consider the Fourier transform of a real signal $x(t) = \overline{x}(t)$:

$$\begin{aligned} X(f) &= \int_{-\infty}^{\infty} x(t) e^{-j2\pi f t}\, dt = \int_{-\infty}^{\infty} x(t)[\cos(2\pi f t) - j\sin(2\pi f t)]\, dt \\ &= X_{\rm r}(f) - jX_{\rm j}(f), \end{aligned} \tag{7.56}$$

where the real part of the spectrum $X_{\rm r}(f)$ is even and the imaginary part $X_{\rm j}(f)$ is odd:

$$X_{\rm r}(f) = \int_{-\infty}^{\infty} x(t)\cos(2\pi ft)\,dt = X_{\rm r}(-f),$$
$$X_{\rm j}(f) = \int_{-\infty}^{\infty} x(t)\sin(2\pi ft)\,dt = -X_{\rm j}(-f). \quad (7.57)$$

We further assume the signal $x(t)$ is either even or odd.

- If $x(t) = x(-t)$ is even, then the integrand $x(t)\sin(2\pi ft)$ in Eq. (7.56) is odd with respect to t and $X_{\rm j}(f) = 0$, but $x(t)\cos(2\pi ft)$ is even and the equation above becomes

$$X(f) = \int_{-\infty}^{\infty} x(t)\cos(2\pi ft)\,dt = 2\int_{0}^{\infty} x(t)\cos(2\pi ft)\,dt = X(-f). \quad (7.58)$$

  This spectrum $X(f)$ is real and even with respect to f. The inverse transform becomes

$$x(t) = \int_{-\infty}^{\infty} X(f)e^{j2\pi ft}\,df = \int_{-\infty}^{\infty} X(f)\cos(2\pi ft)\,df + j\int_{-\infty}^{\infty} X(f)\sin(2\pi ft)\,df$$
$$= 2\int_{0}^{\infty} X(f)\cos(2\pi ft)\,df, \quad (7.59)$$

  owing to the fact that $X(f)\sin(2\pi ft)$ is odd with respect to f, and the second term is zero, but $X(f)\cos(2\pi ft)$ is even. Now we get the cosine transform pair of an even signal $x(t)$:

$$X_{\rm C}(f) = 2\int_{0}^{\infty} x(t)\cos(2\pi ft)\,dt,$$
$$x(t) = 2\int_{0}^{\infty} X_{\rm C}(f)\cos(2\pi ft)\,df. \quad (7.60)$$

  where $X_{\rm C}(f)$ is the cosine transform spectrum of $x(t)$.
- If $x(t) = -x(-t)$ is odd, then the integrand $x(t)\cos(2\pi ft)$ in Eq. (7.56) is odd with respect to t and $X_{\rm r}(f) = 0$, but $x(t)\sin(2\pi ft)$ is even and the equation becomes

$$X(f) = -j\int_{-\infty}^{\infty} x(t)\sin(2\pi ft)\,dt = -j2\int_{0}^{\infty} x(t)\sin(2\pi ft)\,dt = -X(-f). \quad (7.61)$$

  This spectrum $X(f)$ is imaginary and odd with respect to f. The inverse transform becomes

$$x(t) = \int_{-\infty}^{\infty} X(f)e^{j2\pi ft}\,df = \int_{-\infty}^{\infty} X(f)\cos(2\pi ft)\,df + j\int_{-\infty}^{\infty} X(f)\sin(2\pi ft)\,df$$
$$= 2j\int_{0}^{\infty} X(f)\sin(2\pi ft)\,df, \quad (7.62)$$

  owing to the fact that $X(f)\cos(2\pi ft)$ is odd with respective to f and the first term is zero, but $X_{\rm S}(f)\sin(2\pi ft)$ is even. Defining the sine transform as

$X_S(f) = jX(f)$ we get the sine transform pair of an odd signal $x(t)$:

$$X_S(f) = 2\int_0^\infty x(t)\sin(2\pi f t)\,dt,$$
$$x(t) = 2\int_0^\infty X_S(f)\sin(2\pi f t)\,df. \tag{7.63}$$

where $X_S(f)$ is the sine transform spectrum of $x(t)$.

We see that, similar to the Fourier transform, both the sine and cosine transforms also represent a real signal as a linear combination of a set of uncountably infinite basis sinusoids of different frequencies. However, different from the Fourier transform, as here the weighting functions $X_C(f)$ and $X_S(f)$ are real, they only represent the amplitudes of the basis functions but not their phases, which are always zero.

The cosine and sine transforms above are valid only if the signal in question is either even or odd. If the signal is neither even nor odd, but it is known to be zero before $t = 0$, i.e., $x(t) = x(t)u(t)$, we can construct the following even and odd functions

$$x'_e(t) = \begin{cases} x(t) & t \geq 0 \\ x(-t) & t \leq 0 \end{cases} \quad \text{and} \quad x'_o(t) = \begin{cases} x(t) & t > 0 \\ 0 & t = 0 \\ -x(-t) & t < 0 \end{cases} \tag{7.64}$$

so that the cosine and sine transforms can still be applied. Note that $x'_o(0)$ is defined as zero for $x'_o(t)$ to be odd.

7.2.2 From DFT to DCT and DST

The discussion above for continuous signals can be extended to discrete signals of a finite duration. The corresponding cosine and sine transforms are the DCT and DST. However, different from the continuous case, here in the discrete case there is more than one way to construct an even or odd signal based on a set of finite data samples $x[0], \dots, x[N-1]$. By assuming $x[-n] = x[n]$, we can obtain a sequence of $2N-1$ samples that is even with respect to $n = 0$. Alternatively, by assuming $x[-n] = x[n-1]$; i.e., $x[-1] = x[0]$, $x[-2] = x[1]$, and $x[-N] = x[N-1]$, we get a sequence of $2N$ samples that is even with respect to $n = -1/2$. Moreover, there are different ways to assume the periodicity beyond these $2N-1$ or $2N$ data samples. In the following, we will take the second approach to construct a sequence of $2N$ points and assume it is periodic beyond its two ends. The DCT and DST can be derived by applying the DFT to this sequence of $2N$ points.

Given an N-point real signal sequence $x[0], \dots, x[N-1]$, we construct two sequences of $2N$ points,

$$x_e[n] = \begin{cases} x[n] & 0 \leq n \leq N-1 \\ x[-n-1] & -N \leq n \leq -1 \end{cases} \tag{7.65}$$

and

$$x_o[n] = \begin{cases} x[n] & 0 \le n \le N-1 \\ -x[-n-1] & -N \le n \le -1 \end{cases}, \tag{7.66}$$

which are respectively even and odd with respect to $n = -1/2$, as shown in Fig. 7.6. If we shift them to the right by 1/2, or, equivalently, define a new index $n' = n + 1/2$; i.e., $n = n' - 1/2$, then $x_e[n] = x_e[n' - 1/2]$ and $x_o[n] = x_o[n' - 1/2]$ are respectively even and odd with respect to $n' = 0$. These $2N$-point sequences are further assumed to repeat themselves outside the range $-N \le n \le N-1$; i.e., they become periodic with period $2N$:

$$\begin{aligned} x_e[n] &= x_e[n+2N] = x_e[-n-1] = x_e[2N-n-1], \\ x_o[n] &= x_o[n+2N] = -x_o[-n-1] = -x_o[2N-n-1]. \end{aligned} \tag{7.67}$$

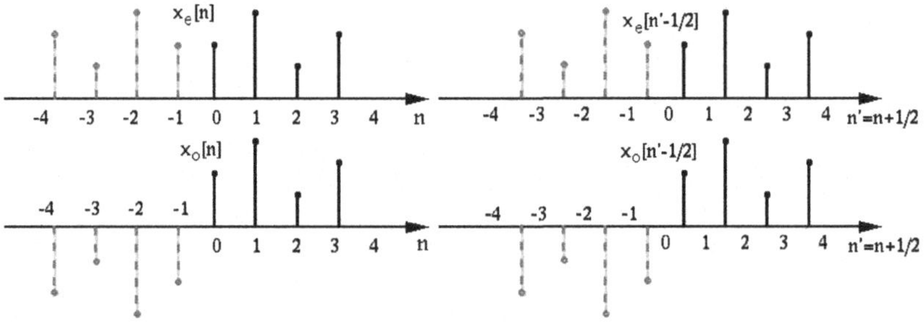

Figure 7.6 Construction of even (top) and odd signals (bottom). Given an $N = 4$ point signal $x[0], \ldots, x[3]$ (black), the even and odd versions can be constructed by including N additional points $x[-1] = \pm x[0], \ldots, x[-4] = \pm x[3]$ (gray). This signal of $2N = 8$ points is either even or odd symmetric with respect to $n = 1/2$. If we define $n' = n + 1/2$, then $x[n] = x[n' - 1/2]$ (from $-N + 1/2 = -3.5$ to $N - 1/2 = 3.5$) are symmetric with respect to $n' = 0$.

Applying the $2N$-point DFT to this constructed signal of $2N$ points, now simply denoted by $x[n]$, we get

$$\begin{aligned} X[k] &= \frac{1}{\sqrt{2N}} \sum_{n'=-N+1/2}^{N-1/2} x\left[n' - \frac{1}{2}\right] e^{-j2\pi n'k/2N} \\ &= \frac{1}{\sqrt{2N}} \sum_{n'=-N+1/2}^{N-1/2} x\left[n' - \frac{1}{2}\right] \cos\left(\frac{2\pi n'k}{2N}\right) \\ &- \frac{j}{\sqrt{2N}} \sum_{n'=-N+1/2}^{N-1/2} x\left[n' - \frac{1}{2}\right] \sin\left(\frac{2\pi n'k}{2N}\right), \qquad k = 0, \ldots, 2N-1. \end{aligned} \tag{7.68}$$

Note that, as shown in Fig. 7.6, $\cos(2\pi n'k/2N)$ and $\sin(2\pi n'k/2N)$ are respectively even and odd with respect to $n' = 0$, while $x[n' - 1/2]$ is either even (top) or odd (bottom) with respect to $n' = 0$. We can now specifically consider the following two cases.

- **DCT**
When $x[n'-1/2] = x_e[n'-1/2] = x_e[-(n'-1/2)]$ is even, the $2N$ terms in the first summation of Eq. (7.68) are even and their sum is twice the sum of either the first or second N terms, while the $2N$ terms in the second summation are odd and their sum is zero. Therefore we have:

$$X[k] = \sqrt{\frac{2}{N}} \sum_{n'=1/2}^{N-1/2} x\left[n' - \frac{1}{2}\right] \cos\left(\frac{2\pi n' k}{2N}\right), \qquad k = 0, \dots, 2N-1. \quad (7.69)$$

Also, as $X[k] = X[-k]$ is even and of period $2N$, we have $X[N+k] = X[N+k-2N] = X[-N+k] = X[N-k]$; i.e., the second N coefficients $X[k]$ for $k = N, \dots, 2N-1$ are redundant and can therefore be dropped. Now the range of the index k in the equation above can be reduced to $k = 0, \dots, N-1$. Replacing n' by $n + 1/2$, we get the DCT:

$$X_C[k] = \sqrt{\frac{2}{N}} \sum_{n=0}^{N-1} x[n] \cos\left(\frac{(2n+1)k\pi}{2N}\right), \qquad k = 0, \dots, N-1. \quad (7.70)$$

Here, $X_C[k]$ is the kth DCT coefficient corresponding to the kth basis function for frequency component $\cos((2n+1)k\pi/2N)$. In particular, when $k = 0$, $X_C[0]$ is proportional to $\sum_{n=0}^{N-1} x[n]$ representing the DC component of the signal.

- **DST**
When $x[n'-1/2] = x_o[n'-1/2] = -x_o[-(n'-1/2)]$ is odd, the $2N$ terms in the first summation of Eq. (7.68) are odd and their sum is zero, while the $2N$ terms in the second summation are even and their sum is twice the sum of either the first or second N terms. Therefore we have:

$$X[k] = \sqrt{\frac{2}{N}} \sum_{n'=1/2}^{N-1/2} x\left[n' - \frac{1}{2}\right] \sin\left(\frac{2\pi n' k}{2N}\right), \qquad k = 0, \dots, 2N-1. \quad (7.71)$$

Also, as $X[k] = -X[-k]$ is odd and of period $2N$, we have $X[N+k] = X[N+k-2N] = X[-N+k] = -X[N-k]$; i.e., the second N coefficients $X[k]$ for $k = N, \dots, 2N-1$ are redundant and can therefore be dropped. Now the range of the index k in the equation above can be reduced to $k = 0, \dots, N-1$. Replacing n' by $n + 1/2$, we get the DST:

$$X_S[k] = \sqrt{\frac{2}{N}} \sum_{n=0}^{N-1} x[n] \sin\left(\frac{(2n+1)k\pi}{2N}\right), \qquad k = 0, \dots, N-1. \quad (7.72)$$

As before, here the spectrum $X_S[k]$ has been redefined to include j. As $X_S[0] = 0$ is always zero, independent of the signal (unlike $X_C[0]$ for the DC component of the signal), the index k in Eq. (7.72) is replaced by $k+1$ to exclude the first term and include an additional term for a high frequency:

$$X_S[k] = \sqrt{\frac{2}{N}} \sum_{n=0}^{N-1} x[n] \sin\left(\frac{(2n+1)(k+1)\pi}{2N}\right), \qquad k = 0, \dots, N-1. \quad (7.73)$$

Note, however, after this modification, the zeroth DST basis function ($k = 0$) becomes $\sin[(2n+1)\pi/2N]$, a function of the time index n, different from the zeroth basis function of the DCT or any other transform, which is a constant. In other words, different from all other orthogonal transforms, the zeroth frequency component $X_S[0]$ of the DST does not represent the DC component or average of the signal.

Comparing the DCT and DST defined above with the DFT considered in Chapter 4 we see the following advantages:

- The DCT and DST are both real transforms without any complex operations needed by the DFT.
- The kth DCT coefficient $X_C[k]$ (Eq. (7.70)) represents a sinusoid of frequency $k/2N$ and the kth DST coefficient $X_S[k]$ (Eq. (7.73)) represents a sinusoid of frequency $(k+1)/2N$, both of which are half of the frequency k/N represented by the kth DFT coefficient $X_F[k]$.
- The highest frequencies represented by the DCT coefficient $X_C[N-1]$ and DST coefficient $X_S[N-1]$ are approximately the same: $f_{\max} = 1/2$ (or period $T = 1/f_{\max} = 2$), also the same as the highest frequency represented by the DFT Eq. (4.167); i.e., all three transforms cover the same frequency range.
- The frequency resolution of the DCT and DST is twice that of the DFT. Each of the N DCT coefficient $X_C[k]$ or DST coefficient $X_S[k]$ corresponds to a different frequency $k/2N$ or $(K+1)/2N$ ($k = 0, \ldots, N-1$), but a pair of two DFT coefficients $X_F[k]$ and $X_F[N-1-k]$ corresponds to one of the $N/2$ different frequencies k/N ($k = 0, \ldots, N/2-1$), as can be seen by comparing Fig. 7.9 for the DCT and DST with Fig. 4.17 for the DFT.
- To perform the Fourier transform on a physical signal, it needs to be truncated to have a finite duration $0 \leq t \leq T$, and then assumed to be periodic beyond T (Figs. 3.1 and 4.14). In this process, an artifactual discontinuity together with some associated high-frequency components will be introduced. However, in the case of the DCT, as an even symmetry (first equation of Eq. (7.67)) is assumed while truncating and imposing periodicity on the time signal, we have $x_e[0] = x_e[2N-1]$; i.e., no discontinuity is introduced and all related artifacts are avoided. However, for the DST based on the assumption of odd symmetry (second equation of Eq. (7.67)), discontinuity is not avoided; i.e., $x_o[0] = -x_o[2N-1] \neq x_o[0]$, same as in the case of the DFT where $x[0] \neq x[N-1]$ in general.

Example 7.4: Figure 7.7 shows two consecutive periods of three time functions on the left, and their DFT coefficients on the right. The first two functions are smooth within the period, while the third one has a discontinuity in the middle of the period. Also, the transitions between the two consecutive periods of the first and third functions are smooth, as $x[0] = x[N-1]$, but there is a discontinuity in the second function at the transition, as $x[0] \neq x[N-1]$. In the

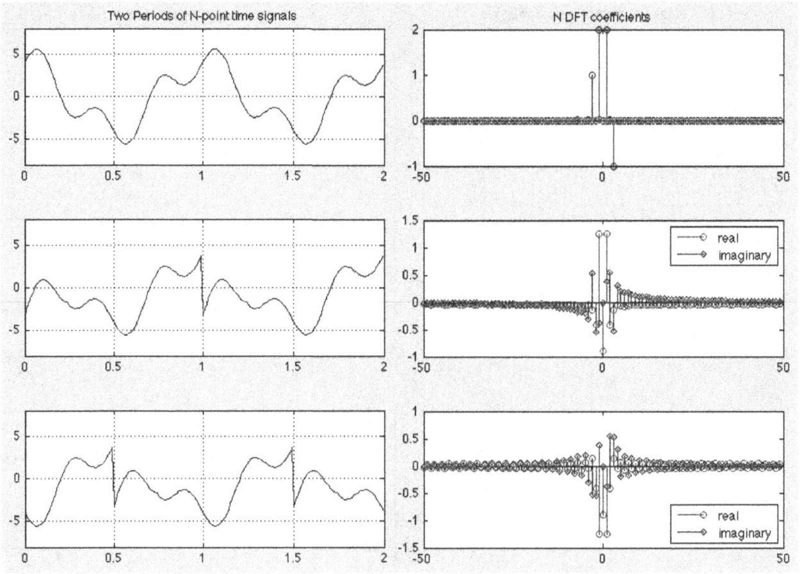

Figure 7.7 Three N-point functions (with two periods shown) and their DFTs. Top: a smooth function composed of two low-frequency sinusoidal components; middle: the same function but slightly modified to have different end points; bottom: shifted version of the second function with a major discontinuity. The DFT spectra of these functions are shown on the right.

frequency domain, corresponding to these discontinuities, some high-frequency components show up in their Fourier spectra.

Next we consider the DCT of these three functions following the discussion above. According to Eq. (7.65), a $2N$-point even function is constructed based on each of these three N-point functions, as shown on the left of Fig. 7.8. As these functions are real and even, their $2N$ DFT spectra are also real and even; therefore, only the N coefficients of the first half of the spectrum need to be kept, as shown on the right of Fig. 7.8, which are essentially also the N DCT coefficients of the original N-point function.

Comparing these three reconstructed $2N$-point even functions and their spectra in the frequency domain, we see that the discontinuity of the second function caused by the different end points is much reduced, but the real discontinuity in the third function remains the same. Consequently, the DCT spectrum of the second function contains few high-frequency components, similar to the spectrum of the first function which is smooth. But the DCT spectrum of the third function contains more significant high-frequency components, owing obviously to the discontinuity in the time domain.

In conclusion, we see that the different end points of a function can cause some significant artifactual high-frequency components in the DFT spectrum. However, this effect can be much reduced in the DCT spectrum; i.e., the DCT is

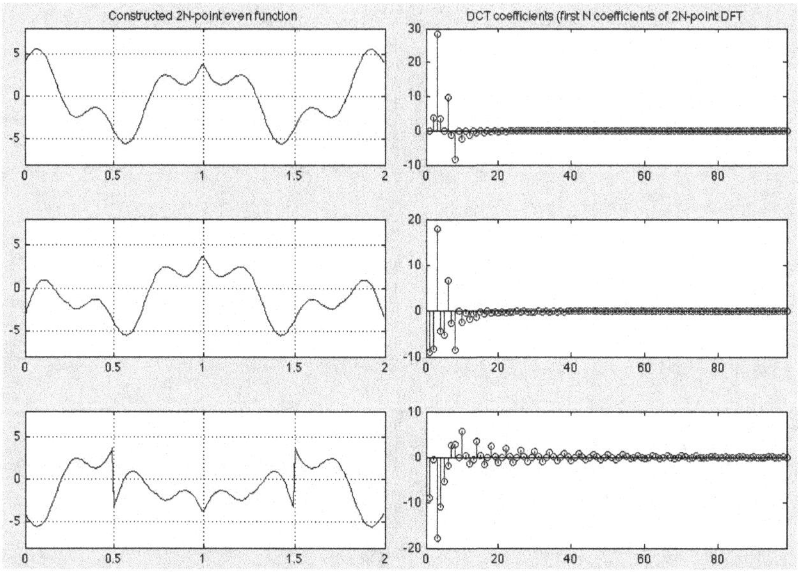

Figure 7.8 The constructed $2N$-point even functions (left) and their DFT spectra (right, showing only the N coefficients in the first half).

insensitive to these kinds of discontinuity caused by the two ending points taking different values.

7.2.3 Matrix forms of DCT and DST

The DCT matrix $\boldsymbol{C}$ and DST matrix $\boldsymbol{S}$ can be constructed respectively by a set of N orthogonal column vectors:

$$\boldsymbol{C} = [\boldsymbol{c}_0 \dots \boldsymbol{c}_{N-1}] = \begin{bmatrix} c[0,0] & \cdots & c[0,N-1] \\ \vdots & \ddots & \vdots \\ c[0,N-1] & \cdots & c[N-1,N-1] \end{bmatrix} \tag{7.74}$$

and

$$\boldsymbol{S} = [\boldsymbol{s}_0 \cdots \boldsymbol{s}_{N-1}] = \begin{bmatrix} s[0,0] & \cdots & s[0,N-1] \\ \vdots & \ddots & \vdots \\ s[0,N-1] & \cdots & s[N-1,N-1] \end{bmatrix}. \tag{7.75}$$

Here, $c[n,k]$ and $s[n,k]$ are the elements in nth row and kth column of $\boldsymbol{C}$ and $\boldsymbol{S}$ $(n,k = 0,1,\cdots,N-1)$, respectively:

$$c[n,k] = \cos\left(\frac{(2n+1)k\pi}{2N}\right), \qquad s[n,k] = \sin\left(\frac{(2n+1)(k+1)\pi}{2N}\right), \tag{7.76}$$

and $\boldsymbol{c}_k$ and $\boldsymbol{s}_k$ are the kth columns of $\boldsymbol{C}$and $\boldsymbol{S}$, respectively:

$$\boldsymbol{c}_k = \left[\cos\left(\frac{k\pi}{2N}\right), \cos\left(\frac{3k\pi}{2N}\right), \cos\left(\frac{5k\pi}{2N}\right), \cdots, \cos\left(\frac{(2N-1)k\pi}{2N}\right)\right]^{\mathrm{T}} \tag{7.77}$$

and

$$\boldsymbol{s}_k = \left[\sin\left(\frac{(k+1)\pi}{2N}\right), \sin\left(\frac{3(k+1)\pi}{2N}\right), \ldots, \sin\left(\frac{(2N-1)(k+1)\pi}{2N}\right)\right]^{\mathrm{T}}, \tag{7.78}$$

representing respectively a sinusoid of frequency $k/2N$ and $(k+1)/2N$.

We now prove that the column vectors of both matrices $\boldsymbol{C}$ and $\boldsymbol{S}$ are orthogonal; i.e., the following hold for all $k, l = 0, \ldots, N-1$ but $l \neq k$:

$$\langle \boldsymbol{c}_k, \boldsymbol{c}_l \rangle = 0, \tag{7.79}$$

$$\langle \boldsymbol{s}_k, \boldsymbol{s}_l \rangle = 0. \tag{7.80}$$

To do so, we need the identity (homework):

$$\sum_{n=0}^{N-1} \cos\left(\frac{(2n+1)k\pi}{2N}\right) = \begin{cases} N & k=0 \\ -N & k=2N \\ 0 & \text{else} \end{cases}. \tag{7.81}$$

First, based on the identity $\cos\alpha\cos\beta = [\cos(\alpha-\beta)+\cos(\alpha+\beta)]/2$, we rewrite Eq. (7.79) as

$$\begin{aligned} \langle \boldsymbol{c}_k, \boldsymbol{c}_l \rangle &= \sum_{n=0}^{N-1} c[n,k]c[n,l] = \sum_{n=0}^{N-1} \cos\left(\frac{(2n+1)k\pi}{2N}\right)\cos\left(\frac{(2n+1)l\pi}{2N}\right) \\ &= \frac{1}{2}\sum_{n=0}^{N-1} \cos\left(\frac{(2n+1)(k-l)\pi}{2N}\right) + \frac{1}{2}\sum_{n=0}^{N-1} \cos\left(\frac{(2n+1)(k+l)\pi}{2N}\right). \end{aligned} \tag{7.82}$$

When $l \neq k$, both terms are zero according to Eq. (7.81); i.e., the column vectors of $\boldsymbol{C}$ are indeed orthogonal. When $l = k$, the first term becomes $N/2$ and the second term is either $N/2$ if $k = 0$ or zero if $k \neq 0$, i.e.,

$$\langle \boldsymbol{c}_k, \boldsymbol{c}_k \rangle = \frac{N}{2} + \frac{1}{2}\sum_{n=0}^{N-1} \cos\left(\frac{(2n+1)2k\pi}{2N}\right) = \begin{cases} N & k=0 \\ N/2 & k \neq 0 \end{cases}. \tag{7.83}$$

For all N columns of $\boldsymbol{C}$ to be normalized, we introduce a scaling factor $a[k]$, defined as

$$a[k] = \begin{cases} \sqrt{1/N} & k=0 \\ \sqrt{2/N} & k \neq 0 \end{cases}, \tag{7.84}$$

so that all columns of the modified version of the DCT matrix, still denoted by $\boldsymbol{C}$, are orthonormal: $\langle \boldsymbol{c}_k, \boldsymbol{c}_l \rangle = \delta[k-l]$.

Next, based on the identity $\sin\alpha\sin\beta = [\cos(\alpha-\beta) - \cos(\alpha+\beta)]/2$, we rewrite Eq. (7.80) as

$$\begin{aligned}\langle \boldsymbol{s}_k, \boldsymbol{s}_l\rangle &= \sum_{n=0}^{N-1} s[n,k]s[n,l] \\ &= \sum_{n=0}^{N-1} \sin\left(\frac{(2n+1)(k+1)\pi}{2N}\right)\sin\left(\frac{(2n+1)(l+1)\pi}{2N}\right) \\ &= \frac{1}{2}\sum_{n=0}^{N-1}\cos\left(\frac{(2n+1)(k-l)\pi}{2N}\right) - \frac{1}{2}\sum_{n=0}^{N-1}\cos\left(\frac{(2n+1)(k+l+2)\pi}{2N}\right).\end{aligned} \tag{7.85}$$

When $l \neq k$, both terms are zero according to Eq. (7.81); i.e., the column vectors of $\boldsymbol{S}$ are indeed orthogonal. When $l = k$, the first term becomes $N/2$ and the second term is either $-N/2$ if $k = N-1$ or zero otherwise, i.e.,

$$\langle \boldsymbol{s}_k, \boldsymbol{s}_k\rangle = \frac{N}{2} - \frac{1}{2}\sum_{n=0}^{N-1}\cos\left(\frac{(2n+1)(2k+2)\pi}{2N}\right) = \begin{cases} N & k = N-1 \\ N/2 & k \neq N-1 \end{cases}. \tag{7.86}$$

For all N columns of $\boldsymbol{S}$ to be normalized, we introduce a scaling factor $b[k]$, defined as

$$b[k] = \begin{cases} \sqrt{2/N} & k \neq N-1 \\ \sqrt{1/N} & k = N-1 \end{cases}, \tag{7.87}$$

so that the columns of the modified version of the DST matrix, still denoted by $\boldsymbol{S}$, are orthonormal: $\langle \boldsymbol{s}_k, \boldsymbol{s}_l\rangle = \delta[k-l]$. As now both $\boldsymbol{C}$ and $\boldsymbol{S}$ are orthonormal; i.e., $\boldsymbol{C}^{\mathrm{T}}\boldsymbol{C} = \boldsymbol{S}^{\mathrm{T}}\boldsymbol{S} = \boldsymbol{I}$, they can be used to define the DCT and DST. Given any N-D signal vector $\boldsymbol{x} = [x[0], \ldots, x[N-1]]^{\mathrm{T}}$, its DCT and DST coefficients can be found simply by matrix multiplication:

$$\boldsymbol{X}_{\mathrm{C}} = \boldsymbol{C}^{\mathrm{T}}\boldsymbol{x} = \begin{bmatrix} \boldsymbol{c}_0^{\mathrm{T}} \\ \vdots \\ \boldsymbol{c}_{N-1}^{\mathrm{T}} \end{bmatrix}\boldsymbol{x}, \qquad \boldsymbol{X}_{\mathrm{S}} = \boldsymbol{S}^{\mathrm{T}}\boldsymbol{x} = \begin{bmatrix} \boldsymbol{s}_0^{\mathrm{T}} \\ \vdots \\ \boldsymbol{s}_{N-1}^{\mathrm{T}} \end{bmatrix}\boldsymbol{x}. \tag{7.88}$$

The kth components $X_{\mathrm{C}}[k]$ of $\boldsymbol{X}_{\mathrm{C}}$ and $X_{\mathrm{S}}[k]$ of $\boldsymbol{X}_{\mathrm{S}}$ are respectively the projections of $\boldsymbol{x}$ onto the kth basis vectors $\boldsymbol{c}_k$ and $\boldsymbol{s}_k$ $(k = 0, \ldots, N-1)$:

$$X_{\mathrm{C}}[k] = \langle \boldsymbol{x}, \boldsymbol{c}_k\rangle = \boldsymbol{x}^{\mathrm{T}}\boldsymbol{c}_k = a[k]\sum_{n=0}^{N-1} x[n]\,\cos\left(\frac{(2n+1)k\pi}{2N}\right), \tag{7.89}$$

$$X_{\mathrm{S}}[k] = \langle \boldsymbol{x}, \boldsymbol{s}_k\rangle = \boldsymbol{x}^{\mathrm{T}}\boldsymbol{s}_k = b[k]\sum_{n=0}^{N-1} x[n]\,\sin\left(\frac{(2n+1)(k+1)\pi}{2N}\right). \tag{7.90}$$

These are Eqs. (7.70) and (7.73) respectively with a scaling factor $a[k] = b[k] = \sqrt{2/N}$, except $a[0] = b[N-1] = 1/\sqrt{N}$; i.e.,

$$a[N-1-k] = b[k] \qquad (k = 0, \ldots, N-1). \tag{7.91}$$

The signal vector $\boldsymbol{x}$ can be reconstructed by the inverse DCT or DST as a linear combination of the corresponding basis:

$$\boldsymbol{x} = \boldsymbol{C}\boldsymbol{X}_\mathrm{C} = [\boldsymbol{c}_0, \ldots, \boldsymbol{c}_{N-1}] \begin{bmatrix} X_\mathrm{C}[0] \\ \vdots \\ X_\mathrm{C}[N-1] \end{bmatrix} = \sum_{k=0}^{N-1} X_\mathrm{C}[k]\boldsymbol{c}_k$$

$$= \boldsymbol{S}\boldsymbol{X}_\mathrm{S} = [\boldsymbol{s}_0, \ldots, \boldsymbol{s}_{N-1}] \begin{bmatrix} X_\mathrm{S}[0] \\ \vdots \\ X_\mathrm{S}[N-1] \end{bmatrix} = \sum_{k=0}^{N-1} X_\mathrm{S}[k]\boldsymbol{s}_k. \tag{7.92}$$

In component form, the nth component $x[n]$ $(n = 0, \ldots, N-1)$ can be found as

$$x[n] = \sum_{k=0}^{N-1} c[n,k]X_\mathrm{C}[k] = \sum_{k=0}^{N-1} X_\mathrm{C}[k]a[k]\cos\left(\frac{(2n+1)k\pi}{2N}\right) \tag{7.93}$$

$$= \sum_{k=0}^{N-1} s[n,k]X_\mathrm{S}[k] = \sum_{k=0}^{N-1} X_\mathrm{S}[k]b[k]\sin\left(\frac{(2n+1)(k+1)\pi}{2N}\right). \tag{7.94}$$

We list below the DCT and DST matrices for $N = 2$, $N = 4$, and $N = 8$ as some specific examples. Here, we use $n = \log_2 N$ as the subscript for the corresponding N-point transform matrices $\boldsymbol{C}_n$ and $\boldsymbol{S}_n$, consistent with the notation used in the following chapter.

When $N = 2$, we have

$$\boldsymbol{C}_1 = \boldsymbol{S}_1 = \frac{1}{\sqrt{2}}\begin{bmatrix} 1 & 1 \\ 1 & -1 \end{bmatrix}. \tag{7.95}$$

This matrix is composed of two row vectors $\boldsymbol{c}_0^\mathrm{T} = [1 \;\; 1]/\sqrt{2}$ and $\boldsymbol{c}_1^\mathrm{T} = [1 \;\; -1]/\sqrt{2}$ and is identical to the two-point DFT matrix considered previously. The DCT of a two-point signal $\boldsymbol{x} = [x[0], x[1]]^\mathrm{T}$ is

$$\boldsymbol{X} = \begin{bmatrix} X[0] \\ X[1] \end{bmatrix} = 0.707\begin{bmatrix} 1 & 1 \\ 1 & -1 \end{bmatrix}\begin{bmatrix} x[0] \\ x[1] \end{bmatrix} = 0.707\begin{bmatrix} x[0]+x[1] \\ x[1]-x[1] \end{bmatrix}. \tag{7.96}$$

The first component $X[0]$ is proportional to the sum $x[0] + x[1]$ of the two samples representing the average or DC component of the signal, and the second component $X[1]$ is proportional to the difference $x[0] - x[1]$ between the two samples. This is also the case for the DFT, as well as all orthogonal transforms when $N = 2$ (as we will see later).

When $N = 4$, we have

$$\boldsymbol{C}_2^\mathrm{T} = \begin{bmatrix} 0.50 & 0.50 & 0.50 & 0.50 \\ 0.65 & 0.27 & -0.27 & -0.65 \\ 0.50 & -0.50 & -0.50 & 0.50 \\ 0.27 & -0.65 & 0.65 & -0.27 \end{bmatrix}, \quad \boldsymbol{S}_2^\mathrm{T} = \begin{bmatrix} 0.27 & 0.65 & 0.65 & 0.27 \\ 0.50 & 0.50 & -0.50 & -0.50 \\ 0.65 & -0.27 & -0.27 & 0.65 \\ 0.50 & -0.50 & 0.50 & -0.50 \end{bmatrix}. \tag{7.97}$$

When $N = 8$ we have

$$C_3^{\mathrm{T}} = \begin{bmatrix} 0.35 & 0.35 & 0.35 & 0.35 & 0.35 & 0.35 & 0.35 & 0.35 \\ 0.49 & 0.42 & 0.28 & 0.10 & -0.10 & -0.28 & -0.42 & -0.49 \\ 0.46 & 0.19 & -0.19 & -0.46 & -0.46 & -0.19 & 0.19 & 0.46 \\ 0.42 & -0.10 & -0.49 & -0.28 & 0.28 & 0.49 & 0.10 & -0.42 \\ 0.35 & -0.35 & -0.35 & 0.35 & 0.35 & -0.35 & -0.35 & 0.35 \\ 0.28 & -0.49 & 0.10 & 0.42 & -0.42 & -0.10 & 0.49 & -0.28 \\ 0.19 & -0.46 & 0.46 & -0.19 & -0.19 & 0.46 & -0.46 & 0.19 \\ 0.10 & -0.28 & 0.42 & -0.49 & 0.49 & -0.42 & 0.28 & -0.10 \end{bmatrix}, \tag{7.98}$$

$$S_3^{\mathrm{T}} = \begin{bmatrix} 0.10 & 0.28 & 0.42 & 0.49 & 0.49 & 0.42 & 0.28 & 0.10 \\ 0.19 & 0.46 & 0.46 & 0.19 & -0.19 & -0.46 & -0.46 & -0.19 \\ 0.28 & 0.49 & 0.10 & -0.42 & -0.42 & 0.10 & 0.49 & 0.28 \\ 0.35 & 0.35 & -0.35 & -0.35 & 0.35 & 0.35 & -0.35 & -0.35 \\ 0.42 & 0.10 & -0.49 & 0.28 & 0.28 & -0.49 & 0.10 & 0.42 \\ 0.46 & -0.19 & -0.19 & 0.46 & -0.46 & 0.19 & 0.19 & -0.46 \\ 0.49 & -0.42 & 0.28 & -0.10 & -0.10 & 0.28 & -0.42 & 0.49 \\ 0.35 & -0.35 & 0.35 & -0.35 & 0.35 & -0.35 & 0.35 & -0.35 \end{bmatrix}. \tag{7.99}$$

The column vectors $\boldsymbol{c}_k$ and $\boldsymbol{s}_k$ $(k = 0, \ldots, N-1)$ of $\boldsymbol{C}$ and $\boldsymbol{S}$ (row vectors of $\boldsymbol{C}^{\mathrm{T}}$ and $\boldsymbol{S}^{\mathrm{T}}$) form an orthonormal basis of space $\mathbb{R}^N$. The N elements of each of the N vectors can also be considered as N samples of the corresponding continuous cosine function $a[k]\cos[(2t+1)k\pi/2N]$ or sine functions $b[k]\sin[(2t+1)(k+1)\pi/2N]$ with progressively higher frequencies, as shown in the first two columns in Fig. 7.9.

Also note that the elements of the first row of $\boldsymbol{S}^{\mathrm{T}}$ for the DST are not the same, unlike either $\boldsymbol{W}$ for the DFT or $\boldsymbol{C}$ for the DCT (or any of the orthogonal transforms to be considered later). In other words, the DC component of the signal is not represented by the DST.

Example 7.5: The DCT and DST coefficients of an $N = 8$ point signal $\boldsymbol{x} = [0, 0, 2, 3, 4, 0, 0, 0]^{\mathrm{T}}$ can be found by a matrix multiplication:

$$\begin{aligned} \boldsymbol{X}_{\mathrm{C}} &= \boldsymbol{C}^{\mathrm{T}}\boldsymbol{x} = [3.18, 0.46, -3.62, -0.70, 1.77, -0.22, -0.42, 1.32]^{\mathrm{T}}, \\ \boldsymbol{X}_{\mathrm{S}} &= \boldsymbol{S}^{\mathrm{T}}\boldsymbol{x} = [4.26, 0.73, -2.72, -0.35, 0.96, -0.84, -0.13, 1.06]^{\mathrm{T}}. \end{aligned} \tag{7.100}$$

The interpretation of these DCT and DST coefficients is much more straightforward than that of the DFT. $X[0]$ represents the DC component or the average of the signal, while the subsequent coefficients $X[k]$ $(k = 1, \ldots, N-1)$ represent the magnitudes of progressively higher frequency components contained in the signal.

The signal is perfectly reconstructed by the inverse DCT or DST as a linear combination of the column vectors of $\boldsymbol{C} = [\boldsymbol{c}_0, \ldots, \boldsymbol{c}_{N-1}]$ or $\boldsymbol{S} = [\boldsymbol{s}_0, \ldots, \boldsymbol{s}_{N-1}]$

as the basis spanning $\mathbb{R}^8$:

$$\boldsymbol{x} = \boldsymbol{C}\boldsymbol{X}_{\mathrm{C}} = \sum_{k=0}^{7} X_{\mathrm{C}}[k]\boldsymbol{c}_k = \boldsymbol{S}\boldsymbol{X}_{\mathrm{S}} = \sum_{k=0}^{7} X_{\mathrm{S}}[k]\boldsymbol{s}_k. \tag{7.101}$$

The reconstruction of the signal by the linear combination of the eight vectors is shown in Fig. 7.9.

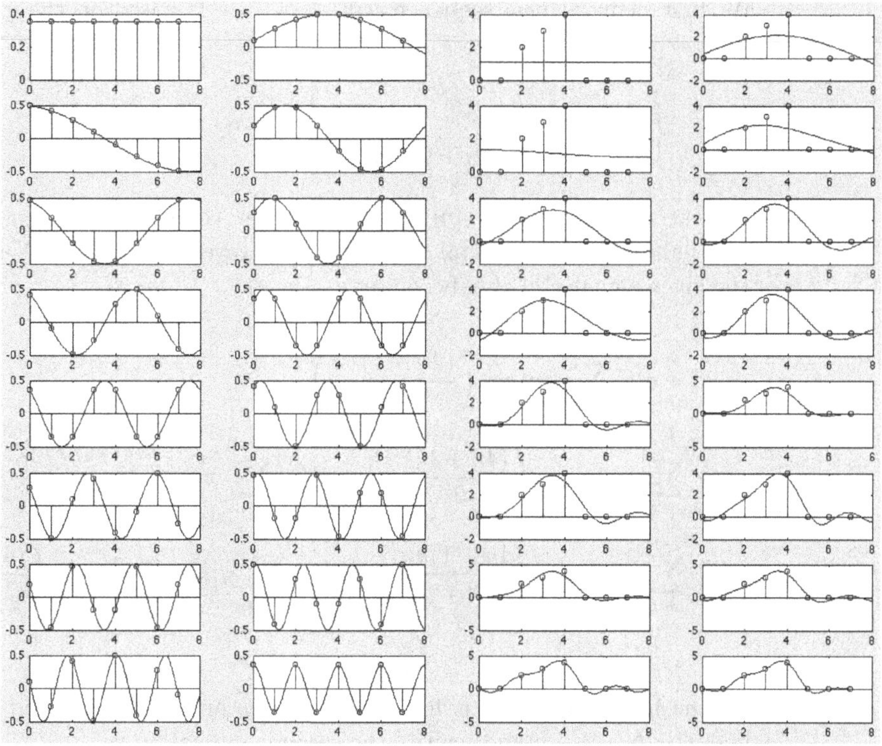

Figure 7.9 Basis functions and vectors of eight-point DCT and DST. The $N = 8$ continuous and discrete basis functions of DCT and DST are shown respectively in the first and second columns; the reconstructions of a discrete signal with progressively more and higher DCT and DST frequency components are shown respectively in the third and fourth columns (Eq. (7.100)). The signal is reconstructed when all N components are used.

In Matlab the forward and inverse DCTs can be carried out by functions `dct` and `idct`, respectively. Also, the forward and inverse DSTs can be carried out by functions `dst` and `idst`, respectively. However, these DST functions are scaled differently. For Parseval's identity to hold, the Matlab forward transform needs to be rescaled: `X=dst(x)/sqrt((length(x)+1)/2)`.

7.2.4 Fast algorithms for the DCT and DST

The computational complexity of both DCT and DST is $O(N^2)$ if implemented as a matrix multiplication ($O(N)$ for each of the N coefficients $X[k]$). However, as the DCT and DST are closely related to the DFT, they can be implemented by the FFT algorithm with complexity $O(N \log_2 N)$. We will first consider the fast algorithm for the DCT and then show that the DST can be carried out based on the fast algorithm for the DCT.

We first define a new sequence $y[0], \ldots, y[N-1]$ based on the given signal $x[0], \ldots, x[N-1]$:

$$\begin{cases} y[n] = x[2n] \\ y[N-1-n] = x[2n+1] \end{cases} \qquad n = 0, \ldots, N/2-1. \tag{7.102}$$

Note that the first half of $y[n]$ contains all even components of $x[n]$, while the second half of $y[n]$ contains all odd ones but in reverse order. The N-point DCT of the given signal $x[n]$ now becomes

$$\begin{aligned} X[k] &= a[k] \sum_{n=0}^{N-1} x[n] \cos\left(\frac{(2n+1)k\pi}{2N}\right) \\ &= a[k] \sum_{n=0}^{N/2-1} x[2n] \cos\left(\frac{(4n+1)k\pi}{2N}\right) + a[k] \sum_{n=0}^{N/2-1} x[2n+1] \cos\left(\frac{(4n+3)k\pi}{2N}\right) \\ &= a[k] \sum_{n=0}^{N/2-1} y[n] \cos\left(\frac{(4n+1)k\pi}{2N}\right) + a[k] \sum_{n=0}^{N/2-1} y[N-1-n] \cos\left(\frac{(4n+3)k\pi}{2N}\right). \end{aligned} \tag{7.103}$$

Here, the first summation is for all even terms and second all odd terms. We define $n' = N-1-n$ and rewrite the second summation as

$$a[k] \sum_{n'=N/2}^{N-1} y[n'] \cos\left(2k\pi - \frac{(4n'+1)k\pi}{2N}\right) = a[k] \sum_{n'=N/2}^{N-1} y[n'] \cos\left(\frac{(4n'+1)k\pi}{2N}\right). \tag{7.104}$$

Now the two summations in the expression of $X[k]$ can be combined to become

$$X[k] = a[k] \sum_{n=0}^{N-1} y[n] \cos\left(\frac{(4n+1)k\pi}{2N}\right). \tag{7.105}$$

We next consider the DFT of $y[n]$:

$$Y[k] = \sum_{n=0}^{N-1} y[n] e^{-j2\pi nk/N}. \tag{7.106}$$

If we multiply both sides by $e^{-jk\pi/2N}$ and take the real part of the result, we get

$$\begin{aligned}\text{Re}[e^{-jk\pi/2N}Y[k]] &= \text{Re}\left[\sum_{n=0}^{N-1} y[n]e^{-j2\pi nk/N}e^{-jk\pi/2N}\right]\\ &= \text{Re}\left[\sum_{n=0}^{N-1} y[n]\left[\cos\left(\frac{(4n+1)k\pi}{2N}\right) - j\,\sin\left(\frac{(4n+1)k\pi}{2N}\right)\right]\right]\\ &= \sum_{n=0}^{N-1} y[n]\cos\left(\frac{(4n+1)k\pi}{2N}\right). \end{aligned} \tag{7.107}$$

As $y[n]$ is real, the second term of the sine function is imaginary and is therefore dropped. Comparing this result with Eq. (7.105) we get

$$X[k] = a[k]\ \text{Re}[\ e^{-jk\pi/2N}Y[k]\ ], \qquad k = 0, \ldots, N-1. \tag{7.108}$$

Now we obtain the fast algorithm for the forward DCT, which can be carried out in the following three steps:

- **Step 1.** Generate a sequence $y[n]$ from the given sequence $x[n]$:

$$\begin{cases} y[n] = x[2n] \\ y[N-1-n] = x[2n+1] \end{cases} \qquad n = 0, \ldots, N/2-1. \tag{7.109}$$

- **Step 2.** Carry out DFT of $y[n]$ by FFT (as $y[n]$ is real, $Y[k]$ is symmetric and only half of the data points need be computed):

$$Y[k] = \mathcal{F}[y[n]] \qquad k = 0, \ldots, N-1. \tag{7.110}$$

- **Step 3.** Obtain DCT $X[k]$ from $Y[k]$:

$$\begin{aligned} X[k] &= [k]\ \text{Re}[\ e^{-jk\pi/2N}Y[k]\ ]\\ &= a[k]\,[Y_\text{r}[k]\cos(k\pi/2N) + Y_\text{j}[k]\sin(k\pi/2N)] \qquad k = 0, \ldots, N-1. \end{aligned} \tag{7.111}$$

  Here, $Y_\text{r}[k]$ and $Y_\text{j}[k]$ are the real and imaginary parts of $Y[k]$, respectively.

Note that the DCT scaling factor $a[k]$ is included in the third step, but no scaling factor (either $1/N$ or $1/\sqrt{N}$) is used during the DFT of $y[n]$.

We now derive the fast algorithm of the inverse DCT. We first consider the real part of the inverse DFT of a sequence $Y[k] = a[k]e^{jk\pi/2N}X[k]$ $(k = 0, \ldots, N-1)$:

$$\begin{aligned}\text{Re}\left[\sum_{k=0}^{N-1} a[k]X[k]e^{jk\pi/2N}e^{j2\pi nk/N}\right] &= \text{Re}\left[\sum_{k=0}^{N-1} a[k]X[k]e^{j(4n+1)k\pi/2N}\right]\\ = \sum_{k=0}^{N-1} a[k]X[k]\cos\left(\frac{(4n+1)k\pi}{2N}\right) &= x[2n] \qquad n = 0, \ldots, N-1. \end{aligned} \tag{7.112}$$

The first half of these N values are the $N/2$ even components $x[2n]$, $(n = 0, \ldots, N/2-1)$. To obtain the odd components, recall that $x[n] = x[2N-n-1]$

(first equation in Eq. 7.67), and we have:

$$x[2n+1] = x[2N-(2n+1)-1] = x[2(N-n-1)] \qquad n = 0, \dots, N/2-1; \tag{7.113}$$

i.e., the $N/2$ odd components are actually the second half ($n = N/2, \dots, N-1$) of the previous equation but in reverse order. Now we have the following three steps for the inverse DCT.

- **Step 1.** Generate a sequence $Y[k]$ from the given DCT coefficients $X[k]$:

$$Y[k] = a[k]X[k]e^{jk\pi/2N}, \qquad k = 0, \dots, N-1. \tag{7.114}$$

- **Step 2.** Carry out the inverse DFT of $Y[n]$ by FFT (Only the real part need be computed):

$$y[n] = \mathrm{Re}[\mathcal{F}^{-1}[Y[k]]]. \tag{7.115}$$

- **Step 3.** Obtain $x[n]'s$ from $y[n]'s$ by

$$\begin{cases} x[2n] = y[n] \\ x[2n+1] = y[N-1-n] \end{cases} \qquad n = 0, \dots, N/2-1. \tag{7.116}$$

Note again that no scaling factor (either $1/N$ or $1/\sqrt{N}$) is used during the inverse DFT of $Y[k]$. Now both the forward and inverse DCT are implemented as a slightly modified DFT which can be carried out by the FFT algorithm with much reduced computational complexity of $O(N\,log_2 N)$.

As the DCT and DST are closely related, the fast DCT algorithm considered above can be readily used for the DST. Specifically, we replace k by $N-1-k$ in Eq. (7.89) and note $a[N-1-k] = b[k]$ (Eq. (7.91)) to get

$$\begin{aligned} & a[N-1-k] \sum_{n=0}^{N-1} x[n] \cos\left(\frac{(2n+1)(N-1-k)\pi}{2N}\right) \\ &= b[k] \sum_{n=0}^{N-1} x[n] \cos\left(\frac{\pi}{2} + n\pi - \frac{(2n+1)(k+1)\pi}{2N}\right) \\ &= b[k] \sum_{n=0}^{N-1} x[n] \sin\left(\frac{(2n+1)(k+1)\pi}{2N} - n\pi\right) \\ &= b[k] \sum_{n=0}^{N-1} x[n](-1)^n \sin\left(\frac{(2n+1)(k+1)\pi}{2N}\right) = Y_{\mathrm{S}}[k]. \end{aligned} \tag{7.117}$$

This is the DST of a signal $y[n] = x[n](-1)^m$. Based on this result, the DST of a signal vector $\boldsymbol{x} = [x[0], \dots, x]N-1]]^{\mathrm{T}}$ can be implemented by the following steps.

- **Step 1.** Negate all odd components of $\boldsymbol{x}$: $y[n] = x[n](-1)^n$ for all $n = 0, \dots, N-1$.
- **Step 2.** Carry out the DCT of $y[n]$ to get $Y_{\mathrm{C}}[k]$.

- **Step 3.** Reverse the order of the DCT coefficients to get the DST coefficients: $X_S[k] = Y_C[N-1-k]$ for all $k = 0, \ldots, N-1$.

The inverse DST can be carried out simply by reversing the steps above:

- **Step 1.** Reverse the order of the DST coefficients: $Y_C[k] = X_S[N-1-k]$ for all $k = 0, \ldots, N-1$.
- **Step 2.** Carry out the inverse DCT of $Y_C[k]$ to get $y[n]$.
- **Step 3.** Negate odd-indexed time samples $x[n] = y[n](-1)^n$ for all $n = 0, \ldots, N-1$.

The C code for the fast algorithms of both DCT and DST is given below. The function `fdct` carries out the DCT (if `inv=0`) to convert a data vector $x[n]$ ($n = 0, \ldots, N-1$) into its DCT coefficients $X[k]$ ($k = 0, \ldots, N-1$). This function is also used for the inverse DCT (if `inv=1`) to reconstruct the signal vector based on its DCT coefficients. This is an in-place algorithm; i.e., the input data will be overwritten by the output. The DST can be implemented by the following function `fdst` based on the function `fdct`. The complexity of both functions is $O(N \log_2 N)$ as they are based on the FFT algorithm.

```
fdct(x,N,inv)     // for forward or inverse DCT
     float *x; int N,inv;
{
  int m,n;
  float a,w, *yr,*yi;
  w=3.14159265/2/N;
  a=sqrt(2.0/N);
  yr=(float *)malloc(N*sizeof(float)); // allocate memory for
  yi=(float *)malloc(N*sizeof(float)); // two temperary vectors
  if (inv) {                           // for IDCT
    for (n=0; n<N; n++) x[n]=x[n]*a;
    x[0]=x[0]/sqrt(2.0);
    for (n=0; n<N; n++) {
      yr[n]=x[n]*cos(n*w);
      yi[n]=x[n]*sin(n*w);
    }
  }                                    // for DCT
  else {
    for (m=0; m<N/2; m++) {
      yr[m]=x[2*m];
      yr[N-1-m]=x[2*m+1];
      yi[m]=yi[N/2+m]=0;
    }
  }
  fft(yr,yi,N,inv);                    // call FFT function
```

```
  if (inv) {                              // for inverse DCT
    for (m=0; m<N/2; m++) {
      x[2*m]=yr[m];
      x[2*m+1]=yr[N-1-m];
    }
  }
  else {                                  // for DCT
    for (n=0; n<N; n++)
      x[n]=cos(n*w)*yr[n]+sin(n*w)*yi[n];
    for (n=0; n<N; n++) x[n]=x[n]*a;
    x[0]=x[0]/sqrt(2.0);
  }
  free(yr); free(yi);
}

fdst(x,N,inv)      // for forward or inverse DST
    float *x; int N,inv;
{
  int n;
  float v;
  if (inv) {                   // inverse DST
    for (n=0; n<N/2; n++)
      { v=x[n]; x[n]=x[N-1-n]; x[N-1-n]=v; }
    fdct(x,N,1);
    for (n=1; n<N; n+=2) x[n]=-x[n];
  }
  else {                       // forward DST
    for (n=1; n<N; n+=2) x[n]=-x[n];
    fdct(x,N,0);
    for (n=0; n<N/2; n++)
      { v=x[n]; x[n]=x[N-1-n]; x[N-1-n]=v; }
  }
}
```

7.2.5 DCT and DST filtering

As a real-valued transform, the computation of the DCT or DST is more straightforward than the DFT filtering. In the discussion below we will mainly consider DCT filtering, as DST filtering is mostly the same.

Example 7.6: The top-left panel of Fig. 7.10 shows a signal (solid curve) with three frequency components: the DC, and two sinusoids at frequencies of 8 Hz and 19 Hz. Moreover, the signal is also contaminated by some white noise (dashed line). The DCT spectrum of the signal is shown in the top-right panel in which the three frequency components are clearly seen (solid curve), together with the DCT spectrum of the noise-contaminated signal (dashed curve), whose energy is spread over a much wider frequency range. The signal is then BP filtered, as shown in the lower right panel of the figure, so that the filtered spectrum only contains the frequency component at 8 Hz. The lower left panel shows the filtered signal in time domain obtained by the inverse DCT. We can see clearly that only the 8-Hz sinusoid remains while all other components in the original signal are filtered out (solid curve), which is compared with the original signal (dashed curve). If we assume this 8-Hz sinusoid is the signal of interest and all other components are interference and noise, then this filtering process has effectively extracted the signal by removing the interference and suppressing the noise.

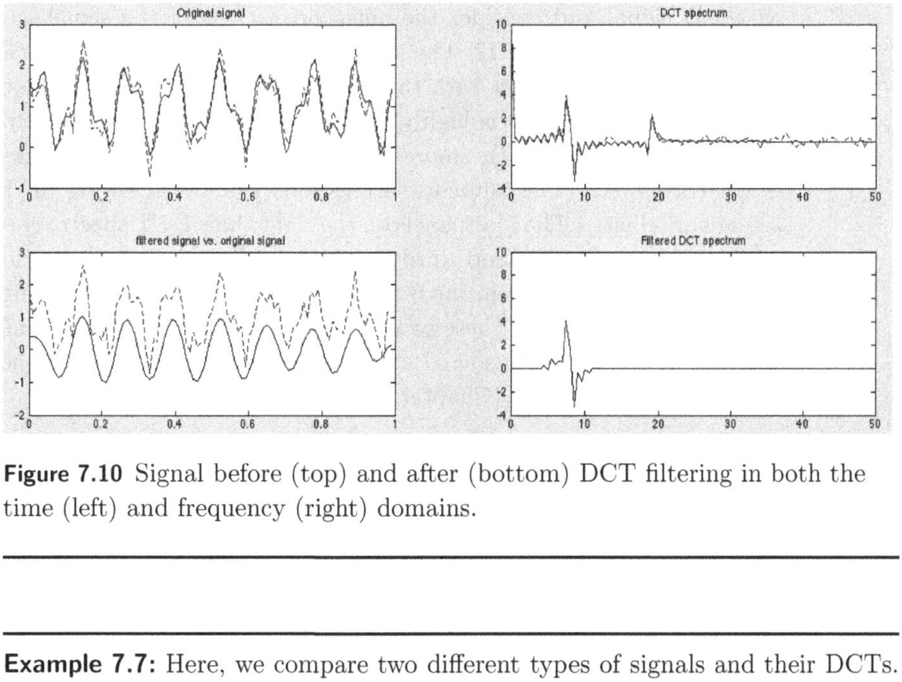

Figure 7.10 Signal before (top) and after (bottom) DCT filtering in both the time (left) and frequency (right) domains.

Example 7.7: Here, we compare two different types of signals and their DCTs. Figure 7.11 shows images of two natural scenes, the clouds on the left and the sand on the right, with very different textures. Specifically, In the cloud image, the value of a pixel is very likely to be similar to those of its neighbors (i.e., they are highly correlated), while in the sand image, the values of neighboring pixels are not likely to be related (i.e., they are much less correlated). Such a difference can be quantitatively described by the autocorrelation of the signal defined before in Eq. (3.119):

$$r_x(t) = \int_{-\infty}^{\infty} x(\tau)x(\tau - t)d\tau = \int_{-\infty}^{\infty} |X(f)|^2 e^{j2\pi f\tau}\, df = \mathcal{F}^{-1}[S_x(f)], \qquad (7.118)$$

where $X(f) = \mathcal{F}[x(t)]$ is the Fourier spectrum of signal $x(t)$ and $S_x(f) = |X(f)|^2$ is the PSD of signal.

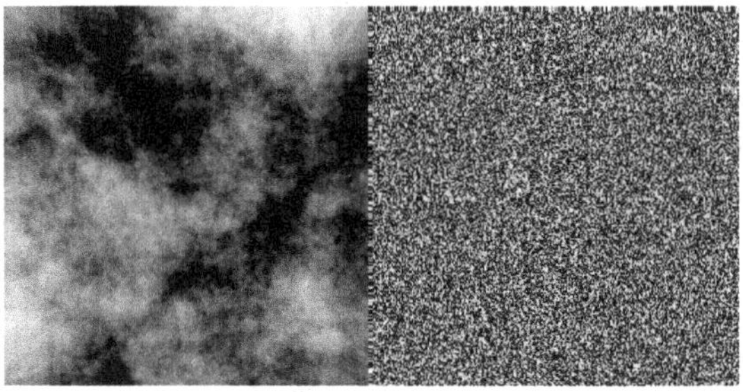

Figure 7.11 Two types of natural scenes: clouds and sand.

To compare the two types of signal, we take one row of each of the two images as a 1-D signal and consider the auto-correlations of the signal as well as its DCT, as shown in Fig. 7.12. The four panels on the left are for the clouds (first) and sand (third) together with their autocorrelation (second and fourth). Note that the signal of clouds is highly correlated, and the closer (smaller t in $r_x(t)$) two signal samples are the more they are correlated, but the signal for sand is not correlated. (The autocorrelations look symmetric owing to the imposed signal periodicity.) The four panels on the right show DCT spectra corresponding to the two signals (first and third) together with their autocorrelations (second and fourth). We see that in the frequency domain the frequency components are hardly correlated at all. These two very different types of signals of high and low correlations will be reconsidered in the future discussion regarding the statistical properties of the signals (Chapter 10).

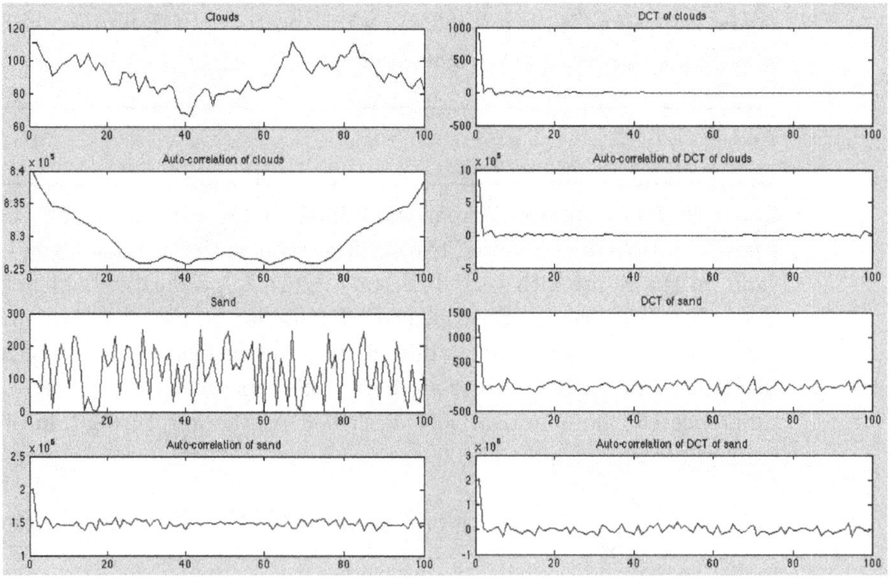

Figure 7.12 Decorrelation of cloud and sand signals.

In general, all natural signals are correlated to different degrees, depending on their specific natures. Most signals are highly correlated, such as the example of clouds, although some exceptions are less so, such as the sand. But in either case, the components in the spectrum of the signal after an orthogonal transform, such as the Fourier or cosine transform, or any other orthogonal transform for that matter, are much less correlated; i.e., all orthogonal transforms tend to decorrelate the signal, as the autocorrelation of a typical signal is significantly reduced in the transform domain.

7.2.6 The 2-D DCT and DST

Here, we consider the DCT and DST filtering of a 2-D signal $x[m, n]$ $(m = 0, \dots, M-1,\ n = 0, \dots, N-1)$, such as an image. In the discussion below we will again mainly consider 2-D DCT filtering, as 2-D DST filtering is mostly the same.

The forward and inverse 2-D DCT are defined respectively as

$$X[k,l] = a[k]a[l] \sum_{n=0}^{N-1} \sum_{m=0}^{M-1} x[m,n] \cos\left(\frac{(2m+1)k\pi}{2M}\right) \cos\left(\frac{(2n+1)l\pi}{2N}\right), \tag{7.119}$$

$$x[m,n] = \sum_{l=0}^{N-1} \sum_{k=0}^{M-1} a[k]a[l]X[k,l] \cos\left(\frac{(2m+1)k\pi}{2M}\right) \cos\left(\frac{(2m+1)l\pi}{2N}\right), \tag{7.120}$$

$$m, k = 0, \dots, M-1, \quad n, l = 0, \dots, N-1.$$

The inverse DCT (second equation) represents the given 2-D signal as a linear combination of a set of MN 2-D basis functions each of size $M \times N$ as a product of two sinusoidal functions in the horizontal and vertical directions. The weighting coefficients $X[k, l]$ can be obtained by the forward DCT. Fig. 7.13 displays a set of $M \times N = 8 \times 8 = 64$ such 2-D basis functions for both the DCT (left) and DST (right).

Similar to the 2-D DFT, the two summations in either the forward DCT in Eq. (7.119) or inverse DCT in Eq. (7.120) can be carried out in two separate steps. First, we can carry out N M-point 1-D DCTs for each of the columns of the 2-D signal array (the inner summation with respect to m in Eq. (7.119), and then carry out M N-point 1-D DCTs for each of the rows of the resulting array after the first step (the outer summation with respect to n in Eq. (7.119). Of course, we can also carry out the row DCTs first and then the column DCTs. In matrix multiplication form, the forward and inverse 2-D DCT can be represented as

$$\begin{cases} \boldsymbol{X} = \boldsymbol{C}_c^{\mathrm{T}} \boldsymbol{x} \boldsymbol{C}_r & \text{(forward)} \\ \boldsymbol{x} = \boldsymbol{C}_c \boldsymbol{X} \boldsymbol{C}_r^{\mathrm{T}} & \text{(inverse)} \end{cases}, \tag{7.121}$$

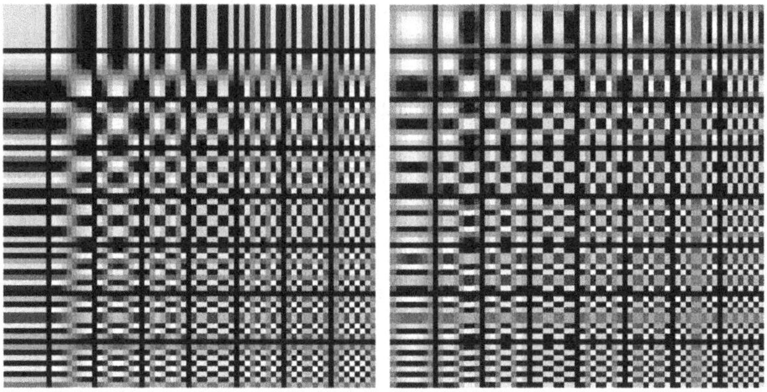

Figure 7.13 The $M \times N = 8 \times 8 = 64$ basis functions $\boldsymbol{B}_{kl}$ of 2-D DCT (left) and DST (right). The DC component is at the top-left corner, and the highest frequency component in both the horizontal and vertical directions is at the lower right corner.

where $\boldsymbol{x}$ and $\boldsymbol{X}$ are both 2-D $M \times N$ matrices representing a 2-D signal and its spectrum, and the pre-multiplication matrix $\boldsymbol{C}_c$ is $M \times M$ for the column transforms, while the post-multiplication matrix $\boldsymbol{C}_r$ is $N \times N$ for the row transforms. The DCT spectrum of a 2-D $M \times N$ real signal (e.g., an image) is also an $M \times N$ real matrix composed of MN DCT coefficients $X[k, l]$ ($k = 0, \dots, M-1, l = 0, \dots, N-1$) representing the magnitudes of the corresponding basis functions.

The DCT matrix $\boldsymbol{C}$ can be expressed in terms of its column vectors and the inverse transform can be written as

$$\begin{aligned}\boldsymbol{x} &= [\boldsymbol{c}_0, \dots, \boldsymbol{c}_{M-1}] \begin{bmatrix} X[0,0] & \cdots & X[0, N-1] \\ \vdots & \ddots & \vdots \\ X[M-1, 0] & \cdots & X[M-1, N-1] \end{bmatrix} \begin{bmatrix} \boldsymbol{c}_0^{\mathrm{T}} \\ \vdots \\ \boldsymbol{c}_{N-1}^{\mathrm{T}} \end{bmatrix} \\ &= \sum_{k=0}^{M-1} \sum_{l=0}^{N-1} X[k, l] \boldsymbol{c}_k \boldsymbol{c}_l^{\mathrm{T}} = \sum_{k=0}^{M-1} \sum_{l=0}^{N-1} X[k, l] \boldsymbol{B}_{kl}. \end{aligned} \tag{7.122}$$

Here, we have defined $\boldsymbol{B}_{kl} = \boldsymbol{c}_k \boldsymbol{c}_l^{\mathrm{T}}$, where $\boldsymbol{c}_k$ is the kth column vector of the $M \times M$ DCT matrix for the row transforms and $\boldsymbol{c}_l$ is the lth column vector of the $N \times N$ DCT matrix for the column transforms. We see that the 2-D signal $\boldsymbol{x}_{M \times N}$ is now expressed as a linear combination of a set of MN 2-D ($M \times N$) DCT basis functions $\boldsymbol{B}_{kl}$ ($k, l = 0, \dots, N-1$), each of which can be obtained from the inverse transform above when all elements of $\boldsymbol{X}$ are zero except $X[k, l] = 1$. When $M = N = 8$, the $8 \times 8 = 64$ such 2-D DCT basis functions are shown in Fig. 7.13. Any 8×8 2-D signal can be expressed as a linear combination of these 64 2-D orthogonal basis functions.

In Eq. (7.122), each basis function $\boldsymbol{B}_{kl}$ is weighted by the klth DCT coefficients $X[k,l]$, the klth component of the 2-D spectrum obtained by the forward transform:

$$\boldsymbol{X} = \boldsymbol{C}^{\mathrm{T}}\boldsymbol{x}\boldsymbol{C} = \begin{bmatrix} \boldsymbol{c}_0^{\mathrm{T}} \\ \vdots \\ \boldsymbol{c}_{M-1}^{\mathrm{T}} \end{bmatrix} \boldsymbol{x}[\boldsymbol{c}_0, \ldots, \boldsymbol{c}_{N-1}]. \tag{7.123}$$

The klth coefficient $X[k,l]$ is

$$\begin{aligned} X[k,l] &= \boldsymbol{c}_k^{\mathrm{T}} \begin{bmatrix} x[0,0] & \cdots & x[0,N-1] \\ \vdots & \ddots & \vdots \\ x[M-1,0] & \cdots & x[M-1,N-1] \end{bmatrix} \boldsymbol{c}_l \\ &= \sum_{m=0}^{M-1}\sum_{n=0}^{N-1} x[m,n]B_{kl}[m,n] = \langle \boldsymbol{x}, \boldsymbol{B}_{kl} \rangle. \end{aligned} \tag{7.124}$$

As in the 2-D DFT case in Eq. (4.272), the coefficient $X[k,l]$ can be found as the projection of the 2-D signal $\boldsymbol{x}$ onto the klth DCT basis function $\boldsymbol{B}_{kl}$.

Example 7.8: An image and its DCT and DST spectra are shown in Fig. 7.14. Different from the DFT spectrum composed of complex DFT coefficients representing the magnitudes and phases for the frequency components, here the spectrum of either the DCT or DST is a real matrix representing the magnitudes of the frequency components all with zero phases.

Figure 7.14 An image (left) and its DCT (middle) and DST (right) spectra.

Various types of filtering, such as LP and HP filtering, can be carried out in the frequency domain by modifying the spectrum of the signal. Fig. 7.15 shows some LP and HP results using two different types of filters: the ideal filter and the Butterworth filter. In the case of an ideal filter, all frequency components farther away from the DC component (top-left corner of the spectrum) than a distance corresponding to the cutoff frequency are suppressed to zero while all other components remain unchanged. The modified spectrum and the resulting LP filtered image after the inverse DCT are shown in the figure at top-left and bottom-left, respectively. Similar to the case of the DFT, some obvious ringing artifacts can be observed in the ideal-filtered image. To avoid this, the

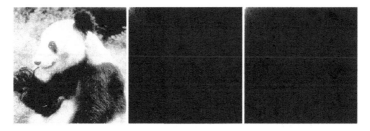

Butterworth filter without sharp edges can be used, as shown by the pair of images second from the left. The same ideal and Butterworth filters can also be used for HP filtering, as shown by the other two pairs of images on the right. Again, note that the ringing artifacts due to the ideal filter are avoided by Butterworth filtering.

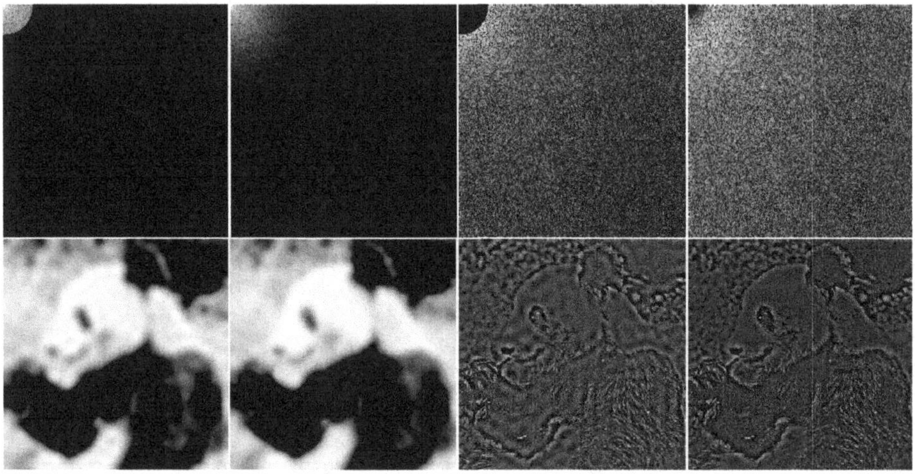

Figure 7.15 LP and HP filtering of an image. Similar to the Fourier transform, DCT also suffers from the ringing artifacts caused by the ideal filters (first and third), which can be avoided by the smooth Butterworth filter.

Example 7.9: The DCT can also be used for data compression, as illustrated in Fig. 7.16. In this case, 90% of the DCT coefficients (corresponding mostly to some high-frequency components) with magnitudes less than a certain threshold value were surprised to zero (black in the image). The image is then reconstructed based on the remaining 10% of the coefficients, but containing over 99.6% of the signal energy. As can be seen in the figure, the reconstructed image, with only 0.4% energy lost, looks very much the same as the original one, except some very fine details corresponding to high-frequency components which were suppressed.

We can throw away 90% of the coefficients but still keep over 99% of the energy only in the frequency domain, not in the spatial domain, owing to the two general properties of all orthogonal transforms: (a) decorrelation of signals and (b) compaction of signal energy. In this example, the effect of energy compaction of the DCT is stronger than that of the DFT discussed before. In general, as a real transform method, the DCT is more widely used in image compression than the DFT. For example, it is used in the most popular image compression standard JPEG (http://en.wikipedia.org/wiki/JPEG).

Figure 7.16 Image compression based on the DCT. An image with its DCT spectrum (left) and the reconstructed image based on 10% of the coefficients containing 99.6% of the total energy (right).

7.3 Homework problems

1. Prove the orthogonality of the discrete Hartley transform given in Eq. (7.27). Hint: consider the trigonometric identities

$$\begin{aligned} \sin(\alpha \pm \beta) &= \sin\alpha\cos\beta \pm \sin\beta\cos\alpha, \\ \cos(\alpha \pm \beta) &= \cos\alpha\cos\beta \mp \sin\beta\sin\alpha, \end{aligned} \tag{7.125}$$

and then use the result of Eq. (1.40).

2. Prove the following relation (Eq. (7.81)):

$$\sum_{n=0}^{N-1} \cos\left(\frac{(2n+1)k\pi}{2N}\right) = \begin{cases} N & k=0 \\ -N & k=2N \\ 0 & 0<k<2N \end{cases} \tag{7.126}$$

Hints: You may find it is helpful to use the identity $\sum_{n=0}^{N-1} x^n = (1 - x^N)/(1-x)$, and to consider the two different cases when k is either even or odd.

3. Let $x[n] = n+1$ $(n = 0, 1, 2, 3)$ be a discrete signal with period $N = 4$. Find its discrete Hartley transform, DCT and DST by matrix multiplication $\boldsymbol{X} = \boldsymbol{H}\boldsymbol{x}$, $\boldsymbol{X} = \boldsymbol{C}^{\mathrm{T}}\boldsymbol{x}$, and $\boldsymbol{X} = \boldsymbol{S}^{\mathrm{T}}\boldsymbol{x}$, respectively, where $\boldsymbol{H}$, $\boldsymbol{C}$, and $\boldsymbol{S}$ are given in Eqs. (7.35) and (7.97). Then carry out the inverse transform also by matrix multiplication $\boldsymbol{x} = \boldsymbol{T}\boldsymbol{X}$, $\boldsymbol{x} = \boldsymbol{C}\boldsymbol{X}$, and $\boldsymbol{x} = \boldsymbol{S}\boldsymbol{X}$ to confirm that the signal is perfectly reconstructed.

4. Develop a Matlab function for the discrete Hartley transform. Apply it to an $N = 8$ sequence $\boldsymbol{x} = [x[0], \ldots, x[7]]^{\mathrm{T}}$ of your choice to obtain its N transform coefficients, then carry out the inverse transform to reconstruct the sequence from these transform coefficients.
5. Understand the C code for the fast DCT algorithm provided in the text and convert it into a Matlab function. Then carry out the forward and inverse DCT using the $N = 8$ sequence chosen for the previous problem. Confirm the perfect reconstruction is achieved.
6. Develop a Matlab function for the DST and repeat the above.
7. Implement the 2-D discrete Hartley transform of the same image used in the homework of Chapter 5 and carry out various types of filtering (LP, HP, etc.) of the image in the transform domain. Then carry out the inverse transform and display the filtered image. Compare the filtering effects with those obtained by the Fourier transform obtained in Chapter 5.
8. Carry out compression of the image used before as shown in Fig. 5.25 by suppressing to zero all frequency components lower than a certain threshold. Obtain the percentage of such suppressed frequency components, and the percentage of lost energy (in terms of signal value squared). (Note that this exercise only serves to illustrate the basic idea of image compression but it is not how image compression is practically done, where those components suppressed need to be recorded as well.)
9. Repeat the two problems above for the DCT.
10. Repeat the same two problems above for the DST.

8 The Walsh-Hadamard, slant, and Haar transforms

In this chapter, we will consider a set of three real orthogonal transforms, namely the Walsh-Hadamard transform (WHT), the slant transform (ST), and the discrete Haar transform (DHT), all of which are defined quite differently from the previously considered transforms all closely related to the Fourier transform based on sinusoidal kernel functions. In fact, the transforms considered here are no longer continuous and smooth in nature, and they can be used to capture some different types of features and components of the signal being transformed.

8.1 The Walsh-Hadamard transform

8.1.1 Hadamard matrix

The Walsh-Hadamard transform matrix can be most conveniently defined based on the concept of *Kronecker product*. The Kronecker product of an m by n matrix $\boldsymbol{A} = [a_{ij}]_{m\times n}$ and a k by l matrix $\boldsymbol{B} = [b_{ij}]_{k\times l}$ is an mk by nl matrix defined as

$$\boldsymbol{A} \otimes \boldsymbol{B} = \begin{bmatrix} a_{11}\boldsymbol{B} & \cdots & a_{1n}\boldsymbol{B} \\ \vdots & \ddots & \vdots \\ a_{m1}\boldsymbol{B} & \cdots & a_{mn}\boldsymbol{B} \end{bmatrix}_{mk\times nl} . \tag{8.1}$$

In general, $\boldsymbol{A}\otimes\boldsymbol{B} \neq \boldsymbol{B}\otimes\boldsymbol{A}$. Now we can define the *Hadamard matrix* recursively as

$$\boldsymbol{H}_1 = \frac{1}{\sqrt{2}}\begin{bmatrix} 1 & 1 \\ 1 & -1 \end{bmatrix}, \tag{8.2}$$

$$\boldsymbol{H}_n = \boldsymbol{H}_1 \otimes \boldsymbol{H}_{n-1} = \frac{1}{\sqrt{2}}\begin{bmatrix} \boldsymbol{H}_{n-1} & \boldsymbol{H}_{n-1} \\ \boldsymbol{H}_{n-1} & -\boldsymbol{H}_{n-1} \end{bmatrix}; \tag{8.3}$$

where $\boldsymbol{H}_n$ is an $N\times N$ matrix with $N = 2^n$. Obviously $\boldsymbol{H}_n = \boldsymbol{H}_n^{\mathrm{T}}$ is symmetric, and, based on the recursion in Eqs. (8.2) and (8.3), we can easily show it is also orthonormal: $\boldsymbol{H}_n^{\mathrm{T}}\boldsymbol{H}_n = \boldsymbol{H}_n\boldsymbol{H}_n = \boldsymbol{I}$. In particular, when $n = 2$, $N = 2^2 = 4$ and

$$\boldsymbol{H}_2 = \boldsymbol{H}_1 \otimes \boldsymbol{H}_1 = \frac{1}{\sqrt{2}}\begin{bmatrix} \boldsymbol{H}_1 & \boldsymbol{H}_1 \\ \boldsymbol{H}_1 & -\boldsymbol{H}_1 \end{bmatrix} = \frac{1}{\sqrt{4}}\begin{bmatrix} 1 & 1 & 1 & 1 \\ 1 & -1 & 1 & -1 \\ 1 & 1 & -1 & -1 \\ 1 & -1 & -1 & 1 \end{bmatrix}. \tag{8.4}$$

When $n = 3$, $N = 2^3 = 8$ and

$$\boldsymbol{H}_3 = \boldsymbol{H}_1 \otimes \boldsymbol{H}_2 = \frac{1}{\sqrt{2}}\begin{bmatrix} \boldsymbol{H}_2 & \boldsymbol{H}_2 \\ \boldsymbol{H}_2 & -\boldsymbol{H}_2 \end{bmatrix} = \frac{1}{\sqrt{8}}\begin{bmatrix} 1 & 1 & 1 & 1 & 1 & 1 & 1 & 1 \\ 1 & -1 & 1 & -1 & 1 & -1 & 1 & -1 \\ 1 & 1 & -1 & -1 & 1 & 1 & -1 & -1 \\ 1 & -1 & -1 & 1 & 1 & -1 & -1 & 1 \\ 1 & 1 & 1 & 1 & -1 & -1 & -1 & -1 \\ 1 & -1 & 1 & -1 & -1 & 1 & -1 & 1 \\ 1 & 1 & -1 & -1 & -1 & -1 & 1 & 1 \\ 1 & -1 & -1 & 1 & -1 & 1 & 1 & -1 \end{bmatrix} \begin{matrix} 0\,0 \\ 1\,7 \\ 2\,3 \\ 3\,4 \\ 4\,1 \\ 5\,6 \\ 6\,2 \\ 7\,5 \end{matrix}. \tag{8.5}$$

The first column to the right of the matrix is for the index number $k = 0, \ldots, N-1$ for the $N = 8$ rows, and the second column is for the *sequency* s of each row, defined as the number of zero-crossings or sign changes in the row. Similar to frequency, sequency also measures the rate of changes or variations in a signal. However, sequency can measure non-periodic signals as well as periodic ones. The conversion between sequency s and index k will be considered later. The $N = 8$ rows in matrix $\boldsymbol{H}_3$ are actually the discrete version of the $N = 8$ continuous functions shown in Fig. 8.1, which can be used as an orthogonal basis to represent any function in the space they span.

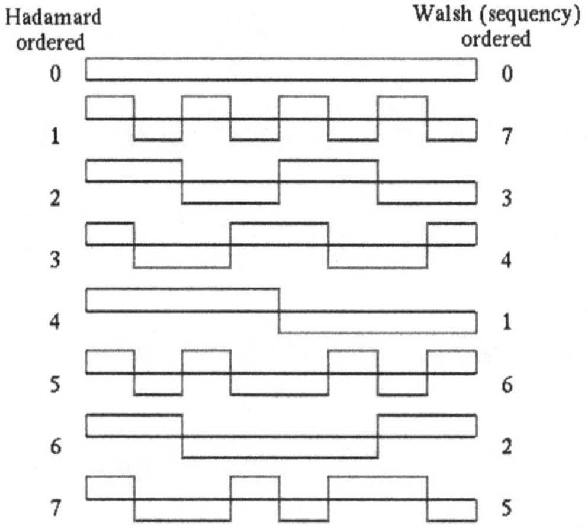

Figure 8.1 The basis functions of the WHT.

Alternatively, a Hadamard matrix $\boldsymbol{H}$ can also be defined in terms of its element $h[k, m]$ in the kth row and mth column as below (for simplicity, the scaling factor

$1/\sqrt{N}$ is neglected for now):

$$h[k,m] = (-1)^{\sum_{i=0}^{n-1} k_i m_i} = \prod_{i=0}^{n-1} (-1)^{k_i m_i} = h[m,k] \qquad k,m = 0,\dots,N-1, \tag{8.6}$$

where

$$k = \sum_{i=0}^{n-1} k_i 2^i = (k_{n-1}k_{n-2}\dots k_1 k_0)_2 \qquad k_i = 0,1, \tag{8.7}$$

$$m = \sum_{i=0}^{n-1} m_i 2^i = (m_{n-1}m_{n-2}\dots m_1 m_0)_2 \qquad m_i = 0,1; \tag{8.8}$$

i.e., $(k_{n-1}k_{n-2}\dots k_1 k_0)_2$ and $(m_{n-1}m_{n-2}\dots m_1 m_0)_2$ are the binary representations of k and m, respectively. Obviously, we need $n = \log_2 N$ bits in these binary representations. For example, when $n = 3$ and $N = 2^n = 8$, the element $h[k,m]$ in row $k = 2 = (010)_2$ and column $m = 3 = (011)_2$ of $\boldsymbol{H}_3$ is $(-1)^{0+1+0} = -1$.

It is easy to show that this alternative definition of the Hadamard matrix is actually equivalent to the previous recursive definition given in Eqs. (8.2) and (8.3). First, when $n = 1$ and $N = 2^n = 2$, the two rows and columns indexed by a single bit of k_0 and m_0, respectively, and the product $k_0 m_0$ of the two bits has four possible values, $0 \times 0 = 0$, $0 \times 1 = 0$, $1 \times 0 = 0$, and $1 \times 1 = 1$; these correspond to the four elements of the matrix; i.e., $h[0,0] = h[0,1] = h[1,0] = (-1)^{k_0 m_0} = (-1)^0 = 1$ and $h[1,1] = (-1)^{k_0 m_0} = (-1)^1 = -1$. This is actually Eq. (8.2).

Next, when n is increased by 1, the size $N = 2^n$ of the matrix is doubled, and one more bit k_{n-1} and m_{n-1} (the most significant bit) is needed for the binary representations of k and m, respectively. The product of these two most significant bits $k_{n-1}m_{n-1}$ determines the four quadrants of the new matrix $\boldsymbol{H}_n$. The first three quadrants (upper left, upper right, and lower left) corresponding to $k_{n-1}m_{n-1} = 0$ are therefore identical to $\boldsymbol{H}_{n-1}$, while the lower right quadrant corresponding to $k_{n-1}m_{n-1} = 1$ is the negation of $\boldsymbol{H}_{n-1}$. This is the recursion in Eq. (8.3).

The Hadamard matrix $\boldsymbol{H}$ is real and symmetric, and also orthogonal:

$$\boldsymbol{H} = \boldsymbol{H}^* = \boldsymbol{H}^{\mathrm{T}} = \boldsymbol{H}^{-1}. \tag{8.9}$$

The orthogonality of $\boldsymbol{H}$ can be proven by induction. This is left for the reader as a homework exercise.

8.1.2 Hadamard-ordered Walsh-Hadamard transform ($\mathrm{WHT_h}$)

The Hadamard matrix can be written in terms of its columns:

$$\boldsymbol{H} = [\boldsymbol{h}_0, \dots, \boldsymbol{h}_{N-1}]. \tag{8.10}$$

As $\boldsymbol{H}$ is an orthogonal matrix, its N vectors are orthonormal:

$$\langle \boldsymbol{h}_k, \boldsymbol{h}_l \rangle = \boldsymbol{h}_k^{\mathrm{T}} \boldsymbol{h}_l = \delta[k-l], \tag{8.11}$$

they form a complete basis that spans the N-D vector space, and the Hadamard matrix $\boldsymbol{H}$ can be used to define an orthogonal transform, called the Hadamard-ordered Walsh-Hadamard transform ($\mathrm{WHT_h}$):

$$\begin{cases} \boldsymbol{X} = \boldsymbol{H}\boldsymbol{x} & \text{(forward)} \\ \boldsymbol{x} = \boldsymbol{H}\boldsymbol{X} & \text{(inverse)} \end{cases}. \tag{8.12}$$

Here, $\boldsymbol{x} = [x[0], \ldots, x[N-1]]^{\mathrm{T}}$ is an N-point signal vector and $\boldsymbol{X} = X[0], \ldots, X[N-1]]^{\mathrm{T}}$ is its WHT spectrum vectors. As $\boldsymbol{H}^{-1} = \boldsymbol{H}$, the forward (first equation) and inverse (second equation) transforms are identical. Also note that the WHT can be carried out by additions and subtractions alone.

The inverse transform ($\mathrm{IWHT_h}$) can be further written as

$$\boldsymbol{x} = [\boldsymbol{h}_0, \ldots, \boldsymbol{h}_{N-1}] \begin{bmatrix} X[0] \\ \vdots \\ X[N-1] \end{bmatrix} = \sum_{k=0}^{N-1} X[k] \boldsymbol{h}_k. \tag{8.13}$$

by which the signal vector $\boldsymbol{x}$ is expressed as a linear combination of the N basis vectors $\boldsymbol{h}_k$, weighted by the WHT coefficients $X[k]$ $(k = 0, \ldots, N-1)$. Taking an inner product with $\boldsymbol{h}_l$ on both sides, we get the WHT coefficient $X[l]$ as the projection of $\boldsymbol{x}$ onto the lth basis vector:

$$X[l] = \langle \boldsymbol{x}, \boldsymbol{h}_l \rangle = \boldsymbol{h}_l^{\mathrm{T}} \boldsymbol{x} \qquad l = 0, \ldots, N-1, \tag{8.14}$$

which is just the component form of the forward WHT:

$$\boldsymbol{X} = \boldsymbol{H}\boldsymbol{x} = \boldsymbol{H}^{\mathrm{T}}\boldsymbol{x} = \begin{bmatrix} \boldsymbol{h}_0^{\mathrm{T}} \\ \vdots \\ \boldsymbol{h}_{N-1}^{\mathrm{T}} \end{bmatrix} \boldsymbol{x}. \tag{8.15}$$

Note that $X[k]$ can also be written as

$$X[k] = \sum_{m=0}^{N-1} h[k,m]x[m] = \frac{1}{\sqrt{N}} \sum_{m=0}^{N-1} x[m] \prod_{i=0}^{n-1} (-1)^{m_i k_i}. \tag{8.16}$$

8.1.3 Fast Walsh-Hadamard transform algorithm

The complexity of WHT implemented as a matrix multiplication $\boldsymbol{X} = \boldsymbol{H}\boldsymbol{x}$ is of course $O(N^2)$. However, similar to the FFT algorithm, we can also derive a fast WHT algorithm with complexity $O(N \log_2 N)$ as shown below. We assume

$n = 3$ and $N = 2^n = 8$, and write the WHT_h of an eight-point signal $\boldsymbol{x}$ as

$$\boldsymbol{X} = \begin{bmatrix} X[0] \\ \vdots \\ X[3] \\ X[4] \\ \vdots \\ X[7] \end{bmatrix} = \boldsymbol{H}_3\boldsymbol{x} = \begin{bmatrix} \boldsymbol{H}_2 & \boldsymbol{H}_2 \\ \boldsymbol{H}_2 & -\boldsymbol{H}_2 \end{bmatrix} \begin{bmatrix} x[0] \\ \vdots \\ x[3] \\ x[4] \\ \vdots \\ x[7] \end{bmatrix}. \tag{8.17}$$

This equation can be separated into two parts. The first half of vector $\boldsymbol{X}$ can be obtained as

$$\begin{bmatrix} X[0] \\ X[1] \\ X[2] \\ X[3] \end{bmatrix} = \boldsymbol{H}_2 \begin{bmatrix} x[0] \\ x[1] \\ x[2] \\ x[3] \end{bmatrix} + \boldsymbol{H}_2 \begin{bmatrix} x[4] \\ x[5] \\ x[6] \\ x[7] \end{bmatrix} = \boldsymbol{H}_2 \begin{bmatrix} x_1[0] \\ x_1[1] \\ x_1[2] \\ x_1[3] \end{bmatrix}, \tag{8.18}$$

where we have defined

$$x_1[i] = x[i] + x[i+4] \qquad i = 0, \ldots, 3. \tag{8.19}$$

Similarly, the second half of vector $\boldsymbol{X}$ can be obtained as

$$\begin{bmatrix} X[4] \\ X[5] \\ X[6] \\ X[7] \end{bmatrix} = \boldsymbol{H}_2 \begin{bmatrix} x[0] \\ x[1] \\ x[2] \\ x[3] \end{bmatrix} - \boldsymbol{H}_2 \begin{bmatrix} x[4] \\ x[5] \\ x[6] \\ x[7] \end{bmatrix} = \boldsymbol{H}_2 \begin{bmatrix} x_1[4] \\ x_1[5] \\ x_1[6] \\ x_1[7] \end{bmatrix}, \tag{8.20}$$

where we have defined

$$x_1[i+4] = x[i] - x[i+4] \qquad i = 0, \ldots, 3. \tag{8.21}$$

What we did above is to convert an eight-point WHT into two four-point WHTs. This process can be carried out recursively. We next rewrite Eq. (8.18) as

$$\begin{bmatrix} X[0] \\ X[1] \\ X[2] \\ X[3] \end{bmatrix} = \begin{bmatrix} \boldsymbol{H}_1 & \boldsymbol{H}_1 \\ \boldsymbol{H}_1 & -\boldsymbol{H}_1 \end{bmatrix} \begin{bmatrix} x_1[0] \\ x_1[1] \\ x_1[2] \\ x_1[3] \end{bmatrix}, \tag{8.22}$$

which can again be separated into two halves. The first half is

$$\begin{aligned} \begin{bmatrix} X[0] \\ X[1] \end{bmatrix} &= \boldsymbol{H}_1 \begin{bmatrix} x_1[0] \\ x_1[1] \end{bmatrix} + \boldsymbol{H}_1 \begin{bmatrix} x_1[2] \\ x_1[3] \end{bmatrix} = \boldsymbol{H}_1 \begin{bmatrix} x_2[0] \\ x_2[1] \end{bmatrix} \\ &= \begin{bmatrix} 1 & 1 \\ 1 & -1 \end{bmatrix} \begin{bmatrix} x_2[0] \\ x_2[1] \end{bmatrix} = \begin{bmatrix} x_2[0] + x_2[1] \\ x_2[0] - x_2[1] \end{bmatrix}, \end{aligned} \tag{8.23}$$

where

$$x_2[i] = x_1[i] + x_1[i+2] \qquad i = 0, 1, \tag{8.24}$$

and

$$X[0] = x_2[0] + x_2[1], \qquad X[1] = x_2[0] - x_2[1]. \tag{8.25}$$

The second half is

$$\begin{aligned}\begin{bmatrix} X[2] \\ X[3] \end{bmatrix} &= \boldsymbol{H}_1 \begin{bmatrix} x_1[0] \\ x_1[1] \end{bmatrix} - \boldsymbol{H}_1 \begin{bmatrix} x_1[2] \\ x_1[3] \end{bmatrix} = \boldsymbol{H}_1 \begin{bmatrix} x_2[2] \\ x_2[3] \end{bmatrix} \\ &= \begin{bmatrix} 1 & 1 \\ 1 & -1 \end{bmatrix} \begin{bmatrix} x_2[2] \\ x_2[3] \end{bmatrix} = \begin{bmatrix} x_2[2] + x_2[3] \\ x_2[2] - x_2[3] \end{bmatrix},\end{aligned} \tag{8.26}$$

where

$$x_2[i+2] = x_1[i] - x_1[i+2] \qquad i = 0, 1, \tag{8.27}$$

and

$$X[2] = x_2[2] + x_2[3], \qquad X[3] = x_2[2] - x_2[3]. \tag{8.28}$$

Similarly, the coefficients $X[4]$ through $X[7]$ in the second half of the transform in Eq. (8.20) can be obtained by the same process. Summarizing the above steps of Eqs. 8.19, 8.21, 8.24, 8.25, 8.27, and 8.28, we get the fast WHT algorithm as illustrated in Fig. 8.2. As the algorithm has $\log_2 N$ stages each of $O(N)$ operations, its complexity is obviously $O(N \log_2 N)$.

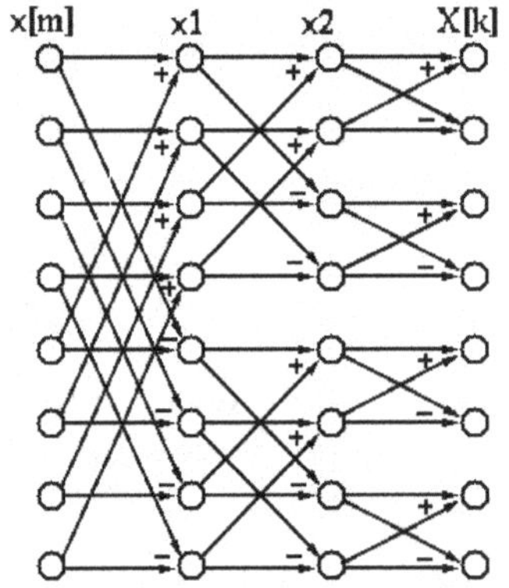

Figure 8.2 The fast WHT algorithm.

8.1.4 Sequency-ordered Walsh-Hadamard matrix (WHT$_\text{w}$)

The rows (or columns) of the WHT matrix $\boldsymbol{H}$, and therefore the elements $X[k]$ of the WHT spectrum $\boldsymbol{X} = \boldsymbol{H}\boldsymbol{x}$, are not arranged in the order of the sequencies, while it is desirable to arrange them according to the sequencies in a low-to-high order, similar to how the DFT coefficients are arranged. To reorder the rows of the Hadamard matrix H according to their sequencies, we first consider the conversion of a given sequency number s into the corresponding row index number k in Hadamard order, in the following three steps:

1. Represent s in binary form:

$$s = (s_{n-1} \dots s_0)_2 = \sum_{i=0}^{n-1} s_i 2^i. \tag{8.29}$$

2. Convert this n-bit binary number to an n-bit Gray code:

$$g = (g_{n-1} \dots g_0)_2, \quad \text{where} \quad g_i = s_i \oplus s_{i+1} \quad i = 0, \dots, n-1. \tag{8.30}$$

Here, $\oplus$ represents exclusive OR of two bits, and $s_n = 0$ is defined as zero.

3. Bit reverse the Gray code bits g_i to get:

$$k_i = g_{n-1-i} = s_{n-1-i} \oplus s_{n-i}. \tag{8.31}$$

Now the row index k can be obtained as

$$k = (k_{n-1}\, k_{n-2} \dots k_1\, k_0)_2 = \sum_{i=0}^{n-1} s_{n-1-i} \oplus s_{n-i} 2^i = \sum_{j=0}^{n-1} s_j \oplus s_{j+1} 2^{n-1-j}, \tag{8.32}$$

where $j = n-1-i$, or equivalently $i = n-1-j$.

For example, when $n = 3$ and $N = 2^3 = 8$ we have

s	0	1	2	3	4	5	6	7
binary	000	001	010	011	100	101	110	111
Gray code	000	001	011	010	110	111	101	100
bit reverse	000	100	110	010	011	111	101	001
k	0	4	6	2	3	7	5	1

(8.33)

Now the sequency-ordered, also called Walsh-ordered, Walsh-Hadamard matrix can be obtained as

$$\boldsymbol{H}_{\text{w}} = \frac{1}{\sqrt{8}} \begin{bmatrix} 1 & 1 & 1 & 1 & 1 & 1 & 1 & 1 \\ 1 & 1 & 1 & 1 & -1 & -1 & -1 & -1 \\ 1 & 1 & -1 & -1 & -1 & -1 & 1 & 1 \\ 1 & 1 & -1 & -1 & 1 & 1 & -1 & -1 \\ 1 & -1 & -1 & 1 & 1 & -1 & -1 & 1 \\ 1 & -1 & -1 & 1 & -1 & 1 & 1 & -1 \\ 1 & -1 & 1 & -1 & -1 & 1 & -1 & 1 \\ 1 & -1 & 1 & -1 & 1 & -1 & 1 & -1 \end{bmatrix} \begin{matrix} 0\;0 \\ 1\;4 \\ 2\;6 \\ 3\;2 \\ 4\;3 \\ 5\;7 \\ 6\;5 \\ 7\;1 \end{matrix}. \tag{8.34}$$

Here, a subscript w is used to indicate the row vectors of this matrix $\boldsymbol{H}$ are sequency ordered (or Walsh-ordered). The two columns to the right of the matrix are the indices of the row vectors in the sequency order (first column) and the original Hadamard order (second column). Note that this sequency-ordered matrix is still symmetric: $\boldsymbol{H}_{\text{w}}^{\text{T}} = \boldsymbol{H}_{\text{w}}$.

Now the sequency-ordered Walsh-Hadamard transform (WHT_{w}) can be carried out as

$$\boldsymbol{X} = \boldsymbol{H}_{\text{w}} \boldsymbol{x}, \tag{8.35}$$

or in component form as

$$X[k] = \sum_{m=0}^{N-1} h_{\mathrm{w}}[k,m]x[m], \tag{8.36}$$

where $h_{\mathrm{w}}[k,m]$ is the element in the kth row and nth column of $\boldsymbol{H}_{\mathrm{w}}$.

The orthogonal basis function shown in Fig. 8.1 can also be rearranged to be sequency-ordered, as shown in Fig. 8.3.

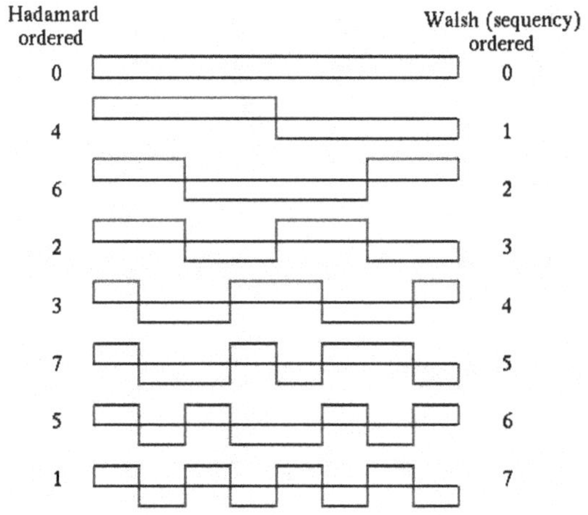

Figure 8.3 The basis functions of the WHT (sequency ordered).

Note that in Matlab the forward and inverse sequency-ordered WHT can be carried out by functions `fwht` and `ifwht`, respectively. However, these WHT functions are scaled differently. For Parseval's identity to hold, the Matlab forward transform needs to be rescaled: `X=fwht(x)*sqrt(length(x))`.

8.1.5 Fast Walsh-Hadamard transform (sequency ordered)

The sequency-ordered Walsh-Hadamard transform $\mathrm{WHT_w}$ can be obtained by first carrying out the fast $\mathrm{WHT_h}$ and then reordering the components of $\boldsymbol{X}$ as shown above. Alternatively, we can use the following fast $\mathrm{WHT_w}$ directly with better efficiency.

Similar to the WHT shown in Eq. (8.16), the sequency-ordered WHT of $x[m]$ can be represented as

$$\begin{aligned} X[k] = \sum_{m=0}^{N-1} h_{\mathrm{w}}[k,m]x[m] &= \sum_{m=0}^{N-1} x[m] \prod_{j=0}^{n-1} (-1)^{(k_{n-1-j}+k_{n-j})m_j} \\ &= \sum_{m=0}^{N-1} x[m] \prod_{i=0}^{n-1} (-1)^{(k_i+k_{i+1})m_{n-1-i}}. \end{aligned} \tag{8.37}$$

Here, $N = 2^n$ and $k_n = 0$. The second equal sign is due to the conversion of index k from Hadamard order to sequency order (Eq. (8.32)). Here we have also defined $i = n-1-j$ and note that $(-1)^{k_i \oplus k_{i+1}} = (-1)^{k_i+k_{i+1}}$, where $m_i, k_i = 0, 1$.

In the following, we assume $n = 3$, $N = 2^3 = 8$, and represent m and k in binary form as $m = (m_2 m_1 m_0)_2$ and $k = (k_2 k_1 k_0)_2$ respectively:

$$m = \sum_{i=0}^{n-1} m_i 2^i = 4m_2 + 2m_1 + m_0, \quad k = \sum_{i=0}^{n-1} k_i 2^i = 4k_2 + 2k_1 + k_0. \tag{8.38}$$

Here, $k_n = k_3 = 0$ by definition. This eight-point WHT_w can be carried out in the following steps.

- As the first step of the algorithm, we rearrange the order of the samples $x[m]$ by bit reversal to define:

$$x_0[4m_0 + 2m_1 + m_2] = x[4m_2 + 2m_1 + m_0] \qquad \text{for } m = 0, \ldots, 7. \tag{8.39}$$

  Now Eq. (8.37) can be written as

$$\begin{aligned} X[k] &= \sum_{m_2=0}^{1} \sum_{m_1=0}^{1} \sum_{m_0=0}^{1} x_0[4m_0 + 2m_1 + m_2] \prod_{i=0}^{2} (-1)^{(k_i + k_{i+1}) m_{n-1-i}} \\ &= \sum_{l_0=0}^{1} \sum_{l_1=0}^{1} \sum_{l_2=0}^{1} x_0[4l_2 + 2l_1 + l_0] \prod_{i=0}^{2} (-1)^{(k_i + k_{i+1}) l_i}. \end{aligned} \tag{8.40}$$

  Here, we have defined $l_i = m_{n-1-i}$.
- Expanding the third summation into two terms for $l_2 = 0$ and $l_2 = 1$, we get

$$\begin{aligned} X[k] &= \sum_{l_0=0}^{1} \sum_{l_1=0}^{1} \prod_{i=0}^{1} (-1)^{(k_i + k_{i+1}) l_i} [x_0[2l_1 + l_0] + (-1)^{k_2 + k_3} x_0[4 + 2l_1 + l_0]] \\ &= \sum_{l_0=0}^{1} \sum_{l_1=0}^{1} \prod_{i=0}^{1} (-1)^{(k_i + k_{i+1}) l_i} x_1[4k_2 + 2l_1 + l_0], \end{aligned} \tag{8.41}$$

  where x_1 is defined as

$$x_1[4k_2 + 2l_1 + l_0] = x_0[2l_1 + l_0] + (-1)^{k_2 + k_3} x_0[4 + 2l_1 + l_0]. \tag{8.42}$$

- Again, expanding the second summation into two terms for $l_1 = 0$ and $l_1 = 1$ we get

$$\begin{aligned} X[k] &= \sum_{l_0=0}^{1} (-1)^{(k_i + k_{i+1}) l_0} [x_1[4k_2 + l_0] + (-1)^{k_1 + k_2} x_1[4k_2 + 2 + l_0]] \\ &= \sum_{l_0=0}^{1} (-1)^{(k_i + k_{i+1}) l_0} x_2[4k_2 + 2k_1 + m_0], \end{aligned} \tag{8.43}$$

  where x_2 is defined as

$$x_2[4k_2 + 2k_1 + l_0] = x_1[4k_2 + l_0] + (-1)^{k_1 + k_2} x_1[4k_2 + 2 + l_0]. \tag{8.44}$$

- Finally, expanding the first summation into two terms for $l_0 = 0$ and $l_0 = 1$, we have

$$X[k] = x_2[4k_2 + 2k_1] + (-1)^{k_0 + k_1} x_2[4k_2 + 2k_1 + 1]. \tag{8.45}$$

Summarizing the above steps, we get the fast WHT_w algorithm composed of the bit reversal and the three equations in Eqs. (8.42), (8.44), and (8.45), as illustrated in Fig. 8.4. In general, the algorithm has $\log_2 N$ stages each with complexity $O(N)$; the total complexity is $O(N \log_2 N)$.

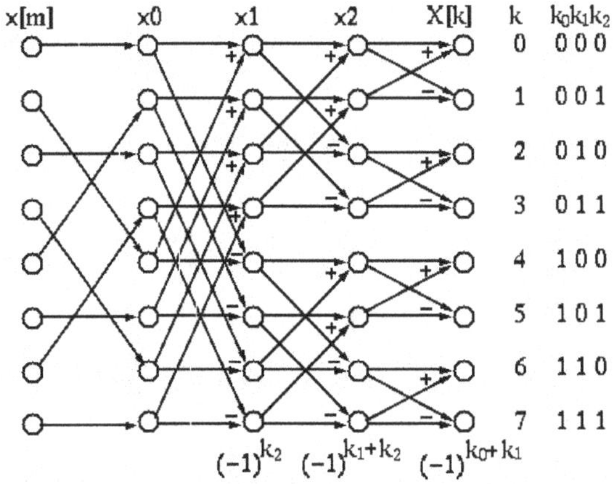

Figure 8.4 The fast WHT algorithm (sequency ordered).

The C code for the fast WHT algorithm is given below. This function `wht` takes a data vector $x[m]$ ($m = 0, \ldots, N-1$) and converts it to WHT coefficients $X[k]$ ($k = 0, \ldots, N-1$), which are Hadamard-ordered if the argument `sequency=0`, or sequency-ordered if `sequency=1`. This is an in-place algorithm; i.e., the input data will be overwritten by the output. The function `wht` can be used for both forward and inverse WHT transforms, as they are identical.

```
wht(float *x, int N, int sequency)
{ int i,j,k,j1,m,n;
  float w,*y,t;
  m=log2f((float)N);
  y=(float *)malloc(N*sizeof(float));
  for (i=0; i<m; i++) {          // for log2 N stages
    n=pow(2,m-1-i);              // length of section
    k=0;
    while (k<N-1) {              // for all sections in a stage
      for (j=0; j<n; j++) {      // for all points in a section
        j1=k+j;
        t=x[j1]+x[j1+n];
        x[j1+n]=x[j1]-x[j1+n];
        x[j1]=t;
      }
      k+=2*n;            // move on to next section
    }
  }
```

```
  w=1.0/sqrt((float)N);
  for (i=0; i<N; i++) x[i]=x[i]*w;
  if (sequency) {      // converting to sequency (Walsh) order
     for (i=0; i<N; i++) { j=h2w(i,m);  y[i]=x[j]; }
     for (i=0; i<N; i++) x[i]=y[i];
  }
  free(y);
}
```

where h2w is a function that converts a sequency index i to Hadamard index j:

```
int h2w(i,m)    // converts a sequency index i to Hadamard index j
     int i,m;
{ int j,k;
  i=i^(i>>1);                          // Gray code
  j=0;
  for (k=0; k<m; ++k)
    j=(j << 1) | (1 & (i >> k));       // bit reversal
  return j;
}
```

Example 8.1: The sequency-ordered WHT of an eight-point signal vector $\boldsymbol{x} = [0,0,2,3,4,0,0,0]^{\mathrm{T}}$ can be obtained by matrix multiplication:

$$\boldsymbol{X} = \boldsymbol{H}_{\mathrm{w}}\boldsymbol{x} = [3.18, 0.35, -3.18, -0.35, 1.77, -1.06, -1.77, 1.06]^{\mathrm{T}}, \qquad (8.46)$$

where $\boldsymbol{H}_{\mathrm{w}}$ is given in Eq. (8.34). The inverse transform (which is identical to the forward transform as $\boldsymbol{H}_{\mathrm{w}}^{-1} = \boldsymbol{H}_{\mathrm{w}}$) represents the signal vector as a linear combination of a set of square waves of different sequencies:

$$\boldsymbol{x} = \boldsymbol{H}_{\mathrm{w}}\boldsymbol{X} = [h_0, \ldots, h_7]\boldsymbol{X} = \sum_{k=0}^{7} X[k]\boldsymbol{h}_k = [0,0,2,3,4,0,0,0]^{\mathrm{T}}. \qquad (8.47)$$

This example is illustrated in Fig. 8.5.

Example 8.2: The WHT and DCT of a set of signals are shown in Fig. 8.6. The original signals are shown in the first and third columns, in comparison with the reconstructions by the inverse transforms based on 20% of the transform coefficients with the greatest magnitudes, while the remaining 80% coefficients are completely removed (suppressed to zero). The DCT and WHT coefficients are shown in the second and fourth columns, respectively.

The reconstruction errors (in percent) depend on the transform method, as well as the specific signal types, as listed in Table 8.1. We see that the two transform

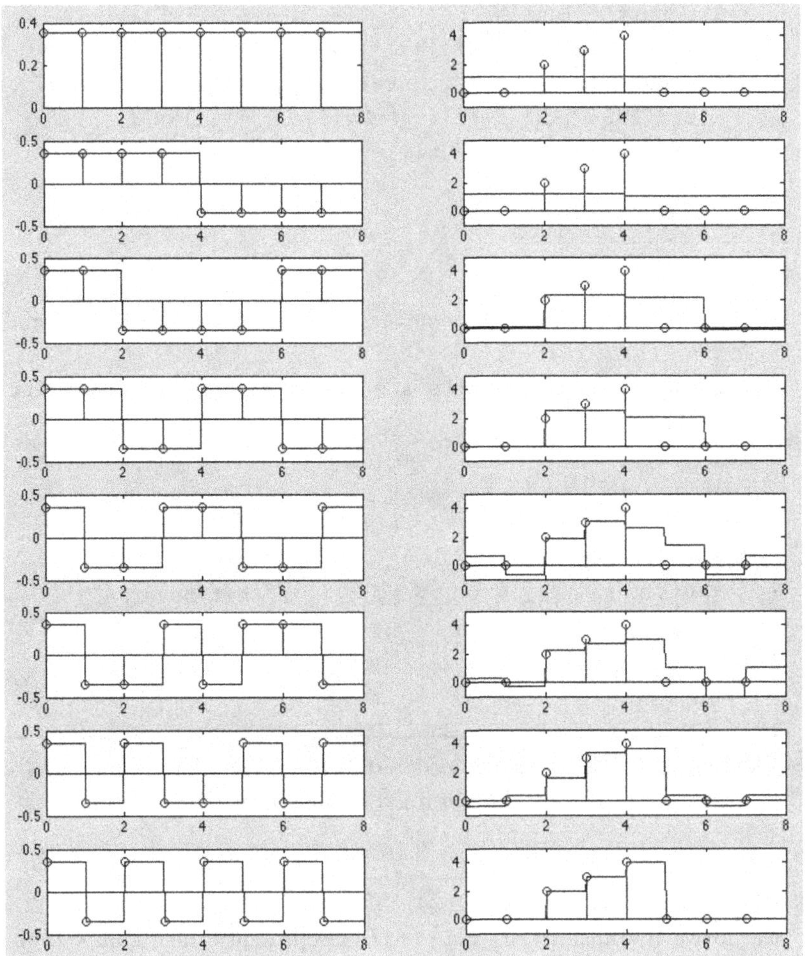

Figure 8.5 The WHT of an eight-point signal. The left column shows the eight basis WHT functions (both continuous and discrete), while the right column shows how a signal can be reconstructed by the inverse WHT (Eq. (8.47)) as a linear combination of these basis functions weighted by WHT coefficients obtained by the forward WHT (Eq. (8.46)). The signal is reconstructed using progressively more components of higher sequencies (from DC component alone to all eight sequency components).

methods are each good for the representation of certain types of signals. For example, the DCT is effective for sinusoidal signals such as in cases 1, 2, and 3, while the WHT is effective for sawtooth and square waves in cases 5 and 7. Note that as the square wave happens to be proportional to one of the basis vectors of the WHT, it can be perfectly represented by a signal WHT coefficient for that basis vector.

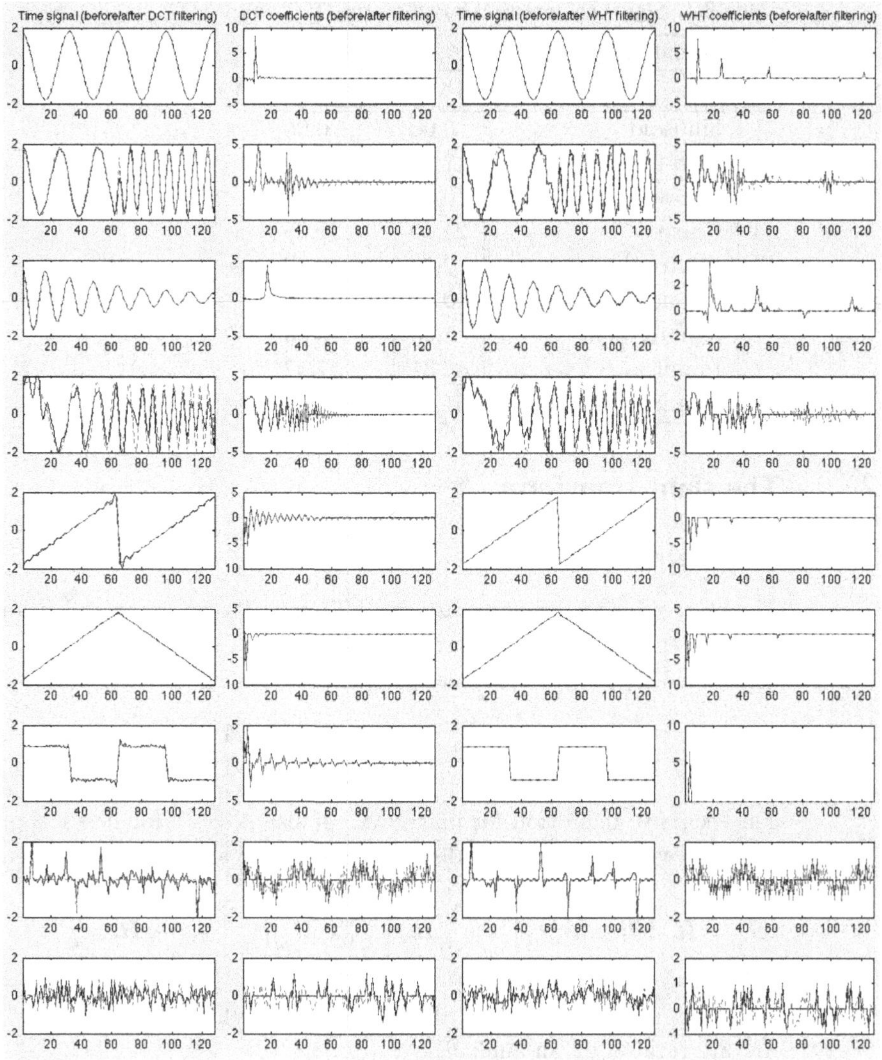

Figure 8.6 Compression of some typical signals by the DCT and WHT. The first and third columns show the time signals compared with their reconstructions based on only 20% of the transform coefficients, as shown in the second and fourth columns for the DCT and WHT, respectively. In both the time and transform domains, the signals before (dashed curves) and after (solid curves) the compression are shown for comparison.

Table 8.1. Signal compression based on the DCT and WHT

	Signal type	Percentage error	
		DCT	WHT
1	Sinusoid	0.00	0.00
2	Two-tune sinusoids	2.23	8.19
3	Decaying sinusoid	0.08	0.47
4	Chirp	24.39	21.55
5	Sawtooth	2.12	0.00
6	Triangle	0.00	0.00
7	Square wave	1.05	0.00
8	Impulses	42.31	47.42
9	Random noise	31.68	31.69

8.2 The slant transform

8.2.1 Slant matrix

The matrix for the slant transform (ST) can also be generated recursively. Initially when $n = 1$, the slant transform matrix of size $N = 2^n = 2$ is identically defined as $\boldsymbol{H}_1$ for the Hadamard matrix (Eq. (8.2)):

$$\boldsymbol{S}_1 = \boldsymbol{S}_1^{\mathrm{T}} = \frac{1}{\sqrt{2}} \begin{bmatrix} 1 & 1 \\ 1 & -1 \end{bmatrix}. \tag{8.48}$$

The recursive definition for matrix $\boldsymbol{S}_n$ of size $N = 2^n$ for $n > 1$ is given below. Here, we will use $\boldsymbol{S}^{\mathrm{T}}$ in the discussion below for a reason to be given later.

$$\boldsymbol{S}_n^{\mathrm{T}} = \boldsymbol{R}_n \left[\boldsymbol{S}_1^{\mathrm{T}} \otimes \boldsymbol{S}_{n-1}^{\mathrm{T}} \right] = \frac{1}{\sqrt{2}} \boldsymbol{R}_n \begin{bmatrix} \boldsymbol{S}_{n-1}^{\mathrm{T}} & \boldsymbol{S}_{n-1}^{\mathrm{T}} \\ \boldsymbol{S}_{n-1}^{\mathrm{T}} & -\boldsymbol{S}_{n-1}^{\mathrm{T}} \end{bmatrix} = \frac{1}{\sqrt{2}} \boldsymbol{R}_n \begin{bmatrix} \boldsymbol{S}_{n-1} & \boldsymbol{S}_{n-1} \\ \boldsymbol{S}_{n-1} & -\boldsymbol{S}_{n-1} \end{bmatrix}^{\mathrm{T}}, \tag{8.49}$$

where $\boldsymbol{R}_n$ is rotation matrix of size $N = 2^n$ by which the $N/4$th row and $N/2$th row are rotated by an angle θ_n:

$$\boldsymbol{R}_n = \begin{bmatrix} 1 & & & & & & & & & \\ & \ddots & & & & & & & & \\ & & 1 & & & & & & & \\ & & & \cos\theta_n & & & -\sin\theta_n & & & \\ & & & & 1 & & & & & \\ & & & & & \ddots & & & & \\ & & & & & & 1 & & & \\ & & & \sin\theta_n & & & \cos\theta_n & & & \\ & & & & & & & 1 & & \\ & & & & & & & & \ddots & \\ & & & & & & & & & 1 \end{bmatrix} \begin{matrix} \\ \\ \\ (2^{n-2} = N/4)\text{th row} \\ \\ \\ \\ (2^{n-1} = N/2)\text{th row} \\ \\ \\ \\ \end{matrix}, \tag{8.50}$$

where

$$\cos\theta_n = \left(\frac{2^{2n-2}-1}{2^{2n}-1}\right)^{1/2} = \sqrt{\frac{3N^2}{4N^2-4}},$$
$$\sin\theta_n = \left(\frac{2^{2n-2}-2^{2n-2}}{2^{2n}-1}\right)^{1/2} = \sqrt{\frac{N^2-4}{4N^2-4}}. \tag{8.51}$$

Note that the trigonometric identity $\sin\theta_n^2 + \cos\theta_n^2 = 1$ is indeed satisfied.

The rotation matrix is obviously orthogonal $\boldsymbol{R}_n^{\mathrm{T}}\boldsymbol{R}_n = \boldsymbol{I}_n$. In particular, if $\theta_n = 0$ then $\boldsymbol{R}_n = \boldsymbol{I}_n$, and Eq. (8.49) for the slant matrix becomes the same as Eq. (8.3) for the Hadamard matrix.

The slant transform matrix $\boldsymbol{S}_n$ is real but not symmetric, and we can show that it is also orthogonal:

$$\boldsymbol{S}_n^{\mathrm{T}} = \boldsymbol{S}_n^{-1}, \quad \text{i.e.,} \quad \boldsymbol{S}_n^{\mathrm{T}}\boldsymbol{S}_n = \boldsymbol{S}_n\boldsymbol{S}_n^{\mathrm{T}} = \boldsymbol{I}_n. \tag{8.52}$$

Similar to the way we prove the orthogonality of the WHT matrix, here the orthogonality of the ST matrix $\boldsymbol{S}$ can also be proven by induction. This is left for the reader as a homework problem.

As the slant matrix $\boldsymbol{S}_n$ is closely related to the Hadamard matrix $\boldsymbol{H}_n$, the sequencies of their corresponding rows are the same. The same re-ordering method given in Eq. (8.33) can be used to rearrange the rows of $\boldsymbol{S}^{\mathrm{T}}$; i.e., the columns of $\boldsymbol{S}$, in ascending order of their sequencies. Based on the recursion of Eq. (8.49) and after conversion to the sequency order, the slant transform matrices of the next two levels for $n = 2$ and $n = 3$ can be obtained as

$$\boldsymbol{S}_2^{\mathrm{T}} = \frac{1}{2}\begin{bmatrix} 1.00 & 1.00 & 1.00 & 1.00 \\ 1.34 & 0.45 & -0.45 & -1.34 \\ 1.00 & -1.00 & -1.00 & 1.00 \\ 0.45 & -1.34 & 1.34 & -0.45 \end{bmatrix} \tag{8.53}$$

$$\boldsymbol{S}_3^{\mathrm{T}} = \frac{1}{\sqrt{8}}\begin{bmatrix} 1.00 & 1.00 & 1.00 & 1.00 & 1.00 & 1.00 & 1.00 & 1.00 \\ 1.53 & 1.09 & 0.65 & 0.22 & -0.22 & -0.65 & -1.09 & -1.53 \\ 1.34 & 0.45 & -0.45 & -1.34 & -1.34 & -0.45 & 0.45 & 1.34 \\ 0.68 & -0.10 & -0.88 & -1.66 & 1.66 & 0.88 & 0.10 & -0.68 \\ 1.00 & -1.00 & -1.00 & 1.00 & 1.00 & -1.00 & -1.00 & 1.00 \\ 1.00 & -1.00 & -1.00 & 1.00 & -1.00 & 1.00 & 1.00 & -1.00 \\ 0.45 & -1.34 & 1.34 & -0.45 & -0.45 & 1.34 & -1.34 & 0.45 \\ 0.45 & -1.34 & 1.34 & -0.45 & 0.45 & -1.34 & 1.34 & -0.45 \end{bmatrix}. \tag{8.54}$$

We make the following observations and comments:

- Unlike the Walsh-Hadamard matrix, the slant matrix $\boldsymbol{S}^{\mathrm{T}} \neq \boldsymbol{S}$ is asymmetric;
- The sequencies of the row vectors in matrix $\boldsymbol{S}^{\mathrm{T}}$ increase from 0 of the first row to $N-1$ of the last one; i.e., these row vectors form a basis containing N basis vectors of space $\mathbb{R}^N$.
- In particular, the second row with sequency of 1 has a negative linear slope, hence the name "slant" matrix.

- In general, we always treat the *column* vectors of an orthogonal transform matrix as the basis vectors of different frequencies/sequencies. This is why we have used $\boldsymbol{S}^{\mathrm{T}}$ in the discussion above, so that the slant transform matrix $\boldsymbol{S}_n = [\boldsymbol{s}_0, \ldots, \boldsymbol{s}_{N-1}]$ is composed of N columns for the N basis vectors $\boldsymbol{s}_n$ $(n = 0, \ldots, N-1)$ of sequency n.
- As the ST and WHT are recursively constructed in a similar manner, their basis vectors bear a similarity, as shown in Fig. 8.7.

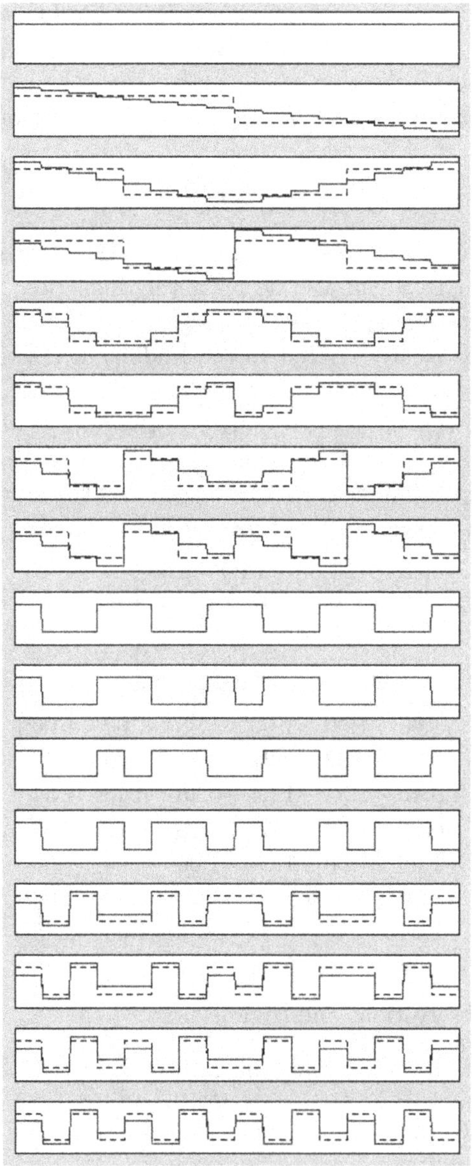

Figure 8.7 Comparison of the basis vectors of slant (solid lines) and Hadamard (dashed lines) transform.

8.2.2 Slant transform and its fast algorithm

Given an orthogonal matrix $\boldsymbol{S}_n$, an orthogonal transform of an N-D vector $\boldsymbol{x}$ can be defined as

$$\boldsymbol{X} = \begin{bmatrix} X[0] \\ \vdots \\ X[N-1] \end{bmatrix} = \boldsymbol{S}^{\mathrm{T}}\boldsymbol{x} = \begin{bmatrix} \boldsymbol{s}_0^{\mathrm{T}} \\ \vdots \\ \boldsymbol{s}_{N-1}^{\mathrm{T}} \end{bmatrix} \boldsymbol{x}, \tag{8.55}$$

or in component form as

$$X[k] = \boldsymbol{s}_k^{\mathrm{T}}\boldsymbol{x} = \langle \boldsymbol{s}_k, \boldsymbol{x} \rangle; \tag{8.56}$$

i.e., $X[k]$ is the projection of the signal vector $\boldsymbol{x}$ onto the kth basis vector $\boldsymbol{s}_k$. The inverse transform reconstructs the signal from its transform coefficients:

$$\boldsymbol{x} = \boldsymbol{S}\boldsymbol{X} = [\boldsymbol{s}_0, \ldots, \boldsymbol{s}_{N-1}] \begin{bmatrix} X[0] \\ \vdots \\ X[N-1] \end{bmatrix} = \sum_{k=0}^{N-1} X[k]\boldsymbol{s}_k. \tag{8.57}$$

Like the Walsh-Hadamard transform, the slant transform also has a fast algorithm with computational complexity of $O(N \log_2 N)$ instead of $O(N^2)$. This algorithm can be explained in the following example of $n = 3$. The slant transform of a vector $\boldsymbol{x}$ of size $N = 2^3 = 8$ is:

$$\boldsymbol{X} = \boldsymbol{S}_3^{\mathrm{T}}\boldsymbol{x} = \frac{1}{\sqrt{2}}\boldsymbol{R}_3 \begin{bmatrix} \boldsymbol{S}_2^{\mathrm{T}} & \boldsymbol{S}_2^{\mathrm{T}} \\ \boldsymbol{S}_2^{\mathrm{T}} & -\boldsymbol{S}_2^{\mathrm{T}} \end{bmatrix} \begin{bmatrix} x[0] \\ \vdots \\ x[3] \\ x[4] \\ \vdots \\ x[7] \end{bmatrix} = \frac{1}{\sqrt{2}}\boldsymbol{R}_3 \begin{bmatrix} \boldsymbol{S}_2^{\mathrm{T}}\boldsymbol{x}_1 \\ \boldsymbol{S}_2^{\mathrm{T}}\boldsymbol{x}_2 \end{bmatrix}, \tag{8.58}$$

where

$$\boldsymbol{x}_1 = \begin{bmatrix} x[0] \\ \vdots \\ x[3] \end{bmatrix} + \begin{bmatrix} x[4] \\ \vdots \\ x[7] \end{bmatrix}, \qquad \boldsymbol{x}_2 = \begin{bmatrix} x[0] \\ \vdots \\ x[3] \end{bmatrix} - \begin{bmatrix} x[4] \\ \vdots \\ x[7] \end{bmatrix}. \tag{8.59}$$

We see that an eight-point slant transform is converted into two four-point slant transforms, each of which can be converted in turn to two two-point transforms. This recursive process is illustrated in the diagram in Fig. 8.8. The three nested boxes (dashed line) represent three levels of recursion for the eight-point, four-point, and two-point transforms, respectively. In general, an N-point transform can be implemented by this algorithm in $\log_2 N$ stages each requiring $O(N)$ operations; i.e., the total complexity is $O(N \log_2 N)$. This algorithm is almost identical to the WHT algorithm shown in Fig. 8.4, except an additional rotation for two of the rows at each level.

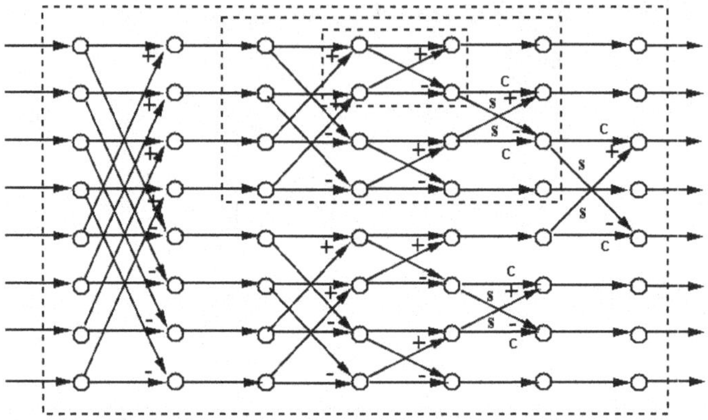

Figure 8.8 A recursive algorithm of fast slant transform. The three nested boxes (dashed line) are for eight-, four-, and two-point transforms, respectively. Letters c and s represent $\cos\theta_n$ and $\sin\theta_n$ for the rotation for each of the transforms (except for $N = 2$).

While the algorithm can be implemented in a manner very similar to the WHT code discussed previously, here we present an alternative implementation based on recursion, which fits the algorithm most naturally.

```
slantf(float *x, int N)
{
  int i,j,k,l,m,n;
  float c,s,u,v,w,*y1,*y2;
  y1=(float*)malloc(N/2 * sizeof(float));
  y2=(float*)malloc(N/2 * sizeof(float));
  if (N==2) {              // 2-point transform
    u=x[0]; v=x[1];
    x[0]=(u+v)/Sqrt2;
    x[1]=(u-v)/Sqrt2;
  }
  else {
    for (n=0; n<N/2; n++) {
      y1[n]=x[n]+x[N/2+n];
      y2[n]=x[n]-x[N/2+n];
    }
    slantf(y1,N/2);        // recursion
    slantf(y2,N/2);
    for (n=0; n<N/2; n++) {
      x[n]=y1[n]/Sqrt2;
      x[N/2+n]=y2[n]/Sqrt2;
    }
    w=4*N*N-4;
```

```
    c=sqrt(3*N*N/w);
    s=sqrt((N*N-4)/w);
    u=x[N/4]; v=x[N/2];
    x[N/4]=c*u-s*v;        // rotation
    x[N/2]=s*u+c*v;
  }
  free(y1); free(y2);
}
```

The inverse transform can be implemented by reversing the steps and operations both mathematically and order-wise in the forward transform:

```
slanti(float *x, int N)
{
  int i,j,k,l,m,n;
  float c,s,u,v,w,*y1,*y2;
  y1=(float*)malloc(N/2 * sizeof(float));
  y2=(float*)malloc(N/2 * sizeof(float));
  if (N==2) {           // 2-point transform
    u=x[0]; v=x[1];
    x[0]=(u+v)/Sqrt2;
    x[1]=(u-v)/Sqrt2;
  }
  else {
    w=4*N*N-4;
    c=sqrt(3*N*N/w);
    s=sqrt((N*N-4)/w);
    u=x[N/4]; v=x[N/2];
    x[N/4]=c*u+s*v;      // rotation
    x[N/2]=c*v-s*u;
    for (n=0; n<N/2; n++) {
      y1[n]=x[n]*Sqrt2;
      y2[n]=x[N/2+n]*Sqrt2;
    }
    slanti(y1,N/2);      // recursion
    slanti(y2,N/2);
    for (n=0; n<N/2; n++) {
      x[n]=(y1[n]+y2[n])/2;
      x[N/2+n]=(y1[n]-y2[n])/2;
    }
  }
  free(y1); free(y2);
}
```

Example 8.3: The slant transform of an eight-point signal vector $\boldsymbol{x} = [0, 0, 2, 3, 4, 0, 0, 0]^{\mathrm{T}}$ can be obtained by matrix multiplication:

$$\boldsymbol{X} = \boldsymbol{S}_3^{\mathrm{T}} \boldsymbol{x} = [3.18, 0.39, -3.64, -0.03, 1.77, -1.06, -0.16, 1.11]^{\mathrm{T}}, \tag{8.60}$$

where $\boldsymbol{S}_3$ is given in Eq. (8.54). The inverse transform will bring the original signal back: $\boldsymbol{x} = \boldsymbol{S}_3 \boldsymbol{X} = [0, 0, 2, 3, 4, 0, 0, 0]^{\mathrm{T}}$.

8.3 The Haar transform

8.3.1 Continuous Haar transform

Similar to the Walsh-Hadamard transform, the Haar transform is yet another orthogonal transform defined by a set of rectangular-shaped basis functions. However, compared to all orthogonal transform methods considered so far, the Haar transform has some unique significance as it is also a special type of the wavelet transforms to be discussed in Chapter 11.

The family of Haar functions $h_k(t)$, $(k = 0, 1, 2, \dots)$ is defined on the interval $0 \le t \le 1$. Except $h_0(t) = 1$, the shape of the kth function $h_k(t)$ for $k > 0$ is determined by two parameters p and q, which are related to k by

$$k = 2^p + q. \tag{8.61}$$

In other words, p and q are uniquely determined so that $2^p \le k$ is the highest power of 2 contained in k, and $q = k - 2^p$ is the remainder. For example, the values of p and q corresponding to $k = 1, \dots, 15$ are as follows:

k	1	2	3	4	5	6	7	8	9	10	11	12	13	14	15
p	0	1	1	2	2	2	2	3	3	3	3	3	3	3	3
q	0	0	1	0	1	2	3	0	1	2	3	4	5	6	7

(8.62)

Now the family of Haar functions can be defined thus:

- when $k = 0$,

$$h_0(t) = 1 \qquad 0 \le t < 1. \tag{8.63}$$

- when $k > 0$, $h_k(t)$ is defined in terms of p and q as

$$h_k(t) = \begin{cases} \sqrt{2}^p & q/2^p \le t < (q+0.5)/2^p \\ -\sqrt{2}^p & (q+0.5)/2^p \le t < (q+1)/2^p \\ 0 & \text{else} \end{cases}. \tag{8.64}$$

The first $N = 8$ Haar functions are shown in Fig. 8.9. We see that the Haar functions $h_k(t)$ for all $k > 0$ contain a single prototype shape composed of a square wave followed by its negative copy, with the two parameters p specifying the magnitude and width (or scale) of the shape and q specifying the position (translation) of the shape. For example, if $k = 5$, then $p = 2$, $q = 1$, and we have:

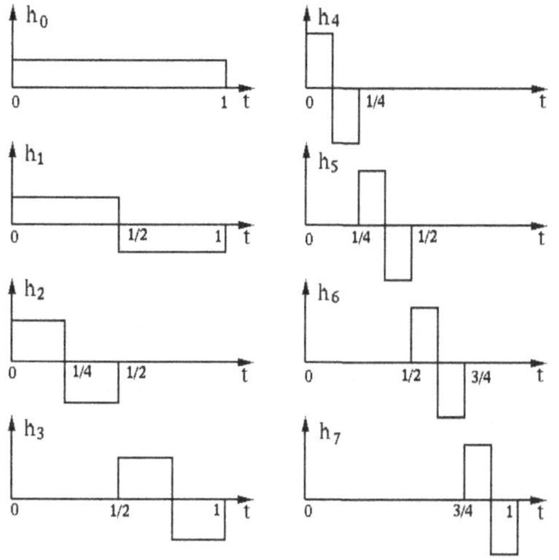

Figure 8.9 The eight basis functions for the Haar transform.

$$h_5(t) = \begin{cases} 2 & 2/8 \leq t < 3/8 \\ -2 & 3/8 \leq t < 4/8 \\ 0 & \text{else} \end{cases} . \tag{8.65}$$

These Haar functions are obviously orthonormal:

$$\langle h_k(t), h_l(t) \rangle = \int_0^1 h_k(t) h_l(t)\, dt = \delta[k-l], \tag{8.66}$$

and they can be used as the basis functions to span a function space over $0 \leq t < 1$. A signal $x(t)$ in this space can be expressed as a linear combination of these Haar functions:

$$x(t) = \sum_{k=0}^{\infty} X[k] h_k(t), \tag{8.67}$$

where the kth coefficient $X[k]$ can be obtained as the projection of $x(t)$ onto the kth basis function $h_k(t)$:

$$X[k] = \langle x(t), h_k(t) \rangle = \int_0^1 x(t) h_k(t)\, dt. \tag{8.68}$$

When $k = 0$, the coefficient

$$X[0] = \int_0^1 x(t) h_0(t)\, dt = \int_0^1 x(t)\, dt \tag{8.69}$$

represents the average or DC component of the signal, like all orthogonal transforms discussed before. When $k > 0$, the coefficient $X[k]$ represents three specific aspects of the signal characteristics:

- certain types of detailed features contained in the signal, in the form of the difference between two consecutive segments of the signal;
- the time interval during which such detailed features occur; and
- the time scale of such features.

For example, a large value (either positive or negative) of the coefficient $X[3]$ for the basis function $h_3(t)$ would indicate that the signal value has some significant variation of the scale of half of its duration in the second half of its duration.

It is interesting to compare the Haar transform with other orthogonal transforms such as the Fourier, cosine, Walsh-Hadamard, and slant transforms discussed before. What all of these transforms, as well as the Haar transform, have in common is that their coefficients represent some types of detail contained in the signal, in terms of different frequencies (Fourier transform and cosine transform), sequencies (Walsh-Hadamard transform), or scales (Haar transform), in the sense that more detailed information is represented by coefficients for higher frequencies, sequencies, or scales. However, none of these transforms is able to indicate when in time such details occur, except the Haar transform, which represents not only the details of different scales, but also their temporal positions. However, we note that this additional capability is gained with the cost of much reduced number of scale levels. All N-point orthogonal transforms can represent N different frequencies/sequencies, but an N-point Haar transform can only represent $\log_2 N$ different scale levels. Owing to such different behaviors the Haar transform is in fact also a special form of the wavelet transform to be discussed in Chapter 11.

8.3.2 Discrete Haar transform

The discrete Haar transform (DHT) is defined based on the family of Haar functions. Specifically, by sampling each of the first N Haar functions $h_k(t)$ ($k = 0, \ldots, N-1$) at time moments $t = n/N$ ($n = 0, \ldots, N-1$), we get N orthogonal vectors. Moreover, if a scaling factor $1/\sqrt{N}$ is included, these vectors become orthonormal:

$$\langle \boldsymbol{h}_k, \boldsymbol{h}_l \rangle = \boldsymbol{h}_k^{\mathrm{T}} \boldsymbol{h}_l = \delta[k-l]. \tag{8.70}$$

These N orthonormal vectors form a basis that spans the N-dimensional vector space, and they form an N by N DHT matrix $\boldsymbol{H}$ (not to be confused with the WHT matrix):

$$\boldsymbol{H} = [\boldsymbol{h}_0, \ldots, \boldsymbol{h}_{N-1}], \qquad \text{or} \qquad \boldsymbol{H}^{\mathrm{T}} = \begin{bmatrix} \boldsymbol{h}_0^{\mathrm{T}} \\ \vdots \\ \boldsymbol{h}_{N-1}^{\mathrm{T}} \end{bmatrix}. \tag{8.71}$$

This is obviously real and orthonormal (but not symmetric):

$$\boldsymbol{H} = \boldsymbol{H}^*, \qquad \boldsymbol{H}^{-1} = \boldsymbol{H}^{\mathrm{T}}, \quad \text{i.e.} \quad \boldsymbol{H}^{\mathrm{T}}\boldsymbol{H} = \boldsymbol{I}. \tag{8.72}$$

The DHT matrices corresponding to $N = 2, 4, 8$ are listed below.

- $N = 2$

$$\boldsymbol{H}_1^{\mathrm{T}} = \frac{1}{\sqrt{2}}\begin{bmatrix} 1 & 1 \\ 1 & -1 \end{bmatrix} = \begin{bmatrix} 0.71 & 0.71 \\ 0.71 & -0.71 \end{bmatrix}. \tag{8.73}$$

This 2×2 DHT matrix is identical to the transform matrices for all other discrete transforms, including DFT, DCT, and WHT. The first row represents the average of the signal, and the second represents the difference between the first and second halves of the signal, as for all transform methods.

- $N = 4$

$$\boldsymbol{H}_2^{\mathrm{T}} = \frac{1}{2}\begin{bmatrix} 1 & 1 & 1 & 1 \\ 1 & 1 & -1 & -1 \\ \sqrt{2} & -\sqrt{2} & 0 & 0 \\ 0 & 0 & \sqrt{2} & -\sqrt{2} \end{bmatrix} = \begin{bmatrix} 0.50 & 0.50 & 0.50 & 0.50 \\ 0.50 & 0.50 & -0.50 & -0.50 \\ 0.71 & -0.71 & 0.00 & 0.00 \\ 0.00 & 0.00 & 0.71 & -0.71 \end{bmatrix}. \tag{8.74}$$

The DCT matrix $\boldsymbol{C}$ and the Walsh-ordered WHT matrix $\boldsymbol{H}_{\mathrm{w}}$ are also listed below for comparison:

$$\boldsymbol{C}^{\mathrm{T}} = \begin{bmatrix} 0.50 & 0.50 & 0.50 & 0.50 \\ 0.65 & 0.27 & -0.27 & -0.65 \\ 0.50 & -0.50 & -0.50 & 0.50 \\ 0.27 & -0.65 & 0.65 & -0.27 \end{bmatrix}, \quad \boldsymbol{H}_{\mathrm{w}}^{\mathrm{T}} = \begin{bmatrix} 0.50 & 0.50 & 0.50 & 0.50 \\ 0.50 & 0.50 & -0.50 & -0.50 \\ 0.50 & -0.50 & -0.50 & 0.50 \\ 0.50 & -0.50 & 0.50 & -0.50 \end{bmatrix}. \tag{8.75}$$

We see that the first rows of all three matrices $\boldsymbol{H}$, $\boldsymbol{C}$, and $\boldsymbol{H}_{\mathrm{w}}$ are identical, representing the DC component of the signal. The elements of their second rows have the same polarities (but of different values), representing the difference between the first and second halves of the signal. However, their third and fourth rows are quite different. For DCT and WHT, these two rows represent progressively higher frequency or sequency components in the signal, but in the case of the DHT these rows represent the same level of details (variations) at a finer scale than the second row, as well as their different temporal locations (either in the first or second half).

- $N = 8$

$$\boldsymbol{H}_3^{\mathrm{T}} = \frac{1}{\sqrt{8}}\begin{bmatrix} 1 & 1 & 1 & 1 & 1 & 1 & 1 & 1 \\ 1 & 1 & 1 & 1 & -1 & -1 & -1 & -1 \\ \sqrt{2} & \sqrt{2} & -\sqrt{2} & -\sqrt{2} & 0 & 0 & 0 & 0 \\ 0 & 0 & 0 & 0 & \sqrt{2} & \sqrt{2} & -\sqrt{2} & -\sqrt{2} \\ 2 & -2 & 0 & 0 & 0 & 0 & 0 & 0 \\ 0 & 0 & 2 & -2 & 0 & 0 & 0 & 0 \\ 0 & 0 & 0 & 0 & 2 & -2 & 0 & 0 \\ 0 & 0 & 0 & 0 & 0 & 0 & 2 & -2 \end{bmatrix} \begin{matrix} 0 \ \varphi_{0,0}(t) \\ 1 \ \psi_{0,0}(t) \\ 2 \ \psi_{1,0}(t) \\ 3 \ \psi_{1,1}(t) \\ 4 \ \psi_{2,0}(t) \\ 5 \ \psi_{2,1}(t) \\ 6 \ \psi_{2,2}(t) \\ 7 \ \psi_{2,3}(t) \end{matrix}. \tag{8.76}$$

It is obvious that the additional four rows represent still more detailed and finer signal variations and their temporal positions at a finer scale than the previous two rows. Note that each row is also labeled as a function ($\varphi_0(t)$ for the first row and $\psi_{p,q}(t)$ for the rest) on the right. The significance of these labelings will be clear in the future when we discuss discrete wavelet transforms.

Now any N-point signal vector $\boldsymbol{x} = [x[0], \ldots, x[N-1]]^{\mathrm{T}}]$ can be expressed as a linear combination of the column vectors $\boldsymbol{h}_k$ $(k = 0, \ldots, N-1)$ of the DHT matrix $\boldsymbol{H}$:

$$\boldsymbol{x} = \boldsymbol{H}\boldsymbol{X} = [\boldsymbol{h}_0, \ldots, \boldsymbol{h}_{N-1}] \begin{bmatrix} X[0] \\ \vdots \\ X[N-1] \end{bmatrix} = \sum_{k=0}^{N-1} X[k]\boldsymbol{h}_k. \tag{8.77}$$

This is the inverse discrete Haar transform (IDHT), where the kth coefficient $X[k]$ for the vector $\boldsymbol{h}_k$ can be obtained as the projection of the signal vector $\boldsymbol{x}$ onto the kth basis vector $\boldsymbol{h}_k$:

$$X[k] = \langle \boldsymbol{x}, \boldsymbol{h}_k \rangle = \boldsymbol{h}_k^{\mathrm{T}} \boldsymbol{x} \qquad (k = 0, \ldots, N-1), \tag{8.78}$$

or in matrix form

$$\boldsymbol{X} = \begin{bmatrix} X[0] \\ \vdots \\ X[N-1] \end{bmatrix} = \boldsymbol{H}^{-1}\boldsymbol{x} = \boldsymbol{H}^{\mathrm{T}}\boldsymbol{x} = \begin{bmatrix} \boldsymbol{h}_0^{\mathrm{T}} \\ \vdots \\ \boldsymbol{h}_{N-1}^{\mathrm{T}} \end{bmatrix} \boldsymbol{x}. \tag{8.79}$$

This is the forward discrete Haar transform (DHT), which can also be obtained by pre-multiplying $\boldsymbol{H}^{-1}$ on both sides of the IDHT equation above. The DHT pair can be written as

$$\begin{cases} \boldsymbol{X} = \boldsymbol{H}^{\mathrm{T}}\boldsymbol{x} & \text{(forward)} \\ \boldsymbol{x} = \boldsymbol{H}\boldsymbol{X} & \text{(inverse)} \end{cases}. \tag{8.80}$$

Example 8.4: The Haar transform coefficients of an eight-point signal $\boldsymbol{x} = [0, 0, 2, 3, 4, 0, 0, 0]^{\mathrm{T}}$ can be obtained by the DHT as

$$\boldsymbol{X} = \boldsymbol{H}^{\mathrm{T}}\boldsymbol{x} = [3.18, 0.35, -2.50, 2.0, 0.0, -0.71, 2.83, 0.0]^{\mathrm{T}}, \tag{8.81}$$

where the eight-point Haar transform matrix is given in Eq. (8.76). Same as in the DCT, WHT, and ST, $X[0] = 3.18$ and $X[1] = 0.35$ represent respectively the sum and difference between the first and second halves of the signal. However, the interpretations of the remaining DHT coefficients are quite different from the DCT and WHT. $X[2] = -2.5$ represents the difference between the first and second quarters in the first half of the signal, while $X[3] = 2$ represents the difference between the third and forth quarters in the second half of the signal. Similarly, $X[4], \ldots, X[7]$ represent the next level of details in terms of the difference between two consecutive eighths of the signal in each of the four quarters of the signal.

The signal vector is reconstructed by the IDHT which expresses the signal as a linear combination of the basis functions, as shown in Eq. (8.77).

8.3.3 Computation of the discrete Haar transform

The computational complexity of an N-point discrete Haar transform implemented as a matrix multiplication is $O(N^2)$. However, a fast algorithm with linear complexity $O(N)$ exists for both DHT and IDHT, as illustrated in Fig. 8.10 for the eight-point DHT transform. The forward transform $\boldsymbol{X} = \boldsymbol{H}_3^{\mathrm{T}} \boldsymbol{x}$ can be written in matrix form as

$$\boldsymbol{X} = \begin{bmatrix} X[0] \\ X[1] \\ X[2] \\ X[3] \\ X[4] \\ X[5] \\ X[6] \\ X[7] \end{bmatrix} = \frac{1}{\sqrt{8}} \begin{bmatrix} 1 & 1 & 1 & 1 & 1 & 1 & 1 & 1 \\ 1 & 1 & 1 & 1 & -1 & -1 & -1 & -1 \\ \sqrt{2} & \sqrt{2} & -\sqrt{2} & -\sqrt{2} & 0 & 0 & 0 & 0 \\ 0 & 0 & 0 & 0 & \sqrt{2} & \sqrt{2} & -\sqrt{2} & -\sqrt{2} \\ 2 & -2 & 0 & 0 & 0 & 0 & 0 & 0 \\ 0 & 0 & 2 & -2 & 0 & 0 & 0 & 0 \\ 0 & 0 & 0 & 0 & 2 & -2 & 0 & 0 \\ 0 & 0 & 0 & 0 & 0 & 0 & 2 & -2 \end{bmatrix} \begin{bmatrix} x[0] \\ x[1] \\ x[2] \\ x[3] \\ x[4] \\ x[5] \\ x[6] \\ x[7] \end{bmatrix}$$

$$= \begin{bmatrix} (1 \;\; 1 \;\; 1 \;\; 1 \;\; 1 \;\; 1 \;\; 1 \;\; 1)/\sqrt{2}^3 \\ (1 \;\; 1 \;\; 1 \;\; 1 \; -1 \; -1 \; -1 \; -1)/\sqrt{2}^3 \\ (1 \;\; 1 \; -1 \; -1 \;\; 0 \;\; 0 \;\; 0 \;\; 0)/\sqrt{2}^2 \\ (0 \;\; 0 \;\; 0 \;\; 0 \;\; 1 \;\; 1 \; -1 \; -1)/\sqrt{2}^2 \\ (1 \; -1 \;\; 0 \;\; 0 \;\; 0 \;\; 0 \;\; 0 \;\; 0)/\sqrt{2} \\ (0 \;\; 0 \;\; 1 \; -1 \;\; 0 \;\; 0 \;\; 0 \;\; 0)/\sqrt{2} \\ (0 \;\; 0 \;\; 0 \;\; 0 \;\; 1 \; -1 \;\; 0 \;\; 0)/\sqrt{2} \\ (0 \;\; 0 \;\; 0 \;\; 0 \;\; 0 \;\; 0 \;\; 1 \; -1)/\sqrt{2} \end{bmatrix} \begin{bmatrix} x[0] \\ x[1] \\ x[2] \\ x[3] \\ x[4] \\ x[5] \\ x[6] \\ x[7] \end{bmatrix}. \tag{8.82}$$

By inspection of this matrix multiplication, we see that each of the last four coefficients $X[4], \ldots, X[7]$ in the second half of vector $\boldsymbol{X}$ can be obtained as the difference between a pair of two signal samples, e.g., $X[4] = (x[0] - x[1])/\sqrt{2}$. Similarly, each of the last two coefficients $X[2]$ and $X[3]$ of the first half of $\boldsymbol{X}$ can be obtained as the difference between two sums of two signal components; e.g., $X[2] = [(x[0] + x[1]) - (x[2] + x[3])]/2$. This process can be carried out recursively as shown on the left of Fig. 8.10, each performing some additions and subtractions on the first half of the data points produced in the previous stage, and in $\log_2 8 = 3$ consecutive stages, the N DHT coefficients $X[0], \ldots, X[7]$ can be obtained. Moreover, if the results of each stage are divided by $\sqrt{2}$, the normalization of the transform can also be taken care of.

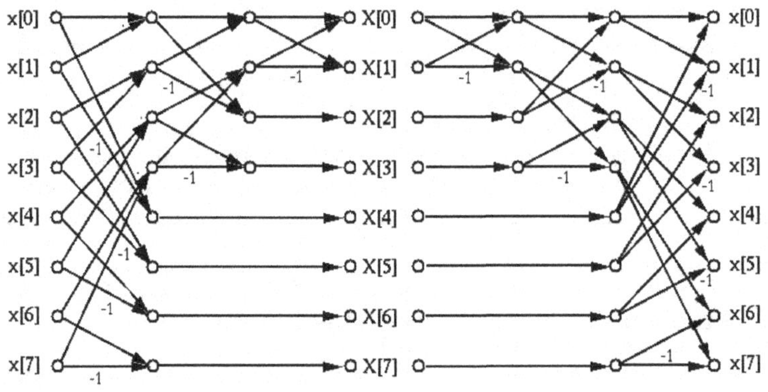

Figure 8.10 The fast Haar transform algorithm. The forward DHT transform shown on the left of the diagram converts the signal $\boldsymbol{x}$ to its DHT coefficients $\boldsymbol{X}$ in the middle, while the inverse transform IDHT shown on the right converts $\boldsymbol{X}$ back to the time domain (reconstruction).

The inverse transform $\boldsymbol{x} = \boldsymbol{H}_3\boldsymbol{X}$ can also be written in matrix form:

$$\begin{bmatrix} x[0] \\ x[1] \\ x[2] \\ x[3] \\ x[4] \\ x[5] \\ x[6] \\ x[7] \end{bmatrix} = \frac{1}{\sqrt{8}} \begin{bmatrix} 1 & 1 & \sqrt{2} & 0 & 2 & 0 & 0 & 0 \\ 1 & 1 & \sqrt{2} & 0 & -2 & 0 & 0 & 0 \\ 1 & 1 & -\sqrt{2} & 0 & 0 & 2 & 0 & 0 \\ 1 & 1 & -\sqrt{2} & 0 & 0 & -2 & 0 & 0 \\ 1 & -1 & 0 & \sqrt{2} & 0 & 0 & 2 & 0 \\ 1 & -1 & 0 & \sqrt{2} & 0 & 0 & -2 & 0 \\ 1 & -1 & 0 & -\sqrt{2} & 2 & 0 & 0 & 2 \\ 1 & -1 & 0 & -\sqrt{2} & 0 & 0 & 0 & -2 \end{bmatrix} \begin{bmatrix} X[0] \\ X[1] \\ X[2] \\ X[3] \\ X[4] \\ X[5] \\ X[6] \\ X[7] \end{bmatrix}$$

$$= \begin{bmatrix} 1 & 1 & 1 & 0 & 1 & 0 & 0 & 0 \\ 1 & 1 & 1 & 0 & -1 & 0 & 0 & 0 \\ 1 & 1 & -1 & 0 & 0 & 1 & 0 & 0 \\ 1 & 1 & -1 & 0 & 0 & -1 & 0 & 0 \\ 1 & -1 & 0 & 1 & 0 & 0 & 1 & 0 \\ 1 & -1 & 0 & 1 & 0 & 0 & -1 & 0 \\ 1 & -1 & 0 & -1 & 0 & 0 & 0 & 1 \\ 1 & -1 & 0 & -1 & 0 & 0 & 0 & -1 \end{bmatrix} \begin{bmatrix} X[0]/\sqrt{2}^3 \\ X[1]/\sqrt{2}^3 \\ X[2]/\sqrt{2}^2 \\ X[3]/\sqrt{2}^2 \\ X[4]/\sqrt{2} \\ X[5]/\sqrt{2} \\ X[6]/\sqrt{2} \\ X[7]/\sqrt{2} \end{bmatrix}. \tag{8.83}$$

By inspection, we see that this matrix multiplication can also be carried out in $\log_2 8 = 3$ stages as shown on the right side of the diagram in Fig. 8.10. Again, the output of each stage needs to be divided by $\sqrt{2}$.

Moreover, from this example of eight-point transform, we can also obtain the computational complexity of the DHT (as well as the fast wavelet transform to be considered in Chapter 11). While the fast algorithms for all orthogonal transforms considered previously have the same complexity $O(N \log_2 N)$, the fast algorithm for DHT shown in Fig. 8.10 is even more efficient, with a complexity

$O(N)$. The proof of this linear complexity can be easily obtained and is left for the reader as a homework problem.

The C code for both the forward and inverse discrete Haar transforms is listed below:

```
dht(x,N,inverse)
     float *x;
     int N,inverse;
{ int i,n;
  float *y,r2=sqrt(2.0);
  y=(float *)malloc(N*sizeof(float));
  if (inverse) {          // inverse DHT
    n=1;
    while(n<N) {
      for (i=0; i<n; i++) {
        y[2*i]  =(x[i]+x[i+n])/r2;
        y[2*i+1]=(x[i]-x[i+n])/r2;
      }
      for (i=0; i<n*2; i++) x[i]=y[i];
      n=n*2;
    }
  }
  else {                  // forward DHT
    n=N;
    while(n>1) {
      n=n/2;
      for (i=0; i<n; i++) {
        y[i]  =(x[2*i]+x[2*i+1])/r2;
        y[i+n]=(x[2*i]-x[2*i+1])/r2;
      }
      for (i=0; i<n*2; i++) x[i]=y[i];
    }
  }
  free(y);
}
```

8.3.4 Filter bank implementation

The fast algorithm of the Haar transform can also be viewed as a special case of the filter bank algorithm for general wavelet transforms, to be discussed in Chapter 11. Here, we briefly discuss such an implementation as a preview of the filter bank idea. To see how this algorithm works, we first consider the convolution of a signal sequence $x[n]$ with some convolution kernel $h[n]$:

$$x'[n] = x[n] * h[n] = \sum_m h[m]x[n-m]. \tag{8.84}$$

In particular, for the Haar transform, we consider four different two-point convolution kernels:

- $h_0[0] = h_0[1] = 1/\sqrt{2}$
- $h_1[0] = 1/\sqrt{2},\ \ h_1[1] = -1/\sqrt{2}$
- $g_0[0] = g_0[1] = 1/\sqrt{2}$
- $g_1[0] = -1/\sqrt{2},\ \ h_1[1] = 1/\sqrt{2}.$

Note that $g_i[n]$ is the time-reversed version of $h_i[n]$ $(i = 0, 1)$; i.e., the order of the elements in the two-point sequence is reversed (the two elements of g_0 and h_0 are identical). Depending on the kernel, the convolution above can be considered as either an LP or HP filter. Specifically, for kernel h_0 (or g_0), we have

$$y[n] = x[n] * h_0[n] = \sum_{m=0}^{1} h_0[m]x[n-m] = \frac{1}{\sqrt{2}}(x[n-1] + x[n]). \tag{8.85}$$

This can be considered as an LP filter, as the output $y[n]$ represents the average of any two consecutive data points $x[n-1]$ and $x[n]$ (corresponding to low frequencies). On the other hand, if the kernel is h_1, then

$$y[n] = x[n] * h_1[n] = \sum_{m=0}^{1} h_0[m]x[n-m] = \frac{1}{\sqrt{2}}(x[n-1] - x[n]). \tag{8.86}$$

This can be considered as an HP filter as the output $y[n]$ represents the difference of the two consecutive data points (corresponding to high frequencies). Finally, if the kernel is g_1, the convolution is also an HP filter:

$$y[n] = x[n] * g_1[n] = \frac{1}{\sqrt{2}}(x[n] - x[n-1]) = -x[n] * h_1[n]. \tag{8.87}$$

Owing to the convolution theorem of the z-transform, these convolutions can also be represented as multiplications in the z-domain:

$$Y(z) = H_i(z)X(z), \quad Y(z) = G_i(z)X(z), \qquad i = 0, 1. \tag{8.88}$$

Now the forward transform of the fast DHT shown on the left of Fig. 8.10 can be considered as a recursion of the following two operations.

- Operation $\mathcal{A}$ (average or approximation): an LP filter implemented as $y[n] = x[n] * h_0[n]$, followed by down-sampling (every other point in $y[n]$ is eliminated).
- Operation $\mathcal{D}$ (difference or detail): an HP filter implemented as $y[n] = x[n] * h_1[n]$, also followed by down-sampling.

For example, operation $\mathcal{A}$ applied to a set of eight-point sequence $x[0], \ldots, x[7]$ will generate a four-point sequence containing $x[0] + x[1]$, $x[2] + x[3]$, $x[4] + x[5]$, and $x[6] + x[7]$ (all divided by $\sqrt{2}$) representing the local average (or approximation) of the signal. When operation $\mathcal{D}$ is applied to the same input, it will generate a different four-point sequence containing $x[0] - x[1]$, $x[2] - x[3]$, $x[4] - x[5]$, and $x[6] - x[7]$ (all divided by $\sqrt{2}$) representing the local difference (or details) of the signal.

In this filter bank algorithm, this pair of operations $\mathcal{A}$ and $\mathcal{D}$ is applied first to the N-point signal $x[n]$ $(n = 0, \ldots, N-1)$, and then recursively to the output of operation $\mathcal{A}$ in the previous recursion. As the data size is reduced by half after each recursion, this process can be carried out $\log_2 N$ times to generate all N transform coefficients. This is the filter bank implementation of the DHT, as illustrated on the left of Fig. 8.11.

The inverse transform of the fast algorithm (right half of Fig. 8.10) can also be viewed as a recursion of two operations.

- Operation $\mathcal{A}$: an LP filter implemented as $y[n] = x[n] * g_0[n]$, applied to the up-sampled version of the data (with a zero inserted between every two consecutive data points, also in front of the first sample and after the last one).
- Operation $\mathcal{D}$: an HP filtered by $y[n] = x[n] * g_1[n]$, applied to the up-sampled input data.

For example, when operation $\mathcal{A}$ is applied to $X[0]$, it will first be up-sampled to become $0, X[0], 0$, which is then convolved with $g_0[n]$ to generate a sequence with two elements $X[0], X[0]$. Also, when operation $\mathcal{D}$ is applied to $X[1]$, it will be up-sampled to become $0, X[1], 0$, which is convolved with $g_1[n]$ to generate a sequence $X[1], -X[1]$. The corresponding elements of these two sequences are then added to generate a new sequence $X[0] + X[1], X[0] - X[1]$. In the next level of recursion, operation $\mathcal{A}$ will be applied to this two-point sequence, while operation $\mathcal{D}$ is applied to the next two data points $X[2], X[3]$, and their outputs, two four-point sequences, are added again. This recursion is also carried out $\log_2 N$ times until all N data points $x[0], \ldots, x[N-1]$ are reconstructed. This is the filter bank implementation of the IDHT, as illustrated on the right of Fig. 8.11.

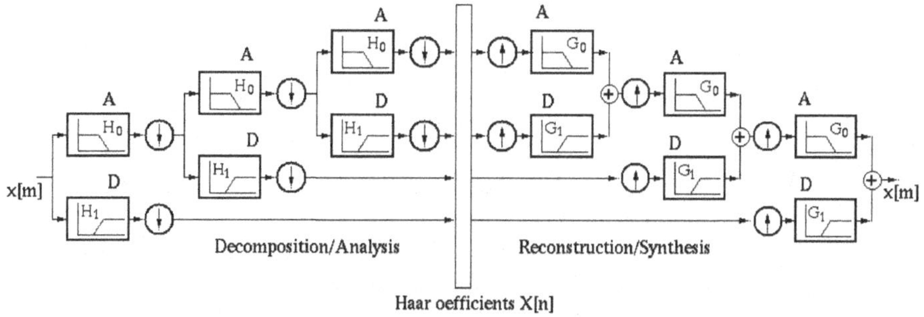

Figure 8.11 Filter bank implementation of DHT. H_0 and G_0 are LP filters and H_1 and G_1 are HP filters. The up and down arrows represent up-sampling and down-sampling, respectively.

8.4 Two-dimensional transforms

As with the discrete Fourier and cosine transforms, all three of the transform methods (Walsh-Hadamard, Slant, and Haar transforms) discussed above can also be applied to a 2-D signal $x[m,n]$ $(m = 0,\ldots,M-1, n = 0,\ldots,N-1)$, such as an image, for purposes such as feature extraction, filtering, and data compression. For convenience, in the following we will represent any of the three orthogonal matrices considered above by a generic orthogonal matrix $\boldsymbol{A}$. The forward and inverse 2-D transform of a 2-D signal are defined respectively as

$$\begin{cases} \boldsymbol{X} = \boldsymbol{A}_c^{\mathrm{T}} \boldsymbol{x} \boldsymbol{A}_r & \text{(forward)} \\ \boldsymbol{x} = \boldsymbol{A}_c \boldsymbol{X} \boldsymbol{A}_r^{\mathrm{T}} & \text{(inverse)} \end{cases}, \tag{8.89}$$

where both the 2-D signal $\boldsymbol{x}$ and its spectrum $\boldsymbol{X}$ are $M \times N$ matrices, and the pre-multiplication matrix $\boldsymbol{A}_c$ is $M \times M$ for the column transforms, while the post-multiplication matrix $\boldsymbol{A}_r$ is $N \times N$ for the row transforms. The inverse transform (second equation) expresses the given 2-D signal $\boldsymbol{x}$ as a linear combination of a set of N^2 2-D basis functions:

$$\begin{aligned} \boldsymbol{x} &= [\boldsymbol{a}_0, \ldots, \boldsymbol{a}_{M-1}] \begin{bmatrix} X[0,0] & \cdots & X[0,N-1] \\ \vdots & \ddots & \vdots \\ X[M-1,0] & \cdots & X[M-1,N-1] \end{bmatrix} \begin{bmatrix} \boldsymbol{a}_0^{\mathrm{T}} \\ \vdots \\ \boldsymbol{a}_{N-1}^{\mathrm{T}} \end{bmatrix} \\ &= \sum_{k=0}^{M-1} \sum_{l=0}^{N-1} X[k,l] \boldsymbol{a}_k \boldsymbol{a}_l^{\mathrm{T}} = \sum_{k=0}^{M-1} \sum_{l=0}^{N-1} X[k,l] \boldsymbol{B}_{kl}, \end{aligned} \tag{8.90}$$

where $\boldsymbol{B}_{kl} = \boldsymbol{a}_k \boldsymbol{a}_l^{\mathrm{T}}$ is the klth 2-D $(M \times N)$ basis function, weighted by the corresponding coefficient $X[k,l]$. Same as in the cases of DFT (Eq. (4.272)) and DCT (Eq. (7.124), this coefficient can be obtained as the projection (inner product) of the 2-D signal $\boldsymbol{x}$ onto the klth 2-D basis function $\boldsymbol{B}_{kl}$:

$$\begin{aligned} X[k,l] &= \boldsymbol{a}_k^{\mathrm{T}} \begin{bmatrix} x[0,0] & \cdots & x[0,N-1] \\ \vdots & \ddots & \vdots \\ x[M-1,0] & \cdots & x[M-1,N-1] \end{bmatrix} \boldsymbol{a}_l \\ &= \sum_{m=0}^{M-1} \sum_{n=0}^{N-1} x[m,n] B_{kl}[m,n] = \langle \boldsymbol{x}, \boldsymbol{B}_{kl} \rangle. \end{aligned} \tag{8.91}$$

When $M = N = 8$, the $8 \times 8 = 64$ such 2-D basis functions corresponding to Walsh-Hadamard (WHT), slant (ST), and Haar (DHT) are shown in Fig. 8.12, in comparison with those of the 2-D DCT. In all four transforms the DC component is at the top-left corner, and the farther away from the corner, the higher are the frequency/sequency contents or scales of details represented. Also note that the spatial positions are represented in the Haar basis.

All of these transform methods can be used for filtering. Fig. 8.13 shows both the LP and HP filtering effects in both the spatial domain and spatial frequency domains for each of the transform methods. We can also see that all of these

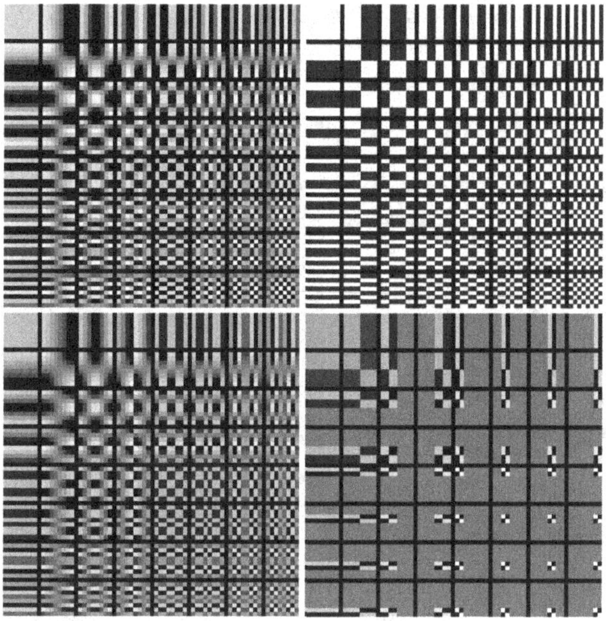

Figure 8.12 The basis functions $\boldsymbol{B}_{kl}$ for the 2-D DCT (top left), WHT (top right), ST (lower left), and DHT (lower right).

transforms have the general property of compacting the signal energy into a small number of low-frequency/sequency/scale components. In the LP filtering examples, only about 1% of the transform coefficients are kept after filtering in the transform domain of the DCT, WHT, ST and DHT, but they carry, respectively, 96.4%, 94.8%, 95.5%, and 93% of the total signal energy. Therefore, all of these transform methods lend themselves to data compression, like the Fourier transform.

As these transform methods are based on different basis functions, they may be suitable for different types of signal. Most obviously, like the DFT, the DCT is based on sinusoidal basis functions and is therefore suitable for representing signals that are smooth in nature. However, it is also possible that in some specific applications other transform methods may be more suitable, as the signals of interest may be more effectively represented by particular types of basis function other than sinusoids. For example, the WHT may be more suitable to use if the signal is of a square-wave nature and may be most effectively represented by a small subset of the WHT basis functions, so that the corresponding transform coefficients may contain most of the signal energy.

Also, we make some special note regarding the Haar transform. As with all other 2-D transforms, the first basis function, the top-left corner in Fig. 8.12, is a constant representing the DC component of the 2-D signal. However, the rest of the basis functions are quite different. Most obviously, the last 16 basis

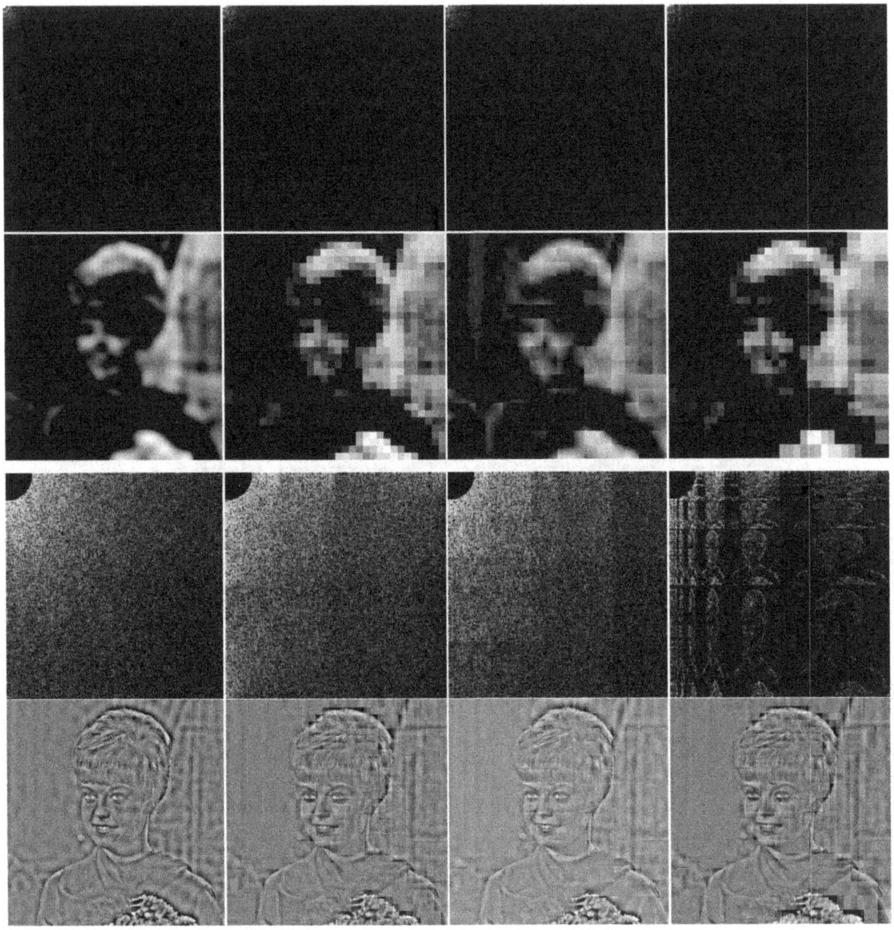

Figure 8.13 LP and HP filtering based on the DCT, WHT, ST, and DHT (from left to right). The filtered spectrum is given in the first (LP) and third (HP) rows; the corresponding filtered image is given directly below the spectrum.

functions in the lower right quarter of the 2-D spectrum represent not only the same (highest) levels of detail in the signal, but also their spatial positions. This contrasts strongly with the spectra of all other transforms, which represent progressively higher spatial frequencies/sequencies (for signal details at different levels) without any indication in terms of their spatial positions. As noted before, this capability of position representation in the spectrum is gained with the cost of a much reduced number of scale levels.

8.5 Homework problems

1. Prove the orthogonality of the WHT matrix $\boldsymbol{H}$ by mathematical induction. First, show that, when $n = 1$ $\boldsymbol{H}_1$ is orthogonal; next, show that if $\boldsymbol{H}_{n-1}$ is orthogonal, then $\boldsymbol{H}_n$ is also orthogonal.
2. Prove the orthogonality of the slant transform matrix $\boldsymbol{S}$ by mathematical induction. As in the previous problem, first show that, when $n = 1$, $\boldsymbol{S}_1$ is orthogonal. Then, based on the assumption that $\boldsymbol{S}_{n-1}$ is orthogonal, show $\boldsymbol{S}_n$ is also orthogonal.
3. Show that computational complexity of the fast algorithm for DHT in Fig. 8.10 is $O(N)$, linear to the size of the data. Hint: follow the analysis of the FFT algorithm, and consider the number of stages in the algorithm and the complexity at each stage.)
4. Understand the C code for the WHT provided in the text and convert it into a Matlab function. Apply it to an $N = 8$ sequence $\boldsymbol{x} = [x[0], \ldots, x[7]]^{\mathrm{T}}$ of your choice to obtain its N transform coefficients, then carry out the inverse transform to reconstruct the sequence from these transform coefficients.
5. Repeat problem 4 for the discrete slant transform ST.
6. Repeat problem 4 for the discrete Haar transform DHT.
7. Implement the 2-D sequency-ordered WHT of the same image used in the homework of Chapter 5 and carry out various types of filtering (LP, HP, etc.) of the image in the transform domain. Then carry out the inverse transform and display the filtered image. Compare the filtering effects with those obtained by the Fourier transform obtained in Chapter 5.
8. Implement the 2-D sequency-ordered WHT of the image used before as shown in Fig. 5.25, and then carry out image compression by suppressing to zero all sequency components lower than a certain threshold. Obtain the percentage of such suppressed sequency components, and the percentage of lost energy (in terms of signal value squared). (Note that this exercise only serves to illustrate the basic idea of image compression but it is not how image compression is practically done, where those components suppressed need to be recorded as well.)
9. Repeat problems 7 and 8 for the discrete slant transform ST.
10. Repeat problems 7 and 8 for the discrete Haar transform DHT.

9 Karhunen-Loève transform and principal component analysis

9.1 Stochastic process and signal correlation

9.1.1 Signals as stochastic processes

In all of our previous discussions, a time signal $x(t)$ is assumed to take a deterministic value $x(t_0)$ at any given moment $t = t_0$. However, in practice, many signals of interest are not deterministic, in the sense that multiple measurements of the same variable may be similar but not identical. While the random nature of such a signal could be caused by some inevitable measurement errors, we also realize that often a variable of certain physical process is affected by a large number of factors too complex to model in terms of how they collectively affect the variable of interest. Consequently, the measured signal appears to be random.

The signal $x(t)$ of a non-deterministic variable can be considered as a *stochastic* or *random process*, of which a time sample $x(t_0)$ at $t = t_0$ is treated as a random variable with a certain probability distribution. In this chapter we will consider a special orthogonal transform that can be applied to such random signals, similar to the way all orthogonal transforms discussed previously are applied to deterministic signals.

Let us first review the following concepts of a stochastic process $x(t)$.

- The *mean function* of $x(t)$ is the expectation of the stochastic process:

$$\mu_x(t) = \int x(t)p(x_t)\,dx = E[x(t)], \tag{9.1}$$

  where $p(x_t)$ is the *probability density function (pdf)* of the variable $x(t)$. If $\mu_x(t) = 0$ for all t, then $x(t)$ is a zero-mean or *centered* stochastic process. As any given process $x(t)$ can be converted into a zero-mean process by simply subtracting the mean from it, $x(t) - \mu_x(t)$, we can always assume a given process $x(t)$ to be centered with a zero mean function without loss of generality.
- The *auto-covariance function* of $x(t)$ is defined as

$$\begin{aligned}\mathrm{Cov}_x(t,\tau) = \sigma_x^2(t,\tau) &= \int\int (x(t) - \mu_x(t))\,(\overline{x}(\tau) - \overline{\mu}_x(\tau))p(x_t, x_\tau)\,dt\,d\tau \\ &= E[(x(t) - \mu_x(t))\,(\overline{x}(\tau) - \overline{\mu}_x(\tau))] = E[x(t)\overline{x}(\tau)] - \mu_x(t)\overline{\mu}_x(\tau),\end{aligned} \tag{9.2}$$

where $p(x_t, x_\tau)$ is the *joint pdf* of $x(t)$ and $x(\tau)$. When $t = \tau$, $\sigma^2(t,t) = \text{Var}_x(t) = E[|x(t)|^2] - \mu_x^2(t)$ becomes the variance of the signal at t. As we can always assume $x(t)$ to be centered with $\mu_x(t) = 0$, the covariance $\sigma_x^2(t,\tau) = E[x(t)\overline{x}(\tau)] = \langle x(t), x(\tau)\rangle$ can be considered as the inner product of the two variables $x(t)$ and $x(\tau)$ (Eq. (2.20) in Chapter 2). In particular, if $\sigma_x^2(t,\tau) = \langle x(t), x(\tau)\rangle = 0$, the two variables are said to be orthogonal to each other.

- The *autocorrelation function* of $x(t)$ is defined as the covariance $\sigma_x^2(t,\tau)$ normalized by $\sigma_x^2(t)$ and $\sigma_x^2(\tau)$:

$$r_x(t,\tau) = \frac{\sigma_x^2(t,\tau)}{\sqrt{\sigma_x^2(t)\,\sigma_x^2(\tau)}} = \frac{\langle x(t), x(\tau)\rangle}{\sqrt{\langle x(t), x(t)\rangle\langle x(\tau), x(\tau)\rangle}}. \tag{9.3}$$

  Owing to the Cauchy-Schwarz inequality (Eq. (2.30)) $|\langle x,y\rangle|^2 \le \langle x,x\rangle\langle y,y\rangle$, we get $|r_x(t,\tau)| \le 1$, and $r_x(t,\tau) = 1$ if $t = \tau$. This result indicates that the similarity between any two different variables $x(t)$ and $x(\tau)$ is no greater than that of a variable $x(t)$ to itself, which is always unity or 100%.

If the joint pdf of the random process $x(t)$ does not change over time, then $x(t)$ is a *stationary process*, and the following hold for any τ:

$$\mu_x(t) = \mu_x(t-\tau), \qquad \sigma_x^2(t,\tau) = \sigma_x^2(t-\tau, 0), \qquad r_x(t,\tau) = r_x(t-\tau,0); \tag{9.4}$$

i.e., the mean function $\mu_x(t) = \mu_x$ becomes a constant. Now the auto-covariance and autocorrelation depend only on the time difference $t-\tau$ and can be written as $\sigma_x^2(t-\tau)$ and $r_x(t-\tau)$, respectively. If the equations above hold but the joint pdf is not necessarily time invariant, then $x(t)$ is said to be a *weak* or *wide-sense stationary (WSS)* process. Moreover, without loss of generality, we can further normalize the signal by a transformation $x'(t) = (x(t) - \mu_x)/\sigma_x^2$ so that its covariance becomes the same as its correlation. In other words, these two functions represent essentially the same characteristics of the signal.

As with a deterministic signal, a random process $x(t)$ can also be truncated and sampled to become a finite set of N random variables $x[n] = x(nt_0)$ $(n = 0, \dots, N-1)$, where $t_0 = 1/F$ is the sampling period and $F = 1/t_0$ is the sampling rate. If the specific sampling rate is not a concern, we could assume $t_0 = F = 1$ for simplicity. The N signal samples can be represented by a random vector $\boldsymbol{x} = [x[0], \dots, x[N-1]]^{\mathrm{T}}$, and, correspondingly, the mean and auto-covariance/autocorrelation functions for a random process become the mean vector and covariance matrix, respectively:

- The *mean vector* of a random vector $\boldsymbol{x}$ is its expectation:

$$\boldsymbol{\mu}_x = E(\boldsymbol{x}) = [\mu[0], \dots, \mu[N-1]]^{\mathrm{T}}, \tag{9.5}$$

  where $\mu[n] = E(x[n])$ is the mean of $x[n]$ $(n = 0, \dots, N-1)$.

- The *covariance matrix* of a random vector $\boldsymbol{x}$ is defined as

$$\boldsymbol{\Sigma}_x = E[(\boldsymbol{x}-\boldsymbol{\mu}_x)(\boldsymbol{x}-\boldsymbol{\mu}_x)^*] = E[\boldsymbol{x}\boldsymbol{x}^*] - \boldsymbol{\mu}_x\boldsymbol{\mu}_x^* = \begin{bmatrix} \sigma_0^2 & \cdots & \sigma_{0(N-1)}^2 \\ \vdots & \ddots & \vdots \\ \sigma_{(N-1)0}^2 & \cdots & \sigma_{N-1}^2 \end{bmatrix}, \tag{9.6}$$

where the element σ_{mn}^2 is the covariance of two random variables $x[m]$ and $x[n]$ $(m, n = 0, \ldots, N-1)$:

$$\sigma_{mn}^2 = E[(x[m]-\mu[m])(\overline{x}[n]-\overline{\mu}[n])] = E(x[m]\overline{x}[n]) - \mu[m]\overline{\mu}[n]. \tag{9.7}$$

As always, we can assume $\mu[n] = 0$ (by trivially subtracting the mean vector from the random vector) and get $\sigma_{mn}^2 = E(x[m]\overline{x}[n]) = \langle x[m], x[n]\rangle$. The nth component on the diagonal is the variance of the nth variable $x[n]$ representing the dynamic energy contained in $x[n]$:

$$\sigma_n^2 = E[|x[n]-\mu[n]|^2] = E(|x[n]|^2) - |\mu[n]|^2. \tag{9.8}$$

This covariance matrix $\boldsymbol{\Sigma}_x^* = \boldsymbol{\Sigma}_x$ is Hermitian and positive definite.
- The *correlation matrix* of a random vector $\boldsymbol{x}$ is defined as

$$\boldsymbol{R}_x = \begin{bmatrix} r_0 & \cdots & r_{0(N-1)} \\ \vdots & \ddots & \vdots \\ r_{(N-1)0} & \cdots & r_{N-1} \end{bmatrix}, \tag{9.9}$$

where the element r_{mn} is the *correlation coefficient* between two random variables $x[m]$ and $x[n]$ defined as the covariance σ_{mn}^2 normalized by σ_m and σ_n:

$$r_{mn} = \frac{\sigma_{mn}^2}{\sqrt{\sigma_m^2\;\sigma_n^2}} = \frac{\langle x[m], x[n]\rangle}{\sqrt{\langle x[n], x[n]\rangle\;\langle x[n], x[n]\rangle}} \qquad m, n = 0, \ldots, N-1, \tag{9.10}$$

where $\langle x[m], x[n]\rangle = E[x[m]\overline{x}[n]]$. Note that r_{mn} measures the similarity between the two variables, and $r_n = 1$ and $|r_{mn}| \le 1$ for all $m \neq n$.

In general, it may not be easy to obtain the true mean vector $\boldsymbol{\mu}_x$ and covariance matrix $\boldsymbol{\Sigma}_x$ of a random vector $\boldsymbol{x}$, as they depend on the joint pdf $p(\boldsymbol{x})$, which is unlikely to be available in practice. However, both $\boldsymbol{\mu}_x$ and $\boldsymbol{\Sigma}_x$ can be estimated if enough samples of the random vector can be obtained. Let $\{\boldsymbol{x}_k, (k = 1, \ldots, K)\}$ be a set of K samples of the N-D random vector $\boldsymbol{x}$, then the mean vector and covariance matrix can be estimated as

$$\hat{\boldsymbol{\mu}}_x = \frac{1}{K}\sum_{k=1}^{K} \boldsymbol{x}_k, \quad \text{and} \quad \hat{\boldsymbol{\Sigma}}_x = \frac{1}{K-1}\sum_{k=1}^{K} (\boldsymbol{x}_k - \boldsymbol{\mu}_x)(\boldsymbol{x}_k - \boldsymbol{\mu}_x)^*, \tag{9.11}$$

where $\boldsymbol{\mu}_x = 0$ can always be assumed to be zero. Moreover, if we define a $K \times N$ matrix $\boldsymbol{D} = [\boldsymbol{x}_1, \ldots, \boldsymbol{x}_K]^{\mathrm{T}}$ composed of the K sample vectors of zero mean as its row vectors, then the estimated covariance matrix can be expressed as

$$\hat{\boldsymbol{\Sigma}}_x = \frac{1}{K-1}[\boldsymbol{D}^{\mathrm{T}}\overline{\boldsymbol{D}}]_{N\times N} = \frac{1}{K-1}[\boldsymbol{x}_1, \dots, \boldsymbol{x}_K]\begin{bmatrix} \boldsymbol{x}_1^* \\ \vdots \\ \boldsymbol{x}_K^* \end{bmatrix} = \frac{1}{K-1}\sum_{k=1}^{K} \boldsymbol{x}_k \boldsymbol{x}_k^* \tag{9.12}$$

9.1.2 Signal correlation

Signal correlation is an important concept in signal processing in general, and in the context of the Karhunen-Loève transform (KLT) in particular. As the measurement of a certain physical system, a signal tends to be smoothly and relatively evenly distributed in either time or space, in the sense that two samples of such a temporal or spatial signal are likely to be similar to each other if they are near to each other, but are less so if they are farther apart; i.e., they tend to be *locally correlated*. For example, given the current temperature as a signal sample $x(t)$, one could predict with reasonable confidence that the next sample $x(t+\tau)$ for the temperature in the near future with a small τ is fairly similar. However, one would be less confident when τ becomes larger. The correlation between two signal samples will eventually diminish when they are so far apart from each other that they are simply not relevant anymore. In other words, the smaller τ is, the larger $r_x(t, t+\tau)$ becomes and vice versa (e.g., the autocorrelation of the clouds in Fig. 7.12).

This common sense experience in everyday life is due to the general phenomenon that the energy associated with a system tends to be distributed smoothly and evenly over both time and space in the physical world governed by the principle of minimum energy and maximum *entropy*, which dictates that in a closed system, concentrated energy tends to disperse over time, and differences in physical quantities (temperature, pressure, density, etc.) tend to even out. Any disruption or discontinuity, typically associated with some kind of energy surge, is a relatively rare and unlikely event.

These signal characteristics of local correlation are reflected in the correlation matrix $\boldsymbol{R}_x$ defined in Eq. (9.9). All elements along the diagonal take the maximum value 1 for self-correlation (always 100%), while any off-diagonal element $|r_{mn}| \leq 1$ for the cross-correlation between two signal samples $x[m]$ and $x[n]$ always takes a smaller value. Moreover, those entries r_{mn} closer to the diagonal (small $|m-n|$) tend to take larger values (close to 1) than those farther away from the diagonal (large $|m-n|$). If the correlation matrix is thought of as a landscape, then there is a ridge along its diagonal along the NW-SE direction.

Based on this observation, a discrete signal can be modeled by a first-order stationary *Markov chain* (see Appendix B), of which the nth random sample $x[n]$ depends only on the previous sample $x[n-1]$ with correlation $0 \leq r \leq 1$. The correlation between any two samples $x[m]$ and $x[n]$ is therefore $r_{mn} = r^{|m-n|}$; i.e., the correlation reduces exponentially as a function of the time interval between them, and the correlation matrix can be written as

$$\boldsymbol{R}_x = \begin{bmatrix} 1 & r & r^2 & \cdots & r^{N-2} & r^{N-1} \\ r & 1 & r & \cdots & r^{N-3} & r^{N-2} \\ r^2 & r & 1 & \cdots & r^{N-4} & r^{N-3} \\ \vdots & \vdots & \vdots & \ddots & \vdots & \vdots \\ r^{N-2} & r^{N-3} & r^{n-4} & \cdots & 1 & r \\ r^{N-1} & r^{N-2} & r^{N-3} & \cdots & r & 1 \end{bmatrix}_{N\times N}. \tag{9.13}$$

This is a Toeplitz matrix with all elements along the diagonal direction being the same. This model of first-order Markov chain will be used later.

To illustrate intuitively the different amount of correlation between two random variables x and y, we consider a set of simple examples shown in Fig. 9.1, where each dot represents an outcome of an experiment in terms of $N = 2$ variable of x and y, with N easily generalized to any value $N > 2$.

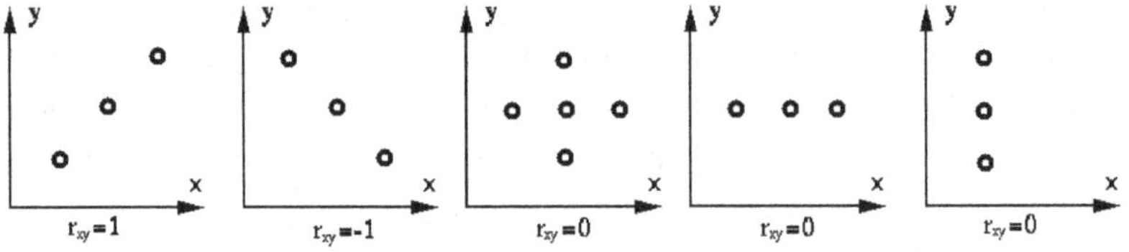

Figure 9.1 Different correlations between x and y. The two variables x and y are positively correlated (first from left), negatively correlated (second), or not correlated (third to fifth).

The different cases shown in the figure can be described by the variances σ_x^2 and σ_y^2 representing the dynamic energy or information contained in the two variables x and y, respectively, and the correlation $r_{xy} = \sigma_{xy}^2/\sqrt{\sigma_x^2\sigma_y^2}$, the covariance σ_{xy}^2 normalized by the variances σ_x^2 and σ_y^2, representing how the two variables are correlated. Specifically, if $r_{xy} > 0$, as in the first case of Fig. 9.1, the two variables are positively correlated; i.e., an increased/decreased x indicates an increased/decreased y and vice versa. On the other hand, if $r_{xy} < 0$, as in the second case of the figure, the two variables are negatively correlated; i.e., an increased/decreased x indicates a decreased/increased y and vice versa. Also, in both cases, the two variables contain the same amount of energy $\sigma_x^2 = \sigma_y^2$, and they are maximally correlated (either positively or negatively) with $|r_{xy}| = 1$; i.e., the information they carry is completely redundant, in the sense that given x we know y and visa versa. If $r_{xy} = 0$ as in the third case in the figure, the two variables are not correlated, each carrying its own independent information. In the last two cases, $\sigma_{xy}^2 = 0$ while either $\sigma_y^2 = 0$ or $\sigma_x^2 = 0$; i.e., one of the two variables contains zero dynamic energy and can therefore be omitted. In such cases the dimension of the data set can be reduced from 2 to 1 without loss any information.

Moreover, we note that by a 45° rotation (an orthogonal transformation) of the coordinate system, as illustrated in Fig. 9.2, the first two cases in Fig. 9.1 can be converted into the last two, in which the two variables are no longer correlated. The energy contained in the 2-D signal is now redistributed in such a way that one of the two variables contains 100% of the dynamic energy, while the other contains none and can therefore be totally omitted. This signal decorrelation by a simple rotation illustrates the very essential reason why orthogonal transforms can be used for data compression.

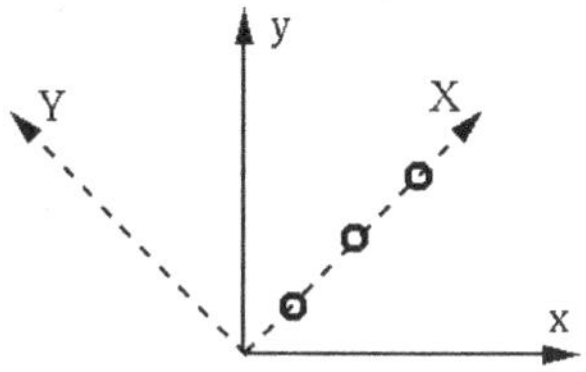

Figure 9.2 Decorrelation by rotation (orthogonal transform).

In general, from the view point of data compression and signal processing, we want to avoid high signal correlation and even energy distribution; therefore, it is desirable to convert the given data set in such a way that (1) the signal components are minimally correlated with least amount of redundancy, and (2) the total energy contained in all signal components is mostly concentrated in a small number of them so that those that carry little energy can be omitted. These properties are commonly desired for many data processing applications such as information extraction, noise reduction, and data compression. We will next consider such a transform method that can achieve these goals optimally.

9.2 Karhunen-Loève transform (KLT)

9.2.1 Continuous KLT

As shown in Eq. (2.106), a deterministic time signal $x(t)$ can be represented by an orthogonal transform as a linear combination of a set of orthogonal basis functions

$$x(t) = \sum_k c[k]\phi_k(t), \tag{9.14}$$

where the coefficients $c[k]$ can be found as the projection of $x(t)$ onto each of the basis functions in Eq. (2.108):

$$c[k] = \langle x(t), \phi_k(t) \rangle = \int x(t)\overline{\phi}_k(t)\, dt. \tag{9.15}$$

On the other hand, according to the Karhunen-Loève theorem (Theorem 2.15), a random signal $x(t)$ as a stochastic process can also be represented in exactly the same form as shown in Eqs. (2.381) and (2.382).

Now we see that Eqs. (9.14) and (9.15) have two different interpretations. If $x(t)$ is a deterministic signal, then its series expansion coefficients $c[k]$ are also deterministic. But if $x(t)$ is a stochastic process, then the coefficients $c[k]$ obtained in Eq. (2.382) become random. In fact, in the series expansion form of Eq. (9.14), the random nature of signal $x(t)$ is reflected by the random coefficients $c[k]$. But, in either case, the orthogonal basis functions $\phi_k(t)$ of the expansion are always deterministic.

Specifically, as discussed in section 2.5, the covariance $\sigma_x^2(t,\tau)$ of a centered stochastic process $x(t)$ is a Hermitian kernel, and the associated integral operator is a self-adjoint and positive definite with real positive eigenvalues λ_k and orthogonal eigenfunctions $\phi_k(t)$. These eigenfunctions form an orthogonal basis that spans a space of all stochastic functions. Any random signal $x(t)$ in the space can be series expanded by Eq. (9.14) with coefficients $c[k]$ given in Eq. (9.15), which can be considered as an orthogonal transformation, the continuous KLT, that converts the given random signal $x(t)$ into a set of coefficients $c[k]$ in the transform domain.

9.2.2 Discrete KLT

We now consider the discrete version of the Karhunen-Loève theorem. When a stochastic process $x(t)$ is truncated and sampled, it becomes a random vector composed of N random variables $\boldsymbol{x} = [x[0], \ldots, x[N-1]]^{\mathrm{T}}$. For convenience and without loss of generality, we will always assume in the following that the signal is centered with $\boldsymbol{\mu}_x = \boldsymbol{0}$, and its covariance matrix is $\boldsymbol{\Sigma}_x = E(\boldsymbol{x}\boldsymbol{x}^*)$ with its mnth element being $\sigma_{mn}^2 = E(x[m]\overline{x}[n]) = \langle x[m], x[n]\rangle$. As $\boldsymbol{\Sigma}_x$ is positive definite and Hermitian, all of its eigenvalues λ_k are real and positive, and its eigenvectors $\boldsymbol{\phi}_k$ $(k = 0, \ldots, N-1)$ form a set of orthogonal basis vectors that span the N-D vector space. Any given N-D random vector in the space can be represented as a linear combination of these basis vectors. This is the discrete Karhunen-Loève theorem.

Let $\boldsymbol{\phi}_k$ $(k = 0, \ldots, N-1)$ be the eigenvector corresponding to the kth eigenvalue λ_k of the covariance matrix $\boldsymbol{\Sigma}_x$; i.e.,

$$\boldsymbol{\Sigma}_x \boldsymbol{\phi}_k = \lambda_k \boldsymbol{\phi}_k \qquad k = 0, \ldots, N-1. \tag{9.16}$$

As $\boldsymbol{\Sigma}_x$ is Hermitian and positive definite, all its eigenvalues $\lambda_k > 0$ are real and positive. Moreover, its N eigenvectors are orthogonal, $\langle \boldsymbol{\phi}_k, \boldsymbol{\phi}_l \rangle = \delta[k-l]$ $(k, l = 0, \ldots, N-1)$, and they form an $N \times N$ unitary matrix $\boldsymbol{\Phi} = [\boldsymbol{\phi}_0, \ldots, \boldsymbol{\phi}_{N-1}]$ satisfying $\boldsymbol{\Phi}^{-1} = \boldsymbol{\Phi}^*$; i.e., $\boldsymbol{\Phi}^*\boldsymbol{\Phi} = \boldsymbol{\Phi}\boldsymbol{\Phi}^* = \boldsymbol{I}$. The N eigenequations in Eq. (9.16) can then be combined to become

$$\boldsymbol{\Sigma}_x \boldsymbol{\Phi} = \boldsymbol{\Sigma}_x [\boldsymbol{\phi}_0, \cdots, \boldsymbol{\phi}_{N-1}] = [\boldsymbol{\phi}_0, \cdots, \boldsymbol{\phi}_{N-1}] \begin{bmatrix} \lambda_0 & \cdots & 0 \\ \vdots & \ddots & \vdots \\ 0 & \cdots & \lambda_{N-1} \end{bmatrix} = \boldsymbol{\Phi}\boldsymbol{\Lambda}, \tag{9.17}$$

where $\mathbf{\Lambda} = \text{diag}(\lambda_0, \ldots, \lambda_{N-1})$ is a diagonal matrix. By pre-multiplying $\mathbf{\Phi}^* = \mathbf{\Phi}^{-1}$ on both sides, the covariance matrix $\mathbf{\Sigma}_x$ is diagonalized:

$$\mathbf{\Phi}^*\mathbf{\Sigma}_x\mathbf{\Phi} = \mathbf{\Phi}^*\mathbf{\Phi}\mathbf{\Lambda} = \mathbf{\Lambda}. \tag{9.18}$$

The discrete KLT of a given random signal vector $\boldsymbol{x}$ can now be defined as

$$\boldsymbol{X} = \begin{bmatrix} X[0] \\ \vdots \\ X[N-1] \end{bmatrix} = \mathbf{\Phi}^*\boldsymbol{x} = \begin{bmatrix} \boldsymbol{\phi}_0^* \\ \vdots \\ \boldsymbol{\phi}_{N-1}^* \end{bmatrix} \boldsymbol{x}, \tag{9.19}$$

where the kth component $X[k]$ of the vector $\boldsymbol{X}$ in the transform domain is the projection of $\boldsymbol{x}$ onto the kth basis vector $\boldsymbol{\phi}_k$:

$$X[k] = \boldsymbol{\phi}_k^*\boldsymbol{x} = \langle \boldsymbol{x}, \boldsymbol{\phi}_k \rangle. \tag{9.20}$$

Pre-multiplying $\mathbf{\Phi}$ on both sides of Eq. (9.19), we get the inverse KLT:

$$\boldsymbol{x} = \mathbf{\Phi}\boldsymbol{X} = [\boldsymbol{\phi}_0, \ldots, \boldsymbol{\phi}_{N-1}] \begin{bmatrix} X[0] \\ \vdots \\ X[N-1] \end{bmatrix} = \sum_{k=0}^{N-1} X[k]\boldsymbol{\phi}_k. \tag{9.21}$$

Equations. (9.21) and (9.19) can be rewritten as a pair of the discrete KLT:

$$\begin{cases} \boldsymbol{X} = \mathbf{\Phi}^*\boldsymbol{x} & \text{(forward)} \\ \boldsymbol{x} = \mathbf{\Phi}\,\boldsymbol{X} & \text{(inverse)} \end{cases}. \tag{9.22}$$

The first equation is the forward transform that gives the random coefficient $X[k]$ as the projection of the random vector $\boldsymbol{x}$ onto the kth deterministic basis vector $\boldsymbol{\phi}_k$ $(k = 0, \ldots, N-1)$, and the second equation is the inverse transform that represents the random vector $\boldsymbol{x}$ as a linear combination of the N eigenvectors $\boldsymbol{\phi}_k$ $(k = 0, \ldots, N-1)$ of $\mathbf{\Sigma}_x$ weighted by the random coefficients $X[k]$. Note that Eqs. (9.21) and (9.19) for the discrete KLT correspond to Eqs. (9.14) and (9.15) for the continuous KLT.

9.2.3 Optimalities of the KLT

As discussed in previous chapters, all orthogonal transforms exhibit to various extents the properties of signal decorrelation and energy compaction. For example, in the frequency domain after the Fourier transform, most of the signal energy is likely to be concentrated in a small number of low-frequency components while little energy is contained in high-frequency components. Moreover, while the signal is typically locally correlated in the time domain, in the sense that the signal value $x[n]$ can be predicted to be similar to the previous one $x[n-1]$, this is no longer the case in the frequency domain, as knowing the value of a frequency component $X[k]$ would provide little information regarding the neighboring components. Other orthogonal transforms have similar effects.

Now we show that, among all orthogonal transforms, the KLT is optimal in terms of signal decorrelation and energy compaction, as stated in Theorem 9.1.

Theorem 9.1. *Let $\boldsymbol{X} = \boldsymbol{\Phi}^*\boldsymbol{x}$ be the KLT of an N-D random signal vector $\boldsymbol{x}$, where $\boldsymbol{\Phi}$ is the eigenvector matrix associated with the covariance matrix $\boldsymbol{\Sigma}_x$ of $\boldsymbol{x}$ (Eq. (9.18)). Then we have the following results.*

1. *The total signal energy $\mathcal{E}_x$ is conserved:*

$$\mathcal{E}_X = \sum_{k=0}^{N-1} E(|X[k]|^2) = tr\boldsymbol{\Sigma}_X = tr\boldsymbol{\Sigma}_x = \sum_{k=0}^{N-1} E(|x[k]|^2) = \mathcal{E}_x. \tag{9.23}$$

2. *The signal is completely decorrelated; i.e., all off-diagonal components of $\boldsymbol{\Sigma}_X$ are zero:*

$$\sigma_{kl}^2 = 0 \qquad \textit{for all } k \neq l. \tag{9.24}$$

3. *The signal energy is maximally compacted:*

$$\mathcal{E}_M(\boldsymbol{\Phi}) \geq \mathcal{E}_M(\boldsymbol{A}), \tag{9.25}$$

where $\mathcal{E}_M(\boldsymbol{A})$ is the energy contained in the first $M < N$ signal components after an arbitrary orthogonal transform $\boldsymbol{X} = \boldsymbol{A}^*\boldsymbol{x}$.

Proof: The first statement is true simply because the trace of the covariance matrix remains the same after any unitary transform:

$$\text{tr}\,\boldsymbol{\Sigma}_X = \text{tr}(\boldsymbol{\Phi}^*\boldsymbol{\Sigma}_x\boldsymbol{\Phi}) = \text{tr}(\boldsymbol{\Phi}^*\boldsymbol{\Phi}\boldsymbol{\Sigma}_x) = \text{tr}\,\boldsymbol{\Sigma}_x, \tag{9.26}$$

where we have used Eq. (A.19). This result is equivalent to Parseval's identity for the property of energy conservation of any orthogonal transform of the deterministic signals.

The second statement is true due to the definition of the KLT by which the covariance matrix $\boldsymbol{\Sigma}_X$ of $\boldsymbol{X} = \boldsymbol{\Phi}^*\boldsymbol{x}$ is diagonalized (Eq. (9.18)):

$$\boldsymbol{\Sigma}_X = E(\boldsymbol{X}\boldsymbol{X}^*) = E[(\boldsymbol{\Phi}^*\boldsymbol{x})(\boldsymbol{\Phi}^*\boldsymbol{x})^*] = \boldsymbol{\Phi}^*E(\boldsymbol{x}\boldsymbol{x}^*)\boldsymbol{\Phi} = \boldsymbol{\Phi}^*\boldsymbol{\Sigma}_x\boldsymbol{\Phi} = \boldsymbol{\Lambda}. \tag{9.27}$$

As all off-diagonal elements $\sigma_{kl}^2 = E(X[k]\overline{X}[l]) = \langle X[k], X[l]\rangle = 0$ $(k \neq l)$, any two different components $X[k]$ and $X[l]$ are indeed decorrelated. The total signal energy is the sum of all eigenvalues (real and positive)

$$\mathcal{E}_X = \text{tr}\,\boldsymbol{\Sigma}_X = \text{tr}\,\boldsymbol{\Lambda} = \sum_{k=0}^{N-1} \lambda_k. \tag{9.28}$$

To prove the third statement, we let $\boldsymbol{A} = [\boldsymbol{a}_0, \ldots, \boldsymbol{a}_{N-1}]$ be an arbitrary unitary matrix satisfying $\boldsymbol{A}^* = \boldsymbol{A}^{-1}$, then the kth element of $\boldsymbol{X} = \boldsymbol{A}^*\boldsymbol{x}$ is $X[k] = \boldsymbol{a}_k^*\boldsymbol{x}$, and the energy contained in the first $M < N$ components in the transform domain is the sum of the first M elements along the diagonal of $\boldsymbol{\Sigma}_X$:

$$\mathcal{E}_M(\boldsymbol{A}) = \sum_{k=0}^{M-1} E(|X[k]|^2) = \sum_{k=0}^{M-1} E(|\boldsymbol{a}_k^* \boldsymbol{x}|^2) = \sum_{k=0}^{M-1} E[(\boldsymbol{a}_k^* \boldsymbol{x})\,(\boldsymbol{a}_k^* \boldsymbol{x})^*]$$
$$= \sum_{k=0}^{M-1} E(\boldsymbol{a}_k^* \boldsymbol{x}\,\boldsymbol{x}^* \boldsymbol{a}_k) = \sum_{k=0}^{M-1} \boldsymbol{a}_k^* E(\boldsymbol{x}\boldsymbol{x}^*)\boldsymbol{a}_k = \sum_{k=0}^{M-1} \boldsymbol{a}_k^* \boldsymbol{\Sigma}_x \boldsymbol{a}_k. \quad (9.29)$$

The task of finding the optimal matrix $\boldsymbol{A}$ that maximizes $\mathcal{E}_M(\boldsymbol{A})$ can be formulated as a constrained optimization problem:

$$\mathcal{E}_M(\boldsymbol{A}) = \sum_{k=0}^{M-1} \boldsymbol{a}_k^* \boldsymbol{\Sigma}_x \boldsymbol{a}_k \to \max$$
$$\text{subject to: } \boldsymbol{a}_k^* \boldsymbol{a}_k = 1 \qquad k = 0, \dots, M-1. \quad (9.30)$$

Here, the constraint $\boldsymbol{a}_k^* \boldsymbol{a}_k = 1$ guarantees that all columns of $\boldsymbol{A}$ are normalized. This problem can be solved by the method of Lagrange multipliers. Specifically, we set to zero the following partial derivative of the modified objective function with respect to $\boldsymbol{a}_l$:

$$\frac{\partial}{\partial \boldsymbol{a}_l}[\mathcal{E}_M(\boldsymbol{A}) - \sum_{k=0}^{M-1} \lambda_k(\boldsymbol{a}_k^* \boldsymbol{a}_k - 1)] = \frac{\partial}{\partial \boldsymbol{a}_l}[\sum_{k=0}^{M-1} (\boldsymbol{a}_k^* \boldsymbol{\Sigma}_x \boldsymbol{a}_k - \lambda_k \boldsymbol{a}_k^* \boldsymbol{a}_k + \lambda_k)]$$
$$= \frac{\partial}{\partial \boldsymbol{a}_l}[\boldsymbol{a}_l^* \boldsymbol{\Sigma}_x \boldsymbol{a}_l - \lambda_l \boldsymbol{a}_l^* \boldsymbol{a}_l] = 2\boldsymbol{\Sigma}_x \boldsymbol{a}_l - 2\lambda_l \boldsymbol{a}_l = \boldsymbol{0}. \quad (9.31)$$

The last equal sign is due to the derivative of a scalar function $f(\boldsymbol{a})$ with respect to its vector argument $\boldsymbol{a}$ (Eq. (A.67)). This equation happens to be the eigenequation of matrix $\boldsymbol{\Sigma}_x$:

$$\boldsymbol{\Sigma}_x \boldsymbol{a}_l = \lambda_l \boldsymbol{a}_l \qquad l = 0, \dots, M-1. \quad (9.32)$$

Comparing this with Eq. (9.16), we see that $\boldsymbol{a}_l = \boldsymbol{\phi}_l$ must be the eigenvectors of $\boldsymbol{\Sigma}_x$; i.e., the optimal transform matrix must be the KLT matrix $\boldsymbol{A} = \boldsymbol{\Phi} = [\boldsymbol{\phi}_0, \dots, \boldsymbol{\phi}_{N-1}]$. The energy contained in the first M components is

$$\mathcal{E}_M(\boldsymbol{\Phi}) = \sum_{k=0}^{M-1} \boldsymbol{\phi}_k^* \boldsymbol{\Sigma}_x \boldsymbol{\phi}_k = \sum_{k=0}^{M-1} \lambda_k, \quad (9.33)$$

where the kth eigenvalue $\lambda_k = E(|X[k]|^2) = \sigma_X^2[k]$ is the average energy contained in the kth component $X[k]$ of $\boldsymbol{X} = \boldsymbol{\Phi}^* \boldsymbol{x}$. This energy $\mathcal{E}_M(\boldsymbol{\Phi})$ is maximized if we choose to keep the M signal components corresponding to the M largest eigenvalues. The fraction of energy contained in the M components is

$$\frac{\mathcal{E}_M(\boldsymbol{\Phi})}{\mathcal{E}_N} = \frac{\sum_{k=0}^{M-1} \lambda_k}{\sum_{k=0}^{N-1} \lambda_k}. \quad (9.34)$$

Q.E.D.

The optimality of energy compaction of the KLT can also be viewed in terms of *Shannon entropy* or simply *entropy*. To understand the concept of entropy, let us first consider a random variable x representing the outcome of a random event. We assume there are in total N possible outcomes each with probability p_k $(k = 0, \ldots, N-1)$ and $\sum_{k=0}^{N-1} p_k = 1$. The uncertainty of a specific outcome x_k can be defined by

$$I(x_k) = \log(1/p_k) = -\log p_k \qquad k = 0, \ldots, N-1. \tag{9.35}$$

In particular, when $p_k = 1$, $I(x_k) = 0$; i.e., a necessary event has zero uncertainty. On the other hand, when $p_k = 0$, $I(x_k) = \infty$; i.e., an impossible event has infinite uncertainty. The entropy of the random event x is its uncertainty defined as the expected uncertainty $I(x_k)$ of any output x_k:

$$H(x) = E[I(x_k)] = -E[\log p_k] = -\sum_{k=0}^{N-1} p_k \log p_k. \tag{9.36}$$

We consider the following two special cases:

- The N outcomes are equally likely; i.e., $p_k = 1/N$ for all $k = 1, \ldots, N$, then $H(x) = \log N$ for the maximum uncertainty.
- All outcomes are impossible except one, e.g., the lth one, which is sure to occur;, i.e., $p_k = 0$ for all $k \neq l$ but $p_l = 1$, then $H(x) = 0$ for the minimum uncertainty.

The specific logarithmic base is unessential, as the entropies corresponding to different bases are equivalent up to a scaling factor. The unit of entropy H is a *bit* if the base is 2, or *nat* (or *nit*) if the natural logarithm base $e = 2.71828$ is used. The two units are related by a scaling factor of ln 2.

When a certain amount of information regarding the outcome of a random event x is gained, its uncertainty may be reduced from $H(x)$ to $H'(x)$, and the reduction $I(x) = H(x) - H'(x)$ can be used as a quantitative measurement of the amount of information gained. In particular, if the outcome of the event is completely known, the uncertainty is reduced from $H(x)$ to $H'(x) = 0$; i.e., the entropy $H(x)$ also represents the total amount of information contained in the random variable x.

The concept of entropy defined in Eq. (9.36) can be used to measure quantitatively how well the signal energy is concentrated among its N components. To do so, we treat the energy distribution among all N components $x[n]$ of a signal $\boldsymbol{x}$ as the probability distribution of the N possible outcomes of a random event. Typically, the energy of a signal is relatively evenly distributed among all signal components; i.e, the uncertainty H is large. But after certain orthogonal transform (e.g., the DFT or DCT), the energy is redistributed so that most of it is compacted into a small number of components (e.g., the low-frequency components); i.e., the uncertainty H is reduced. As the KLT is optimal in the sense that it maximally compacts signal energy into a small number of signal components, it minimizes the entropy H.

From the data compression point of view, the signal energy distribution measured by entropy H is also indicative of by how much the data can be compressed, for example, the optimal *Huffman coding* compression method. This is an entropy encoding algorithm that assigns variable code lengths to a set of N signal symbols according to their probabilities p_k. The optimality is achieved by always assigning shorter code to more probable symbols so that the average code length is minimized. As can be seen in one of the homework problems, the average code length is closely related to the signal entropy. Therefore, for the purpose of data compression, it is always desirable to carry out a certain orthogonal transform by which the signal energy is compacted and its entropy reduced, so that shorter average code will result to achieve better compression effect.

9.2.4 Geometric interpretation of the KLT

The property of optimal energy compaction of the KLT can also be viewed in terms of the information contained in the signal. We assume a signal vector composed of a set of N real random variables $\boldsymbol{x} = [x[0], \dots, x[N-1]]^{\mathrm{T}}$ has a normal joint pdf (Eq. (B.36))

$$p(\boldsymbol{x}) = N(\boldsymbol{x}, \boldsymbol{\mu}_x, \boldsymbol{\Sigma}_x) = \frac{1}{(2\pi)^{N/2} \, |\boldsymbol{\Sigma}_x|^{1/2}} \exp\left[-\frac{1}{2}(\boldsymbol{x}-\boldsymbol{\mu}_x)^{\mathrm{T}} \boldsymbol{\Sigma}_x^{-1} (\boldsymbol{x}-\boldsymbol{\mu}_x)\right]. \tag{9.37}$$

As always, we also assume, without loss of generality, $\boldsymbol{\mu}_x = \mathbf{0}$. The shape of the normal distribution in the N-D space given in Eq. (9.37) can be represented by an iso-value hyper-surface in the space determined by

$$N(\boldsymbol{x}, \boldsymbol{\mu}_x, \boldsymbol{\Sigma}_x) = c, \tag{9.38}$$

where the constant can be so chosen so that

$$(\boldsymbol{x}-\boldsymbol{\mu}_x)^{\mathrm{T}} \boldsymbol{\Sigma}_x^{-1} (\boldsymbol{x}-\boldsymbol{\mu}_x) = \boldsymbol{x}^{\mathrm{T}} \boldsymbol{\Sigma}_x^{-1} \boldsymbol{x} = 1. \tag{9.39}$$

As $\boldsymbol{\Sigma}_x$ is positive definite, this quadratic equation represents a hyper-ellipsoid in the N-D space, whose spatial orientation is totally determined by $\boldsymbol{\Sigma}_x$.

After the KLT the signal vector $\boldsymbol{x}$ becomes $\boldsymbol{X} = \boldsymbol{\Phi}^{\mathrm{T}} \boldsymbol{x}$ which is completely decorrelated with a diagonalized covariance matrix (Eq. (9.27)):

$$\boldsymbol{\Sigma}_X = \boldsymbol{\Lambda} = \begin{bmatrix} \lambda_0 & 0 & \cdots & 0 \\ 0 & \lambda_1 & \cdots & 0 \\ \vdots & \vdots & \ddots & \vdots \\ 0 & 0 & \cdots & \lambda_{N-1} \end{bmatrix} = \begin{bmatrix} \sigma_X^2[0] & 0 & \cdots & 0 \\ 0 & \sigma_X^2[1] & \cdots & 0 \\ \vdots & \vdots & \ddots & \vdots \\ 0 & 0 & \cdots & \sigma_X^2[N-1] \end{bmatrix}. \tag{9.40}$$

Substituting $\boldsymbol{x} = \boldsymbol{\Phi}\boldsymbol{X}$ into the quadratic equation Eq. (9.39), we get

$$\begin{aligned} \boldsymbol{x}^{\mathrm{T}} \boldsymbol{\Sigma}_x^{-1} \boldsymbol{x} &= \boldsymbol{X}^{\mathrm{T}} \boldsymbol{\Phi}^{\mathrm{T}} \boldsymbol{\Sigma}_x^{-1} \boldsymbol{\Phi} \boldsymbol{X} = \boldsymbol{X}^{\mathrm{T}} \boldsymbol{\Sigma}_X^{-1} \boldsymbol{X} \\ &= \boldsymbol{X}^{\mathrm{T}} \boldsymbol{\Lambda}^{-1} \boldsymbol{X} = \sum_{k=0}^{N-1} \frac{X^2[k]}{\lambda_k} = \sum_{k=0}^{N-1} \frac{X^2[k]}{\sigma_X^2[k]} = 1. \end{aligned} \tag{9.41}$$

This is the equation of a standard hyper-ellipsoid with its N semi-axes being $\sqrt{\lambda_k} = \sigma_X[k]$. We see that the KLT can be interpreted geometrically in terms of the following effects.

- The coordinate system of the N-D space is rotated in such a way that it is now aligned with the eigenvectors $\boldsymbol{\phi}_k$ $(k = 0, \dots, N-1)$ of $\boldsymbol{\Sigma}_x$.
- The semi-principal axes of the hyper-ellipsoid representing the distribution $N(\boldsymbol{x}, \boldsymbol{\mu}_x, \boldsymbol{\Sigma}_x)$ are in parallel with the new coordinates $\boldsymbol{\phi}_k$.
- The lengths of these semi-principal axes are the square root of the corresponding eigenvalue $\sqrt{\lambda_k}$ $(k = 0, \dots, N-1)$.

Given the normal pdf in Eq. (9.37), the uncertainty, or the amount of information contained in this signal, can be measured in terms of the entropy defined in Eq. (9.36):

$$H(\boldsymbol{x}) = -E[\ln p(\boldsymbol{x})] = \frac{N}{2}\ln 2\pi + \frac{1}{2}\ln|\boldsymbol{\Sigma}_x| + \frac{1}{2}E[\boldsymbol{x}^{\mathrm{T}}\boldsymbol{\Sigma}_x^{-1}\boldsymbol{x}]. \tag{9.42}$$

According to Eq. (A.43), the second term can be further written as

$$\frac{1}{2}\ln|\boldsymbol{\Sigma}_x| = \frac{1}{2}\ln\left(\prod_{k=0}^{N-1}\lambda_k\right) = \frac{1}{2}\sum_{k=0}^{N-1}\ln\lambda_k, \tag{9.43}$$

and according to Eq. (A.19), the last term (a scalar) can be further written as

$$\begin{aligned}\frac{1}{2}E[tr(\boldsymbol{x}^{\mathrm{T}}\boldsymbol{\Sigma}_x^{-1}\boldsymbol{x})] &= \frac{1}{2}E[tr(\boldsymbol{\Sigma}_x^{-1}\boldsymbol{x}\boldsymbol{x}^{\mathrm{T}})]\\ = \frac{1}{2}tr(\boldsymbol{\Sigma}_x^{-1}E[\boldsymbol{x}\boldsymbol{x}^{\mathrm{T}}]) &= \frac{1}{2}tr(\boldsymbol{\Sigma}_x^{-1}\boldsymbol{\Sigma}_x) = \frac{1}{2}tr\boldsymbol{I} = \frac{N}{2}.\end{aligned} \tag{9.44}$$

Substituting these two terms back into Eq. (9.42) we get

$$H(\boldsymbol{x}) = \frac{N}{2}(\ln 2\pi + 1) + \frac{1}{2}\sum_{k=0}^{N-1}\ln\lambda_k. \tag{9.45}$$

If, for the purpose of data compression, we want to keep only $M < N$ out of the N variables with minimum information loss, we can first take the KLT $\boldsymbol{X} = \boldsymbol{\Phi}^*\boldsymbol{x}$, and then keep the M components of $\boldsymbol{X}$ corresponding to the M greatest eigenvalues λ_k. By doing so, the entropy $H(\boldsymbol{x})$ in Eq. (9.42) is maximized; i.e., maximum information is preserved. This is the same conclusion as stated in Theorem 9.1.

These properties can be most conveniently visualized when $N = 2$, as illustrated in Fig. 9.3. Here, a signal $\boldsymbol{x} = [x_0, x_1]^{\mathrm{T}}$ is originally represented under the standard basis vectors $\boldsymbol{e}_0$ and $\boldsymbol{e}_1$:

$$\boldsymbol{x} = \begin{bmatrix} x_0 \\ x_1 \end{bmatrix} = x_0\boldsymbol{e}_0 + x_1\boldsymbol{e}_1 = x_0\begin{bmatrix} 1 \\ 0 \end{bmatrix} + x_2\begin{bmatrix} 0 \\ 1 \end{bmatrix}. \tag{9.46}$$

The quadratic equation in Eq. (9.39) representing the 2-D normal distribution of the signal $\boldsymbol{x}$ can be written as

$$\boldsymbol{x}^{\mathrm{T}}\boldsymbol{\Sigma}_x^{-1}\boldsymbol{x} = [x_0, x_1]\begin{bmatrix} a & b/2 \\ b/2 & c \end{bmatrix}\begin{bmatrix} x_0 \\ x_1 \end{bmatrix} = ax_0^2 + bx_0x_1 + cx_1^2 = 1, \tag{9.47}$$

where we have assumed

$$\boldsymbol{\Sigma}_x^{-1} = \begin{bmatrix} a & b/2 \\ b/2 & c \end{bmatrix}. \tag{9.48}$$

As $\boldsymbol{\Sigma}_x$ is positive definite and so is $\boldsymbol{\Sigma}_x^{-1}$, we have $\left|\boldsymbol{\Sigma}_x^{-1}\right| = ac - b^2/4 > 0$; i.e., the quadratic equation above represents an ellipse (instead of any other quadratic curves, such as a hyperbola or parabola) centered at the origin (or at $\boldsymbol{\mu}_x$ if it is not zero). As shown in Fig. 9.3, the two signal components x_0 and x_1 are maximally correlated with $r_{01} = 1$ and contain equal amount of energy $\sigma_{x_0}^2 = \sigma_{x_1}^2$; i.e., the energy is evenly distributed among both components.

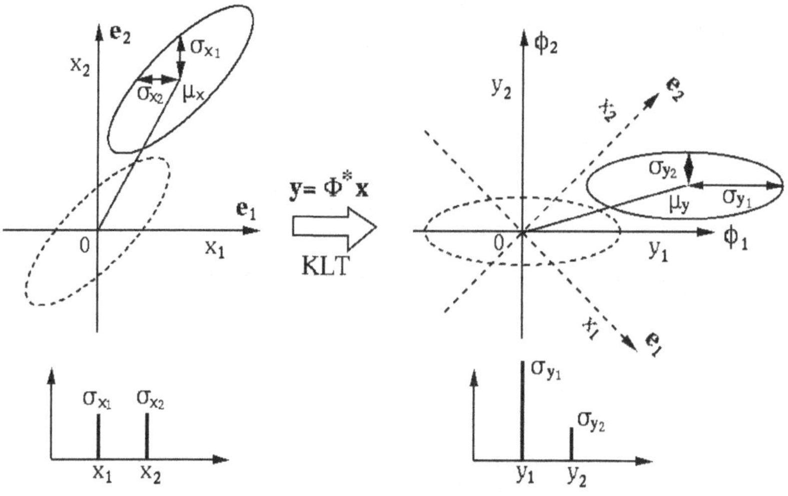

Figure 9.3 Geometric interpretation of KLT $\boldsymbol{y} = \boldsymbol{\Phi}^{\mathrm{T}}\boldsymbol{x}$.

Now a 2-D KLT $\boldsymbol{y} = \boldsymbol{\Phi}^{\mathrm{T}}\boldsymbol{x}$ is carried out in three stages: (1) subtract the mean $\boldsymbol{\mu}_x$ from $\boldsymbol{x}$ so that it is centered, (2) carry out the rotation $\boldsymbol{y} = \boldsymbol{\Phi}^{\mathrm{T}}\boldsymbol{x}$, and (3) add back the mean vector in the rotated space $\boldsymbol{\mu}_y = \boldsymbol{\Phi}^{\mathrm{T}}\boldsymbol{\mu}_x$. After the KLT, the signal is represented by two new basis vectors $\boldsymbol{\phi}_0$ and $\boldsymbol{\phi}_1$, which are just rotated versions of $\boldsymbol{e}_0$ and $\boldsymbol{e}_1$. In this space spanned by $\boldsymbol{\phi}_0$ and $\boldsymbol{\phi}_1$, the ellipse representing the joint probability density $p(\boldsymbol{x})$ becomes standardized with major semi-axis $\sqrt{\lambda_0} = \sigma_{X_0}$ and minor semi-axis $\sqrt{\lambda_1} = \sigma_{X_1}$, in parallel with the new basis vectors $\boldsymbol{\phi}_0$ and $\boldsymbol{\phi}_1$, respectively.

We see that after the KLT, the two components y_0 and y_1 are completely decorrelated with $r_{01} = 0$; and $\lambda_0 > \lambda_1$, indicating that the energy is maximally compacted into y_0 while y_1 contains minimal energy. We also note that this KLT rotation is optimal in terms of both signal decorrelation and energy compaction, as no other rotation can do any better in these regards.

9.2.5 Principal component analysis (PCA)

Owing to its optimality of signal decorrelation and energy compaction, the KLT can be used to reduce the dimensionality of a given data set while preserving maximum signal energy/information in information extraction and data compression. The signal components $X[k]$ after the KLT are called the *principal components*, and the data analysis method based on the KLT transform is called *principal component analysis (PCA)*, which is widely used in a wide variety of fields. Specifically the PCA can be carried out in the following steps:

1. Estimate the mean vector $\boldsymbol{\mu}_x$ of the given random signal vector $\boldsymbol{x}$. Subtract $\boldsymbol{\mu}_x$ from $\boldsymbol{x}$ so that it becomes centered with zero mean.
2. Estimate the covariance matrix $\boldsymbol{\Sigma}_x$ of the centered signal.
3. Find all N eigenvalues and sort then in descending order:

$$\lambda_0 \geq \ldots \geq \lambda_{N-1}. \tag{9.49}$$

4. Determine a reduced dimensionality $M < N$ so that the fraction of energy contained $\sum_{n=0}^{M-1} \lambda_n / \sum_{n=0}^{N-1} \lambda_n$ is no less than a predetermined threshold (e.g., 99%).
5. Construct an $N \times M$ transform matrix composed of the M eigenvectors corresponding to the M largest eigenvalues $\lambda_0, \ldots, \lambda_{M-1}$ of $\boldsymbol{\Sigma}_x$:

$$\boldsymbol{\Phi}_M = [\boldsymbol{\phi}_0, \ldots, \boldsymbol{\phi}_{M-1}]_{N \times M}, \tag{9.50}$$

and carry out the KLT based on this $\boldsymbol{\Phi}_M$:

$$\boldsymbol{X}_M = \begin{bmatrix} X[0] \\ \vdots \\ X[M-1] \end{bmatrix}_{M \times 1} = \boldsymbol{\Phi}_M^* \boldsymbol{x} = \begin{bmatrix} \boldsymbol{\phi}_0^* \\ \vdots \\ \boldsymbol{\phi}_{M-1}^* \end{bmatrix}_{M \times N} \begin{bmatrix} x[0] \\ \vdots \\ x[N-1] \end{bmatrix}_{N \times 1}, \tag{9.51}$$

where the kth element of $\boldsymbol{X}_M$ is $X[k] = \boldsymbol{\phi}_k^* \boldsymbol{x} = \langle \boldsymbol{x}, \boldsymbol{\phi}_k \rangle$. As the dimension M of $\boldsymbol{X}_M$ is less than the dimension N of $\boldsymbol{x}$, data compression is achieved. This is a lossy compression method with an energy loss $\sum_{k=M}^{N-1} \lambda_k$. But as λ_k's in the numerator summation are the smallest eigenvalues, the error is minimum.
6. Carry out the analysis needed in the M-D space, and, if needed, use the inverse KLT to reconstruct the signal (e.g., for compression):

$$\hat{\boldsymbol{x}} = \boldsymbol{\Phi}_M \boldsymbol{X}_M = \boldsymbol{\Phi}_M \boldsymbol{\Phi}_M^* \boldsymbol{x}, \tag{9.52}$$

or in component form:

$$\hat{\boldsymbol{x}} = \begin{bmatrix} \hat{x}[0] \\ \vdots \\ \hat{x}[N-1] \end{bmatrix} = [\boldsymbol{\phi}_0 \cdots \boldsymbol{\phi}_{M-1}] \begin{bmatrix} X[0] \\ \vdots \\ X[M-1] \end{bmatrix} = \sum_{k=0}^{M-1} X[k]\boldsymbol{\phi}_k \quad (9.53)$$

$$= [\boldsymbol{\phi}_0 \cdots \boldsymbol{\phi}_{M-1}] \begin{bmatrix} \boldsymbol{\phi}_0^* \\ \vdots \\ \boldsymbol{\phi}_{M-1}^* \end{bmatrix} \boldsymbol{x} = \left[\sum_{k=0}^{M-1} \boldsymbol{\phi}_k \boldsymbol{\phi}_k^* \right]_{N \times N} \boldsymbol{x}. \quad (9.54)$$

Equation (9.53) indicates that the reconstruction $\hat{\boldsymbol{x}}$ is a linear combination of the first M of the N eigenvectors that span the N-D space, and Eq. (9.54) indicates that $\hat{\boldsymbol{x}}$ is a linear transformation of $\boldsymbol{x}$ by an $N \times N$ matrix formed as the sum of the M outer products $\boldsymbol{\phi}_k \boldsymbol{\phi}_k^*$ ($k = 0, \ldots, M-1$). In particular when $M = N$, this matrix becomes $\boldsymbol{\Phi}_N \boldsymbol{\Phi}_N^* = \boldsymbol{I}_{N \times N}$ and $\hat{\boldsymbol{x}} = \boldsymbol{x}$ is a perfect reconstruction.

9.2.6 Comparison with other orthogonal transforms

To illustrate the optimality of the KLT in terms of the two desirable properties of signal decorrelation and energy compaction discussed above, we compare its performance with a set of orthogonal transforms considered in previous chapters including the identity transform IT (no transform), DCT, WHT, SLT, and DHT in the following examples, using two images of different characteristics shown in Fig. 9.4, an image of clouds on the left and another image of sand on the right.

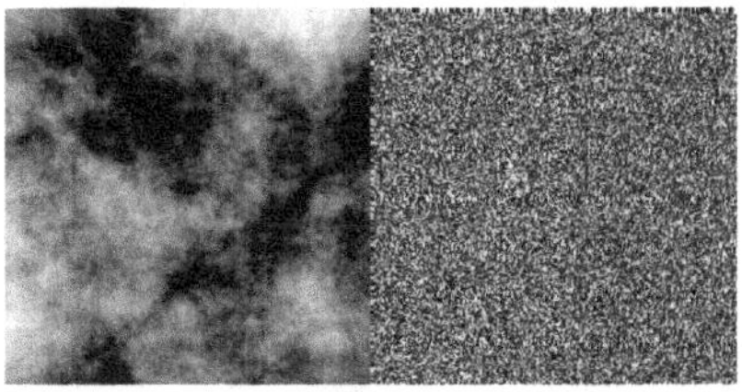

Figure 9.4 Images of clouds and sand.

We first carry out a generic orthogonal transform $\boldsymbol{X} = \boldsymbol{A}^{\mathrm{T}}\boldsymbol{x}$, where $\boldsymbol{A}^{\mathrm{T}} = \boldsymbol{A}^{-1}$ is an orthogonal matrix representing each one of the orthogonal transforms, and $\boldsymbol{x}$ is each one of the row vectors of the image, treated as an instantiation of a random vector $\boldsymbol{x}$. We then compare the covariance matrix $\boldsymbol{\Sigma}_x$ of the original vector $\boldsymbol{x}$ with the covariance matrix $\boldsymbol{\Sigma}_X = \boldsymbol{A}^{\mathrm{T}}\boldsymbol{\Sigma}_x\boldsymbol{A}$ (Eq. (B.34)) after the transform to see the performance of the transform method in terms of how well it decorrelates the signal and compacts its energy.

The covariance matrices of the cloud and sand images after each of the transforms are shown in image form in Figs. 9.5 and 9.6, respectively. The intensities of the image pixels representing the $N \times N$ covariance matrix elements are rescaled by a non-linear mapping $y = x^{0.3}$ for the very low values to be visible as well as the high values.

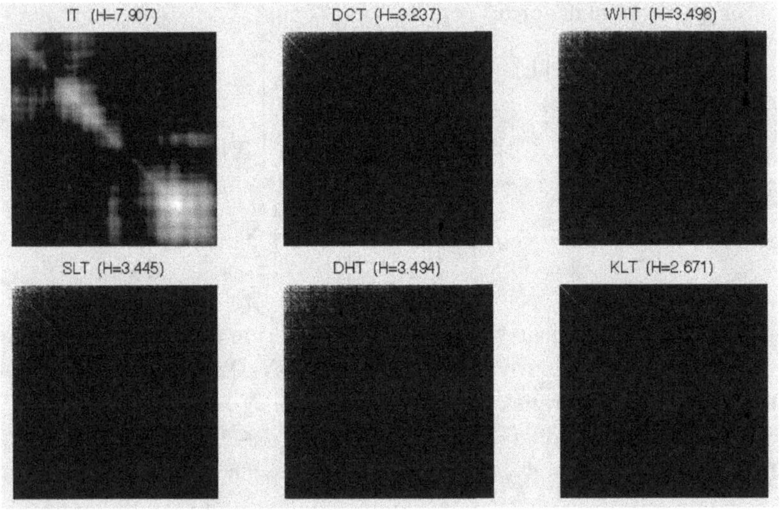

Figure 9.5 Covariance matrices of cloud image after various transforms.

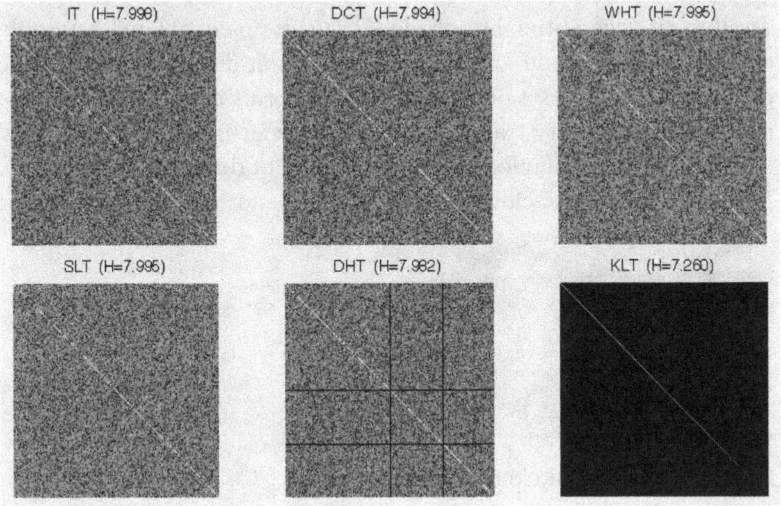

Figure 9.6 Covariance matrices of sand image after various transforms.

In the top left panel of Fig. 9.5 showing the covariance matrix of the original signal without any transform (or IT), some very bright areas exist off the diagonal, indicating that a significant number of signal components are highly correlated. We can also observe a general trend that the elements around the diagonal are brighter than those farther away from the diagonal, indicating the fact that neighboring signal components tend to be more correlated than those that are farther away from each other. In the next few panels of the figure showing the covariance matrix after each of a set of orthogonal transforms (the DCT, WHT, SLT, DHT, and KLT), the values of the off-diagonal elements are much reduced, as the signal components are significantly decorrelated after the transform. In particular, in the lower right panel showing the covariance matrix after the KLT, all off-diagonal elements become zero; i.e., the signal components are completely decorrelated.

The effect of energy compaction is also represented in the images by the brightness of the elements along the diagonal, which is reduced gradually from top left to bottom right. This effect can be more clearly seen in Fig. 9.7 showing the profile of the diagonal of the covariance matrices, the variances of the N signal components after each of the transform methods. We note that the dashed curve representing the energy distribution without any transform (or IT) is mostly flat; i.e., the signal energy is relatively evenly distributed among all signal components. The remaining curves of energy distribution after each of the transforms all show some steep descent (high on the left and low on the right), indicating that the signal energy is greatly compacted with most energy concentrated in a small number of signal components (corresponding to mainly low frequencies). In particular, the solid curve corresponding to the KLT has the steepest descent representing the optimal energy compaction. As in Fig. 9.5, here a non-linear mapping $y = x^{0.3}$ is used for low values to be visible as well as the high ones.

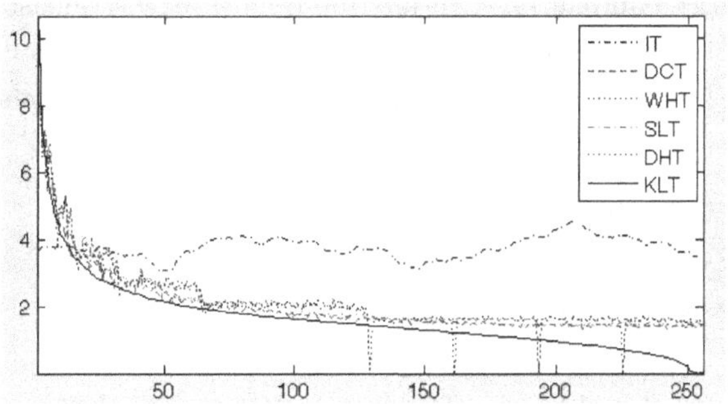

Figure 9.7 Energy distribution after various transforms (clouds).

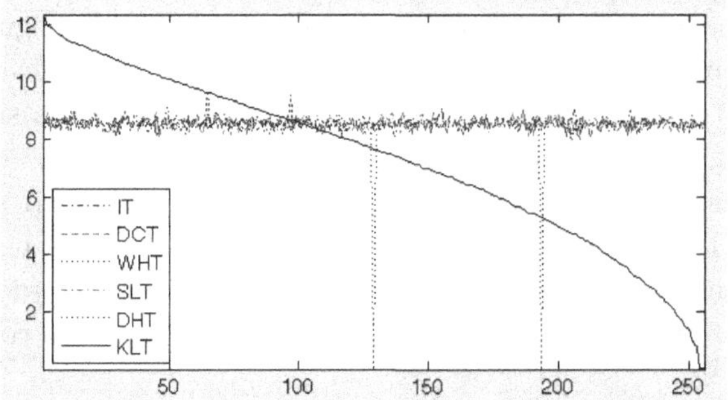

Figure 9.8 Energy distribution after various transforms (sand).

The same analysis is also carried out to the image of sand (right in Fig. 9.4), which has a drastically different texture from the image of clouds (left). This is because the color of a grain of sand is irrelevant to that of the neighboring grains; i.e., the signal components are much less correlated than those in the image of clouds in the previous case. Consequently, in the covariance matrix of the original signal (IT) shown in the top-left panel of Fig. 9.6, all off-diagonal elements look random with relatively low values, indicating that the pixels are not correlated. Also, we no longer see the trend of brighter pixels around the diagonal as observed in the covariance matrix for the image of clouds. Moreover, we see that all covariance matrices after the transforms shown in the other panels of the figure look very similar to the first one, indicating that the signal is hardly decorrelated by these transforms, except for the last covariance matrix in the lower right panel after the optimal transform of KLT, by which the signal is completely decorrelated, as indicated by the diagonalized covariance matrix.

The profiles of the diagonals of the covariance matrices for the sand signal are also plotted in Fig. 9.8, showing how the signal energy is distributed among all signal components before and after various orthogonal transforms. We see that none of the transform methods is able to further compact signal energy, except for the optimal KLT, by which the signal is maximally compacted, as shown by the solid curve.

The effect of energy compaction can also be quantitatively measured by the entropies of the energy distribution profiles in Figs. 9.7 and 9.8, as listed in the table below for different transform methods applied to both the signals of clouds and sand.

Signal	None (IT)	DCT	WHT	SLT	DHT	KLT
Clouds	7.907	3.237	3.496	3.445	3.494	2.671
Sand	7.998	7.894	7.965	7.995	7.982	7.260

We see that for the signal of clouds with significant correlation, all orthogonal transforms perform well in terms of energy compaction as the entropy is significantly reduced after each transform, and the optimal KLT achieves the minimum entropy slightly lower than others. However, for the signal of sand with low correlation, most transform methods have very limited effect of energy compaction, except the KLT with the minimum entropy which is significantly lower than those of all other transforms.

The different energy compaction effects achieved by the IT (no transform), DCT and KLT for the cloud signal are also illustrated in the table below, which lists the number (and percentage) of signal components needed in order to keep a certain percentage of the total signal energy (information) for data compression.

Transform	90%	95%	99%	100%
None (IT)	209 (82%)	230 (90%)	250 (98%)	256 (100%)
DCT	10 (4%)	22 (9%)	97 (38%)	256 (100%)
KLT	7 (3%)	13 (5%)	55 (21%)	256 (100%)

For example, if it is tolerable to lose 5% of the signal energy, then we need to keep, out of the total $N = 256$ signal components, 230 (90% of data) without any transform, 22 (9% of data) after the DCT, but only 13 (5% of data) after the KLT. In other words, using the optimal KLT, we can achieve a data compression rate of $13/256 \approx 5\%$, by keeping only 5% of the data containing 95% of the signal energy.

Two observations can be made based on the two examples above.

- All orthogonal transforms tend to decorrelate a natural signal and compact its energy, and KLT does it optimally. Typically, after an orthogonal transform, consecutive signal components in the transform domain are much less correlated, and the signal energy tends to be compacted into a small number of signal components. For example, after the DFT or DCT, two consecutive frequency components in the spectrum are not likely to be correlated, and most of the signal energy is concentrated in a small number of low-frequency components as well as the DC component, while most of the high-frequency components carry little energy. These are essentially the reasons why orthogonal transforms are widely used in data processing.
- The general claim that orthogonal transforms can significantly reduce signal correlation and compact its energy is based on the implicit assumption that the signal in question is continuous and smooth, due to the nature of the underlying physics in most applications. However, this assumption may not be valid in some unlikely cases, such as the image of sands. In fact, the effects of the orthogonal transform in terms of signal decorrelation and energy compaction depend heavily on the nature of the specific signal at hand. Very limited effect may be achieved if the signal in question is not highly correlated to start with.

Although the KLT is optimal in terms of signal decorrelation and energy compaction, some other orthogonal transforms, such as the DCT, can achieve very similar effects to that of the KLT, and are still widely used for two reasons. First, by definition the KLT transform is for random signals and it depends on the specific data being analyzed. The transform matrix $\boldsymbol{\Phi} = [\boldsymbol{\phi}_0, \dots, \boldsymbol{\phi}_{N-1}]$ is composed of the eigenvectors of the covariance matrix $\boldsymbol{\Sigma}_x$ of the signal $\boldsymbol{x}$, which can be estimated only when enough data are available. Second, the computational cost of the KLT transform is much higher than other orthogonal transforms. The computational complexity of the eigenvalue problem of the N-D covariance matrix is $O(N^3)$, while the complexity for any other orthogonal transform based on a predetermined transform matrix is no worse than $O(N^2)$. Moreover, fast algorithms with complexity $O(N \log_2 N)$ exist for most transforms such as DFT, DCT, and WHT. For these reasons, the DFT, DCT or some other transforms may be the preferred method in many applications. The KLT can be used when the covariance matrix of the data can be estimated and computational cost is not critical. Also the KLT as the optimal transform can be used to serve as a standard against which all other transform methods can be compared and evaluated.

9.2.7 Approximation of the KLT by the DCT

Here we show that the KLT can be approximated by the DCT if the signal is locally correlated and can therefore be modeled by a first-order Markov process with Toeplitz correlation matrix $\boldsymbol{R}$ (Eq. (9.13)). Specifically, we will show that when the correlation of the Markov process approaches one, its KLT transform approaches the DCT. The proof is a two-step process: (1) find the KLT matrix for the Markov process by solving the eigenvalue problem of its correlation matrix $\boldsymbol{R}$; (2) let $r \rightarrow 1$ and show the KLT matrix approaches the DCT matrix.

The KLT matrix of a first-order Markov process is the eigenvector matrix $\boldsymbol{\Phi} = [\boldsymbol{\phi}_0, \ldots, \boldsymbol{\phi}_{N-1}]$ of the Toeplitz correlation matrix $\boldsymbol{R}$:

$$\boldsymbol{R}\boldsymbol{\Phi} = \boldsymbol{\Phi}\boldsymbol{\Lambda}; \qquad \text{i.e.,} \qquad \boldsymbol{\Phi}^{\mathrm{T}}\boldsymbol{R}\boldsymbol{\Phi} = \boldsymbol{\Lambda}. \tag{9.55}$$

As $\boldsymbol{R}$ is symmetric (self-adjoint), all λ_n are real and all $\boldsymbol{\phi}_n$ are orthogonal. Also, it can be shown[1] that $\boldsymbol{\Phi}$ and $\boldsymbol{\Lambda}$ of the Toeplitz correlation matrix $\boldsymbol{R}$ take the following forms:

- The nth eigenvalue is

$$\lambda_n = \frac{1-r}{1-2r\cos\omega_n + r^2} \qquad n = 0, \ldots, N-1. \tag{9.56}$$

- The mth element ϕ_{mn} of the nth eigenvector $\boldsymbol{\phi}_n = [\ldots, \phi_{mn}, \ldots]^{\mathrm{T}}$ is

$$\phi_{mn} = \left(\frac{2}{N+\lambda_n}\right)^{1/2} \sin\left(\omega_n\left(m - \frac{N-1}{2}\right) + (n+1)\frac{\pi}{2}\right). \tag{9.57}$$

- In the above, ω_n $(n = 0, \ldots, N-1)$ are the N real roots of the equation

$$\tan(N\omega) = -\frac{(1-r^2)\sin\omega}{(1+r^2)\cos\omega - 2r}. \tag{9.58}$$

The proof for these expressions is lengthy and therefore omitted.

Next, we consider the three expressions given above when $r \rightarrow 1$. First, Eq. (9.58) simply becomes

$$\tan(N\omega) = 0. \tag{9.59}$$

Solving this for ω we get

$$\omega_n = \frac{n\pi}{N}. \tag{9.60}$$

However, when $n = 0$, $\omega_0 = 0$ and $\cos\omega_0 = 1$, and Eq. (9.59) becomes an indeterminate form 0/0. But applying L'Hopital's rule twice yields:

$$\lim_{\omega\rightarrow 0} \tan(N\omega) = \lim_{\omega\rightarrow 0} \frac{0}{2\cos\omega} = 0; \tag{9.61}$$

[1] Ray, W.D. and Driver, R.M., Further decomposition of the Karhunen-Loève series representation of a stationary process, *IEEE Transactions on Information Theory*, 16(6), 1970, 663-668.

i.e., $\omega_0 = 0$ is still a valid root for Eq. (9.58). Having found $\omega_n = n\pi/N$ for all $0 \le n \le N-1$, we can further find the eigenvalues λ_n in Eq. (9.56) when $r \to 1$. For $n > 0$, $\omega_n \neq 0$ and $\cos\omega_n \neq 1$, we have

$$\lambda_n = \lim_{r\to 1} \frac{1-r}{1-2r\cos\omega_n + r^2} = 0 \qquad 1 \le n \le N-1. \tag{9.62}$$

We also get $\lambda_0 = N$ by noting that the second equation in Eq. (9.55) is a similarity transformation of $\boldsymbol{R}$ which conserves its trace

$$\operatorname{tr}\boldsymbol{R} = N = \operatorname{tr}\boldsymbol{\Lambda} = \sum_{n=0}^{N-1} \lambda_n = \lambda_0. \tag{9.63}$$

We can now find the elements ϕ_{mn} in the eigenvector $\boldsymbol{\phi}_n$. For all $n > 0$, we have $\lambda_n = 0$ and $\omega_n = n\pi/N$, and Eq. (9.57) becomes

$$\begin{aligned}\phi_{mn} &= \sqrt{\frac{2}{N}}\sin\left(\frac{n\pi}{N}(m - \frac{N-1}{2}) + (n+1)\frac{\pi}{2}\right) = \sqrt{\frac{2}{N}}\sin\left(\frac{n\pi}{2N}(2m+1) + \frac{\pi}{2}\right)\\ &= \sqrt{\frac{2}{N}}\cos\left(\frac{n\pi}{2N}(2m+1)\right) \qquad (0 \le m \le N-1, 1 \le n \le N-1).\end{aligned} \tag{9.64}$$

When $n = 0$, $\omega_0 = 0$ and $\lambda_0 = N$, and Eq. (9.57) becomes:

$$\phi_{m0} = \sqrt{\frac{1}{N}}\sin\left(\frac{\pi}{2}\right) = \sqrt{\frac{1}{N}} \qquad 0 \le m \le N-1. \tag{9.65}$$

This happens to be precisely the DCT matrix derived in section 7.2.3, and we can therefore conclude that the KLT of a first-order Markov process approaches the DCT when $r \to 1$.

However, we note that the result above cannot be extended to the limit of $r = 1$, as when $r = 1$ all elements of $\boldsymbol{R}$ become 1, and its eigenvectors are no longer unique. In fact, the column vectors of any other orthogonal transform matrix $\boldsymbol{A}$ are the eigenvectors of this all-1 matrix $\boldsymbol{R}$:

$$\boldsymbol{A}^{\mathrm{T}}\boldsymbol{R}\boldsymbol{A} = \boldsymbol{\Lambda} = \operatorname{diag}[N, 0, \ldots, 0]; \tag{9.66}$$

i.e.,

$$\boldsymbol{a}_m^{\mathrm{T}}\boldsymbol{R}\boldsymbol{a}_n = \begin{cases} N & m = n = 0 \\ 0 & \text{else} \end{cases} \tag{9.67}$$

To see this, we note that the first column $\boldsymbol{a}_0$ of any orthogonal transform matrix $\boldsymbol{A} = [\boldsymbol{a}_0, \ldots, \boldsymbol{a}_{N-1}]$ (DFT, WHT, as well as DCT, except DST) is always composed of N constants $1/\sqrt{N}$ (representing the DC component), and as all other columns $\boldsymbol{a}_n$ $(n > 0)$ are orthogonal to $\boldsymbol{a}_0$, they all sum up to zero:

$$\langle \boldsymbol{a}_n, \boldsymbol{a}_0 \rangle = \boldsymbol{a}_n^{\mathrm{T}}\boldsymbol{a}_0 = \frac{1}{\sqrt{N}}\sum_{m=0}^{N-1} a[m, n] = 0. \tag{9.68}$$

As a result, all elements of matrix $\boldsymbol{A}^{\mathrm{T}}\boldsymbol{R}\boldsymbol{A}$ in Eq. (9.66) are zero:

$$\boldsymbol{a}_m^{\mathrm{T}}\boldsymbol{R}\boldsymbol{a}_n = \boldsymbol{a}_m^{\mathrm{T}} \begin{bmatrix} 1 & \cdots & 1 \\ \vdots & \ddots & \vdots \\ 1 & \cdots & 1 \end{bmatrix} \boldsymbol{a}_n = \boldsymbol{a}_m^{\mathrm{T}} \begin{bmatrix} 0 \\ \vdots \\ 0 \end{bmatrix} = 0 \qquad m \neq 0 \text{ or } n \neq 0, \tag{9.69}$$

except when $m = n = 0$, when the top-left element is

$$\boldsymbol{a}_0^{\mathrm{T}}\boldsymbol{R}\boldsymbol{a}_0 = \frac{1}{N}[1, \ldots, 1] \begin{bmatrix} 1 & \ldots & 1 \\ \vdots & \ddots & \vdots \\ 1 & \ldots & 1 \end{bmatrix} \begin{bmatrix} 1 \\ \vdots \\ 1 \end{bmatrix} = N. \tag{9.70}$$

The approximation of the KLT of a first-order Markov process by the DCT can also be seen from another point of view. It can be shown that the N DCT basis vectors, the column vectors of the DCT matrix $\boldsymbol{C} = [\boldsymbol{c}_0, \ldots, \boldsymbol{c}_{N-1}]$, are the eigenvectors of the tridiagonal matrix of the following form (independent of the parameter α):

$$\boldsymbol{Q} = \begin{bmatrix} 1-\alpha & -\alpha & 0 & \cdots & 0 \\ -\alpha & 1 & \ddots & \ddots & \vdots \\ 0 & \ddots & \ddots & \ddots & 0 \\ \vdots & \ddots & \ddots & 1 & -\alpha \\ 0 & \cdots & 0 & -\alpha & 1-\alpha \end{bmatrix}; \tag{9.71}$$

i.e., $\boldsymbol{C}^{\mathrm{T}}\boldsymbol{Q}\boldsymbol{C} = \boldsymbol{M}$, where $\boldsymbol{M} = \mathrm{diag}(\mu_0, \ldots, \mu_{N-1})$ is a diagonal matrix composed of N eigenvalues of $\boldsymbol{Q}$.

On the other hand, it can also be shown that the inverse of the correlation matrix $\boldsymbol{R}$ of a first-order Markov process given in Eq. (9.13) takes the form

$$\boldsymbol{R}^{-1} = \frac{1}{\beta} \begin{bmatrix} 1-r\alpha & -\alpha & 0 & \cdots & 0 \\ -\alpha & 1 & \ddots & \ddots & \vdots \\ 0 & \ddots & \ddots & \ddots & 0 \\ \vdots & \ddots & \ddots & 1 & -\alpha \\ 0 & \cdots & 0 & -\alpha & 1-r\alpha \end{bmatrix}, \tag{9.72}$$

where

$$\alpha = \frac{r}{1+r^2}, \qquad \beta = \frac{1-r^2}{1+r^2}. \tag{9.73}$$

Based on Eq. (9.55), we have $\boldsymbol{\Phi}^{\mathrm{T}}\boldsymbol{R}^{-1}\boldsymbol{\Phi} = \boldsymbol{\Lambda}^{-1}$. It is therefore clear that $\lim_{r\to 1}\boldsymbol{\Phi} = \boldsymbol{C}$; i.e., when $r \to 1$, the N KLT basis vectors, the column vectors of the eigenvector matrix $\boldsymbol{\Phi} = [\boldsymbol{\phi}_0, \ldots, \boldsymbol{\phi}_{N-1}]$ of $\boldsymbol{R}$ (same as that of $\boldsymbol{R}^{-1}$), can be approximated by the DCT basis vectors. Note, again, that the approximation breaks down when $r = 1$ and $\beta = 0$.

As an example, Fig. 9.9 shows the first eight of the $N = 128$ basis vectors of the KLT of a Markov process with $r = 0.9$, in comparison with the corresponding DCT basis vectors. Note that the KLT basis vectors match those of the DCT very closely and the similarity will increase when $r \to 1$. Also note that, as an eigenvector of $\boldsymbol{R}$, a KLT vector ϕ_n, can be either positive or negative; i.e., the corresponding transform coefficients of the KLT and DCT may have opposite polarity. However, this does not affect the transform, as the reconstructed signal will still be the same. Also shown in the left and right panels of Fig. 9.10 are the 3-D plots of the covariance matrices after the KLT and DCT of a Markov process. We see that the two transforms are very similar in terms of the energy compaction and signal decorrelation. The performances of the DCT are almost as good as the optimal KLT.

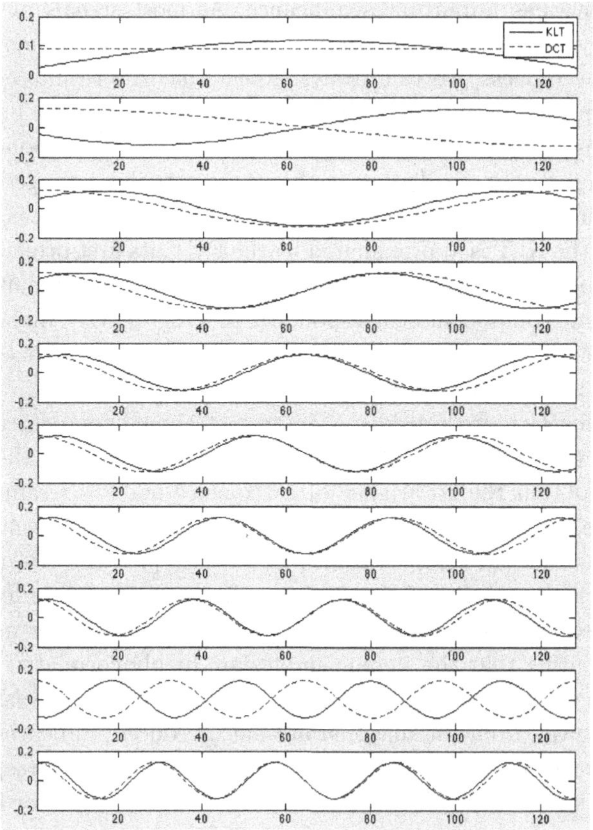

Figure 9.9 Comparison of the first eight basis vectors of the DCT and KLT of a first-order Markov process.

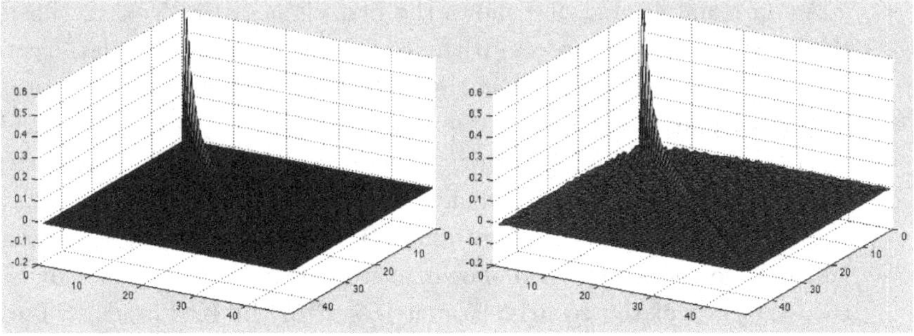

Figure 9.10 The 3-D plots of the covariance matrices after the KLT (left) and DCT (right).

The result above has important significance. As most signals of interest in practice are likely to be locally correlated and can therefore be modeled by a first-order Markov process, we can always expect that the results of the DCT are close to the optimal transform of the KLT. Furthermore, as the basis vectors of the KLT are the eigenvectors of the signal covariance $\mathbf{\Sigma}_x$ corresponding to the eigenvalues arranged in descending order, they are actually arranged according the energy contained in the signal components (represented by the eigenvalues). Consequently, as the KLT is approximated by the DCT, its first principal component corresponding to the DC component contains the largest amount of energy, and the subsequent components corresponding to progressively higher frequencies in the DCT contain progressively lower energy. This approximation is valid in general for all locally correlated signals.

To illustrate this fact, we consider a dataset of annual temperatures in the Los Angeles area collected over the period of 1878–1997, shown in the top panel of Fig. 9.11. To obtain the covariance of a sequence of $N = 8$ samples of the data, we truncate the signal into a set of segments of N samples each, and treat these segments as random samples from a stochastic process. We next obtain the $N \times N$ covariance matrix of the data, as shown in the lower left panel of the figure. We see that the elements around the diagonal of the matrix have high values, indicating that the signal samples are highly correlated when they are close to each other (taken within a short duration), but the values of the elements farther away from the diagonal are much reduced, indicating that the signal samples are much less correlated when they are far apart (separated by a long period of time). This behavior can be modeled by a first-order Markov chain of N points whose covariance is shown in the lower right panel of the same figure (correlation between two consecutive samples assumed to be $r = 0.5$), which looks similar to the covariance of the actual signal, in the sense that the correlation is gradually reduced between signal samples when they are farther apart.

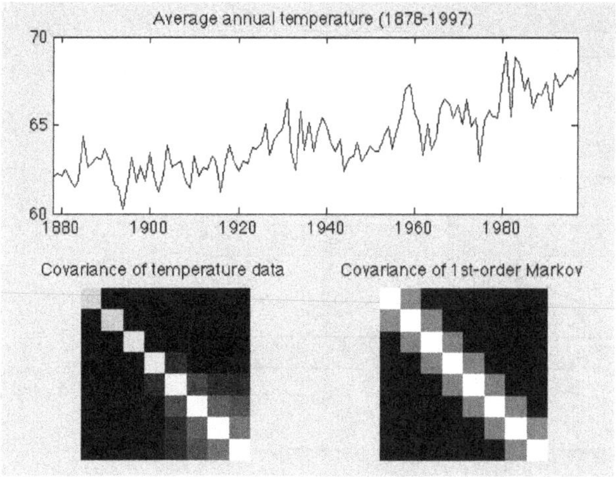

Figure 9.11 Covariances of natural signal and first-order Markov chain.

We can further compare the KLT of the signal and its approximation by the DCT. The KLT matrix is composed of the N eigenvectors of the signal covariance, shown in the panels on the left of Fig. 9.12, which are compared with the eigenvectors of the covariance of the Markov model (solid curves) shown in the panels on the right of the figure. These two sets of curves look similar in terms of the general wave forms and their frequencies (not necessarily in the same order). Moreover, the row vectors of the DCT transform matrix are also plotted (dashed curves) together with the eigenvectors based on the Markov model. Comparing all three sets of curves, we see that they match very closely (except for different polarity in some cases), indicating the fact that indeed a correlated signal can be modeled by a first-order Markov process, and its KLT can be approximated by the DCT.

We can make the following observations based on this example.

- The temperature-time function, as one of the weather parameters representing a natural process, confirms the general assumption that the correlation between signal samples tends to decay as they are farther apart.
- The signal correlation can be indeed closely modeled by a first-order Markov chain model with a correlation r and the only parameter.
- The eigenvectors of the covariance matrix above can be closely matched by the row vectors of the DCT matrix.
- The KLT transform of typical natural signals can be approximately carried out as a DCT.
- In particular, the first eigenvector $\boldsymbol{\phi}_0$, corresponding to the largest eigenvalue, is approximated by the first row of the DCT matrix composed of all constants, representing the first principal component $y_0 = \langle \boldsymbol{x}, \boldsymbol{\phi}_0 \rangle = \boldsymbol{\phi}_0^* \boldsymbol{x}$, which is the average (DC component) of all elements in signal $\boldsymbol{x}$.

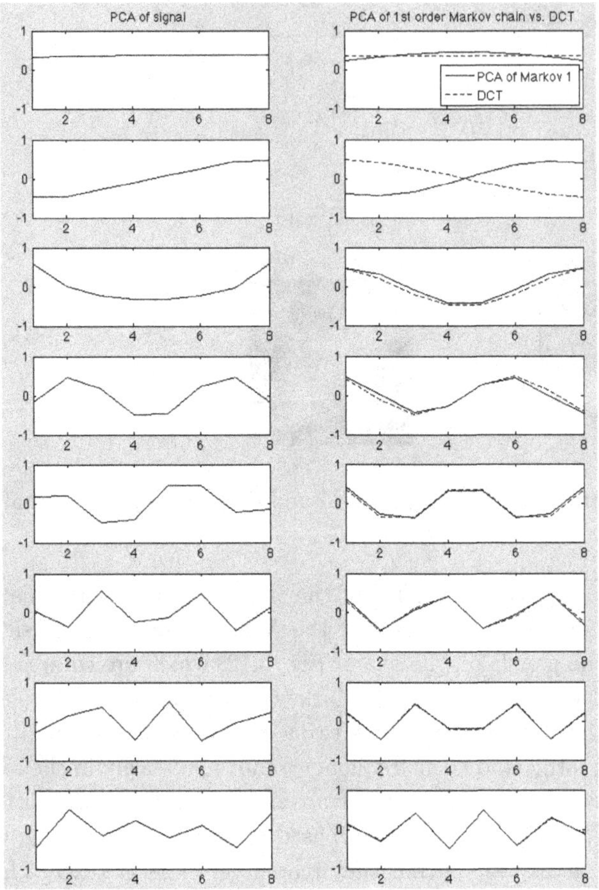

Figure 9.12 The eigenvectors of the covariance matrix of the signal (left), in comparison with the eigenvectors of the covariance matrix of its Markov model and the row vectors of the DCT (right).

9.3 Applications of the KLT

9.3.1 Image processing and analysis

The KLT can be carried out on a set of N images for various purposes such as feature extraction and data compression. There are two alternative ways to carry out the KLT on the N images, depending on how a random signal is defined. First, an N-D vector can be formed by N pixels each taken at the same position (e.g., ith row and jth column) from one of the N images. The number of such vectors is obviously the total number of pixels in each image, assumed to be K, and they form a K by N matrix $\boldsymbol{D}$, whose covariance matrix can be estimated as (Eq. (9.12))

$$\hat{\boldsymbol{\Sigma}}_x = \frac{1}{K-1}[\boldsymbol{D}^{\mathrm{T}}\boldsymbol{D}]_{N\times N}. \tag{9.74}$$

Alternatively, a K-D vector can be formed by concatenating the rows (or columns) of each of the N images, and each of these vectors from the N images can be treated as a sample of a K-D random vector, represented by a column of $\boldsymbol{D}$ defined above, or a row of $\boldsymbol{D}^{\mathrm{T}}$, and the covariance matrix can be estimated as

$$\hat{\boldsymbol{\Sigma}}'_x = \frac{1}{N-1}[\boldsymbol{D}\boldsymbol{D}^{\mathrm{T}}]_{K\times K}. \tag{9.75}$$

We can show that the eigenvalue problems of these two different covariance matrices are equivalent. First, assume the eigenequations for $\boldsymbol{D}^{\mathrm{T}}\boldsymbol{D}$ and $\boldsymbol{D}^{\mathrm{T}}\boldsymbol{D}$ are

$$\boldsymbol{D}^{\mathrm{T}}\boldsymbol{D}\boldsymbol{\phi} = \lambda\boldsymbol{\phi} \qquad \boldsymbol{D}\boldsymbol{D}^{\mathrm{T}}\boldsymbol{\psi} = \mu\boldsymbol{\psi}. \tag{9.76}$$

Pre-multiplying $\boldsymbol{D}^{\mathrm{T}}$ on both sides of the second equation we get

$$\boldsymbol{D}^{\mathrm{T}}\boldsymbol{D}[\boldsymbol{D}^{\mathrm{T}}\boldsymbol{\psi}] = \mu[\boldsymbol{D}^{\mathrm{T}}\boldsymbol{\psi}]. \tag{9.77}$$

This is actually the first eigenequation with the same eigenvalue $\mu = \lambda$ and eigenvector $\boldsymbol{D}^{\mathrm{T}}\boldsymbol{\psi}$, which is the same as $\boldsymbol{\phi}$, when both are normalized. The two covariance matrices $\boldsymbol{\Sigma}_x$ and $\boldsymbol{\Sigma}'_x$ have the same rank $R = \min(N, K)$ (if D is not degenerate) and therefore the same number of non-zero eigenvalues. Consequently, the KLT can be carried out based on either matrix with the same effects in terms of the signal decorrelation and energy compaction. As the number of pixels in the image is typically much greater than the number of images, $K > N$, we will take the first approach above to treat the same pixels from all N images as a sample of an N-D random signal vector and carry out the KLT based on the $N \times N$ covariance matrix $\hat{\boldsymbol{\Sigma}}_x$. We can now carry out the KLT to each of the K N-D vectors $\boldsymbol{x}$ for a pixel of the N images to obtain another N-D vector $\boldsymbol{X} = \boldsymbol{\Phi}^*\boldsymbol{x}$ for the same pixel of a set of N *eigen-images*, as shown in Fig. 9.13. Owing to the nature of the KLT, most of the energy/information contained in the N images, representing the variations among all N images, is now concentrated in the first few eigen-images, while the remaining eigen-images can be omitted without losing much energy/information. This is the foundation for various KLT-based image compression and feature extraction algorithms. The subsequent operations such as image recognition and classification can all be carried out in a much lower dimensional space.

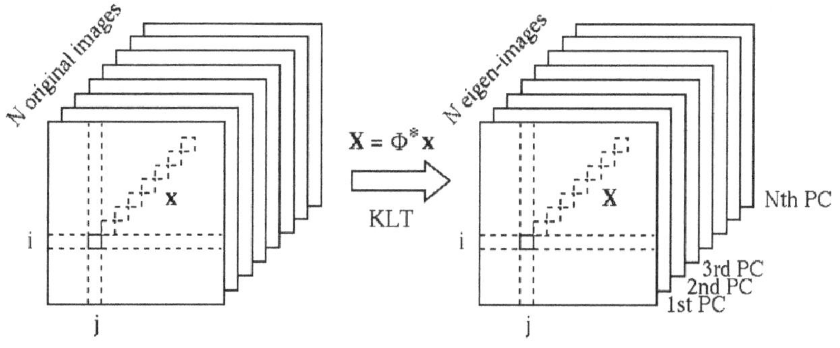

Figure 9.13 KLT of a set of images.

Example 9.1: In *remote sensing* , images of the surface of the Earth (or other planets) are taken by orbiting satellites, for various studies in fields such as geology, geography, and agriculture. The camera system on the satellite has a set of N sensors each sensitive to a different wavelength band in the visible and infrared range of the electromagnetic spectrum. Depending on the number of sensors N, the image data collected are either multi-spectral ($N < 10$) or hyperspectral (N is up to 200 or more). For instance, the $N = 210$ bands of the HYDICE (Hyperspectral Digital Imagery Collection Experiment) data cover the wavelength range from 400 to 2500 nm with 10 nm separation between two neighboring bands. In this example, we choose 20 bands separated by 100 nm from a set of HYDICE image data (Lincoln Memorial, Washington, DC), [2] as shown in Fig. 9.14 (top). We see that the images corresponding to neighboring wavelength bands are often similar to each other; i.e., they are highly correlated with much redundancy. (Obviously in the complete HYDICE data, the 210 bands separated by 10 nm are even more highly correlated.) When the KLT is carried out on these $N = 20$ dimensional vectors (each for a pixel in the images), the resulting $N = 20$ PCA images are obtained as shown in Fig. 9.14 (bottom). Two observations can be made. First, after the KLT, the images are completely decorrelated. The PCA images all look different, each carrying its independent information. Second, the signal energy is highly compacted into the first few PCA images, as also seen in Fig. 9.15 and Table 9.1 for the energy distributions before and after the KLT. The data can be compressed by keeping only the first three PCA components (15% of data) containing 98.5% of the total energy/information.

Table 9.1. Energy distribution before and after KLT

Component	0	1	2	3	4	5	6	7	8	9
Before KLT	3.7	3.7	3.7	3.9	5.0	5.4	6.8	6.9	7.5	8.3
After KLT	70.6	23.3	4.6	0.6	0.5	0.1	0.1	0.1	0.1	.04
Cont'd	10	11	12	13	14	15	16	17	18	19
	7.5	5.7	6.4	4.9	2.9	3.8	3.7	3.3	2.3	4.7
	0.02	0.02	0.02	0.01	0.01	0.01	0.0	0.0	0.0	

[2] Credit to the School of Electrical and Computer Engineering, ITaP and LARS, Purdue University.

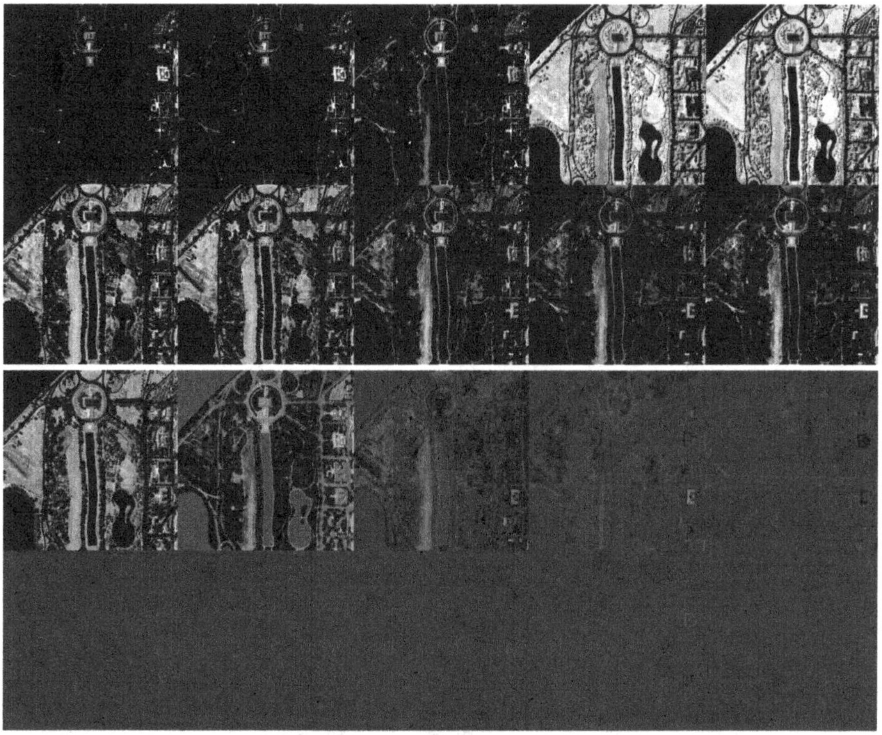

Figure 9.14 Ten out of the 210 spectral bands of the HYDICE image data (Lincoln Memorial, Washington DC). Top: before the KLT; bottom: after the KLT.

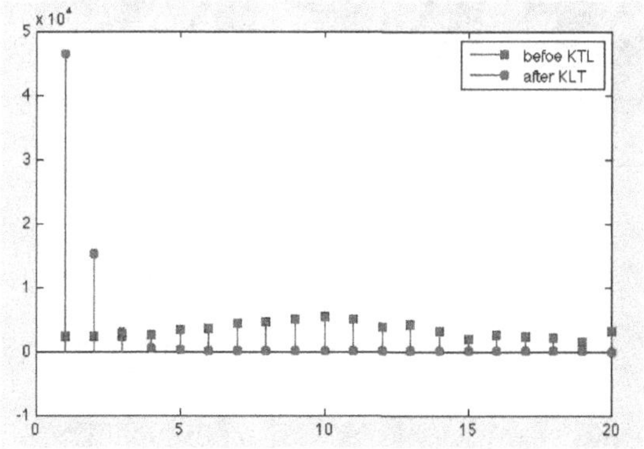

Figure 9.15 Signal energy distributions among 20 signal components before and after the KLT.

Example 9.2: A sequence of $N = 8$ frames of a video of a moving escalator and their eigen-images are shown respectively in the upper and lower parts of Fig. 9.16. The covariance matrix and the energy distribution among the eight components plot both before and after the KLT are shown in Fig. 9.17. We see that due to the local correlation, the covariance matrix before the KLT (left) does indeed resemble the correlation matrix $\boldsymbol{R}$ of a first-order Markov model (bottom right in Fig. 9.12), and the covariance matrix after the KLT (middle) is completely decorrelated and its energy highly compacted, as also clearly shown in the comparison of the energy distribution before and after the KLT (right). Also as shown in Eq. (9.18), the KLT basis, the set of all $N = 8$ eigenvectors of the signal covariance matrix, is very much similar to the DCT basis, indicating that the DCT with a fast algorithm would produce almost the same results as the KLT. Moreover, it is interesting to observe that the first eigen-image (left panel of the third row of Fig. 9.16 represents mostly the static scene of the image frames corresponding to the main variations in the image (carrying most of the energy), while the subsequent eigen-images represent mostly the motion in the video, the variation between the frames. For example, the motion of the people riding on the escalator is mostly reflected by the first few eigen-images following the first one, while the motion of the escalator stairs is mostly reflected in the subsequent eigen-images.

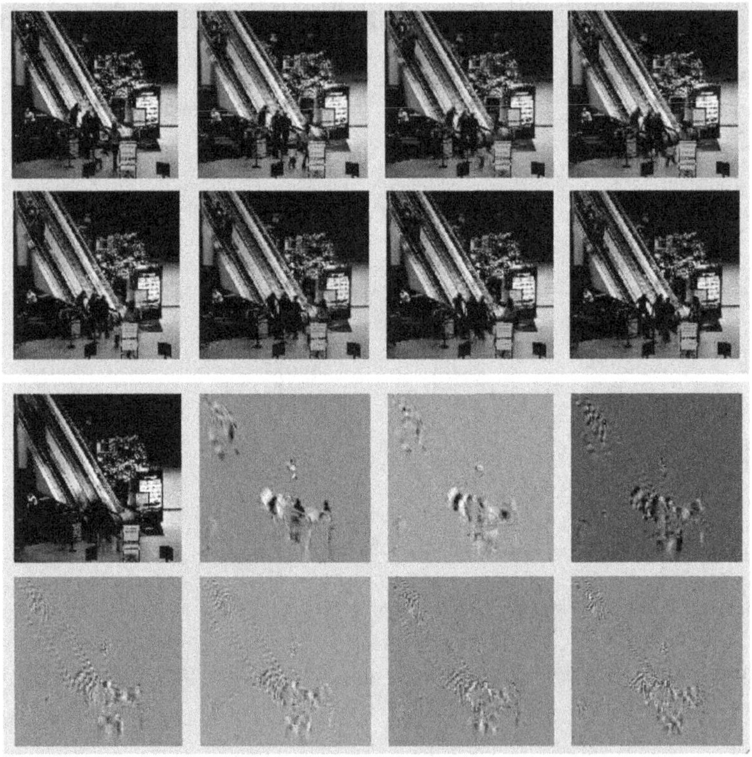

Figure 9.16 Video frames (top) and the eigen-images (bottom).

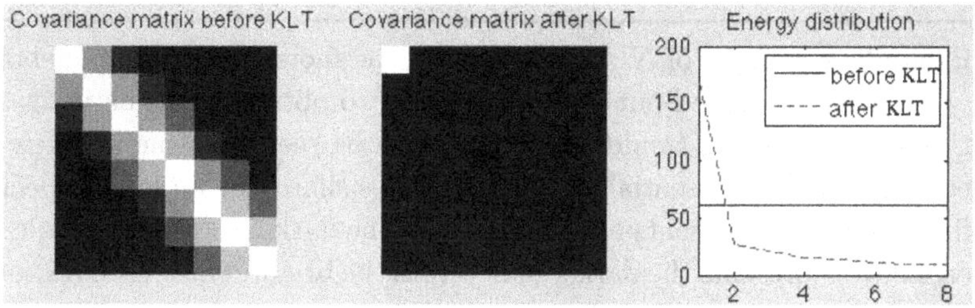

Figure 9.17 Covariance matrix before and after the KLT. The covariance matrices before and after the KLT are shown in image form (left and middle), while the energy distributions among the N components before and after the transform are also plotted (right).

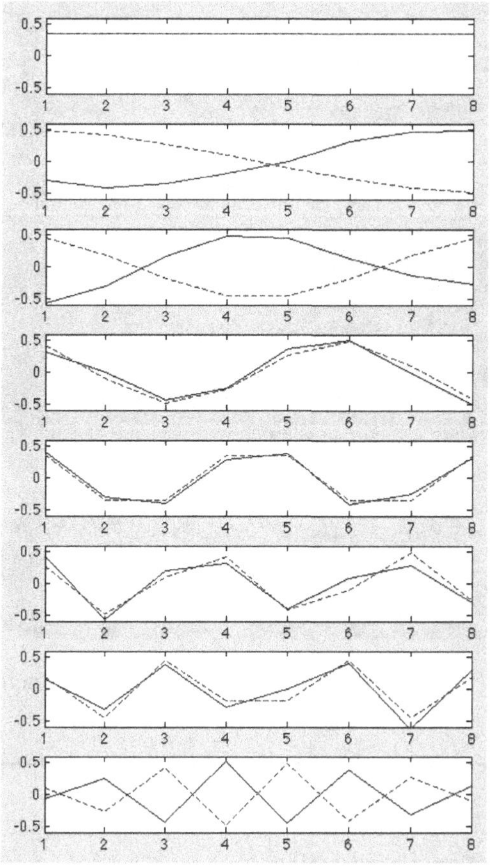

Figure 9.18 KLT basis vectors compared with the DCT basis. The basis vectors of the KLT of the video frames closely resemble the DCT basis vectors (may have opposite polarities).

Example 9.3: A set of $N = 20$ face images is shown in the top panel of Fig. 9.19. [3] The KLT is carried out on these images to obtain the eigen-images, called in this case *eigenfaces* (middle panel). It can be seen that the first few eigenfaces capture the most essential common features shared by all faces. Specifically, the first eigenface represents a generic face in the dark background, while the second eigenface represents the darker hair versus the brighter face. The rest of the eigenfaces represent some other features with progressively less significance. Table 9.2 shows the percentage of energy contained in the eigenfaces. The faces are then reconstructed based on 95% of the total information, as shown in the bottom panel in Fig. 9.19. The method of eigenfaces can be used in facial recognition and classification.

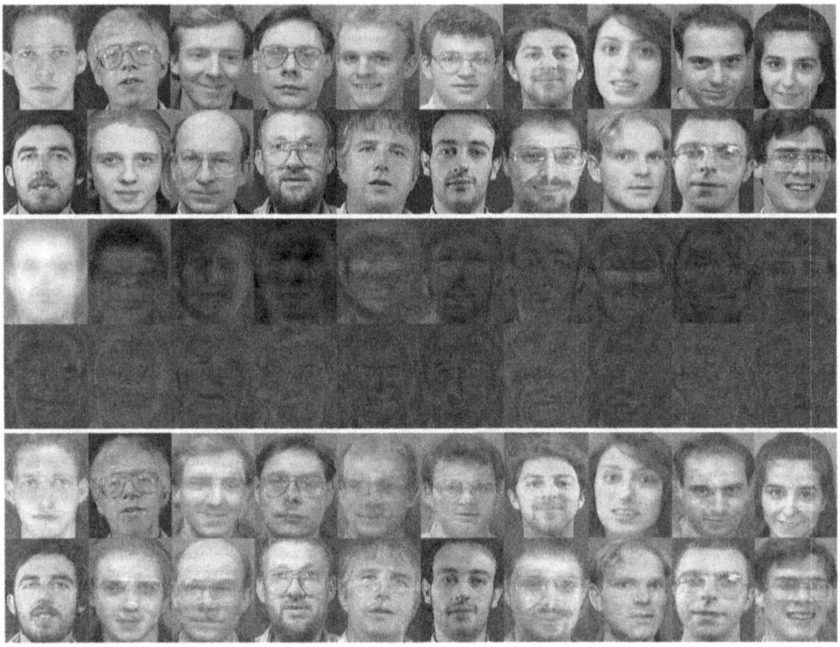

Figure 9.19 Original faces (top), eigenfaces (middle), and reconstructed faces (bottom).

9.3.2 Feature extraction for pattern classification

In the field of machine learning, pattern classification/recognition is a general method that classifies a set of objects of interest into different *categories* or *classes* and recognizes any given object as a member of one of these classes.

[3] Credit to AT&T Laboratories, Cambridge.

Table 9.2. Energy contained in the eigenfaces

# of components	1	2	3	4	5	6	7	8
% energy contained	48.5	11.6	6.1	4.6	3.8	3.7	2.6	2.5
Accumulative	48.5	60.1	66.2	70.8	74.6	78.3	81.0	83.5
Cont'd	9	10	11	12	13	14	15	16
	1.9	1.9	1.8	1.6	1.5	1.4	1.3	1.2
	85.4	87.3	89.0	90.7	92.2	93.6	94.9	96.1
Cont'd	17	18	19	20				
	1.1	1.1	0.9	0.8				
	97.2	98.2	99.2	100.0				

Specifically, each object is represented as an N-D vector, known as a *pattern*, based on a set of N*features* that can be observed and measured to characterize the object. Then a pattern classification algorithm can be carried out in this N-D *feature space* to classify all pattern vectors in it. The classification is therefore essentially the partitioning of the feature space into a set of regions each corresponding to one particular class. A given pattern is classified to the class corresponding to the region in which it resides. There are, in general, two types of pattern classification algorithm, depending on whether certain *a priori* knowledge or information regarding the classes is available. An algorithm is *supervised* if it is based on the assumed availability of a set of patterns with known classes, called *training samples.* When such training samples cannot be obtained, an *unsupervised* algorithm has to be used. If the number of features N is large, especially if the N features are not all pertinent to the representation of the classes of interest, a process called *feature extraction* is needed to find a set of $M < N$ features to form a much lower dimensional feature space in which the classification can be more effectively and efficiently carried out. These M features can be either directly chosen from the N original features, or they can be generated based on N original features.

For example, in the hyperspectral remote sensing image data, at each pixel position a set of N values corresponding to the same number of bands of wavelengths form a pattern vector in the N-D feature space, representing the spectral signature of the surface material covered by the pixel. All such pattern vectors in the feature space can then be classified according to the different types of the surface material of interest, such as vegetation (e.g., different types of crops and forests), water bodies (e.g., oceans, lakes, rivers), soil and rock (e.g., different types of mineral), snow and ice, desert, man-made objects (e.g., pavement, roads, and buildings). For example, the spectral signatures of four different ground cover types (water, grass, tree, and building roof) of the HYDICE image data used in Example 9.1 are shown in Fig. 9.20.

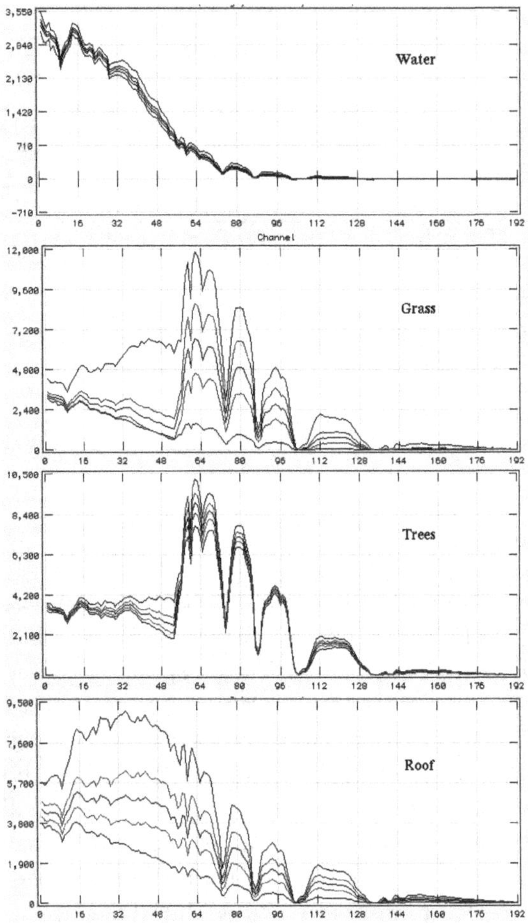

Figure 9.20 Spectral signatures of four ground cover types. The five curves are for the maximum, mean plus standard deviation, mean, mean minus standard deviation and minimum, respectively.

For another example of image recognition, some objects given in image form, such as the 26 letters of the English alphabet or the 10 digits from 0 to 9, may need to be recognized. Extracting from the image a set of relevant features representative of the patterns may be difficult, as it requires specific knowledge regarding the objects of interest. A more straightforward way of representing such image objects is to simply use all N pixels in the image (e.g., $N = 256$ for 16×16 images), arranged as an N-D vector pattern obtained by concatenating its rows or columns of the image.

A challenge in both examples above is that the number of features N is large ($N = 210$ or $N = 256$), and not all of them are necessarily pertinent to the classification of the specific classes of interest. In such cases, we need to carry out the feature extraction as a pre-processing stage to find a set of $M < N$ features most relevant to the subsequent classification. Owing to the property of optimal energy compaction stated in Theorem 9.1, the KLT $\boldsymbol{X} = \boldsymbol{\Phi}^* \boldsymbol{x}$ can be applied to generate such a set of M new features as the linear combination of the N original features. However, it may no longer be proper for the KLT matrix $\boldsymbol{\Phi}$ to be based on the covariance matrix $\boldsymbol{\Sigma}_x$ in Eq. (9.17), which represents the variations among all pattern vectors in the data. Instead, the KLT matrix $\boldsymbol{\Phi}$ here needs to be based on some different matrix that reflects more specifically the differences between the classes to be distinguished.

Let $\{\boldsymbol{x}_i^{(k)}, (i = 1, \dots, n_k)\}$ be a set of n_k N-D vectors for the training samples known to belong to class k, where $k = 1, \dots, K$ for each of the K classes. Based on these training samples we can define the following *scatter matrices*.

- *Scatter matrix* (same as the covariance matrix) of the kth class for the variation or scatteredness within the class:

$$\boldsymbol{S}_k = \frac{1}{n_k} \sum_{i=1}^{n_k} (\boldsymbol{x}_i^{(k)} - \boldsymbol{m}_k)(\boldsymbol{x}_i^{(k)} - \boldsymbol{m}_k)^{\mathrm{T}} \qquad k = 1, \dots, K, \tag{9.78}$$

  where $\boldsymbol{m}_k$ is the mean vector of the kth class:

$$\boldsymbol{m}_k = \frac{1}{n_k} \sum_{i=1}^{n_k} \boldsymbol{x}_i^{(k)} \qquad k = 1, \dots, K. \tag{9.79}$$

- *Within-class scatter matrix* for the within-class scatteredness of all K classes:

$$\boldsymbol{S}_w = \sum_{k=1}^{K} p_k \boldsymbol{S}_k = \frac{1}{n} \sum_{k=1}^{K} n_k \boldsymbol{S}_k, \tag{9.80}$$

  where $n = \sum_{k=1}^{K} n_k$ is the total number of training samples of all K classes and $p_k = n_k/n$.

- *Between-class scatter matrix* for the separability, or the variation between all K classes:

$$\boldsymbol{S}_b = \sum_{k=1}^{K} p_k (\boldsymbol{m}_k - \boldsymbol{m})(\boldsymbol{m}_k - \boldsymbol{m})^{\mathrm{T}}, \tag{9.81}$$

  where $\boldsymbol{m}$ is the mean vector of all n training samples of all K classes:

$$\boldsymbol{m} = \frac{1}{n} \sum_{\boldsymbol{x}} \boldsymbol{x} = \frac{1}{n} \sum_{k=1}^{k} n_k \frac{1}{n_k} \sum_{i=1}^{n_k} \boldsymbol{x}_i^{(k)} = \sum_{k=1}^{k} p_k \boldsymbol{m}_k. \tag{9.82}$$

- *Total scatter matrix* (same as the covariance matrix) of all n samples of the K classes for the total variation or scatteredness among them:

$$\begin{aligned}
\boldsymbol{S}_t &= \frac{1}{n} \sum_{\boldsymbol{x}} (\boldsymbol{x} - \boldsymbol{m})(\boldsymbol{x} - \boldsymbol{m})^{\mathrm{T}} \\
&= \frac{1}{n} \sum_{k=1}^{K} \sum_{i=1}^{n_k} (\boldsymbol{x}_i^{(k)} - \boldsymbol{m}_k + \boldsymbol{m}_k - \boldsymbol{m})(\boldsymbol{x}^{(k)} - \boldsymbol{m}_k + \boldsymbol{m}_k - \boldsymbol{m})^{\mathrm{T}} \\
&= \frac{1}{n} \sum_{k=1}^{K} \sum_{i=1}^{n_k} (\boldsymbol{x}_i^{(k)} - \boldsymbol{m}_k)(\boldsymbol{x}_i^{(k)} - \boldsymbol{m}_k)^{\mathrm{T}} + \frac{1}{n} \sum_{k=1}^{K} \sum_{i=1}^{n_k} (\boldsymbol{m}_k - \boldsymbol{m})(\boldsymbol{m}_k - \boldsymbol{m})^{\mathrm{T}} \\
&= \boldsymbol{S}_w + \boldsymbol{S}_b.
\end{aligned} \tag{9.83}$$

The second to last equal sign is due to the fact that

$$\sum_{k=1}^{K}\sum_{i=1}^{n_k}(\boldsymbol{x}_i^{(k)} - \boldsymbol{m}_k)(\boldsymbol{m}_k - \boldsymbol{m})^{\mathrm{T}} = 0. \tag{9.84}$$

The relation $\boldsymbol{S}_t = \boldsymbol{S}_w + \boldsymbol{S}_b$ in Eq. (9.83) indicates the fact that the total scatteredness $\boldsymbol{S}_t$ of the n samples is due to the contributions of the total within-class scatteredness $\boldsymbol{S}_w$ and the total between-class scatteredness $\boldsymbol{S}_b$, as one would intuitively expect.

Now we can carry out the KLT based on the between-class scatter matrix $\boldsymbol{S}_b$, so that after the transform most of the information specifically representing the separability of the K classes (for different surface materials in remote sensing or letters/digits in character recognition) will be compacted into a small number of $M < N$ components. The classification/recognition can then be carried out in the resulting M-D feature space containing most of the information relevant to the classification (separability) with much reduced computational complexity, by a particular classification algorithm. As a simple example, we could classify a given pattern $\boldsymbol{x}$ to the class with minimum distance $D(\boldsymbol{x}, \boldsymbol{m}_k)$ between its mean and the pattern $\boldsymbol{x}$:

$$\boldsymbol{x} \text{ belongs to class } k \text{ iff } \quad D(\boldsymbol{x}, \boldsymbol{m}_k) \leq D(\boldsymbol{x}, \boldsymbol{m}_l) \quad \text{for all } l = 1, \ldots, K. \tag{9.85}$$

Another application of the KLT is data visualization. It may be desirable to be able to intuitively assess the data by visualizing how the data points are distributed in the N-D feature space. However, visualization is obviously impossible when $N > 3$. In such cases, the KLT transform based on the overall covariance matrix of the data can be used to project the data points from the original N-D space to a 2- or 3-D space in which most of the information characterizing the distribution of the data points in the feature space is conserved for visualization.

Example 9.4: Consider the classification of the 10 digits from 0 to 9, each written multiple times by students in a class, in the form of a 16×16 image, as shown in the top-left panel of Fig. 9.21. Each pattern can be simply represented by the $N = 256 = 16 \times 16$ pixels in the image, which can be converted to N-D vectors obtained by concatenating the rows of its image. Based on $\boldsymbol{S}_b$ representing the separability of the 10 classes, the KLT can be carried out. The energy distribution plots both before and after the KLT are shown in the two right panels in Fig. 9.21. Different from the KLT based on the covariance matrix of the data as discussed previously, the KLT here is based on the between-class scatter matrix $\boldsymbol{S}_b$, and consequently the energy in question represents specifically the separability information most pertinent to the classification of the 10 digits. From the distribution plots, we see that, before the KLT, the energy is relatively evenly distributed throughout most of the 256 pixels, with high local correlation in the same row (each corresponding to one of the 16 peaks in the plot), but after the

KLT, the energy is highly compacted into the first nine principal components, while the remaining $256 - 9 = 247$ components contain little energy and therefore can be omitted. The classification can then be carried out in the $M = 9$ dimensional feature space with much reduced computational cost. Also, in order to visualize the information contained in the 9-D space used in the classification, we can carry out the inverse KLT to reconstruct the images based on these nine components (Eq. (9.52)), as shown in the bottom-left panel of the figure. We see that these images contain most of the information pertinent to the classification, in that the within-class variation is minimized while the between-class variation is maximized.

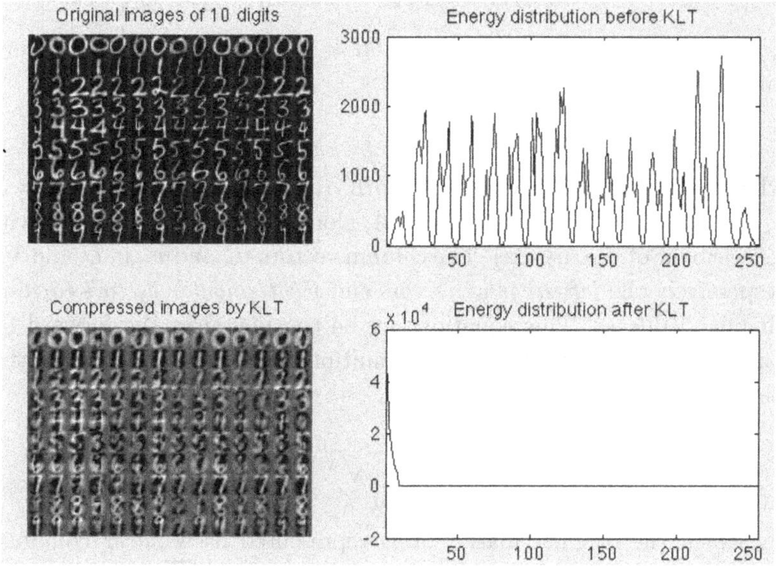

Figure 9.21 The KLT of image pattern classification based on the between-class scatter matrix

9.4 Singular value decomposition transform

9.4.1 Singular value decomposition

The *singular value decomposition (SVD)* of an $M \times N$ matrix $\boldsymbol{A}$ of rank $R \leq \min(M, N)$ is based on the following eigenvalue problems of an $M \times M$ matrix $\boldsymbol{AA}^*$ and an $N \times N$ matrix $\boldsymbol{A}^*\boldsymbol{A}$:

$$\boldsymbol{AA}^*\boldsymbol{u}_n = \lambda_n \boldsymbol{u}_n, \qquad \boldsymbol{A}^*\boldsymbol{A}\boldsymbol{v}_n = \lambda_n \boldsymbol{v}_n, \qquad n = 1, \ldots, R. \tag{9.86}$$

As the rank of $\boldsymbol{A}$ is R, there exist only R non-zero eigenvalues. Also, as both $\boldsymbol{AA}^*$ and $\boldsymbol{A}^*\boldsymbol{A}$ are self-adjoint (symmetric if real), their eigenvalues λ_n are real and their eigenvectors $\boldsymbol{u}_n$ and $\boldsymbol{v}_n$ are orthogonal

$$\boldsymbol{u}_m^* \boldsymbol{u}_n = \boldsymbol{v}_m^* \boldsymbol{v}_n = \delta[m-n], \tag{9.87}$$

and they form two unitary (orthogonal if real) matrices $\boldsymbol{U} = [\boldsymbol{u}_1, \dots, \boldsymbol{u}_N]_{M\times M}$ and $\boldsymbol{V} = [\boldsymbol{v}_1, \dots, \boldsymbol{v}_N]_{N\times N}$ that satisfy

$$\boldsymbol{U}\boldsymbol{U}^* = \boldsymbol{U}^*\boldsymbol{U} = \boldsymbol{I}_{M\times M} \qquad \text{and} \qquad \boldsymbol{V}\boldsymbol{V}^* = \boldsymbol{V}^*\boldsymbol{V} = \boldsymbol{I}_{N\times N}. \tag{9.88}$$

Both $\boldsymbol{A}\boldsymbol{A}^*$ and $\boldsymbol{A}^*\boldsymbol{A}$ can therefore be diagonalized by $\boldsymbol{U}$ and $\boldsymbol{V}$ respectively:

$$\begin{aligned} \boldsymbol{U}^*(\boldsymbol{A}\boldsymbol{A}^*)\boldsymbol{U} &= \boldsymbol{\Lambda}_{M\times M} = \text{diag}[\lambda_1, \dots, \lambda_R, 0, \dots, 0], \\ \boldsymbol{V}^*(\boldsymbol{A}^*\boldsymbol{A})\boldsymbol{V} &= \boldsymbol{\Lambda}_{N\times N} = \text{diag}[\lambda_1, \dots, \lambda_R, 0, \dots, 0]. \end{aligned} \tag{9.89}$$

The *SVD theorem* states that the $M \times N$ matrix $\boldsymbol{A}$ can be diagonalized by $\boldsymbol{U}$ and $\boldsymbol{V}$:

$$\boldsymbol{U}^*\boldsymbol{A}V = \boldsymbol{\Lambda}^{1/2} = \text{diag}[\sqrt{\lambda_1}, \dots, \sqrt{\lambda_R}] = \text{diag}[s_1, \dots, s_R], \tag{9.90}$$

where $\boldsymbol{\Lambda}$ is an $M \times N$ matrix with R non-zero elements $s_n = \sqrt{\lambda_n}$ ($n = 1, \dots, R$), called *singular values* of $\boldsymbol{A}$, along the diagonal (starting with the top-left element of the matrix). The column vectors $\boldsymbol{u}_k$ and $\boldsymbol{v}_k$ in $\boldsymbol{U}$ and $\boldsymbol{V}$ are called respectively the *left-singular vectors* and *right-singular vectors* corresponding to singular value s_k. This equation can be considered as the forward SVD transform. Pre-multiplying $\boldsymbol{U}$ and post-multiplying $\boldsymbol{V}^*$ on both sides of the equation above, we get the inverse transform:

$$\boldsymbol{A} = \boldsymbol{U}\Lambda^{1/2}\boldsymbol{V}^* = \sum_{k=1}^{R} \sqrt{\lambda_i}[\boldsymbol{u}_k \boldsymbol{v}_k^*] = \sum_{k=1}^{R} s_k [\boldsymbol{u}_k \boldsymbol{v}_k^*], \tag{9.91}$$

by which the original matrix $\boldsymbol{A}$ is represented as a linear combination of R matrices $[\boldsymbol{u}_k \boldsymbol{v}_k^*]$ weighted by the singular values $\sqrt{\lambda_k}$ ($k = 1, \dots, R$). We can rewrite both the forward and inverse SVD transform as a pair of forward and inverse transforms:

$$\begin{cases} \boldsymbol{\Lambda}^{1/2} = \boldsymbol{U}^*\boldsymbol{A}\boldsymbol{V} \\ \boldsymbol{A} = \boldsymbol{U}\boldsymbol{\Lambda}^{1/2}\boldsymbol{V}^*. \end{cases} \tag{9.92}$$

Given the SVD of an $M \times N$ matrix $\boldsymbol{A} = \boldsymbol{U}\boldsymbol{\Lambda}^{1/2}\boldsymbol{V}^*$, its pseudo-inverse can be found to be

$$\boldsymbol{A}^- = \boldsymbol{V}\boldsymbol{\Lambda}^{-1/2}\boldsymbol{U}^*, \tag{9.93}$$

where both $\boldsymbol{A}^-$ and $\boldsymbol{\Lambda}^{-1/2}$ are $N \times M$ matrices, and $\boldsymbol{\Lambda}^{-1/2}$ is the pseudo-inverse of $\boldsymbol{\Lambda}$ composed of the reciprocals $1/s_k = 1/\sqrt{\lambda_k}$ of the R singular values along the diagonal.

Example 9.5: Consider the SVD of the following 3×2 matrix ($M = 3 > N = 2$).

$$\begin{aligned} \boldsymbol{A} &= \begin{bmatrix} 2 & 3 \\ 4 & 1 \\ 3 & 2 \end{bmatrix} = \boldsymbol{U}\boldsymbol{\Lambda}^{1/2}\boldsymbol{V}^T \\ &= \begin{bmatrix} 0.53 & 0.75 & 0.41 \\ 0.63 & -0.67 & 0.41 \\ 0.58 & 0.04 & -0.82 \end{bmatrix} \begin{bmatrix} \sqrt{39.17} & 0.00 \\ 0.00 & \sqrt{3.83} \\ 0.00 & 0.00 \end{bmatrix} \begin{bmatrix} 0.84 & -0.54 \\ 0.54 & 0.84 \end{bmatrix}^{\mathrm{T}}. \end{aligned} \tag{9.94}$$

The pseudo-inverse of $\boldsymbol{A}$ is

$$\begin{aligned} \boldsymbol{A}^- &= \boldsymbol{V}\boldsymbol{\Lambda}^{-1/2}\boldsymbol{U}^{\mathrm{T}} \\ &= \begin{bmatrix} 0.84 & -0.54 \\ 0.54 & 0.84 \end{bmatrix} \begin{bmatrix} 1/\sqrt{39.17} & 0.00 & 0.00 \\ 0.00 & 1/\sqrt{3.83} & 0.00 \end{bmatrix} \begin{bmatrix} 0.53 & 0.75 & 0.41 \\ 0.63 & -0.67 & 0.41 \\ 0.58 & 0.04 & -0.82 \end{bmatrix}^{\mathrm{T}} \\ &= \frac{1}{30} \begin{bmatrix} -4 & 8 & 2 \\ 11 & -7 & 2 \end{bmatrix}. \end{aligned} \tag{9.95}$$

This result can be verified by $\boldsymbol{A}^-\boldsymbol{A} = \boldsymbol{I}_{2\times 2}$.

Next, we let

$$\boldsymbol{B} = \boldsymbol{A}^{\mathrm{T}} = \begin{bmatrix} 2 & 4 & 3 \\ 3 & 1 & 2 \end{bmatrix}. \tag{9.96}$$

Taking the transpose on both sides of Eq. (9.94) we get the SVD of a 2×3 matrix ($M = 2 < N = 3$)

$$\begin{aligned} \boldsymbol{B} = \boldsymbol{A}^{\mathrm{T}} &= \begin{bmatrix} 2 & 4 & 3 \\ 3 & 1 & 2 \end{bmatrix} = \boldsymbol{V}[\boldsymbol{\Lambda}^{1/2}]^{\mathrm{T}}\boldsymbol{U}^{\mathrm{T}} \\ &= \begin{bmatrix} 0.84 & -0.54 \\ 0.54 & 0.84 \end{bmatrix} \begin{bmatrix} \sqrt{39.17} & 0.00 & 0.00 \\ 0.00 & \sqrt{3.83} & 0.00 \end{bmatrix} \begin{bmatrix} 0.53 & 0.75 & 0.41 \\ 0.63 & -0.67 & 0.41 \\ 0.58 & 0.04 & -0.82 \end{bmatrix}^{\mathrm{T}}. \end{aligned} \tag{9.97}$$

The pseudo-inverse of $\boldsymbol{B}$ is

$$\begin{aligned} \boldsymbol{B}^- &= \boldsymbol{U}\boldsymbol{\Lambda}^{-1/2}\boldsymbol{V}^{\mathrm{T}} \\ &= \begin{bmatrix} 0.53 & 0.75 & 0.41 \\ 0.63 & -0.67 & 0.41 \\ 0.58 & 0.04 & -0.82 \end{bmatrix} \begin{bmatrix} 1/\sqrt{39.17} & 0.00 \\ 0.00 & 1/\sqrt{3.83} \\ 0.00 & 0.00 \end{bmatrix} \begin{bmatrix} 0.84 & -0.54 \\ 0.54 & 0.84 \end{bmatrix}^{\mathrm{T}} \\ &= \frac{1}{30} \begin{bmatrix} -4 & 11 \\ 8 & -7 \\ 2 & 2 \end{bmatrix}. \end{aligned} \tag{9.98}$$

This result can be verified by $\boldsymbol{B}\boldsymbol{B}^- = \boldsymbol{I}_{2\times 2}$.

The $M \times N$ matrix $\boldsymbol{A}$ can be considered as any linear transformation that converts a vector $\boldsymbol{x} \in \mathbb{C}^N$ to another vector $\boldsymbol{y} = \boldsymbol{A}\boldsymbol{x} \in \mathbb{C}^M$. Then the SVD $\boldsymbol{A} = \boldsymbol{U}\boldsymbol{\Lambda}^{1/2}\boldsymbol{V}^*$ indicates that this linear transformation can be carried out as

$$\boldsymbol{y} = \boldsymbol{A}\boldsymbol{x} = \boldsymbol{U}\boldsymbol{\Lambda}^{1/2}\boldsymbol{V}^*\boldsymbol{x} \tag{9.99}$$

in the following three steps.

1. Rotate vector $\boldsymbol{x}$ by the unitary matrix $\boldsymbol{V}^*$:
$$\boldsymbol{y}_1 = \boldsymbol{V}^*\boldsymbol{x}. \tag{9.100}$$
2. Scale $\boldsymbol{y}_1$ by a factor of $s_k = \sqrt{\lambda_k}$ $(k = 1, \dots, R)$:
$$\boldsymbol{y}_2 = \boldsymbol{\Lambda}^{1/2}\boldsymbol{y}_1 = \boldsymbol{\Lambda}^{1/2}\boldsymbol{V}^*\boldsymbol{x}. \tag{9.101}$$
3. Rotate vector $\boldsymbol{y}_2$ by the unitary matrix $\boldsymbol{U}$:
$$\boldsymbol{y} = \boldsymbol{U}\boldsymbol{y}_2 = \boldsymbol{U}\boldsymbol{\Lambda}^{1/2}\boldsymbol{V}^*\boldsymbol{x} = \boldsymbol{A}\boldsymbol{x}. \tag{9.102}$$

Example 9.6: A linear transformation $\boldsymbol{y} = \boldsymbol{A}\boldsymbol{x}$ with

$$\boldsymbol{A} = \boldsymbol{U}\boldsymbol{\Lambda}^{1/2}\boldsymbol{V}^{\mathrm{T}} = \begin{bmatrix} 0.87 & 0.50 \\ -0.50 & 0.87 \end{bmatrix} \begin{bmatrix} 3 & 0 \\ 0 & 2 \end{bmatrix} \begin{bmatrix} 0.71 & -0.71 \\ 0.71 & 0.71 \end{bmatrix}^{\mathrm{T}} = \begin{bmatrix} 2.54 & -1.13 \\ 0.16 & 2.29 \end{bmatrix} \tag{9.103}$$

can be implemented in three steps: rotation counter-clockwise by 45°, scaling in the horizontal and vertical directions by factors 3 and 2 respectively, and rotation clockwise by 30°. Applying this linear transformation to a triangle with vertices

$$\boldsymbol{x}_1 = \begin{bmatrix} 0 \\ 0 \end{bmatrix} \quad \boldsymbol{x}_2 = \begin{bmatrix} 0 \\ 1 \end{bmatrix} \quad \boldsymbol{x}_3 = \begin{bmatrix} 1 \\ 0 \end{bmatrix}, \tag{9.104}$$

we get the three new vertices

$$\boldsymbol{y}_1 = \begin{bmatrix} 0 \\ 0 \end{bmatrix} \quad \boldsymbol{y}_2 = \begin{bmatrix} -1.13 \\ 2.29 \end{bmatrix} \quad \boldsymbol{y}_3 = \begin{bmatrix} 2.54 \\ 0.16 \end{bmatrix}. \tag{9.105}$$

This linear transformation and its three component steps are shown in Fig. 9.22.

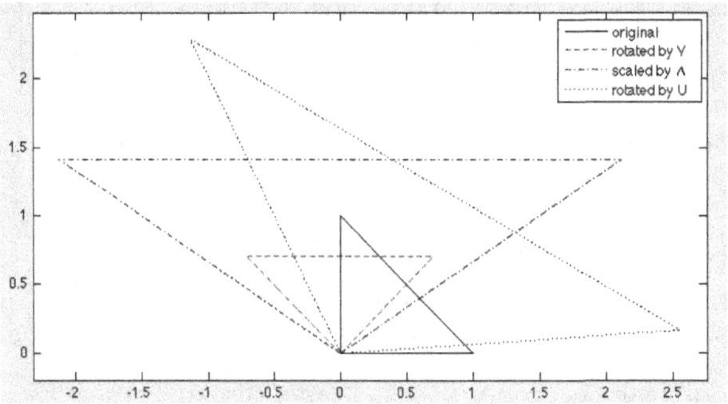

Figure 9.22 A linear transformation of a triangle can be decomposed into three steps: counter-clockwise rotation by 45°, scaling in horizontal and vertical directions by factors 3 and 2 respectively, and clockwise rotation by 30°.

The matrix $\boldsymbol{A}$ can be considered as a 2-D signal, such as an image. By the forward SVD transform, this image can be represented by the R singular values $s_k = \sqrt{\lambda_k}$ as the coefficients for the R components or eigen-images $[\boldsymbol{u}_k \boldsymbol{v}_k^*]$ $(k = 1, \ldots, R)$. On the other hand, by the inverse SVD transform, the signal $\boldsymbol{A}$ can be reconstructed as a linear combination of these eigen-images weighted by the singular values, as shown in Eq. (9.91).

As with all orthogonal transforms discussed previously, the SVD transform also conserves the signal energy. The total energy contained in $\boldsymbol{A}$ is simply the sum of the energy contained in each of its $M \times N$ elements, which is equal to the trace of either $\boldsymbol{A}\boldsymbol{A}^*$ and $\boldsymbol{A}^*\boldsymbol{A}$:

$$\mathcal{E} = \sum_{m=1}^{M} \sum_{n=1}^{N} |a_{mn}|^2 = \text{tr}(\boldsymbol{A}\boldsymbol{A}^*) = \text{tr}(\boldsymbol{A}^*\boldsymbol{A}). \tag{9.106}$$

Moreover, as the trace is conserved by an orthogonal transform, we take the trace on both sides of Eq. (9.89) to get

$$\begin{aligned}
\text{tr}[\boldsymbol{U}^*(\boldsymbol{A}\boldsymbol{A}^*)\boldsymbol{U}] &= \text{tr}(\boldsymbol{A}\boldsymbol{A}^*) = \text{tr}\,\boldsymbol{\Lambda} = \sum_{k=1}^{R} \lambda_k, \\
\text{tr}[\boldsymbol{V}^*(\boldsymbol{A}^*\boldsymbol{A})\boldsymbol{V}] &= \text{tr}(\boldsymbol{A}^*\boldsymbol{A}) = \text{tr}\,\boldsymbol{\Lambda} = \sum_{k=1}^{R} \lambda_k.
\end{aligned} \tag{9.107}$$

This result indicates that the energy contained in the signal $\boldsymbol{A}$ is the same as the sum of all singular values squared representing the signal energy in the transform domain after the SVD transform.

We can further show that the *degrees of freedom (DOFs)*, the number of independent variables in the representation of the signal, are also conserved by the SVD transform, indicating that the signal information is conserved. For simplicity, we assume $M = N = R$ and the DOFs of $\boldsymbol{A}_{N \times N}$ are N^2. After the transform, the signal is represented in terms of $\boldsymbol{U}$, $\boldsymbol{V}$, and $\boldsymbol{\Lambda}$. Obviously the DOFs of $\boldsymbol{\Lambda}$ are N. We can also show that the DOFs of both $\boldsymbol{U}$ and $\boldsymbol{V}$ are $(N^2 - N)/2$. The DOFs of the first column with N elements are $N - 1$ due to the constraint of normalization, and the DOFs of the second column are $N - 2$ due to the constraints of being orthogonal to the first one as well as being normalized. In general, the DOFs of a column are always one fewer than that of the previous one, and the total DOFs of all N vectors of $\boldsymbol{U}$ are

$$(N-1) + (N-2) + \cdots + 1 = N(N-1)/2 = (N^2 - N)/2. \tag{9.108}$$

The same is true for $\boldsymbol{V}$. Together with the DOFs of N for Λ, the total DOF in the transform domain are $2\,(N^2 - N)/2 + N = N^2$, same as that of $\boldsymbol{A}$ before the SVD transform.

9.4.2 Application in image compression

The SVD transform has various applications, including image processing and analysis. We now consider how it can be used for data compression. For simplicity we consider an $N \times N$ real image matrix $\boldsymbol{A}$. Image compression can be achieved by using only the first $M < N$ eigen-images of $\boldsymbol{A}$ in Eq. (9.91):

$$\boldsymbol{A}_M = \sum_{k=1}^{M} \sqrt{\lambda_k} \boldsymbol{u}_k \boldsymbol{v}_k^{\mathrm{T}}. \tag{9.109}$$

The energy contained in $\boldsymbol{A}_k$ is

$$\begin{aligned}
\mathrm{tr}[\boldsymbol{A}_M^{\mathrm{T}} \boldsymbol{A}_M] &= \mathrm{tr}\left[\sum_{k=1}^{M} \sqrt{\lambda_k} \boldsymbol{v}_k \boldsymbol{u}_k^{\mathrm{T}}\right] \left[\sum_{l=1}^{M} \sqrt{\lambda_l} \boldsymbol{u}_l \boldsymbol{v}_l^{\mathrm{T}}\right] \\
&= \mathrm{tr}\left[\sum_{k=1}^{M} \left(\sum_{l=1}^{M} \sqrt{\lambda_k} \sqrt{\lambda_l} \boldsymbol{v}_k \boldsymbol{u}_k^{\mathrm{T}} \boldsymbol{u}_l \boldsymbol{v}_l^{\mathrm{T}}\right)\right] = \mathrm{tr}\left[\sum_{k=1}^{M} \lambda_k \boldsymbol{v}_k \boldsymbol{v}_k^{\mathrm{T}}\right] \\
&= \sum_{k=1}^{M} \lambda_k \, \mathrm{tr}\,[\boldsymbol{v}_k \boldsymbol{v}_k^{\mathrm{T}}] = \sum_{k=1}^{M} \lambda_k \boldsymbol{v}_k^{\mathrm{T}} \boldsymbol{v}_k = \sum_{k=1}^{M} \lambda_k.
\end{aligned}$$

The fraction of energy contained in the compressed image $\boldsymbol{A}_M$ is $\sum_{k=1}^{M} \lambda_k / \sum_{k=1}^{N} \lambda_k$. Obviously, if we use the M components corresponding to the M largest eigenvalues, the energy contained in $\boldsymbol{A}_M$ is maximized.

Next we consider the compression rate in terms of the DOFs of $\boldsymbol{A}_M$. The DOFs in the M orthogonal vectors $\{\boldsymbol{u}_1, \ldots, \boldsymbol{u}_M\}$ are

$$(N-1) + (N-2) + \cdots + (N-M) = NM - M(M+1)/2. \tag{9.110}$$

The same is true for $\{\boldsymbol{v}_1, \ldots, \boldsymbol{v}_M\}$. Including the DOFs of M in $\{\lambda_1, \ldots, \lambda_M\}$, we get the total DOFs:

$$2NM - M(M+1) + M = 2NM - M^2, \tag{9.111}$$

and the compression ratio is

$$\frac{2NM - M^2}{N^2} = \frac{2M}{N} - \frac{M^2}{N^2} \approx \frac{2M}{N}. \tag{9.112}$$

We consider a specific example of the image of Lenna of size $N = 128$ as shown in Fig. 9.23 (left) together with its SVD matrices $\boldsymbol{U}$ and $\boldsymbol{V}$ (middle and right). The singular values $s_i = \sqrt{\lambda_i}$ in descending order and the energy λ_i contained are also plotted respectively in the top and bottom panels of Fig. 9.24. The reconstructed images using different M of the SVD eigen-images are shown in Fig. 9.25. The top two rows show the SVD eigen-images corresponding to the 10 largest singular values (first row), and the corresponding reconstructions as the partial sums (second row). The bottom two rows show the rest of the eigen-images and the corresponding reconstructions, with M increased by 10 ($M = 10, 20, 30, \ldots, 100$).

We see that the reconstructed images approximate the original image progressively more closely as M is increased to include more eigen-images in the partial sum. This effect can be quantitatively explained by the energy distribution over the total 128 SVD components, shown in the lower panel of Fig. 9.24. The distribution curve is obtained by simply squaring the singular-value curve in the top panel so that it represents the energy contained in each of the eigen-images. As most of the signal energy is contained in the first few SVD components, all eigen-images for $M > 20$ in the third row contain little information, correspondingly, the reconstructed images in the fourth row closely approximate the original image, which is perfectly reconstructed only if all $M = N = 128$ eigen-images are used.

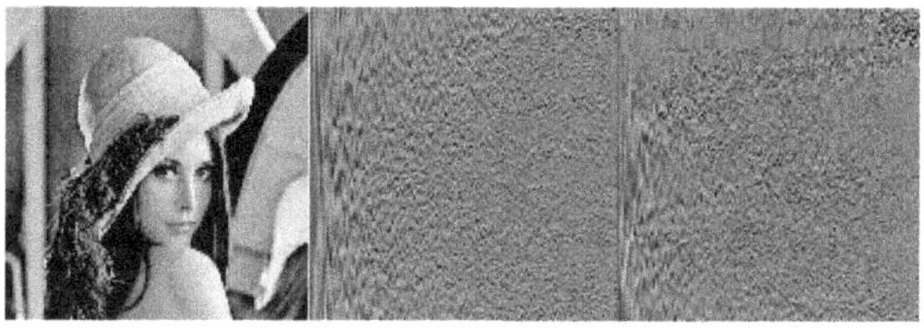

Figure 9.23 Original image (left), matrices $\boldsymbol{U}$ (middle) and $\boldsymbol{V}$ (right).

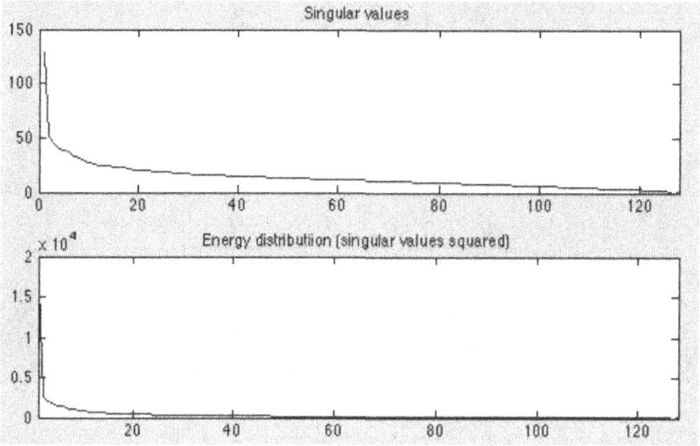

Figure 9.24 Singular values $s_i = \sqrt{\lambda_i}$ (top) and their energy distribution λ_i (bottom).

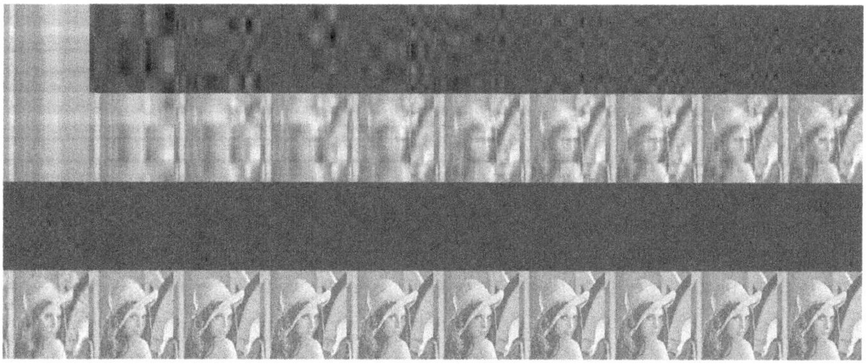

Figure 9.25 Top two rows: the SVD components and the corresponding partial reconstructions with $M = 1, \ldots, 10$. Bottom two rows: same as before but with $M = 10, 20, \ldots, 100$.

9.5 Homework problems

1. An experiment concerning two random variables x and y is carried out $K = 3$ times with different outcomes as listed in the tables given below. Calculate their correlation r_{xy} based on the estimated means and covariances:

$$\hat{\mu}_x = \frac{1}{K}\sum_{k=1}^{K} x^{(k)}, \qquad \hat{\mu}_y = \frac{1}{K}\sum_{k=1}^{K} y^{(k)}. \tag{9.113}$$

$$\hat{\sigma}_{xy}^2 = \frac{1}{K-1}\sum_{k=1}^{K} x^{(k)} y^{(k)} - \hat{\mu}_x \hat{\mu}_y, \qquad \hat{r}_{xy} = \frac{\hat{\sigma}_{xy}^2}{\sqrt{\hat{\sigma}_x^2\ \hat{\sigma}_y^2}}. \tag{9.114}$$

(a)

k	1st	2nd	3rd
$x^{(k)}$	1	2	3
$y^{(k)}$	1	2	3

(9.115)

(b)

k	1st	2nd	3rd
$x^{(k)}$	2	4	6
$y^{(k)}$	3	6	9

(9.116)

(c)

k	1st	2nd	3rd
$x^{(k)}$	1	2	3
$y^{(k)}$	3	2	1

(9.117)

(d)

k	1st	2nd	3rd
$x^{(k)}$	1	2	3
$y^{(k)}$	2	2	2

(9.118)

(e)

k	1st	2nd	3rd
$x^{(k)}$	2	2	2
$y^{(k)}$	1	2	3

(9.119)

(f)

k	1st	2nd	3rd	4th	5th
$x^{(k)}$	1	2	2	2	3
$y^{(k)}$	2	1	2	3	2

(9.120)

2. In the 2-D normal distribution in Eq. (9.47), let $a = c = 5$ and $b = 8$.

(a) Find the two eigenvalues λ_0 and λ_1 and their corresponding eigenvectors $\boldsymbol{\phi}_0$ and $\boldsymbol{\phi}_1$.
(b) Find the KLT matrix $\boldsymbol{\Phi} = [\boldsymbol{\phi}_0 \boldsymbol{\phi}_1]$. What kind of rotation does it represent? Carry out the KLT rotation $\boldsymbol{y} = \boldsymbol{\Phi}^{\mathrm{T}}\boldsymbol{x}$ so that $\boldsymbol{y} = [y[0], y[1]]^{\mathrm{T}}$ can be expressed in terms of $\boldsymbol{x} = [x[0], x[1]]^{\mathrm{T}}$. Find $\boldsymbol{\Sigma}_y$.
(c) Give the quadratic equation associated with a 2-D normal distribution of $\boldsymbol{y}$ after the KLT. Confirm this is an equation of an ellipse and find the major and minor semi-axes.

3. Consider a set of $K = 9$ data points in an $N = 2$-D space:

$$\boldsymbol{x} = \begin{bmatrix} 1 & 2 & 3 & 4 & 5 & 2 & 3 & 3 & 4 \\ 1 & 2 & 3 & 4 & 5 & 3 & 2 & 4 & 3 \end{bmatrix}. \tag{9.121}$$

Do the following by hand or using Matlab (or any other computer tools).
(a) Plot the data points on a 2-D plane to visualize the data.
(b) Find the mean vector and covariance matrix of these $K = 9$ data points.
(c) Find $\boldsymbol{\Sigma}_x$'s eigenvalues λ_i and corresponding normalized eigenvectors $\boldsymbol{\phi}_i$ $(i = 0, 1)$, form an orthogonal KLT matrix $\boldsymbol{\Phi} = [\boldsymbol{\phi}_0\ \boldsymbol{\phi}_1]$ by the two eigenvectors.
(d) Carry out KLT of the original data $\boldsymbol{X} = \boldsymbol{\Phi}^{\mathrm{T}}\boldsymbol{x}$.
(e) Find the mean vector and covariance matrix of $\boldsymbol{X}$ in the KLT transform domain.
(f) Verify that the total signal energy (trace of the covariance matrix) is conserved. If one of the two dimensions of $\boldsymbol{X}$ corresponding to the smaller eigenvalue is dropped, what is the percentage of energy remaining?
(g) Re-plot the $K = 9$ data points $\boldsymbol{X}$ in the KLT domain spanned by $\boldsymbol{\phi}_0$ and $\boldsymbol{\phi}_1$.

4. Repeat the problem above with the same data set augmented with four additional points

$$\boldsymbol{x} = \begin{bmatrix} 1 & 2 & 3 & 4 & 5 & 2 & 3 & 3 & 4 & 2 & 4 & 1 & 5 \\ 1 & 2 & 3 & 4 & 5 & 3 & 2 & 4 & 3 & 4 & 2 & 5 & 1 \end{bmatrix}. \tag{9.122}$$

5. Carry out SVD in Matlab (or any other programming language) of the following $M = 2$ by $N = 4$ matrix:

$$\boldsymbol{A} = \begin{bmatrix} 2 & 1 & 2 & 4 \\ 4 & 3 & 3 & 1 \end{bmatrix}. \tag{9.123}$$

(a) Find $\boldsymbol{U}$, $\boldsymbol{V}$ and the singular values.
(b) Verify that $\boldsymbol{A} = \boldsymbol{U}\boldsymbol{\Sigma}^{1/2}\boldsymbol{V}^{\mathrm{T}}$.
(c) Find the pseudo-inverse $\boldsymbol{A}^{-} = \boldsymbol{V}\boldsymbol{\Sigma}^{-1/2}\boldsymbol{U}^{\mathrm{T}}$.
(d) Verify that $\boldsymbol{A}\boldsymbol{A}^{-} = \boldsymbol{I}$.

6. Repeat the problem above for $\boldsymbol{B} = \boldsymbol{A}^{\mathrm{T}}$.
7. Develop code in Matlab or any other programming language to implement the following:

(a) Use the Matlab function "rand" or "normrnd" to generate a set of $K = 1000$ samples of an $N = 8$ dimensional random vector $\boldsymbol{x}$. Find the mean vector $\boldsymbol{m}_x$ and covariance matrix $\boldsymbol{\Sigma}_x$. Observe the diagonal and off-diagonal elements of the covariance matrix and explain what you have observed. Justify that $\boldsymbol{\Sigma}_x$ can be modeled by an identity matrix $c\boldsymbol{I}$ with some constant c.

(b) Generate the $N \times N$ covariance matrix $\boldsymbol{R}$ of a first-order Markov process as given in Eq. (9.13) with some r such as $r = 0.9$, and then design an $N \times N$ transform matrix $\boldsymbol{A}$ (not orthogonal) so that the signal after the transform $\boldsymbol{y} = \boldsymbol{A}\boldsymbol{x}$ becomes Markov in the sense that its covariance given below can also be modeled by Eq. (9.13):

$$\boldsymbol{\Sigma}_y = E(\boldsymbol{y}\boldsymbol{y}^{\mathrm{T}}) = E(\boldsymbol{A}\boldsymbol{x}\boldsymbol{x}^{\mathrm{T}}\boldsymbol{A}^{\mathrm{T}}) = \boldsymbol{A}E(\boldsymbol{x}\boldsymbol{x}^{\mathrm{T}})\boldsymbol{A}^{\mathrm{T}} = \boldsymbol{A}\boldsymbol{\Sigma}_x\boldsymbol{A}^{\mathrm{T}}. \quad (9.124)$$

Hint: Assume $\boldsymbol{\Sigma}_x = \boldsymbol{I}$ and consider using SVD as given in Eq. (9.92).

(c) Carry out transform $\boldsymbol{y} = \boldsymbol{A}\boldsymbol{x}$ and verify that $\boldsymbol{\Sigma}_y$ is indeed Toeplitz-like. Then carry out both the KLT and DCT to $\boldsymbol{y}$:

$$\boldsymbol{z}_{KLT} = \boldsymbol{\Phi}^{\mathrm{T}}\boldsymbol{y}, \quad \text{and} \quad \boldsymbol{z}_{DC} = \boldsymbol{C}^{\mathrm{T}}\boldsymbol{y}. \quad (9.125)$$

(d) Compare the 3-D plots of the covariance matrices of both $\boldsymbol{z}_{KLT}$ and $\boldsymbol{z}_{DCT}$ to convince yourself that they are very similar to each other. Plot each of the N columns of $\boldsymbol{\Phi}$ and those of $\boldsymbol{C}$ to convince yourself that they also look very similar to each other.

(e) The steps above can be repeated for larger values of N.

8. A signal composed of N symbols (e.g., values of signal samples before or after an orthogonal transform) can be encoded by the optimal Huffman coding with minimum total code length, which can be carried out in the following algorithm.

   - Estimate the probability p_k of the kth outcome ($k = 0, \ldots, N-1$) and sort them in descending order. Here, we assume $N = 2^n$ for convenience. Set $M = N$.
   - The forward path (left to right): Replace the two lowest probabilities by their sum. Set $M = M - 1$. Resort the M probabilities. Repeat this step until $M = 2$.
   - Backward path (right to left): Add a bit (0 or 1) to the binary code of the two probabilities newly emerging. Set $M = M + 1$. Repeat this step until $M = N$.

   For example, consider $N = 2^2 = 4$ symbols A, B, C, and D with probabilities $p_A = 0.4$, $p_B = 0.3$, $p_C = 0.2$, and $p_D = 0.1$:

	p_k	code	p_k	code	p_k	code
A	0.4	1	0.4	1	0.6	0
B	0.3	00	0.3	00	0.4	1
C	0.2	010	0.3	01		
D	0.1	011				

The average code length (number of bits) is:

$$L = 0.4 \times 1 + 0.3 \times 2 + 0.2 \times 3 + 0.1 \times 3 = 1.9 \text{ bits}$$

and the uncertainty is

$$H = -0.4 \log_2 0.4 - 0.3 \log_2 0.3 - 0.2 \log_2 0.2 - 0.1 \log_2 0.1 = 1.846 \text{ bits.}$$

Now carry out the Huffman encoding to each of the following cases and compare the average code length with the uncertainty.
(a) $p_A = 0.5$, $p_B = 0.5$, $p_C = 0.0$, and $p_D = 0.0$;
(b) $p_A = 0.9$, $p_B = 0.1$, $p_C = 0.0$, and $p_D = 0.0$;
(c) $p_A = 0.8$, $p_B = 0.1$, $p_C = 0.06$, and $p_D = 0.04$;
(d) $p_A = 0.25$, $p_B = 0.25$, $p_C = 0.25$, and $p_D = 0.25$.

9. Consider the classification of $K = 2$ gender classes using height and weight as $N = 2$ features.
   (a) Based on the assumed average weight of 65 kg and height of 162 cm for a group of K_f female students, and 80 kg and 175 cm for a group of K_m male students on a co-ed campus, generate two sets of normally distributed 2-D data points $\boldsymbol{x} = [x_1, x_2]^{\mathrm{T}}$ (x_1 and x_2 are the weight and height, respectively) with standard deviation of 8 for both features and both genders.
   Hint: you can use Matlab function `normrnd(mu,sigma)` to generate a set of random numbers of normal distribution with mean `mu` and variance `sigma`.
   (b) In the N-D feature space, different classification algorithms can be carried out, such as the following:
   * Minimum distance: a pattern $\boldsymbol{x}$ is classified to class ω_i if

$$||\boldsymbol{x} - \boldsymbol{m}_i||^2 \le ||\boldsymbol{x} - \boldsymbol{m}_j||^2 \quad \text{for all } j = 1, \ldots, C. \qquad (9.126)$$

   * Bayes method: a pattern $\boldsymbol{x}$ is classified to class ω_i if it is most likely to belong to the class; i.e.,

$$P(\omega_i/\boldsymbol{x}) \ge P(\omega_j/\boldsymbol{x}) \quad \text{for all } j = 1, \ldots, C, \qquad (9.127)$$

   where the likelihood is defined below according to Bayes formula:

$$P(\omega_c/\boldsymbol{x}) = \frac{p(\boldsymbol{x}/\omega_c)P(\omega_c)}{p(\boldsymbol{x})} \propto p(\boldsymbol{x}/\omega_c)P(\omega_c), \qquad (9.128)$$

   and $P(\omega_c)$ is the *a priori* probability for any randomly chosen pattern to belong to class ω_c, and $P(\omega_c/\boldsymbol{x})$ is the *a posteriori* probability for a specific pattern $\boldsymbol{x}$ to belong to the class. The denominator $p(\boldsymbol{x})$ is a distribution of all patterns independent of their classes, which can be dropped as it is the same for all classes.

   Specifically in this problem, we have $P(\omega_f) = K_f/(K_f + K_m)$, $P(\omega_m) = K_m/(K_f + K_m)$, and $p(\boldsymbol{x}/\omega_f)$ and $p(\boldsymbol{x}/\omega_m)$ can be assumed to be normal and their means and variances can be estimated respectively by the K_f patterns $\boldsymbol{x} \in \omega_f$ and the K_m patterns $\boldsymbol{x} \in \omega_m$ of known class, called *training samples* in practice.

Apply both minimum distance and Bayes' methods to classify the patterns and compare the results in terms of error rates defined as the percentage of misclassified patterns.

(c) Convert the pattern vectors in the $N = 2$-D feature space into 1-D vectors by each of the following methods:
 * Use the first feature x_1 for weight only (drop x_2 for height).
 * Use the second feature x_2 for height only (drop x_1 for weight).
 * Use the KLT method to generate a new feature $y = ax_1 + bx_2$ as a linear combination of both x_1 and x_2.

 Carry out classification in each of the three 1-D feature spaces and compare the results in terms of error rates.

10. Based on the provided $N = 20$ (filename "DC0" through "DC19") out of the 210 wavelength bands of the HYDICE image data (Lincoln Memorial, Washington, DC), carry out the supervised classification based on the spectral signatures of a set of $K = 4$ typical ground cover material types in the region (water surface, lawn areas, trees, and building roof) in the following two steps:

 (a) **Training:** For each of the K classes of interest; e.g., water surface, pick a set of pixels in the image known to belong to the class (called *training samples*). Find the mean vector $\boldsymbol{m}_k$ $(k = 1, \ldots, K)$ of the training samples for each class. You could use the following areas for the four training classes:

Ground	Area 1		Area 2	
type	Rows	Columns	Rows	Columns
Water	230–390	10–40	360–400	10–50
Grass	400–420	150–170	390–410	290–300
Trees	200–230	80–110	240–270	85–105
Roofs	512–517	13–49	700–710	207–233

 (b) **Classification:** For each pixel $\boldsymbol{x}$ in the image, find all K N-D Euclidean distances $D(\boldsymbol{x}, \boldsymbol{m}_l)$ $(l = 1, \ldots, K)$, and classify the pixel to the kth class if $D(\boldsymbol{x}, \boldsymbol{m}_k)$ is minimum among all K distances.

 Next, use the KLT-based feature extraction method discussed in Subsection 9.3.2 to generate a set of M new features that conserve 99% of the information (now in terms of the separability between the four specific classes of interest), and then carry out the supervised classification in this M-D feature space. Finally, compare the classification results of the two parts.

11. Ten handwritten digits from 0 to 9 are provided in an image DigitsClaaes.gif, which is composed of 10 sets of 225 subimages for each of the 10 digits. As each digit is represented as a 16×16 image, we can consider these patterns as vectors in an $N = 256$ feature space. Now carry out the KLT-based feature extraction as discussed in Example 9.4 to obtain an $M = 9$-D feature space. Then carry out the classification of these $225 \times 10 = 2250$ patterns using the minimum distance method used in the previous problem.

10 Continuous- and discrete-time wavelet transforms

10.1 Why wavelet?

10.1.1 Short-time Fourier transform and Gabor transform

In Chapter 3, we learned that a signal can be represented as either a time function $x(t)$ as the amplitude of the signal at any given moment t, or, alternatively and equivalently, as a spectrum $X(f) = \mathcal{F}[x(t)]$ representing the magnitude and phase of the frequency component at any given frequency f. However, no information in terms of the frequency contents is explicitly available in the time domain, and no information in terms of the temporal characteristics of the signal is explicitly available in the frequency domain. In this sense, neither $x(t)$ in the time domain nor $X(f)$ in the frequency domain provides complete description of the signal. In other words, we can have either temporal or spectral locality regarding the information contained in the signal, but never both at the same time.

To address this dilemma, the *short-time Fourier transform (STFT)*, also called *windowed Fourier transform*, can be used. The signal $x(t)$ to be analyzed is first truncated by a window function $w(t)$ before it is Fourier transformed to the frequency domain. As all frequency components in the spectrum are known to be contained in the signal segment inside this particular time window, certain temporal locality in the frequency domain is achieved.

We first consider a simple rectangular window with width T:

$$w_{\rm r}(t) = \begin{cases} 1 & 0 < t < T \\ 0 & \text{else} \end{cases}. \tag{10.1}$$

If a particular segment $\tau < t < \tau + T$ of the signal $x(t)$ is of interest, the signal is first multiplied by the window $w_{\rm r}(t)$ shifted by τ, and then Fourier transformed to get:

$$X_{\rm r}(f, \tau) = \mathcal{F}[x(t) w_{\rm r}(t - \tau)] = \int_{-\infty}^{\infty} x(t) w_{\rm r}(t - \tau) e^{-j2\pi ft}\, dt = \int_{\tau}^{\tau+T} x(t) e^{-j2\pi ft}\, dt. \tag{10.2}$$

Based on the time-shift and frequency convolution properties of the Fourier transform, the spectrum of this windowed signal can also be expressed as

$$X_{\rm r}(f, \tau) = X(f) * [W_{\rm r}(f) e^{-2\pi f\tau}], \tag{10.3}$$

where $W_r(f) = \mathcal{F}[w_r(t)]$ is the Fourier transform of the rectangular window $w_r(t)$. We see that the temporal locality in the frequency domain is gained at the expenses of the severe distortion of the STFT spectrum $X_r(f)$ due to the convolution with the ringing sinc function $W_r(f) = \mathcal{F}[w_r(t)]$ of the rectangular window. This distortion could be reduced if a smooth window such as a bell-shaped Gaussian function is used:

$$w_g(t) = e^{-\pi(t/\sigma)^2}, \tag{10.4}$$

where the parameter σ controls the width of the window. The spectrum of the Gaussian window is also a Gaussian function (Eq. 3.171):

$$W_g(f) = \mathcal{F}[w_g(t)] = \sigma\, e^{-\pi(\sigma f)^2}. \tag{10.5}$$

Now the spectrum of the signal windowed by a Gaussian (shifted by τ) is:

$$X_g(f,\tau) = \mathcal{F}[x(t)w_g(t-\tau)] = \int_{-\infty}^{\infty} x(t)e^{-(t-\tau)^2/\sigma^2} e^{-j2\pi ft}\, dt. \tag{10.6}$$

This Fourier transform of the Gaussian windowed signal is called the *Gabor transform* of the signal .

The original time signal can be obtained by the inverse Gabor transform. Multiplying $e^{j2\pi f\tau}$ on both sides of the equation, and then integrating with respect to f, we get

$$\begin{aligned}
&\int_{-\infty}^{\infty} X_g(f,\tau)e^{j2\pi f\tau}\, df = \int_{-\infty}^{\infty} \left[\int_{-\infty}^{\infty} x(t)e^{-(t-\tau)^2/\sigma^2} e^{-j2\pi ft}\, dt\right] e^{j2\pi f\tau}\, df \\
&= \int_{-\infty}^{\infty} x(t)e^{-(t-\tau)^2/\sigma^2} \left[\int_{-\infty}^{\infty} e^{-j2\pi ft} e^{j2\pi f\tau}\, df\right] dt = \int_{-\infty}^{\infty} x(t)e^{-(t-\tau)^2/\sigma^2}\, \delta(t-\tau)\, dt \\
&= x(\tau).
\end{aligned} \tag{10.7}$$

Similar to the case of rectangular windowing in Eq. (10.3), the Gabor spectrum in Eq. (10.6) can also be written as

$$X_g(f,\tau) = [W_g(f)e^{-j2\pi f\tau}] * X(f). \tag{10.8}$$

As before, the Gabor spectrum $X_g(f,\tau)$ in Eq. (10.8) is a blurred version of the true Fourier spectrum $X(f)$, although the ringing artifact caused by the rectangular window is avoided.

10.1.2 The Heisenberg uncertainty

In general the STFT method, based on either rectangular or Gaussian windowing, suffers from a profound difficulty, namely, the increased time locality results necessarily in a decreased frequency locality, as the resolution of the STFT spectrum, a blurred version of the true Fourier spectrum $X(f)$, is much reduced due to the convolution in Eqs. (10.3) or (10.8). For example, in the case of the Gabor transform, as the width $1/\sigma$ of $W_g(f)$ in the frequency domain is inversely proportional to the width σ of $w_g(t)$ in the time domain, a narrower time window $w_g(t)$ for higher temporal resolution will necessarily cause a wider $W_g(f)$ and thereby a more blurred Gabor spectrum $X_g(f)$.

This issue could also be illustrated if we further assume the truncated signal repeats itself outside a finite window of width T; i.e., the signal $x(t+T) = x(t)$ becomes periodic. Correspondingly, its spectrum becomes discrete, composed of an infinite set of coefficients $X[k]$ each for one of the frequency components $e^{j2\pi kt/T}$ ($k = 0, \pm 1, \pm 2, \ldots$). Obviously this discrete spectrum contains no information in the gap of $f_0 = 1/T$ between any two consecutive components $X[k]$ and $X[k+1]$. Moreover, the higher temporal resolution we achieve by reducing T, the lower frequency resolution will result due to the larger gap $f_0 = 1/T$ in the frequency domain. We see that it is fundamentally impossible to have the complete information of a given signal in both the time and frequency domains at the same time, as increasing the resolution in one domain will necessarily reduce that in the other, due to the *Heisenberg uncertainty* discussed in Chapter 3 (Eq. (3.185)).

The STFT approach also has another drawback. The window width is fixed throughout the analysis, even though there may be a variety of different signal characteristics of interest with varying time scales. For example, the signal may contain some random, irregular and sparse spikes, or bursts of rapid oscillation, which can be localized only if a very narrow time window is used. On the other hand, there may be some totally different features in the signal, such as slowly changing drifts and trends, which can be captured only if the time window has much greater width. It would be impossible for the STFT method with a fixed window width to detect and represent all of these different types of signal characteristics of interest.

In summary, if the signal is stationary and its characteristics of interest do not change much over time, then the Fourier transform may be sufficient for the analysis of the signal in terms of characterizing these features in the frequency domain. However, in many applications it is the transitory or non-stationary aspects of the signal such as drifts, trends, and abrupt changes that are of most concern and interest, but the Fourier analysis is unable to detect and characterize such features in the frequency domain.

In order to overcome these limitations of the Fourier analysis and to gain localized information in both the frequency and time domains, a different kind of transform, called the *wavelet transform*, can be used. This method can be viewed as a trade-off between the time and frequency domains. Unlike the Fourier transform which converts a signal between the time (or space) and frequency domains, the coefficients of the wavelet transform represent signal details of different scales (corresponding to different frequencies in the Fourier analysis), and also their temporal (or spatial) locations. Information contained in different scale levels reflects the signal characteristics of different scales.

The discussion above can be summarized by the *Heisenberg box (or Heisenberg cell)* illustrated in Fig. 10.1, which illustrates the issue of resolution and locality in both time and frequency in the Fourier transform, STFT, and wavelet transform.

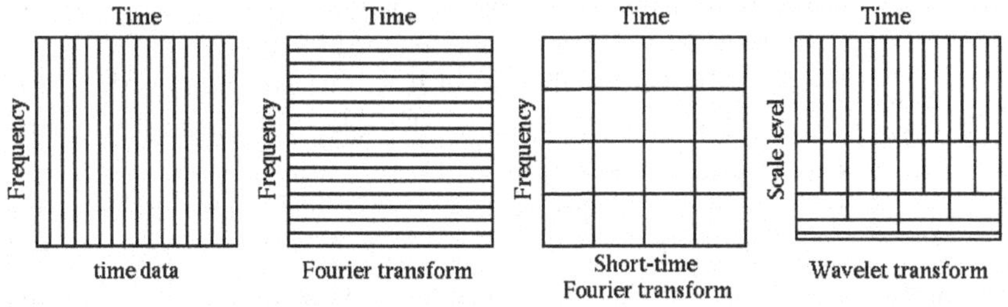

Figure 10.1 Heisenberg box: comparisons of temporal and frequency locality in Fourier and wavelet transforms.

The panel on the left is the time signal with full time resolution (temporal locality) but zero frequency resolution (frequency locality). The second panel represents its Fourier spectrum with full frequency resolution but zero temporal resolution. The third panel is the STFT whose temporal and frequency localities are inversely proportional to each other. In fact, this STFT method can be considered as a trade-off between the first two cases depending on the fixed window width. The last panel on the right represents the wavelet transform with varying scale levels and the corresponding time resolutions. At a low scale level (less detailed information corresponding to low frequencies) the window size is large, while at a high scale level (more signal details corresponding to high frequencies) the window size is small. In other words, local information in both the time and frequency domains can be represented in this transform scheme.

10.2 Continuous-time wavelet transform (CTWT)

10.2.1 Mother and daughter wavelets

All continuous orthogonal transforms previously discussed, such as the Fourier transform, are integral transforms that can be expressed as an inner product of the signal $x(t)$ and a transform kernel function $\phi_f(t)$:

$$X(f) = \langle x(t), \phi_f(t) \rangle = \int x(t)\overline{\phi}_f(t)\, dt. \tag{10.9}$$

Here, the family of the kernel functions $\phi_f(t)$ corresponding to different f form an orthogonal basis that spans the vector space in which the signal $x(t)$ resides. For example, in the case of the Fourier transform, a member of the kernel function family is a complex exponential $\phi_f(t) = e^{j2\pi ft}$ corresponding to a parameter f representing a specific frequency.

Similarly, the *continuous-time wavelet transform (CTWT)* is also an integral transform based on a set of kernel functions, sometimes referred to as the *daugh-*

ter wavelets, all derived from a *mother wavelet* $\psi(t)$ that should satisfy the following conditions:

- $\psi(t)$ has a compact support; i.e., $\psi(t) \neq 0$ only inside a bounded range $a < t < b$.
- $\psi(t)$ has a zero mean:

$$\int_{-\infty}^{\infty} \psi(t)\, dt = 0; \quad \text{i.e.} \quad \Psi(f)\Big|_{f=0} = \Psi(0) = 0, \tag{10.10}$$

  where $\Psi(f) = \mathcal{F}[\psi(t)]$ is the Fourier spectrum of $\psi(t)$. In other words, the DC component of the mother wavelet is zero.
- $\psi(t) \in \mathcal{L}^2$ is square-integrable; i.e.,

$$\int_{-\infty}^{\infty} |\psi(t)|^2\, dt < \infty. \tag{10.11}$$

- $\psi(t)$ can be normalized so that:

$$||\psi(t)||^2 = \int_{-\infty}^{\infty} |\psi(t)|^2\, dt = 1. \tag{10.12}$$

Intuitively, a mother wavelet $\psi(t)$ needs to satisfy two conditions. First, it is non-zero only within a finite range (first condition); i.e., it is "small." Second, it has a zero mean (second condition); i.e., it is a "wave" that takes both positive and negative values around zero. In other words, $\psi(t)$ is a small wave, hence the name "wavelet". Obviously this is essentially different from all other continuous orthogonal transforms such as the Fourier and cosine transforms, whose kernel functions are sinusoidal waves (infinite waves) defined over the entire time axis.

Based on the mother wavelet, a family of kernel functions $\psi_{s,\tau}(t)$, the daughter wavelets, can be generated by scaling and translating the mother wavelet by s and τ, respectively:

$$\psi_{s,\tau}(t) = \frac{1}{\sqrt{s}}\psi\left(\frac{t-\tau}{s}\right), \tag{10.13}$$

where τ is the time translation ($\tau > 0$ for right shift and $\tau < 0$ for left shift) and $s > 0$ is a scaling factor ($s > 1$ for expansion and $s < 1$ for compression). Unlike the kernel function $\phi_f(t) = e^{j2\pi ft}$ of the Fourier transform with only one parameter f for frequency, the CTWT kernel $\psi_{s,\tau}(t)$ has two parameters, τ and s for translation and scaling, respectively. This is the reason why the wavelet transform is capable of representing localized information in the time domain as well as in different scale levels (corresponding to different frequencies), while the Fourier transform is only capable of representing localized frequency information.

The factor $1/\sqrt{s}$ is included in the wavelet $\psi_{s,\tau}(t)$ so that it is also normalized as the mother wavelet, independent of the scaling factor s:

$$\begin{aligned} ||\psi_{s,\tau}(t)||^2 &= \langle \psi_{s,\tau}(t), \psi_{s,\tau}(t) \rangle = \frac{1}{s}\int_{-\infty}^{\infty} \left|\psi(\frac{t-\tau}{s})\right|^2 dt \\ &= \frac{1}{s}\int_{-\infty}^{\infty} |\psi(t')|^2\, s\, dt' = ||\psi(t)||^2 = 1. \end{aligned} \tag{10.14}$$

Here, we have assumed $t' = (t-\tau)/s$ and therefore $dt' = dt/s$.

10.2.2 The forward and inverse wavelet transforms

Given a mother wavelet $\psi(t)$, we can derive all of her daughter wavelets $\psi_{s,\tau}(t)$ for different s and τ, and then define the CTWT of a time signal $x(t)$ as an integral transform:[1]

$$X(s,\tau) = \mathcal{W}[x(t)] = \langle x(t), \psi_{s,\tau}(t)\rangle = \int_{-\infty}^{\infty} x(t)\overline{\psi}_{s,\tau}(t)\, dt$$
$$= \frac{1}{\sqrt{s}}\int_{-\infty}^{\infty} x(t)\overline{\psi}\left(\frac{t-\tau}{s}\right) dt = x(\tau) \star \psi_{s,0}(\tau). \tag{10.15}$$

We see that the CTWT of $x(t)$ is actually the correlation of the signal $x(t)$ and the wavelet function $\psi_{s,0}(t) = \psi(t/s)/\sqrt{s}$. If we take the Fourier transform on both sides of the CTWT $X(s,\tau)$ above, while treating τ as the time variable and s as a parameter, we get the Fourier spectrum of the CTWT of $x(t)$ (correlation property of the Fourier transform Eq. (3.117)):

$$\hat{X}(s,f) = \mathcal{F}[X(s,\tau)] = \mathcal{F}[\mathcal{W}[x(t)]] = X(f)\,\overline{\Psi}_{s,0}(f), \tag{10.16}$$

where $X(f) = \mathcal{F}[x(t)]$ and $\Psi_{s,0}(f) = \mathcal{F}[\psi_{s,0}(t)]$ are the Fourier spectra of the signal $x(t)$ and the wavelet $\psi_{s,0}(t)$, respectively. Note that here we have to use a hat in addition to a capital letter X to denote the result obtained by applying two different transforms (CTWT followed by CTFT) consecutively to a signal $x(t)$. This will be the only deviation from our convention of representing the transform of a signal $x(t)$ by a capital letter $X(f)$.

Given the Fourier spectrum of the mother wavelet $\Psi(f) = \mathcal{F}[\psi(t)]$, we can find the spectrum of a daughter wavelet $\psi_{s,\tau}(t)$ (time-shift and scaling properties of the Fourier transform Eqs. (3.111) and (3.107)):

$$\Psi_{s,\tau}(f) = \mathcal{F}[\psi_{s,\tau}(t)] = \mathcal{F}\left[\frac{1}{\sqrt{s}}\psi\left(\frac{t-\tau}{s}\right)\right] = \sqrt{s}\Psi(sf)e^{-j2\pi f\tau}. \tag{10.17}$$

In particular when $\tau = 0$, we have:

$$\Psi_{s,0}(f) = \sqrt{s}\Psi(sf)e^{-j2\pi f\tau}\Big|_{\tau=0} = \sqrt{s}\Psi(sf). \tag{10.18}$$

Now we see that the CTWT can also be obtained by taking the inverse Fourier transform of $\hat{X}(s,f)$ in Eq. (10.16):

$$X(s,\tau) = \mathcal{F}^{-1}[\hat{X}(s,f)] = \mathcal{F}^{-1}[X(f)\,\overline{\Psi}_{s,0}(f)] = \sqrt{s}\int_{-\infty}^{\infty} X(f)\overline{\Psi}(sf)e^{j2\pi f\tau}\, df. \tag{10.19}$$

[1] In the wavelet literature different notations have been used for the CTWT of a signal $x(t)$, such as $CWT_x(s,\tau)$ and $Wx(s,\tau)$. However, here we simply use the capitalized letter $X(s,\tau) = \mathcal{W}[x(t)]$ to represent the CTWT of $x(t)$, consistent with the convention used for all orthogonal transforms considered in previous chapters, such as $X(f) = \mathcal{F}[x(t)]$ for the Fourier transform of $x(t)$.

The time signal $x(t)$ can be reconstructed from its CTWT transform $X(s,\tau)$ by the inverse wavelet transform:

$$\begin{aligned} x(t) = \mathcal{W}^{-1}[X(s,\tau)] &= \frac{1}{C_\psi}\int_0^\infty \int_{-\infty}^\infty X(s,\tau)\psi_{s,\tau}(t)\, d\tau\, \frac{ds}{s^2} \\ &= \frac{1}{C_\psi}\int_0^\infty \frac{1}{\sqrt{s}}\int_{-\infty}^\infty X(s,\tau)\psi\left(\frac{t-\tau}{s}\right) d\tau\, \frac{ds}{s^2}, \end{aligned} \tag{10.20}$$

where C_ψ is defined as

$$C_\psi = \int_0^\infty \frac{|\Psi(s)|^2}{s} ds < \infty. \tag{10.21}$$

This inequality, referred to as the *admissibility condition*, is necessary for the inverse CTWT to exist. Note that for this condition to hold, we must have $\Psi(f)\big|_{f=0} = \Psi(0) = 0$, one of the conditions specified before (Eq. (10.10)). Consequently, Eq. (10.19) will produce the same result for different $X(0)$, as it is always multiplied by $\Psi(0) = 0$. In other words, the CTWT is insensitive to the DC component $X(0)$ of the signal $x(t)$.

Now we prove that the signal $x(t)$ can indeed be reconstructed by the inverse CTWT given in Eq. (10.20). We first multiply $\Psi_{s,0}(f)/s^2$ on both sides of Eq. (10.16) and integrate with respect to s to get:

$$\int_0^\infty \hat{X}(s,f)\Psi_{s,0}(f)\, \frac{ds}{s^2} = X(f)\int_0^\infty |\Psi_{s,0}(f)|^2\, \frac{ds}{s^2} = X(f)\int_0^\infty \frac{|\Psi(sf)|^2}{s}\, ds. \tag{10.22}$$

The last equal sign is due to Eq. (10.18). The integral on the right-hand side can be further written as

$$\int_0^\infty \frac{|\Psi(sf)|^2}{s}\, ds = \int_0^\infty \frac{|\Psi(sf)|^2}{sf}\, d(sf) = \int_0^\infty \frac{|\Psi(s')|^2}{s'}\, ds' = C_\psi, \tag{10.23}$$

where we have assumed $s' = sf$, and the last equal sign is due to the definition of C_ψ in Eq. (10.21). Now we can solve Eq. (10.22) for $X(f)$ to get

$$X(f) = \frac{1}{C_\psi}\int_0^\infty \hat{X}(s,f)\ \Psi_{s,0}(f)\, \frac{ds}{s^2}. \tag{10.24}$$

Taking the inverse Fourier transform on both sides we get the inverse CTWT in Eq. (10.20):

$$\begin{aligned} x(t) = \mathcal{F}^{-1}[X(f)] &= \frac{1}{C_\psi}\int_0^\infty \mathcal{F}^{-1}[\hat{X}(s,f)\ \Psi_{s,0}(f)]\, \frac{ds}{s^2} \\ &= \frac{1}{C_\psi}\int_0^\infty X(s,t) * \psi_{s,0}(t)\, \frac{ds}{s^2} \\ &= \frac{1}{C_\psi}\int_0^\infty \frac{1}{\sqrt{s}}\int_{-\infty}^\infty X(s,\tau)\psi\left(\frac{t-\tau}{s}\right) d\tau\, \frac{ds}{s^2}. \end{aligned} \tag{10.25}$$

Here, we have used the convolution theorem of the Fourier transform (Eq. (3.122)).

The result of Eq. (10.23) also indicates an interesting fact as a side product. For any given function $f(x)$, in this case $|\Psi(f)|^2$, the integral of its scaled version $f(sx)/s$ over all scale s is a constant independent of x; i.e., a constant function over the entire domain of x. This result has some important significance, as we will see later in the future discussion of the discrete-time wavelet transform (DTWT).

In summary, both the forward and inverse CTWTs in Eqs. (10.15) and (10.20) can be written as the following CTWT pair:

$$x(s,\tau) = \mathcal{W}[x(t)] = \frac{1}{\sqrt{s}} \int_{-\infty}^{\infty} x(t)\overline{\psi}\left(\frac{t-\tau}{s}\right) dt$$
$$x(t) = \mathcal{W}^{-1}[X(s,\tau)] = \frac{1}{C_\psi} \int_0^{\infty} \frac{1}{\sqrt{s}} \int_{-\infty}^{\infty} X(s,\tau)\psi\left(\frac{t-\tau}{s}\right) d\tau \frac{ds}{s^2}. \tag{10.26}$$

The forward CTWT in the first equation converts a 1-D signal $x(t)$ into a 2-D function $X(s,\tau)$ of s for scale and τ for translation, while the inverse CTWT in the second equation reconstructs the signal from $X(s,\tau)$. The CTWT transform has some essential differences compared with all previously considered orthogonal transforms such as the Fourier transform. First, the Fourier spectrum $X(f) = \mathcal{F}[x(t)]$ is a 1-D function of frequency f, but the CTWT $X(s,\tau) = \mathcal{W}[x(t)]$ is a 2-D function of two variables s and τ. Second, the CTWT is not an orthogonal transform, as its kernel functions, the daughter wavelets, $\psi_{s,\tau}(t)$, are not orthogonal to each other. Owing to such differences, the CTWT representation of a 1-D signal is necessarily redundant. It can be used for signal filtering, as to be seen later, but it is not suitable for data compression.

10.3 Properties of the CTWT

In the discussion below we will always assume $X(s,\tau) = \mathcal{W}[x(t)]$ and $Y(s,\tau) = \mathcal{W}[y(t)]$.

- **Linearity**

$$\mathcal{W}[ax(t) + by(t)] = a\mathcal{W}[x(t)] + b\mathcal{W}[y(t)] = aX(s,\tau) + bY(s,\tau). \tag{10.27}$$

The wavelet transform of a function $x(t)$ is simply an inner product of the function with a kernel function $\psi_{s,\tau}(t)$ (Eq. 10.15). Therefore, owing to the linearity of the inner product in the first variable, the wavelet transform is also linear.

- **Time shift**

$$\mathcal{W}[x(t-t')] = X(s,\tau-t'). \tag{10.28}$$

The proof is left for the reader as a homework problem.

- **Time scaling**

$$\mathcal{W}[x(t/a)] = \sqrt{a}X(s/a,\tau/a). \tag{10.29}$$

The proof is left for the reader as a homework problem.

- **Localization**
Let the center and width of a mother wavelet $\psi(t)$ be $t = t_0$ and Δt in the time domain and those of its spectrum be $\Psi(f)$ f_0 and Δf in the frequency domain, respectively. Then the center and width of a scaled and translated daughter wavelet $\psi_{s,\tau}(t) = \psi((t-\tau)/s)/\sqrt{s}$ are

$$t_{0,s,\tau} = st_0 + \tau, \qquad \text{and} \qquad \Delta t_{s,\tau} = s\Delta t, \tag{10.30}$$

and, according to the time/frequency scaling property (Eq. (3.107)), the center and width of its spectrum $\Psi_{s,\tau}(f) = \sqrt{s}\Psi(sf)e^{-j2\pi f\tau}$ (Eq. (10.17)) are

$$f_{0,s,\tau} = \frac{1}{s} f_0, \qquad \Delta f_{s,\tau} = \frac{1}{s}\Delta f. \tag{10.31}$$

We can now make two observations.
  - The product of the widths of the wavelet function $\psi_{s,\tau}(t)$ in the time domain and its spectrum $\Psi_{s,\tau}(f)$ in the frequency domain is constant, independent of s and τ:

$$\Delta t_{s,\tau}\ \Delta f_{s,\tau} = s\Delta t\ \frac{1}{s}\Delta f = \Delta t\ \Delta f. \tag{10.32}$$

  - The spectrum $\Psi_{s,\tau}(f)$ of the wavelet function can be considered as a BP filter with a quality factor Q (Eq. (5.67)), here defined as the ratio of its bandwidth and the center frequency:

$$Q = \frac{\Delta f_{s,\tau}}{f_{0,s,\tau}} = \frac{\Delta f}{f_0}; \tag{10.33}$$

  i.e., the quality factor Q of the filter is constant, independent of the scaling factor s.
- **Multiplication theorem**
Corresponding to the multiplication theorem (Eq. (3.105)) for the Fourier transform $\langle x(t), y(t)\rangle = \langle X(f), Y(f)\rangle$, where $X(f) = \mathcal{F}[x(t)]$ and $Y(f) = \mathcal{F}[y(t)]$, a similar theorem also exists for the wavelet transform. However, as the CTWT $X(s,\tau)$ is a function of two variables s and τ, we first need to define the inner product of two CTWTs as:

$$\langle X(s,\tau), Y(s,\tau)\rangle = \int_0^\infty \int_{-\infty}^\infty X(s,\tau)\overline{Y}(s,\tau)\, d\tau\, \frac{ds}{s^2}. \tag{10.34}$$

The multiplication theorem states:

$$\langle x(t), y(t)\rangle = \frac{1}{C_\psi}\langle X(s,\tau), Y(s,\tau)\rangle. \tag{10.35}$$

To prove this theorem, we substitute the CTWTs of two functions $x(t)$ and $y(t)$ (Eq. (10.19))

$$X(s,\tau) = \mathcal{W}[x(t)] = \sqrt{s}\int_{-\infty}^\infty X(f)\overline{\Psi}(sf)e^{j2\pi f\tau}\, df$$

$$Y(s,\tau) = \mathcal{W}[y(t)] = \sqrt{s}\int_{-\infty}^\infty Y(f)\overline{\Psi}(sf)e^{j2\pi f\tau}\, df \tag{10.36}$$

into the inner product defined above and get

$$
\begin{aligned}
\langle X(s,\tau), Y(s,\tau)\rangle &= \int_0^\infty \int_{-\infty}^\infty X(s,\tau)\overline{Y}(s,\tau)\,d\tau\,\frac{ds}{s^2} \\
&= \int_0^\infty \int_{-\infty}^\infty \left[\int_{-\infty}^\infty X(f)\overline{\Psi}(sf)e^{j2\pi f\tau}\,df\right]\left[\int_{-\infty}^\infty \overline{Y}(f')\Psi(sf')e^{-j2\pi f'\tau}\,df'\right] d\tau\,\frac{ds}{s} \\
&= \int_0^\infty \int_{-\infty}^\infty \int_{-\infty}^\infty \left[X(f)\overline{Y}(f')\overline{\Psi}(sf)\Psi(sf')\int_{-\infty}^\infty e^{j2\pi(f-f')\tau}\,d\tau\right] df'\,df\,\frac{ds}{s} \\
&= \int_0^\infty \int_{-\infty}^\infty \int_{-\infty}^\infty X(f)\overline{Y}(f')\overline{\Psi}(sf)\Psi(sf')\delta(f-f')\,df'\,df\,\frac{ds}{s} \\
&= \int_{-\infty}^\infty X(f)\overline{Y}(f)\left[\int_0^\infty \frac{|\Psi(sf)|^2}{s}\,ds\right] df = C_\psi \int_{-\infty}^\infty X(f)\overline{Y}(f)\,df \\
&= C_\psi\langle X(f), Y(f)\rangle = C_\psi\langle x(t), y(t)\rangle,
\end{aligned}
\tag{10.37}
$$

where, again, C_ψ is given in Eq. (10.21). In particular, when $y(t) = x(t)$, we have

$$
||x(t)||^2 = ||X(f)||^2 = \frac{1}{C_\psi}\langle X(s,\tau), X(s,\tau)\rangle = \frac{1}{C_\psi}\int_0^\infty \int_{-\infty}^\infty |X(s,\tau)|^2\,d\tau\,\frac{ds}{s^2}.
\tag{10.38}
$$

This is Parseval's theorem of the CTWT, where $|X(s,\tau)|^2$ is the signal energy distribution in the 2-D wavelet transform domain spanned by s and τ.

- **Non-orthogonality**

  All previously considered orthogonal transforms represent a given signal in terms of a set of orthogonal basis functions or vectors that span the vector space in which the signal resides. For example, in the Fourier transform $X(f) = \mathcal{F}[x(t)] = \langle x(t), \phi_f(t)\rangle$, the basis functions $\phi_f(t) = e^{j2\pi ft}$ (for all f) are orthogonal:

$$
\langle \phi_f(t), \phi_{f'}(t)\rangle = \int_{-\infty}^\infty \phi_f(t)\overline{\phi}_{f'}(t)\,dt = 0 \qquad (f \neq f'),
\tag{10.39}
$$

  indicating that they are uncorrelated with zero redundancy. In other words, the kernel function $\phi_f(t)$ at every single point f in the transform domain makes its unique contribution to the representation of the time signal in the inverse transform $x(t) = \int X(f)\phi_f(t)\,df = \int \langle x(t), \phi_f(t)\rangle \phi_f(t)\,df$.

  However, this is no longer the case for the CTWT, which converts a 1-D time signal $x(t)$ to a 2-D function $X(s,\tau) = \mathcal{W}[x(t)] = \langle x(t), \psi_{s,\tau}(t)\rangle$ defined over the half plane $-\infty < \tau < \infty$ and $s > 0$. Redundancy exists in this 2-D transform domain (s,τ) in terms of the information needed for the reconstruction of the time signal $x(t)$. The redundancy between any two points (s,τ) and (s',τ') in the transform domain can be measured by the *reproducing kernel*, defined as the inner product of the two kernel functions (basis functions) $\psi_{s,\tau}(t)$ and $\psi_{s',\tau'}(t)$:

$$
K(s,\tau,s',\tau') = \langle \psi_{s,\tau}(t), \psi_{s',\tau'}(t)\rangle = \int_{-\infty}^\infty \psi_{s,\tau}(t)\overline{\psi}_{s',\tau'}(t)\,dt.
\tag{10.40}
$$

Unlike Eq. (10.39) for any orthogonal transform, the inner product above is not zero in general. This is a major difference between the non-orthogonal CTWT and all orthogonal transforms. This reproducing kernel can be considered as the correlation between two kernel functions $\psi_{s,\tau}(t)$ and $\psi_{s',\tau'}(t)$, representing the redundancy between them.
Let $X(s,\tau) = \mathcal{W}[x(t)] = \langle x(t), \psi_{s,\tau}(t)\rangle \neq 0$ be the CTWT at point (s,τ). Then the CTWT at another point (s',τ') is

$$X(s',\tau') = \langle x(t), \psi_{s',\tau'}(t)\rangle = \int_{-\infty}^{\infty} x(t)\overline{\psi}_{s',\tau'}(t)\,dt. \tag{10.41}$$

Substituting the reconstructed $x(t)$ by the inverse CTWT (Eq. (10.20)) into this equation we get

$$\begin{aligned} X(s',\tau') &= \int_{-\infty}^{\infty}\left[\frac{1}{C_\psi}\int_0^{\infty}\int_{-\infty}^{\infty} X(s,\tau)\psi_{s,\tau}(t)\,d\tau\,\frac{ds}{s^2}\right]\overline{\psi}_{s',\tau'}(t)\,dt \\ &= \frac{1}{C_\psi}\int_0^{\infty}\int_{-\infty}^{\infty} X(s,\tau)\left[\int_{-\infty}^{\infty}\psi_{s,\tau}(t)\overline{\psi}_{s',\tau'}(t)\,dt\right]d\tau\,\frac{ds}{s^2} \\ &= \frac{1}{C_\psi}\int_0^{\infty}\int_{-\infty}^{\infty} K(s,\tau,s',\tau')X(s,\tau)\,d\tau\,\frac{ds}{s^2}. \end{aligned} \tag{10.42}$$

Consider two cases. First, if $K(s,\tau,s',\tau') = 0$ for all points (s,τ); i.e., $\psi_{s',\tau'}(t)$ at point (s',τ') is not correlated with $\psi_{s,\tau}(t)$ at any other point (s,τ), then $X(s',\tau') = 0$; i.e., it does not contribute to the representation of the signal in the inverse CTWT (Eq. (10.20)). Second, if $K(s,\tau,s',\tau') \neq 0$ for some points (s,τ), then $X(s',\tau') \neq 0$ does contribute to the representation of the signal. However, as it is a linear combination of all other $X(s,\tau) \neq 0$ (weighted by $K(s,\tau,s',\tau')$), its contribution is redundant.

10.4 Typical mother wavelet functions

Throughout the previous discussion of the wavelet transform, the mother wavelet function is not specifically defined. Here, we consider some commonly used mother wavelets.

- **Shannon wavelet**
  The Shannon wavelet can be more conveniently defined in the frequency domain as an ideal BP filter:

$$\Psi(f) = \begin{cases} 1 & f_1 < |f| < f_2 \\ 0 & \text{else} \end{cases}. \tag{10.43}$$

By inverse Fourier transform we get the Shannon wavelet in the time domain:

$$\begin{aligned} \psi(t) &= \mathcal{F}^{-1}[\Psi(f)] = \int_{-\infty}^{\infty}\Psi(f)e^{j2\pi ft}\,df = \int_{-f_2}^{-f_1} e^{j2\pi ft}\,df + \int_{f_1}^{f_2} e^{j2\pi ft}\,df \\ &= \frac{1}{\pi t}[\sin(2\pi f_2 t) - \sin(2\pi f_1 t)]. \end{aligned} \tag{10.44}$$

The Shannon wavelet and its spectrum are shown in Fig. 10.2. Obviously, this wavelet has very good frequency locality but poor temporal locality. However, this wavelet has some significance in the discussion of an algorithm for the reconstruction of the time signal, to be considered later.

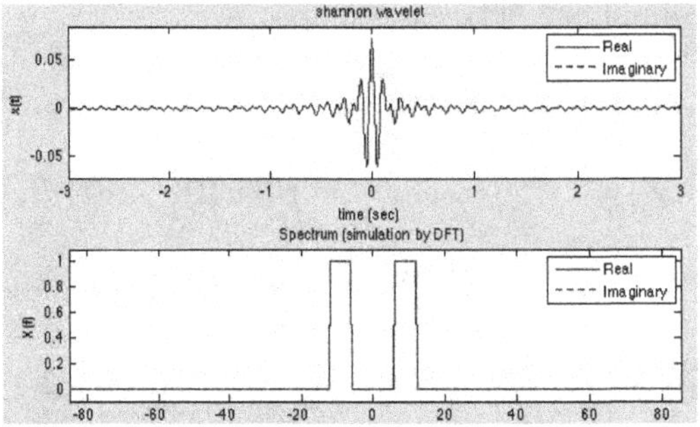

Figure 10.2 Shannon wavelet (top) and its spectrum (bottom).

- **Morlet wavelet**

The Morlet wavelet is a complex exponential $e^{j\omega_0 t}$ modulated by a normalized Gaussian function $e^{-t^2/2}/\sqrt{2\pi}$:

$$\psi(t) = \frac{1}{\sqrt{2\pi}} e^{j\omega_0 t} e^{-t^2/2} = \frac{1}{\sqrt{2\pi}} \left[\cos(\omega_0 t) e^{-t^2/2} + j \sin(\omega_0 t) e^{-t^2/2} \right]. \quad (10.45)$$

According to the frequency shift property of the Fourier transform (Eq. (3.112)), the spectrum of the Morlet wave is another Gaussian function shifted by $-\omega_0$:

$$\Psi(\omega) = \mathcal{F}[\psi(t)] = \int_{-\infty}^{\infty} \psi(t) e^{-j\omega t} \, dt = \frac{1}{\sqrt{2\pi}} \int_{-\infty}^{\infty} e^{-t^2/2} e^{j(\omega-\omega_0)t} \, dt$$
$$= e^{-(\omega-\omega_0)^2/2} = e^{-(2\pi(f-f_0))^2/2}. \quad (10.46)$$

The Morlet wavelet and its spectrum are shown in Fig. 10.3. Note that when $\omega_0 = 0$, $\Psi(0) = e^{-\omega_0^2/2} > 0$, violating the admissibility condition. However, if ω_0 is large enough; e.g., $f_0 = 1$ Hz or $\omega_0 = 2\pi$, $\Psi(0) = e^{-6.28^2/2} = 2.7 \times 10^{-9}$ is small enough to be neglected. As the Fourier spectrum $\Psi(\omega)$ of the Morlet wavelet is zero when $\omega < 0$, it is an analytic signal according to the definition discussed in Chapter 3.

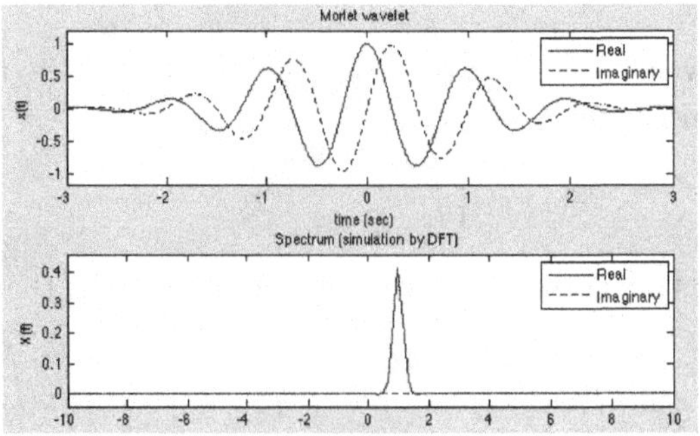

Figure 10.3 Morlet wavelet (top) and its spectrum (bottom).

- **Derivative of Gaussian (DoG)**
  This wavelet is the first-order derivative of a normalized Gaussian function $g(t) = e^{-\pi(t/a)^2}/a$:
  $$\psi(t) = \frac{d}{dt}g(t) = \frac{d}{dt}\left[\frac{1}{a}e^{-\pi(t/a)^2}\right] = -\frac{2\pi t}{a^3}e^{-\pi(t/a)^2}. \tag{10.47}$$
  Note that the Gaussian function is normalized
  $$\int_{-\infty}^{\infty} g(t)\,dt = 1, \tag{10.48}$$
  and the parameter a is related to the standard deviation σ by $a = \sqrt{2\pi\sigma^2}$. The Fourier transform of this derivative of Gaussian can be easily found according to the time derivative property of the Fourier transform (Eq. 3.127) to be
  $$\Psi(f) = \mathcal{F}[\psi(t)] = j2\pi f t\; e^{-\pi(af)^2}. \tag{10.49}$$
- **Marr wavelet (Mexican hat)**
  The Marr wavelet is the negative version of the second derivative of the Gaussian function $g(t) = e^{-\pi(t/a)^2}/a$:
  $$\psi(t) = -\frac{d^2}{dt^2}g(t) = -\frac{d}{dt}\left[-\frac{2\pi t}{a^3}e^{-\pi(t/a)^2}\right] = \frac{2\pi}{a^3}\left(1 - \frac{2\pi t}{a^2}\right)e^{-\pi(t/a)^2}. \tag{10.50}$$
  If we let $a = \sqrt{2\pi\sigma^2}$, the Gaussian function $g(t)$ is normalized, $\int g(t)\,dt = 1$, and the Marr wavelet becomes
  $$\psi(t) = \frac{1}{\sqrt{2\pi}\sigma^3}\left(1 - \frac{t^2}{\sigma^2}\right)e^{-t^2/2\sigma^2}. \tag{10.51}$$
  The Marr wavelet function is also referred to as the Mexican hat function owing to its waveform. The Fourier transform of the Gaussian function is also Gaussian (Eq. (3.170)):
  $$\mathcal{F}\left[\frac{1}{a}e^{-\pi(t/a)^2}\right] = e^{-\pi(af)^2}, \tag{10.52}$$
  and according to the time derivative property of the Fourier transform (Eq. (3.127)), we get the spectrum of the Marr wavelet
  $$\Psi(f) = \mathcal{F}[\psi(t)] = -(j2\pi f t)^2 e^{-\pi(af)^2} = 4\pi^2 f^2 e^{-\pi(af)^2}. \tag{10.53}$$
  The Marr wavelet and its Fourier transform are shown in Fig. 10.4.

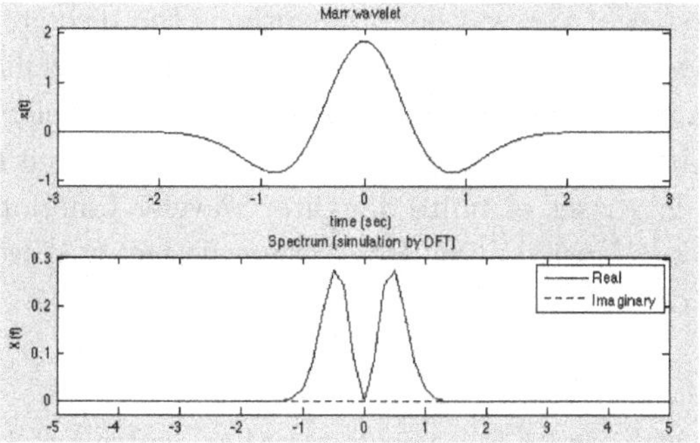

Figure 10.4 Marr wavelets of different scale levels and their spectra.

- **Difference of Gaussians**

  As the name suggests, this wavelet is simply the difference between two Gaussian functions with different parameters $a_1 > a_2$ (representing the variance):

$$\psi(t) = g_1(t) - g_2(t) = \frac{1}{a_1} e^{-\pi(t/a_1)^2} - \frac{1}{a_2} e^{-\pi(t/a_2)^2}. \tag{10.54}$$

  The spectrum of this function is the difference between the spectra of the two Gaussian functions, which are also Gaussian:

$$\Psi(f) = G_1(f) - G_2(f) = e^{-\pi(a_1 f)^2} - e^{-\pi(a_2 f)^2}. \tag{10.55}$$

  Note that $\Psi(0) = 0$ as required. As can be seen in Fig. 10.5, the difference of Gaussians looks very much like the second derivative of Gaussian (Marr) wavelet, and both functions could be abbreviated as DoG. But note that they are two different types of function.

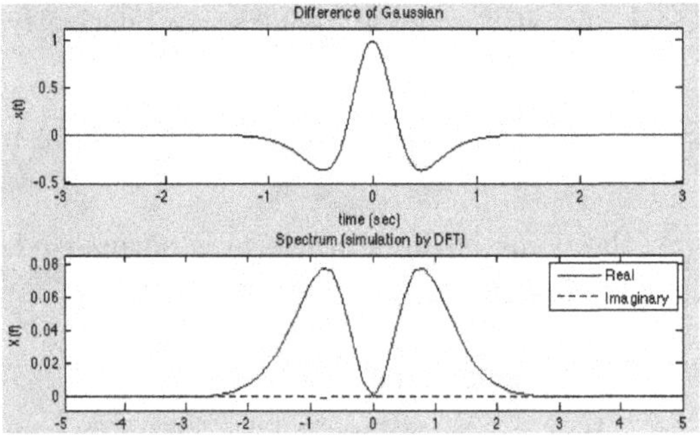

Figure 10.5 Difference of Gaussians and its spectrum.

10.5 Discrete-time wavelet transform (DTWT)

10.5.1 Discretization of wavelet functions

In order to actually obtain the wavelet transform of a real time signal in practice, we need to discretize both the signal and the wavelet functions and the resulting discrete version of the wavelet transform is the discrete-time wavelet transform (DTWT), which can be carried out numerically. Specifically, we need not only sample both the time signal $x(t)$ and the mother wavelet function $\psi(t)$ to get a set of finite samples $x[n]$ and $\psi[n]$ $(n = 0, \dots, N-1)$, but also discretize the scale factor s to get a set of finite daughter wavelet functions of different scales $\psi_{s_l,0}[n] = \psi[n/s_l]$ $(l = 1, \dots, S)$. Here, the scale factor s_l is defined as an exponential function of the scale index l:

$$s_l = s_0 2^{l/r} = s_0(2^{1/r})^l, \tag{10.56}$$

where s_0 is the base scale and r is a parameter that controls the total number of scale levels $S = r\log_2(N/s_0)$.

Having discretized the time signal and the mother wavelet, we can also obtain their DFT coefficients $X[k] = \mathcal{F}[x[n]]$ and $\Psi_{s_l,0}[k] = \mathcal{F}[\psi_{s_l,0}[n]]$ (with s_l treated as a parameter) ($k = 0, \ldots, N-1$). When the mother wavelet function $\psi[n]$ is scaled by $s_l > 1$, it is expanded in the time domain to become $\psi_{s_l,0}[n] = \psi[n/s_l]$, and its spectrum is compressed in the frequency domain to become $\Psi_{s_l,0}[k] = \Psi[s_l k]$. When $l = 1$, the mother wavelet is scaled minimally by a factor $s_0 r^{1/r}$, but, when $l = S$, it is maximally expanded by a factor of $s_l = s_0 2^{S/r} = s_0 2^{\log_2(N/s_0)} = N$, and its N-point Fourier spectrum $\Psi_{s_l,0}[k] = \Psi[Nk]$ is maximally compressed to become a single point. Moreover, if $r > 1$, the base of the exponent is reduced from 2 to $2^{1/r} < 2$ for a finer scale resolution with a smaller step size between two consecutive scale levels. For example, when $r = 2$, the base of the exponent in Eq. (10.56) is reduced from 2 to $\sqrt{2} = 1.442$, and the total number of scale levels is correspondingly doubled and the scale resolution is increased. Particularly when $s_0 = r = 1$, we have $s_l = 2^l$, and the corresponding transform is called the *dyadic wavelet transform*.

The exponentially scaled Shannon, Morlet, and Marr wavelets are shown in Figs. 10.6, 10.7, and 10.8.

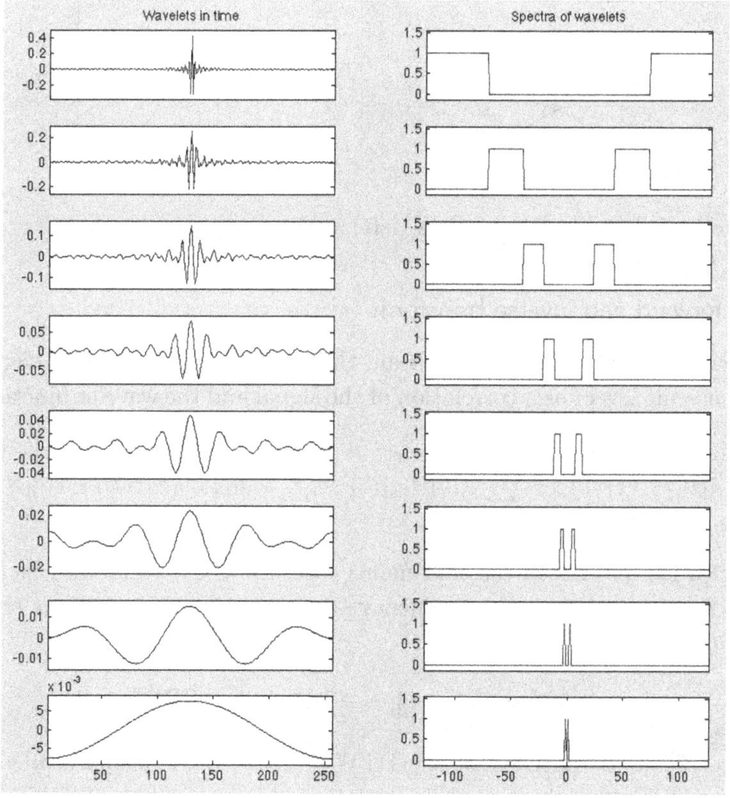

Figure 10.6 The Shannon wavelets (left) and their spectra (right).

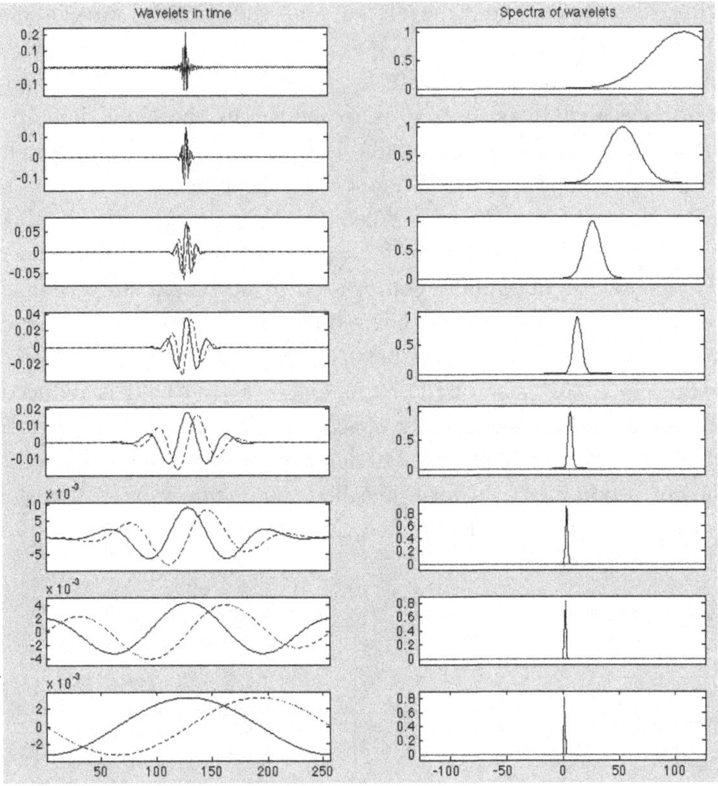

Figure 10.7 The Morlet wavelets (left) and their spectra (right).

10.5.2 The forward and inverse transform

Following Eq. (10.15), we can obtain the DTWT coefficients of a discrete signal $x[n]$ at scale level l as a correlation of the signal and the wavelet function $\psi_{s_l,0}[n]$:

$$X[l,n] = \mathcal{W}[x[n]] = \sum_{n=0}^{N-1} x[n]\overline{\psi}_{s_l,l}[n] = \sum_{n=0}^{N-1} x[n]\overline{\psi}_{s_l,0}[n-l] = x[n] \star \psi_{s_l,0}[n]. \tag{10.57}$$

As with Eq. (10.16) in the continuous case, the DTWT can also be carried out as a multiplication in the frequency domain (with scale index l treated as a parameter):

$$\hat{X}[l,k] = \mathcal{F}[X[l,n]] = \mathcal{F}[\mathcal{W}[x[n]] = X[k]\,\overline{\Psi}_{s_l,0}[k], \tag{10.58}$$

where $\hat{X}[l,k]$ is the DFT of the DTWT $X[l,n]$ of the signal $x[n]$. Taking the inverse DFT on both sides of the equation above we get the DTWT in the time domain:

$$X[l,n] = \mathcal{F}^{-1}\left[\hat{X}[l,k]\right] = \mathcal{F}^{-1}\left[X[k]\,\overline{\Psi}_{s_l,0}[k]\right]. \tag{10.59}$$

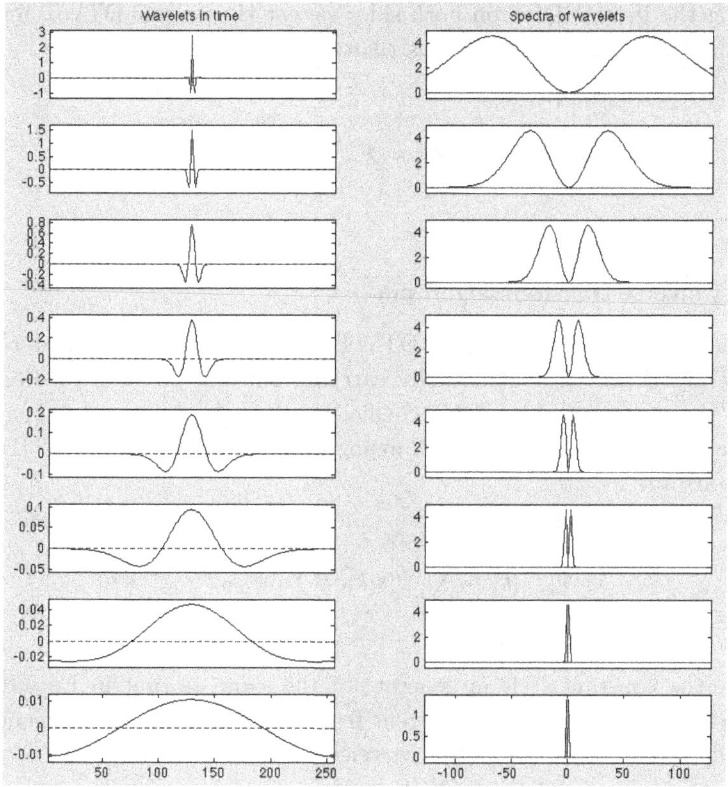

Figure 10.8 The Marr wavelets (left) and their spectra (right).

The inverse DTWT can also be more conveniently obtained in the frequency domain, similar to the derivation of the inverse transform in Eq. 10.25 for the continuous case. We first multiply both sides of Eq. (10.58) by $\Psi_{s_l,0}[k]$ and then sum both sides over all scale levels to get:

$$\sum_{l=1}^{S} \hat{X}[l,k]\Psi_{s_l,0}[k] = \sum_{l=1}^{S}[X[k]\overline{\Psi}_{s_l,0}[k]]\ \Psi_{s_l,0}[k] = X[k]\ \sum_{l=1}^{S} |\Psi_{s_l,0}[k]|^2. \quad (10.60)$$

But according to Eq. (10.23), the summation of the daughter wavelets squared over all scales is a constant; i.e., in the discrete case we have

$$\sum_{l=1}^{S} |\Psi_{s_l,0}[k]|^2 = C. \quad (10.61)$$

Now the above equation becomes

$$X[k] = \frac{1}{C}\sum_{l=1}^{S} \hat{X}[l,k]\Psi_{s_l,0}[k]. \quad (10.62)$$

Taking the inverse DFT on both sides we get the inverse DTWT by which the original time signal $x[n]$ is reconstructed:

$$x[n] = \mathcal{F}^{-1}[X[k]] = \mathcal{F}^{-1}\left[\frac{1}{C}\sum_{l=1}^{S}\hat{X}[l,k]\Psi_{s_l,0}[k]\right]. \tag{10.63}$$

10.5.3 A fast inverse transform algorithm

We now show that the inverse DTWT can be more conveniently obtained by a fast algorithm without actually carrying out Eq. (10.63). To do so, we first show that the sum of the DFT coefficients $\Psi_{s_l,0}[k] = \mathcal{F}[\psi_{s_l,0}[n]]$ (for all $n \neq 0$) of the daughter wavelets over all exponential scales $s_l = s_0(2^{1/r})^l$ (Eq. (10.56)) is a constant:

$$\sum_{l=1}^{S}\Psi_{s_l,0}[k] = \sum_{l=1}^{S}\Psi[s_l k] = \sum_{l=1}^{S}\Psi[s_0 2^{l/r}\,k] = C \tag{10.64}$$

where the constant C is in general not the same as that in Eq. (10.61). This equation holds for all k for different frequency components, independent of the specific waveform of the mother wavelet.

To prove Eq. (10.64), we first consider the corresponding situation in the continuous case, the integral of an arbitrary function $\Psi(f)$ scaled exponentially by a factor $s = b^u$:

$$\int_{-\infty}^{\infty}\Psi(b^u f)du = \int_{-\infty}^{\infty}\Psi(sf)d(\log_b s) = \frac{1}{\ln b}\int_0^{\infty}\Psi(sf)\frac{ds}{s}$$
$$= \frac{1}{\ln b}\int_0^{\infty}\frac{\Psi(sf)}{sf}d(sf) = \frac{1}{\ln b}\int_0^{\infty}\frac{\Psi(s')}{s'}ds' = C. \tag{10.65}$$

Here, we have assumed $s' = sf$, and that the integral converges to some constant. This result is independent of the variable f; i.e. the integral of all exponentially scaled versions of any function $\Psi(f)$ is a constant over the entire domain f of the function, irrespective of the specific waveform of the function. As a discrete approximation of the integral in Eq. (10.65), the summation in Eq. (10.64) should also converge to a constant, so long as the resolution of the different scales is high enough (large enough value for parameter r). For example, as shown in Fig. 10.9, the spectra of the exponentially scaled Morlet and Marr wavelets do indeed sum up to a constant over the frequency f. Note that Eq. (10.64) still holds if we take the complex conjugate on both sides; i.e., $\overline{\Psi}_{s_l,0}[k]$ also add up to a constant $\sum_{l=1}^{S}\overline{\Psi}_{s_l,0}[k] = C$. Also note that the DFTs of most typical wavelets are real $\overline{\Psi}_{s_l,0}[k] = \Psi_{s_l,0}[k]$.

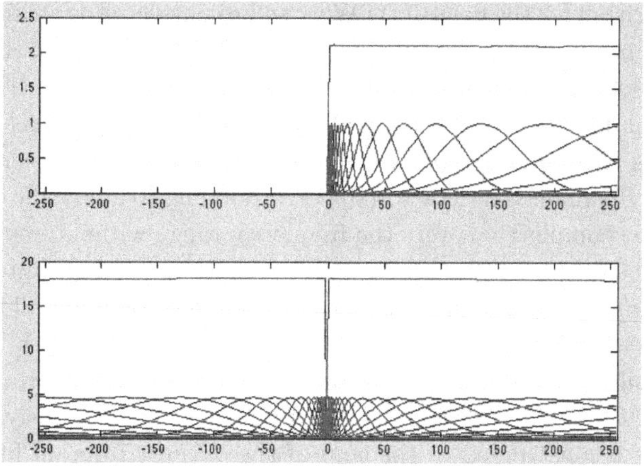

Figure 10.9 Summations of the spectra of Morlet wavelets (top) and Marr (Mexican hat) wavelets (bottom).

We are now ready to consider the fast algorithm for the inverse DTWT. Specifically we will show that the inverse DTWT can be carried out simply by summing all the DTWT coefficients obtained by Eq. (10.59):

$$\sum_{l=1}^{S} X[l,n] = \sum_{s=1}^{S} \mathcal{F}^{-1}\left[X[k]\overline{\Psi}_{s_l,0}[k]\right] = \sum_{l=1}^{S}\left[\sum_{k=0}^{N-1} X[k]\overline{\Psi}_{s_l,0}[k]e^{j2\pi nk/N}\right]$$
$$= \sum_{k=0}^{N-1} X[k]\left[\sum_{l=1}^{S}\overline{\Psi}_{s_l,0}[k]\right] e^{j2\pi nk/N} = C\sum_{k=0}^{N-1} X[k]e^{j2\pi nk/N} = C\,x[n], \quad (10.66)$$

where $C = \sum_{l=1}^{S}\overline{\Psi}_{s_l,0}[k]$ according to Eq. (10.64), and we note the last sign above is due to the inverse DFT. Now the original time signal can be trivially obtained from its DTWT coefficients:

$$x[n] = \frac{1}{C}\sum_{l=1}^{S} X[l,n] = \frac{\sum_{l=1}^{S} X[l,n]}{\sum_{l=1}^{S}\overline{\Psi}_{s_l,0}[k]}. \quad (10.67)$$

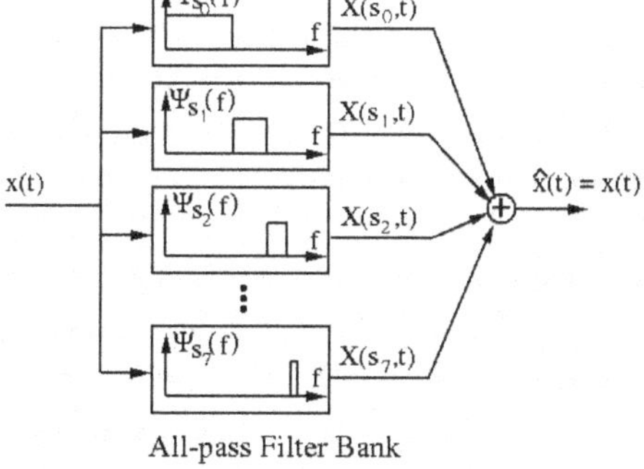

Figure 10.10 All-pass filter bank composed of BP wavelets.

This fast algorithm for the inverse DTWT can be considered as an all-pass filter bank illustrated in Fig. 10.10. We first consider the DTWT based on the Shannon dyadic wavelet, which is an ideal BP filter in the frequency domain (Eq. (10.43)) that preserves all information of the signal inside the passing band $\Delta f = f_2 - f_1$, while suppresses all frequency components outside to zero. Moreover, as shown in Fig. 10.6, the Shannon wavelets $\Psi_{s_l}(f)$ corresponding to all dyadic scales form a filter bank that completely covers the frequency range without any overlap or gap; i.e., Eq. (10.64) is indeed satisfied. Collectively these ideal BP filters form an all-pass filter bank with a constant frequency response throughout all frequencies except at $f = 0$ where $\Psi_{s_l,0}[0] = 0$ for all $l = 1, \ldots, S$ (Eq. (10.10)), as required by the admissibility condition. The outputs of these BP filters are simply the DTWT coefficients $X[l, n]$ carrying all signal information. Obviously, the signal can be perfectly reconstructed as the sum of the outputs from all filters in the filter bank, as indicated in Eq. (10.67).

The wavelet transform can therefore be represented by the all-pass filter bank shown in Fig. 10.10. The forward transform is implemented as the BP filtering process by which the DTWT coefficients $X(s_l, \tau)$ for different scales s_l and translation τ are produced, and the inverse transform is implemented as the summation of the outputs of these BP filters by which the time signal is perfectly reconstructed.

The Shannon wavelets assumed in the discussion above can be generalized to any other wavelet function, such as the Morlet and Marr wavelets. Although as BP filters they are overlapped, they still form an all-pass filter bank with constant gain over the entire frequency range due to Eq. (10.64), as shown in Fig. 10.9. The information contained in the signal is preserved collectively by all BP filters in the filter bank, and the signal can be reconstructed simply by summing their outputs.

Example 10.1: The wavelet transform of a sawtooth time signal of $N = 128$ samples is shown in Fig. 10.11. Here, we choose to use the Morlet wavelets of $S = 8$ different scale levels, corresponding to the same number of BP filters. These wavelets $\psi_{s_l}(t)$ in the time domain and their spectra $\Psi_{s_l}(f)$ in the frequency domain have already been shown in Fig. 10.7. The DTWT coefficients $X[l, n]$ corresponding to different scale levels s_l are shown on the left of Fig. 10.11, and their partial sums as the reconstructions of the signal are shown on the right, where the lth panel is the partial sum of the DTWT coefficients of the first l scale levels. We see that the approximation of the original sawtooth signal $x[n]$ improves progressively as more scale levels are included, until eventually a perfect reconstruction of the signal is obtained when all S scale levels are used.

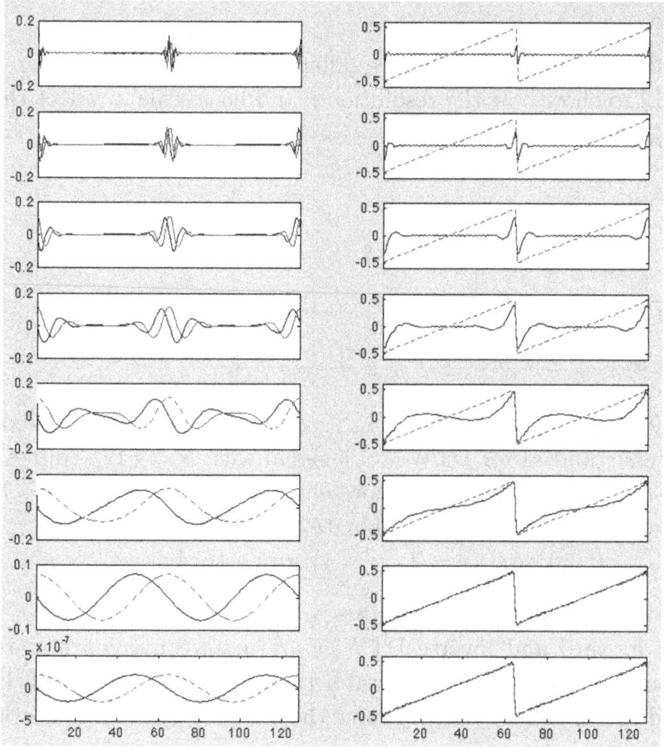

Figure 10.11 The reconstruction of a sawtooth signal (right) as the sum of its DTWT (left). The DTWT coefficients over $S = 8$ scale levels are shown on the left (solid and dashed curves for the real and imaginary parts), while the partial sums of the DTWT coefficients of l scale levels are shown on the right (solid curves), compared with the original signal (dashed curves).

10.6 Wavelet transform computation

Here, we give a few segments of C code for the implementation of both the forward and inverse DTWT discussed above.

- **Generation of S scale levels**

```
r=2;                          // scale resolution
s0=1;                         // base scale
S=r*log2((float)N/s0);        // number of scale levels
scale=alloc1df(S);            // allocate memory for S scales
for (l=0; l<S; l++) {
  scale[l]=s0*pow(2.0,(l+1)/r); // lth scale s_l
}
```

The scales corresponding to three different sets of parameters are plotted in Fig. 10.12 to show how the resolution r and base scale s_0 affect the scales s_l.

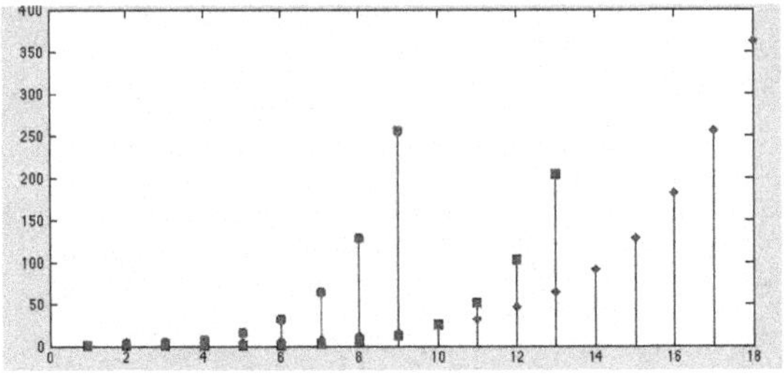

Figure 10.12 Scales s_l versus $l = 1, \ldots, S$ corresponding to different parameters r and s_0 for DTWT of a signal with $N = 512$ samples. The circles, squares, and diamonds represent $S = 9$, $S = 13$, and $S = 18$ scales, corresponding to parameter combinations $(r = 1,\ s_0 = 1)$, $(r = 1,\ s_0 = 0.05)$, and $(r = 2,\ s_0 = 1)$, respectively.

- **Generation of wavelet functions**
  As both forward and inverse DTWTs are more conveniently carried out in the frequency domain, the spectra of the wavelet functions will be specified and used in the code. First, we show the code for generating Morlet wavelets of S scales:

```
f0=0.6;                        // wavelet parameter
for (l=0; l<S; l++) {          // for all S scale levels
  for (n=0; n<N; n++) {        // for all N frequencies
    v=2*Pi*(scale[l]*((float)(n-N/2)/N)-f0); // DC in middle
    waver[l][n]=exp(-v*v/2); // spectrum (real)
    wavei[l][n]=0;             // spectrum (imaginary)
  }
}
```

Here, waver and wavei are two 2-D arrays for the real and imaginary parts of the wavelet spectrum for N samples (frequencies) and S scales. As an example, the code below generates the Marr wavelets based on Eq. (10.53):

```
for (l=0; l<S; l++) {          // for all S scale levels
  for (n=0; n<N; n++) {        // for all N frequencies
    v=a*scale[l]*(n-N/2)/N;    // DC in middle
    waver[l][n]=4*Pi*Pi*v*v*exp(-Pi*v*v); // spectrum (real)
    wavei[l][n]=0;             // spectrum (imaginary)
  }
}
```

- **The forward DTWT**
  Here, we assume the real and imaginary parts of the time signal are stored in two $N \times 1$ arrays `xr` and `xi`, respectively, and the real and imaginary parts of the DTWT of the time signal are stored in two $S \times N$ arrays `Xr` and `Xi` for wavelet coefficients of S scales and N time translations:

```
dft(xr,xi,N,0);                  // DFT of signal
for (l=0; l<S; l++) {            // for all S scale levels
  for (n=0; n<N; n++) {          // for all N frequencies
    Xr[l][n]=xr[n]*waver[l][n]+xi[n]*wavei[l][n];
    Xi[l][n]=xi[n]*waver[l][n]-xr[n]*wavei[l][n];
  }
  dft(Xr[l],Xi[l],N,1);          // inverse DFT, back to time
}
```

- **The inverse DTWT**
  Here, we only give the code for the inverse DTWT based on Eq. (10.63). The code based on Eq. (10.67) is trivial and not listed. Again, the real and imaginary parts of the DTWT coefficients are stored in the two $S \times N$ arrays `Xr` and `Xi`, and the real and imaginary parts of the reconstructed time signal are in two $N \times 1$ arrays `yr` and `yi`, respectively.

```
for (n=0; n<N; n++) {
  yr[n]=yi[n]=0;                 // initialization
  for (l=0; l<S; l++) {          // for all S scale levels
    dft(Xr[l],Xi[l],N,0);        // DFT of DTWT coefficients
    for (n=0; n<N; n++) {
      yr[n]=yr[n]+Xr[l][n]*waver[l][n]-Xi[l][n]*wavei[l][n];
      yi[n]=yi[n]+Xr[l][n]*wavei[l][n]+Xi[l][n]*waver[l][n];
    }
    dft(yr,yi,N,1);              // inverse DFT back to time
  }
}
```

A set of typical signals and their DTWT transforms based on both the Marr and Morlet wavelets are shown in Fig. 10.13 in image forms. These signals include sinusoids and their combinations, a chirp signal (a sinusoid with continuously changing frequency), square, sawtooth, and triangle waves, impulse train, and random noise.

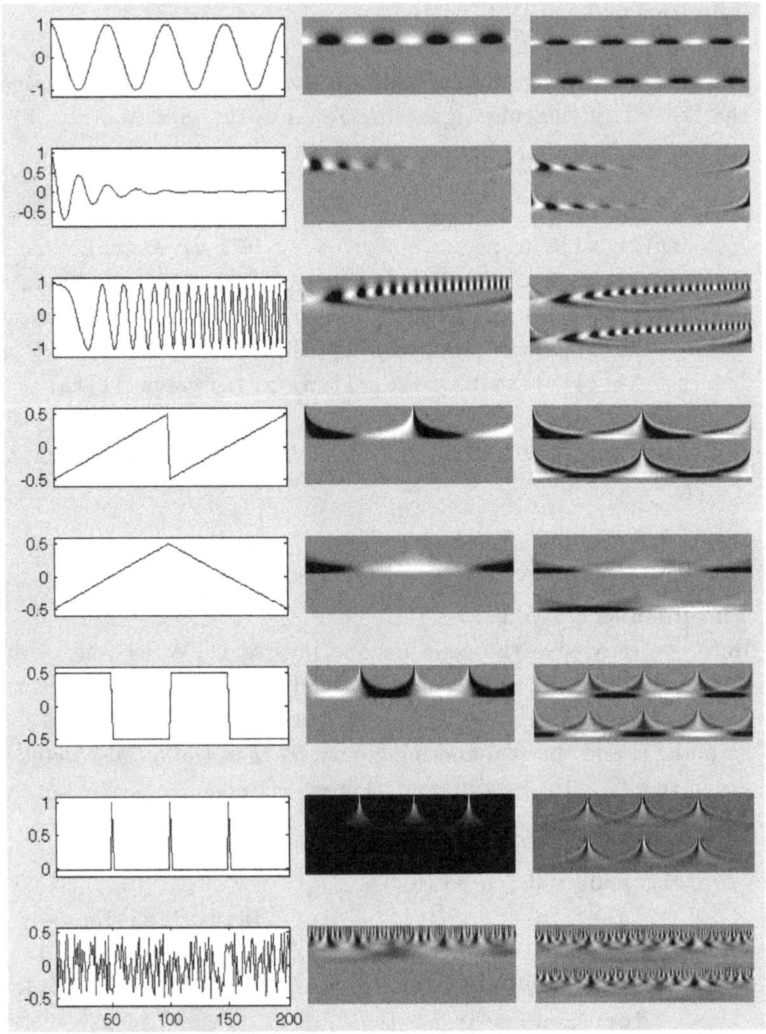

Figure 10.13 Typical signals (left) and their real DTWT based on the Marr wavelets (middle) and complex DTWT based on Morlet wavelets. The real and imaginary parts of the Morlet DTWT are shown in the upper and lower parts, respectively.

10.7 Filtering based on wavelet transform

Similar to Fourier filtering (LP, HP, BP, etc.) that takes place in the frequency domain, various wavelet filtering can also be carried out in the transform domain where the wavelet coefficients $X[l, n]$ are modified to achieve certain desired effects for purposes such as noise reduction and information extraction. Here, we consider a set of examples that illustrate the filtering effects based on the wavelet transform in comparison with those based on the Fourier transform.

Example 10.2: The monthly Dow Jones Industrial Average (DJIA) index as a time function and its Fourier spectrum are plotted in the top two panels of Fig. 10.14. The LP filtered Fourier spectrum is plotted in panel 3. Similar LP filtering is also carried out based on the wavelet transform (Morlet), as shown in Fig. 10.15. The LP filtered data obtained by both the Fourier and wavelet transforms are re-plotted as the solid and dashed curves respectively in panel 4 of Fig. 10.14, in comparison with the original one as the dotted curve. We see that the LP filtered curves by both transform methods are very similar to each other, and, as expected, they are both much smoother than the original one.

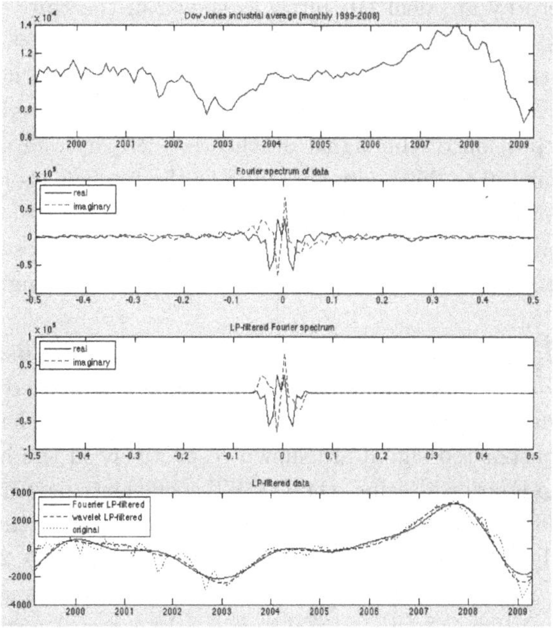

Figure 10.14 The four panels are, respectively, the DJIA index (1999 – 2008), its Fourier spectrum, the LP filtered spectrum, and the LP filtered data by both the Fourier and wavelet transforms.

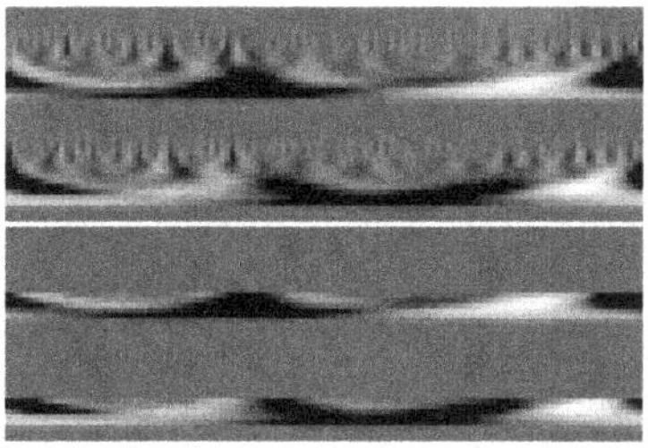

Figure 10.15 LP filtering of DJIA data based on Morlet wavelet transform. The DTWT coefficients before and after LP filtering are shown respectively in the top and bottom panels. The coefficients suppressed to zero appear gray.

Example 10.3: A chirp is a sinusoidal signal whose frequency is monotonically and continuously changing, either linearly or exponentially. Here we compare the filtering effects of an exponential chirp based on both the Fourier transform and the wavelet transform. As the frequency changes over time, it may seem that filtering out a certain frequency should only affect the signal locally in the time segment corresponding to the frequency removed. However, this is not actually the case if the filtering is carried out in the Fourier domain.

A chirp and its Fourier spectrum are shown respectively in the first and second panels of Fig. 10.16. Then, certain frequency components in the spectrum are suppressed to zero by an ideal BP filter, as shown in the third panel. The signal is then reconstructed by the inverse Fourier transform, as shown in the bottom panel. Note that although only the frequency components within a relatively narrow band are suppressed, the entire time signal is affected, including the slowly changing portion of the signal on the very left, as well as the time interval (roughly from 150 to 250) corresponding to the frequencies suppressed. This is due to the nature of the Fourier transform that the frequency information is extracted from the entire time span of the signal, and those frequency components that are suppressed also contribute to the slowly changing portion of the signal as well.

On the other hand, the filtering based on the wavelet transform demonstrates some different effect, as shown in Fig. 10.17, where the same chirp and its DTWT coefficients are shown in the top two panels, and the filtering in the transform domain and the reconstructed signal are shown respectively in the bottom two panels. Similar to the Fourier filtering, the DTWT coefficients here inside a certain band of scale levels are suppressed to zero. However, different from the Fourier filtering, only a local portion (also roughly from 150 to 250) of the reconstructed signal corresponding to the suppressed scale levels is significantly affected, while the waveforms of the signal outside the interval remain mostly the same. This very different filtering effect reflects the fact that the wavelet transform possesses temporal locality as well as frequency (scale levels) locality.

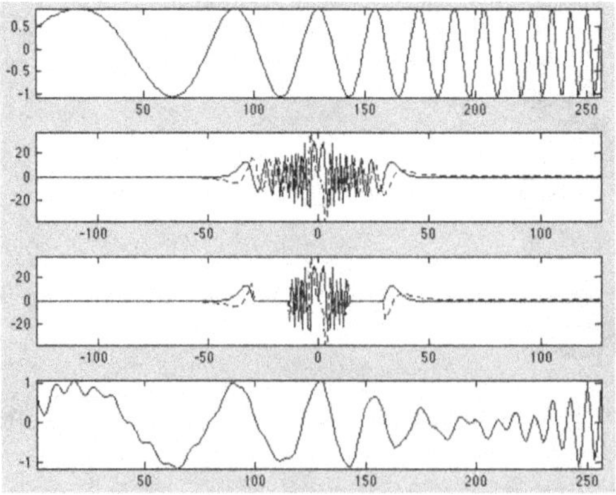

Figure 10.16 Fourier filtering of chirp signal.

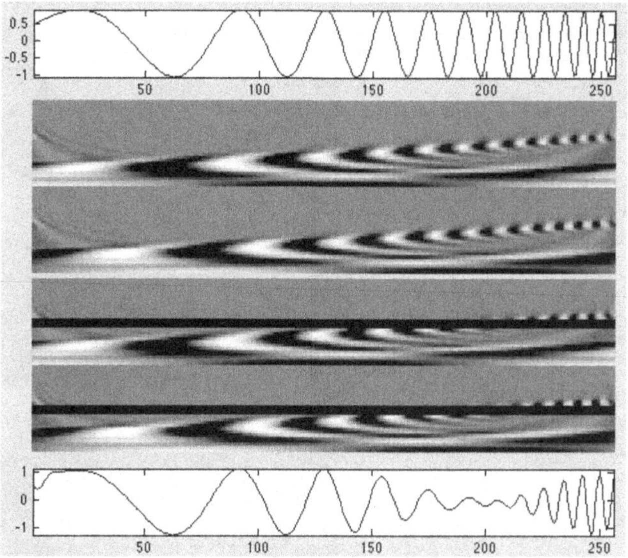

Figure 10.17 CTWT filtering of chirp signal.

Example 10.4: One weakness of the Fourier transform is that it is insensitive to non-stationary characteristics in the signal because the frequency information is extracted from the entire signal duration without temporal locality. Here, we consider a signal before and after it is contaminated by some spiky noise, as shown on the top and bottom panels on the left of Fig. 10.18, and the corresponding Fourier spectra shown on the right. As we can see, the spiky noise has a very wide energy distribution spreading over the entire spectrum; i.e., all frequency components of the signal are affected by the noise. In particular, some of the weaker frequency components in the signal are completely overwhelmed by the noise, and it is obvious that separating the noise from the signal by Fourier filtering is extremely difficult.

This problem of noise removal can be addressed by wavelet filtering, as shown in Fig. 10.19. The original signal and its reconstructions after HP and LP filtering are shown respectively in the top, middle, and bottom panels on the left, while the corresponding wavelet coefficients are shown on the right. We see that it is now possible to separate the noise from the signal by wavelet filtering, due obviously to the temporal locality of the wavelet transform. The spiky noise is separated out by HP filtering (middle left), while the signal is reasonably recovered after LP filtering (lower left).

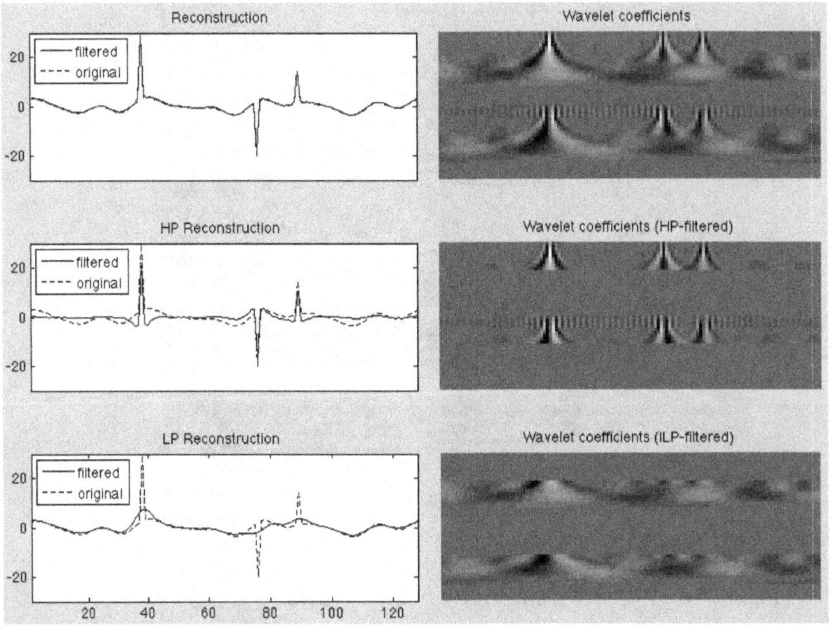

Figure 10.19 Separation of the signal and noise by wavelet filtering. The original signal with spiky noise and its reconstructions after HP and LP filtering are shown respectively in the top, middle, and bottom panels on the left, while their wavelet coefficients are shown in the corresponding panels on the right.

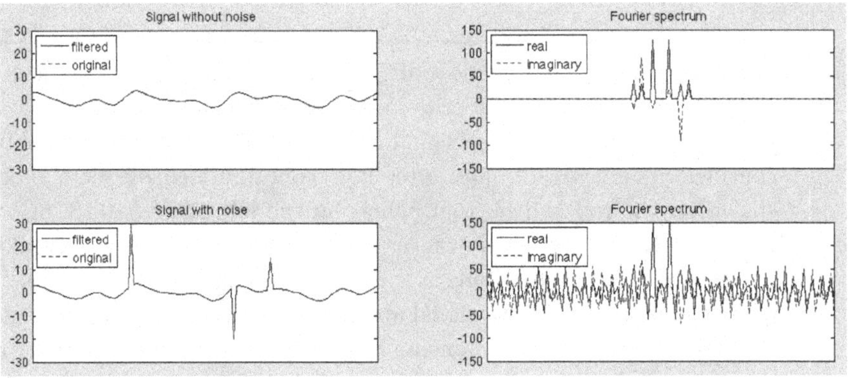

Figure 10.18 A noise-contaminated signal and its Fourier spectrum.

Example 10.5: The annual average temperature in the Los Angeles area from 1878 to 1997 (NOAA National Weather Service Center in the USA) is shown in the top panel of Fig. 10.20 (solid curve). The data clearly show an upward trend of the annual temperature, with a 5.57°F total rise over the 120 years, with an average annual increase of 0.0464° F.

The upward drift in the data can be removed in the time domain. We first find the linear regression of the curve in terms of the slope and the intercept representing the trend, and then subtract it from the data. The result is shown as the dashed curve in the top panel of Fig. 10.20. We next consider if and how this could also be done by filtering in either the Fourier or wavelet transform domain.

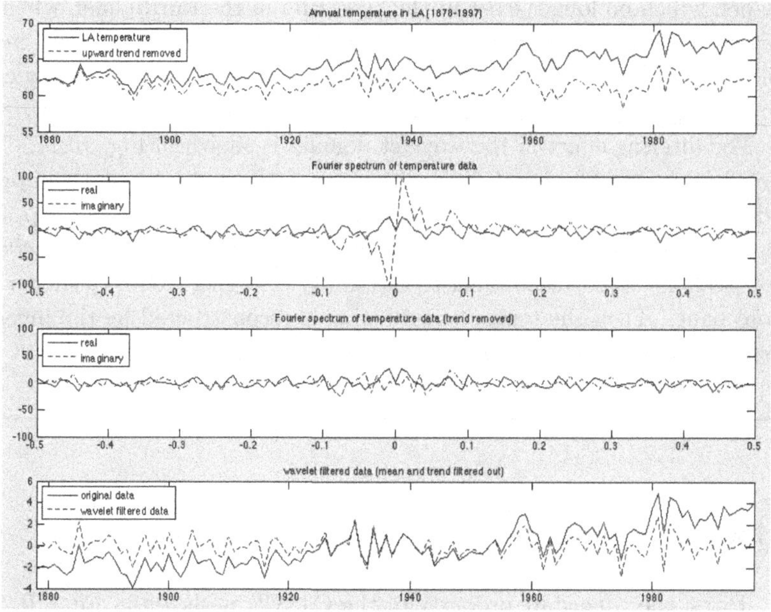

Figure 10.20 Annual temperature in LA area (1878–1997).

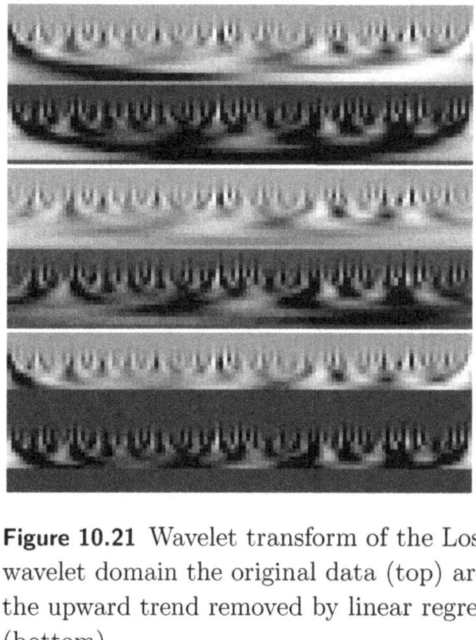

Figure 10.21 Wavelet transform of the Los Angeles temperature data. In the wavelet domain the original data (top) are compared with the same data with the upward trend removed by linear regression (middle), and by LP filtering (bottom).

The Fourier spectra of the temperature data with and without the upward drift are shown in the second and third panel of Fig. 10.20. We see that their real parts are the same, but their imaginary parts differ significantly at the low-frequency region as the upward trend is an odd function, represented by both the positive and negative peaks in the imaginary part of the spectrum in the second panel, which no longer exist in the spectrum in the third panel, when this trend is removed. It is difficult to separate out the slowly changing trend from the rest of the signal by filtering in the frequency domain, as their frequency components are mixed.

The filtering effect in the wavelet domain is shown in Fig. 10.21. The wavelet coefficients of the signal before and after the removal of the upward trend (detected by linear regression) are shown respectively in the top and middle panels. Also LP filtering is carried out by suppressing the wavelet coefficients of low scale levels corresponding to the slowly changing trend, as shown in the bottom panel. Then the temperature signal is reconstructed by the inverse wavelet transform, as shown in the bottom panel of Fig. 10.20. We see that indeed the upward trend is removed by wavelet filtering.

10.8 Homework problems

1. Prove the time shift property of the CTWT as shown in Eq. (10.28).
2. Prove the time scaling property of the CTWT as shown in Eq. (10.29).
3. Show that if the center and width of a compactly supported mother wavelet function $\psi(t)$ are respectively t_0 and Δt, then those of a daughter wavelet $\psi_{s,\tau}(t) = \psi((t-\tau)/s)/\sqrt{s}$ are as shown in Eq. (10.30).
4. Develop an m-file in Matlab to implement the DTWT algorithm for both the forward and inverse transform. Generate the DTWT of the eight signals in Fig. 10.13 based on first the Morlet wavelets and then the Marr wavelet (Mexican hat).
5. Generate the following two signals in Matlab with $f_1 = 5$ and $f_2 = 25$:
   - First,
   $$x_1[n] = \cos(2\pi n f_1/N) + \cos(2\pi n f_2/N). \tag{10.68}$$
   - Second, composed of two halves of sinusoids of different frequencies
   $$\begin{aligned} x_2[n] &= \cos(2\pi n f_1/N) \qquad n = 0, \ldots, N/2-1, \\ x_2[n] &= \cos(2\pi n f_2/N) \qquad n = N/2, \ldots, N-1. \end{aligned} \tag{10.69}$$

   For the purpose of separating the two frequencies f_1 and f_2 contained in both signals $x_1[n]$ and $x_2[n]$, design a two-channel filter bank composed of two filters so that they each output one of the two frequencies. Carry out this approach based on both Fourier filtering and wavelet filtering.

6. As seen in the text, the wavelet transform can achieve locality in both the temporal and frequency domains, which is desirable in representing, detecting, and possibly removing, if so desired, certain temporal signal features that are either local (such as irregular spikes) or non-stationary (such as long-term effects of trend or non-periodic frequency change). Obtain datasets of your own choice that contain such characteristics and carry out filtering to separate such features with the rest of the signal in both the Fourier frequency domain and wavelet transform domain. Compare the filtering effects of both methods.
7. Repeat Example 10.3 using Marr wavelets.
8. Repeat Example 10.4 using Marr wavelets.
9. Repeat Example 10.5 using Marr wavelets.

11 Multiresolution analysis and discrete wavelet transform

In Chapter 10 we considered the CTWT that converts a signal $x(t)$ in the 1-D time domain into a 2-D function $X(s,\tau)$ in the transform domain, based on the kernel functions $\psi_{s,\tau}(t)$, which are non-orthogonal and redundant. Now we will consider the concept of multiresolution analysis (MRA), also called multiscale approximation (MSA), based on which various orthogonal and bi-orthogonal wavelets can be constructed as bases that span the function space $L^2(\mathbb{R})$, as in the case for all the orthogonal transforms discussed before. The discrete implementation of this method is called the *discrete wavelet transform (DWT)*, not to be confused with the DTWT previously discussed.

11.1 Multiresolution analysis (MRA)

11.1.1 Scale spaces

We can discretize both parameters s and τ in the wavelet function $\psi_{s,\tau}(t)$ defined in Eq. (10.13) in a dyadic manner so that it becomes:

$$\psi_{j,k}(t) = \frac{1}{\sqrt{2^{-j}}}\psi\left(\frac{t-2^{-j}k}{2^{-j}}\right) = 2^{j/2}\psi(2^j t - k) \qquad j,k \in \mathbb{Z} = \{\dots,-1,0,1,\dots\}. \tag{11.1}$$

The mother wavelet $\psi(t)$ is either expanded (dilated) if $j < 0$, or compressed if $j > 0$. In either case, it is also translated by an integer amount in time to the right if $k > 0$ or to the left if $k < 0$. While constructing the specific mother wavelet function $\psi(t)$, we can further impose the orthogonality requirement so that all wavelets $\psi_{j,k}(t)$ are orthogonal with respect not only to integer translation (in terms of k) but also the dyadic scaling (in terms of j). In other words, at any given scale level j, these wavelets form an orthogonal basis that spans a space at the level, and all bases across different scale levels are also orthogonal to each other. In the following, we will develop the theory for the construction of such a set of orthogonal wavelet basis functions across different scale levels.

Definition: An MRA is a sequence of nested *scale spaces* $V_j \subset L^2(\mathbb{R})$,

$$\{0\} = V_{-\infty} \subset \cdots \subset V_{-2} \subset V_{-1} \subset V_0 \subset V_1 \subset V_2 \subset \cdots \subset V_\infty = L^2(\mathbb{R}), \tag{11.2}$$

that satisfies the following conditions:

- Completeness. The union of the nested spaces is the entire function space and their intersection is a set containing 0 as its only member:

$$\cup_{j\in\mathbb{Z}}V_j = V_\infty = L^2(\mathbb{R}), \qquad \cap_{j\in\mathbb{Z}}V_j = V_{-\infty} = \{0\}. \tag{11.3}$$

- Self-similarity in scale.

$$x(t) \in V_0 \quad \text{iff} \quad x(2^j t) \in V_j, \quad j \in \mathbb{Z}. \tag{11.4}$$

- Self-similarity in translation.

$$x(t) \in V_0 \quad \text{iff} \quad x(t-k) \in V_0, \quad k \in \mathbb{Z}. \tag{11.5}$$

- Existence of a function $\theta(t) \in V_0$ so that the family $\{\theta(t-k)\,|(k \in \mathbb{Z})\}$ is a Riesz basis (linearly independent frame, Definition 2.25) that spans V_0.

$$V_0 = \text{span}(\theta(t-k),\ k \in \mathbb{Z}). \tag{11.6}$$

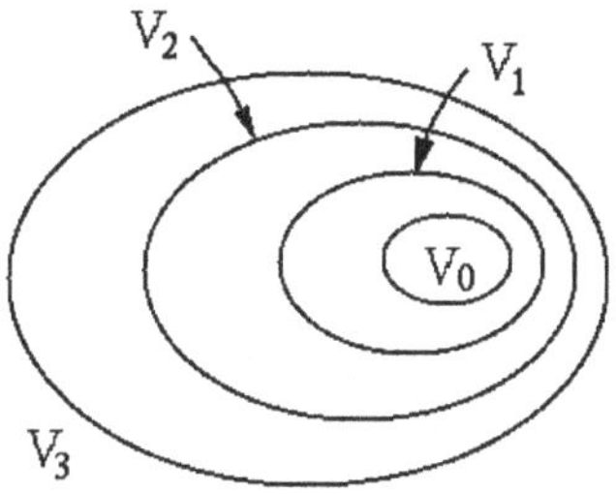

Figure 11.1 The nested V_j spaces for MRA.

The self-similarities in scale and translation can be combined if we relate Eq. (11.4) to Eq. (11.5) with t replaced by $2^j t$:

$$x(t) \in V_0 \quad \text{iff} \quad x(2^j t) \in V_j \quad \text{iff} \quad x(2^j t - k) \in V_j. \tag{11.7}$$

If we define another function $y(t) = x(2^j t)$, then the two self-similarities above can be expressed as

$$y(t) \in V_j \quad \text{iff} \quad y(t - 2^{-j}k) \in V_j; \tag{11.8}$$

i.e., any function in V_j translated by $2^{-j}k$ is still in V_j. This sequence of nested scale spaces is illustrated in Fig. 11.1.

The significance of the scale spaces V_j $(j \in \mathbb{Z})$ is that any given function $x(t) \in L^2(\mathbb{R})$ can be approximated in any one of these spaces V_j with different resolutions or levels of details, and the greater j, the better the approximations. Owing to the dyadic scaling, the resolution of V_{j+1} is twice that of V_j. We further consider the following two cases with $j > 0$.

- Space V_j is spanned by basis $\theta(2^j t)$ which is 2^j times narrower than $\theta(t)$ that spans V_0. Therefore, it is capable of representing smaller scale or more detailed information in a signal $x(t)$; i.e., $V_0 \subset V_j$. In particular, when $j \to \infty$, a basis function of V_∞ is maximally compressed to become an impulse function and the space a family of such basis function spans becomes the entire $L^2(\mathbb{R})$ in which any details in a signal can be represented as (Eq. (1.6)):

$$\int_{-\infty}^{\infty} x(\tau)\delta(t-\tau)\,d\tau = x(t) \in L^2(\mathbb{R}). \tag{11.9}$$

- Space V_{-j} is spanned by basis $\theta(2^{-j}t)$ which is 2^j times wider than $\theta(t)$ that spans V_0, it can therefore only represent larger scale or less detailed information in a signal $x(t)$; i.e., $V_{-j} \subset V_0$. In particular, when $j \to \infty$, the basis function of $V_{-\infty}$ is expanded to have an infinite width but zero height, a constant 0 for all t, and the corresponding space becomes $\{0\}$, containing 0 as its only member.

Based on the Riesz basis $\theta(t) \in V_0$, a *father wavelet* $\phi(t)$ can be constructed in the frequency domain as

$$\Phi(f) = \mathcal{F}[\phi(t)] = \frac{\Theta(f)}{[\sum_k |\Theta(f-k)|^2]^{1/2}}, \qquad k \in \mathbb{Z}, \tag{11.10}$$

where $\Theta(f) = \mathcal{F}[\theta(t)]$ is the Fourier spectrum of $\theta(t)$.

Now we show that the father wavelet so defined is orthogonal to itself shifted by any integer amount k; i.e.,

$$\langle \phi(t-k), \phi(t) \rangle = \int_{-\infty}^{\infty} \phi(t-k)\overline{\phi}(t)\,dt = \delta[k], \qquad k \in \mathbb{Z}. \tag{11.11}$$

As the inner product is actually the autocorrelation of $\phi(t)$ evaluated at $t = k$, the equation above can be expressed as the product of the autocorrelation $r_\phi(\tau)$ and an impulse train with unity interval:

$$\int_{-\infty}^{\infty} \phi(t-\tau)\overline{\phi}(t)\,dt\Big|_{\tau=k\in\mathbb{Z}} = r_\phi(\tau)\Big|_{\tau=k\in\mathbb{Z}} = r_\phi(\tau)\sum_{k\in\mathbb{Z}} \delta(\tau-k) = \delta[k]. \tag{11.12}$$

This product in the time domain corresponds to a convolution in the frequency domain:

$$|\Phi(f)|^2 * \sum_{k\in\mathbb{Z}} \delta(f-k) = \sum_{k\in\mathbb{Z}} |\Phi(f-k)|^2 = 1, \tag{11.13}$$

where $|\Phi(f)|^2 = \mathcal{F}[r_\phi(t)]$ (Eq. (3.119)) and $\sum_{k\in\mathbb{Z}} \delta(f-k) = \mathcal{F}[\sum_{k\in\mathbb{Z}} \delta(\tau-k)]$ (Eq. (3.173)). We see that this equation is indeed satisfied by $\Phi(f)$ constructed in Eq. (11.10), and consequently the orthogonality of Eq. (11.11) in the time domain is also satisfied. Now the father wavelet $\phi(t) = \mathcal{F}^{-1}[\Phi(f)]$ can be used to form an orthogonal basis to span V_0:

$$V_0 = \text{span}(\phi(t-k),\ k \in \mathbb{Z}). \tag{11.14}$$

The result in Eq. (11.11) for V_0 can be generalized to any V_j by replacing t by $2^j t$ in the equation to get

$$\int_{-\infty}^{\infty} \phi(2^j t-k)\overline{\phi}(2^j t)\,d(2^j t) = \int_{-\infty}^{\infty} \sqrt{2^j}\phi(2^j t-k)\sqrt{2^j}\overline{\phi}(2^j t)\,dt$$
$$= \langle \phi_{j,k}(t), \phi_{j,0}(t) \rangle = \delta[k]. \tag{11.15}$$

Here $\phi_{j,k}(t)$ is a set of *scaling functions* defined as

$$\phi_{j,k}(t) = \sqrt{2^j}\phi(2^j t - k) = 2^{j/2}\phi(2^j t - k) \in V_j \qquad k \in \mathbb{Z}, \tag{11.16}$$

which can be used as an orthogonal basis to span V_j:

$$V_j = \text{span}(\phi_{j,k}(t),\ k \in \mathbb{Z}). \tag{11.17}$$

In particular, when $j = 0$, $\phi_{0,k}(t) = \phi(t-k)$ and the expression above becomes Eq. (11.14). Now any $x(t) \in V_j$ can be represented in terms of the scaling functions as

$$x(t) = \sum_{k\in\mathbb{Z}} \langle x(t), \phi_{j,k}(t)\rangle \phi_{j,k}. \tag{11.18}$$

The scaling functions $\phi_{j,k}(t)$ in space V_j are also related to those in other levels. Specifically, $\phi(t) \in V_0 \subset V_1$ can be expressed in terms of the orthogonal basis $\phi_{1,k}(t) = \sqrt{2}\phi(2t-k) \in V_1$:

$$\phi(t) = \sum_{k\in\mathbb{Z}} h_0[k]\phi_{1,k}(t) = \sqrt{2}\sum_{k\in\mathbb{Z}} h_0[k]\phi(2t-k), \tag{11.19}$$

where the coefficients $h_0[k]$ can be found as the projection of $\phi(t)$ onto the kth basis function $\phi_{1,k}(t) = \sqrt{2}\phi(2t-k)$:

$$h_0[k] = \langle \phi(t), \sqrt{2}\phi(2t-k)\rangle = \sqrt{2}\int_{-\infty}^{\infty} \phi(t)\overline{\phi}(2t-k)\,dt. \tag{11.20}$$

The relationship between V_0 and V_1 can be further generalized to V_j and V_{j+1}. Replacing t by $2^j t - l$ in Eq. (11.19), we get

$$\phi(2^j t - l) = \sqrt{2}\sum_{k\in\mathbb{Z}} h_0[k]\phi(2(2^j t - l) - k) = \sqrt{2}\sum_{k\in\mathbb{Z}} h_0[k]\phi(2^{j+1}t - (2l+k)). \tag{11.21}$$

But due to Eq. (11.16), the above can be written as

$$\phi_{j,l}(t) = \phi(2^j t - l) = \sum_{k\in\mathbb{Z}} h_0[k]\phi_{j+1,2l+k}(t) = \sum_{k'\in\mathbb{Z}} h_0[k'-2l]\phi_{j+1,k'}(t), \tag{11.22}$$

where we have assumed $k' = 2l + k$. Comparing this equation with a discrete convolution $y[l] = h[l] * x[l] = \sum_k h[l-k]x[k]$, we see that it can be considered as a convolution under two conditions: (1) the coefficients are time reversed; (2) the output is down-sampled. In other words, the equation actually describes a discrete FIR filter with $h_0[k]$ as its impulse response, called a *scaling filter*, followed by a down-sampler. As the resolution of the output $\phi_{j,l}(t) \in V_j$ is lower than that of the input $\phi_{j+1,k'}(t) \in V_{j+1}$, this scaling filter is a LP filter.

This filtering process can also be described in the frequency domain. Taking the Fourier transform of Eq. (11.19), we get

$$\begin{aligned}\Phi(f) &= \int_{-\infty}^{\infty} \phi(t)e^{-j2\pi ft}\,dt = \sqrt{2}\sum_{k\in\mathbb{Z}} h_0[k]\int_{-\infty}^{\infty}\phi(2t-k)e^{-j2\pi ft}\,dt \\ &= \sqrt{2}\sum_{k\in\mathbb{Z}} h_0[k]\int_{-\infty}^{\infty}\phi(t')e^{-j2\pi f(t'+k)/2}\,d\left(\frac{t'}{2}\right) \\ &= \frac{1}{\sqrt{2}}\sum_{k\in\mathbb{Z}} h_0[k]e^{-jk\pi f}\int_{-\infty}^{\infty}\phi(t')e^{-j2\pi ft'/2}\,dt' \\ &= \frac{1}{\sqrt{2}}H_0\left(\frac{f}{2}\right)\Phi\left(\frac{f}{2}\right), \end{aligned} \tag{11.23}$$

where $t' = 2t - k$, and $H_0(f)$ is the DTFT spectrum of the discrete impulse response $h_0[k]$; i.e., the frequency response function of the scaling filter:

$$H_0(f) = \mathcal{F}[h_0[k]] = \sum_{k\in\mathbb{Z}} h_0[k]e^{-j2k\pi f}. \tag{11.24}$$

Note that as the time gap between neighboring samples of $h_0[k]$ is $t_0 = 1$ (sampling frequency $F = 1/t_0 = 1$, $H_0(f \pm 1) = H_0(f)$ is periodic with period of $F = 1$; i.e., $H(f \pm 1) = H(f)$ and $H_0(f + 1/2) = H_0(f - 1/2)$.

Equation (11.23) can be further expanded recursively:

$$\begin{aligned}\Phi(f) &= \frac{1}{\sqrt{2}}H_0\left(\frac{f}{2}\right)\left[\frac{1}{\sqrt{2}}H_0\left(\frac{f}{4}\right)\Phi\left(\frac{f}{4}\right)\right] = \cdots \\ &= \prod_{j=1}^{\infty}\frac{1}{\sqrt{2}}H_0\left(\frac{f}{2^j}\right)\Phi(0) = \prod_{j=1}^{\infty}\frac{1}{\sqrt{2}}H_0\left(\frac{f}{2^j}\right). \end{aligned} \tag{11.25}$$

The last equal sign is based on the assumption that $\phi(t)$ is normalized; i.e., its DC component is unity:

$$\Phi(0) = \int_{-\infty}^{\infty}\phi(t)e^{-j2\pi ft}\,dt\Big|_{f=0} = \int_{-\infty}^{\infty}\phi(t)\,dt = 1. \tag{11.26}$$

The summation index in the discussion above always takes values in the set of integers; e.g., $k \in \mathbb{Z}$. For simplicity, In the following we will only specify the summation index without explicitly showing the limits.

Example 11.1: Consider a father function defined as

$$\phi(t) = \begin{cases} 1 & 0 < t < 1 \\ 0 & \text{else} \end{cases}. \tag{11.27}$$

This is a square impulse which is indeed orthogonal to itself translated by any integer k:

$$\langle\phi(t), \phi(t-k)\rangle = \int_{-\infty}^{\infty}\phi(t)\phi(t-k)\,dt = \delta[k], \qquad k \in \mathbb{Z}. \tag{11.28}$$

Based on this father function, we can construct a set of scaling functions $\phi_{0,k}(t)$ that spans V_0. Any function $x(t) \in L^2(\mathbb{R})$ can be approximated in V_0:

$$x(t) \approx \sum_k c_k \phi_{0,k}(t) = \sum_k c_k \phi(t-k). \tag{11.29}$$

Replacing t in $\phi_{0,k}(t) = \phi(t-k)$ by $2^j t$ and including a normalization factor $2^{j/2}$, we get another set of orthonormal functions:

$$\phi_{j,k}(t) = 2^{j/2}\phi(2^j t - k) \qquad k \in \mathbb{Z}. \tag{11.30}$$

As $\phi(t) = 1$ when its argument satisfies $0 < t < 1$, we also have $\phi(2^j t - k) = 1$ when $0 < 2^j t - k < 1$; i.e.,

$$\frac{k}{2^j} < t < \frac{k}{2^j} + \frac{1}{2^j}. \tag{11.31}$$

We see that $\phi_{j,k}(t)$ is a rectangular impulse of height $2^{j/2} = \sqrt{2^j}$ and width $1/2^j$, and it is shifted k times its width. Obviously, these functions are also orthonormal and they span space V_j:

$$\langle \phi_{j,k}(t), \phi_{j,l}(t) \rangle = \delta[k-l] \qquad k, l \in \mathbb{Z}. \tag{11.32}$$

The basic ideas above are illustrated in Fig. 11.2. The first two panels show two scaling functions $\phi(t) = \phi_{0,0}(t)$ and $\phi_{0,1}(t) = \phi(t-1)$, both in V_0; the next two panels show another two scaling functions $\phi_{1,0}(t) = \sqrt{2}\phi(2t)$ and $\phi_{1,1}(t) = \sqrt{2}\phi(2t-1)$ in V_1. Panel 5 shows a function $x(t) \in V_1$ represented as a linear combination of the scaling functions $\phi_{1,k}(t)$:

$$x(t) = 0.5\phi_{1,0}(t) + \phi_{1,1}(t) - 0.25\phi_{1,4}(t). \tag{11.33}$$

Finally, panel 6 shows that a scaling function $\phi_{0,0}(t) \in V_0$ represented as a linear combination of the basis functions $\phi_{1,k}(t) \in V_1$ (Eq. (11.22)):

$$\phi_{0,l}(t) = h_0[0]\phi_{1,2l}(t) + h_0[1]\phi_{1,2l+1}(t) = \frac{1}{\sqrt{2}}\phi_{1,2k}(t) + \frac{1}{\sqrt{2}}\phi_{1,2k+1}(t), \tag{11.34}$$

where the coefficients $h_0[0] = h_0[1] = 1/\sqrt{2}$ are obtained according to Eq. (11.20). The ideas illustrated in this example are valid in general if the square impulses are replaced by any family of functions with compact support; i.e., the functions are non-zero only over a finite duration.

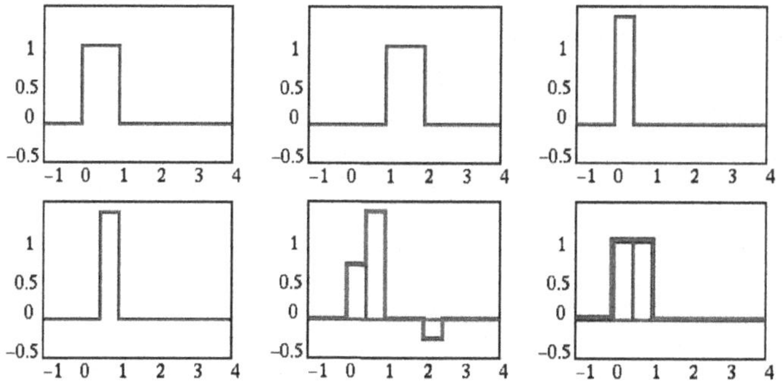

Figure 11.2 The basis functions that span scale spaces and some functions they represent.

11.1.2 Wavelet spaces

Previously we constructed a sequence of nested scale spaces $V_j \subset V_{j+1}$ in which a given function $x(t) \in L^2(\mathbb{R})$ can be approximated at different scale levels; i.e., the approximation in V_{j+1} contains more detailed information in the signal than that in V_j. In other words, certain functions in V_{j+1} but not representable in V_j are contained in the difference space $W_j = V_{j+1} - V_j$, called the *wavelet space.* As $W_j \subset V_{j+1}$, $V_j \subset V_{j+1}$, and $W_j \cap V_j = \{0\}$, W_j is the complementary space of V_j; i.e., V_{j+1} is the direct sum of V_j and W_j, this relationship can be carried out recursively:

$$V_{j+1} = W_j \oplus V_j = W_j \oplus W_{j-1} \oplus V_{j-1} = \cdots . \tag{11.35}$$

This result indicates that the approximation of $x(t)$ in V_j can be improved by including more detailed information in W_j, so that $x(t)$ is now approximated in $V_{j+1} = W_j \oplus V_j$ of higher resolution. As can be seen from Fig. 11.3, this improvement can be continued if we start at an arbitrary initial level such as V_0, and keep including more detailed information contained in W_j when $j \to \infty$, so that the signal can be ever more precisely approximated:

$$\left[\oplus_{j=0}^{\infty} W_j\right] \oplus V_0 = L^2(\mathbb{R}). \tag{11.36}$$

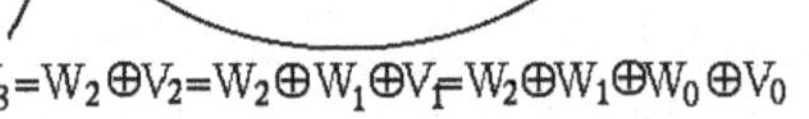

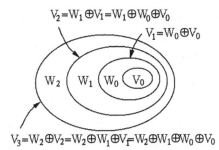

Figure 11.3 The nested V_j and W_j spaces for MRA.

As with the scale space V_0 that is spanned by a set of orthogonal scaling functions $\phi_{0,k} = \phi(t-k)$ derived from a father wavelet $\phi(t)$, we assume here the wavelet space W_0 is also spanned by a set of orthogonal wavelet functions $\psi(t-k)$ that are derived from a *mother wavelet* $\psi(t)$. Similar to Eq. (11.11), these wavelet functions are required to be orthogonal to themselves shifted by any integer amount:

$$\langle \psi(t-k), \psi(t) \rangle = \int_{-\infty}^{\infty} \psi(t-k)\overline{\psi}(t)\, dt = \delta[k] \qquad k \in \mathbb{Z}, \tag{11.37}$$

and they span the space W_0:

$$W_0 = \text{span}(\psi(t-k),\ k \in \mathbb{Z}). \tag{11.38}$$

Moreover, the mother and father wavelets are required to be orthogonal to each other with any integer shift:

$$\langle \phi(t-k), \psi(t) \rangle = \int_{-\infty}^{\infty} \phi(t-k)\overline{\psi}(t)\,dt = 0 \qquad k \in \mathbb{Z}. \tag{11.39}$$

Following the same process for the derivation of Eq. (11.13) from Eq. (11.11) for the scaling functions, we can also represent the orthogonality of Eqs. (11.37) and (11.39) in the frequency domain as

$$\sum_k |\Psi(f-k)|^2 = 1, \tag{11.40}$$

$$\sum_k \Phi(f-k)\overline{\Psi}(f-k) = 0. \tag{11.41}$$

This result in W_0 can be generalized to space W_j. Replacing t by $2^j t$ in Eq. (11.37) we get

$$\int_{-\infty}^{\infty} \psi(2^j t - k)\overline{\psi}(2^j t)\,d(2^j t) = \int_{-\infty}^{\infty} \sqrt{2^j}\psi(2^j t - k)\sqrt{2^j}\overline{\psi}(2^j t)\,dt$$
$$= \langle \psi_{j,k}(t), \psi_{j,0}(t) \rangle = \delta[k], \tag{11.42}$$

where we have defined a set of orthogonal *wave functions* $\psi_{j,k}(t)$ as

$$\psi_{j,k}(t) = \sqrt{2^j}\psi(2^j t - k) = 2^{j/2}\psi(2^j t - k) \in W_j \qquad k \in \mathbb{Z}, \tag{11.43}$$

which can be used as an orthogonal basis to span W_j:

$$W_j = \text{span}(\psi_{j,k}(t),\ k \in \mathbb{Z}). \tag{11.44}$$

Moreover, these wavelet functions $\psi_{j,k}(t)$ are further required to be orthogonal to the scaling functions $\phi_{j,k}(t)$ as well as themselves:

$$\langle \psi_{j,k}(t), \psi_{i,l}(t) \rangle = \delta[i-j]\delta[k-l] \tag{11.45}$$

$$\langle \phi_{j,k}(t), \psi_{j,l}(t) \rangle = 0. \tag{11.46}$$

Consequently, spaces W_j and V_j, spanned respectively by $\psi_{j,k}(t)$ and $\phi_{j,l}(t)$, are orthogonal; i.e., $W_j \perp V_j$. Moreover, as $V_j = W_{j-1} \oplus V_{j-1}$, it follows that $W_j \perp V_{j-1}$ and $W_j \perp W_{j-1}$; i.e., the wavelet functions $\psi_{j,k}(t)$ are orthogonal with respect to j for different scale levels as well as to k for different integer translations in each scale level. Furthermore, since all wavelet spaces W_j are spanned by $\psi_{j,k}(t)$, the entire function space $L^2(\mathbb{R}) = \oplus_j W_j$ is also spanned by these orthogonal wavelet functions:

$$L^2(\mathbb{R}) = \text{span}(\psi_{j,k}(t),\ j, k \in \mathbb{Z}). \tag{11.47}$$

Similar to the representation of the father wavelet $\phi(t) \in V_0 \in V_1$ in Eq. (11.19), the mother wavelet $\psi(t) \in V_0 \in V_1$ can also be expressed as a linear combination of the basis $\phi_{1,k}(t) = \sqrt{2}\phi(2t-k)$ in V_1:

$$\psi(t) = \sum_k h_1[k]\phi_{1,k}(t) = \sqrt{2}\sum_k h_1[k]\phi(2t-k). \tag{11.48}$$

where the coefficients $h_1[k]$ can be found as the projection of $\psi(t)$ onto the kth basis function $\psi_{1,k}(t)$. These coefficients $h_1[k]$ must be related in some way to the coefficients $h_0[k]$ in order for the mother $\psi(t)$ and father wavelet $\phi(t)$ to be orthogonal as required, as to be discussed later.

We replace t by $2^j t - l$ in the equation above to get

$$\psi(2^j t - l) = \sqrt{2}\sum_k h_1[k]\phi(2(2^j t - l) - k) = \sqrt{2}\sum_k h_1[k]\phi(2^{j+1}t - (2l + k))$$
$$= \sqrt{2}\sum_{k'} h_1[k' - 2l]\phi(2^{j+1}t - k'), \tag{11.49}$$

where $k' = 2l + k$. Owing to Eq. (11.16), we have

$$\phi(2^{j+1}t - k) = 2^{-(j+1)/2}\phi_{j+1,k}(t). \tag{11.50}$$

Substituting this into the equation above we get

$$\psi_{j,l}(t) = 2^{j/2}\psi(2^j t - l) = \sum_k h_1[k - 2l]\phi_{j+1,k}(t). \tag{11.51}$$

Similar to Eq. (11.22) for the scaling functions $\phi_{j,l}(t)$, under the two conditions that the coefficients $h_1[k]$ are reversed in time and the output is down-sampled, Eq. (11.51) also describes a discrete FIR filter, called a *wavelet filter* with $h_1[k]$ as the impulse response, followed by a down-sampler. The input $\phi_{j+1,k}(t) \in V_{j+1}$ of the wavelet filter is the same as the scaling filter, but the output $\psi_{j,l}(t) \in W_j$ contains the high resolution contents of the input in V_{j+1} not represented by the output $\phi_{j,l}(t) \in V_j$ of the scaling filter; i.e., this wavelet filter is an HP filter.

This filtering process can also be described in the frequency domain. Taking the Fourier transform on both sides of Eq. (11.48) and following the steps in Eq. (11.23) for the scaling functions, we get

$$\Psi(f) = \mathcal{F}[\psi(t)] = \sqrt{2}\sum_k h_1[k]\mathcal{F}[\phi(2t - k)]$$
$$= \frac{1}{\sqrt{2}}\sum_k h_1[k]e^{-jk\pi}\Phi\left(\frac{f}{2}\right) = \frac{1}{\sqrt{2}}H_1\left(\frac{f}{2}\right)\Phi\left(\frac{f}{2}\right), \tag{11.52}$$

where $H_1(f) = \mathcal{F}[h_1[k]]$ is the frequency response function of the wavelet filter:

$$H_1(f) = \sum_k h_1[k]e^{-j2k\pi f}. \tag{11.53}$$

Note again $H_1(f \pm 1) = H_1(f)$ is periodic with period 1 and $H_1(f + 1/2) = H_1(f - 1/2)$.

As in Eq. (11.25), the wavelet filter can also be recursively expanded to become

$$\Psi(f) = \frac{1}{\sqrt{2}}H_1\left(\frac{f}{2}\right)\prod_{j=2}^{\infty}\frac{1}{\sqrt{2}}H_0\left(\frac{f}{2^j}\right). \tag{11.54}$$

In order to satisfy the admissibility condition (Eq. (10.21)), the DC component of the wavelet $\psi(t)$ is required to be zero (Eq. (10.10)):

$$\Psi(0) = \int_{-\infty}^{\infty} \psi(t)e^{-j2\pi ft}\,dt\Big|_{f=0} = \int_{-\infty}^{\infty} \psi(t)\,dt = 0. \tag{11.55}$$

The LP scaling filter and the HP wavelet filter followed by a down-sampler described respectively in Eqs. (11.22) and (11.51) are illustrated in the frequency domain in Fig. 11.4, where the input $\mathbf{\Phi}_{j+1,k}(f)$ is filtered by the scale and wavelet filters and then down-sampled (denoted by the down-arrow) to produce $\mathbf{\Phi}_{j,k}(f)$ and $\mathbf{\Psi}_{j,k}(f)$, respectively. Moreover, this filtering-down-sampling process can be further carried out recursively when the output $\phi_{j,k}(t)$ of the scaling filter is taken as the input of the scale and wavelet filters of the next level to produce $\mathbf{\Phi}_{j-1,k}(t)$ and $\mathbf{\Psi}_{j-1,k}$,(f), as shown on the left of Fig. 11.12, to be considered later.

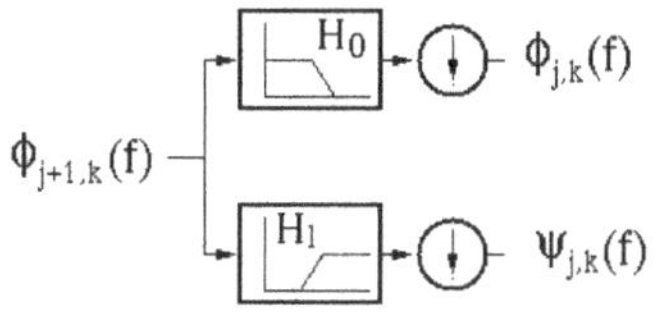

Figure 11.4 Scaling and wavelet filters in the frequency domain.

11.1.3 Properties of the scaling and wavelet filters

Here, we consider a set of properties required of the LP scaling filter and HP wavelet filter. Specifically the coefficients $h_0[k]$ and $h_1[k]$, or the frequency response functions $H_0(f)$ and $H_1(f)$, of these filters have to satisfy a set of conditions for their outputs, the scaling and wavelet functions $\phi(t)$ and $\psi(t)$ to orthogonal as discussed previously. These required properties of the scaling and wavelet filters will be used in the design of these filters by which the wavelet transform is actually carried out.

1. Normalization in the time domain.

$$\frac{1}{\sqrt{2}}\sum_k h_0[k] = 1. \tag{11.56}$$

We integrate both sides of Eq. (11.19) with respect to t to get

$$\int_{-\infty}^{\infty} \phi(t)\,dt = \sqrt{2}\sum_k h_0[k]\int_{-\infty}^{\infty} \phi(2t-k)\,dt = \sum_k h_0[k]\frac{1}{\sqrt{2}}\int_{-\infty}^{\infty} \phi(t')\,dt', \tag{11.57}$$

where we have assumed $t' = 2t - k$; i.e., $t = (t' - k)/2$. Dividing both sides by $\int_{-\infty}^{\infty} \phi(t)\,dt \neq 0$, we get Eq. (11.56).

2. Normalization in the frequency domain.

$$H_0(0) = \sqrt{2}, \quad H_1(0) = 0. \tag{11.58}$$

These can be easily obtained by letting $f = 0$ in Eqs. (11.23) and (11.52), and noting $\Phi(0) = 1$ (Eq. (11.26)) and $\Psi(0) = 0$ (Eq. (11.55)). Equivalently, we have

$$\sum_k h_1[k] = 0, \quad \sum_k h_0[k] = \sqrt{2}, \tag{11.59}$$

which can also be easily shown by letting $f = 0$ in Eqs. (11.24) and (11.53) and applying the results $H_0(0) = \sqrt{2}$ and $H_1(0) = 0$ above.

3. Orthogonalities of scaling and wavelet functions (time domain).
   Previously we considered the required orthogonalities of the scaling functions (Eq. (11.15)), the wavelet functions (Eq. (11.42)), and between the scaling and wavelet functions (Eq. (11.46)). Now we show that these orthogonalities can also be represented in terms of the scaling and wavelet filters $h_0[k]$ and $h_1[k]$:

$$\sum_k h_0[k]\overline{h}_0[k-2n] = \delta[n],$$
$$\sum_k h_1[k]\overline{h}_1[k-2n] = \delta[n],$$
$$\sum_k h_0[k]\overline{h}_1[k-2n] = 0. \tag{11.60}$$

In particular, when $n = 0$, we have

$$\sum_k |h_0[k]|^2 = 1, \quad \sum_k |h_1[k]|^2 = 1. \tag{11.61}$$

Proof: Substituting Eq. (11.22) into Eq. (11.15) (and replacing k by l), we get

$$\begin{aligned}\delta[l] &= \langle \phi_{j,l}(t), \phi_{j,0}(t) \rangle = \int_{-\infty}^{\infty} \phi_{j,l}(t)\overline{\phi}_{j,0}(t)\, dt \\ &= \sum_k \sum_{k'} h_0[k-2l]\overline{h}_0[k'] \int_{-\infty}^{\infty} \phi_{j+1,k}(t)\overline{\phi}_{j+1,k'}(t)\, dt \\ &= \sum_k \sum_{k'} h_0[k-2l]\overline{h}_0[k]\delta[k-k'] = \sum_k h_0[k-2l]\overline{h}_0[k].\end{aligned} \tag{11.62}$$

In the same manner, we can also prove the second equation in Eq. (11.60) for $h_1[k]$ by substituting Eq. (11.51) into Eq. (11.42), and the third equation for both $h_0[k]$ and $h_1[k]$ by substituting both Eqs. (11.22) and (11.51) into Eq. (11.46).

4. Orthogonalities of scaling and wavelet functions (frequency domain).
   Previously we considered the orthogonalities of the scaling functions (Eqs. (11.11) and (11.13)), of the wavelet functions (Eqs. (11.37) and (11.40)), and between the scaling and wavelet functions (Eqs. (11.39) and (11.41)). Now we further show that these orthogonalities can also be represented in terms of the scaling and wavelet filters $H_0(f)$ and $H_1(f)$ in the frequency domain.

$$|H_0(f)|^2 + \left|H_0\left(f+\frac{1}{2}\right)\right|^2 = 2,$$
$$|H_1(f)|^2 + \left|H_1\left(f+\frac{1}{2}\right)\right|^2 = 2,$$
$$H_0(f)\overline{H}_1(f) + H_0\left(f+\frac{1}{2}\right)\overline{H}_1\left(f+\frac{1}{2}\right) = 0. \tag{11.63}$$

Proof: Substituting Eq. (11.23) into Eq. (11.13), we get

$$\sum_k \left|H_0\left(\frac{f-k}{2}\right)\right|^2 \left|\Phi\left(\frac{f-k}{2}\right)\right|^2 = 2. \tag{11.64}$$

We then separate the even and odd terms in the summation to get

$$\sum_k \left|H_0\left(\frac{f-2k}{2}\right)\right|^2 \left|\Phi\left(\frac{f-2k}{2}\right)\right|^2 + \sum_k \left|H_0\left(\frac{f-(2k+1)}{2}\right)\right|^2 \left|\Phi\left(\frac{f-(2k+1)}{2}\right)\right|^2 = 2. \tag{11.65}$$

But as $H_0(f \pm k) = H_0(f)$ is periodic and due to Eq. (11.13), the above can be written as

$$\left|H_0\left(\frac{f}{2}\right)\right|^2 \sum_k \left|\Phi\left(\frac{f}{2}-k\right)\right|^2 + \left|H_0\left(\frac{f+1}{2}\right)\right|^2 \sum_k \left|\Phi\left(\frac{f+1}{2}-k\right)\right|^2$$
$$= \left|H_0\left(\frac{f}{2}\right)\right|^2 + \left|H_0\left(\frac{f}{2}+\frac{1}{2}\right)\right|^2 = 2. \tag{11.66}$$

Replacing $f/2$ by f, we complete the proof. The second equation in Eq. (11.63) for $H_1(f)$ can be proven in the same way by substituting Eq. (11.52) into Eq. (11.40). To prove the third equation in Eq. (11.63) involving both $H_0(f)$ and $H_1(f)$, we substitute Eqs. (11.23) and (11.52) into Eq. (11.41) to get

$$\sum_k H_0\left(\frac{f-k}{2}\right)\Phi\left(\frac{f-k}{2}\right)\overline{H}_1\left(\frac{f-k}{2}\right)\overline{\Phi}\left(\frac{f-k}{2}\right)$$
$$= \sum_k H_0\left(\frac{f-k}{2}\right)\overline{H}_1\left(\frac{f-k}{2}\right)\left|\Phi\left(\frac{f-k}{2}\right)\right|^2 = 0. \tag{11.67}$$

We then separate the even and odd terms in the summation to get

$$\sum_k H_0\left(\frac{f-2k}{2}\right)\overline{H}_1\left(\frac{f-2k}{2}\right)\left|\Phi\left(\frac{f-2k}{2}\right)\right|^2 + \sum_k H_0\left(\frac{f-(2k+1)}{2}\right)\overline{H}_1\left(\frac{f-(2k+1)}{2}\right)\left|\Phi\left(\frac{f-(2k+1)}{2}\right)\right|^2 = 0. \tag{11.68}$$

Replacing $f/2$ by f' and noting that $H_i(f \pm k) = H_i(f)$ $(i = 1, 2)$, we get

$$H_0(f')\overline{H}_1(f') \sum_k |\Phi(f'-k)|^2$$
$$+H_0\left(f'-\frac{1}{2}\right)\overline{H}_1\left(f'-\frac{1}{2}\right)\sum_k \left|\Phi\left(f'-k-\frac{1}{2}\right)\right|^2 = 0. \quad (11.69)$$

As both summations are equal to unity (Eq. (11.13)), The proof is complete

In our discussion above, the discrete scaling and wavelet filters are represented in the frequency domain by their DTFT $H_0(f)$ and $H_1(f)$ (Eqs. (11.24) and (11.53)), respectively. Alternatively, these filters can also be represented in the Z-domain as

$$H_0(z) = \sum_k h_0[k]z^{-k}, \qquad H_1(z) = \sum_k h_1[k]z^{-k}, \quad (11.70)$$

which are also used in much of the wavelet literature. When $H_0(z)$ and $H_1(z)$ are evaluated along the unit circle $|z| = 1$; i.e., $z = e^{j2\pi f}$, they become the same as $H_0(f)$ and $H_1(f)$. In particular, corresponding to $f = 0$ and $f + 1/2$, we have respectively $z = e^0 = 1$ and $e^{j2\pi(f+1/2)} = -e^{j2\pi f} = -z$. Now the normalization and orthogonality properties considered above can also be represented in the Z-domain as

$$H_0(1) = \sqrt{2}, \qquad H_1(1) = 0, \quad (11.71)$$
$$|H_0(z)|^2 + |H_0(-z)|^2 = 2, \quad (11.72)$$
$$|H_1(z)|^2 + |H_1(-z)|^2 = 2, \quad (11.73)$$
$$H_0(z)\overline{H}_1(z) + H_0(-z)\overline{H}_1(-z) = 0. \quad (11.74)$$

11.1.4 Relationship between scaling and wavelet filters

We now show the scaling filter $H_0(f)$ and wavelet filter $H_1(f)$ can be related by

$$H_1(f) = e^{-j2\pi f}\,\overline{H}_0\left(f-\frac{1}{2}\right); \quad \text{i.e.} \quad H_0(f) = e^{j2\pi f}\,\overline{H}_1\left(f-\frac{1}{2}\right). \quad (11.75)$$

We can easily verify that all required conditions in Eq. (11.63) are satisfied by $H_0(f)$ and $H_1(f)$ related in Eq. (11.75); i.e., the scaling and wavelet functions generated by the filters $H_0(f)$ and $H_1(f)$ related by Eq. (11.75) are indeed orthogonal to themselves with integer translation, and they are also orthogonal to each other with integer translation and across different scale levels. First, given $H_0(f)$ (or $H_1(f)$) that satisfies the first (or second) equation in Eq. (11.63), the corresponding $H_1(f)$ (or $H_0(f)$) given in Eq. (11.75) will satisfy the second (or first) one. Second, substituting $H_1(f)$ in Eq. (11.75) into the third equation in Eq. (11.63) we see that it indeed holds:

$$H_0(f)e^{j2\pi f}H_0\left(f-\frac{1}{2}\right) + H_0\left(f-\frac{1}{2}\right)e^{j2\pi(f+1/2)}H_0(f)$$
$$= H_0(f)\,e^{j2\pi f}H_0\left(f-\frac{1}{2}\right) - H_0\left(f-\frac{1}{2}\right)e^{j2\pi f}H_0(f) = 0. \quad (11.76)$$

This relationship in Eq. (11.75) between $H_0(f)$ and $H_1(f)$ in the frequency domain can be converted into the time domain by taking the inverse Fourier transform on both sides of the equation and applying the time shift, modulation, and complex conjugate properties of the DTFT (Eqs. (4.33), (4.46), and (4.29)):

$$h_1[k] = \mathcal{F}^{-1}\left[-e^{-j2\pi f}\,\overline{H}_0\left(f-\frac{1}{2}\right)\right] = (-1)^k\,\overline{h}_0[1-k]. \tag{11.77}$$

The actual wavelet function $\psi(t) \in V_0$, therefore, can be obtained by substituting these coefficients into Eq. (11.48):

$$\psi(t) = \sqrt{2}\sum_k h_1[k]\phi(2t-k) = \sqrt{2}\sum_k (-1)^k\,\overline{h}_0[1-k]\phi(2t-k). \tag{11.78}$$

We can verify that this wavelet function $\psi(t)$ is indeed orthogonal to its integer translations $\psi(t-l)$ for all $l \in \mathbb{Z}$; i.e., $\langle\psi(t-l),\psi(t)\rangle = \delta[l]$:

$$\begin{aligned}
\langle\psi(t-l),\psi(t)\rangle &= \int_{-\infty}^{\infty}\psi(t-l)\overline{\psi}(t)\,dt \\
&= 2\sum_{k'}\sum_k(-1)^{k+k'}\overline{h}_0[1-k]h_0[1-k']\int_{-\infty}^{\infty}\phi(2(t-l)-k')\overline{\phi}(2t-k)\,dt \\
&= 2\sum_k\sum_m(-1)^{m+k}\overline{h}_0[1-k]h_0[1-m+2l]\int_{-\infty}^{\infty}\phi(2t-m)\overline{\phi}(2t-k)\,dt \\
&\qquad (\text{where } m = 2l+k') \\
&= \sum_k\sum_m(-1)^{m+k}\overline{h}_0[1-k]h_0[1-m+2l]\delta[m-k] \\
&= \sum_k \overline{h}_0[1-k]h_0[1-k+2l] = \delta[l].
\end{aligned} \tag{11.79}$$

Here, we have used the fact that $\phi_{1,k}(t)$ are orthonormal (Eq. (11.15)), and the last equal sign is due to Eq. (11.60).

Replacing t by 2^jt-k, we obtain the wavelet functions $\psi_{j,k}(t) = \psi(2^jt-k)$ that span W_j.

Example 11.2: The scaling function $\phi(t)$ considered in the previous example is a square impulse with unit height and width, and the coefficients are $h_0[0] = h_0[1] = 1/\sqrt{2}$. Now, based on Eq. (11.77) the coefficients for the wavelet functions $\psi_{1,k}(t)$ can be obtained as

$$\begin{aligned}
h_1[0] &= (-1)^0 h_0[1-0] = h_0[1] = 1/\sqrt{2}, \\
h_1[1] &= (-1)^1 h_0[1-1] = -h_0[0] = -1/\sqrt{2},
\end{aligned} \tag{11.80}$$

and the wavelet function is

$$\psi(t) = \sum_l h_1[l]\sqrt{2}\phi[2t-l] = \phi(2t) - \phi(2t-1) = \begin{cases} 1 & 0 \le t < 1/2 \\ -1 & 1/2 \le t < 1 \\ 0 & \text{else} \end{cases}. \tag{11.81}$$

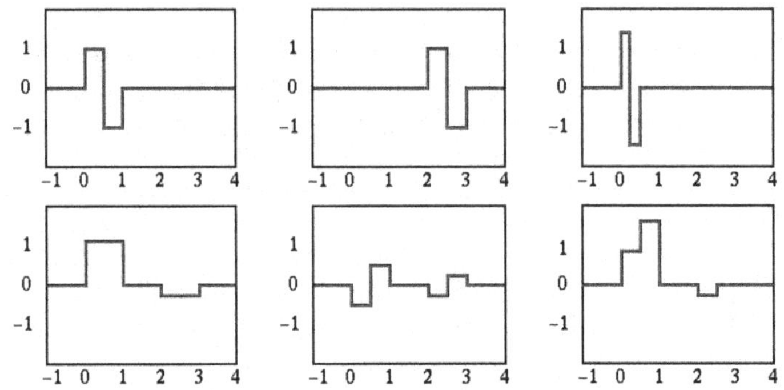

Figure 11.5 Basis functions that span wavelet spaces (top) and some functions they represent.

The first two panels of Fig. 11.5 show two of the wavelet functions $\psi(t) = \psi_{0,0}(t)$ and $\psi_{0,2}(t) = \psi(t-2)$ in W_0; the third panel shows a wavelet function $\psi_{1,0}(t) = \sqrt{2}\psi(2t)$ in W_1. The fourth panel shows a function in V_0 spanned by $\phi_{0,k}(t)$, and the fifth panel shows a function in W_0 spanned by $\psi_{0,k}(t)$, which cannot be represented in V_0. The sixth panel shows the sum of these two functions in $V_1 = V_0 \oplus W_0$, which can be represented by $\phi_{1,k}(t)$ spanning V_0, or, equivalently, by $\phi_{0,k}(t)$ and $\psi_{0,k}(t)$.

11.1.5 Wavelet series expansion

A signal $x(t)$ can be approximated in any scale space V_j spanned by a set of scaling functions $\phi_{j,k}(t)$ as the orthogonal basis. For example, when $j = 0$ the approximation in V_0 is

$$x(t) \approx \sum_k c_{0,k}\phi(t-k) = \sum_k c_{0,k}\phi_{0,k}(t), \tag{11.82}$$

where the *approximation coefficients* $c_{0,k}$ can be found as the projections of the signal onto the corresponding basis vector:

$$c_{0,k} = \langle x(t), \phi_{0,k}(t)\rangle = \int x(t)\overline{\phi}_{0,k}(t)\,dt \qquad \text{(for all } k\text{)}. \tag{11.83}$$

Moreover, the signal can be ever more precisely approximated if progressively more detailed information contained in wavelet space W_j spanned by $\psi_{j,k}(t)$ is included when $j \to \infty$ (Eq. (11.36)):

$$\begin{aligned} x(t) &= \sum_k c_{0,k}\phi_{0,k}(t) + \sum_{j=0}^{\infty}\sum_k d_{j,k}\psi_{j,k}(t) \\ &= \sum_k \langle x(t), \phi_{0,k}(t)\rangle\phi_{0,k}(t) + \sum_{j=0}^{\infty}\sum_k \langle x(t), \psi_{j,k}(t)\rangle\psi_{j,k}(t), \end{aligned} \tag{11.84}$$

where $d_{j,k}$, called the *detail coefficients*, can be found as

$$d_{j,k} = \langle x(t), \psi_{j,k}(t) \rangle = \int x(t)\overline{\psi}_{j,k}(t)\, dt \quad \text{(for all } k \text{ and } j > 0). \tag{11.85}$$

Equation (11.84) is the *wavelet series expansion* of the signal $x(t)$, corresponding to Eq. (3.5) of the Fourier series expansion considered in Chapter 3.

Example 11.3: Here we use the Haar wavelets to approximate the following continuous function $x(t)$ defined over the period $0 \le t < 1$, as shown in panel 1 of Fig. 11.6:

$$x(t) = \begin{cases} t^2 & 0 \le t < 1 \\ 0 & \text{else.} \end{cases} \tag{11.86}$$

First note that each individual space (V_0, W_0, W_1, ...) is spanned by a different number of basis functions. For example, spaces V_0 and W_0 are spanned by only one basis function, whereas space W_1 is spanned by two basis functions, and space W_2 is spanned by four (Fig. 8.9).

We can choose to start at scale level $j = 0$. According to Eqs. (11.83) and (11.85), the approximation and wavelet coefficients can be obtained as

$$\begin{aligned}
c_0(0) &= \int_0^1 t^2 \varphi_{0,0}(t)\, dt = \int_0^1 t^2(t)\, dt = \frac{1}{3}, \\
d_0(0) &= \int_0^1 t^2 \psi_{0,0}(t)\, dt = \int_0^{0.5} t^2(t)\, dt - \int_{0.5}^1 t^2(t)\, dt = -\frac{1}{4}, \\
d_1(0) &= \int_0^1 t^2 \psi_{1,0}(t)\, dt = \int_0^{0.25} \sqrt{2}t^2(t)\, dt - \int_{0.25}^{0.5} t^2\sqrt{2}(t)\, dt = -\frac{\sqrt{2}}{32}, \\
d_1(1) &= \int_0^1 t^2 \psi_{1,1}(t)\, dt = \int_{0.5}^{0.75} \sqrt{2}t^2(t)\, dt - \int_{0.75}^1 t^2\sqrt{2}(t)\, dt = -\frac{3\sqrt{2}}{32}.
\end{aligned} \tag{11.87}$$

Therefore, the wavelet series expansion of the function $x(t)$ is

$$x(t) = \frac{1}{3}\phi_{0,0}(t) + \left[-\frac{1}{4}\psi_{0,0}(t)\right] + \left[-\frac{\sqrt{2}}{32}\psi_{1,0}(t) - \frac{3\sqrt{2}}{32}\psi_{1,1}(t)\right] + \cdots. \tag{11.88}$$

The first two coefficients are for $\phi_{0,0}(t) \in V_0$ and $\psi_{0,0}(t) \in W_0$, respectively, as shown in panels 2 and 3 in Fig. 11.6, and their weighted sum is the approximation of the function in space $V_1 = V_0 \oplus W_0$ as shown in panel 4. The last two coefficients are for $\psi_{1,0}(t)$ and $\psi_{1,1}(t)$ both in space W_1, and their weighted sum is the approximation of the function in W_1 as shown in panel 5. Then in space $V_2 = V_1 \oplus W_1$ the function $x(t)$ can be approximated as the sum of the approximations in V_1 and W_1 as shown in panel 6. This process can be carried out further by including progressively more detailed information in wavelet spaces $W_2, W_3, \ldots W_j$ as $j \to \infty$.

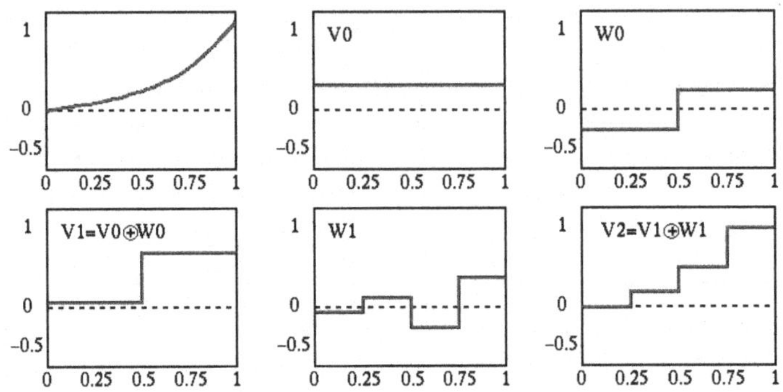

Figure 11.6 Wavelet approximation of a function.

The definition of the MRA requires the existence of a Riesz basis (not necessarily an orthogonal basis) that spans space V_0; i.e., the MRA may be a biorthogonal MRA. . In this case, a dual function exists corresponding to each scaling or wavelet function. Specifically at the jth level of such a biorthogonal MRA, corresponding to the scaling function $\phi_{j,k}(t)$ and wavelet function $\psi_{j,k}(t)$, there exist respectively a dual scaling function $\tilde{\phi}_{j,k}(t)$ and a dual wavelet function $\tilde{\psi}_{j,k}(t)$ so that

$$\begin{aligned}\langle \phi_{j,k}(t), \tilde{\phi}_{j,l}(t)\rangle &= \delta[k-l],\\ \langle \psi_{j,k}(t), \tilde{\psi}_{i,l}(t)\rangle &= \delta[i-j]\delta[k-l],\\ \langle \phi_{j,k}(t), \tilde{\psi}_{i,l}(t)\rangle &= \langle \psi_{j,k}(t), \tilde{\phi}_{i,l}(t)\rangle = 0.\end{aligned} \tag{11.89}$$

As with $\phi_{j,k}(t)$ and $\psi_{j,k}(t)$, which span respectively V_j and W_j satisfying $V_j \oplus W_j = V_{j+1}$, the dual scaling and wavelet functions $\tilde{\phi}_{j,k}(t)$ and $\tilde{\psi}_{j,k}(t)$ respectively span $\tilde{V}_j$ and $\tilde{W}_j$ satisfying $\tilde{V}_j \oplus \tilde{W}_j = \tilde{V}_{j+1}$. Note, however, as these basis functions are in general not orthogonal, V_j and W_j are not an orthogonal complement of each other in V_{j+1}; neither are $\tilde{V}_j$ and $\tilde{W}_j$ in $\tilde{V}_{j+1}$.

In this case, the wavelet series expansion in Eq. (11.84) becomes

$$\begin{aligned}x(t) &= \sum_k \langle x(t), \phi_{0,k}(t)\rangle \tilde{\phi}_{0,k}(t) + \sum_{j=0}^{\infty}\sum_k \langle x(t), \psi_{j,k}(t)\rangle \tilde{\psi}_{j,k}(t)\\ &= \sum_k \langle x(t), \tilde{\phi}_{0,k}(t)\rangle \phi_{0,k}(t) + \sum_{j=0}^{\infty}\sum_k \langle x(t), \tilde{\psi}_{j,k}(t)\rangle \psi_{j,k}(t).\end{aligned} \tag{11.90}$$

11.1.6 Construction of scaling and wavelet functions

To carry out the wavelet transform of a given signal, the scaling function $\phi(t)$ and the wavelet functions $\psi(t)$ need to be specifically determined. In general, this is a design process which can be carried out in one of the following ways:

- specify $\phi(t)$ and $\psi(t)$ in the time domain;
- specify their spectra $\Phi(f)$ and $\Psi(f)$ in the frequency domain;
- specify the corresponding filter coefficients $h_0[k]$ and $h_1[k]$ in the time domain;
- specify the corresponding filter frequency response functions $H_0(f)$ and $H_1(f)$ in the frequency domain.

In the following we will consider these different methods. Keep in mind that it is desirable for the scaling and wavelet functions to have good localities in both the time and frequency domains. Ideally, they should be *compactly supported*, i.e., they are non-zero only within a finite domain.

- **Haar wavelets**

  Construct the scaling and wavelet functions by the following steps.

  1. Choose the scaling function $\phi(t)$ satisfying Eq. (11.11)
  $$\langle \phi(t-k), \phi(t) \rangle = \delta[k], \tag{11.91}$$
  or $\Phi(f)$ satisfying Eq. (11.13)
  $$\sum_{k \in \mathbb{Z}} |\Phi(f-k)|^2 = 1. \tag{11.92}$$
  For the Haar transform, we simply choose the scaling function as
  $$\phi(t) = \begin{cases} 1 & 0 \le t < 1 \\ 0 & \text{else} \end{cases}. \tag{11.93}$$
  2. Find the scaling coefficients $h_0[k]$ based on Eq. (11.20),
  $$h_0[k] = \langle \phi(t), \sqrt{2}\phi(2t-k) \rangle, \tag{11.94}$$
  or $H_0(f)$ according to Eq. (11.23),
  $$H_0(f) = \sqrt{2}\,\frac{\Phi(2f)}{\Phi(f)}. \tag{11.95}$$
  For the Haar transform, we have
  $$\begin{aligned} h_0[k] &= \sqrt{2}\int_{-\infty}^{\infty} \phi(t)\phi(2t-k)\,dt = \sqrt{2}\int_0^1 \phi(2t-k)\,dt \\ &= \frac{1}{\sqrt{2}}\int_0^2 \phi(t'-k)\,dt' = \frac{1}{\sqrt{2}}\begin{cases} 1 & k = 0, 1 \\ 0 & \text{else} \end{cases}. \end{aligned} \tag{11.96}$$
  3. Find the wavelet coefficients $h_1[k]$ according to Eq. (11.77)
  $$h_1[k] = (-1)^k\, \overline{h}_0[1-k], \tag{11.97}$$
  or $H_1(f)$ according to Eq. (11.75)
  $$H_1(f) = e^{-j2\pi f}\, \overline{H}_0\left(f - \frac{1}{2}\right). \tag{11.98}$$
  For the Haar transform, we have
  $$h_1[k] = (-1)^k h_0[1-k] = \frac{1}{\sqrt{2}}\begin{cases} 1 & k = 0 \\ -1 & k = 1 \\ 0 & \text{else} \end{cases}. \tag{11.99}$$

4. Find wavelet function $\psi(t)$ according to Eq. (11.78)

$$\psi(t) = \sqrt{2}\sum_k (-1)^k\, \overline{h}_0[1-k]\phi(2t-k), \tag{11.100}$$

or $\Psi(f)$ according to Eq. (11.52)

$$\Psi(f) = \frac{1}{\sqrt{2}} H_1(\frac{f}{2})\Phi(\frac{f}{2}). \tag{11.101}$$

For the Haar transform, we have:

$$\psi(t) = h_1[0]\phi_{1,0}(t) + h_1[1]\phi_{1,1}(t) = \begin{cases} 1 & 0 \le t < 1/2 \\ -1 & 1/2 \le t < 1 \\ 0 & \text{else} \end{cases}. \tag{11.102}$$

Based on $\phi(0) = \phi_{0,0}(t)$ and $\psi(0) = \psi_{0,0}(t)$, all other $\psi_{j,k}(t)$ can be obtained, as the rows in the Haar matrix in Eq. (8.76).
The Haar scaling function $\phi(t)$ and the first few Haar wavelet functions $\psi_{j,k}(t)$ are shown in Fig. 11.7. Obviously, they have perfect temporal locality. However, similar to the ideal filter discussed before, the drawback of the Haar wavelets is their poor frequency locality, due obviously to their sinc-like spectra $\Phi(f)$ and $\Psi(f)$ caused by the sharp corners of the rectangular time window in both $\phi(t)$ and $\psi(t)$, as shown in Fig. 11.8.

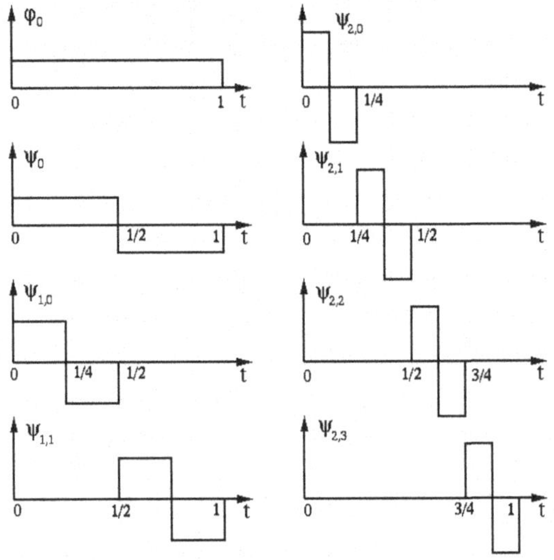

Figure 11.7 Haar scaling and wavelet functions.

- **Meyer wavelets**
  Here we construct a wavelet with good locality in both the time and frequency domains by avoiding sharp discontinuities in both domains. We start in the frequency domain by considering the spectrum $\Phi(f)$ of the scaling function $\phi(t)$. First define a function for the smooth transition from 0 to 1 and then use it to define a smooth frequency window. Specifically, consider the third-order

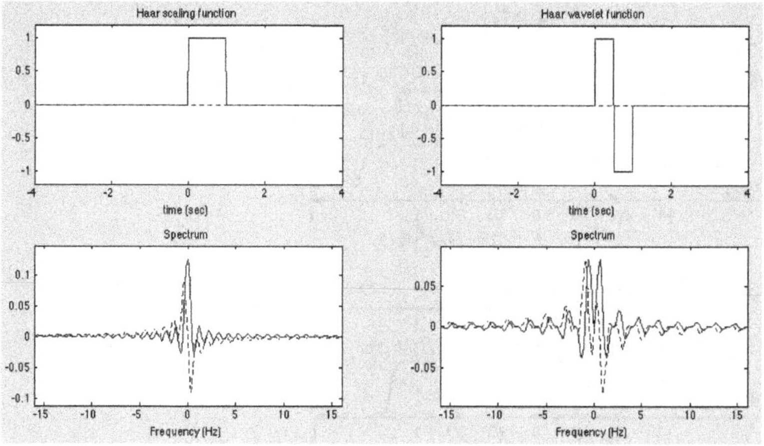

Figure 11.8 Haar scaling and wavelet functions (top) and their spectra (bottom). The real and imaginary parts of the spectra are shown respectively by solid and dashed curves.

polynomial shown in Fig. 11.9(a),

$$\nu(f) = \begin{cases} 0 & f < 0 \\ 3f^2 - 2f^3 & 0 \leq f \leq 1 \\ 1 & 1 < f \end{cases}, \tag{11.103}$$

and define $\Phi(f)$ as

$$\Phi(f) = \begin{cases} \sqrt{\nu(2+3f)} & f \leq 0 \\ \sqrt{\nu(2-3f)} & f \geq 0 \end{cases}. \tag{11.104}$$

Here, the function $3f^2 - 2f^3$ is chosen so that $\nu(1/2) = 1/2$ and $\nu(f) + \nu(1-f) = 1$, in order to satisfy the orthogonality in Eq. (11.13). (Other functions such as $10f^3 - 15f^4 + 6f^5$ satisfying the same conditions could also be used.) As shown in Fig. 11.9(b), $\Phi^2(f) = 1$ when $|f| \leq 1/3$, $\Phi^2(f) = 0$ when $2/3 \leq |f| < 1$, and $\phi^2(f) + \phi^2(f \pm 1) = 1$ during the transition interval $1/3 < |f| < 2/3$, where the two neighboring copies of $\Phi(f)$ overlap; i.e., Eq. (11.13) is indeed satisfied.

Given $\Phi(f)$, we next find the scaling filter $H_0(f)$ based on $H_0(f) = \sqrt{2}\Phi(2f)/\Phi(f)$ (Eq. (11.23)), where $\Phi(2f)$, a compressed version of $\Phi(f)$, is zero for all $|f| > 1/3$. When $|f| < 1/3$, $\Phi(f) = 1$ and $\Phi(2f)/\Phi(f) = \Phi(2f)$. Also, as $H_0(f \pm 1) = H_0(f)$ is periodic, it can be obtained as

$$H_0(f) = \sum_k \Phi(2(f-k)) = \sum_k \Phi(2f - 2k). \tag{11.105}$$

These functions $\Phi(f)$, $\Phi(2f)$, and $H_0(f)$ are shown in Fig. 11.9(b), (c), and (d), respectively.

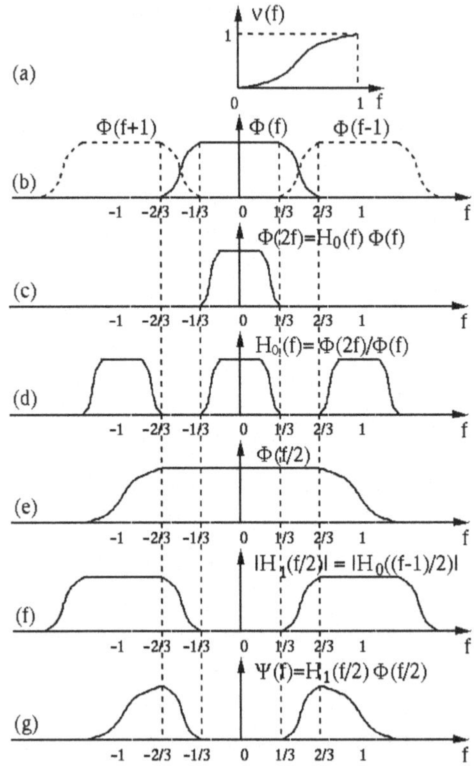

Figure 11.9 Construction of Meyer scaling and wavelet functions.

Given $H_0(f)$, we can find $H_1(f)$ based on Eq. (11.75):

$$H_1(f) = e^{-j2\pi f}\overline{H}_0(f - \frac{1}{2}) = e^{-j2\pi f} \sum_k \Phi(2f - 2k - 1), \tag{11.106}$$

and then $\Psi(f)$ based on Eq. (11.52):

$$\begin{aligned}\Psi(f) &= \frac{1}{\sqrt{2}} H_1\left(\frac{f}{2}\right) \Phi\left(\frac{f}{2}\right) = \frac{1}{\sqrt{2}} e^{-j\pi f} H_0\left(\frac{f-1}{2}\right) \Phi\left(\frac{f}{2}\right) \\ &= \frac{1}{\sqrt{2}} e^{-j\pi f} \sum_k \Phi(f - 2k - 1) \Phi\left(\frac{f}{2}\right) \\ &= \begin{cases} 0 & |f| < 1/3 \\ -\frac{1}{\sqrt{2}} e^{-j2\pi f} \Phi(f-1) & 1/3 < |f| < 2/3 \\ -\frac{1}{\sqrt{2}} e^{-j2\pi f} \Phi(f/2) & 2/3 < |f| < 4/3 \\ 0 & 4/3 < |f| \end{cases}.\end{aligned} \tag{11.107}$$

These functions $\Phi(f/2)$, $H_1(f/2)$, and $\Psi_0(f)$ are shown in Fig. 11.9(e), (f), and (g), respectively.

Finally, the scaling function $\phi(t)$ and wavelet function $\psi(t)$ can be obtained by the inverse Fourier transform of $\Phi(f)$ and $\Psi(f)$ respectively, as shown in Fig. 11.10, and the coefficients for the scaling and wavelet filters can be found by the inverse DTFT:

$$h_i[k] = \mathcal{F}^{-1}[H_i(f)] = \int_0^1 H_i(f)e^{-j2\pi kf}\,df \qquad i = 0, 1, \quad k \in \mathbb{Z}. \quad (11.108)$$

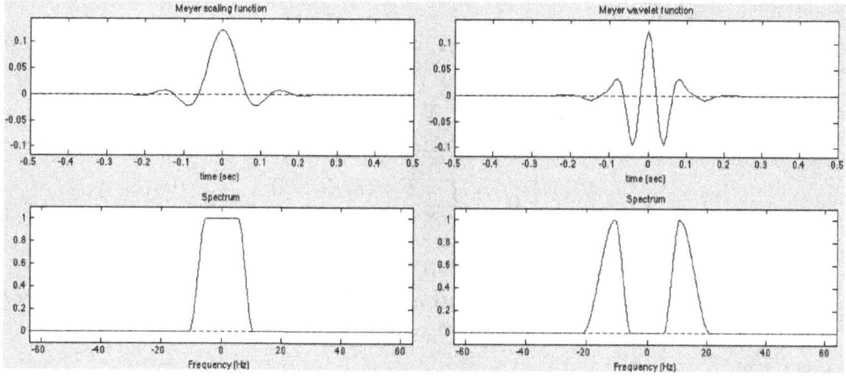

Figure 11.10 Meyer scaling and wavelet functions (top) and their spectra (bottom). The imaginary parts are shown as dashed curves.

The Matlab code segment for generating the Mayer wavelets is given below:

```
N=1024;  % total number of samples
M=N/8;   % size of period
Phi=zeros(1,N);
Psi=zeros(1,N);
for i=1:N
    f=abs(i-N/2-1);
    if f<M/3
        Phi(i)=1;
    elseif f<2*M/3
        Phi(i)=nu(2-f/(M/3));  % Meyer scaling function
    end
    if (f>M/3 & f<2*M/3)
        Psi(i)=nu(f/(M/3)-1);
    elseif (f>2*M/3 & f<4*M/3)
        Psi(i)=nu(2-f/(2*M/3));
    end
end
phi=fftshift(ifft(fftshift(Phi)));
psi=fftshift(ifft(fftshift(Psi)));
```

where

```
function y = nu(f)
  y=3*f^2-2*f^3;
end
```

- **Daubechies' wavelets**
  In addition to the temporal locality (ideally with compact support), it is also desirable for a wavelet function $\psi(t)$ to have a high number of *vanishing moments*, so that a signal can be effectively represented by the wavelet transform.
  To understand this point, we first need to understand the concepts of vanishing moments and *regular functions* . The number of vanishing moments of a wavelet function $\psi(t)$ is N if all of its moments lower than N are zero:

$$\langle t^n, \psi(t) \rangle = \int_{-\infty}^{\infty} t^n \psi(t)\, dt = 0 \qquad 0 \le n < N. \tag{11.109}$$

  Also, a function $x(t)$ is regular if it can be approximated by a polynomial $p(t) = \sum_{n=0}^{M} c_n t^n$ around any t. When this signal is represented in space W_0 spanned by the wavelet basis $\psi_{0,k} = \psi(t-k)$:

$$x(t) = \sum_k d_{0,k}\psi_{0,k}(t) = \sum_k d_{0,k}\psi(t-k), \tag{11.110}$$

  then the coefficient

$$d_{0,k} = \langle x(t), \psi(t-k) \rangle \approx \sum_{n=0}^{M} c_n \langle t^n, \psi(t-k) \rangle \tag{11.111}$$

  is zero if $N > M$. We see that the greater number of vanishing moments N, the more coefficients $d_{0,k}$ in the wavelet expansion may become zero (or small enough to be ignored). The same argument can also be made for the higher scale levels of $j > 0$. Owing to this result of much reduced transform coefficients, the signal can be more effectively represented. This is obviously desirable in various applications such as data compression.
  Daubechies' wavelets are widely used owing to their two favorite features: (1) they are compactly supported, and (2) they have the maximal number of vanishing moments for a given support. The derivation of Daubechies' wavelets is based on the normalization and orthogonality properties of $H_0(f)$ given in Eqs. (11.58) and (11.63). But for convenience and conciseness we will use the alternative expressions of these properties in Eqs. (11.71) – (11.74) in terms of $H_0(z)$ in the z-domain (with $z = e^{j2\pi f}$).
  We let $z = 1$ in Eq. (11.72) and note $H_0(1) = \sqrt{2}$ to get

$$|H_0(1)|^2 + |H_0(-1)|^2 = 2 + |H_0(-1)|^2 = 2, \tag{11.112}$$

  from which we get $H_0(-1) = 0$. We further see that $z = -1$ is a root of the polynomial $H_0(z) = \sum_k h_0[k]z^{-k}$; i.e., it must have a factor $(1+z^{-1})^N$ for some N, and therefore can be written in the following form:

$$H_0(z) = (1+z^{-1})^N Q(z), \tag{11.113}$$

Here, $Q(z)$ is a polynomial of z^{-1}. Daubechies proved[1] that the minimum degree of $Q(z)$ is $N-1$; i.e., $H_0(z)$ is a polynomial of order $N+N-1=2N-1$ containing $2N$ terms $h_0[k]z^{-k}$ ($k=0,\dots,2N-1$, assuming the filter is causal with $h_0[k]=0$ for all $k<0$), and the scaling and wavelet functions $\phi(t)$ and $\psi(t)$ corresponding such a $H_0(z)$ are compactly supported. Specifically, $\phi(t)\neq 0$ for $0\le t\le 2N-1$ and $\psi(t)\neq 0$ for $-(N-1)\le t\le N$, and the wavelet function has the maximum number of vanishing moments N given the compact support of length $2N$.

Here, we consider the three cases when $N=1$, $N=2$, and $N=3$.

- $N=1$ (Daubechies 2 or D2, same as the Haar transform):
  The order of $Q(z)$ is $N-1=0$; i.e., $Q(z)=c$ is a constant and $H_0(z)=c(1+z^{-1})$. But as $H_0(1)=c2=\sqrt{2}$, we get $c=1/\sqrt{2}$ and $h_0[0]=h_0[1]=1/\sqrt{2}$; i.e., this is the Haar scaling filter already considered above.
- $N=2$ (Daubechies 4 or D4):
  The order of $Q(z)$ is $N-1=1$ and

$$H_0(z)=(1+z^{-1})^2Q(z)=(1+z^{-1})^2(c_0+c_1z^{-1}). \tag{11.114}$$

The two coefficients c_0 and c_1 can be obtained by using Eqs. (11.71) through (11.74) as constraining equations. We first evaluate $H_0(z)$ above at $z=1$ to get (Eq. (11.71)):

$$H_0(1)=4(c_0+c_1)=\sqrt{2},\quad \text{i.e.}\quad c_1+c_2=\frac{\sqrt{2}}{4}. \tag{11.115}$$

We next evaluate $H_0(z)$ and $H_0(-z)$ at $z=j$ to get

$$\begin{aligned} H_0(j)&=(1-j)^2(c_0-jc_1)=-2(jc_0+c_1),\\ H_0(-j)&=(1+j)^2(c_0+jc_1)=2(jc_0-c_1). \end{aligned} \tag{11.116}$$

Substituting these into Eq. (11.72) we get

$$|H_0(j)|^2+|H_0(-j)|^2=8(c_0^2+c_1^2)=2;\quad \text{i.e.,}\quad c_0^2+c_1^2=\frac{1}{4}. \tag{11.117}$$

Solving Eqs. (11.115) and (11.117), we get $c_{0,1}=(1\pm\sqrt{3})/4\sqrt{2}$ and

$$\begin{aligned} H_0(z)&=\frac{1}{4\sqrt{2}}(1+z^{-1})^2[(1+\sqrt{3})+(1-\sqrt{3})z^{-1}]\\ &=\frac{1}{4\sqrt{2}}[(1+\sqrt{3})+(3+\sqrt{3})z^{-1}+(3-\sqrt{3})z^{-2}+(1-\sqrt{3})z^{-3}], \end{aligned} \tag{11.118}$$

[1] Daubechies, I., *Ten Lectures on Wavelets (CBMS-NSF Regional Conference Series in Applied Mathematics)*, Society for Industrial and Applied Mathematics, 1992.

and the four Daubechies scaling filter coefficients are:

$$\begin{aligned} h_0[0] &= \frac{1+\sqrt{3}}{4\sqrt{2}} = 0.482\,962\,1, \\ h_0[1] &= \frac{3+\sqrt{3}}{4\sqrt{2}} = 0.836\,516\,3, \\ h_0[2] &= \frac{3-\sqrt{3}}{4\sqrt{2}} = 0.224\,143\,9, \\ h_0[3] &= \frac{1-\sqrt{3}}{4\sqrt{2}} = -0.129\,409\,5. \end{aligned} \tag{11.119}$$

The corresponding wavelet coefficients can be obtained according to Eq. (11.77) $h_1[k] = (-1)^k h_0[1-k]$ as

$$\begin{aligned} h_1[1] &= -h_0[0] = -0.482\,962\,1, \\ h_1[0] &= h_0[1] = 0.836\,516\,3, \\ h_1[-1] &= -h_0[2] = -0.224\,143\,9, \\ h_1[-2] &= h_0[3] = -0.129\,409\,5. \end{aligned} \tag{11.120}$$

- $N = 3$ (Daubechies 6 or D6):
  The order of $Q(z)$ is $N-1 = 2$ and

$$H_0(z) = (1+z^{-1})^3 Q(z) = (1+z^{-1})^3 (c_0 + c_1 z^{-1} + c_2 z^{-2}). \tag{11.121}$$

  Here again, the three coefficients c_0, c_1, and c_2 can be obtained by using the normalization and orthogonality conditions given in Eqs. (11.71) through (11.74) as the constraining equations. Similar to the case of $N = 2$, we can find the $2N = 6$ coefficients of the scaling filter as

$$\begin{aligned} h_0[0] &= \left[1+\sqrt{10}+\sqrt{5+2\sqrt{10}}\right]/16\sqrt{2} = 0.332\,670\,6, \\ h_0[1] &= \left[5+\sqrt{10}+3\sqrt{5+2\sqrt{10}}\right]/16\sqrt{2} = 0.806\,891\,5, \\ h_0[2] &= \left[10-2\sqrt{10}+2\sqrt{5+2\sqrt{10}}\right]/16\sqrt{2} = 0.459\,877\,5, \\ h_0[3] &= \left[10-2\sqrt{10}-2\sqrt{5+2\sqrt{10}}\right]/16\sqrt{2} = -0.135\,011\,0, \\ h_0[4] &= \left[5+\sqrt{10}-3\sqrt{5+2\sqrt{10}}\right]/16\sqrt{2} = -0.085\,441\,3, \\ h_0[5] &= \left[1+\sqrt{10}-\sqrt{5+2\sqrt{10}}\right]/16\sqrt{2} = 0.035\,226\,3. \end{aligned} \tag{11.122}$$

The corresponding wavelet coefficients $h_1[k]$ can be found by Eq. (11.77). No analytical expression exists for either the scaling function $\phi(t)$ or the wavelet function $\psi(t)$. However, once the coefficients $h_0[k]$ and $h_1[k]$ for the scaling and wavelet filters are available, $\phi(t)$ and $\psi(t)$ can be iteratively constructed by Eqs. (11.19) and (11.48) (or Eqs. (11.25) and (11.54) in the frequency domain) based on the initial D2 or Haar scaling function (Eq. (11.93)).

Such a construction is implemented in the Matlab function given below, by which the Daubechies scaling function $\phi(t)$ and wavelet function $\psi(t)$ are iteratively constructed. The resulting scaling and wavelet functions of the first six iterations are shown in Fig. 11.11. The waveforms of the scaling and wavelet functions of order $N > 3$ can be similarly obtained, and they indeed become smoother as order N increases.

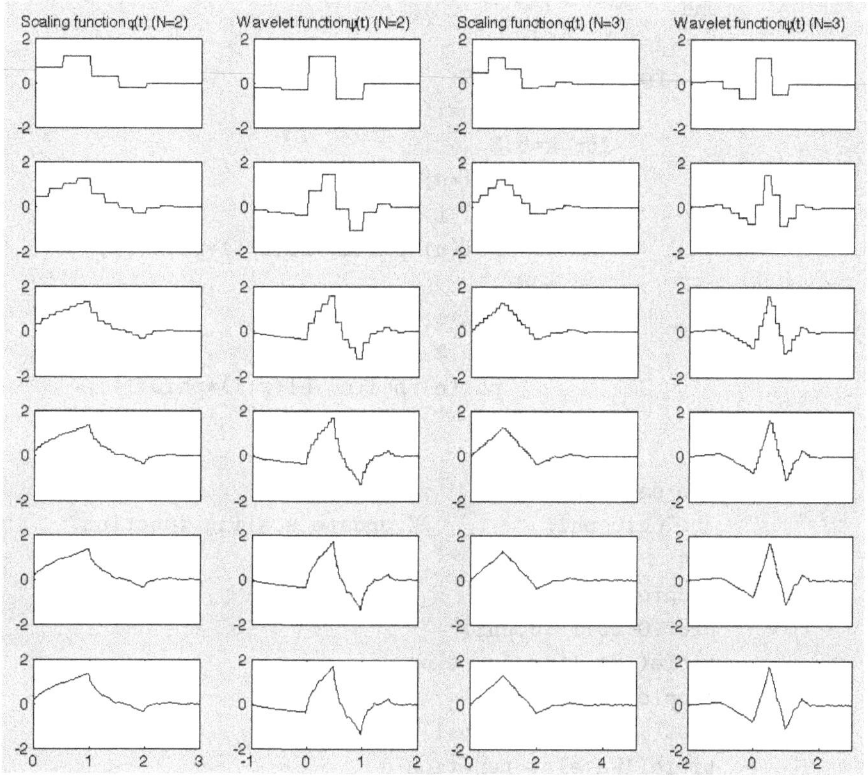

Figure 11.11 Iterative approximations of Daubechies' scaling and wavelet functions. The scaling and wavelet functions $\phi(t)$ and $\psi(t)$ for $N = 2$ are shown in the two columns on the left, while those for $N = 3$ are shown in the two columns on the right. The six rows show the first six intermediate results of the iteration based on Eqs. (11.19) and (11.48).

```
function daubechies
T=3;                    % time period in second
s=64;                   % sampling rate: s samples/second
t0=1/s;                 % sampling period
N=T*s;                  % total number of samples
K=4;                    % length of coefficient vector
r3=sqrt(3);
h0=[1+r3 3+r3 3-r3 1-r3]/4;  % Daubechies coefficients
h1=fliplr(h0);          % time reversal of h0
```

```
h1(2:2:K)=-h1(2:2:K); % negate odd terms
phi=zeros(1,N);       % scaling function
psi=zeros(1,N);       % wavelet function
phi0=zeros(1,N);
for i=1:s
    phi0(i)=1;        % initialize scaling function
end
for j=1:log2(s);
    for n=1:N
        phi(n)=0; psi(n)=0;
        for k=0:3
            l=2*n-k*s;
            if (l>0 & l<=N)
                phi(n)=phi(n)+h0(k+1)*phi0(l);
            end
            l=2*n-k*s;
            if (l>0 & l<=N)
                psi(n)=psi(n)+h1(k+1)*phi0(l);
            end
        end
    end
    phi0=phi;         % update scaling function
end
subplot(2,1,1)
plot(0:t0:T-t0,phi)
title('Scaling function');
subplot(2,1,2)
plot(-1:t0:T-1-t0,psi);
title('Wavelet function')
```

11.2 Discrete wavelet transform (DWT)

11.2.1 Discrete wavelet transform (DWT)

To numerically carry out the wavelet series expansion of a signal $x(t)$ as shown in Eq. (11.84), both the scaling function $\phi_{0,k}(t)$ and wavelet function $\psi_{j,k}(t)$ as well as the signal $x(t)$ need to be discretized so that they are all represented as N-D vectors $\boldsymbol{\phi}_{0,k}$, $\boldsymbol{\psi}_{j,k}$, and $\boldsymbol{x}$, composed respectively of $\phi_{j,k}[n]$, $\psi_{j,k}[n]$, and $x[n]$ as the nth components. Owing to the dyadic scaling, there are in total $J = \log_2 N$ scale levels (N assumed to be a power of 2 for convenience). Also, as the data size is 2^j at each level $j = 0, \ldots, J-1$, there are 2^j possible translations $k = 0, \ldots, 2^j - 1$. In particular, at the highest scale level $j = J-1$, there is only $2^0 = 1$ sample at the lowest scale level $j = 0$ and $2^{J-1} = N/2$ samples.

Now the wavelet expansion becomes the DFT by which the discrete signal $\boldsymbol{x} = [x[0], \ldots, x[N-1]]^{\mathrm{T}}$ is represented as a weighted sum in the scaling and wavelet spaces spanned by the orthogonal basis vectors $\boldsymbol{\phi}_{0,k}$ and $\boldsymbol{\psi}_{j,k}$:

$$\boldsymbol{x} = \langle \boldsymbol{x}, \boldsymbol{\phi}_{0,0} \rangle \boldsymbol{\phi}_{0,0} + \sum_{j=0}^{J-1} \sum_{k=0}^{2^j-1} \langle \boldsymbol{x}, \boldsymbol{\psi}_{j,k} \rangle \boldsymbol{\psi}_{j,k}, \tag{11.123}$$

which can also be represented in component form:

$$x[n] = X_\phi[0,0]\phi_{0,0}[n] + \sum_{j=0}^{J-1} \sum_{k=0}^{2^j-1} X_\psi[j,k]\psi_{j,k}[n] \qquad n = 0, \ldots, N-1. \tag{11.124}$$

This is the inverse DWT, by which the signal $\boldsymbol{x}$ is reconstructed from its DWT *approximation coefficient* $X_\phi[0,0]$ and *detail coefficients* $X_\psi[j,k]$, which can be found as the projections of the signal vector onto the corresponding basis vectors, similar to the case of wavelet series expansion in Eqs. (11.83) and (11.85):

$$X_\phi[0,0] = \langle \mathbf{x}, \boldsymbol{\phi}_{0,0} \rangle = \sum_{n=0}^{N-1} x[n]\overline{\phi}_{0,0}[n], \tag{11.125}$$

$$X_\psi[j,k] = \langle \mathbf{x}, \boldsymbol{\psi}_{j,k} \rangle = \sum_{n=0}^{N-1} x[n]\overline{\psi}_{j,k}[n] \qquad j = 0, \ldots, J-1,\ k = 0, \ldots, 2^j - 1. \tag{11.126}$$

These equations are the forward DWT by which the DWT coefficients are obtained, including $X_\phi[0,0]$ and $X_\psi[j,k]$ for all $J = \log_2$ scale levels ($j = 0, \ldots, J-1 = \log_2 N - 1$) each with 2^j integer translations ($k = 0, \ldots, 2^j - 1$). As there are in total $1 + \sum_{j=0}^{J-1} 2^j = 2^J = N$ coefficients, we can arrange them as an N-D vector in the DWT domain, as shown later in Fig. 11.13 for $N = 2^3 = 8$, just like the coefficients of all discrete orthogonal transforms considered in previous chapters.

At the lowest level when $j = 0$, the signal is simply approximated by its average represented by $\phi_{0,0}[n] = 1$. However, it is not always necessary to start the approximation process from this lowest scale level. On the other hand, at the highest possible level when $j = J$ (not part of the DFT in Eqs. (11.124) or (11.126)), the full resolution is achieved in V_J, where the signal is simply represented by all of its N original samples $x[n]$ ($n = 0, \ldots, N-1$).

As with all other discrete orthogonal transforms considered in previous chapters, the DWT also represents a discrete signal in terms of its DWT coefficients. (Note, however, different from all previous transforms, the DWT coefficients represent different translations as well as different scales, while the coefficients of other transforms, such as the DFT and DCT, only represent different frequencies.) In the DWT domain, various signal processing operations, such as filtering, noise reduction, feature extraction and data compression, can be carried out. The inverse DWT transform can then be carried out to reconstruct the signal back in the time domain.

Example 11.4: When $N = 4$, the discrete Haar scaling and wavelet functions are given as the rows of the following matrix (Eq. (8.74)):

$$\begin{bmatrix} 1 & 1 & 1 & 1 \\ 1 & 1 & -1 & -1 \\ \sqrt{2} & -\sqrt{2} & 0 & 0 \\ 0 & 0 & \sqrt{2} & -\sqrt{2} \end{bmatrix} \begin{matrix} \phi_{0,0}[n] \\ \psi_{0,0}[n] \\ \psi_{1,0}[n] \\ \psi_{1,1}[n] \end{matrix}. \tag{11.127}$$

Given a discrete signal $\mathbf{x} = [x[0], \ldots, x[N-1]]^{\mathrm{T}} = [1, 4, -3, 0]^{\mathrm{T}}$, the DWT coefficients can be found by Eqs. (11.125) and (11.126). The coefficient in V_0 is

$$X_\phi[0,0] = \frac{1}{2}\sum_{m=0}^{3} x[n]\phi_{0,0}[n] = \frac{1}{2}[1 \cdot 1 + 4 \cdot 1 - 3 \cdot 1 + 0 \cdot 1] = 1. \tag{11.128}$$

The coefficient in W_0 is

$$X_\psi[0,0] = \frac{1}{2}\sum_{m=0}^{3} x[n]\psi_{0,0}[n] = \frac{1}{2}[1 \cdot 1 + 4 \cdot 1 - 3 \cdot (-1) + 0 \cdot (-1)] = 4. \tag{11.129}$$

The two coefficients in W_1 are

$$X_\psi[1,0] = \frac{1}{2}\sum_{m=0}^{3} x[n]\psi_{1,0}[n] = \frac{1}{2}[1 \cdot \sqrt{2} + 4 \cdot (-\sqrt{2}) - 3 \cdot 0 + 0 \cdot 0] = -1.5\sqrt{2}, \tag{11.130}$$

$$X_\psi[1,1] = \frac{1}{2}\sum_{m=0}^{3} x[n]\psi_{1,0}[n] = \frac{1}{2}[1 \cdot 0 + 4 \cdot 0 - 3 \cdot \sqrt{2} + 0 \cdot (-\sqrt{2})] = -1.5\sqrt{2}. \tag{11.131}$$

Or in matrix form we have

$$\begin{bmatrix} 1 \\ 4 \\ -1.5\sqrt{2} \\ -1.5\sqrt{2} \end{bmatrix} = \frac{1}{2} \begin{bmatrix} 1 & 1 & 1 & 1 \\ 1 & 1 & -1 & -1 \\ \sqrt{2} & -\sqrt{2} & 0 & 0 \\ 0 & 0 & \sqrt{2} & -\sqrt{2} \end{bmatrix} \begin{bmatrix} 1 \\ 4 \\ -3 \\ 0 \end{bmatrix}. \tag{11.132}$$

Now the four-point discrete signal can be expressed as a linear combination of these basis functions:

$$x[n] = \frac{1}{2}[X_\phi[0,0]\phi_{0,0}[n] + C_\psi[0,0]\psi_{0,0}[n] + X_\phi[1,0]\psi_{1,0}[n] + X_\phi[1,1]\psi_{1,1}[n]\,]$$
$$n = 0, \ldots, 3, \tag{11.133}$$

or in matrix form as

$$\begin{bmatrix} 1 \\ 4 \\ -3 \\ 0 \end{bmatrix} = \frac{1}{2} \begin{bmatrix} 1 & 1 & \sqrt{2} & 0 \\ 1 & 1 & -\sqrt{2} & 0 \\ 1 & -1 & 0 & \sqrt{2} \\ 1 & -1 & 0 & -\sqrt{2} \end{bmatrix} \begin{bmatrix} 1 \\ 4 \\ -1.5\sqrt{2} \\ -1.5\sqrt{2} \end{bmatrix}. \tag{11.134}$$

This is the inverse DWT.

11.2.2 Fast wavelet transform (FWT)

The total number of operations in Eq. 11.126 for the forward DWT (or Eq. (11.124) for the inverse DWT) is proportional to the product of the vector length N and number of integer shifts $\sum_{j=0}^{J-1} 2^j = N$; i.e., the computational complexity is $O(N^2)$. For example, when the Haar transform as a DWT is implemented as a matrix multiplication in Eq. (8.80), its complexity is obviously $O(N^2)$. Now we will consider Mallat's fast wavelet transform (FWT) algorithm for the DWT with a linear complexity of $O(N)$ (as we have already seen in the case of the discrete Haar transform in section 8.3.3).

A given N-D signal vector $\boldsymbol{x} = [x[0], \ldots, x[N-1]]^{\mathrm{T}}$ can be represented in any scale space V_j spanned by orthogonal basis $\boldsymbol{\phi}_{j,k}$ or wavelet space W_j spanned by orthogonal basis $\boldsymbol{\psi}_{j,k}$ $(j = 0, \ldots, J-1)$, in terms of the following coefficients:

$$X_\phi[j,k] = \sum_{n=0}^{N-1} x[n]\overline{\phi}_{j,k}[n], \qquad X_\psi[j,k] = \sum_{n=0}^{N-1} x[n]\overline{\psi}_{j,k}[n]. \tag{11.135}$$

Note that these equations are the same as the forward DWT in Eqs. (11.125) (when $j = k = 0$) and (11.126). Moreover, owing to the recursive relationships of $\phi_{j,l}(t)$ and $\psi_{j,l}(t)$ (Eqs. (11.22) and (11.51)), these equations can both be expressed in terms of the coefficients $X_\phi[j+1,k]$ at the next higher scale level:

$$\begin{aligned} X_\phi[j,k] &= \sum_{n=0}^{N-1} x[n] \sum_l \overline{h}_0[l-2k]\overline{\phi}_{j+1,l}[n] = \sum_l \overline{h}_0[l-2k] \sum_{n=0}^{N-1} x[n]\overline{\phi}_{j+1,l}[n] \\ &= \sum_l \overline{h}_0[l-2k] X_\phi[j+1,l], \end{aligned} \tag{11.136}$$

$$\begin{aligned} X_\psi[j,k] &= \sum_{n=0}^{N-1} x[n] \sum_l \overline{h}_1[l-2k]\overline{\phi}_{j+1,l}[n] = \sum_l \overline{h}_1[l-2k] \sum_{n=0}^{N-1} x[n]\phi_{j+1,l}[n] \\ &= \sum_l \overline{h}_1[l-2k] X_\phi[j+1,l]. \end{aligned} \tag{11.137}$$

These operations can be carried out recursively until the highest scale level at $j+1 = J$ is reached. The corresponding space V_J is spanned by $\phi_{J,k}[n] = \delta[k-n]$ as the standard basis, and the signal $\boldsymbol{x}$ is simply represented by all of its N samples:

$$X_\phi[J,k] = \sum_{n=0}^{N-1} x[n]\phi_{J,k}[n] = \sum_{n=0}^{N-1} x[n]\delta[k-n] = x[k] \qquad k = 0, \ldots, N-1. \tag{11.138}$$

Comparing Eqs. (11.136) and (11.137) with the discrete convolution (Eq. (4.152)),

$$y[k] = h[k] * x[k] = \sum_{n=0}^{N-1} x[n]h[k-n], \tag{11.139}$$

we see that both DWT coefficients $X_\phi[j,k]$ and $X_\psi[j,k]$ at the jth scale level can be obtained from the coefficients $X_\phi[j+1,k]$ at the $(j+1)$th scale level by:

- convolution with time-reversed $\overline{h}_0[k]$ and $\overline{h}_1[k]$;
- subsampling to keep every other sample in the convolution.

Equations (11.136) and (11.137) can, therefore, be considered as a filtering process:

$$X_\phi[j,k] = \overline{h}_0[-l] * X_\phi[j+1,l]\big|_{l=2k},$$
$$X_\psi[j,k] = \overline{h}_1[-l] * X_\phi[j+1,l]\big|_{l=2k}. \tag{11.140}$$

(Note that this operation of convolution followed by subsampling for the DWT coefficients is the same as in that in Eqs. (11.22) and (11.51) for the scaling and wavelet functions.) This filtering process can be implemented as either a convolution in the time domain or, equivalently, as a multiplication in the frequency domain.

Now we see that both $X_\phi[j,k]$ and $X_\psi[j,k]$ for all $j < J$ can be obtained by filtering $X_\phi[j+1,k]$ of the next higher scale level $j+1$, which in turn can be obtained from a still higher level $j+2$, and this recursion can be carried out until the highest level $j = J$ is reached, where $X_\phi[J,k] = x[k]$ are simply the N signal samples originally given. Based on this recursion, the forward DWT in Eqs. (11.125) and (11.126) can be implemented by the *analysis filter bank* shown on the left-hand side of Fig. 11.12, by which all N DWT coefficients in Eqs. (11.125) and (11.126) can be generated, as represented by the vertical bar in the middle of the figure. This is the FWT algorithm. As the data size is halved by the subsampling of each iteration, the total computational complexity of the FWT is linear:

$$O\left(N + \frac{N}{2} + \frac{N}{4} + \frac{N}{8} + \cdots + 1\right) = O(N). \tag{11.141}$$

The right-hand side of Fig. 11.12 is for the inverse DWT by which the signal is to be reconstructed from its DWT coefficients, to be discussed next. Same as all orthogonal transforms considered before, for an N-D signal vector $\boldsymbol{x} = [x[0], \ldots, [x[N-1]]^{\mathrm{T}}$, there are also N DWT coefficients in the transform domain which can be arranged as an N-D vector, same as the N-D spectrum vector of the DFT or DCT, as shown in Fig. 11.13.

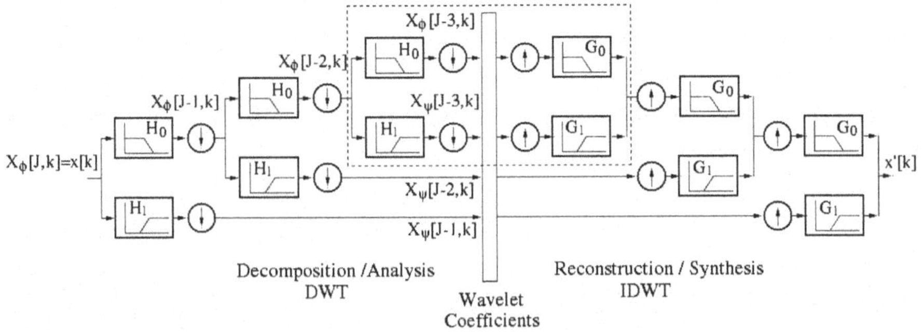

Figure 11.12 Filter banks for both forward and inverse DWT. Inside the dashed box is the building block, the two-channel decomposition-reconstruction filter bank system.

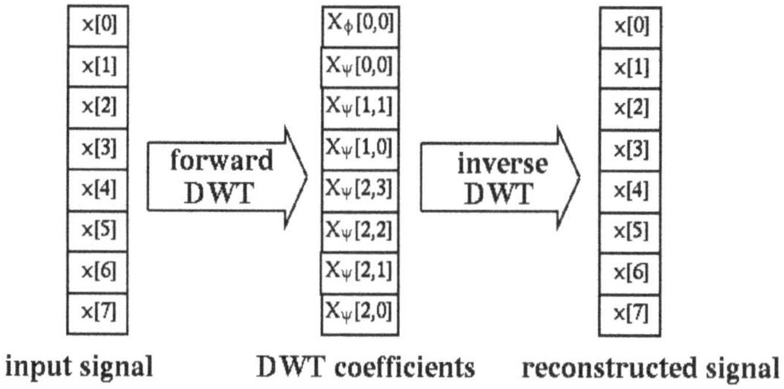

Figure 11.13 Vector representations of the forward and inverse DWT ($N = 8$).

11.3 Filter bank implementation of DWT and inverse DWT

11.3.1 Two-channel filter bank and inverse DWT

The complexity of the inverse DWT in Eq. (11.124) for signal reconstruction is $O(N^2)$, as mentioned above. Here we consider the fast inverse DWT that can be carried out as a sequence of filtering operations with the same linear complexity $O(N)$ for the forward DWT in the *synthesis filter bank*, as illustrated on the right-hand side of Fig. 11.12. In the following we will derive the theory needed for the design of the filters G_0 and G_1 in the synthesis filter bank.

The DWT filter bank shown in Fig. 11.12 can be considered as a hierarchical structure composed of a set of basic *two-channel filter banks*, which in turn is composed of two pairs of filters, the analysis and synthesis filter banks, as shown inside the dashed box in Fig. 11.12 and also Fig. 11.14. The analysis bank contains a LP filter, represented by $h_0[n]$ or $H_0(f) = \mathcal{F}[h_0[n]]$, that takes input $x[n]$ and generates output $a[n]$ (approximation), and an HP filter, represented by $h_1[n]$ or $H_1(f) = \mathcal{F}[h_1[n]]$, that takes the same input $x[n]$ and generates output $d[n]$ (detail). Each of these filters is followed by a down-sampler. The synthesis bank also contains a pair of filters represented respectively by $g_0[n]$ or $G_0(f) = \mathcal{F}[g_0[n]]$ and $g_1[n]$ or $G_1(f) = \mathcal{F}[g_1[n]]$, each proceeded by an upsampler. We have already considered filters $h_0[n]$ and $h_1[n]$ of the analysis filter bank, and will now concentrate on the design of $g_0[n]$ and $g_1[n]$, so that the sum of their outputs, the output $\hat{\boldsymbol{x}}$ of the synthesis filter bank, can be identical to the input $\boldsymbol{x}$ (with possibly some delay). Once this perfect reconstruction is achieved by the basic two-channel filter bank at this lowest level, it can also be achieved recursively at each of the next higher levels in the entire filter bank in Fig. 11.12.

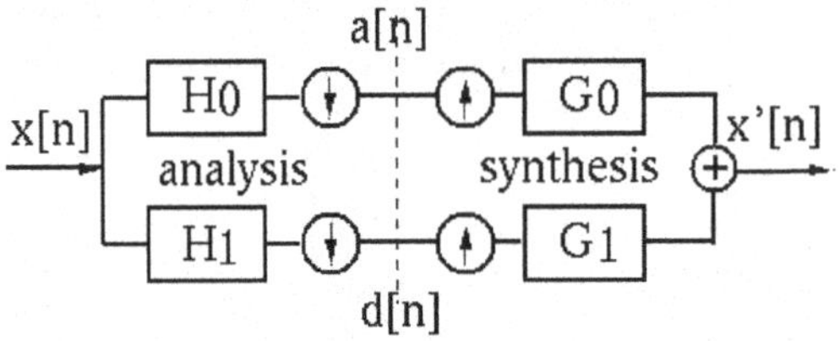

Figure 11.14 Two-channel filter bank.

As in Eqs. (11.136) and (11.137), the outputs $a[k]$ and $d[k]$ of the two analysis filters of the two-channel filter bank can be written as

$$a[k] = \sum_n h_0[n-2k]x[n], \qquad d[k] = \sum_n h_1[n-2k]x[n], \tag{11.142}$$

which can be considered as the inner products of vectors $\boldsymbol{x} = [\ldots, x[n], \ldots]^{\mathrm{T}}$ and $\boldsymbol{h}_i(k) = [\ldots, h_i[n-2k], \ldots]^{\mathrm{T}}$ $(i = 0, 1)$:

$$a[k] = \langle \boldsymbol{x}, \boldsymbol{h}_0(k) \rangle, \qquad d[k] = \langle \boldsymbol{x}, \boldsymbol{h}_1(k) \rangle. \tag{11.143}$$

The output $\hat{x}[n]$ of the two-channel filter bank can be written as

$$\hat{x}[n] = \sum_k a[k]g_0[n-2k] + \sum_k d[k]g_1[n-2k], \qquad \text{for all } n, \tag{11.144}$$

or in vector form:

$$\begin{aligned} \hat{\boldsymbol{x}} &= \sum_k a[k]\boldsymbol{g}_0(k) + \sum_k d[k]\boldsymbol{g}_1(k) \\ &= \sum_k \langle \boldsymbol{x}, \boldsymbol{h}_0(k) \rangle \boldsymbol{g}_0(k) + \sum_k \langle \boldsymbol{x}, \boldsymbol{h}_1(k) \rangle \boldsymbol{g}_1(k). \end{aligned} \tag{11.145}$$

where $\boldsymbol{g}_0(k)$ and $\boldsymbol{g}_1(k)$ are two vectors composed of the time-reversed version of the synthesis filter coefficients $g_i[n-2k]$. Our goal here is to design the two filters $g_0[n]$ and $g_1[n]$ in the synthesis filter bank so that its output $\hat{x}[n] = x[n]$ is a perfect reconstruction of the input for the original signal. The derivation can be carried out in either the time or frequency domain based on the DTFT or Z-transform (with $z = e^{j2\pi f}$ evaluated along the unit circle). Here we choose to use the DTFT approach, although the Z-transform is also used in some literature. Note again that all DTFT spectra are periodic with period 1; e.g., $H_0(f \pm 1) = H_0(f)$ and $H_0(f + 1/2) = H_0(f - 1/2)$.

According to the down-sampling property of the DTFT (Eq. (4.47) for $k = 2$), the subsampled outputs $a[n]$ of $H_0(f)$ and $d[n]$ of $H_1(f)$, when given the same input $x[n]$, can be expressed in the frequency domain as

$$A(f) = \frac{1}{2}\left[H_0\left(\frac{f}{2}\right)X\left(\frac{f}{2}\right) + H_0\left(\frac{f+1}{2}\right)X\left(\frac{f+1}{2}\right)\right], \tag{11.146}$$

$$D(f) = \frac{1}{2}\left[H_1\left(\frac{f}{2}\right)X\left(\frac{f}{2}\right) + H_1\left(\frac{f+1}{2}\right)X\left(\frac{f+1}{2}\right)\right]. \tag{11.147}$$

Next, according to the upsampling property of the DTFT (Eq. (4.53)), the overall output of the two-channel filter bank can be expressed as

$$\begin{aligned}\hat{X}(f) &= G_0(f)A(2f) + G_1(f)D(2f)\\ &= \frac{1}{2}[G_0(f)H_0(f) + G_1(f)H_1(f)]\,X(f)\\ &+\frac{1}{2}\left[G_0(f)H_0\left(f+\frac{1}{2}\right) + G_1(f)H_1\left(f+\frac{1}{2}\right)\right]\,X\left(f+\frac{1}{2}\right).\end{aligned} \tag{11.148}$$

For perfect reconstruction we need to have $X(f) = \hat{X}(f)$; i.e., the coefficient of the first term of $X(f)$ should be 1 (or a pure delay) and that of the second term of $X(f+1/2)$ is zero:

$$\begin{cases} G_0(f)H_0(f) + G_1(f)H_1(f) = 2 \\ G_0(f)H_0(f+\frac{1}{2}) + G_1(f)H_1(f+\frac{1}{2}) = 0 \end{cases}. \tag{11.149}$$

These two equations can be written in matrix form as

$$\begin{bmatrix} H_0(f) & H_1(f) \\ H_0(f+\frac{1}{2}) & H_1(f+\frac{1}{2}) \end{bmatrix} \begin{bmatrix} G_0(f) \\ G_1(f) \end{bmatrix} = \mathbf{H}(f) \begin{bmatrix} G_0(f) \\ G_1(f) \end{bmatrix} = \begin{bmatrix} 2 \\ 0 \end{bmatrix}, \tag{11.150}$$

where $\boldsymbol{H}$ is a 2×2 matrix defined as

$$\mathbf{H}(f) = \begin{bmatrix} H_0(f) & H_1(f) \\ H_0(f+\frac{1}{2}) & H_1(f+\frac{1}{2}) \end{bmatrix}, \qquad \mathbf{H}^{-1}(f) = \frac{1}{\Delta(f)} \begin{bmatrix} H_1(f+\frac{1}{2}) & -H_1(f) \\ -H_0(f+\frac{1}{2}) & H_0(f) \end{bmatrix}, \tag{11.151}$$

where $\Delta(f)$ is the determinant of $\mathbf{H}(f)$:

$$\Delta(f) = H_0(f)H_1\left(f+\frac{1}{2}\right) - H_0\left(f+\frac{1}{2}\right)H_1(f). \tag{11.152}$$

Note that

$$\Delta\left(f+\frac{1}{2}\right) = -\Delta(f). \tag{11.153}$$

Solving Eq. (11.150) we get

$$\begin{bmatrix} G_0(f) \\ G_1(f) \end{bmatrix} = \mathbf{H}^{-1}(f) \begin{bmatrix} 2 \\ 0 \end{bmatrix} = \frac{1}{\Delta(f)} \begin{bmatrix} H_1(f+\frac{1}{2}) & -H_1(f) \\ -H_0(f+\frac{1}{2}) & H_0(f) \end{bmatrix} \begin{bmatrix} 2 \\ 0 \end{bmatrix}; \tag{11.154}$$

i.e.,

$$G_0(f) = \frac{2}{\Delta(f)}H_1\left(f+\frac{1}{2}\right), \qquad G_1(f) = \frac{-2}{\Delta(f)}H_0\left(f+\frac{1}{2}\right). \tag{11.155}$$

Replacing f by $f+\frac{1}{2}$ in the second equation for $G_1(f)$ we get

$$G_1\left(f+\frac{1}{2}\right) = \frac{2}{\Delta(f)}H_0(f). \tag{11.156}$$

Dividing the two sides of this equation by those of the first equation in Eq. (11.155) we get

$$G_1\left(f+\frac{1}{2}\right)H_1\left(f+\frac{1}{2}\right)=G_0(f)H_0(f), \tag{11.157}$$

which can also be written as

$$G_1(f)H_1(f)=G_0\left(f+\frac{1}{2}\right)H_0\left(f+\frac{1}{2}\right). \tag{11.158}$$

This equation can be substituted back into the two equations in Eq. (11.149) in different ways to get the following four conditions for perfect reconstruction:

$$\begin{aligned}
G_0(f)H_0(f)+G_0\left(f+\frac{1}{2}\right)H_0\left(f+\frac{1}{2}\right)&=2,\\
G_1(f)H_1(f)+G_1\left(f+\frac{1}{2}\right)H_1\left(f+\frac{1}{2}\right)&=2,\\
G_1(f)H_0(f)+G_1\left(f+\frac{1}{2}\right)H_0\left(f+\frac{1}{2}\right)&=0,\\
G_0(f)H_1(f)+G_0\left(f+\frac{1}{2}\right)H_1\left(f+\frac{1}{2}\right)&=0.
\end{aligned} \tag{11.159}$$

Comparing these four equations with Eq. (11.63) required of $H_0(f)$ and $H_1(f)$ (orthogonalities of the scaling and wavelet functions), we see that if we let

$$G_0(f)=\overline{H}_0(f),\qquad G_1(f)=\overline{H}_1(f), \tag{11.160}$$

then all four equations in Eq. (11.159) hold; i.e., the condition for a perfect reconstruction is satisfied. Moreover, applying the DTFT property in Eq. (4.31) to these two relations in the frequency domain we get the following in the time domain:

$$g_0[n]=\overline{h}_0[-n],\qquad g_1[n]=\overline{h}_1[-n]. \tag{11.161}$$

In other words, the perfect reconstruction can be achieved by the synthesis filters if their coefficients are the complex conjugate and time-reversed (conjugate mirror) version of coefficients of the analysis filters.

We also note that the four equations in Eq. (11.159) are actually the down- and upsampled versions of $G_0(f)H_0(f)$, $G_1(f)H_1(f)$, $G_1(f)H_0(f)$, and $G_0(f)H_1(f)$ (recall Eq. (4.56)), and they correspond to the following four down-sampled convolutions in the time domain:

$$\begin{aligned}
g_0[2n]*h_0[2n]&=\sum_k h_0[k]g_0[2n-k]=\delta[n],\\
g_1[2n]*h_1[2n]&=\sum_k h_1[k]g_1[2n-k]=\delta[n],\\
g_1[2n]*h_0[2n]&=\sum_k h_1[k]g_0[2n-k]=0,\\
g_0[2n]*h_1[2n]&=\sum_k h_0[k]g_1[2n-k]=0.
\end{aligned} \tag{11.162}$$

Comparing these four time convolutions with Eq. (11.60), we reach the same conclusion as above: the condition for perfect reconstruction is satisfied if the coefficients of the synthesis filters satisfy $g_0[n] = \overline{h}_0[-n]$ and $g_1[n] = \overline{h}_1[-n]$.

To see how the two-channel filter bank can actually be implemented, we list below the Matlab code, which carries out first the analysis filtering for signal decomposition with $H_0(f) = \mathcal{F}[h_0[k]]$ (filtering coefficients $h_0[k]$ provided as input) and $H_1(f)$ (Eq. 11.75), and then the synthesis filtering for signal reconstruction with $G_0(f)$ and $G_1(f)$ (Eq. (11.160)).

```
function y=TwoChannelFilterBank(x,h)
   h=h/norm(h);                   % normalize h
   K=length(h);                   % length of filter (K<N)
   N=length(x);                   % length of signal vector
   h0=zeros(1,N); h0(1:K)=h;      % analysis filter H0
   H0=fft(h0);
   for k=0:N-1
      l=mod(k-N/2,N);             % rotation by 1/2
      H1(k+1)=exp(-j*2*pi*k/N)*conj(H0(l+1)); % analysis filter H1
   end
   G0=conj(H0); G1=conj(H1);      % synthesis filters G0 and G1:
   % Decomposition by analysis filters:
   A=fft(x);                      % input
   d=ifft(A.*H1);                 % filtering to get d (detail)
   a=ifft(A.*H0);                 % filtering to get a (approximation)
   d=d(1:2:length(d));            % downsampling d
   a=a(1:2:length(a));            % downsampling a
   % Reconstruction by synthesis filters:
   a=upsample(a,2);               % upsampling a
   d=upsample(d,2);               % upsampling d
   a=ifft(fft(a).*G0);            % filtering of a
   d=ifft(fft(d).*G1);            % filtering of d
   y=a+d;                         % perfect reconstruction of x
end
```

Here the input `x` is the signal vector, and input `h` is a vector containing the filtering coefficients $h_0[k]$. For example, $h = [1\ 1]$ for D2 (Haar), or $h = [0.4830\ 0.8365\ 0.2241\ \ -0.1294]$ for D4. The output `y` is a perfect reconstruction of the input `x`. Also note that here both the decomposition and reconstruction are implemented as multiplications in the frequency domain, although they can also be equivalently carried out by circular convolutions in the time domain.

Having obtained the two-channel filter bank in Fig. 11.14 capable of perfect reconstruction, we can use it as the building block to construct the filter bank in Fig. 11.12, by which the input signal is perfectly reconstructed at the output. Note that the iteration of the DWT on the left of the figure can be terminated at any scale level before reaching the lowest possible scale level (top level in the figure), depending on the actual signal processing need, as the data can always be perfectly reconstructed from any level by the inverse DWT on the right.

The Matlab code for both the forward DWT for signal decomposition and the inverse DWT for signal reconstruction is listed below. The algorithm is basically a recursion of the operations in the two-channel filter bank shown above. The input of the forward DWT function includes a vector x for the signal to be DWT transformed and another vector h for the father wavelet coefficients $h_0[k]$, and the output is a vector w for the DWT coefficients. Note that the size $N = 2^n$ of the data vector x is assumed to be a power of two for convenience. Note that, unlike the fast algorithms of all previously considered orthogonal transforms (except DHT) of complexity of $O(N \log_2 N)$, all containing an inner loop that is carried out $\log_2 N$ times, here the number of iterations is smaller than $\log_2 N$ when the length of the filter h is greater than two (except D2 or Haar when h has only two non-zero components). In general, the iteration in the DWT does not have to be always carried out to the lowest possible scale level.

```
function w=mydwt(x,h)
    K=length(h);
    if K>N
        error('K should be less than N');          % assume N>K
    end
    N=length(x);
    n=log2(N);
    if n~=int16(n)
        error('Length of data x should be power of 2');
    end
    h=h/norm(h);                                    % normalize h
    h0=zeros(1,N); h0(1:K)=h; H0=fft(h0);           % scaling function
    for k=0:N-1
        l=mod(k-N/2,N);                             % rotation by 1/2
        H1(k+1)=exp(-j*2*pi*k/N)*conj(H0(l+1)); % wavelet function
    end
    a=x;
    n=length(a);
    w=[];
    while n>=K
        A=fft(a);
        d=real(ifft(A.*H1));          % convolution d=a*h1
        a=real(ifft(A.*H0));          % convolution a=a*h0
        d=d(2:2:n);                   % downsampling d
        a=a(2:2:n);                   % downsampling a
        H0=H0(1:2:length(H0));        % subsampling H0
        H1=H1(1:2:length(H1));        % subsampling H1
```

```
        w=[d,w];                    % concatenate DWT coefficients
        n=n/2;
    end
    w=[a w];                        % residual in scale space V_0
end
```

The input of the inverse DWT function includes a vector w for the DWT coefficients and a vector h for the father wavelet coefficients $h_0[k]$, and the output is a vector y for the reconstructed signal x.

```
function y=myidwt(w,h)
    K=length(h);
    N=length(w);
    n=log2(N);
    h=h/norm(h);                   % normalize h
    h0=zeros(1,N); h0(1:K)=h; H0=fft(h0);
    for k=0:N-1
        l=mod(k-N/2,N);
        H1(k+1)=exp(-j*2*pi*k/N)*conj(H0(l+1));
    end
    G0=conj(H0); G1=conj(H1);  % synthesis filters
    i=0;
    while 2^i<K
       i=i+1;            % starting scale based on filter length
    end
    n=2^(i-1);
    a=w(1:n);
    while n<N
        d=w(n+1:2*n);              % get detail
        a=upsample(a,2,1);         % upsampling a
        d=upsample(d,2,1);         % upsampling d
        if n==1
           a=a'; d=d';             % upsampling 1x1 is column vector
        end
        n=2*n;                     % signal size is doubled
        A=fft(a).*G0(1:N/n:N); % convolve a with subsampled G0
        D=fft(d).*G1(1:N/n:N); % convolve d with subsampled G1
        a=real(ifft(A));
        d=real(ifft(D));
        a=a+d;
    end
    y=a;
end
```

Example 11.5: The DWT of an eight-point signal vector $\boldsymbol{x} = [0, 0, 2, 3, 4, 0, 0, 0]^{\mathrm{T}}$ can be obtained by the code above. Depending on the wavelet functions used, different DWT coefficients will be generated. When the Haar wavelets are used, the output is exactly the same as Eq. (8.81) obtained by the discrete Haar transform:

$$\boldsymbol{X} = \boldsymbol{H}^{\mathrm{T}}\boldsymbol{x} = [3.18, 0.35, -2.50, 2.0, 0.0, -0.71, 2.83, 0.0]^{\mathrm{T}}. \tag{11.163}$$

But when Daubechies' wavelets are used, we get a different set of DWT coefficients:

$$\boldsymbol{X} = [0.91, 3.60, -1.84, 2.65, 0.84, -0.65, 1.93, 0.000]^{\mathrm{T}}. \tag{11.164}$$

In either case, the signal is perfectly reconstructed by the inverse DWT.

11.3.2 Two-dimensional DWT

Similar to all orthogonal transforms previously discussed, the DWT can also be extended to a 2-D transform that can be applied to 2-D signals such as an image. To do so, we first extend the 1-D two-channel filter bank shown in Fig. 11.14 to a basic 2-D filter bank, as shown in Fig. 11.15, where the left half is the analysis filter bank for signal decomposition and the right half is the synthesis filter bank for signal reconstruction. The input of the analysis filter bank is a 2-D array treated as the coefficients $X_\phi[j]$ at the previous scale level j. We first carry out both LP and HP filtering corresponding to H_0 and H_1, respectively, on each of the N columns of this array (vertical filtering), and then, after down-sampling, we carry out the same filtering on the rows of the resulting array (horizontal filtering). The outcomes of this two-stage filtering process are four sets of coefficients at the next lower scale level $j-1$, including $W_\phi[j-1]$, LP-filtered in both vertical and horizontal directions, $W_\psi^h[j-1]$, LP-filtered vertically but HP-filtered horizontally, $W_\psi^v[j-1]$, HP-filtered vertically but LP-filtered horizontally, and $W_\psi^d[j-1]$, HP-filtered in both directions. These four sets of coefficients, each one-quarter of the original size of the input 2-D array, are stored as the upper-left, upper-right, lower-left, and lower-right quarters of a 2-D array, respectively. As in the 1-D case, the synthesis filter bank on the right of Fig. 11.15 reverses the process to generate a perfect reconstruction of the input signal as the output. An example of the decomposition and reconstruction carried out by this 2-D two-channel filter bank is shown in Fig. 11.16.

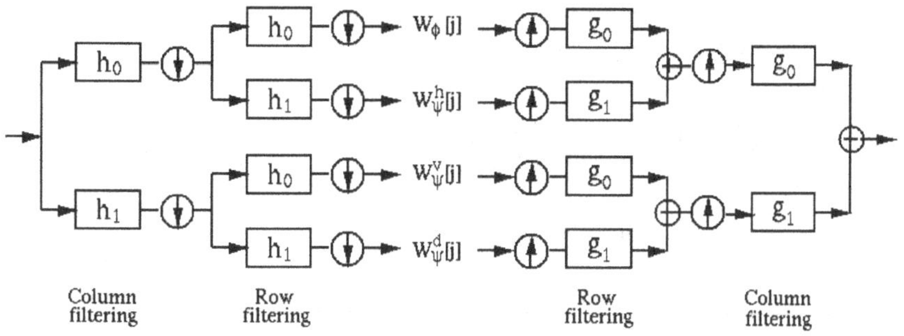

Figure 11.15 A 2-D two-channel filter bank.

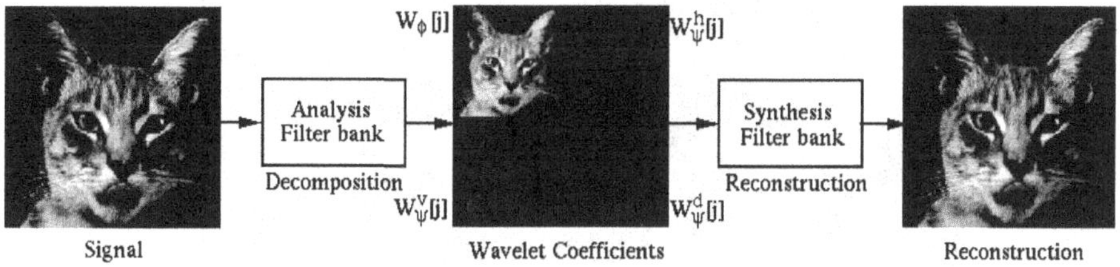

Figure 11.16 Signal decomposition and reconstruction by 2-D two-channel filter bank.

This two-stage filtering-down-sampling operation can be applied to $W_\phi[j-1]$, one of the four sets of coefficients that is LP-filtered in both directions and stored in the top-left quarter of the array, to generate the four sets of coefficients at the next lower scale level $j-2$, as illustrated in Fig. 11.17. Moreover, similar to the hierarchical process shown in Fig. 11.12, this process can be carried out recursively, until, if needed, the lowest possible scale level is reached. If the input data are an $N \times N$ 2-D array, then the 2-D DWT coefficients at any scale level, including the final and lowest level, is also an $N \times N$ matrix. For example, the 2-D DWT coefficients obtained at each of four consecutive iterations of the 2-D DWT recursion are shown in Fig. 11.18. Same as in the case of 1-D DWT, the 2-D DWT iteration can terminate at any of these scale levels, at which the data can always be perfectly reconstructed by the inverse transform.

Note that the 2-D array composed of the DWT coefficients is similar to the spectrum of most 2-D orthogonal transforms (except DFT), such as the DCT, in the sense that the coefficients around the top-left and lower-right corners represent respectively the signal components of the lowest and highest scale levels, corresponding to the lowest and highest frequency components in the DCT. The 2-D DWT coefficients can therefore be filtered (HP, LP, BP, etc.), similar to the filtering of the 2-D DCT spectrum.

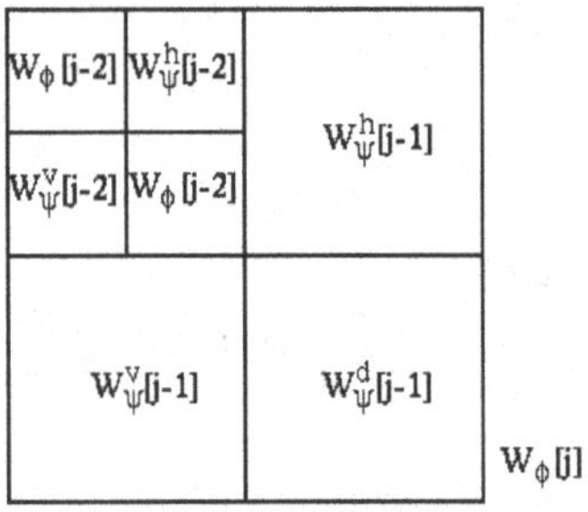

Figure 11.17 Recursion of the 2-D discrete wavelet transform.

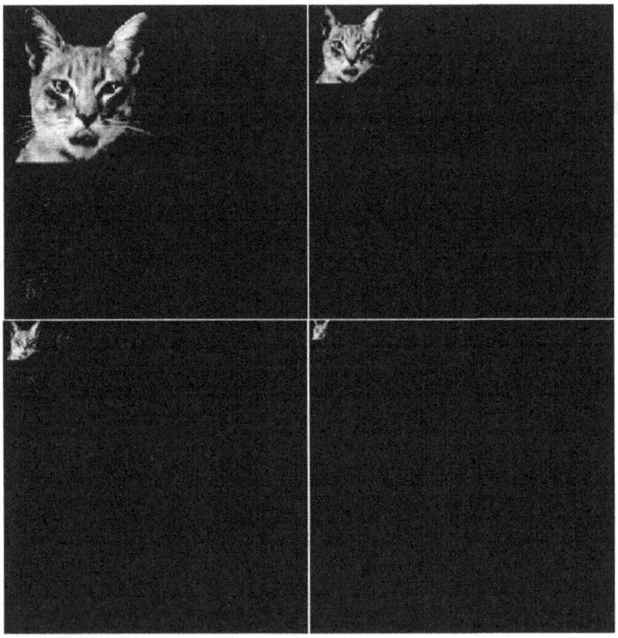

Figure 11.18 The 2-D DWT coefficients obtained at four consecutive stages.

The Matlab code for both the forward and inverse 2-D DWTs is listed below. The input of the forward DWT function includes a 2-D array `x` for the signal, such as an image, and a vector `h` for the father wavelet filter coefficients $h_0[k]$, and the output is a 2-D array `w` of the same size as the input array for the DWT coefficients.

```
function w=dwt2d(x,h)
   K=length(h);
   [M,N]=size(x);
   if M~=N
      error('Input should be a square array');
   end
```

```
if K>N
   error('Data size should be larger than size of filter');
end
n=log2(N);
if n~=int16(n)
   error('Length of data x should be power of 2');
end
h0=zeros(1,N);
h0(1:K)=h;
H0=fft(h0);
for k=0:N-1
   l=mod(k-N/2,N);
   H1(k+1)=-exp(-j*2*pi*k/N)*conj(H0(l+1));
end
a=x;
w=zeros(N);
n=length(a);
while n>=K
   t=zeros(n,n);
   for k=1:n                          % for all n columns
      A=fft(double(a(:,k)));          % get the kth column
      D=real(ifft(A.*H1'));           % convolution d=a*h1
      A=real(ifft(A.*H0'));           % convolution a=a*h0
      t(:,k)=[A(2:2:n); D(2:2:n)];    % save filtered column
   end
   for k=1:n                          % for all n rows
      A=fft(t(k,:));                  % get the kth row
      D=real(ifft(A.*H1));            % convolution d=a*h1
      A=real(ifft(A.*H0));            % convolution a=a*h0
      t(k,:)=[A(2:2:n) D(2:2:n)];     % save filtered row
   end
   w(1:n,1:n)=t;                      % concatenate coefficients
   H0=H0(1:2:length(H0));             % subsampling H0
   H1=H1(1:2:length(H1));             % subsampling H1
   n=n/2;                             % size of the next level
   a=t(1:n,1:n);                      % up-left quarter as input
end
```

The inputs of the inverse DWT function include a 2-D array w for the 2-D DWT coefficients and a vector h for the father wavelet coefficients $h_0[k]$, and the output is a 2-D array y for the reconstruction of the input data array.

```
function y=idwt2d(w,h)
   K=length(h);
```

```
N=length(w);
n=log2(N);
h=h/norm(h);                % normalize h
h0=zeros(1,N); h0(1:K)=h; H0=fft(h0);
for k=0:N-1
   l=mod(k-N/2,N);
   H1(k+1)=-exp(-j*2*pi*k/N)*conj(H0(l+1));
end
G0=conj(H0); G1=conj(H1); % synthesis filters
i=0;
while 2^i<K
   i=i+1;           % starting scale based on filter length
end
n=2^(i-1);                  % signal size of initial scale
y=w;
t=y(1:n,1:n);
while n<N
   g0=G0(1:N/(2*n):N);
   g1=G1(1:N/(2*n):N);
   for k=1:n                            % filtering n rows
      % rows in top half:
      a=upsample(y(k,1:n),2,1);         % approximate
      d=upsample(y(k,n+1:2*n),2,1);     % detail
      A=fft(a).*g0;                     % convolve a with G0
      D=fft(d).*g1;                     % convolve d with G1
      y(k,1:2*n)=real(ifft(A)+ifft(D));
      % rows in bottom half:
      a=upsample(y(n+k,1:n),2,1);       % approximate
      d=upsample(y(n+k,n+1:2*n),2,1);   % detail
      A=fft(a).*g0;                     % convolve a with G0
      D=fft(d).*g1;                     % convolve d with G1
      y(n+k,1:2*n)=real(ifft(A)+ifft(D));
   end
   for k=1:2*n                          % filtering 2n columns
      a=upsample(y(1:n,k),2,1);         % top half
      d=upsample(y(n+1:2*n,k),2,1);     % bottom half
      A=fft(a).*g0';                    % convolve a with G0
      D=fft(d).*g1';                    % convolve d with G1
      y(1:2*n,k)=real(ifft(A)+ifft(D))/2;
   end
   n=n*2;
end
```

11.4 Applications in filtering and compression

Example 11.6: Consider a set of signals, denoted by $\boldsymbol{x}$, as shown in the first and third columns (dashed curves) of Fig. 11.19, and their DCT and DWT (Daubechies D6) coefficients, generically denoted by $\boldsymbol{X} = \mathcal{T}[\boldsymbol{x}]$, as shown in the second and fourth columns of the figure. Compression is then carried out in both the DCT and DWT domains by suppressing to zero a certain percentage (80% in this case) of the transform coefficients with lowest magnitudes. The compressed coefficients, denoted by $\boldsymbol{X}'$, are shown as the solid curves in the second and fourth columns, in comparison with the original ones (dashed curves). Finally, the signals are reconstructed by the inverse transforms of the modified coefficients to get $\boldsymbol{x}' = \mathcal{T}^{-1}[\boldsymbol{X}']$, shown as the solid curves in the first and third columns, in comparison with the original signals.

The performance of the DCT and DWT when used for compression can be evaluated in terms of both energy loss and signal error. As $||\boldsymbol{x}||^2 = ||\boldsymbol{X}||^2$ and $||\boldsymbol{x}'||^2 = ||\boldsymbol{X}'||^2$ (Parseval's identity), the percentage energy loss due to the compression can be found in either the time or transform domain as

$$\frac{||\boldsymbol{X}||^2 - ||\boldsymbol{X}'||^2}{||\boldsymbol{X}||^2} = \frac{||\boldsymbol{x}||^2 - ||\boldsymbol{x}'||^2}{||\boldsymbol{x}||^2}. \tag{11.165}$$

On the other hand, the percentage signal error caused by the compression can also be defined as

$$\frac{||\boldsymbol{x} - \boldsymbol{x}'||^2}{||\boldsymbol{x}||^2}. \tag{11.166}$$

It can be shown (see homework) that the signal error happens to be the same as energy loss, $||\boldsymbol{x} - \boldsymbol{x}'||^2 = ||\boldsymbol{x}||^2 - ||\boldsymbol{x}'||^2$, in this case.

The signal error depends on the transform method used, as well as the specific signal, as listed in Table 11.1. We see that the DCT and DWT are each good at representing certain types of signals. For example, the DCT is effective for the sinusoidal signals such as in cases 1, 2, and 3, while the DWT is effective for non-periodic and spiky signals such as in cases 4, 5, and 8. Note in particular that the DWT is especially effective at representing irregular spiky signals. Also, neither transform method can represent the random noise as it is close to a white noise with energy relatively evenly distributed over all components in either the DCT or DWT domain. It is also interesting to compare the errors with D4 and D6 wavelets in cases 1 and 7. When compared with D4, D6 performs better in case 1 of a smooth sinusoid, but worse in case 7 of a square wave. In general, as D6 is smoother than D4, it is more effective than D4 to represent smooth signals but less so for signals with discontinuities.

Table 11.1. Signal compression based on the DCT and DWT

	Signal type	Percentage error		
		DCT	D4	D6
1	Sinusoid	0.00	0.56	0.11
2	Two-tune sinusoids	2.23	9.57	10.17
3	Decaying sinusoid	0.08	4.00	2.01
4	Chirp	24.39	16.64	14.99
5	Sawtooth	2.12	0.00	0.16
6	Triangle	0.00	0.00	0.00
7	Square wave	1.05	0.31	1.82
8	Impulses	42.31	1.90	3.86
9	Random noise	35.82	41.01	40.83

Example 11.7: A piecewise linear signal (first row in Fig. 11.20) is contaminated by some random noise (second row). Two different types of filtering are then applied to remove the noise as much as possible, based on the DWT, (first four rows in the figure) as well as the DCT (last four rows) for comparison.

- LP filtering is first carried out to remove the lower 7/8 of the coefficients after either the DCT or DWT, and then the filtered signal is reconstructed by the inverse transform, as shown in the third and seventh rows, respectively. While the high-frequency noise is significantly reduced, the original signal is also distorted due to removal of the high-frequency or high scale-level components in the signal.
- Thresholding filtering is then carried out to remove all transform coefficients with values lower than a threshold (0.2 in this example), as shown in the fourth and eighth rows for the DCT and DWT filtering, respectively. We see that filtering based on the DWT removes more noise than that based on the DCT, due to the fact that in the transform domain, the signal is better separated from the noise by the DWT, while they are completely mixed together in the DCT spectrum. Comparing the first and second rows on the right for the DCT coefficients, we see that the high-frequency components of the signal are mixed those of the noise, while the same comparison of the fifth and sixth rows for the DWT coefficients shows that the signal components have more concentrated energy compared to those of the noise, allowing them to be better separated. Further comparison of the DWT and DCT representation of this specific piece-wise linear signal in the first and fifth rows reveals that the signal can be much more efficiently represented by the DWT rather than the DCT, as many fewer coefficients are needed to in the representation by the DWT, also indicating that better compression rate can be achieved by the DWT.

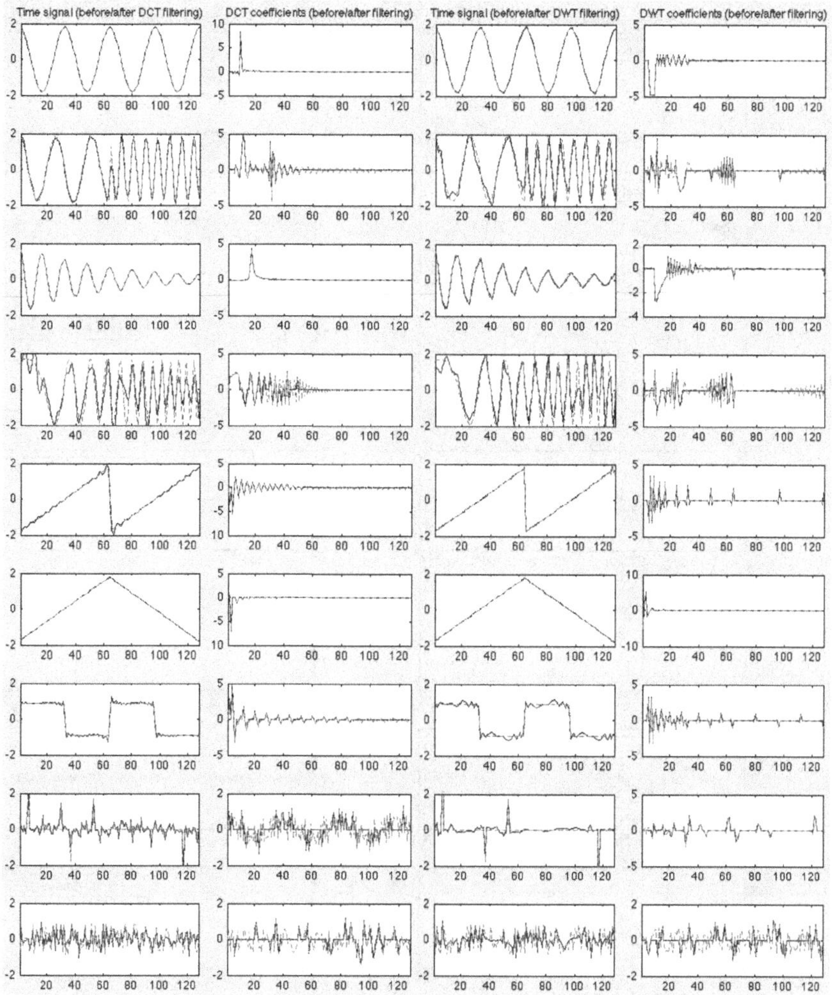

Figure 11.19 Compression of some typical signals by DCT (left)and DWT (right). The first and third columns show the time signals compared with their reconstructions based on only 20% of the transform coefficients, as shown in the second and fourth columns for the DCT and DWT, respectively. In both the time and transform domains, the signals before (dashed curves) and after (solid curves) the compression are shown for comparison.

Example 11.8: In Fig. 11.21, the first two rows show the images of Lenna, both the original (first row) and contaminated by white noise (second row), together with their 2-D DCT (middle) and DWT (right) spectra. The third row shows the noise and its DCT and DWT spectra. We see that the noise is indeed white as its energy is relatively evenly distributed over the entire frequency domain. To remove the noise, two different types of filtering are carried out in the transform

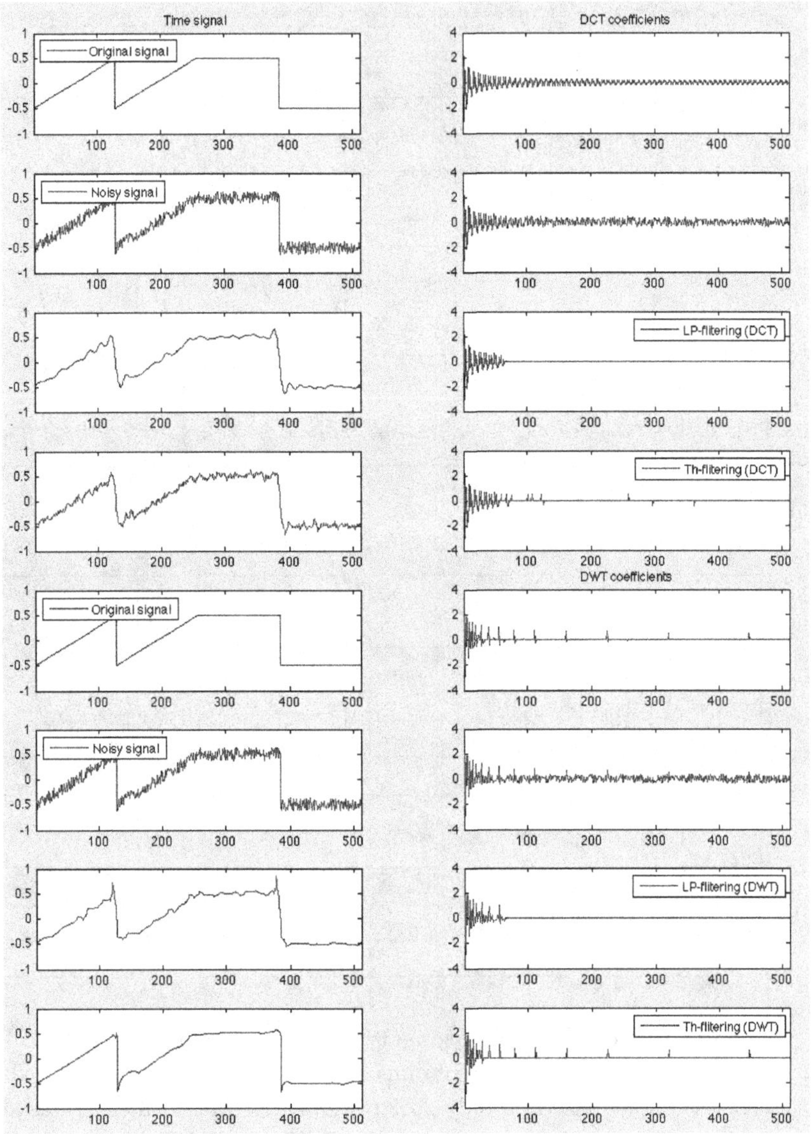

Figure 11.20 Signal filtering based on DCT and DWT. The time signals are shown on the left while the corresponding transform (DWT and DCT) coefficients are shown on the right. The top four rows are for the DCT while the bottom four are for the DWT. A non-linear mapping $y = x^{0.6}$ is used to plot the coefficients in the transform domain for the low values to be better seen.

domains, as shown in Fig. 11.22. First, we use an ideal LP filter to remove all frequencies higher than a given cutoff frequency; i.e., farther than a specified distance away from the the DC component (top-left corner) in the spectrum. Then the image is reconstructed based on the filtered spectrum. The results

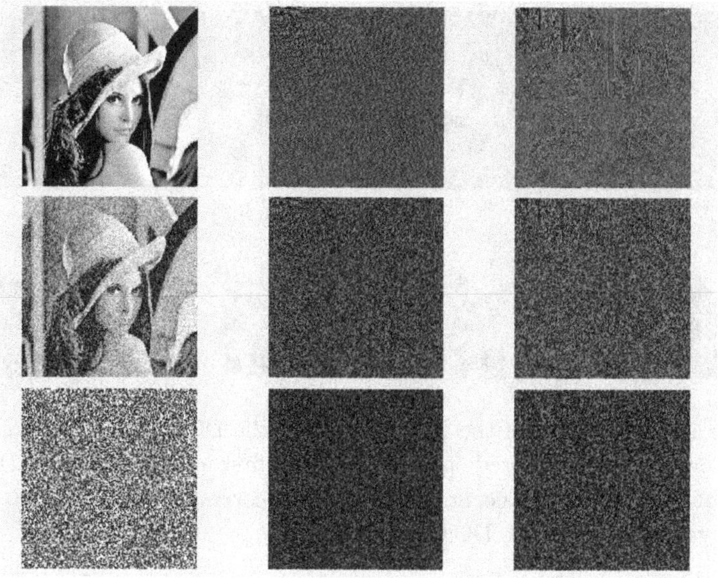

Figure 11.21 The Lenna image and its DCT and DWT spectra. The first row shows the image and its DCT and DWT spectra, while the second row shows the same but contaminated by white noise and its DCT and DWT spectra.

are shown in columns 1 for DCT and 3 for DWT. Next, we remove 98% of the frequency components to keep only the remaining 2% components carrying maximum possible energy, as shown in columns 2 for DCT and 4 for DWT. By visual inspection we see that the DWT-filtered image is obviously better in terms of both image details and remaining amount of noise.

Example 11.9: The same Lenna image is transformed by each of seven different 2-D transform methods considered throughout the book, the DFT, DCT, DST, WHT, SLT, and DHT, as well as the DWT, generically denoted by $\mathcal{T}$, resulting in seven spectra $\boldsymbol{X} = \mathcal{T}[\boldsymbol{x}]$, as shown on the left in Fig. 11.23. Moreover, for the purpose of compression, 99.5% of the coefficients in each of the spectra $\boldsymbol{X}$ is suppressed with only 0.5% of the coefficients of the greatest magnitudes kept (1 to 200 compression rate). The modified spectra, denoted by $\boldsymbol{Y}$, are shown in the middle of the figure. Then the corresponding inverse transform is carried out to reconstruct the image as $\boldsymbol{y} = \mathcal{T}^{-1}[\boldsymbol{X}]$, as shown on the right of the figure. The compression results based on these different transform methods can be evaluated both subjectively by visual inspection and numerically by the relative error. As in the 1-D compression considered in Example 11.6, the energy loss due to the

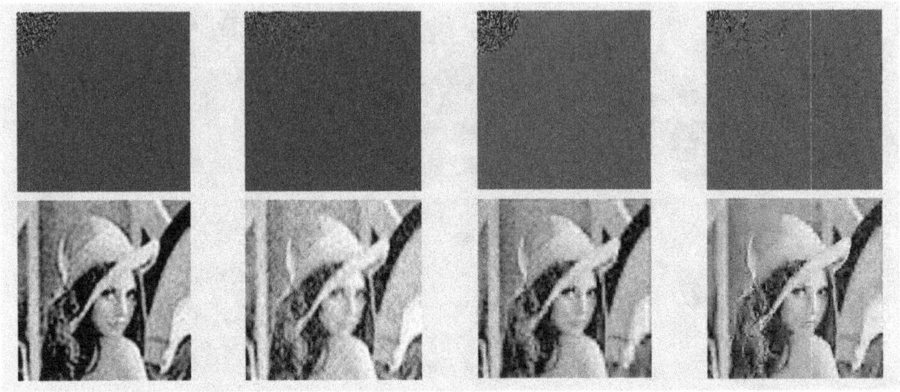

Figure 11.22 Filtering of the Lenna image in the DCT and DWT domains. The first row shows the filtered spectra by DCT (first and second) and DWT (third and fourth), while the second row shows the corresponding images reconstructed by inverse DCT and DWT.

Table 11.2. Image compression based on different orthogonal transforms

	Percentage error		
Transform method	Panda	Cat	Lenna
DFT	1.84	9.24	2.52
DCT	1.47	7.69	2.19
DST	2.60	8.06	3.07
WHT	2.19	10.98	2.99
SLT	1.78	9.46	2.57
DHT	2.13	9.76	2.63
DWT	1.73	7.59	2.38

compression is the same as the signal error:

$$\frac{||\boldsymbol{X}||^2 - ||\boldsymbol{Y}||^2}{||\boldsymbol{X}||^2} = \frac{||\boldsymbol{x}||^2 - ||\boldsymbol{y}||^2}{||\boldsymbol{x}||^2} = \frac{||\boldsymbol{x} - \boldsymbol{y}||^2}{||\boldsymbol{x}||^2}. \tag{11.167}$$

The same compression is also carried out for two other images, namely a cat and a panda. The compression results in terms of the signal error are summarized in Table 11.2, from which we see that the error depends on the specific transform method used as well as the data being processed. For all three images, the DCT and DWT always have the lowest error among all methods. Moreover, based on visual inspection of the compressed images, we see that the compressed image reconstructed by the DWT method always looks the best, even when its error is slightly higher than that of the DCT.

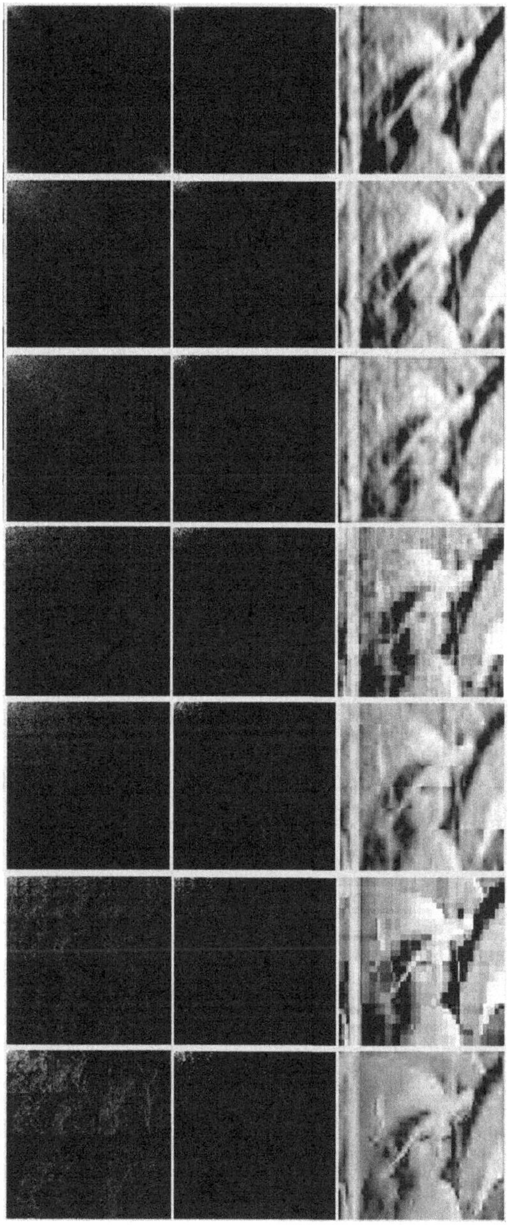

Figure 11.23 Image compression based on seven different transform methods including, from top down, DFT, DCT, DST, WHT, SLT, DHT, and DWT. The spectra and their compressed versions are shown on the left and middle columns (A non-linear mapping $y = x^{0.3}$ is applied for the coefficients of low magnitude to be visible. Also, as the DFT is a complex transform, the spectra shown here are for the magnitudes of the transform coefficients). The reconstructed signal based on the compressed spectra are shown in the right column.

11.5 Homework problems

1. Consider the Haar wavelet transform as illustrated in Examples 11.1 and 11.2.
   (a) Verify that all properties of the scaling and wavelet filters (Eqs. (11.56), (11.60), (11.58), and (11.63)) are satisfied by the scaling and wavelet filters of the Haar transform.
   (b) Based on Eqs. (11.27) and (11.81), find the spectra $\Phi(f) = \mathcal{F}[\phi(t)]$ and $\Psi(f) = \mathcal{F}[\psi(t)]$.
   (c) Verify that Eqs. (11.23) and (11.52) hold.
2. Reconsider the two-channel filter bank shown in Fig. 11.14 but now using the z-transform as the analysis tool. Design the filter for perfect reconstruction (PR) by following the steps below.
   (a) Show that the output of the two-channel filter is

$$\begin{aligned}\hat{X}(z) &= G_0(z)A(z^2) + G_1(z)D(z^2) \\ &= \frac{1}{2}[G_0(z)H_0(z) + G_1(z)H_1(z)]\, X(z) \\ &+\frac{1}{2}[G_0(z)H_0(-z) + G_1(z)H_1(-z)]\, X(-z). \qquad (11.168)\end{aligned}$$

   (b) For the two-channel filter bank to achieve perfect reconstruction, its output $\hat{x}[n]$ has to be identical to the input $x[n]$, with a delay of m samples, i.e., $\hat{x}[n] = x[n-m]$, or $\hat{X}(z) = X(z)z^{-m}$ in the z-domain. Given filters $H_0(z)$ and $H_1(z)$, find $G_0(z)$ and $G_1(z)$ for perfect reconstruction.
   Hint: for PR we let: $G_0(z)H_0(z) + G_1(z)H_1(z) = 2z^{-m}$ and $G_0(z)H_0(-z) + G_1(z)H_1(-z) = 0$.
   (c) For convenience, we set

$$\Delta(z) = H_0(z)H_1(-z) - H_0(-z)H_1(z) = 2z^{-m}. \qquad (11.169)$$

   Show that $G_0(z)$ and $G_1(z)$ obtained above can be expressed as

$$G_0(z) = H_1(-z), \qquad G_1(z) = -H_0(-z). \qquad (11.170)$$

   What do these relationships mean in the time domain in terms of the filter coefficients $g_0[n]$ and $g_1[n]$ given $h_0[n]$ and $h_1[n]$? Hint: Consider Eq. (6.212)). Is the function $\Delta(z)$ given in Eq. (11.169) even, odd, or neither? Is m an even or odd number?
3. Obtain the coefficients of the four-tap Daubechies filters by following the steps below.
   (a) Derive the following identity:

$$\begin{aligned}1 &= [\cos^2(\pi f) + \sin^2(\pi f)]^3 \\ &= \cos^6(\pi f) + 3\cos^4(\pi f)\sin^2(\pi f) + 3\sin^2(\pi f + \pi/2)\cos^4(\pi f + \pi/2) \\ &+ \cos^6(\pi f + \pi/2). \qquad (11.171)\end{aligned}$$

(b) Define

$$|H_0(f)|^2 = 2[\cos^6(\pi f) + 3\cos^4(\pi f)\sin^2(\pi f)]. \tag{11.172}$$

Show both the normalization and orthogonality properties of a scaling filter given in Eqs. (11.58) and (11.63) are satisfied by this $H_0(f)$; i.e., it can indeed be used as a scaling filter, as the notation suggested.

(c) Find $H_0(f)$ by taking square root of $|H_0(f)|^2$, which can be written as

$$\begin{aligned}|H_0(f)|^2 &= 2\cos^4(\pi f)\left[(\cos(\pi f))^2 + (\sqrt{3}\sin(\pi f))^2\right]\\ &= 2\cos^4(\pi f)\,|\cos^2(\pi f) + j\sqrt{3}\sin^2(\pi f)|^2.\end{aligned} \tag{11.173}$$

Express the result in the form of a third order polynomial of $e^{-j2\pi kf}$. Verify the four coefficients are indeed the coefficients for the Daubechies scaling filter of $N = 2$.

4. Obtain the coefficients of the four-tap Daubechies filters for the two-channel filter bank with perfect reconstruction by following the steps below.

(a) Define $Q(z) = a_0 + a_1 z^{-1} + a_2 z^{-2}$ (coefficients a_0, a_1, and a_2 to be determined) and choose

$$H_0(z)G_0(z) = (1 + z^{-1})^{2N} Q(z) = (1 + z^{-1})^4 Q(z), \tag{11.174}$$

where we have chosen $N = 2$. Write $H_0(z)G_0(z)$ and $H_0(-z)G_0(-z)$ as a polynomial of z^{-1}, and show $\Delta(z)$ in Eq. (11.169) can be written as

$$\Delta(z) = 2(4a_0 + a_1)z^{-1} + 2(4a_0 + 6a_1 + 4a_2)z^{-3} + 2(a_1 + 4a_2)z^{-5} = 2z^{-m}. \tag{11.175}$$

(b) Determine the coefficients a_0, a_1, and a_2 by choosing to keep only the term of z^{-3} in $\Delta(z)$ above; i.e.,

$$4a_0 + a_1 = 0, \quad 4a_0 + 6a_1 + 4a_2 = 1, \quad a_1 + 4a_2 = 0. \tag{11.176}$$

Solve these equations to find a_0, a_1, a_2, and show

$$Q(z) = a_0 + a_1 z^{-1} + a_2 z^{-2} = \frac{z^{-1}}{16}(-z + 4 - z^{-1}). \tag{11.177}$$

(c) Show that the term $-z + 4 - z^{-1}$ in $Q(z)$ obtained above can be factored to become

$$-z + 4 - z^{-1} = (a + bz)(a + bz^{-1}). \tag{11.178}$$

Find the two coefficients a and b.

(d) Given the coefficients a and b, show that $Q(z)$ can be written as

$$Q(z) = \frac{z^{-1}}{32}[(1 + \sqrt{3}) + (1 - \sqrt{3})z][(1 + \sqrt{3}) + (1 - \sqrt{3})z^{-1}], \tag{11.179}$$

and $H_0(z)G_0(z)$ can be written as the product

$$\left[\frac{(1+z^{-1})^2(1+\sqrt{3})+(1-\sqrt{3})z^{-1}}{4\sqrt{2}}\right]\left[\frac{z^{-3}(1+z)^2(1+\sqrt{3})+(1-\sqrt{3})z}{4\sqrt{2}}\right], \tag{11.180}$$

which is actually a product of the following two third order polynomials of z^{-1}:

$$H_0(z) = \frac{1}{4\sqrt{2}}[(1+\sqrt{3})+(3+\sqrt{3})z^{-1}+(3-\sqrt{3})z^{-2}+(1-\sqrt{3})z^{-3}],$$
$$G_0(z) = \frac{1}{4\sqrt{2}}[(1-\sqrt{3})+(3-\sqrt{3})z^{-1}+(3+\sqrt{3})z^{-2}+(1+\sqrt{3})z^{-3}].$$

As $H_0(z) = \sum_n h_0[n]z^{-n}$, we see that the coefficients $h_0[n]$ are

$$h_0[0] = \frac{1+\sqrt{3}}{4\sqrt{2}}, \quad h_0[1] = \frac{3+\sqrt{3}}{4\sqrt{2}}, \quad h_0[2] = \frac{3-\sqrt{3}}{4\sqrt{2}}, \quad h_0[3] = \frac{1-\sqrt{3}}{4\sqrt{2}}, \tag{11.181}$$

the same as those given in Eq. (11.119).

(e) Find $H_1(z)$ and $G_1(z)$ according to Eq. (11.170). These four filters $H_0(z)$, $G_0(z)$, $H_1(z)$, and $G_1(z)$ form an orthonormal filter bank with perfect reconstruction and leads to the Daubechies D_4 wavelets.

5. Obtain the six coefficients $h_0[k]$ $(k = 0, \ldots, 5)$ of the Daubechies scaling filter of order $N = 3$. Verify that they are the same as those given in Eq. (11.122). Revise the Matlab code provided to construct the scaling and wavelet functions $\phi(t)$ and $\psi(t)$ of order $N = 3$.
6. Prove that the energy loss in Eq. (11.165) and signal error in Eq. (11.166) in Example 11.6 are the same; i.e.,

$$||\boldsymbol{X}||^2 - ||\boldsymbol{Y}||^2 = ||\boldsymbol{x}||^2 - ||\boldsymbol{y}||^2 = ||\boldsymbol{x}-\boldsymbol{y}||^2. \tag{11.182}$$

<u>Hint:</u> As $\boldsymbol{X} = \mathcal{T}[\boldsymbol{x}]$ is an orthogonal transform, we have $\langle \boldsymbol{x}, \boldsymbol{y}\rangle = \langle \boldsymbol{X}, \boldsymbol{Y}\rangle$. Also, the compression in the transform domain can be expressed as $Y[n] = c_n X[n]$ $(n = 0, \ldots, N-1)$ where $c_n = 1$ if the nth coefficient $Y[n] = X[n]$ is kept during the compression, or $c_n = 0$ if it is suppressed to zero.

7. Compress each of the signals in Example 11.6 by suppressing 90% of the coefficients after each one of the orthogonal transform methods discussed throughout the book, including DFT, DCT, DST, WHT, DHT, as well as DWT. Evaluate these methods quantitatively and qualitatively in terms of
   - Percentage of signal energy contained in the remaining 10% of the transform coefficients;
   - Percentage error between reconstructed signal and the original;
   - Subjective comparison of the reconstructed signal and the original.
8. Repeat the previous problem on a set of different images of your choice; evaluate all of the orthogonal transform methods with the same quantitative and qualitative criteria.

Appendices

A Review of linear algebra

A.1 Basic definitions

- **Matrix**

An $m \times n$ matrix $\boldsymbol{A} \in \mathbb{R}^{m \times n}$ or $\mathbb{C}^{m \times n}$ is an array of m rows and n columns

$$\boldsymbol{A} = \begin{bmatrix} a_{11} & a_{12} & \cdots & a_{1n} \\ a_{21} & a_{22} & \cdots & a_{2n} \\ \vdots & \vdots & \ddots & \vdots \\ a_{m1} & a_{m2} & \cdots & a_{mn} \end{bmatrix}_{m \times n}, \tag{A.1}$$

where $a_{ij} \in \mathbb{R}$ or $\mathbb{C}$ is the element in the ith (first index) row and jth (second index) column. In particular,

– if $m = n$, $\boldsymbol{A}$ becomes a square matrix;
– if $m = 1$, $\boldsymbol{A}$ becomes an n-dimensional $(1 \times n)$ row vector;
– if $n = 1$, $\boldsymbol{A}$ becomes an m-dimensional $(m \times 1)$ column vector.

Throughout the book, a vector $\boldsymbol{a}$ is always assumed to be a column vector, unless specified otherwise. Sometimes it is convenient to express a matrix in terms of its column vectors

$$\boldsymbol{A} = [\boldsymbol{a}_1, \ldots, \boldsymbol{a}_n], \tag{A.2}$$

where $\boldsymbol{a}_j$ $(j = 1, \ldots, n)$ is an m-dimensional column vector:

$$\boldsymbol{a}_j = \begin{bmatrix} a_{1j} \\ a_{2j} \\ \vdots \\ a_{mj} \end{bmatrix}. \tag{A.3}$$

The ith row is an n-dimensional row vector $[a_{i1}\ a_{i2}\ \cdots\ a_{in}]$.

- **Transpose and conjugate transpose**

The *transpose* of an $m \times n$ matrix $\boldsymbol{A}$, denoted by $\boldsymbol{A}^{\mathrm{T}}$, is an $n \times m$ matrix obtained by swapping elements a_{ij} and a_{ji} for all $i, j \in \{1, \ldots, n\}$. In other words, the jth column of $\boldsymbol{A}$ becomes the jth row of $\boldsymbol{A}^{\mathrm{T}}$, and at the same time, the ith row of $\boldsymbol{A}$ becomes the ith column of $\boldsymbol{A}^{\mathrm{T}}$:

$$\boldsymbol{A}^{\mathrm{T}} = [\boldsymbol{a}_1, \boldsymbol{a}_2, \ldots, \boldsymbol{a}_n]^{\mathrm{T}} = \begin{bmatrix} \boldsymbol{a}_1^{\mathrm{T}} \\ \boldsymbol{a}_2^{\mathrm{T}} \\ \vdots \\ \boldsymbol{a}_n^{\mathrm{T}} \end{bmatrix} = \begin{bmatrix} a_{11} & a_{21} & \cdots & a_{m1} \\ a_{12} & a_{22} & \cdots & a_{m2} \\ \vdots & \vdots & \ddots & \vdots \\ a_{1n} & a_{2n} & \cdots & a_{mn} \end{bmatrix}_{n\times m}, \tag{A.4}$$

where $\boldsymbol{a}_j$ is the jth column of $\boldsymbol{A}$ and its transpose $\boldsymbol{a}_j^{\mathrm{T}}$ is the jth row of $\boldsymbol{A}^{\mathrm{T}}$:

$$\boldsymbol{a}_j^{\mathrm{T}} = \begin{bmatrix} a_{1j} \\ a_{2j} \\ \vdots \\ a_{nj} \end{bmatrix}^{\mathrm{T}} = [a_{1j}, a_{2j}, \ldots, a_{nj}]. \tag{A.5}$$

Here are some important properties related to transpose:

$$(\boldsymbol{A}^{\mathrm{T}})^{\mathrm{T}} = \boldsymbol{A}, \quad (\boldsymbol{A}\boldsymbol{B})^{\mathrm{T}} = \boldsymbol{B}^{\mathrm{T}}\boldsymbol{A}^{\mathrm{T}}. \tag{A.6}$$

The *conjugate transpose* of an $m \times n$ complex matrix $\boldsymbol{A}$, denoted by $\boldsymbol{A}^*$, is the complex conjugate of its transpose; i.e.,

$$\boldsymbol{A}^* = \overline{\boldsymbol{A}^{\mathrm{T}}} = \overline{\boldsymbol{A}}^{\mathrm{T}}; \tag{A.7}$$

i.e., the element in the ith row and jth column of $\boldsymbol{A}^*$ is the complex conjugate of the element in the jth row and ith column of $\boldsymbol{A}$. We obviously have

$$(\boldsymbol{A}^*)^* = \boldsymbol{A}, \quad (\boldsymbol{A}\boldsymbol{B})^* = \boldsymbol{B}^*\boldsymbol{A}^*. \tag{A.8}$$

- **Identity matrix**
  The *identity matrix* $\boldsymbol{I}$ is a special $n \times n$ square matrix with all elements being zero except those along the main diagonal which are 1:

$$\boldsymbol{I} = \mathrm{diag}[1, \ldots, 1] = \begin{bmatrix} 1 & 0 & \cdots & 0 \\ 0 & 1 & \cdots & 0 \\ \vdots & \vdots & \ddots & \vdots \\ 0 & 0 & \cdots & 1 \end{bmatrix}_{n\times n}. \tag{A.9}$$

The identity matrix can also be expressed in terms of its column vectors:

$$\boldsymbol{I} = [\boldsymbol{e}_1, \ldots, \boldsymbol{e}_i, \ldots, \boldsymbol{e}_n], \tag{A.10}$$

where $\boldsymbol{e}_i$ $(i = 1, \ldots, n)$ is an n-dimensional column vector with all elements equal to zero except the ith one, which is 1:

$$\boldsymbol{e}_i = [e_{1i}, \ldots, e_{ni}]^{\mathrm{T}} = [0, \ldots, 0, 1, 0, \ldots, 0]^{\mathrm{T}}; \tag{A.11}$$

i.e., the $e_{ij} = 0$ for all $i \neq j$ and $e_{ii} = 1$ for all $i = 1, \ldots, n$.

- **Scalar multiplication**
A matrix $\boldsymbol{A}$ can be multiplied by a scalar c to get

$$c\boldsymbol{A} = c \begin{bmatrix} a_{11} & a_{12} & \cdots & a_{1n} \\ a_{21} & a_{22} & \cdots & a_{2n} \\ \vdots & \vdots & \ddots & \vdots \\ a_{m1} & a_{m2} & \cdots & a_{mn} \end{bmatrix} = \begin{bmatrix} ca_{11} & ca_{12} & \cdots & ca_{1n} \\ ca_{21} & ca_{22} & \cdots & ca_{2n} \\ \vdots & \vdots & \ddots & \vdots \\ ca_{m1} & ca_{m2} & \cdots & ca_{mn} \end{bmatrix}. \tag{A.12}$$

- **Dot product**
The *dot product*, also called the *inner product*, of two real column vectors $\boldsymbol{x} = [x_1, \ldots, x_n]^{\mathrm{T}}$ and $\boldsymbol{y} = [y_1, \ldots, y_n]^{\mathrm{T}}$ is defined as

$$\boldsymbol{x} \cdot \boldsymbol{y} =< \boldsymbol{x}, \boldsymbol{y} >= \boldsymbol{x}^{\mathrm{T}}\overline{\boldsymbol{y}} = \boldsymbol{y}^*\boldsymbol{x} = [x_1, \ldots, x_n] \begin{bmatrix} \overline{y}_1 \\ \vdots \\ \overline{y}_n \end{bmatrix} = \sum_{i=1}^{n} x_i \overline{y}_i, \tag{A.13}$$

where $\overline{u + jv} = u - jv$ is the complex conjugate of $u + jv$. If the inner product of $\boldsymbol{x}$ and $\boldsymbol{y}$ is zero, then the two vectors are said to be *orthogonal*, denoted by $\boldsymbol{x} \perp \boldsymbol{y}$. In particular, when $\boldsymbol{x} = \boldsymbol{y}$, we have

$$< \boldsymbol{x}, \boldsymbol{x} >= ||\boldsymbol{x}||^2 = \sum_{i=1}^{n} x_i \overline{x}_i = \sum_{i=1}^{n} |x_i|^2 > 0, \tag{A.14}$$

where

$$||\boldsymbol{x}|| = \sqrt{\sum_{i=1}^{n} |x_i|^2} \tag{A.15}$$

is called the *norm* of $\boldsymbol{x}$. If $||\boldsymbol{x}|| = 1$, $\boldsymbol{x}$ is *normalized.*

- **Matrix multiplication**
The product of an $m \times k$ matrix $\boldsymbol{A}$ and a $k \times n$ matrix $\boldsymbol{B}$ is

$$\boldsymbol{A}_{m \times k} \boldsymbol{B}_{k \times n} = \boldsymbol{C}_{m \times n}, \tag{A.16}$$

where the element in the ith row and jth column of $\boldsymbol{C}$ is the dot product of the ith row vector of $\boldsymbol{A}$ and the jth column of $\boldsymbol{B}$:

$$c_{ij} = [a_{i1}, \ldots, a_{ik}] \begin{bmatrix} b_{k1} \\ \vdots \\ b_{kn} \end{bmatrix} = \sum_{l=1}^{k} a_{il} b_{lj}. \tag{A.17}$$

For this multiplication to be possible, the number of columns of $\boldsymbol{A}$ must be equal to the number of rows of $\boldsymbol{B}$, so that the dot product can be carried out. Otherwise, the two matrices cannot be multiplied.

- **Trace**
The *trace* of $\boldsymbol{A}$ is defined as the sum of the elements along the main diagonal:

$$\mathrm{tr}(\boldsymbol{A}) = \sum_{i=1}^{n} a_{ii}. \tag{A.18}$$

Here are some properties of the trace:

$$\mathrm{tr}(\boldsymbol{A} + \boldsymbol{B}) = \mathrm{tr}\,\boldsymbol{A} + \mathrm{tr}\,\boldsymbol{B}, \quad \mathrm{tr}(c\boldsymbol{A}) = c\ \mathrm{tr}\,\boldsymbol{A}, \quad \mathrm{tr}(\boldsymbol{AB}) = \mathrm{tr}(\boldsymbol{BA}). \tag{A.19}$$

- **Rank**
  If none of a set of vectors can be expressed as a linear combination of the rest of the vectors, then these vectors are *linearly independent.* The *rank* of a matrix $\boldsymbol{A}$, denoted by rank $\boldsymbol{A}$, is the maximum number of linearly independent columns of $\boldsymbol{A}$, which is the same as the maximum number of linearly independent rows. Obviously the rank of an $m \times n$ matrix is no larger than the smaller of m and n:
  $$\text{rank}\, \boldsymbol{A} \leq \min(m, n). \tag{A.20}$$
  If the equation holds, matrix $\boldsymbol{A}$ has a *full rank.*
- **Determinant**
  The *determinant* of an $n \times n$ matrix $\boldsymbol{A}$, denoted by $\det \boldsymbol{A}$ or $|\boldsymbol{A}|$, is a scalar that can be recursively defined as
  $$|\boldsymbol{A}| = \det \boldsymbol{A} = \sum_{j=1}^{n} (-1)^{j+1}\, a_{1j}\, \det \boldsymbol{A}_{1j}, \tag{A.21}$$
  where $\boldsymbol{A}_{1j}$ is an $(n-1) \times (n-1)$ matrix obtained by deleting the first row and jth column of $\boldsymbol{A}$, and the determinant of a 1×1 matrix is $\det(a) = a$. If $\boldsymbol{A}$ is not a full rank matrix, its determinant is 0. In particular, when $n = 2$,
  $$\det \begin{bmatrix} a & b \\ c & d \end{bmatrix} = ad - bc, \tag{A.22}$$
  and when $n = 3$,
  $$\det \begin{bmatrix} a & b & c \\ d & e & f \\ g & h & i \end{bmatrix} = a \det \begin{bmatrix} e & f \\ h & i \end{bmatrix} - b \det \begin{bmatrix} d & f \\ g & i \end{bmatrix} + c \det \begin{bmatrix} d & e \\ g & h \end{bmatrix}$$
  $$= aei - afh - bdi + bfg + cdh - ceg = (aei + bfg + cdh) - (gec + hfa + idb). \tag{A.23}$$
  Here are some important properties of the determinant ($\boldsymbol{A}$ and $\boldsymbol{B}$ are square matrices):
  $$\det(\boldsymbol{AB}) = \det(\boldsymbol{BA}) = \det \boldsymbol{A}\, \det \boldsymbol{B},$$
  $$\det(\boldsymbol{A}^{\mathrm{T}}) = \det \boldsymbol{A}, \qquad \det(c\boldsymbol{A}) = c^n \det \boldsymbol{A}. \tag{A.24}$$
- **Inverse matrix**
  If $\boldsymbol{A}$ is an $n \times n$ square matrix and there exists another $n \times n$ matrix $\boldsymbol{B}$ so that $\boldsymbol{AB} = \boldsymbol{BA} = \boldsymbol{I}$, then $\boldsymbol{B} = \boldsymbol{A}^{-1}$ is the *inverse* of $\boldsymbol{A}$, which can be obtained by:
  $$\boldsymbol{A}^{-1} = \frac{1}{\det \boldsymbol{A}} \begin{bmatrix} c_{11} & c_{12} & \cdots & c_{1n} \\ c_{21} & c_{22} & \cdots & c_{2n} \\ \vdots & \vdots & \ddots & \vdots \\ c_{n1} & c_{n2} & \cdots & c_{nn} \end{bmatrix}^{\mathrm{T}} = \frac{1}{\det \boldsymbol{A}} \begin{bmatrix} c_{11} & c_{21} & \cdots & c_{n1} \\ c_{12} & c_{22} & \cdots & c_{n2} \\ \vdots & \vdots & \ddots & \vdots \\ c_{1n} & c_{2n} & \cdots & c_{nn} \end{bmatrix} \tag{A.25}$$
  where c_{ij} is the ij-th *cofactor* defined as
  $$c_{ij} = (-1)^{i+j}\, \det \boldsymbol{\mu}_{ij}, \tag{A.26}$$

where $\boldsymbol{\mu}_{ij}$ is an $(n-1) \times (n-1)$ *minor matrix* obtained by removing the ith row and jth column of $\boldsymbol{A}$. Obviously, if $\boldsymbol{A}$ is not a full rank matrix, $\det \boldsymbol{A} = 0$, then $\boldsymbol{A}^{-1}$ does not exist. The following statements are equivalent:

- $\boldsymbol{A}$ is invertible; i.e., inverse matrix $\boldsymbol{A}^{-1}$ exists.
- $\boldsymbol{A}$ is full rank.
- $\det \boldsymbol{A} \neq 0$.
- All column and row vectors are linearly independent.
- All eigenvalues of $\boldsymbol{A}$ are non-zero (to be discussed later).

These are some basic properties related to the inverse of a matrix $\boldsymbol{A}$:

$$(\boldsymbol{A}^{-1})^{-1} = \boldsymbol{A}, \quad (c\boldsymbol{A})^{-1} = \frac{1}{c}\boldsymbol{A}^{-1}, \quad (\boldsymbol{A}\boldsymbol{B})^{-1} = \boldsymbol{B}^{-1}\boldsymbol{A}^{-1}, \quad (\boldsymbol{A}^{-1})^{\mathrm{T}} = (\boldsymbol{A}^{\mathrm{T}})^{-1}. \tag{A.27}$$

- **Pseudo-inverse matrix**

  Let $\boldsymbol{A}$ be an $m \times n$ matrix. If $m \neq n$, then $\boldsymbol{A}$ does not have an inverse. However, we can find its *pseudo-inverse* $\boldsymbol{A}^{-}$, an $n \times m$ matrix, as shown below.

  - If $\boldsymbol{A}$ has more rows than columns; i.e., $m > n$, then

  $$\boldsymbol{A}^{-} = (\boldsymbol{A}^{*}\boldsymbol{A})^{-1}\boldsymbol{A}^{*}. \tag{A.28}$$

  We can verify that $\boldsymbol{A}^{-}\boldsymbol{A} = \boldsymbol{I}$:

  $$\boldsymbol{A}^{-}\boldsymbol{A} = (\boldsymbol{A}^{*}\boldsymbol{A})^{-1}\boldsymbol{A}^{*}\boldsymbol{A} = \boldsymbol{I}_{n \times n}. \tag{A.29}$$

  But $\boldsymbol{A}\boldsymbol{A}^{-} \neq \boldsymbol{I}$.

  - If $\boldsymbol{A}$ has more columns than rows; i.e., $m < n$, then

  $$\boldsymbol{A}^{-} = \boldsymbol{A}^{*}(\boldsymbol{A}\boldsymbol{A}^{*})^{-1}. \tag{A.30}$$

  We can verify that $\boldsymbol{A}\boldsymbol{A}^{-} = \boldsymbol{I}$:

  $$\boldsymbol{A}\boldsymbol{A}^{-} = \boldsymbol{A}\boldsymbol{A}^{*}(\boldsymbol{A}\boldsymbol{A}^{*})^{-1} = \boldsymbol{I}_{m \times m}. \tag{A.31}$$

  But $\boldsymbol{A}^{-}\boldsymbol{A} \neq \boldsymbol{I}$.

  Note that the pseudo-inverses in Eq. (A.28) $(m > n)$ and Eq. (A.30) $(m < n)$ are essentially the same. Assume $\boldsymbol{A}$ has more rows than columns $(m > n)$, then another matrix defined as $\boldsymbol{B} = \boldsymbol{A}^{*}$ has more columns than rows. Taking the conjugate transpose on both sides of Eq. (A.28), we get

  $$(\boldsymbol{A}^{-})^{*} = [(\boldsymbol{A}^{*}\boldsymbol{A})^{-1}\boldsymbol{A}^{*}]^{*} = \boldsymbol{A}(\boldsymbol{A}^{*}\boldsymbol{A})^{-1} = (\boldsymbol{A}^{*})^{-}; \tag{A.32}$$

  i.e.,

  $$\boldsymbol{B}^{-} = \boldsymbol{B}^{*}(\boldsymbol{B}\boldsymbol{B}^{*})^{-1}, \tag{A.33}$$

  which is the same as Eq. (A.30). We can also show that $(\boldsymbol{A}^{-})^{-} = \boldsymbol{A}$. If $m > n$, then we have

  $$\begin{aligned}(\boldsymbol{A}^{-})^{-} &= [(\boldsymbol{A}^{*}\boldsymbol{A})^{-1}\boldsymbol{A}^{*}]^{-} = [(\boldsymbol{A}^{*}\boldsymbol{A})^{-1}\boldsymbol{A}^{*}]^{*}\left[(\boldsymbol{A}^{*}\boldsymbol{A})^{-1}\boldsymbol{A}^{*}[(\boldsymbol{A}^{*}\boldsymbol{A})^{-1}\boldsymbol{A}^{*}]^{*}\right]^{-1} \\ &= \boldsymbol{A}(\boldsymbol{A}^{*}\boldsymbol{A})^{-1}\left[(\boldsymbol{A}^{*}\boldsymbol{A})^{-1}\boldsymbol{A}^{*}\boldsymbol{A}(\boldsymbol{A}^{*}\boldsymbol{A})^{-1}\right]^{-1} \\ &= \boldsymbol{A}(\boldsymbol{A}^{*}\boldsymbol{A})^{-1}(\boldsymbol{A}^{*}\boldsymbol{A}) = \boldsymbol{A}.\end{aligned} \tag{A.34}$$

  Similarly. we can show the same is true if $m < n$. In particular, when $m = n$, $\boldsymbol{A}$ is invertible and the pseudo-inverse in either Eq. (A.28) or Eq. (A.30) becomes the regular inverse $\boldsymbol{A}^{-} = \boldsymbol{A}^{-1}$.

A.2 Eigenvalues and eigenvectors

For any $n \times n$ matrix $\boldsymbol{A}$, if there exists an $n \times 1$ vector $\boldsymbol{\phi}$ and a scalar λ satisfying

$$\boldsymbol{A}_{n\times n}\boldsymbol{\phi}_{n\times 1} = \lambda\boldsymbol{\phi}_{n\times 1}, \tag{A.35}$$

then λ and $\boldsymbol{\phi}$ are called the *eigenvalue* and *eigenvector* of $\boldsymbol{A}$, respectively. To obtain λ, we rewrite the above equation as

$$(\lambda\boldsymbol{I} - \boldsymbol{A})\boldsymbol{\phi} = \boldsymbol{0}. \tag{A.36}$$

This homogeneous algebraic equation system of n equations for n unknowns, the elements in vector $\boldsymbol{\phi}$has non-zero solutions if and only if

$$\det(\lambda\boldsymbol{I} - \boldsymbol{A}) = 0. \tag{A.37}$$

This nth-order equation of λ is the *characteristic equation* of the matrix $\boldsymbol{A}$, which can be solved to get n solutions, the n eigenvalues $\{\lambda_1, \ldots, \lambda_n\}$ of $\boldsymbol{A}$. Substituting each λ_i back into the equation system in Eq. (A.36), we can obtain the non-zero solution, the eigenvector $\boldsymbol{\phi}_i$ corresponding to λ_i:

$$\boldsymbol{A}\boldsymbol{\phi}_i = \lambda_i\boldsymbol{\phi}_i, \qquad i = 1, \ldots, n. \tag{A.38}$$

Putting all n such equations together, we get

$$\boldsymbol{A}\left[\boldsymbol{\phi}_1, \ldots, \boldsymbol{\phi}_n\right] = \left[\lambda_1\boldsymbol{\phi}_1, \ldots, \lambda_n\boldsymbol{\phi}_n\right] = \left[\boldsymbol{\phi}_1, \ldots, \boldsymbol{\phi}_n\right]\begin{bmatrix} \lambda_1 & 0 & \cdots & 0 \\ 0 & \lambda_2 & \cdots & 0 \\ \vdots & \vdots & \ddots & \vdots \\ 0 & 0 & \cdots & \lambda_n \end{bmatrix}. \tag{A.39}$$

Defining

$$\boldsymbol{\Phi} = \left[\boldsymbol{\phi}_1, \ldots, \boldsymbol{\phi}_n\right], \quad \text{and} \quad \boldsymbol{\Lambda} = \text{diag}\left[\lambda_1, \ldots, \lambda_n\right], \tag{A.40}$$

we can write the equation above in a more compact form:

$$\boldsymbol{A}\boldsymbol{\Phi} = \boldsymbol{\Phi}\boldsymbol{\Lambda} \quad \text{or} \quad \boldsymbol{\Phi}^{-1}\boldsymbol{A}\boldsymbol{\Phi} = \boldsymbol{\Lambda}. \tag{A.41}$$

The trace and determinant of $\boldsymbol{A}$ can be obtained in terms of its eigenvalues

$$\text{tr}\,\boldsymbol{A} = \sum_{k=1}^{n} \lambda_k, \tag{A.42}$$

$$\det\boldsymbol{A} = \prod_{k=1}^{n} \lambda_k. \tag{A.43}$$

$\boldsymbol{A}^{\mathrm{T}}$ has the same eigenvalues and eigenvectors as $\boldsymbol{A}$:

$$\boldsymbol{A}^{\mathrm{T}}\boldsymbol{\phi}_i = \lambda_i\boldsymbol{\phi}_i \qquad i = 1, \ldots, n. \tag{A.44}$$

$\boldsymbol{A}^m$ has the same eigenvectors as $\boldsymbol{A}$, but its eigenvalues are $\{\lambda_1^m, \ldots, \lambda_n^m\}$:

$$\boldsymbol{A}^m\boldsymbol{\phi}_i = \lambda_i^m\boldsymbol{\phi}_i \qquad i = 1, \ldots, n. \tag{A.45}$$

In particular, when $m = -1$, the eigenvalues of $\boldsymbol{A}^{-1}$ are $\{1/\lambda_1, \ldots, 1/\lambda_n\}$:

$$\boldsymbol{A}^{-1}\boldsymbol{\phi}_i = \frac{1}{\lambda_i}\boldsymbol{\phi}_i \qquad i = 1, \ldots, n. \tag{A.46}$$

A Hermitian matrix $\boldsymbol{A}$ is *positive definite*, denoted by $\boldsymbol{A} > 0$, if and only if for any non-zero $\boldsymbol{x} = [x_1, \ldots, x_n]^{\mathrm{T}}$, the quadratic form $\boldsymbol{x}^* \boldsymbol{A} \boldsymbol{x}$ is greater than zero:

$$\boldsymbol{x}^* \boldsymbol{A} \boldsymbol{x} > 0. \tag{A.47}$$

In particular, if we let $\boldsymbol{x} = \boldsymbol{\phi}_i$ be the eigenvector corresponding to the ith eigenvalue λ_i, then the above becomes

$$\boldsymbol{\phi}_i^* \boldsymbol{A} \boldsymbol{\phi}_i = \lambda_i \boldsymbol{\phi}_i^* \boldsymbol{\phi}_i > 0, \tag{A.48}$$

as $\boldsymbol{\phi}_i^* \boldsymbol{\phi}_i = ||\boldsymbol{\phi}_i||^2 > 0$, we know $\lambda_i > 0$ for all $i = 1, \ldots, n$; i.e., $\boldsymbol{A} > 0$ if and only if all of its eigenvalues are greater than zero. Also, as the eigenvalues of $\boldsymbol{A}^{-1}$ are $1/\lambda_i$, $i = (1, \ldots, n)$, we have $\boldsymbol{A} > 0$ if and only if $\boldsymbol{A}^{-1} > 0$.

A.3 Hermitian matrix and unitary matrix

A matrix $\boldsymbol{A}$ is *Hermitian* if it is equal to its *conjugate transpose*:

$$\boldsymbol{A} = \overline{\boldsymbol{A}}^{\mathrm{T}} = \overline{\boldsymbol{A}^{\mathrm{T}}} = \boldsymbol{A}^*. \tag{A.49}$$

In particular if a Hermitian matrix $\overline{\boldsymbol{A}} = \boldsymbol{A}$ is real, then it is *symmetric* $\boldsymbol{A} = \boldsymbol{A}^{\mathrm{T}}$. All eigenvalues λ_i of a Hermitian matrix are real, and all eigenvectors $\boldsymbol{\phi}_i$ corresponding to distinct eigenvalues are orthogonal. If the eigenvectors are normalized with unit norm, then they are *orthonormal* (both orthogonal and normalized):

$$< \boldsymbol{\phi}_i, \boldsymbol{\phi}_j >= \delta[i-j] \qquad i, j = 1, \ldots, n. \tag{A.50}$$

A matrix $\boldsymbol{A}$ is *unitary* if its conjugate transpose is equal to its inverse:

$$\boldsymbol{A}^* = \boldsymbol{A}^{-1}; \qquad \text{i.e.} \qquad \boldsymbol{A}^* \boldsymbol{A} = \boldsymbol{A} \boldsymbol{A}^* = \boldsymbol{I}. \tag{A.51}$$

When a unitary matrix is real $\boldsymbol{A} = \overline{\boldsymbol{A}}$, then it is *orthogonal* $\boldsymbol{A}^{\mathrm{T}} = \boldsymbol{A}^{-1}$. The absolute values of all eigenvalues (may be complex) of a unitary matrix are $|\lambda_i| = 1$; i.e. they lie on the unit circle centered at zero in the complex plane. The determinant of a unitary matrix $\boldsymbol{A}$ is $\det \boldsymbol{A} = \prod_{k=1}^n \lambda_k = pm1$.

Let $\boldsymbol{\Lambda} = \mathrm{diag}[\lambda_1, \ldots, \lambda_n]$ and $\boldsymbol{\Phi} = [\boldsymbol{\phi}_1, \ldots, \boldsymbol{\phi}_n]$ be the eigenvalue and eigenvector matrices of a Hermitian matrix $\boldsymbol{A}^* = \boldsymbol{A}$. If all columns $\boldsymbol{\phi}_i$ of $\boldsymbol{\Phi}$ are orthonormal, then $\boldsymbol{\Phi}$ is unitary satisfying:

$$\boldsymbol{\Phi}^{-1} = \boldsymbol{\Phi}^*; \quad \text{i.e.} \quad \boldsymbol{\Phi}\boldsymbol{\Phi}^* = \boldsymbol{\Phi}^*\boldsymbol{\Phi} = \boldsymbol{I}, \tag{A.52}$$

and the eigenequation of the Hermitian matrix $\boldsymbol{A}$ can be written as:

$$\boldsymbol{A}\boldsymbol{\Phi} = \boldsymbol{\Phi}\boldsymbol{\Lambda}; \tag{A.53}$$

i.e.,

$$\boldsymbol{\Phi}^{-1}\boldsymbol{A}\boldsymbol{\Phi} = \boldsymbol{\Phi}^*\boldsymbol{A}\boldsymbol{\Phi} = \boldsymbol{\Lambda}, \quad \text{or} \quad \boldsymbol{A} = \boldsymbol{\Phi}\boldsymbol{\Lambda}\boldsymbol{\Phi}^{-1} = \boldsymbol{\Phi}\boldsymbol{\Lambda}\boldsymbol{\Phi}^*. \tag{A.54}$$

From the first equation above we see that the Hermitian matrix $\boldsymbol{A}$ can be diagonalized by its unitary eigenvector matrix $\boldsymbol{\Phi}$. From the second equation we see that the matrix $\boldsymbol{A}$ can be decomposed to be expressed as

$$\boldsymbol{A} = \boldsymbol{\Phi}\boldsymbol{\Lambda}\boldsymbol{\Phi}^* = [\boldsymbol{\phi}_1, \ldots, \boldsymbol{\phi}_n] \begin{bmatrix} \lambda_1 & \cdots & 0 \\ \vdots & \ddots & \vdots \\ 0 & \cdots & \lambda_n \end{bmatrix} \begin{bmatrix} \boldsymbol{\phi}_1^* \\ \vdots \\ \boldsymbol{\phi}_n^* \end{bmatrix} = \sum_{i=1}^{n} \lambda_i \boldsymbol{\phi}_i \boldsymbol{\phi}_i^*. \tag{A.55}$$

Based on any unitary matrix $\boldsymbol{A} = [\boldsymbol{a}_1 \ldots, \boldsymbol{a}_n]$ (where the ith column vector is $\boldsymbol{a}_k = [a_{1k}, \ldots, a_{nk}]^{\mathrm{T}}$), a *unitary transform* of a vector $\boldsymbol{x} = [x_1, \ldots, x_n]^{\mathrm{T}}$ can be defined as:

$$\begin{cases} \boldsymbol{y} = \begin{bmatrix} y_1 \\ \vdots \\ y_n \end{bmatrix} = \boldsymbol{A}^{-1}\boldsymbol{x} = \boldsymbol{A}^*\boldsymbol{x} = \begin{bmatrix} \boldsymbol{a}_1^* \\ \vdots \\ \boldsymbol{a}_n^* \end{bmatrix} \boldsymbol{x} \\ \boldsymbol{x} = \begin{bmatrix} x_1 \\ \vdots \\ x_n \end{bmatrix} = \boldsymbol{A}\boldsymbol{y} = \begin{bmatrix} \boldsymbol{a}_1 & \ldots & \boldsymbol{a}_n \end{bmatrix} \begin{bmatrix} y_1 \\ \vdots \\ y_n \end{bmatrix} = \sum_{j=1}^{n} y_j \, \boldsymbol{a}_j \end{cases}. \tag{A.56}$$

The first and second equations are respectively the forward and inverse transforms. In particular, when $\boldsymbol{A} = \overline{\boldsymbol{A}}$ is real, $\boldsymbol{A}^{-1} = \boldsymbol{A}^{\mathrm{T}}$ is an orthogonal matrix and the corresponding transform is an *orthogonal transform.*

The forward transform can also be written in component form:

$$y_j = <\boldsymbol{x}, \boldsymbol{a}_j> = \boldsymbol{x}^{\mathrm{T}}\overline{\boldsymbol{a}} = \boldsymbol{a}_j^*\boldsymbol{x} = \sum_{i=1}^{n} x_i \overline{a}_{ij}, \qquad j = 1, \ldots, n, \tag{A.57}$$

where the transform coefficient $y_i = \boldsymbol{a}_i^*\boldsymbol{x}$ represents the projection of $\boldsymbol{x}$ onto the ith column vector $\boldsymbol{a}_i$ of the transform matrix $\boldsymbol{A}$. The *inverse transform* can also be written as:

$$\boldsymbol{x} = \sum_{j=1}^{n} y_j \, \boldsymbol{a}_j, \qquad \text{or in component form as} \qquad x_i = \sum_{j=1}^{n} a_{ij} \, y_j \qquad i = 1, \ldots, n. \tag{A.58}$$

By this transform, vector $\boldsymbol{x}$ is represented as a linear combination (weighted sum) of the n column vectors $\boldsymbol{a}_1, \boldsymbol{a}_2, \ldots, \boldsymbol{a}_n$ of matrix $\boldsymbol{A}$. Geometrically, $\boldsymbol{x}$ is a point in the n-dimensional space spanned by these n orthonormal basis vectors. Each coefficient y_i is the coordinate in the ith dimension, which can be obtained as the projection of $\boldsymbol{x}$ onto the corresponding basis vector $\boldsymbol{a}_i$.

A unitary (orthogonal) transform $\boldsymbol{y} = \boldsymbol{A}\boldsymbol{x}$ can be interpreted geometrically as the rotation of the vector $\boldsymbol{x}$ about the origin, or equivalently, the representation of the same vector in a rotated coordinate system. A unitary (orthogonal) transform $\boldsymbol{y} = \boldsymbol{A}\boldsymbol{x}$ does not change the vector's length:

$$||\boldsymbol{y}||^2 = \boldsymbol{y}^*\boldsymbol{y} = (\boldsymbol{A}^*\boldsymbol{x})^*(\boldsymbol{A}^*\boldsymbol{x}) = \boldsymbol{x}^*\boldsymbol{A}\boldsymbol{A}^*\boldsymbol{x} = \boldsymbol{x}^*\boldsymbol{x} = ||\boldsymbol{x}||^2, \tag{A.59}$$

as $\boldsymbol{A}\boldsymbol{A}^* = \boldsymbol{A}\boldsymbol{A}^{-1} = \boldsymbol{I}$. This is the Parseval's relation. If $\boldsymbol{x}$ is interpreted as a signal, then its length $||\boldsymbol{x}||^2 = ||\boldsymbol{y}||^2$ represents the total energy or information contained in the signal, which is preserved during any unitary transform.

A.4 Toeplitz and circulant matrices

A square matrix is called a *Toeplitz matrix* if any element a_{mn} is equal to its lower-right neighbor $a_{m+1\,n+1}$; i.e., every diagonal of the matrix is composed of the same value. For example, the following matrix is a Toeplitz matrix:

$$A_T = \begin{bmatrix} a & b & c & d & e & f \\ g & a & b & c & d & e \\ h & g & a & b & c & d \\ i & h & g & a & b & c \\ j & i & h & g & a & b \\ k & j & i & h & g & a \end{bmatrix}. \tag{A.60}$$

An $N \times N$ Toeplitz matrix can be formed by a sequence $\dots a_{-2}, a_{-1}, a_0, a_1, a_2, \dots$:

$$A_T = \begin{bmatrix} a_0 & a_1 & a_2 & \cdots & a_{N-3} & a_{N-2} & a_{N-1} \\ a_{-1} & a_0 & a_1 & \cdots & a_{N-4} & a_{N-3} & a_{N-2} \\ a_{-2} & a_{-1} & a_0 & \cdots & a_{N-5} & a_{N-4} & a_{N-3} \\ \vdots & \vdots & \vdots & \ddots & \vdots & \vdots & \vdots \\ a_{3-N} & a_{4-N} & a_{5-N} & \cdots & a_0 & a_1 & a_2 \\ a_{2-N} & a_{3-N} & a_{4-N} & \cdots & a_{-1} & a_0 & a_1 \\ a_{1-N} & a_{2-N} & a_{3-N} & \cdots & a_{-2} & a_{-1} & a_0 \end{bmatrix}. \tag{A.61}$$

In particular, if the sequence is periodic, $x_n = x_{n+N}$ with period N, then the Toeplitz matrix above becomes a *circulant matrix*, composed of N rows each rotated one element to the right relative to the previous row:

$$A_T = \begin{bmatrix} a_0 & a_1 & a_2 & \cdots & a_{N-3} & a_{N-2} & a_{N-1} \\ a_{N-1} & a_0 & a_1 & \cdots & a_{N-4} & a_{N-3} & a_{N-2} \\ a_{N-2} & a_{N-1} & a_0 & \cdots & a_{N-5} & a_{N-4} & a_{N-3} \\ \vdots & \vdots & \vdots & \ddots & \vdots & \vdots & \vdots \\ a_3 & a_4 & a_5 & \cdots & a_0 & a_1 & a_2 \\ a_2 & a_3 & a_4 & \cdots & a_{N-1} & a_0 & a_1 \\ a_1 & a_2 & a_3 & \cdots & a_{N-2} & a_{N-1} & a_0 \end{bmatrix}. \tag{A.62}$$

When the period N of the sequence is increased to approach infinity $N \to \infty$, the periodic sequence approaches aperiodic, correspondingly, the circulant matrix asymptotically becomes a Toeplitz matrix.

A.5 Vector and matrix differentiation

Let $\boldsymbol{x} = [x_1, \dots, x_n]^{\mathrm{T}}$ be an n-D vector composed of n variables x_k $(k = 1, \dots, n)$. A vector differentiation operator is defined as

$$\frac{d}{d\boldsymbol{x}} = \left[\frac{\partial}{\partial x_1}, \dots, \frac{\partial}{\partial x_n}\right]^{\mathrm{T}}, \tag{A.63}$$

which can be applied to any scalar function $f(\boldsymbol{x})$ to find its derivative with respect to its variable argument $\boldsymbol{x}$:

$$\frac{d}{d\boldsymbol{x}} f(\boldsymbol{x}) = \left[\frac{\partial f}{\partial x_1}, \ldots, \frac{\partial f}{\partial x_n} \right]^{\mathrm{T}}. \tag{A.64}$$

Vector differentiation has the following properties:

-
$$\frac{d}{d\boldsymbol{x}} (\boldsymbol{b}^{\mathrm{T}} \boldsymbol{x}) = \frac{d}{d\boldsymbol{x}} (\boldsymbol{x}^{\mathrm{T}} \boldsymbol{b}) = \boldsymbol{b}. \tag{A.65}$$

-
$$\frac{d}{d\boldsymbol{x}} (\boldsymbol{x}^{\mathrm{T}} \boldsymbol{x}) = 2\boldsymbol{x}. \tag{A.66}$$

-
$$\frac{d}{d\boldsymbol{x}} (\boldsymbol{x}^{\mathrm{T}} \boldsymbol{A}\boldsymbol{x}) = 2\boldsymbol{A}\boldsymbol{x}, \qquad \text{where } \boldsymbol{A} = \boldsymbol{A}^{\mathrm{T}} \tag{A.67}$$

To show the third one, we consider the kth element of the vector ($k = 1, \ldots, n$):

$$\frac{\partial}{\partial x_k} (\boldsymbol{x}^T \boldsymbol{A}\boldsymbol{x}) = \frac{\partial}{\partial x_k} \sum_{i=1}^{n} \sum_{j=1}^{n} a_{ij} x_i x_j = \sum_{i=1}^{n} a_{ik} x_i + \sum_{j=1}^{n} a_{kj} x_j = 2 \sum_{i=1}^{n} a_{ik} x_i. \tag{A.68}$$

Note that here we have used the assumption $\boldsymbol{A}^{\mathrm{T}} = \boldsymbol{A}$. Putting all n elements in vector form, we get Eq. (A.67). In particular, when $\boldsymbol{A} = \boldsymbol{I}$, we obtain Eq. (A.66). More specially, when $n = 1$, we get this familiar derivative in the scalar case:

$$\frac{d}{dx} (ax^2) = \frac{d}{dx} (xax) = 2ax. \tag{A.69}$$

Let $\boldsymbol{A} = [a_{ij}]_{m \times n}$ ($i = 1, \ldots, m,\ j = 1, \ldots, n$) be an $m \times n$ matrix. A matrix differentiation operator is defined as

$$\frac{d}{d\boldsymbol{A}} = \begin{bmatrix} \frac{\partial}{\partial a_{11}} & \cdots & \frac{\partial}{\partial a_{1n}} \\ \vdots & \ddots & \vdots \\ \frac{\partial}{\partial a_{m1}} & \cdots & \frac{\partial}{\partial a_{mn}} \end{bmatrix}, \tag{A.70}$$

which can be applied to any scalar function $f(\boldsymbol{A})$ to find its derivative with respect to its matrix argument $\boldsymbol{A}$:

$$\frac{d}{d\boldsymbol{A}} f(\boldsymbol{A}) = \begin{bmatrix} \frac{\partial}{\partial a_{11}} f(\boldsymbol{A}) & \cdots & \frac{\partial}{\partial a_{1n}} f(\boldsymbol{A}) \\ \vdots & \ddots & \vdots \\ \frac{\partial}{\partial a_{m1}} f(\boldsymbol{A}) & \cdots & \frac{\partial}{\partial a_{mn}} f(\boldsymbol{A}) \end{bmatrix}. \tag{A.71}$$

In particular when $f(\boldsymbol{A}) = \boldsymbol{u}^{\mathrm{T}} \boldsymbol{A}\boldsymbol{v}$, where $\boldsymbol{u}$ and $\boldsymbol{v}$ are $m \times 1$ and $n \times 1$ constant vectors, respectively, we have

$$\frac{\partial}{\partial a_{ij}} [\boldsymbol{u}^{\mathrm{T}} \boldsymbol{A}\boldsymbol{v}] = \frac{\partial}{\partial a_{ij}} \left[\sum_{i=1}^{m} \sum_{j=1}^{n} u_i a_{ij} v_j \right] = u_i v_j, \qquad i = 1, \ldots, m,\ j = 1, \ldots, n; \tag{A.72}$$

i.e.,

$$\frac{d}{d\boldsymbol{A}} (\boldsymbol{u}^{\mathrm{T}} \boldsymbol{A}\boldsymbol{v}) = \boldsymbol{u}\boldsymbol{v}^{\mathrm{T}}. \tag{A.73}$$

B Review of random variables

B.1 Random variables

- **Random experiment and its sample space**
  A *random experiment* is a procedure that can be carried out repeatedly with a random outcome generated each time. The *sample space* Ω of the random experiment is a set containing all of its possible outcomes. Ω may be finite, countable infinite, or uncountable. For example, "Randomly pick a card from a deck of cards labeled 0, 1, 2, 3, and 4" is a random experiment. The sample space is a set of all of the possible outcomes: $\Omega = \{0, 1, 2, 3, 4\}$.
- **Random events**
  An *event* $A \subset \Omega$ is a subset of the sample space Ω. A can be an empty set $\emptyset$, a proper subset (e.g., a single outcome), or the entire sample space Ω. Event A occurs if the outcome is a member of A.
  The *event space* $\mathcal{F}$ is set of events. If Ω is finite and countable, then $\mathcal{F} = \text{Pow}(\Omega)$ is the power set of Ω (a set of all possible subsets of Ω). But if Ω is infinite or uncountable, $\mathcal{F}$ is a σ-algebra on Ω satisfying the following:
  - $\Omega \in \mathcal{F}$ (or $\emptyset \in \mathcal{F}$).
  - closed to countable unions: if $A_i \in \mathcal{F}$ $(i = 1, 2, \ldots)$, then $\cup_i A_i \in \mathcal{F}$;
  - closed to complements: if $A \in \mathcal{F}$, then $\overline{\Omega} = \Omega - A \in \mathcal{F}$.

  The ordered pair $(\Omega, \mathcal{F})$ is called a *measurable space*. The concept of σ-algebra is needed to introduce a probability measure for all events in $\mathcal{F}$. For example, $\mathcal{F} = \{\emptyset, \{0, 1, 2\}, \{2, 3\}, \Omega = \{0, 1, 2, 3, 4\}\}$.
- **Probability**
  The *probability* is a measure on $\mathcal{F}$. Probability of any event $A \in \mathcal{F}$ is a function $P(A)$ from A to a real value in the range $[0, 1]$, satisfying the following:
  - $0 \le P(A) \le 1$ for all $A \in \mathcal{F}$.
  - $P(\emptyset) = 0$, and $P(\Omega) = 1$.
  - $P(A \cup B) = P(A) + P(B)$ if $A \cap B = \emptyset$ for all $A, B \in \mathcal{F}$.

  For example, "The randomly chosen card has a number smaller than 3" is a random event, which is represented by a subset $A = \{0, 1, 2\} \subset \Omega$. The probability of this event A is $P(A) = 3/5$. Event A occurs if the outcome ω is one of the members of A, $\omega \in A$, e.g., 2.
- **Probability Space**
  The triple $(\Omega, \mathcal{F}, P)$ is called the *probability space*.

- **Random variables**
  A random variable $x(\omega)$ is a complex-valued (or real-valued as a special case) function $x : \Omega \to \mathbb{R}$ that maps every outcome $\omega \in \Omega$ into a complex number x. Formally, the function $x(\omega)$ is a random variable if

$$\{\omega : x(\omega) \leq r\} \in \mathcal{F}, \quad \forall r \in \mathbb{R}. \tag{B.1}$$

  Random variables x can be either continuous or discrete.
- **Cumulative distribution function**
  The *cumulative distribution function* of a random variable x is defined as

$$F_x(u) = P(x < u), \tag{B.2}$$

  and we have $F_x(\infty) = 1$ and $F_x(-\infty) = 0$.
- **Density function**
  The *density function* of a random variable x is defined by

$$p_x(x) = \frac{d}{du} F_x(u); \quad \text{i.e.,} \quad F_x(u) = \int_{-\infty}^{u} p_x(x)\, dx. \tag{B.3}$$

  We have

$$P(a \leq x < b) = F_x(b) - F_x(a) = \int_{a}^{b} p_x(x)\, dx. \tag{B.4}$$

  In particular,

$$P(x < \infty) = F_x(\infty) - F_x(-\infty) = \int_{-\infty}^{\infty} p_x(x)\, dx = 1. \tag{B.5}$$

  The subscript of p_x can be dropped if no confusion will be caused.
- **Discrete random variables**
  If a random variable x can only take one of a set of N values $\{x_n \quad n = 1, \ldots, N\}$, then its *probability distribution* is

$$P(x = x_n) = p_n \quad (n = 1, \ldots, N), \tag{B.6}$$

  where

$$0 \leq p_n \leq 1, \quad \text{and} \quad \sum_{i=n}^{N} p_n = 1. \tag{B.7}$$

  The cumulative distribution function is

$$F_x(\xi) = P(x < \xi) = \sum_{x_n < \xi} p_n. \tag{B.8}$$

- **Expectation**
  The *expectation* is the mathematical mean of a random variable x. If x is continuous,

$$\mu_x = E(x) = \int_{-\infty}^{\infty} x\, p(x)\, dx. \tag{B.9}$$

If x is discrete,

$$\mu_x = E(x) = \sum_{n=1}^{N} x_n p_n. \tag{B.10}$$

- **Variance**

  The *variance* represents the statistical variability of a random variable x. If x is continuous,

$$\sigma_x^2 = \mathrm{Var}(x) = E[|x - \mu_x|^2] = \int_{-\infty}^{\infty} |x - \mu_x|^2 p(x)\, dx. \tag{B.11}$$

  If x is discrete,

$$\sigma_x^2 = \mathrm{Var}(x) = E[|x - \mu_x|^2] = \sum_{n=1}^{N} |x_n - \mu_x|^2 p_n. \tag{B.12}$$

  We also have

$$\begin{aligned} \sigma_x^2 &= \mathrm{Var}(x) = E(|x - \mu_x|^2) = E[(x - \mu_x)\overline{(x - \mu_x)}] \\ &= E(|x|^2) - \mu_x E(\overline{x}) - E(x)\overline{\mu} + |\mu_x|^2 = E(|x|^2) - |\mu_x|^2. \end{aligned} \tag{B.13}$$

  The *standard deviation* of x is defined as

$$\sigma_x = \sqrt{\mathrm{Var}(x)}. \tag{B.14}$$

- **Normal (Gaussian) distribution**

  A random variable x has a *normal distribution* if its density function is

$$p(x) = N(x, \mu, \sigma) = \frac{1}{\sqrt{2\pi\sigma^2}} e^{-\frac{(x-\mu)^2}{\sigma^2}}. \tag{B.15}$$

  It can be shown that

$$\int_{-\infty}^{\infty} N(x, \mu, \sigma)\, dx = 1, \tag{B.16}$$

$$E(x) = \int_{-\infty}^{\infty} x\, N(x, \mu, \sigma)\, dx = \mu, \tag{B.17}$$

  and

$$\mathrm{Var}(x) = \int_{-\infty}^{\infty} (x - \mu)^2 N(x, \mu, \sigma)\, dx = \sigma^2. \tag{B.18}$$

B.2 Multivariate random variables

- **Multivariate random variables**

  A set of N *multivariate random variables* can be represented as a *random vector* $\boldsymbol{x} = [x_1, \ldots, x_N]^{\mathrm{T}}$. When a *stochastic* or *random process* (to be discussed later) $x(t)$ is sampled, it can be represented as a random vector $\boldsymbol{x}$.

- **Joint distribution function and density function**

  The *joint distribution function* of a random vector $\boldsymbol{x}$ is defined as

$$F_{\boldsymbol{x}}(u_1, \ldots, u_N) = P_{\boldsymbol{x}}(x_1 < u_1, \ldots, x_n < u_N)$$
$$= \int_{-\infty}^{u_1} \cdots \int_{-\infty}^{u_n} p_{\boldsymbol{x}}(x_1, \ldots, x_N)\, dx_1 \ldots dx_N = \int_{-\infty}^{\boldsymbol{u}} p(\boldsymbol{x})\, d\boldsymbol{x}, \quad \text{(B.19)}$$

  where $p(\boldsymbol{x}) = p_{\boldsymbol{x}}(x_1, \ldots, x_N)$ is the *joint density function* of the random vector $\boldsymbol{x}$.
- **Mean vector**

  The *expectation* or *mean* of random variable x_n is defined as

$$\mu_n = E(x_n) = \int_{-\infty}^{\infty} \cdots \int_{-\infty}^{\infty} x_n\, p_{\boldsymbol{x}}(x_1, \ldots, x_N)\, dx_1 \ldots dx_N. \quad \text{(B.20)}$$

  The *mean vector* of random vector $\boldsymbol{x}$ is defined as

$$\boldsymbol{\mu}_x = E(\boldsymbol{x}) = \int_{-\infty}^{\infty} \boldsymbol{x}\, p(\boldsymbol{x})\, d\boldsymbol{x} = [E(x_1), \ldots, E(x_N)]^{\mathrm{T}} = [\mu_1, \ldots, \mu_N]^{\mathrm{T}}, \quad \text{(B.21)}$$

  which can be interpreted as the center of gravity of an N-D object with $p_{\boldsymbol{x}}(x_1, \ldots, x_N)$ being the density function.
- **Covariance matrix**

  The *variance* of random variable x_n measures its variability and is defined as

$$\sigma_n^2 = \mathrm{Var}(x_n) = E[|x_n - \mu_n|^2] = E(|x_n|^2) - |\mu_n|^2$$
$$= \int_{-\infty}^{\infty} \cdots \int_{-\infty}^{\infty} |x_n - \mu_n|^2\, p_{\boldsymbol{x}}(x_n, \ldots, x_N)\, dx_1 \ldots dx_N. \quad \text{(B.22)}$$

  The *covariance* of x_m and x_n $(m, n = 1, \ldots, N)$ measures their similarity and is defined as

$$\sigma_{mn}^2 = \mathrm{Cov}(x_m, x_n) = E[(x_m - \mu_m)\overline{(x_n - \mu_n)}]$$
$$= \int_{-\infty}^{\infty} \cdots \int_{-\infty}^{\infty} (x_m - \mu_m)\overline{(x_n - \mu_n)}\, p_{\boldsymbol{x}}(x_1, \ldots, x_N)\, dx_1 \ldots dx_N. \quad \text{(B.23)}$$

  Note that

$$\sigma_{mn}^2 = E[(x_m - \mu_m)\overline{(x_n - \mu_n)}] = E(x_m\overline{x}_n) - E(x_m)\overline{\mu}_n - \mu_m E(\overline{x}_n) + \mu_m\overline{\mu}_n$$
$$= E(x_m\overline{x}_n) - \mu_m\overline{\mu}_n. \quad \text{(B.24)}$$

  The *covariance matrix* of a random vector $\boldsymbol{x}$ is defined as

$$\boldsymbol{\Sigma}_x = \int_{-\infty}^{\infty} (\boldsymbol{x} - \boldsymbol{\mu}_x)(\boldsymbol{x} - \boldsymbol{\mu}_x)^* p(\boldsymbol{x})\, d\boldsymbol{x}$$
$$= E[(\boldsymbol{x} - \boldsymbol{\mu}_x)(\boldsymbol{x} - \boldsymbol{\mu}_x)^*] = E(\boldsymbol{x}\boldsymbol{x}^*) - \boldsymbol{\mu}_x\boldsymbol{\mu}_x^*$$
$$= \begin{bmatrix} \sigma_{11}^2 & \cdots & \sigma_{1N}^2 \\ \vdots & \ddots & \vdots \\ \sigma_{N1}^2 & \cdots & \sigma_{NN}^2 \end{bmatrix}. \quad \text{(B.25)}$$

When $m = n$, $\sigma_n^2 = E(|x_n|^2) - |\mu_n|^2$ is the variance of x_n, which can be interpreted as the amount of information, or energy, contained in the nth component x_n of the signal $\boldsymbol{x}$. Therefore the total information or energy contained in $\boldsymbol{x}$ is

$$\text{tr } \boldsymbol{\Sigma}_x = \sum_{n=1}^{N} \sigma_n^2. \tag{B.26}$$

Obviously, $\boldsymbol{\Sigma}$ is symmetric as $\sigma_{mn}^2 = \sigma_{nm}^2$. Moreover, it can be shown that $\boldsymbol{\Sigma}$ is also *positive definite*, and all its eigenvalues $\lambda_n > 0$ $(n = 1, \ldots, N)$ are positive and we have

$$\text{tr } \boldsymbol{\Sigma}_x = \sum_{n=1}^{N} \lambda_n > 0 \qquad \det \boldsymbol{\Sigma}_x = \prod_{n=1}^{N} \lambda_n > 0. \tag{B.27}$$

- **Correlation coefficient**
  The covariance σ_{mn}^2 of two random variables x_m and x_n represents the statistical similarity between them. If $\sigma_{mn}^2 > 0$, x_m and x_n are positively correlated; if $\sigma_{mn}^2 < 0$, they are negatively correlated, if $\sigma_{mn}^2 = 0$, they are *uncorrelated* or *decorrelated*. The normalized covariance is called the *correlation coefficient*:

$$r_{mn} = \frac{\sigma_{mn}^2}{\sigma_m \sigma_n} = \frac{E(x_m \overline{x}_n) - \mu_m \overline{\mu}_n}{\sqrt{E(|x_m|^2) - |\mu_m|^2}\sqrt{E(|x_n|^2) - |\mu_n|^2}}. \tag{B.28}$$

  The correlation coefficient $-1 \le r_{mn} \le 1$ measures the similarity between the two random variables x_m and x_n. They are either positively correlated if $r_{mn} > 0$, negatively correlated if $r_{mn} < 0$, or uncorrelated if $r_{mn} = 0$. The correlation matrix of a random vector is therefore defined as

$$\boldsymbol{R} = \begin{bmatrix} r_{11} & \cdots & r_{1N} \\ \vdots & \ddots & \vdots \\ r_{N1} & \vdots & r_{NN} \end{bmatrix}. \tag{B.29}$$

  Obviously, all elements $r_{nn} = 1$ $(n = 1, \ldots, N)$ along the main diagonal of $\boldsymbol{R}$ are 1, and all off-diagonal elements $|r_{mn}| < 1$ $(m \neq n)$.
- **Correlation and independence**
  A set of N random variables x_n $(n = 1, \ldots, N)$ are independent if and only if

$$p(\boldsymbol{x}) = p_{\boldsymbol{x}}(x_1, \ldots, x_N) = p(x_1)\, p(x_2) \ldots p(x_N). \tag{B.30}$$

  Two random variables x_m and x_n are uncorrelated if $r_{mn} = 0$; i.e.,

$$\sigma_{mn}^2 = E(x_m \overline{x}_n) - \mu_m \overline{\mu}_n = 0, \qquad \text{or} \qquad E(x_m \overline{x}_n) - \mu_m \overline{\mu}_n. \tag{B.31}$$

  Obviously, if x_m and x_n are independent, we have $E(x_m \overline{x}_n) = E(x_m) E(\overline{x}_n) = \mu_m \overline{\mu}_n$ and they are uncorrelated. However, if they are uncorrelated, they are not necessarily independent, unless they are normally distributed.

A random vector $\boldsymbol{x} = [x_1, \ldots, x_N]^{\mathrm{T}}$ is uncorrelated or decorrelated if $r_{mn} = 0$ for all $m \neq n$, and both its covariance $\boldsymbol{\Sigma}$ and correlation $\boldsymbol{R}$ become diagonal matrices with only non-zero σ_n^2 $(n = 1, \ldots, N)$ on its diagonal.

- **Mean and covariance under unitary transforms**
  If the inverse of a matrix is the same as its conjugate transpose: $\boldsymbol{A}^{-1} = \boldsymbol{A}^*$, then it is a unitary matrix. Given any unitary matrix $\boldsymbol{A}$, an orthogonal transform of a random vector $\boldsymbol{x}$ can be defined as

$$\begin{cases} \boldsymbol{X} = \boldsymbol{A}^*\boldsymbol{x} \\ \boldsymbol{x} = \boldsymbol{A}\boldsymbol{X} \end{cases}. \tag{B.32}$$

The mean vector $\boldsymbol{\mu}_X$ and the covariance matrix $\boldsymbol{\Sigma}_X$ of $\boldsymbol{X}$ are related to the $\boldsymbol{\mu}_x$ and $\boldsymbol{\Sigma}_x$ of $\boldsymbol{x}$ by

$$\boldsymbol{\mu}_X = E(\boldsymbol{X}) = E(\boldsymbol{A}^*\boldsymbol{x}) = \boldsymbol{A}^*E(\boldsymbol{x}) = \boldsymbol{A}^*\boldsymbol{\mu}_x, \tag{B.33}$$

$$\begin{aligned} \boldsymbol{\Sigma}_X &= E(\boldsymbol{X}\boldsymbol{X}^*) - \boldsymbol{\mu}_X\boldsymbol{\mu}_X^* = E(\boldsymbol{A}^*\boldsymbol{x}\boldsymbol{x}^*\boldsymbol{A}) - \boldsymbol{A}^*\boldsymbol{\mu}_x\boldsymbol{\mu}_x^*\boldsymbol{A} \\ &= \boldsymbol{A}^*E(\boldsymbol{x}\boldsymbol{x}^*)\boldsymbol{A} - \boldsymbol{A}^*\boldsymbol{\mu}_x\boldsymbol{\mu}_x^*\boldsymbol{A} = \boldsymbol{A}^*[E(\boldsymbol{x}\boldsymbol{x}^*) - \boldsymbol{\mu}_x\boldsymbol{\mu}_x^*]\boldsymbol{A} \\ &= \boldsymbol{A}^*\boldsymbol{\Sigma}_x\boldsymbol{A}. \end{aligned} \tag{B.34}$$

The unitary transform does not change the trace of $\boldsymbol{\Sigma}$:

$$\mathrm{tr}\ \boldsymbol{\Sigma}_X = \mathrm{tr}\ \boldsymbol{\Sigma}_x, \tag{B.35}$$

which means the total amount of energy or information contained in $\boldsymbol{x}$ is not changed after a unitary transform $\boldsymbol{X} = \boldsymbol{A}^*\boldsymbol{x}$ (although the distribution of energy among the components may be changed).

- **Normal distribution**
  The density function of a normally distributed random vector $\boldsymbol{x}$ is

$$p(\boldsymbol{x}) = N(\boldsymbol{x}, \boldsymbol{\mu}_x, \boldsymbol{\Sigma}_x) = \frac{1}{(2\pi)^{n/2}\,|\boldsymbol{\Sigma}_x|^{1/2}} \exp\left[-\frac{1}{2}(\boldsymbol{x} - \boldsymbol{\mu}_x)^{\mathrm{T}}\boldsymbol{\Sigma}_x^{-1}(\boldsymbol{x} - \boldsymbol{\mu}_x)\right]. \tag{B.36}$$

When $n = 1$, $\boldsymbol{\Sigma}_x$ and $\boldsymbol{\mu}_x$ become σ_x and μ_x, respectively, and the density function becomes single variable normal distribution. To find the shape of a normal distribution, consider the iso-value hyper-surface in the N-D space determined by the equation

$$N(\boldsymbol{x}, \boldsymbol{\mu}_x, \boldsymbol{\Sigma}_x) = c_0, \tag{B.37}$$

where c_0 is a constant. This equation can be written as

$$(\boldsymbol{x} - \boldsymbol{\mu}_x)^{\mathrm{T}}\boldsymbol{\Sigma}_x^{-1}(\boldsymbol{x} - \boldsymbol{\mu}_x) = c_1, \tag{B.38}$$

where c_1 is another constant related to c_0, $\boldsymbol{\mu}_x$, and $\boldsymbol{\Sigma}_x$. This equation represents a hyper-ellipsoid in the N-D space. The center and spatial distribution of this ellipsoid are determined by $\boldsymbol{\mu}_x$ and $\boldsymbol{\Sigma}_x$, respectively.

In particular, when $\boldsymbol{x} = [x_1, \dots, x_N]^{\mathrm{T}}$ is decorrelated; i.e., $\sigma^2_{mn} = 0$ for all $m \neq n$, $\boldsymbol{\Sigma}_x$ becomes a diagonal matrix

$$\boldsymbol{\Sigma}_x = \mathrm{diag}[\sigma_1^2, \dots, \sigma_N^2] = \begin{bmatrix} \sigma_1^2 & 0 & \cdots & 0 \\ 0 & \sigma_2^2 & \cdots & 0 \\ \vdots & \vdots & \ddots & \vdots \\ 0 & 0 & \cdots & \sigma_N^2 \end{bmatrix}, \tag{B.39}$$

and the equation $N(\boldsymbol{x}, \boldsymbol{\mu}_x, \boldsymbol{\Sigma}_x) = c_0$ can be written as

$$(\boldsymbol{x} - \boldsymbol{\mu}_x)^{\mathrm{T}} \boldsymbol{\Sigma}_x^{-1} (\boldsymbol{x} - \boldsymbol{\mu}_x) = \sum_{n=1}^{N} \frac{(x_n - \mu_n)^2}{\sigma_n^2} = c_1, \tag{B.40}$$

which represents a standard ellipsoid with all its axes in parallel with those of the coordinate system.

- **Estimation of $\boldsymbol{\mu}_x$ and $\boldsymbol{\Sigma}_x$**
When $p(\boldsymbol{x}) = p(x_1, \dots, x_n)$ is not known, $\boldsymbol{\mu}_x$ and $\boldsymbol{\Sigma}_x$ cannot be found by their definitions, but they can be estimated if a set of K outcomes ($\boldsymbol{x}^{(k)}$, $k = 1, \dots, K$) of the random experiment can be observed. Then the mean vector can be estimated as

$$\hat{\boldsymbol{\mu}}_x = \frac{1}{K} \sum_{k=1}^{K} \boldsymbol{x}^{(k)}; \tag{B.41}$$

i.e., the nth element of $\hat{\boldsymbol{\mu}}_x$ is estimated as

$$\hat{\mu}_n = \frac{1}{K} \sum_{k=1}^{K} x_n^{(k)}, \quad (n = 1, \dots, N), \tag{B.42}$$

where $x_n^{(k)}$ is the nth element of the kth outcome $\boldsymbol{x}^{(k)}$. The covariance matrix $\boldsymbol{\Sigma}_x$ can be estimated as

$$\hat{\boldsymbol{\Sigma}}_x = \frac{1}{K-1} \sum_{k=1}^{K} (\boldsymbol{x}^{(k)} - \hat{\boldsymbol{\mu}}_x)(\boldsymbol{x}^{(k)} - \hat{\boldsymbol{\mu}}_x)^{\mathrm{T}} = \frac{1}{K-1} \sum_{k=1}^{K} \boldsymbol{x}^{(k)} \boldsymbol{x}^{(k)T} - \hat{\boldsymbol{\mu}}_x \hat{\boldsymbol{\mu}}_x^{\mathrm{T}}; \tag{B.43}$$

i.e., the mnth element of $\hat{\boldsymbol{\Sigma}}_x$ is

$$\hat{\sigma}_{mn} = \frac{1}{K-1} \sum_{k=1}^{K} (x_m^{(k)} - \mu_m)(x_n^{(k)} - \mu_n) = \frac{1}{K-1} \sum_{k=1}^{K} x_m^{(k)} x_n^{(k)} - \hat{\mu}_m \hat{\mu}_n. \tag{B.44}$$

Note that for the estimation of the covariance to be unbiased; i.e., $E(\hat{\boldsymbol{\Sigma}}_x) = \boldsymbol{\Sigma}_x$, the coefficient $1/(K-1)$, instead of $1/K$, needs to be used. Obviously, this makes little difference when the number of samples K is large.

B.3 Stochastic models

A physical signal can be modeled as a time function $x(t)$ which takes a complex value (or real value as a special case) $x(t_0)$ at each time moment $t = t_0$. This value may be either deterministic or random with a certain probability distribution. In the latter case the time function is called a *stochastic process* or *random process*.

Recall that a random variable $x(\omega)$ is a function that maps the outcomes $\omega \in \Omega$ in the sample space Ω of a random experiment to a real number between 0 and 1. Here, a stochastic process can be considered as a function $x(\omega, t)$ of two arguments of time t as well as the outcome $\omega \in \Omega$.

If the mean and covariance functions of a random process $x(t)$ do not change over time; i.e.,

$$\mu_x(t) = \mu_x(t-\tau), \qquad R_x(t,\tau) = R_x(t-\tau) \qquad \Sigma_x(t,\tau) = \Sigma_x(t-\tau), \tag{B.45}$$

then $x(t)$ is a *stationary process*, in the weak or wide sense (*weak-sense* or *wide-sense stationarity*). If the probability distribution of $x(t)$ does not change over time, it is said to have *strict or strong stationarity.* We will only consider stationary processes.

- The *mean function* of $x(t)$ is the expectation, defined as

$$\mu_x(t) = E[x(t)]. \tag{B.46}$$

  If $\mu_x(t) = 0$ for all t, then $x(t)$ is a zero-mean or centered stochastic process, which can be easily obtained by subtracting the mean function $\mu_x(t)$ from the original process $x(t)$. If the stochastic process is stationary, then $\mu_x(t) = \mu_x$ is a constant.
- The *auto-covariance function* of $x(t)$ is defined as

$$\begin{aligned} \sigma_x^2(t,\tau) &= \text{Cov}[x(t), x(\tau)] = E[(x(t)-\mu_x(t))(x(\tau)-\mu_x(\tau))] \\ &= E[x(t)x(\tau)] - \mu_x(t)\mu_x(\tau). \end{aligned} \tag{B.47}$$

  If the stochastic process is stationary, then $\sigma_x^2(t) = \sigma_x^2(\tau) = \sigma_x^2$, $\mu_x(t) = \mu_x(\tau) = \mu_x$, and $\sigma_x^2(t,\tau) = \sigma_x^2(t-\tau)$, then the above can be expressed as

$$\sigma_x^2(t-\tau) = E[(x(t)-\mu_x(t))(x(\tau)-\mu_x(\tau))] = E[x(t)x(\tau)] - \mu_x^2. \tag{B.48}$$

- The *autocorrelation function* of $x(t)$ is defined as

$$r_x(t,\tau) = \frac{\sigma_x^2(t,\tau)}{\sigma_x(t)\sigma_x(\tau)}. \tag{B.49}$$

  If the stochastic process is stationary, then $\sigma_x^2(t) = \sigma_x^2(\tau) = \sigma_x^2$, and $\sigma_x^2(t,\tau) = \sigma_x^2(t-\tau)$, the above can be expressed as

$$r_x(t-\tau) = \frac{\sigma_x^2(t-\tau)}{\sigma_x^2}. \tag{B.50}$$

- When two stochastic processes $x(t)$ and $y(t)$ are of interest, then their *cross-covariance* and *cross-correlation functions* are defined respectively as

$$\begin{aligned} \sigma_{xy}^2(t,\tau) &= \text{Cov}[x(t), y(\tau)] = E[(x(t)-\mu_x(t))(y(\tau)-\mu_y(\tau))] \\ &= E[x(t)y(\tau)] - \mu_x(t)\mu_y(\tau), \end{aligned} \tag{B.51}$$

  and

$$r_{xy}(t,\tau) = \frac{\sigma_{xy}^2(t,\tau)}{\sigma_x(t)\sigma_y(\tau)}. \tag{B.52}$$

When only one stochastic process $x(t)$ is concerned, $\mu_x(t)$ and σ_x^2 can be simply referred to as its mean and covariance. If a stochastic process $x(t)$ has a zero mean; i.e., $\mu_x(t) = 0$ for all t, then it is said to be centered. Any stochastic process can be centered by a simple subtraction:

$$x'(t) = x(t) - \mu_x(t), \tag{B.53}$$

so that $\mu_{x'} = 0$. Without loss of generality, any stochastic process can be assumed to be centered. In this case, its covariance becomes

$$\sigma_x^2 = E[x^2(t)]. \tag{B.54}$$

A *Markov process* $x(t)$ is a particular type of stochastic process whose future values depend only on its present value, but independent of any past values. In other words, the probability of any future value conditioned on present and all past values is equal to the probability conditioned only on the present value:

$$P(x(t+h) = y|x(s) = \xi(s),\ \forall s \le t) = Pr[x(t+h) = y|x(t) = \xi(t)] \quad \forall h > 0. \tag{B.55}$$

When a stochastic process is sampled it becomes a time sequence of random variables $x[n]$ $(n = 1, \dots, N)$, which can be represented by a random vector $\boldsymbol{x} = [x[0], \dots, x[N-1]]^{\mathrm{T}}$. A *Markov chain* is defined as

$$\begin{aligned} &P(x[n] = y|x[m] = \xi[m],\ \forall m < n) \\ &= P(x[n] = y|x[n-m] = \xi[n-m],\ m = 1, \dots, k); \end{aligned} \tag{B.56}$$

i.e., the value $x[n]$ depends only on the k prior values. In particular, when $k = 1$, this is a first-order Markov chain:

$$P(x[n] = y|x[m] = \xi[m],\ \forall m < n) = P(x[n] = y|x[n-1] = \xi[n-1]). \tag{B.57}$$

Let $-1 < r < 1$ be the correlation coefficient between any two consecutive values $x[n]$ and $x[n-1]$ of a stationary first-order Markov chain of size N, then the correlation matrix is

$$\boldsymbol{R}_x = \begin{bmatrix} 1 & r & r^2 & \cdots & r^{N-1} \\ r & 1 & r & \cdots & r^{N-2} \\ r^2 & r & 1 & \cdots & r^{N-3} \\ \vdots & \vdots & \vdots & \ddots & \vdots \\ r^{N-1} & r^{N-2} & r^{N-3} & \cdots & 1 \end{bmatrix}_{N \times N}. \tag{B.58}$$

We see that the correlation between two variables $x[m]$ and $x[n]$ is $\rho^{|m-n|}$, which decays exponentially as a function of the distance $|m-n|$ between the two variables. This matrix $\boldsymbol{R}$ is a Toeplitz matrix.

Moreover, when $k = 0$, we get a memoryless zero-order Markov chain of which any value $x[n]$ is a random variable independent of any other value $x[m]$. In other words, all elements of the chain are totally decorrelated; i.e., $r_{mn} = \delta[m-n]$, and the correlation matrix is the identity matrix $\boldsymbol{R} = \boldsymbol{I} = \mathrm{diag}(1, \dots, 1)$. Also, let σ^2 be the variance of any $x[n]$ of a stationary zero-order Markov chain, then the covariance matrix is

$$\boldsymbol{\Sigma}_x = \begin{bmatrix} \sigma^2 & 0 & 0 & \cdots & 0 \\ 0 & \sigma^2 & 0 & \cdots & 0 \\ 0 & 0 & \sigma^2 & \cdots & 0 \\ \vdots & \vdots & \vdots & \ddots & \vdots \\ 0 & 0 & 0 & \cdots & \sigma^2 \end{bmatrix} = \sigma^2 \boldsymbol{I}. \tag{B.59}$$

Bibliography

Ahmed, N. and Rao, K.R. (1975) *Orthogonal Transforms for Digital Signal Processing*, Springer-Verlag

Bracewell, R.N. (2000) *The Fourier Transform and Its Applications*, McGraw Hill

Brigham, E.O. (1988) *The Fast Fourier Transform and Its Applications*, Prentice Hall

Britanak, V., Yip, P.C., and Rao, K.R. (2007) *Discrete Cosine and Sine Transforms*, Academic Press

Christersen, O. (2003) *An Introduction to Frames and Riesz Bases*, Birkhäuser

Hirsch, F. and Lacombe, G. (1999) *Elements of Functional Analysis*, Graduate Texts in Mathematics, Vol. 192, Springer

Jain, A.K. (1989) *Fundamentals of Digital Image Processing*, Prentice Hall

Jolliffe, I.T. (2002) *Principal Component Analysis*, 2nd ed., Springer

Loève, M. (1978) *Probability Theory II*, 4th ed., Graduate Texts in Mathematics, Vol. 46, Springer-Verlag

Mallat, S. (1998) *A Wavelet Tour of Signal Processing*, Academic Press

Oppenheim, A.V. and Willsky A.S. (1997) *Signals and Systems*, 2nd ed., Prentice Hall

Oppenheim, A.V. and Schafer, R.W. (1975) *Digital Signal Processing*, Prentice Hall

Poularikas, A.D. and Seely, S. (1991) *Signals and Systems*, PWS-KENT Publishing Company

Rao, K.R. (ed.) (1985) *Discrete Transforms and Their Applications*, Van Nostrand Reinhold

Rao, K.R. and Yip, P.C. (2001) *The Transform and Data Compression Handbook*, CRC Press LLC

Rao, K.R., Kim, D.N., and Hwang, J.J. (2010) *Fast Fourier Transform – Algorithms and Applications*, (Signals and Communication Technology), Springer

Roman, S. (2008). *Advanced Linear Algebra (Graduate Texts in Mathematics)*, 3rd ed., Vol. 135, Springer

Strang, G. and Nguyen, T. (1996) *Wavelets and Filter Banks*, Wellesley-Cambridge Press

Vetterli, M. and Kovacevic, J. (1995) *Wavelets and Subband Coding*, Prentice Hall

Index

Printed in the United States
By Bookmasters